Roderich Groß · Andreas Kolling
Spring Berman · Emilio Frazzoli
Alcherio Martinoli · Fumitoshi Matsuno
Melvin Gauci
Editors

Distributed Autonomous Robotic Systems

The 13th International Symposium

Springer

Editors
Roderich Groß
Department of Automatic Control and Systems Engineering
University of Sheffield
Sheffield
UK

Andreas Kolling
Department of Automatic Control and Systems Engineering
University of Sheffield
Sheffield
UK

Spring Berman
School for Engineering of Matter, Transport and Energy (SEMTE)
Arizona State University
Tempe, AZ
USA

Emilio Frazzoli
Massachusetts Institute of Technology
Cambridge, MA
USA

Alcherio Martinoli
ENAC, IIE, DIAL
École Polytechnique Fédérale de Lausanne (EPFL)
Lausanne
Switzerland

Fumitoshi Matsuno
Department of Mechanical Engineering and Science
Kyoto University
Kyoto
Japan

Melvin Gauci
Wyss Institute for Biologically Inspired Engineering
Harvard University Wyss Institute for Biologically Inspired
Cambridge, MA
USA

ISSN 2511-1256 ISSN 2511-1264 (electronic)
Springer Proceedings in Advanced Robotics
ISBN 978-3-030-10300-2 ISBN 978-3-319-73008-0 (eBook)
https://doi.org/10.1007/978-3-319-73008-0

Printed on acid-free paper

This Springer imprint is published by the registered company Springer International Publishing AG part of Springer Nature
The registered company address is: Gewerbestrasse 11, 6330 Cham, Switzerland

Springer Proceedings in Advanced Robotics 6

Series editors

Editorial Advisory Board

More information about this series at http://www.springer.com/series/15556

Foreword

Robots! Robots on Mars and in oceans, in hospitals and homes, in factories and schools; robots fighting fires, making goods and products, saving time and lives. Robots today are making a considerable impact from industrial manufacturing to healthcare, transportation, and exploration of the deep space and sea. Tomorrow, robots will become pervasive and touch upon many aspects of modern life.

The *Springer Tracts in Advanced Robotics (STAR)* was launched in 2002 with the goal of bringing to the research community the latest advances in the robotics field based on their significance and quality. During the past 15 years, the STAR series has featured publication of both monographs and edited collections. Among the latter, the proceedings of thematic symposia devoted to excellence in robotics research, such as ISRR, ISER, FSR, and WAFR, have been regularly included in STAR.

The expansion of our field as well as the emergence of new research areas has motivated us to enlarge the pool of proceedings in the STAR series in the past few years. This has ultimately led to launching a sister series in parallel to STAR. The *Springer Proceedings in Advanced Robotics (SPAR)* is dedicated to the timely dissemination of the latest research results presented in selected symposia and workshops.

This volume of the SPAR series brings the proceedings of the thirteenth edition of the DARS symposium on Distributed Autonomous Robotic Systems, whose proceedings have been previously published within STAR. This symposium took place at the Natural History Museum in London from November 7th to 9th, 2016. The volume edited by Roderich Groß, Andreas Kolling, Spring Berman, Emilio Frazzoli, Alcherio Martinoli, Fumitoshi Matsuno, and Melvin Gauci contains 47 scientific contributions organized in seven chapters. This collection focuses on robotic exploration, modular and swarm robotics, multi-robot control, estimation, planning, and applications.

From its excellent technical program to its warm social interaction, DARS culminates with this unique reference on the current developments and new advances in distributed autonomous robotic systems—a genuine tribute to its contributors and organizers!

Naples, Italy
Stanford, CA, USA
November 2017

Bruno Siciliano
Oussama Khatib
SPAR Editors

Preface

These proceedings contain the papers presented at DARS 2016, the 13th International Symposium on Distributed Autonomous Robotic Systems, which was held at the Natural History Museum in London, UK, from November 7th to 9th, 2016. The goal of DARS is to provide a forum for scientific advances in the theory and practice of distributed autonomous robotic systems. Distributed robotics is an interdisciplinary and rapidly growing area, combining research in computer science, communication and control systems, and electrical and mechanical engineering. Distributed robotic systems can autonomously solve complex problems while operating in highly unstructured real-world environments. They are expected to play a major role in addressing future societal needs, for example, by improving environmental impact assessment, food supply, transportation, manufacturing, security, and emergency and rescue services.

Building upon previous editions, the symposium presented a strong and varied technical program. We received a record 120 paper submissions—a testament to the thriving and growing nature of the field. The review process was overseen by the Program Chairs. Each paper was reviewed by at least three reviewers. Moreover, each paper received a final evaluation by a Program Chair. We would like to thank all members of the Program Committee as well as the additional referees for their diligent and constructive reviews—a crucial element for upholding the high technical standard of DARS. The review process yielded 47 papers to be included in the symposium, corresponding to an acceptance rate of 39%. Of the 47 papers, 30 papers were presented orally, and 17 papers were presented as posters. The method of presentation was chosen not only based on the quality of each paper, but also on content in order to ensure a well-balanced oral track, which is of interest to most of the attendees. Additionally, the papers for oral presentation were divided into seven thematic areas, namely Distributed Coverage and Exploration, Multi-Robot Control, Multi-Robot Estimation, Multi-Robot Planning, Modular Robots and Smart Materials, Swarm Robotics, and Multi-Robot Systems in Applications. All 47 accepted papers are included in these proceedings.

The program also featured four invited keynote addresses by researchers who are making a lasting contribution to science and robotics: "Material-Integrated

Intelligence for Robot Autonomy" by Nikolaus Correll (University of Colorado Boulder, USA), "Coordination, Cooperation and Collaboration in Multi-Robot Systems" by Vijay Kumar (University of Pennsylvania, USA), "Go to the Bee and Be Wise: Swarm Engineering Inspired by House-Hunting Honeybees" by James Marshall (University of Sheffield, UK), and "Robust Human Control of Multi-Robot Swarms" by Katia Sycara (Carnegie Mellon University, USA). The abstracts of these four keynote addresses are included in the proceedings.

This edition of DARS included three awards: Best Paper, Best Application Paper, and Best Poster. The awards committee was chaired by Michael Rubenstein (Northwestern University, USA) and included Melvin Gauci (Harvard University, USA), Sabine Hauert (Bristol University, UK), and Bahar Haghighat (EPFL, Switzerland). For the Best Paper award, the Program Chairs nominated six papers as finalists from among all the accepted papers based on the reports and award nominations by the referees, as well as on the revised contributions included in the digital proceedings. The final decision also took into account the presentation quality at the symposium. The Best Paper award went to "Robust Coordinated Aerial Deployments for Theatrical Applications Given Online User Interaction via Behavior Composition" by Elen Cappo et al. The Best Application Paper award was sponsored by the Institution of Engineering and Technology. All orally-presented papers were eligible for this award, and the decision took into account the degree to which the work addressed problems of practical implementation, and the quality of the presentation. This award went to "Multi-Swarm Infrastructure for Swarm Versus Swarm Experimentation" by Duane Davis et al. All posters were eligible for the Best Poster award, and the decision was based on the quality of the work and the poster presentation. This award went to "Vertex: A New Distributed Underwater Robotic Platform for Environmental Monitoring" by Felix Schill et al.

We would like to thank everyone involved in making DARS 2016 a success, including VICON Motion Systems (DARS 2016 Platinum Sponsor), RS Components (DARS 2016 Gold Sponsor), the Advisory Committee, the Program Committee and additional referees, the Organizing Committee, and all the authors of all submitted papers. Finally, we would like to thank the local organization team, in particular Ana Macintosh and Stefan M. Trenkwalder.

Sheffield, UK
July 2017

Roderich Groß
Andreas Kolling
Spring Berman
Emilio Frazzoli
Alcherio Martinoli
Fumitoshi Matsuno
Melvin Gauci

Organization

General Chair
Roderich Groß, The University of Sheffield, UK

General Co-Chair
Andreas Kolling, iRobot, USA

Technical Program Co-Chairs
Spring Berman, Arizona State University, USA
Emilio Frazzoli, MIT, USA
Alcherio Martinoli, EPFL, Switzerland
Fumitoshi Matsuno, Kyoto University, Japan

Publication Chair
Melvin Gauci,
Harvard University, USA

Publicity Chair
Sabine Hauert, University of Bristol, UK

Local Organization Team
Louise A. Caffrey, The University of Sheffield, UK
Ana MacIntosh, The University of Sheffield, UK
Stefan M. Trenkwalder, The University of Sheffield, UK

Advisory Committee
Hajime Asama, University of Tokyo, Japan
Marcelo H. Ang, National University of Singapore, Singapore
Tamio Arai, University of Tokyo, Japan
Raja Chatila, UPMC, France
Gregory S. Chirikjian, Johns Hopkins University, USA
Young-Jo Cho, ETRI, Republic of Korea
Nak Young Chong, JAIST, Japan
Nikolaus Correll, University of Colorado Boulder, USA

Rüdiger Dillmann, KIT, Germany
Toshio Fukuda, Nagoya University, Japan
Maria Gini, University of Minnesota, USA
M. Ani Hsieh, University of Pennsylvania, USA
Alcherio Martinoli, EPFL, Switzerland
Francesco Mondada, EPFL, Switzerland
Lynne E. Parker, University of Tennessee, USA

Program Committee

William Agassounon, Textron Defense Systems Inc., USA
Antonio P. Aguiar, University of Porto, Portugal
Rachid Alami, LAAS-CNRS, France
Javier Alonso-Mora, MIT, USA
Francesco Amigoni, Polytechnic University of Milan, Italy
Marcelo H. Ang, National University of Singapore, Singapore
Adrian Arfire, EPFL, Switzerland
Ryo Ariizumi, Nagoya University, Japan
Filippo Arrichiello, University of Cassino and Southern Lazio, Italy
Masoud Asadpour, University of Tehran, Iran
Shun-ichi Azuma, Kyoto University, Japan
Nicola Basilico, University of Milan, Italy
Meysam Basiri, EPFL, Switzerland
Jacob Beal, Raytheon BBN Technologies, USA
Kostas Bekris, Rutgers University, USA
Gerardo Beni, University of California, Riverside, USA
Sarah Bergbreiter, University of Maryland, USA
Navneet Bhalla, Harvard University, USA
Subhrajit Bhattacharya, University of Pennsylvania, USA
Mauro Birattari, Université Libre de Bruxelles, Belgium
Nicolas Bredeche, Pierre and Marie Curie University, France
Andreas Breitenmoser, University of Southern California, USA
Zack J. Butler, Rochester Institute of Technology, USA
Stefano Carpin, University of California, Merced, USA
Luiz Chaimowicz, Federal University of Minas Gerais, Brazil
Han-Lim Choi, KAIST, Republic of Korea
Anders L. Christensen, University Institute of Lisbon, Portugal
Timothy H. Chung, DARPA, USA
Brian Coltin, Carnegie Mellon University, USA
Nikolaus Correll, University of Colorado Boulder, USA
Jorge Cortés, University of California, San Diego, USA
Raffaello D'Andrea, ETH Zurich, Switzerland
Philip Dames, Temple University, USA
Karthik Dantu, SUNY Buffalo, USA

Prithviraj Dasgupta, University of Nebraska Omaha, USA
Carrick Detweiler, University of Nebraska Lincoln, USA
Gianni A. Di Caro, IDSIA USI, Switzerland
Rüdiger Dillmann, KIT, Germany
Dimos Dimarogonas, KTH, Sweden
Clare Dixon, University of Liverpool, USA
Marco Dorigo, Université Libre de Bruxelles, Belgium
Takahiro Endo, Kyoto University, Japan
William C. Evans, Google, USA
Alessandro Farinelli, University of Verona, Italy
Eliseo Ferrante, University of Leuven, Belgium
Rafael Fierro, The University of New Mexico, USA
Robert Fitch, The University of Sydney, Australia
Ryusuke Fujisawa, Hachinohe Institute of Technology, Japan
Rui Fukui, University of Tokyo, Japan
Simon Garnier, New Jersey Institute of Technology, USA
Andrea Gasparri, Roma Tre University, Italy
Melvin Gauci, Harvard University, USA
Veysel Gazi, Istanbul Kemerburgaz University, Turkey
Katie Genter, University of Texas Austin, USA
Maria Gini, University of Minnesota, USA
Heiko Hamann, University of Paderborn, Germany
Kiyohiko Hattori, NICT, Japan
Sabine Hauert, University of Bristol, UK
Tomohisa Hayakawa, Tokyo Institute of Technology, Japan
Geoffrey Hollinger, Oregon State University, USA
Satoshi Hoshino, Utsunomiya University, Japan
Jonathan P. How, MIT, USA
M. Ani Hsieh, University of Pennsylvania, USA
Hiroyuki Iizuka, Hokkaido University, Japan
Volkan Isler, University of Minnesota, USA
Yoshiaki Katada, Setsunan University, Japan
Takashi Kawakami, Hokkaido University of Science, Japan
Mirko Kovac, Imperial College London, UK
Masao Kubo, National Defense Academy, Japan
Daisuke Kurabayashi, Tokyo Institute of Technology, Japan
Haruhisa Kurokawa, AIST, Japan
Konstantinos J. Kyriakopoulos, National Technical University of Athens, Greece
Dongjun Lee, Seoul national University, Republic of Korea
Somchaya Liemhetcharat, Uber Advanced Technologies Center, USA
Pedro U. Lima, University of Lisbon, Portugal
Ali Marjovi, EPFL, Switzerland
Lino Marques, University of Coimbra, Portugal
Fulvio Mastrogiovanni, University of Genova, Italy

Nathan Michael, Carnegie Mellon University, USA
Dejan Milutinovic, University of California, Santa Cruz, USA
Melanie Moses, The University of New Mexico, USA
Masaaki Nagahara, Kyoto University, Japan
Radhika Nagpal, Harvard University, USA
Toru Namerikawa, Keio University, Japan
Nils Napp, SUNY Buffalo, USA
Daniele Nardi, Sapienza University of Rome, Italy
Keitaro Naruse, University of Aizu, Japan
Iñaki Navarro, EPFL, Switzerland
Giuseppe Notarstefano, University of Salento, Italy
Michael Novitzky, Georgia Institute of Technology, USA
Shinsuke Oh-hara, University of Yamanashi, Japan
Kazuhiro Ohkura, Hiroshima University, Japan
Derek Paley, University of Maryland, USA
Lucia Pallottino, University of Pisa, Italy
Antonio Pascoal, University of Lisbon, Portugal
Marco Pavone, Stanford University, USA
José Pereira, EPFL, Switzerland
Kirstin H. Petersen, Cornell University, USA
Hemma Philamore, University of Bristol, UK
Luciano C. A. Pimenta, Federal University of Minas Gerais, Brazil
Carlo Pinciroli, Ecole Polytechnique de Montréal, Canada
Amanda Prorok, University of Pennsylvania, USA
Subramanian Ramamoorthy, The University of Edinburgh, UK
Andreagiovanni Reina, The University of Sheffield, UK
Ioannis Rekleitis, University of South Carolina, USA
Paolo Remagnino, Kingston University London, UK
Alessandro Renzaglia, LAAS CNRS, France
Paolo Robuffo Giordano, IRISA/INRIA Rennes, France
Michael Rubenstein, Northwestern University, USA
Lorenzo Sabattini, University of Modena and Reggio Emilia, Italy
Brian Sadler, US Army Research Laboratory, USA
Erol Sahin, Middle East Technical University, Turkey
Kazunori Sakurama, Tottori University, Japan
Ketan Savla, University of Southern California, USA
Thomas Schmickl, University of Graz, Austria
Mac Schwager, Stanford University, USA
Iman Shames, University of Melbourne, Australia
Dylan A. Shell, Texas A&M University, USA
Wei-Min Shen, University of Southern California, USA
Tomohiro Shirakawa, National Defense Academy of Japan, Japan
Stephen L. Smith, University of Waterloo, Canada
Paolo Stegagno, Cornell University, USA

Kasper Stoy, IT University of Copenhagen, Denmark
Ken Sugawara, Tohoku Gakuin University, Japan
Ikuo Suzuki, Kitami Institute of Technology, Japan
Keiki Takadama, The University of Electrocommunications, Japan
Herbert G. Tanner, University of Delaware, USA
Danilo Tardioli, University Center of Defense, Spain
Guy Theraulaz, Paul Sabatier University and CNRS, France
Jonathan Timmis, University of York, UK
Vito Trianni, ISTC CNR, Italy
Elio Tuci, Aberystwyth University, UK
Kazuki Umemoto, Kanagawa University, Japan
Richard T. Vaughan, Simon Fraser University, Canada
Rodrigo Ventura, University of Lisbon, Portugal
Richard Voyles, Purdue University, USA
Justin Werfel, Harvard University, USA
Kazuaki Yamada, Toyo University, Japan
Masahito Yamamoto, Hokkaido University, Japan
Toshiyuki Yasuda, Hiroshima University, Japan
Jingjin Yu Rutgers, University, USA
Ikemoto Yusuke, Meijo University, Japan
Uwe R. Zimmer, Australian National University, Australia

Additional Referees

Charuvahan Adhivarahan
Jacopo Banfi
Florian Berlinger
Barbara Bruno
Levi DeVries
Sedat Dogru
Kevin Eckenhoff
Elizabeth Esterly
Andres Faina
Boris Gromov
Bahar Haghighat
Christoph Hintz
Lucas Janson
Matthew Kelly
Yara Khaluf
Jose Marcio Luna
Massimo Mecella
Michael Otte
Alyssa Pierson
Ragesh K. Ramachandran
Daniel Selvaratnam

Sara Spedicato
Khalil Taheri
Stefan M. Trenkwalder
Constantinos Vrohidis
Jonathan West
Sean Wilson
Indrajeet Yadav
Dingjiang Zhou
Saeed Ahmadizadeh
Cenk Baykal
Dimitris Boskos
Alessio Capitanelli
Krishna Doddapaneni
Miguel Duarte
Iñaki Esnaola
Mark Fabbro
Jorge Gomes
Meng Guo
Shahab Heshmati-Alamdari
Frank Imeson
Aris Kanellopoulos
Monroe Kennedy
Ganesh Kumar
Yoshiyuki Matsumura
Ivano Notarnicola
Cammy Peterson
Hasan Poonawala
Philipp Schillinger
Wenceslao Shaw-Cortez
Adam Stager
Andrea Testa
Andrea Vanzo
Zijian Wang
Michael Whitzer
Peter Wurman
Michael Zavlanos

Contents

Part I Distributed Coverage and Exploration

A Probabilistic Topological Approach to Feature Identification Using a Stochastic Robotic Swarm 3
Ragesh K. Ramachandran, Sean Wilson and Spring Berman

Communication-Restricted Exploration for Search Teams 17
Elizabeth A. Jensen, London Lowmanstone and Maria Gini

From Ants to Birds: A Novel Bio-Inspired Approach to Online Area Coverage 31
Luca Giuggioli, Idan Arye, Alexandro Heiblum Robles and Gal A. Kaminka

Information Based Exploration with Panoramas and Angle Occupancy Grids 45
Daniel Mox, Anthony Cowley, M. Ani Hsieh and C. J. Taylor

Multirobot Persistent Patrolling in Communication-Restricted Environments 59
Marta Romeo, Jacopo Banfi, Nicola Basilico and Francesco Amigoni

Part II Multi-Robot Control

A Comparative Study of Collision Avoidance Algorithms for Unmanned Aerial Vehicles: Performance and Robustness to Noise 75
Steven Roelofsen, Denis Gillet and Alcherio Martinoli

A Decentralized Control Strategy for Resilient Connectivity Maintenance in Multi-robot Systems Subject to Failures 89
Cinara Ghedini, Carlos H. C. Ribeiro and Lorenzo Sabattini

Chase Your Farthest Neighbour 103
Rotem Manor and Alfred M. Bruckstein

OuijaBots: Omnidirectional Robots for Cooperative Object Transport with Rotation Control Using No Communication 117
Zijian Wang, Guang Yang, Xuanshuo Su and Mac Schwager

Persistent Multi-robot Formations with Redundancy 133
Alyxander Burns, Bernd Schulze and Audrey St. John

Triangular Networks for Resilient Formations 147
David Saldaña, Amanda Prorok, Mario F. M. Campos and Vijay Kumar

Part III Multi-Robot Estimation

Construction of Optimal Control Graphs in Multi-robot Systems 163
Gal A. Kaminka, Ilan Lupu and Noa Agmon

Decision-Making Accuracy for Sensor Networks with Inhomogeneous Poisson Observations 177
Chetan D. Pahlajani, Indrajeet Yadav, Herbert G. Tanner and Ioannis Poulakakis

Distributed Laplacian Eigenvalue and Eigenvector Estimation in Multi-robot Systems 191
Mehran Zareh, Lorenzo Sabattini and Cristian Secchi

Distributed Object Characterization with Local Sensing by a Multi-robot System 205
Golnaz Habibi, Sándor P. Fekete, Zachary Kingston and James McLurkin

Optical Wireless Communications for Heterogeneous DARS 219
Patricio J. Cruz, Christoph Hintz, Jonathan West and Rafael Fierro

Part IV Multi-Robot Planning

Bundling Policies for Sequential Stochastic Tasks in Multi-robot Systems 237
Changjoo Nam and Dylan A. Shell

Decomposition of Finite LTL Specifications for Efficient Multi-agent Planning 253
Philipp Schillinger, Mathias Bürger and Dimos V. Dimarogonas

Informative Path Planning and Mapping with Multiple UAVs in Wind Fields 269
Doo-Hyun Cho, Jung-Su Ha, Sujin Lee, Sunghyun Moon and Han-Lim Choi

Multi-robot Informative and Adaptive Planning for Persistent Environmental Monitoring 285
Kai-Chieh Ma, Zhibei Ma, Lantao Liu and Gaurav S. Sukhatme

The Effectiveness Index Intrinsic Reward for Coordinating Service Robots 299
Yinon Douchan and Gal A. Kaminka

United We Move: Decentralized Segregated Robotic Swarm Navigation 313
Fabrício R. Inácio, Douglas G. Macharet and Luiz Chaimowicz

Part V Modular Robots and Smart Materials

A Rule Synthesis Algorithm for Programmable Stochastic Self-assembly of Robotic Modules 329
Bahar Haghighat and Alcherio Martinoli

Distributed Adaptive Locomotion Learning in ModRED Modular Self-reconfigurable Robot 345
Ayan Dutta, Prithviraj Dasgupta and Carl Nelson

Distributed Camouflage for Swarm Robotics and Smart Materials 359
Yang Li, John Klingner and Nikolaus Correll

Evo-Bots: A Simple, Stochastic Approach to Self-assembling Artificial Organisms 373
Juan A. Escalera, Matthew J. Doyle, Francesco Mondada and Roderich Groß

Geometrical Study of a Quasi-spherical Module for Building Programmable Matter 387
Benoît Piranda and Julien Bourgeois

HyMod: A 3-DOF Hybrid Mobile and Self-Reconfigurable Modular Robot and its Extensions 401
Christopher Parrott, Tony J. Dodd and Roderich Groß

Network Characterization of Lattice-Based Modular Robots with Neighbor-to-Neighbor Communications 415
André Naz, Benoît Piranda, Thadeu Tucci, Seth Copen Goldstein and Julien Bourgeois

Part VI Swarm Robotics

Decentralized Progressive Shape Formation with Robot Swarms 433
Carlo Pinciroli, Andrea Gasparri, Emanuele Garone and Giovanni Beltrame

Discovery and Exploration of Novel Swarm Behaviors Given Limited Robot Capabilities 447
Daniel S. Brown, Ryan Turner, Oliver Hennigh and Steven Loscalzo

Effects of Spatiality on Value-Sensitive Decisions Made by Robot Swarms 461
Andreagiovanni Reina, Thomas Bose, Vito Trianni and James A. R. Marshall

Emergence and Inhibition of Synchronization in Robot Swarms 475
Fernando Perez-Diaz, Stefan M. Trenkwalder, Rüdiger Zillmer and Roderich Groß

Evolving Behaviour Trees for Swarm Robotics 487
Simon Jones, Matthew Studley, Sabine Hauert and Alan Winfield

Evolving Group Transport Strategies for e-Puck Robots: Moving Objects Towards a Target Area 503
Muhanad H. Mohammed Alkilabi, Aparajit Narayan, Chuan Lu and Elio Tuci

From Formalised State Machines to Implementations of Robotic Controllers 517
Wei Li, Alvaro Miyazawa, Pedro Ribeiro, Ana Cavalcanti, Jim Woodcock and Jon Timmis

Human Responses to Stimuli Produced by Robot Swarms - the Effect of the Reality-Gap on Psychological State 531
Gaëtan Podevijn, Rehan O'Grady, Carole Fantini-Hauwel and Marco Dorigo

Localization of Inexpensive Robots with Low-Bandwidth Sensors 545
Shiling Wang, Francis Colas, Ming Liu, Francesco Mondada and Stéphane Magnenat

Modelling Mood in Co-operative Emotional Agents 559
Joe Collenette, Katie Atkinson, Daan Bloembergen and Karl Tuyls

Programmable Self-disassembly for Shape Formation in Large-Scale Robot Collectives 573
Melvin Gauci, Radhika Nagpal and Michael Rubenstein

Towards Differentially Private Aggregation of Heterogeneous Robots 587
Amanda Prorok and Vijay Kumar

Part VII Multi-Robot Systems in Applications

Construction Planning for a Modularized Rail Structure: Type Selection of Rail Structure Modules and Dispatch Planning of Constructor Robots 605
Rui Fukui, Yuta Kato, Gen Kanayama, Ryo Takahashi and Masayuki Nakao

Distributed Convolutional Neural Networks for Human Activity Recognition in Wearable Robotics 619
Dana Hughes and Nikolaus Correll

Formation Control of a Drifting Group of Marine Robotic Vehicles . . . 633
Nicholas R. Rypkema and Henrik Schmidt

Multi-swarm Infrastructure for Swarm Versus Swarm Experimentation 649
Duane T. Davis, Timothy H. Chung, Michael R. Clement and Michael A. Day

Robust Coordinated Aerial Deployments for Theatrical Applications Given Online User Interaction via Behavior Composition 665
Ellen A. Cappo, Arjav Desai and Nathan Michael

Vertex: A New Distributed Underwater Robotic Platform for Environmental Monitoring 679
Felix Schill, Alexander Bahr and Alcherio Martinoli

Author Index 695

Abstracts of Invited Keynote Presentations

Material-Integrated Intelligence for Robot Autonomy

Prof. Nikolaus Correll, University of Colorado Boulder, USA

Advances in miniature electronics, distributed algorithms and manufacturing technology have enabled a new generation of smart composites that tightly integrate sensing, actuation, computation and communication. Such "robotic materials" are inspired by multifunctional natural structures such as the skin of the cuttlefish that can change its color and patterning, bird wings that can change their shape, or the human skin that provides tactile sensing at high dynamic range. I will describe a series of recent results that best illustrate the benefits of material integrated computation: high-bandwidth sensing for texture recognition and localization in artificial skins, distributed optimization for controlling shape change, distributed classification for recognizing gestures drawn onto a modular facade, and feedback control of soft robotic actuators. I will then describe current challenges in robotic grasping and manipulation, and demonstrate how robotic materials can provide critical sensing and control during a series of manipulation tasks with applications to warehouse automation, manufacturing and lab automation.

Coordination, Cooperation, and Collaboration in Multi-Robot Systems

Prof. Vijay Kumar, University of Pennsylvania, USA

The central challenge in multi-robot systems lies in the synthesis of collective behaviors which enable group performance that exceeds the ability of individuals. We explore three different paradigms for collective behaviors. At a fundamental level, coordination is beneficial when individuals are confronted with a task that they can complete but can do so more efficiently as a group. Cooperation refers to the ability of robots to accomplish tasks they could not have completed on their own. Collaboration is useful for groups with different types of robots with diverse capabilities and tasks which cannot be completed with a single type of robot. This talk will discuss biological inspiration for these paradigms, mathematical frameworks, and resilience in collective behaviors with applications to ground and aerial robots.

Go to the Bee and Be Wise: Swarm Engineering Inspired by House-Hunting Honeybees

Prof. James A. R. Marshall, The University of Sheffield, UK

Distributed autonomous systems are likely to become increasingly important for robotics and other applications, due to their potential for resilience, scalability,and flexibility. However, designing grouplevel behaviors that are implemented by simple individual-level rules operating with local information is an inherently hard problem, and guaranteeing properties of that behavior is even harder. For example, search techniques and formal methods applied to swarms both rapidly fall foul of the curse of dimensionality as number of agents increase. However natural selection has successfully designed such systems repeatedly, and tools from the natural sciences have rigorously described the behaviour of very large systems of interacting components. In this talk, I will recount how observations of house-hunting honeybees led to the design of a new class of distributed decision-making algorithm, and its deployment on hundreds of small and simple robots. Rather than simply imitating nature, however, the algorithm's principled development requires the integration of concepts and techniques from areas as diverse as behavioural ecology and statistical physics.

Robust Human Control of Multi-Robot Swarms

Prof. Katia Sycara, Carnegie Mellon University, USA

As robotic platforms become cheaper and more reliable, multirobot deployment becomes possible and desirable. Since complete robot autonomy for these deployments is not yet possible, the presence of a human operator is necessary. Multiple human studies have shown that cognitive limitations prevent effective human control of multi-robot systems of tens of robots. Another difficulty is that many different types of human interactions may be necessary to maintain and control multi-robot systems. Additionally, the coordination scheme of multiple robots can vary which has consequences on the operator's difficulty of control. We have developed a characterization of human-robot tasks, and appropriate human robot interaction modes, based on the task's cognitive complexity of control. This scheme helps explicate the forms of control likely to be needed and the demands they pose on human operators. This talk will present two lines of research following from this characterization. The first evaluates the potential for using scheduling techniques to improve the performance of systems in which operators must attend to multiple independently operating robots. The second presents challenges and results pertaining to human control of autonomously cooperating robotic swarms.

Part I
Distributed Coverage and Exploration

A Probabilistic Topological Approach to Feature Identification Using a Stochastic Robotic Swarm

Ragesh K. Ramachandran, Sean Wilson and Spring Berman

Abstract This paper presents a novel automated approach to quantifying the topological features of an unknown environment using a swarm of robots with local sensing and limited or no access to global position information. The robots randomly explore the environment and record a time series of their estimated position and the covariance matrix associated with this estimate. After the robots' deployment, a point cloud indicating the free space of the environment is extracted from their aggregated data. Tools from topological data analysis, in particular the concept of persistent homology, are applied to a subset of the point cloud to construct barcode diagrams, which are used to determine the numbers of different types of features in the domain. We demonstrate that our approach can correctly identify the number of topological features in simulations with zero to four features and in multi-robot experiments with one to three features.

Keywords Unlocalized robotic swarm · Stochastic robotics · Mapping
GPS-denied environments · Topological data analysis · Algebraic topology

1 Introduction

Many potential applications for robotic swarms, such as environmental monitoring, exploration, disaster response, search-and-rescue, and mining, will require the robots to operate in uncertain environments. Constraints on the robots' onboard power may preclude the use of GPS and inter-robot communication, and even if the robots are

R. K. Ramachandran · S. Wilson · S. Berman (✉)
School for Engineering of Matter, Transport and Energy, Arizona State University, Tempe, AZ, USA
e-mail: spring.berman@asu.edu

S. Wilson
e-mail: sean.t.wilson@asu.edu

R. K. Ramachandran
e-mail: rageshkr@asu.edu

R. Groß et al. (eds.), *Distributed Autonomous Robotic Systems*, Springer Proceedings in Advanced Robotics 6,
https://doi.org/10.1007/978-3-319-73008-0_1

equipped with localization devices, they may be deployed in GPS-denied environments (e.g., indoors or underground). Despite these constraints, the robots may be required to map their environment in order to perform desired tasks. For instance, the robots may need to identify target payloads to transport or obstacles and hazardous regions to avoid. Since the robots will have limited sensing and computational capabilities, it would not be feasible to implement existing techniques such as occupancy grid mapping [24], simultaneous localization and mapping (SLAM) [21, 24], and Probability Hypothesis Density (PHD) filtering [27] to address this problem.

As an initial step toward constructing a map with metric information, we present an automated method for computing the number of topological features in an unknown domain from data obtained by a swarm of inexpensive robots with local sensing, no inter-robot communication, and limited or no access to global position information. The features represent obstacles or other regions of interest that robots do not pass through. The data consist of robots' position estimates and the covariance matrices of these estimates, recorded by the robots during random exploration of the domain. The robots collect this data autonomously and independently during their deployment, without relying on input from a supervisory agent. We assume that after a set period of time, the robots navigate to an easily identifiable landmark (e.g., a beacon), where they transfer this data to a central computer. The computer then processes the data from the entire swarm to extract a point cloud that covers the domain's free space and applies tools from *Topological Data Analysis* (TDA), namely *persistent homology*, to identify the numbers of different types of topological features. Our approach scales with the number of robots and is robust to the failure of a small portion of the swarm.

Although topological mapping has been extensively studied, TDA has only recently been applied in robotics for environmental characterization. For a scenario with a single robot, [7] presents a method for topological SLAM that encodes the topology of the environment in a generalized Voronoi graph. Few works address the problem of mapping an environment using a robotic swarm with limited sensing, no inter-robot communication, and no global localization. In [19], we presented an optimal control approach to mapping a GPS-denied environment with a robotic swarm using a partial differential equation model of the swarm population dynamics. This strategy works best when the domain contains only a few sparsely distributed features, whereas the approach presented here can be applied to domains that are more densely populated with features. In [20], the authors propose an algorithm that covers the free space of the environment with robots and then constructs an approximate generalized Voronoi graph of the covered region. This algorithm requires the robots to communicate with a central server that commands their actions. In contrast, our approach does not require a centralized decision maker during the robots' operation. Alternatively, [13] obtains a simplicial approximation of a region of interest as a topological map using dual pairs of nerves that are constructed using relevant visibility and observation covers. Contrary to our strategy, [13] requires the robots to have the ability to detect and maintain a record of landmarks in the domain, such as obstacle corners and edges. The mapping approach in [8] is similar to ours in that it generates a point cloud of the domain's free region and uses persistent homology to compute topological features in the environment. However, unlike our strategy, this approach requires each robot to have an identification label that can be recognized by other robots.

The paper is structured as follows. Section 2 introduces the tools of TDA that are used in our methodology. Section 3 presents the problem statement and describes assumptions about the robot capabilities and motion model. Our approach for extracting topological features of the domain from the robots' data is discussed in Sect. 4. Sects. 5 and 6 validate our approach with simulations and multi-robot experiments, respectively. Finally, Sect. 7 concludes the paper and proposes future work.

2 Background

Topological Data Analysis (TDA) [5] is an emerging field that aims to provide algorithmic and mathematical tools for studying topological and geometric attributes of data. The fundamental idea underlying TDA is that data has an inherent *shape* that encodes important information regarding the connectivity of the data and yields insight into its global structure. TDA exploits the mathematical framework of *algebraic topology* [15], especially the concept of persistent homology [9], to characterize the topological structure of data. In many applications, data is obtained as a *point cloud* consisting of noisy samples of an intensity map in a Euclidean space. Prominent topological features of a point cloud can be computed using TDA and presented in the form of compact representations such as *persistence diagrams* [10] and *barcode diagrams* [11]. TDA has been extensively applied to problems in computer vision and image processing [23], sensor networks [6, 14], robotics [4, 18], localization [22], and map comparison [3].

We provide a brief introduction to persistent homology, which is central to our mapping methodology. More detailed treatments of the associated theory and computations are given in [10, 16, 28]. Persistent homology is a method of analyzing the correlation of homological information gathered across different scales. This technique enables the identification of topological features that are present over a large range of scales, as opposed to those which are only temporarily present (short-scale features). Homology is a robust tool that facilitates the study of *global* attributes of spaces and functions from *local* computations on noisy data. A topological space $\mathbf{T}$ can be associated with a collection of vector spaces called *homology groups*, denoted by $H_k(\mathbf{T}), k = 0, 1, 2, ..., dim(\mathbf{T}) - 1$, each of which encodes a particular topological feature of $\mathbf{T}$. In persistent homology, these features are characterized using *Betti numbers*, which are the ranks of the homology groups. These numbers are topological invariants. The kth Betti number of $\mathbf{T}$, denoted by β_k, is the rank of $H_k(\mathbf{T})$ and represents the number of independent k-dimensional cycles in $\mathbf{T}$. For example, if $\mathbf{T} \subset \mathbb{R}^2$, then β_0 is the number of connected components in $\mathbf{T}$ and β_1 is the number of holes in $\mathbf{T}$. If $\mathbf{T} \subset \mathbb{R}^3$, then β_0, β_1, and β_2 are the numbers of connected components, tunnels, and voids in $\mathbf{T}$, respectively.

In a typical TDA application, a finite set of samples from a space $\mathbf{M}$ is available. These samples, along with the metric associated with $\mathbf{M}$, comprise the point cloud

C of the space. In TDA, the metric is used to map **C** onto a collection of simplices called a simplicial complex. Simplices are combinatorial objects constructed from the subsets of **C**. A k-simplex $\sigma = [v_0, v_1, ..., v_k]$ is an ordered list of $k+1$ elements $\{v_0, v_1, ..., v_k\} \in \mathbf{C}$, called vertices. The simplicial complex provides a discrete representation of the underlying topological space using a combinatorial structure that can be represented algebraically using linear operators (matrices). It is this combinatorial structure that permits us to develop algorithms for homological computation. There are various ways to build a simplicial complex from a point cloud. The simplest way is to choose a parameter $\delta > 0$ and add a k-simplex to the simplicial complex if every vertex in the simplex is within a distance δ from every other. The simplicial complex constructed in this manner is called the Vietoris–Rips complex [12] or Rips complex for short, often denoted as $Rips(\mathbf{C}, \delta)$.

For large datasets, the number of simplices in the simplicial complex can be enormous, making the computations highly inefficient. We reduce the computational requirements by choosing a subset of the point cloud consisting of *landmark points*, denoted by $\mathbf{L} \subset \mathbf{C}$, as vertices for the Rips complex. These landmark points were selected using a greedy inductive selection process called a sequential max-min algorithm [1]. In order to compute persistent Betti numbers, we require a *filtration*, defined as a family of $Rips(\mathbf{C}, \delta)$ parametrized by δ such that $Rips(\mathbf{C}, \delta_1) \subseteq Rips(\mathbf{C}, \delta_2)$ for all $\delta_1 > 0$, $\delta_2 > 0$ where $\delta_1 \leq \delta_2$.

The persistent topological features of **T** over multiple values of δ can be identified using a *barcode diagram*, which is a graphical representation of $H_k(\mathbf{T})$ in terms of the homology generators. A barcode plots a set of horizontal line segments on a graph whose x-axis spans a range of δ values and whose y-axis depicts an arbitrary ordering of homology generators. The numbers of arrows in the barcode for dimension 0 and dimension 1 indicate the numbers of connected components and features in the domain, respectively. A barcode diagram can be computed automatically using algorithms that find the homology generators of the homology that is constructed on a point cloud. Figure 1 illustrates a barcode diagram that is obtained from an example point cloud.

3 Problem Statement

We consider a scenario in which N robots are deployed into a bounded, unknown, GPS-denied 2D environment in order to collect data that can be used to determine the number of topological features in the domain. The robots have local sensing capabilities and can identify features and other robots at distances within their sensing range to perform collision avoidance maneuvers. Each robot is equipped with a compass and wheel encoders, which enable it to estimate its position and orientation with uncertainty.

The robots perform correlated random walks in the domain, avoiding features and other robots. During its motion, each robot estimates its position in a global reference frame using its onboard odometry and a Kalman filter. At fixed time intervals, the

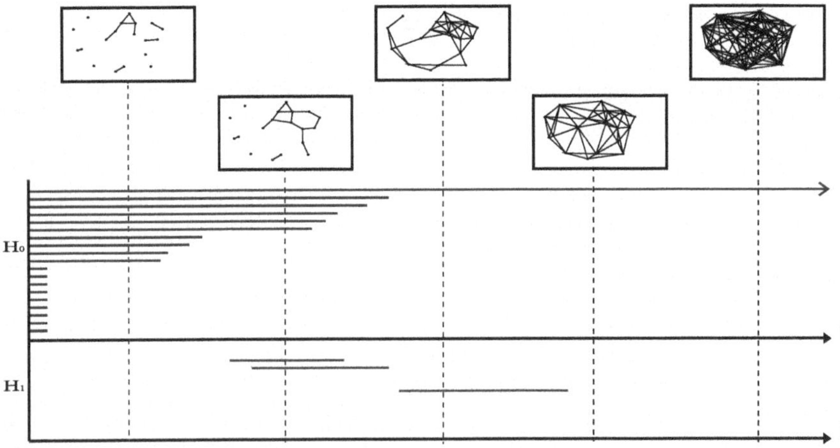

Fig. 1 An example barcode diagram of a filtration formed from a Rips complex. $\beta_k(\delta_i)$ is the number of horizontal segments in the barcode for $H_k(\mathbf{T})$ that intersect the dashed line at $\delta = \delta_i$

robot records its estimated position and the covariance matrix corresponding to the uncertainty of the estimate. After a time span T, all robots travel to a common location where their stored data is retrieved and processed. We assume that T is sufficiently large for the robots to thoroughly cover the domain and that the robots have sufficient memory to store the data that they obtain during their deployment.

The robots follow the motion model described in [25]. Each robot has a constant translational speed v and an orientation $\theta(t)$ at time t with respect to a global frame. We define a robot's velocity vector at time t as $\mathbf{V}(t) = [v_x(t),\ v_y(t)]^T = [v\ \cos(\theta(t)),\ v\ \sin(\theta(t))]^T$ and its position vector as $\mathbf{X}(t) = [x(t),\ y(t)]^T$. The displacement of a robot over a time step Δt is given by

$$\mathbf{X}(t + \Delta t) = \mathbf{X}(t) + \mathbf{V}(t)\Delta t + \mathbf{W}(t), \tag{1}$$

where $\mathbf{W}(t) \in \mathbb{R}^2$ is a vector of independent, zero-mean normal random variables that are generated at time t to model the randomness in the robot's motion due to sensor and actuator noise. At the beginning of a time step, each robot generates a random number between 0 and 1. If this number is below a predefined threshold p_{th}, the robot randomly chooses a new $\theta(t) \in [-\pi, \pi]$. At time $t = 0$, the start of a deployment, each robot is assigned the parameters v and p_{th} and obtains accurate measurements of its position $\mathbf{X}(0)$ and orientation $\theta(0)$.

We consider two types of scenarios. In **Type I** scenarios, robots receive accurate estimates of their global positions when they are close to the boundary of the domain. For example, robots on the exterior of a building will have access to GPS measurements that are unavailable to robots inside. In **Type II** scenarios, robots do not receive global position updates anywhere in the domain, which may for instance be located underground or underwater.

4 Feature Extraction Methodology

During a deployment, the data that robot $j \in \{1, ..., N\}$ obtains at time $t_k \in [0, T]$, $k \in \{1, ..., K\}$, consists of the element $d_k^j = \{\mu_k^j, \Sigma_k^j\}$, where $\mu_k^j \in \mathbb{R}^2$ is the mean of the robot's estimate of its (x, y) position at time t_k, and $\Sigma_k^j \in \mathbb{R}^{2\times 2}$ is the covariance matrix of its position estimate at this time. In this section, we present a three-step methodology for extracting the topological features of the domain from this data.

In the first step, we discretize the domain into a high-resolution uniform grid, as in the occupancy grid mapping algorithms described in [25]. Let m_i denote the grid cell with index $i \in \{1, ..., M\}$ and $\mathbf{M} = \{m_i\}$ denote the set of all grid cells. The goal of this step is to use the robots' data to assign each grid cell m_i a probability p_i^f of being *free*, or unoccupied by a topological feature. Toward this end, we compute p_{ijk}, the probability that robot j occupied grid cell m_i at time t_k, for all robots, cells, and measurement times. This probability is obtained by numerically integrating the Gaussian distribution with mean μ_k^j and covariance matrix Σ_k^j over the region $[x_i^l, x_i^u] \times [y_i^l, y_i^u]$ occupied by the cell:

$$p_{ijk} = \int_{y_i^l}^{y_i^u} \int_{x_i^l}^{x_i^u} \mathcal{N}(\mu_k^j, \Sigma_k^j)\, dx\, dy \tag{2}$$

Next, we assign a score $s_i \in [0, \infty)$ to each grid cell m_i according to the formula

$$s_i = \sum_{j=1}^{N} \sum_{k=1}^{K} \log\left(\frac{1}{1 - p_{ijk}}\right) \tag{3}$$

We rescale each score s_i to a value $s_i^C \in [0, C]$, where C is chosen such that the value of $1 - \exp(C)^{-1}$ is close to one. This rescaling improves numerical stability when converting the scores to probabilities, especially values near zero and one. Finally, the probability of each grid cell being free is computed as $p_i^f = 1 - \exp(s_i^C)^{-1}$.

In the second step, we extract a point cloud $\mathbf{C}$ and select a subset $\mathbf{L}$ of these points as landmark points. The point cloud is constructed by sampling a dense, uniformly random set of points from the domain and rejecting those points that are inside a grid cell m_i for which p_i^f is below a given threshold (i.e., there is a high probability of cell m_i being occupied by a feature). In this work, we set the threshold heuristically. Landmark points are selected from the point cloud using the sequential max-min algorithm [1]. This algorithm initially chooses a random point in $\mathbf{C}$ as the first landmark. Given a set of $i - 1$ landmarks denoted by $\mathbf{L}_{i-1}$, the algorithm selects the ith landmark as the point $c \in \mathbf{C}$ that maximizes the function $d(c, \mathbf{L}_{i-1}) = \min\{\|c - l\| \;\; l \in \mathbf{L}_{i-1}\}$. The landmarks chosen in this manner tend to cover the point cloud.

Finally, we use the landmark points to construct a filtration using the tools discussed in Sect. 2, and we extract barcode diagrams from this filtration. We chose the Rips complex as a basis for constructing the filtration [11] and used the MATLAB-based JavaPlex package [2] to perform all persistent homology computations and generate the barcodes. We computed persistent homology only for dimensions zero and one, since higher dimensions are not relevant for our application.

5 Simulations

We applied the methodology described in Sect. 4 to estimate the number of topological features in the simulated environments shown in Fig. 2 for both **Type I** and **Type II** scenarios. The simulations were coded in Python, and all other computations were performed in MATLAB. The simulated swarm consisted of 30 point robots, each with a sensing radius of 5 cm, an average speed of $v = 20$ cm/s, and $p_{th} = 0.2$. The robots explored a 200 cm × 200 cm domain over a time period $T = 200$ s. At the start of each simulation, the robots were placed at random locations near the domain boundary. The robots dispersed throughout the domain according to the model Eq. 1, where the covariance matrix of the random variables in $\mathbf{W}(t)$ was set to be a diagonal matrix with 0.1 on the diagonal. Upon encountering a feature, the domain boundary, or another robot within a distance of 5 cm, a robot would randomly choose a different direction to avoid a collision. After each simulated swarm deployment, we randomly sampled 16,000 points from the domain, extracted a point cloud $\mathbf{C}$ by using a threshold of 0.2 for p_i^f, and selected a set $\mathbf{L}$ of 1,000 landmark points from $\mathbf{C}$. The maximum filtration value (maximum value of δ) used for the barcode computation was heuristically chosen to be 3Δ, where $\Delta = \max\{d(c, \mathbf{L}) : c \in \mathbf{C}\}$.

Figures 3, 4, 5, 6, 7, 8, 9 and 10 plot the outputs of the different steps of our methodology for both **Type I** and **Type II** scenarios: contour plots of p_i^f (Figs. 3 and 4), point clouds (Figs. 5 and 6), landmark points (Figs. 7 and 8), and barcode diagrams (Figs. 9 and 10). The contour plots of p_i^f in the **Type I** scenarios are more accurate than the plots in the **Type II** scenarios, in the sense that they display higher probabilities of free space in areas that are actually unoccupied by features. However, the plots in the **Type II** scenarios do correctly estimate very low probabilities of free space in areas that are occupied by features. The barcode arrows for both **Type I** and **Type II** scenarios give the correct numbers of connected components and features for

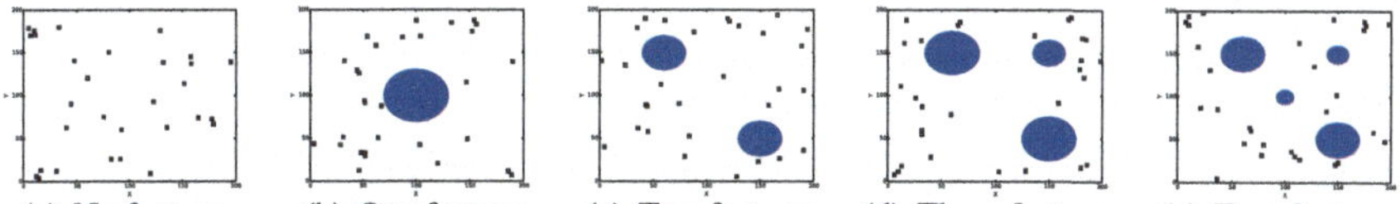

(a) No feature (b) One feature (c) Two features (d) Three features (e) Four features

Fig. 2 Snapshots of a simulated swarm moving through different domains

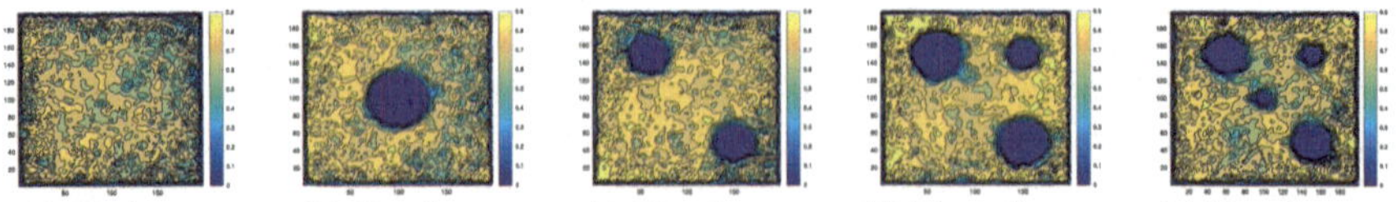

(a) No feature (b) One feature (c) Two features (d) Three features (e) Four features

Fig. 3 Contour plots of p_i^f, the probability that grid cell m_i is free, over all grid cells of discretized domains in **Type I** scenarios. Colorbar values range from 0 to 0.9

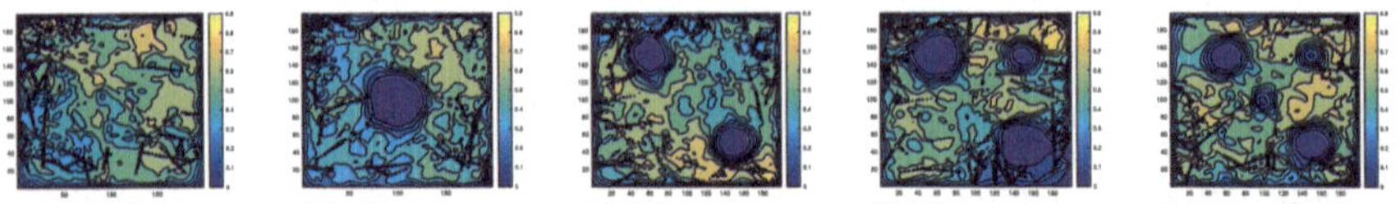

(a) No feature (b) One feature (c) Two features (d) Three features (e) Four features

Fig. 4 Contour plots of p_i^f, the probability that grid cell m_i is free, over all grid cells of discretized domains in **Type II** scenarios. Colorbar values range from 0 to 0.9

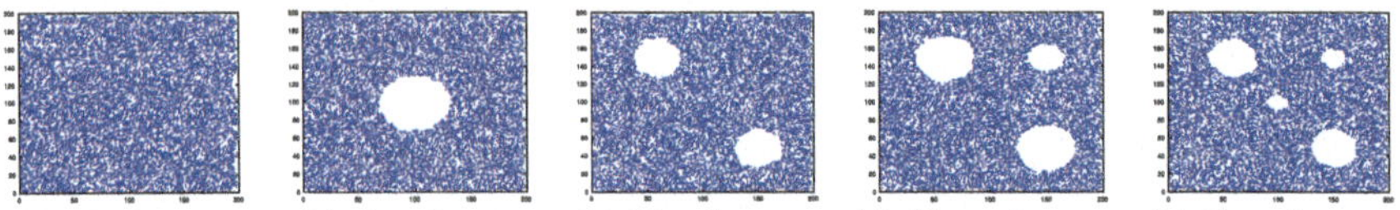

(a) No feature (b) One feature (c) Two features (d) Three features (e) Four features

Fig. 5 Point clouds computed over domains in **Type I** scenarios

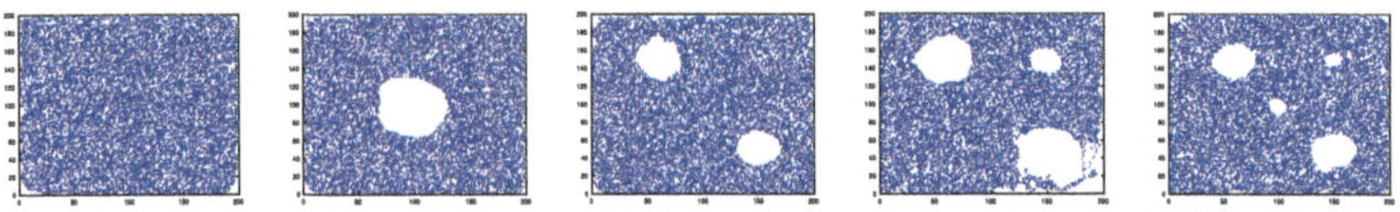

(a) No feature (b) One feature (c) Two features (d) Three features (e) Four features

Fig. 6 Point clouds computed over domains in **Type II** scenarios

each simulated environment. These results show that our methodology can accurately extract topological features even when the robots do not receive accurate estimates of their global positions.

We also examined the effect of the quantity of robot position data on the accuracy of our approach for **Type II** scenarios. Larger quantities of robot data can be obtained by extending the time period T of the swarm deployment or by deploying a larger number of robots, N. We ran simulations with 30 robots over the four domains shown in Fig. 2 with deployment times T that varied from 40 to 240 s, at intervals of 20 s. At the end of each deployment, the number of topological features was computed from the resulting barcode diagram. Figure 11 plots the computed number of features in each domain for every value of T. The figure shows that the correct number of

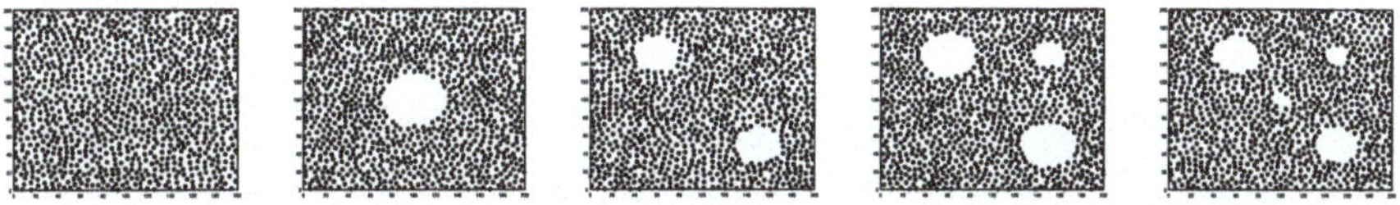

(a) No feature (b) One feature (c) Two features (d) Three features (e) Four features

Fig. 7 Landmark points selected over domains in **Type I** scenarios

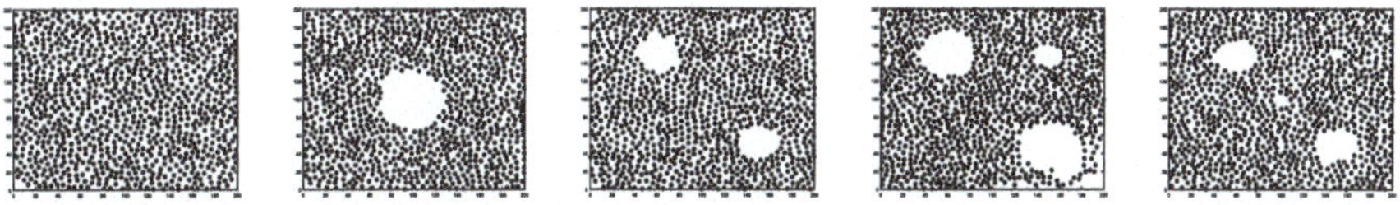

(a) No feature (b) One feature (c) Two features (d) Three features (e) Four features

Fig. 8 Landmark points selected over domains in **Type II** scenarios

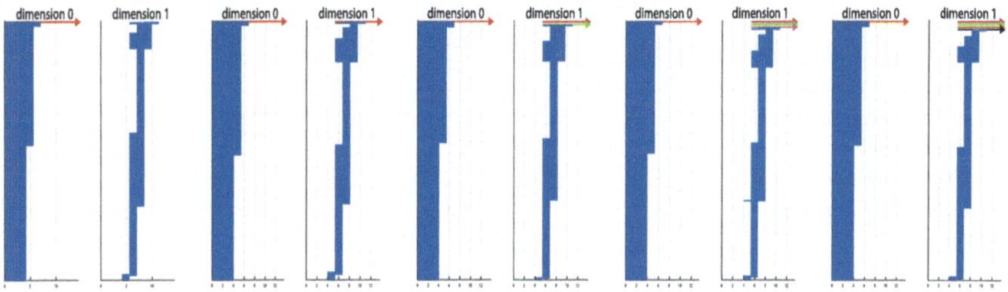

(a) No feature (b) One feature (c) Two features (d) Three features (e) Four features

Fig. 9 Barcodes computed for domains in **Type I** scenarios

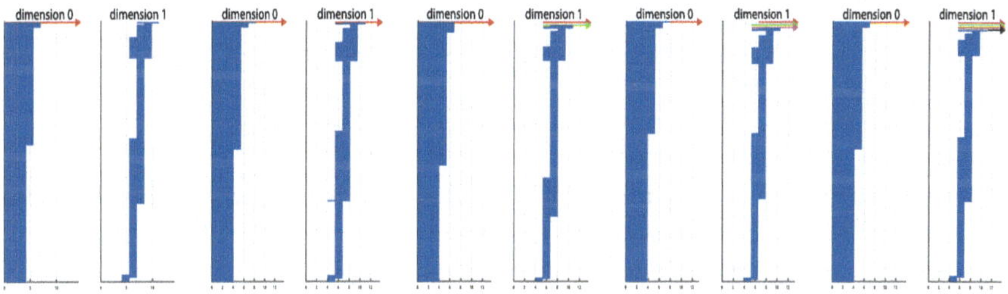

(a) No feature (b) One feature (c) Two features (d) Three features (e) Four features

Fig. 10 Barcodes computed for domains in **Type II** scenarios

features is identified in each domain when $T \geq 100$ s. For shorter deployments, the robots do not always cover a sufficiently large area of the domain for their recorded position data to yield an accurate count of the number of features. Hence, Fig. 11 shows that the shortest possible time period over which the swarm should be deployed in the simulated scenarios is $T \in (80\text{ s}, 100\text{ s}]$. We also ran simulations over the four domains in Fig. 2 with robot population sizes $N \in \{5, 10, 20, 30, 40, 50\}$ and $T = 200$ s and computed the numbers of connected components and topological

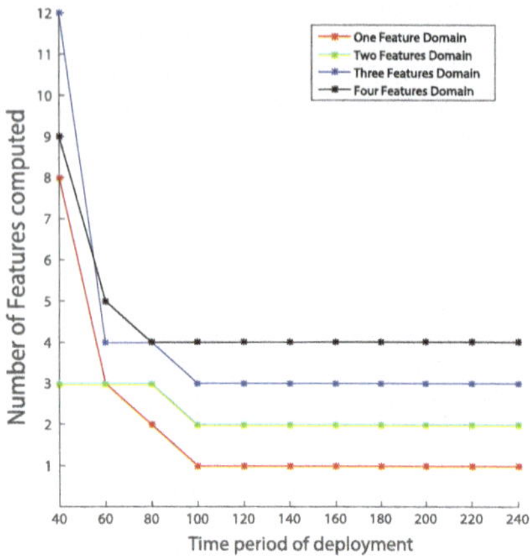

Fig. 11 Computed number of features versus swarm deployment time period T (in seconds) for simulations with 30 robots on the domains shown in Fig. 2 for **Type II** scenarios

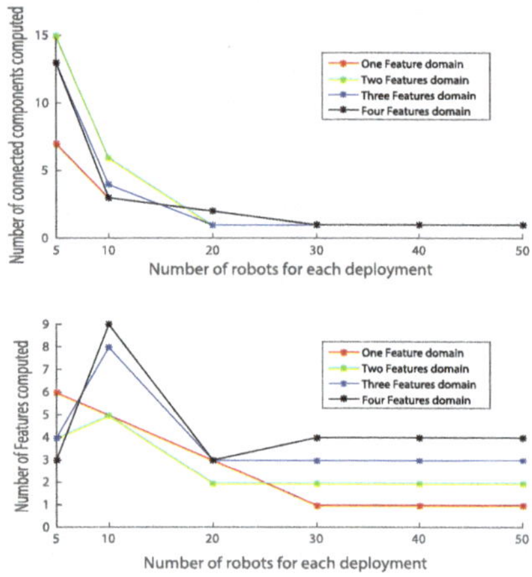

Fig. 12 Computed numbers of connected components (*top*) and features (*bottom*) versus number of robots N for simulations with $T = 200$ s on the domains shown in Fig. 2 for **Type II** scenarios

features in each domain. Figure 12 plots these numbers for every value of N and shows that they are accurate when data is obtained by $N \geq 30$ robots. In practice, such simulations can be performed to determine estimates of the minimum values of T and N that will yield accurate counts of the number of features in an environment.

We also note that the effectiveness of our approach depends on the degree of uncertainty in the robots' position estimates, as quantified by the covariance matrices associated with the position data. A large covariance indicates a highly uncertain position estimate and results in a low probability p_i^f of the corresponding grid cell being free. This low value reduces the likelihood that the central computer will misidentify the grid cell as being free (i.e., a possible location for a robot) if it is already known to be occupied by an obstacle. In addition, since covariances in robot positions will increase over time, newly acquired position data will not result in significant changes in the value of p_i^f.

6 Experimental Results

In addition to simulations, we validated our methodology through experiments with four Pheeno mobile robot platforms [26] in a **Type II** scenario with one to three features. The robots were initially placed at random locations in a 1.5 × 2.1 meter rectangular arena that was bounded by wooden walls, as shown in Fig. 13. The robots were controlled to move at 10 cm/s with an avoidance radius of 10 cm. Whenever a robot detected a feature, wall, or another robot, it avoided a collision by moving according to a specular reflection from the detected object and then continued in a straight line. The robots were marked with 2D binary identification tags to enable real-time tracking of their positions and orientations by an overhead camera (Microsoft Life Cam, resolution of 1920 × 1080 pixels). A control computer broadcast each robot's initial state $\mathbf{x} = [x, y, \phi]^T$ over WiFi, where x and y are the robot's position coordinates in the arena and ϕ is its heading. Each robot used an Extended Kalman Filter (EKF) to estimate its state at intervals of 200 ms. This state was updated according to a kinematic unicycle model and a measurement state vector, $\mathbf{z} = [\Delta d_e, \Delta\phi_e, \phi_c]^T$, where Δd_e is the encoders' measurement of the linear distance traveled, $\Delta\phi_e$ is the change in heading angle measured by the encoders, and ϕ_c is the orientation of the robot in the global frame measured by the compass. The state error covariance matrix **P**, process covariance matrix **Q**, and measurement covariance matrix **R** were set to $\mathbf{P} = diag(0.2, 0.2, 0.1)$, $\mathbf{Q} = diag(2, 2, 4)$, and $\mathbf{R} = diag(0.1, 5, 0.4)$ These matrices were chosen to favor the robot's measurements over the kinematic motion model. The initial state estimate covariance was chosen to reflect errors in tag placement on the robots and camera discretization error. The EKF was implemented on Pheeno's Arduino Pro Mini microcontroller (3.3V 8MHz), while the state data and covariance matrices were stored onboard its Raspberry Pi 2 Model B.

The results in Fig. 14a–d confirm that our methodology correctly extracts one connected component and two topological features from the robots' data after being deployed in the environment in Fig. 13b. The plots in Figs. 15 and 16 show that given a sufficiently long deployment time T and a sufficiently large number of robots N, our approach produces accurate counts of the numbers of connected components and features in environments with one, two, and three features. There is a trade-off between the robots' deployment time and the reliability of their position data, since the EKF state estimates will drift due to the robots' wheel slip and sensor noise. These factors cause the covariances of the position estimates to eventually grow larger than the environment and thus yield no useful information for mapping. This uncertainty can be reduced in a **Type 1** scenario by correcting the drift with direct GPS measurements or with estimates of global position using local measurements of known objects in the environment. From our experiments, it is evident that larger numbers of robots yield more accurate mapping results, since there is a higher chance of robots exploring small gaps between features before the covariances of their position data grow too large to provide useful information.

(a) (b) (c)

Fig. 13 The experimental arena with four Pheeno robots and (a) one feature, (b) two features, or (c) three features. At the start of the experiment, the control computer identifies the robots' positions and orientations, indicated by the red dots and cyan lines, from the robots' 2D binary identification tags. This identification is done using the thresholding, boxpoint, and contouring OpenCV libraries on a Windows computer

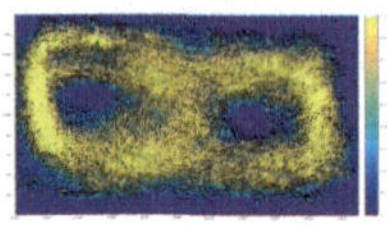

(a) Contour plot of p_i^f; colorbar values range from 0 to 0.9

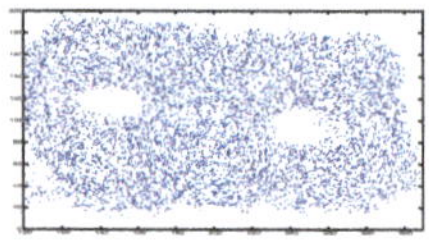

(b) Point cloud

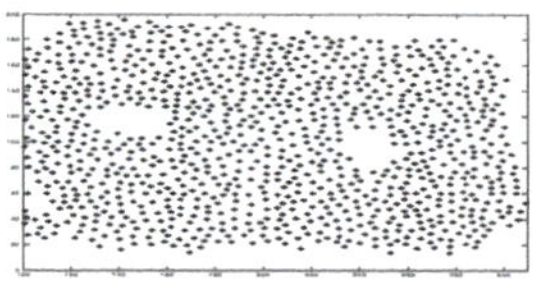

(c) Landmark points

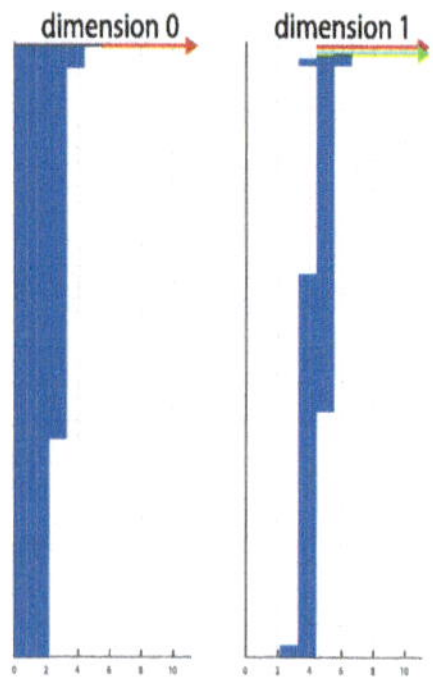

(d) Barcode Diagram

Fig. 14 Experimental results from a **Type II** environment containing two objects

Fig. 15 Computed numbers of connected components (*top*) and features (*bottom*) versus swarm deployment time period T (in seconds) for experiments with four robots on the domains shown in Fig. 13 for **Type II** scenarios

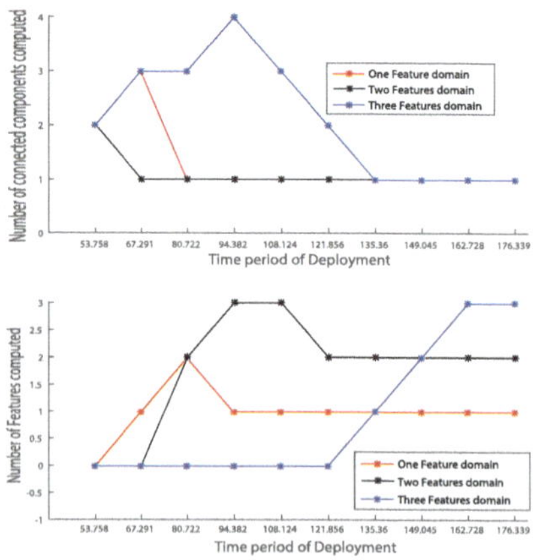

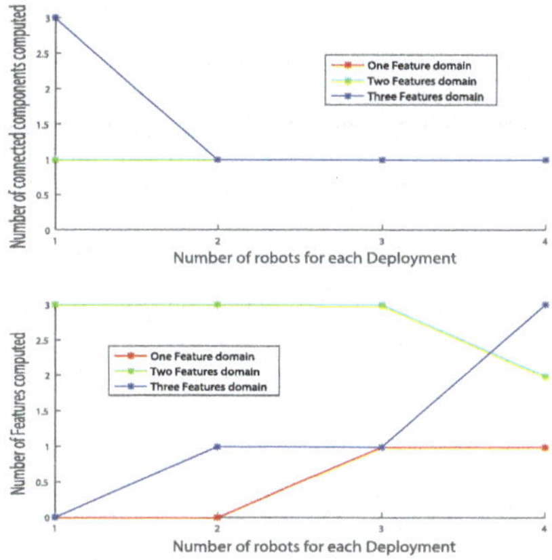

Fig. 16 Computed numbers of connected components (*top*) and features (*bottom*) versus number of robots N for experiments with $T = 180$ s on the domains shown in Fig. 13 for **Type II** scenarios

7 Conclusion

We have formulated a new approach to identifying the numbers of topological features in an unknown domain by applying tools from Topological Data Analysis (TDA) to data collected by a robotic swarm. The proposed methodology was shown to be effective through simulations on different domains and experiments with mobile robots. We note that the point cloud generated in our procedure is embedded with a metric, making it a *metric space*. In future work, we will extend our approach to incorporate the metric in order to construct a metric map of the unknown environment. This can be done using the techniques of manifold learning [17], which focus on identifying manifolds from a point cloud. In addition, we will employ TDA techniques to compute the optimal threshold of p_i^f for extracting the point cloud.

Acknowledgements R.K.R. thanks Dr. Subhrajit Bhattacharya for his valuable input on this work. This work was supported by NSF Award CMMI-1363499 and DARPA Young Faculty Award D14AP00054.

References

1. Adams, H., Carlsson, G.: On the nonlinear statistics of range image patches. SIAM J. Imaging Sci. **2**(1), 110–117 (2009)
2. Adams, H., Tausz, A., Vejdemo-Johansson, M.: JavaPlex: a research software package for persistent (co)homology. In: Mathematical Software – ICMS, pp. 129–136. Springer, Berlin (2014)
3. Ahmed, M., Fasy, B.T., Wenk, C.: Local persistent homology based distance between maps. In: ACM SIGSPATIAL International Conference on Advances in Geographic Information Systems (SIGSPATIAL), pp. 43–52 (2014)
4. Bhattacharya, S., Ghrist, R., Kumar, V.: Persistent homology for path planning in uncertain environments. IEEE Trans. Robot. **31**(3), 578–590 (2015)
5. Carlsson, G.: Topology and data. Bull. Am. Math. Soc. **46**(2), 255–308 (2009)

6. Chintakunta, H., Krim, H.: Distributed localization of coverage holes using topological persistence. IEEE Trans. Signal Process. **62**(10), 2531–2541 (2014)
7. Choset, H., Nagatani, K.: Topological simultaneous localization and mapping (SLAM): toward exact localization without explicit localization. IEEE Trans. Robot. Autom. **17**, 125–137 (2001)
8. Dirafzoon, A., Bozkurt, A., Lobaton, E.J.: Dynamic topological mapping with biobotic swarms (2015). CoRR arXiv:1507.03206
9. Edelsbrunner, H., Harer, J.L.: Persistent homology - a survey. Contemp. Math. **453**, 257–282 (2008)
10. Edelsbrunner, H., Harer, J.L.: Computational Topology: An Introduction. American Mathematical Society, Providence, RI (2010)
11. Ghrist, R.: Barcodes: the persistent topology of data. Bull. Am. Math. Soc. **45**(1), 61–75 (2008)
12. Ghrist, R.: Elementary Applied Topology, 1st edn. Createspace, USA (2014)
13. Ghrist, R., Lipsky, D., Derenick, J., Speranzon, A.: Topological landmark-based navigation and mapping. University of Pennsylvania, Department of Mathematics, Technical Report Vol. 8 (2012)
14. Ghrist, R., Muhammad, A.: Coverage and hole-detection in sensor networks via homology. In: International Symposium on Information Processing in Sensor Networks (IPSN) (2005)
15. Hatcher, A.: Algebraic Topology. Cambridge University Press, New York (2002)
16. Kaczynski, T., Mischaikow, K., Mrozek, M.: Computational Homology. Springer, New York (2004)
17. Ma, Y., Fu, Y.: Manifold Learning Theory and Applications, 1st edn. CRC Press Inc., Boca Raton (2011)
18. Pokorny, F.T., Stork, J.A., Kragic, D.: Grasping objects with holes: a topological approach. In: IEEE International Conference on Robotics and Automation (ICRA), pp. 1100–1107 (2013)
19. Ramachandran, R.K., Elamvazhuthi, K., Berman, S.: An optimal control approach to mapping GPS-denied environments using a stochastic robotic swarm. In: International Symposium on Robotics Research (ISRR) (2015)
20. Ramaithitima, R., Whitzer, M., Bhattacharya, S., Kumar, V.: Automated creation of topological maps in unknown environments using a swarm of resource-constrained robots. IEEE Robot. Autom. Lett. **1**(2), 746–753 (2016)
21. Robertson, P., Angermann, M., Krach, B.: Simultaneous localization and mapping for pedestrians using only foot-mounted inertial sensors. In: International Conference on Ubiquitous Computing (Ubicomp), pp. 93–96 (2009)
22. Robinson, M.: Topological Signal Processing. Springer, Berlin (2014)
23. Skraba, P., Ovsjanikov, M., Chazal, F., Guibas, L.: Persistence-based segmentation of deformable shapes. In: IEEE Computer Society Conference on Computer Vision and Pattern Recognition - Workshops, pp. 45–52 (2010)
24. Thrun, S.: A probabilistic online mapping algorithm for teams of mobile robots. Int. J. Robot. Res. **20**(5), 335–363 (2001)
25. Thrun, S., Burgard, W.: Probabilistic Robotics. Intelligent Robotics and Autonomous Agents. MIT, Cambridge (2005)
26. Wilson, S., Gameros, R., Sheely, M., Lin, M., Dover, K., Gevorkyan, R., Haberland, M., Bertozzi, A., Berman, S.: Pheeno, a versatile swarm robotic research and education platform. IEEE Robot. Autom. Lett. **1**(2), 884–891 (2016)
27. Zajic, T., Ravichandran, R.B., Mahler, R.P.S., Mehra, R.K., Noviskey, M.J.: Joint tracking and identification with robustness against unmodeled targets. Signal Process. Sens. Fusion Target Recognit. XII **5096**, 279–290 (2003)
28. Zomorodian, A.J., Ablowitz, M.J., Davis, S.H., Hinch, E.J., Iserles, A., Ockendon, J., Olver, P.J.: Topology for Computing (Cambridge Monographs on Applied and Computational Mathematics). Cambridge University Press, New York, NY (2005)

Communication-Restricted Exploration for Search Teams

Elizabeth A. Jensen, London Lowmanstone and Maria Gini

Abstract Exploring an unknown environment comes with risks and complications, and in some cases an environment may be too dangerous for humans to explore, but immediate exploration is critical, as in the aftermath of an earthquake. Robots, however, can be deployed to seek out points of interest and report back to the waiting human operators. One aspect of a disaster scenario is that communication is often more limited than we are accustomed to in everyday life, so these robots cannot rely on having constant contact with the outside world, or even with all other robots in the environment. In this paper, we present two algorithms for a small team of robots to explore an unknown environment, and use both simulation and experiments with physical robots to demonstrate the algorithms' performance. We provide proofs of correctness and guarantee full coverage of the environment, even with attrition.

1 Introduction

When a search and rescue team arrives on the scene of a disaster, it may be prevented from entering immediately due to the instability of the environment. Earthquakes may have aftershocks well after the main quake ends, and fires may have burned through the support structure of a building, making it too dangerous for humans to enter. Many researchers have developed robots or the means to use them in an initial exploration, to map points of interest such as the location of survivors or weak supports, and send this information to the human rescuers waiting outside [2, 24].

While this is necessary, many of these previous avenues of research have considered using only one robot for the exploration, or a pairing of a ground robot and an

E. A. Jensen (✉) · L. Lowmanstone · M. Gini
University of Minnesota, Minneapolis, MN, USA
e-mail: ejensen@cs.umn.edu

L. Lowmanstone
e-mail: london4@comcast.net

M. Gini
e-mail: gini@cs.umn.edu

R. Groß et al. (eds.), *Distributed Autonomous Robotic Systems*,
Springer Proceedings in Advanced Robotics 6,
https://doi.org/10.1007/978-3-319-73008-0_2

aerial robot to reach more locations. Instead, we are interested in algorithms which allow teams of robots to enter and explore unknown and dangerous environments, leveraging greater numbers of robots to more quickly cover the environment, while providing guarantees of full coverage even in the face of individual failures. However, we are also aware that a common feature of a disaster scenario is that communication may be more limited than we are accustomed to in everyday life [27], so these robots cannot rely on having constant contact with the outside world, or even with each other inside the environment. We therefore focused on developing algorithms which can function under various communication restrictions, such as chemical or line-of-sight means of communicating between robots.

Our primary contributions in this work are two distributed algorithms for exploration using small teams of robots. The innovation in these algorithms comes from how the robots disperse into and subsequently explore the environment, even with communication restrictions. We provide proofs that the algorithms will achieve full coverage of the environment, return all functioning robots to the entry point, and that points of interest are marked in such a way that the human rescuers can go directly to those points when the environment is deemed safe for them to enter. We demonstrate the algorithms' functionality through simulations and experiments using physical robots.

2 Related Work

A multi-robot system has several advantages over a single robot, including cost, efficiency and robustness [5, 11]. While a single robot can be designed to efficiently complete its task, it may then be suitable for only a small set of tasks, and if even a small part fails, the robot may be unable to complete the desired task. In contrast, a multi-robot system comprised of smaller, individually less-capable robots, with several types to complete various parts of the task, can still accomplish their goal even if some robots fail.

There are multiple methods for a team of robots to explore an unknown environment. Gage [10] proposed three types of coverage. Blanket coverage provides simultaneous coverage of the entire area. Barrier coverage sets up a complete perimeter around an area. In sweep coverage, the robots make a pass over the environment and ensure every point has been seen by at least one robot, but don't stay in any one location, instead moving progressively through the environment.

Most coverage algorithms are focused on surveillance, and thus aim to achieve either blanket or barrier coverage. However, the number of robots required to achieve either can be prohibitively large. In contrast, sweep coverage can be accomplished with a small team, down to a single robot, if necessary, should all other robots fail [9]. Thus, in our approach, we use an exploration algorithm in which the team of robots completes a single sweep of the environment to locate points of interest that can be relayed to the search and rescue team.

An additional consideration is whether the system should be centralized or distributed. In a centralized system, a controller issues instructions and keeps the group coordinated. Stump et al. [29], used a robot as a base station, while the other robots formed a communication bridge as they moved into the unknown region. Similarly, Rekleitis et al. [26] used one robot as a stationary beacon for another robot, thus reducing odometry errors. The centralized approach also has the advantage of making it easy to create a global map, which can later be used to direct other agents, human or robot [4, 28, 31]. However, a centralized approach fails completely if the controller fails, and does not scale well.

A distributed approach, on the other hand, is inherently more scalable and can also take better advantage of the robustness of having multiple robots. Each robot is responsible for its own movements and data collection, and relies on only local neighbors for coordinating exploration and dispersion. It may seem that the robots are working together on a global scale, but in actuality the decisions are made individually on a local scale. Information can be passed throughout the group, similar to the communication bridge [29], but using broadcast messages rather than point-to-point messages. An alternative approach is to allow the robots to separate to explore distant frontiers, and work towards each other to build a map of the environment [15]. This allows a single sweep through the environment, but also relies on precise measurements from the robots' sensors to fuse the maps together into a global whole.

Fusing the maps together is often made easier by using a grid-based approach to the coverage problem [27, 30]. Grid-based approaches can be quite effective in ensuring full coverage of an environment, and are particularly well suited to smaller, indoor spaces [11], but the size of the grid cells can require longer travel distances to achieve full coverage. In contrast, a graph based approach allows the robots more latitude in choosing where to explore [3, 18], and decomposes the environment in a way that is more tuned to that particular environment, but with less structure, there can also be more overlap in the robots' sensor coverage. Brass et al. [3] are able to provide optimal paths for two robots using their algorithms for exploration of undirected graphs, particularly in environments such as office buildings or caves. In our work, we use a graph-based approach to help provide flexibility in how the robots disperse.

Ma and Yang [21] show that the most efficient dispersion of mobile nodes is triangular, producing the maximal overall coverage with minimal overlap or gap. The dispersion formation is achieved through the nodes' local communication, in which they determine distance and bearing to their neighbors. Lee et al. [18] use online distributed triangulation to break the environment into discrete areas, which the robots can then monitor and update in both coverage and patrolling tasks. Liu et al. [19] have shown that repeated location updates can lead to better coverage over time. Similar approaches by Howard et al. [12] and Cortes et al. [7] used potential fields and gradient descent, respectively, to disperse the nodes. In simulation, these methods successfully spread the nodes throughout the environment to achieve blanket coverage, but a sufficient number of robots may not be available outside of simulation.

Another method is to model distributed multi-robot algorithms on insect behavior. The robots have very little individual ability, but can communicate with local neighbors and arrange themselves according to a desired dispersion pattern. McLurkin and Smith [23] have developed robots and several algorithms for dispersion and exploration in indoor environments. Their algorithms rely on the robots maintaining connectivity in order to perform correctly. These algorithms are similar to those in [7, 12, 19], but allow for greater variability in the dispersion pattern, including clusters and perimeter formations. Additional work based on insect behavior includes pheromone-based algorithms [1, 16, 22, 25], which rely on items placed in the environment for communication and navigation.

Dirafzoon et al. [8] provide an overview of many sensor network coverage algorithms which can be applied to multi-robot systems as well. However, many of these rely on individual robots knowing the distance and bearing of other robots around them, which requires more sophisticated sensors. For example, Kurazume and Hirose [17] developed an algorithm in which the team of robots was split into two groups, one of which remained stationary as landmarks while the other moved, and then they traded roles. On the other hand, research has shown that a team of robots can disperse into an unknown environment using only wireless signal intensity to guide the dispersion [14, 20]. This method allows the use of simple robots, without the need to carry a heavy payload of sensors, so that the robots can run longer and explore further. Smaller, simpler robots are also less expensive, so more robots can be acquired for a task. Overall, a distributed system allows an individual robot to work independently, while also sharing data with neighbors as necessary. It also does not rely on long distance communication.

3 Communication-Restricted Exploration

Our primary objective is for our algorithms to achieve full exploration of an unknown environment using a team of robots. Our algorithms aim for sweep coverage, and thus can function with a single robot, if needed (due to availability or attrition). Our distributed approach takes advantage of the robustness inherent in having multiple robots, and is not impeded by the communication restrictions, since only local communication is required. Lastly, as in insect-based algorithms, the robots carry and drop off beacons (such as ZigBee motes or RFID tags) to provide longer lasting trails and information to mobile agents that may pass by later.

We assume that the robots can detect and avoid obstacles, communicate with each other in some manner (wi-fi, line-of-sight, chemical, etc.), and can carry and drop off beacons. We also assume that the specifics of the environment are currently unknown, even if pre-disaster information, such as a map, is available.

Our algorithms use the communication signal intensity to direct the robots' movements, keeping them linked as a group during the entire exploration. This provides the benefit of reducing both the likelihood of robots getting lost and the possibility that part of the environment will be overlooked. Our innovation lies in making the algorithms independent of the type of communication used, while still making the robot team capable of achieving full coverage in an efficient manner even with some attrition.

3.1 Algorithm Details

In the Rolling Dispersion Algorithm (RDA) the robots are either *explorers*, which move into the frontier, or *sentries*, which maintain a return path to the entrance. Each robot uses connectivity with its neighbors and distance to nearby obstacles to choose which of the following behaviors it will execute on each iteration of the algorithm. We show the finite state machine, using the initials of the behaviors for the node labels in Fig. 1.

Avoid Collisions: Use proximity sensors to avoid obstacles.
Disperse: Move towards open space and the frontier, away from neighbors.
Follow Path: Accept request and follow the path to the requesting robot.
Guard: Stay in place and act as a sentry for other robots.
Retract: Dead-end reached, drop a beacon and return to the frontier.
Seek Connection: Re-establish communication with the rest of the group.

Definition 1 An exploring robot is in a dead-end when every direction in which it might move is towards an obstacle, be it a wall, a beacon, or another robot.

The robots initially disperse to the furthest extent of their communication range. When the robots can no longer move apart without losing communication with another robot, they call for reinforcements, which leap-frog their way to the frontier, leaving behind beacons to mark the path to the entrance and any unexplored regions as necessary. When robots encounter a dead-end (see Definition 1), they drop off a beacon to mark the area as explored, and retract to the previous intersection before moving to a new frontier. This retraction process is repeated for every robot along that path until all robots have moved on to the frontier, leaving the entire path marked as explored by beacons. The direction of the dispersion and exploration is primarily informed by the wireless signal intensities between agents, as in [20], though the individual robots also make decisions based on their proximity sensors to avoid collisions. When there are no more paths left to explore, the robots will retract back to the entry and we can then guarantee that all parts of the environment have been explored.

The Sweep Exploration Algorithm (SEA) is based on RDA, but is intended for use in scenarios with much more restrictive communication, such as chemical signals, or line-of-sight using a camera and color LEDs. With such restrictions, it is critical to reduce the number and size of messages to be able to ensure full exploration, so the robots decide their next action based on the states of their neighbors. There are two states used only by robots, one state used only by beacons, and five states used by both.

Branch: Robot or beacon marking an intersection.
Call Path: Robot or beacon on the path to the unexplored frontier.
Explorer: Robot moving along a path to answer a call or failure notification.
Failure Path: Robot or beacon along the path leading to a location where an agent failed and needs to be replaced.
Repel: Beacon marking an area as explored, preventing repeated exploration.
Retract Path: Robot or beacon that has received notice of a retracting robot on its way.
Retractor: Robot has reached a dead-end, dropped a beacon to mark the explored area and is in the process of retracting to the nearest branch.
Sentry: Robot or beacon marking a path where there are no active notifications (call, failure, or retract).

However, this also means that only one robot can be moving at a time, or the messages get mixed up and parts of the environment may be missed. Therefore, instead of the robots initially dispersing in any direction, as in RDA, they travel one at a time down a single path, until it is completely explored, and then retract and explore a new path. We show the finite state machine for robots using SEA and give the transitions in Fig. 1. The nodes are labeled with the initials of the states.

A robot in the Explorer state will move away from the rest of the robots into the unexplored frontier, while a robot in the Retractor state will return along the path of robots and beacons to the previous intersection (see Definition 2) and then become an Explorer again. A robot in any of the other states will remain in place, marking the path and providing other robots with the status of those robots further down the path (passing along calls for additional Explorers, or notices of retraction or failures). Robots in intersections use the state Branch to alert the Retractors that they have reached the point where there is again a frontier to explore.

Definition 2 Intersections are defined as locations where either there is an obstacle creating multiple paths (such as a corner), or where the robot could move in perpendicular directions without obstruction.

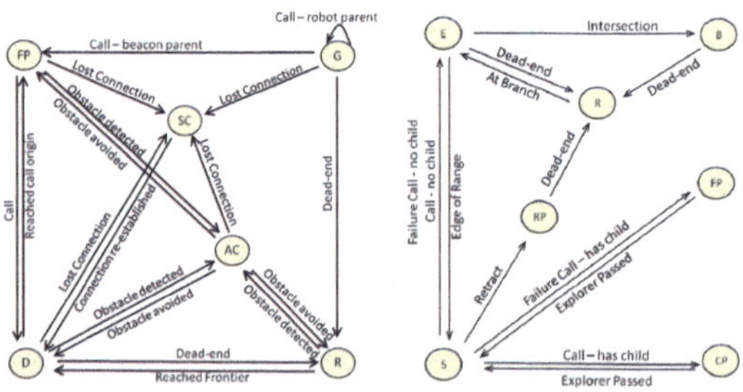

Fig. 1 (left) Rolling dispersion algorithm finite state machine. (right) Sweep exploration algorithm finite state machine

3.2 Algorithm Correctness

We present here formal proofs that the robots running our algorithms will, even with communication restrictions, complete the exploration without missing any point in the environment, will not end up in infinite loops (so that the robots exit when done), and can succeed in these goals even with robot and beacon failures.

Though SEA has more restrictions on communication than RDA, both algorithms operate in the same manner at their core, so the following properties and proofs apply to both algorithms. The differences in the algorithms shows in how the robots move (multiple or one at a time), and how much information is shared between robots.

Lemma 1 *The algorithms avoid unnecessarily repeated exploration.*

Proof We will do this proof in two parts: first assuming that the beacons do not fail and then assuming that they may fail. In either case we will prove by contradiction that the algorithms will avoid unnecessary repeated exploration.

First, assume that the robots explore an area that has been previously explored. This produces an immediate contradiction because, when an exploring robot reaches a dead-end it drops a beacon to mark the explored area. Any re-explorations are prevented by the presence of these beacons.

Second, in the case when a beacon marking an explored area fails, the area around that beacon becomes unmarked. If the failed beacon is surrounded by beacons marking the path as explored, then no robot will reach that area, so it will not be re-explored. If, however, the beacon was bordering unexplored regions, then the robots will have to re-explore the now unmarked area until reaching a dead-end, and then once again mark the area as explored, preventing future unnecessary visits. The important caveat here is that we need to assume a finite number of beacon failures, otherwise robots would re-mark areas infinitely, which would lead to other areas not being explored or the robots not returning to the entrance to report the completed exploration. Therefore, given a finite number of beacon failures, which is a reasonable assumption, we again derive a contradiction. □

Lemma 2 *The algorithms avoid infinite loops.*

Proof We will again consider two cases in this proof: first assuming that the beacons do not fail; and then assuming that they may fail. In both cases, we will prove by contradiction that the algorithms will not get stuck in infinite loops.

First, assume that a robot is repeatedly exploring a loop in the environment. This is immediately a contradiction of Lemma 1 because the point at which the exploring robot first closed the loop (by reaching a previously explored location), it would have detected it was in a dead-end, and dropped a beacon to mark the area as explored. That beacon will break the loop, since the robots will treat it as an impassable obstacle no matter the direction from which they approach.

Second, in the case where beacons can fail, the area surrounding a beacon's location becomes unmarked. If there are still surrounding beacons marking the area as explored, then there will be no effect on the robots' exploration. If the beacon is neighbors with a robot or the path to the frontier or entrance, then the area will be re-explored, as in Lemma 1, but will again be stopped when a dead-end is once again located. We must assume that there will be a finite number of beacon failures, which is a reasonable assumption, so we again derive a contradiction. □

Lemma 3 *The algorithms achieve full coverage with a single robot.*

Proof Assume we have one robot and an infinite number of beacons. In both algorithms, the robot will advance, leaving beacons to mark the return path and areas that have been explored. The explore/retract behaviors are the same as Depth-First search, which is complete in a finite search space when repeated states and loops are avoided. By Lemmas 1 and 2, we have proven that our algorithms avoid repeated states and loops. Our environment is finite. Thus, our algorithms will achieve full coverage with only one robot. □

Theorem 1 *The algorithms will achieve full coverage of the environment with multiple robots, and all functional robots will return to the entrance.*

Proof We prove by induction that the algorithms function correctly when multiple robots are used.

Base Case: The base case is that the algorithms achieve full exploration when only one robot explores the environment. The proof is given in Lemma 3.

Induction Step: Assuming that the algorithms achieve full exploration with k robots and the remaining functional robots return to base, we want to prove that the algorithms achieve full exploration when 1 more robot is added. Suppose that there are $k+1$ robots, then we need to show that: (1) the robots will not continuously explore overlapping areas, and (2) that the robots will not miss an area because they lost contact with the other agents, and (3) that no robot will be stranded.

First, in Lemma 1, we have already shown that there will not be unnecessary repeated exploration of an area, so long as the beacons remain active. Every robot that approaches the area will detect the beacons and move to a different area, and this remains true no matter how many robots are added.

Second, both algorithms enforce the restriction that the robots remain in contact with at least one other agent at all times, and when that connection is lost, the robots will immediately stop exploring and retreat in order to reconnect. This is required for ensuring complete coverage, because the connectivity means that robots will not miss any area, and will not re-explore areas previously covered.

Third, the connectivity keeps the robots from being stranded in the environment, because only the exploring robot in a dead-end will mark an area as explored. The retraction step then brings the robot back into the group before the next robot marks an area as explored, so that no robot is left behind or trapped. In addition, the fact that the robots explore a path and then retract to the previous intersection, and then repeat the process, similar to Depth-First search, means that all the still functional robots will eventually retract back to the entrance when the exploration is complete. Once again, adding another robot to the team does not change this functionality.

Thus, with $k+1$ robots, the algorithms achieve full exploration, and the remaining functional robots return to the entrance when the exploration is complete. □

3.3 Algorithm Properties

In addition to the previous proofs, there are several important properties of the algorithms. Using multiple robots reduces the individual load on each robot, but the coordination adds costs in location visits and number of messages.

We can represent the environment as a graph, in which nodes representing locations that are separated by the distance of the communication range, in order to keep it to a finite number of nodes. We show a possible graph as an overlay on the simulation environment map in Fig. 2, though this depends on how the robots move about the environment, so subsequent runs may lead to different locations for the intersections and dead-ends. With a single robot, each node is visited at most twice (leaf nodes are visited only once), assuming there are no failures. With multiple robots, each intermediate node *n* is visited at most *2n'* times, where *n'* is the number of nodes beyond node *n* on that path, and that number is bounded by the number of robots, in the case where the total number of robots is less than *n'*.

In RDA, the robots must send many messages to confirm that they are still within communication range of each other, since multiple robots can move at the same time. This is quite costly in terms of bandwidth, processing, and power consumption, and may not be feasible with some kinds of communication. If we use chemical signals, flooding the environment with those chemicals will cause us to lose new messages in the old ones. But in restricting the communication in SEA, we lose the ability for multiple robots to move at the same time. However, it does put bounds on the number of messages being sent, which makes the coordination easier as well. In SEA, we require only eight message types to complete the exploration, which correspond directly to the eight states listed previously. Because of this, the robots can complete the exploration with very little required in terms of information sharing, which is often one of the more expensive aspects of multi-robot systems.

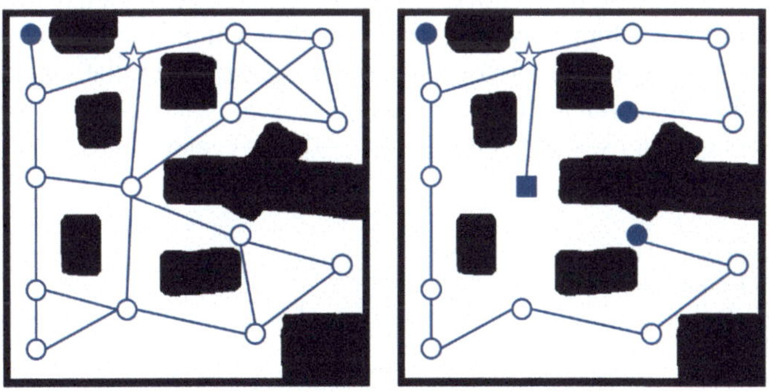

Fig. 2 (left) Cave-like environment used in simulations, showing overlay graph. (right) One possible exploration. The star marks the start, the square marks a location where exploration was begun, then left to finish a different path, resulting in the filled circles becoming dead-ends

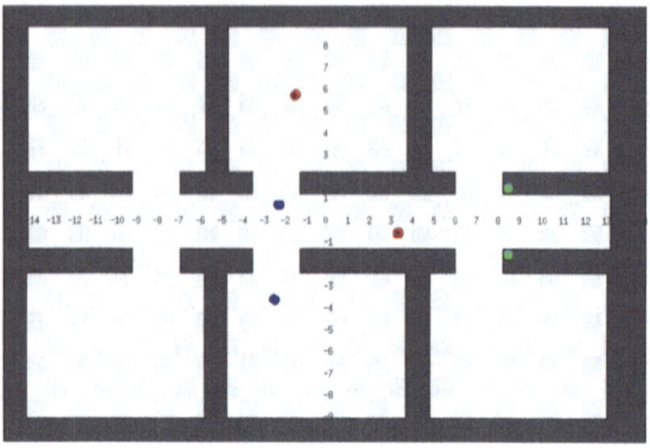

Fig. 3 A simple environment with RDA exploration partially completed. The robots (red and blue) are moving right to left, and have dropped two beacons (green) so far

4 Simulation Results

We have conducted experiments in Player/Stage and ROS/Stage, using the same robot models and movement/sensor attributes, in order to test the viability of our algorithms. The testing environments, shown in Figs. 2 and 3, provide two very different types of spaces–one very open, with several large obstacles at varying intervals and in non-uniform shapes that leave wide open areas and potential loops, and one a series of rooms off a single long hallway. The robots start from the same location for each run in each environment. The robots choose their directions with some randomness to avoid hitting each other, especially at the start of each run, so results are averaged over 15 runs of each environment and number of robots.

Figure 4 shows the rate of coverage for the RDA and SEA algorithms in the cave environment. While in most cases the simulations show fairly steady increase in the percentage of coverage, the RDA with 5 robots does show a plateau, due to the fact that the robots initially disperse in all directions, and then must wait for others to leap-frog out to the frontier. The SEA algorithm does not have this plateau because the robots explore along only a single path at a time. The experiments using SEA start with a higher initial coverage, due to the fact that the algorithms use different initial dispersion methods, and using the exact same cluster at the start led to many robot collisions. Accounting for that difference, and comparing times from

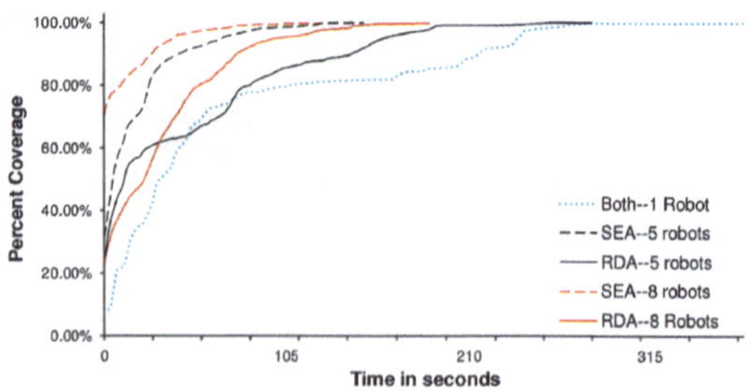

Fig. 4 Average time to full exploration for simulations using 1, 5 and 8 robots with each algorithm in the cave environment simulation

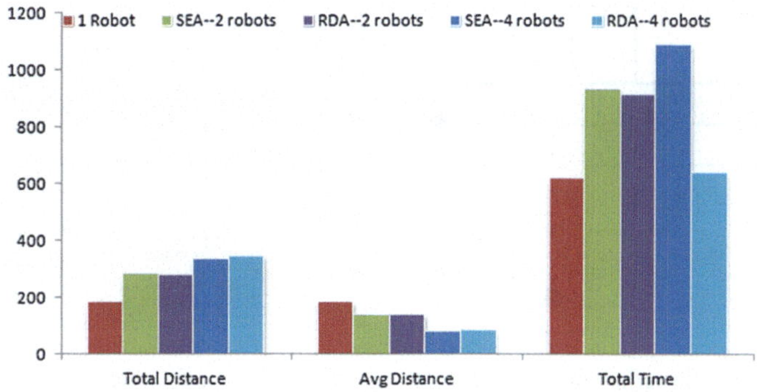

Fig. 5 Total distance, average distance per robot, and total time for simulations using 1, 2, and 4 robots with each algorithm in the office environment

the same starting percentage to full coverage, SEA is 1.35 times faster at achieving full coverage than RDA. SEA outperforms RDA in environments where there are long paths, because SEA fully explores only one path at a time, while RDA will attempt to explore many paths simultaneously, but will run into delays when robots have to be pulled from other paths to fully explore long paths.

In the office building environment, with its long corridor and single rooms off of it, both methods performed the same for 1 and 2 robots (shown in Fig. 5), because with only two robots, they are constantly leap-frogging around each other to continue exploring. However, with 4 robots RDA took significantly less time, even though the robots traveled approximately the same total distance as the 4 robots running SEA. While the robots using RDA could explore in multiple directions simultaneously, the robots using SEA had wait for each individual path and room to be cleared, and then had to move back to the corridor one at a time.

For our experiments, we used the Scribbler robots with the IPRE Fluke [13] attachment. We made use of the camera to read beacons and IR sensors to locate intersections. Due to the limitations of the robots, only SEA could be implemented. However, an advantage to using the Scribblers in testing is that we will be able to run experiments with 10–30 robots in the future. The Clearpath Jackal [6] would be a more appropriate robot for search and rescue operations.

5 Experimental Results

Our experimental environment has a more structured lay-out, similar to a house or apartment floor-plan, with small doorways and multiple rooms and hallways, shown in Fig. 6. We ran the experiments with 1, 2 and 3 robots, which was the maximum number of robots that could maneuver given the size of the environment. We measured the time to complete the exploration, number of messages sent, and distance traveled

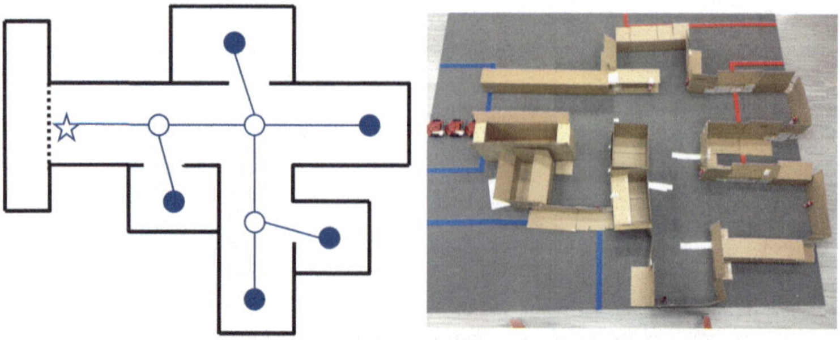

Fig. 6 (left) Overlay graph of experimental set-up (dotted line denotes start area). The star is the first point in the environment, and the filled circles mark dead-ends. (right) Experimental set-up with robots in start and cardboard walls

(see Fig. 7). SEA does not perform well in this environment due to the restrictions on movement. The short paths mean that by the time additional robots arrive, the first has completed the branch and moved onward, leading to significant overlap in areas covered.

6 Conclusion

We have presented two distributed algorithms for multi-robot exploration of unknown environments. Both algorithms make use of signal intensity to direct the movement of the robots, and beacons are used to mark explored areas in the environment in addition to creating a trail to the entrance and other points of interest within the environment. We have provided formal proofs that each of the algorithms will allow

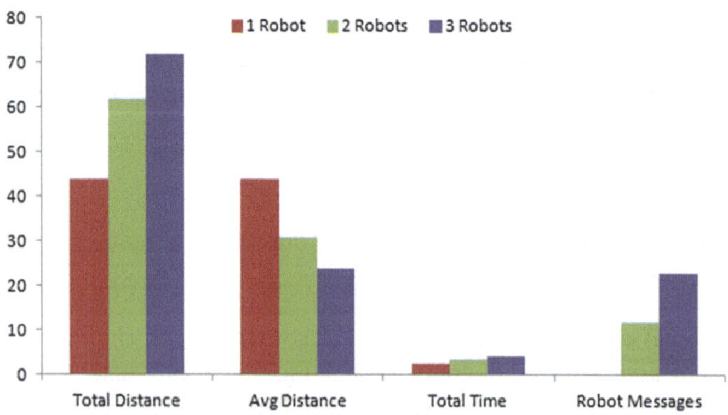

Fig. 7 Experimental results runs with 1, 2, or 3 robots using SEA

a team of robots to fully explore the environment, so long as at least one member of the robot team is still functional at the end of the exploration.

In future work, we plan to test the algorithms in other types of environments, including larger ones with many loops, open areas that create the potential for loops, and varying path lengths. We will also include human interaction during the exploration and rescue stages.

References

1. Batalin, M.A., Sukhatme, G.S.: The design and analysis of an efficient local algorithm for coverage and exploration based on sensor network deployment. IEEE Trans. Robot. **23**(4), 661–675 (2007). https://doi.org/10.1109/TRO.2007.903809
2. Birk, A., Carpin, S.: Rescue robotics - a crucial milestone on the road to autonomous systems. Adv. Robot. **20**(5), 595–605 (2006)
3. Brass, P., Cabrera-Mora, F., Gasparri, A., Xiao, J.: Multirobot tree and graph exploration. Trans. Robot. **27**(4), 707–717 (2011). https://doi.org/10.1109/TRO.2011.2121170
4. Burgard, W., Moors, M., Stachniss, C., Schneider, F.E.: Coordinated multi-robot exploration. IEEE Trans. Robot. **21**(3), 376–386 (2005). https://doi.org/10.1109/TRO.2004.839232
5. Choset, H.: Coverage for robotics – a survey of recent results. Ann. Math. Artif. Intell. **31**, 113–126 (2001). https://doi.org/10.1023/A:1016639210559
6. ClearpathRobotics: Jackal ugv. Technical report (2015)
7. Cortes, J., Martinez, S., Karatas, T., Bullo, F.: Coverage control for mobile sensing networks. IEEE Trans. Robot. Autom. **20**(2), 243–255 (2004). https://doi.org/10.1109/TRA.2004.824698
8. Dirafzoon, A., Emrani, S., Salehizadeh, S.M.A., Menhaj, M.B.: Coverage control in unknown environments using neural networks. Artif. Intell. Rev. 237–255 (2012)
9. Fazli, P., Davoodi, A., Pasquier, P., Mackworth, A.K.: Complete and robust cooperative robot area coverage with limited range. In: IEEE/RSJ Int'l Conference on Intelligent Robots and Systems (IROS), pp. 5577–5582. IEEE (2010)
10. Gage, D.W.: Command control for many-robot systems. In: 19th Annual AUVS Technical Symposium, pp. 22–24 (1992)
11. Galceran, E., Carreras, M.: A survey on coverage path planning for robotics. Robot. Auton. Syst. **61**(12), 1258–1276 (2013)
12. Howard, A., Mataric, M.J., Sukhatme, G.S.: Mobile sensor network deployment using potential fields: A distributed, scalable solution to the area coverage problem. In: Proceedings of the International Symposium on Distributed Autonomous Robotic Systems, pp. 299–308 (2002)
13. IPRE: Myro hardware. Technical report (2011)
14. Jensen, E.A., Gini, M.: Rolling dispersion for robot teams. In: Proceedings Int'l Joint Conference on Artificial Intelligence (IJCAI), pp. 2473–2479 (2013)
15. Khawaldah, M.A., Nuchter, A.: Enhanced frontier-based exploration for indoor environment with multiple robots. Adv. Robot. **29**(10), 657–669 (2015)
16. Koenig, S., Liu, Y.: Terrain coverage with ant robots: a simulation study. In: Proceedings Fifth Int'l Conference on Autonomous Agents, pp. 600–607. ACM, New York, NY, USA (2001). http://doi.acm.org/10.1145/375735.376463
17. Kurazume, R., Hirose, S.: An experimental study of a cooperative positioning system. Auton. Robot. **00**, 43–52 (2000)
18. Lee, S.K., Fekete, S.P., McLurkin, J.: Structured triangulation in multi-robot systems: Coverage, patrolling, voronoi partitions, and geodesic centers. Int. J. Robot. Res. **35**(10), 1234–1260 (2016)

19. Liu, B., Brass, P., Dousse, O., Nain, P., Towsley, D.: Mobility improves coverage of sensor networks. In: MobiHoc '05: Proceedings 6th ACM Int'l Symposium on Mobile ad hoc Networking and Computing, pp. 300–308. ACM, New York, NY, USA (2005). http://doi.acm.org/10.1145/1062689.1062728
20. Ludwig, L., Gini, M.: Robotic swarm dispersion using wireless intensity signals. In: Proceedings Int'l Symposium on Distributed Autonomous Robotic Systems (DARS), pp. 135–144 (2006)
21. Ma, M., Yang, Y.: Adaptive triangular deployment algorithm for unattended mobile sensor networks. IEEE Trans. Comput. **56**(7), 946–958 (2007). https://doi.org/10.1109/TC.2007.1054
22. Mamei, M., Zambonelli, F.: Pervasive pheromone-based interaction with rfid tags. ACM Trans. Auton. Adapt. Syst. **2**(2), 4 (2007). https://doi.org/10.1145/1242060.1242061
23. McLurkin, J., Smith, J.: Distributed algorithms for dispersion in indoor environments using a swarm of autonomous mobile robots. In: Proceedings Int'l Symposium on Distributed Autonomous Robotic Systems (DARS) (2004)
24. Murphy, R.R.: Disaster Robotics. The MIT Press (2014)
25. O'Hara, K.J., Walker, D.B., Balch, T.R.: Physical path planning using a pervasive embedded network. IEEE Trans. Robot. **24**(3), 741–746 (2008). https://doi.org/10.1109/TRO.2008.919303
26. Rekleitis, I., Dudek, G., Milios, E.: Multi-robot exploration of an unknown environment, efficiently reducing the odometry error. In: Proceedings Int'l Joint Conference on Artificial Intelligence (IJCAI), vol. 2, pp. 1340–1345. Morgan Kaufmann Publishers, Inc., Nagoya, Japan (1997)
27. Rooker, M.N., Birk, A.: Multi-robot exploration under the constraints of wireless networking. Control Eng. Pract. **15**(4), 435–445 (2007)
28. Stachniss, C., Burgard, W.: Exploring unknown environments with mobile robots using coverage maps. In: Proceedings Int'l Joint Conference on Artificial Intelligence (IJCAI) (2003)
29. Stump, E., Jadbabaie, A., Kumar, V.: Connectivity management in mobile robot teams. In: Proceedings IEEE Int'l Conference on Robotics and Automation (ICRA), pp. 1525–1530 (2008). https://doi.org/10.1109/ROBOT.2008.4543418
30. Viet, H.H., Dang, V.H., Choi, S.: Bob: an online coverage approach for multi-robot systems. Appl. Intell. **42**(2), 157–173 (2015)
31. Wurm, K.M., Stachniss, C., Burgard, W.: Coordinated multi-robot exploration using a segmentation of the environment. In: Proceedings IEEE/RSJ Int'l Conference on Intelligent Robots and Systems, pp. 1160–1165 (2008). https://doi.org/10.1109/IROS.2008.4650734

From Ants to Birds: A Novel Bio-Inspired Approach to Online Area Coverage

Luca Giuggioli, Idan Arye, Alexandro Heiblum Robles and Gal A. Kaminka

Abstract Online coverage path planning is a canonical multi-robot task, where the objective is to minimize the time it takes for robots to visit every point in an unknown area. Two general major approaches have been explored in the literature: a stigmergic approach, inspired by *ant behavior*, relies on active marking of the environment. In contrast, the *collaborative approach* relies instead on localization, memory of positions, and global communications. In this paper, we report on a new approach, inspired by territorial *bird chirping*, which borrows from both previous approaches: it relies on localization and memory, but not on global communications. We provide a detailed analytic and empirical evaluation of this model.

1 Introduction and Background

Coverage path planning is a canonical robotics task, with many applications such as environmental monitoring, surveillance, exploration, and search [5]. In *online coverage*, one or more robots is to move inside an unknown target area, such that every point in the area is visited by one or more robots, often with the secondary objective of minimizing the time for such full coverage [8]. For multiple robots, two major approaches emerge for online coverage: A *stigmergic* [25] *approach*, which relies

L. Giuggioli (✉)
Bristol Centre for Complexity Sciences, Department of Engineering Mathematics and School of Biological Sciences, University of Bristol, Bristol BS81TH, UK
e-mail: luca.giuggioli@bristol.ac.uk

I. Arye · G. A. Kaminka
The MAVERICK Group, Computer Science Department, Bar Ilan University, 5290002 Ramat Gan, Israel
e-mail: idanarye@gmail.com

G. A. Kaminka
e-mail: galk@cs.biu.ac.il

A. Heiblum Robles
Department of Engineering Mathematics, University of Bristol, Bristol BS81TH, UK
e-mail: alex.heiblum@bristol.ac.uk

R. Groß et al. (eds.), *Distributed Autonomous Robotic Systems*,
Springer Proceedings in Advanced Robotics 6,
https://doi.org/10.1007/978-3-319-73008-0_3

on environmental marking by the robots to direct their motion towards uncovered territory, and a *collaborative approach*, which relies on global communications to have robots explicitly—and remotely—coordinate their actions.

The stigmergic approach is often thought to be inspired by *ants*, though it is used by some mammals as well [7], e.g., foxes [20]. Here, robots mark visited points as they move around, simultaneously reading previously left markings. Robots move away from points marked by others [10, 17, 27]. This causes them to divide the environment into territories, each maintained by a single robot. Clear benefits of this approach include simplicity of the control algorithm (a random walk), and the fact there is no need for localization or memory of markings; robots use the markings themselves to identify locations visited by themselves or others. Unfortunately, coverage is often redundant, and relies strongly on the duration of the markers existence; moreover, building robots with actual marker reading and writing mechanisms is quite difficult in practice [10].

The collaborative approach is often associated with artificial methods (though it could just as easily be inspired by human teamwork). Here robots communicate with each other (in most studies, regardless of their distance), and divide up the area between them, e.g., [1, 11, 13]. The coverage time can be minimized using such collaborative algorithms. However, this approach requires not only localization (to identify current position) and navigation (to move to agreed-upon new locations), but also memory (to store visited locations and future positions), communications that allow task assignment, and most importantly, a shared coordinate system that allows a shared understanding of the regions to be divided between the robots. Satisfying these requirements in practice is again difficult. While memory and localization can perhaps be fairly easily had (some commercial vacuum cleaners now employ SLAM), establishing global communications, and a shared coordinate system in an unknown area, with unknown initial poses, is a difficult challenge [18, 24].

There have been surprisingly few attempts at addressing these challenges. Rekleities et al. have worked on coverage with range-limited communications [21], yet still rely on a shared coordinate systems. Batalin et al. have replaced the need for communications and localization with the need to sense others remotely, distinguishing robots from other objects in the environment [2]. Rutishauser et al. examine collaborative algorithms that display graceful degradation (to random motions) when positional, sensory, and communication failures accumulate [22], and are therefore less reliant on explicit collaboration. Durham et al. have discussed a related approach to ours, for offline coverage, whereby robots that meet exchange information by pairwise gossip communications, to statically partition a known an area between them [4]. In contrast, our online coverage approach only requires robots to detect each other, but no information exchange is necessary.

In this paper we propose a novel online coverage approach that is inspired by the territorial behaviour of higher organisms in particular certain species of birds [12] and mammals [23]. The key idea is that when two robots meet, they detect each other (in birds, this is done by chirping a challenge which is then countered), then remember the location of the encounter and treat it as a border landmark. Robots using this approach are assumed to have localization and memory, but do not need global communications or a shared coordinate system. Moreover, they utilize the simple-to-implement random walk algorithm for their motion. We provide a comprehensive mathematical and empirical analysis of this approach. Specifically, we analyze its characteristics and determine its efficiency in terms of coverage time as a function of its control parameters.

2 The Memory-Based Territorial Exclusion Model

We are given a set of N identical circular robots of radius r with a radially uniform detection distance $d \geq r$ that move in continuous space with speed of magnitude $|\overrightarrow{v}|$ within an arena of size A with periodic boundary conditions (toroidal domain). Every time a robot detects another, they both remember the location (in their own coordinate system), and move away from it. They mark the location in memory, and then use this to avoid the location if they run into it again. Thus upon detecting another robot or remembering a mark, the robots turn away.

Algorithm 1 SINGLE- STEP

Require: Current position $\overrightarrow{p}$
1: Randomly choose λ, ϑ
2: $\overrightarrow{\ell} \leftarrow$RANDOMWALK($\lambda$, ϑ)
3: **if** A robot is detected in position $\overrightarrow{b}$ within d
OR
Remembered location $\overrightarrow{b}$ is within d **then**
4: $\quad \overrightarrow{\ell} \leftarrow \overrightarrow{b}$
5: Move to new position $\overrightarrow{\ell}$
6: If robot or mark detected, REPEL.

The position of each robot is updated following Algorithm 1, which controls a single motion step of size d at most. First (line 2), a new motion vector $\overrightarrow{\ell}$ is generated by calling a correlated random walk procedure with parameters λ, ϑ described below (Sect. 2.1). Then, the robot considers whether another robot is detected along the motion vector, within the detection distance d (line 3). If so, the robot remembers the location (line 4) and sets a revised, shortened motion vector to it, $\overrightarrow{b}$, reaching the detected robot or mark by not moving beyond them. Alternatively, if a previously marked location is retrieved from memory, it is used to set $\overrightarrow{\ell}$ (line 5–6). The robot then moves along $\overrightarrow{\ell}$. If it encounters a robot or remembers a mark in the new location, it repels towards the opposite direction, following the exclusion rule described in Sect. 2.2.

2.1 Selecting a Motion Vector

The robots possess a degree of persistence in their motion and move as correlated random walkers (see e,g. [3]). At one extreme each new step is uncorrelated with the direction of the previous step and their movement is random. At the other extreme the direction of movement is always identical to the one at the preceding step and their movement is ballistic. At each time step, each robot randomly selects a step size λ and an angle ϑ. The step size is sampled from an exponential distribution of mean size equal to half the size of the robot diameter. The new angular direction is selected relative to the previous movement direction. Except at the start of the simulations when the choice of angle is completely random, each robot selects from a distribution of turning angles $f(\theta)$. This distribution is a (symmetric) wrapped Cauchy distribution, that is a Cauchy distribution $\mathscr{C}(x) = \rho\left[\pi(\rho^2 + x^2)\right]^{-1}$, which is wrapped around the origin. For values between $-\pi \leq \theta \leq \pi$ [16]. The parameter ρ is called the concentration parameter and represents the mean cosine of the distribution $\langle\cos(\theta)\rangle = \int_{-\pi}^{\pi} \cos(\theta) f(\theta)\, d\theta = \rho$. In other words it indicates the 'strength' by which a robot would go forward at each step. Integration and inversion of $f(\theta)$ allows to sample from a uniform distribution U between 0 and 1 and obtain θ angles distributed according to $f(\theta)$ via $\theta = 2 \arctan\left\{\frac{1-\rho}{1+\rho} \tan\left[\pi\left(U - \frac{1}{2}\right)\right]\right\}$.

In the limit $\rho \to 0$ the distribution $f(\theta)$ reduces to a uniform one between $-\pi$ and π, while in the limit $\rho \to 1$ the distribution $f(\theta)$ reduces to a Dirac delta distribution. In the former case, sampling turning angles from a uniform distribution means that the robot moves at random, whereas in the latter limit the robot moves ballistically, i.e. always going straight (except upon encountering others). For intermediate values, the robot moves as a so-called correlated random walker with the degree of persistence determined by ρ. For computational simplicity the random step length for which a robot is initially selected to move is rounded down to its first integer. The call to Algorithm RANDOMWALK represents the computation of a new motion vector based on the sampled parameters λ, ϑ.

2.2 Repelling Away from a Collision or a Remembered Mark

Two robots detect each other whenever their distance becomes equal to d. Upon detection the two robots will mark the positions where such detection has occurred and retreat from each other. The retreat represents an exclusion interaction at the moment of detection, but the detection position is also a virtual mark is remembered by the robots for the duration T (called memory time). Neither of the two robots will cross the virtual mark. As each robot may interact with any neighbour that comes within a detection distance, the number of active marks that a robot has with any other members of the population fluctuates over time and depends on the various parameters of the model.

Consider a robot R_1 executing Algorithm 1. Assume that it is in collision course with another robot, robot R_j, which is considered static initially. At the predicted location of collision, robot R_1 marks the terrain at location $\vec{b}$. Robot R_1's location now gets updated by accounting for the presence of the virtual mark at $\vec{b}$ with robot j fixed. We assume symmetrical detection, i.e., Robot R_j now also remember a virtual mark at $\vec{b}$, though in its own coordinate system.

The geometry of the collision is as follows (see Fig. 1). After the mark location at $\vec{b}$ is established, robot 1 moves up to $\vec{a}$ when it collides with the virtual mark at position $\vec{b}$. The mark now acts as a virtual barrier since robot 1's velocity gets reflected as if a barrier tangent to the robot's detection circle was present at $\vec{b}$. Formally the reflected velocity vector $\vec{v}\,'$ changes from the old velocity $\vec{v}$ through the relation

$$\vec{v}\,' = \vec{v} - 2\left\langle \vec{v}, \widehat{\vec{a}\,\vec{b}} \right\rangle \cdot \widehat{\vec{a}\,\vec{b}}, \tag{1}$$

where $\widehat{\vec{a}\,\vec{b}}$ is the normalized vector from $\vec{a}$ to $\vec{b}$ and $\langle \vec{z}, \vec{z}\,' \rangle$ represents the scalar product between $\vec{z}$ and $\vec{z}\,'$, that is the projection of $\vec{z}$ along $\vec{z}\,'$. As the reflection of the velocity is performed only when the robot is moving towards the virtual barrier, the exclusion rule applies only when $\left\langle \vec{v}, \widehat{\vec{a}\,\vec{b}} \right\rangle > 0$. All of this computation is carried out by the REPEL algorithm.

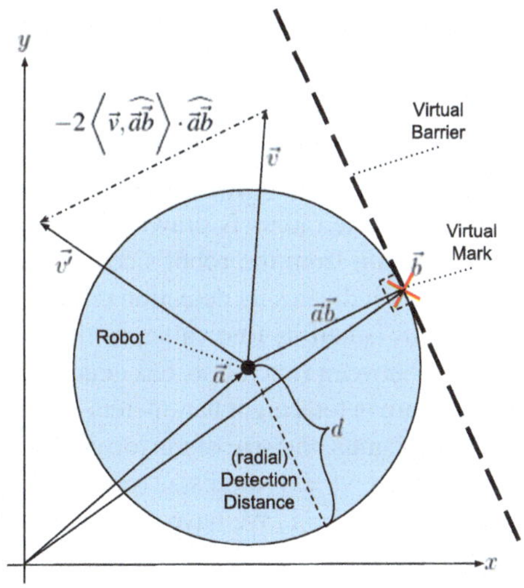

Fig. 1 Representation of the robot exclusion (repel) rule whereby one agent changes its direction upon encountering a virtual barrier

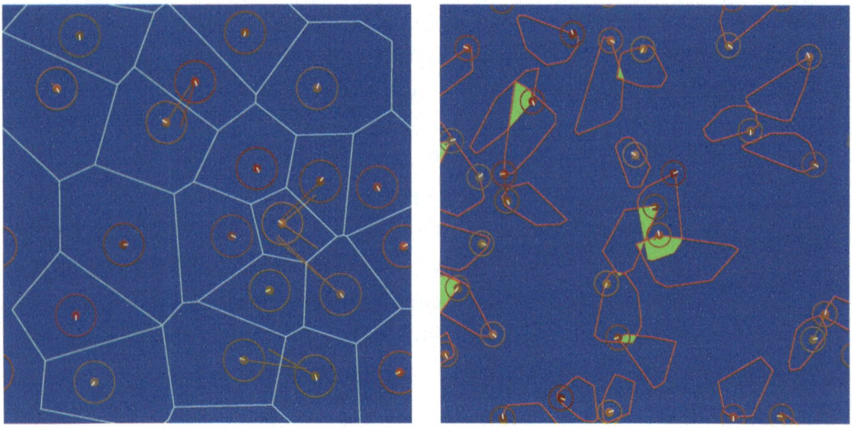

(a) A snap-shot of the simulation with Voronoi tessellation displayed.

(b) A snap-shot of the simulation with a visualization of the home range size of each robot.

Fig. 2 Visualization of the simulator with Voronoi partitioning in panel (**a**) and home range size in panel (**b**). The circles around each robot represent half the size of the robot's detection distance, while the long lines centered on a robot indicate the locations of the virtual marks. The direction of motion of each robot is shown by a small radial vector pointing outward from each robot. Although home ranges and Voronoi tessellation are correctly computed also for the agents that are close to the 'edges' of the toroidal domain, they are not visualized here. The home range size are computed with a minimum convex polygon estimate over 50 time steps. Overlaps between the estimated home ranges are colored in green

3 Simulating the Behavior of Robotic Birds

The simulator was developed in Java using the MASON simulation framework (https://cs.gmu.edu/~eclab/projects/mason/). The simulator uses MASON's visualization facilities to draw the moving robots. It portraits each robot as a small disk of radius r. The radius is drawn in a different color to indicate the current movement direction from the robot's centre to the radius tip. A larger circle around the robot's disk indicates its detection circle. For ease of visualization the detection circle's radius is half its true value, so that the touching of two circles indicates that a collision between two robots has occurred.

There are in total eight parameters in the model. Those that we have kept fixed are, in arbitrary units, the size of the toroidal arena $A = 100 \times 100$, the magnitude of the robots' speed $|\vec{v}| = 1$, their size (robot radius $r = 1$) and the mean of the distribution of step length $\lambda = 1$. We have changed the remaining four parameters consisting of the number of robots $1 \leq N \leq 100$, the detection distance $2 \leq d \leq 30$ (arb. units), the memory time $0 \leq T \leq 100$ (arb. units), and the random walk persistence $0 \leq \rho \leq 1$.

Additionally, the simulation draws lines from each robot to the virtual marks that were generated when other robots were detected. These lines remain visible for an amount of time T and show the mark locations where the robot may collide with.

To get a perception of the global patterns we display at each time step the Voronoi tessellation [19] of the arena and the size and locations of the boundaries of the so-called agent home range (see e.g. [6]) that gives an indication of where a robot has been over a prescribed amount of time. Figure 2 shows the Voronoi partitioning and the outer boundaries of the home ranges measured by computing the minimum convex polygon (MCP) (see e.g. [28]) of a robot's localizations. Although other more sophisticated procedures exist (see e.g. [14]), MCP is sufficient for our purpose.

To mimic experimental conditions where no global clock exists that causes all the elements of the system to update their state at the same time [9], we use the following asynchronous update scheme. At the start of a simulation the N robots are numbered. At each time step robots are displaced sequentially from 1 to N. Within one simulation step, say robot R_1 is selected, then its position is updated considering all other robots as static. When R_2 is then selected, R_1's updated position is accounted for in moving R_2, and so on with all other robots. A discrete simulation step is considered elapsed once all the N robot positions have been updated. This sequential scheme has also the advantage of reducing the computational costs of dealing with multi-agent detections if synchronous updating were to be used.

3.1 Measured Outputs

For each parameter set of the model the measured outputs are obtained by running 1000 simulations starting from the same initial condition (the number of simulations for each figure is specified in the caption). The outputs that aim at giving an indication of the degree of spatial heterogeneity of the system are of two kinds. One kind is instantaneous and is obtained by observing the locations of each robot at a given time and averaging over simulation runs. The instantaneous measurements include the size of each Voronoi cell, the distance with respect to Voronoi neighbours, the number of active virtual marks both within each Voronoi cell and within Voronoi neighbours. The time-integrated measurements require accumulation of the data over each simulation run for a certain period of time and include the robot home range size, the spatial overlap and the coverage time. Outputs of the simulations are represented via the values of the mean, the standard deviation and the coefficient of variation (CV), that is the ratio of the standard deviation divided by the mean.

Whenever a simulation run starts, the initial directions of the robots are randomized and the robots are placed in an hexagonal pattern. To do so it calculates the maximal circle radius for packing N circle in the toroidal arena of size A, and chooses initial placements for the robots as if they were circles of that radius. After the initial placement, to 'thermalize' the initial configuration the simulation is run for a burning time $t_b = 100\,T$.

4 Results

Given our interest in proposing a new coverage algorithm, we have focused on the analysis of the memory-based territorial exclusion model rather than on a comparison between the various algorithms employing stigmergic or collaborative approaches. We do so by characterizing the heterogeneity of the spatial arrangement of the robots and computing coverage times.

The degree of spatial heterogeneity of the emerging spatial segregation patterns is obtained by studying the variability of the Voronoi partitioning. In Fig. 3 we display CV of the Voronoi tile size at time t_b as a function of detection radius and memory time. The CV of Voronoi partitioning is a measure of the relative strength of the fluctuations in the Voronoi cells among each robot and is mainly dependent on the detection distance d. For a given T the robots are allowed to wander more throughout space the smaller is d. Voronoi tiles with varying shape and size appear more frequently the smaller is the detection distance as indicated by the increase in CV while d is decreased in panel (a) and (b). The dependence of the spatial heterogeneity of the robot position on T for a fixed d is in general very minor, as shown in panel (a) with 100 robots. However, this is not the case anymore when one uses a small number of robots for which a decrease in T reduces the movement constraints and allows for more variability between the positions of the robots (and the resulting Voronoi tiles) as displayed in the top right part of panel (b). The associated heat maps with $(d, T) = (3, 15)$ and $(d, T) = (8, 15)$ in panels (a) and (b), which represent the spatial occupation probability where a single agent have been, also gives some indication of a wider range and variability in the sizes of the Voronoi cells when a smaller number of robots is being deployed.

A way to measure the efficiency of the swarm algorithm is to estimate the coverage time (CT) of the domain A for different choice of detection distance d and number

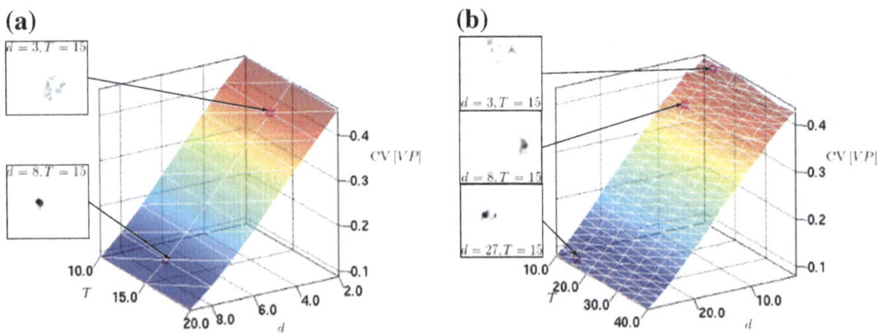

Fig. 3 Coefficient of variation of Voronoi cell size as a function of the memory time and detection distance for 100 robots in panel **(a)** and 10 robots animals in panel **(b)**. Variability is evaluated through 1000 simulation runs for each parameter combination. Heat maps of one simulation run that indicate where one robot (selected at random) has been over 1000 time steps are also plotted for different combinations of values of d and T. The concentration parameter ρ is set to 0

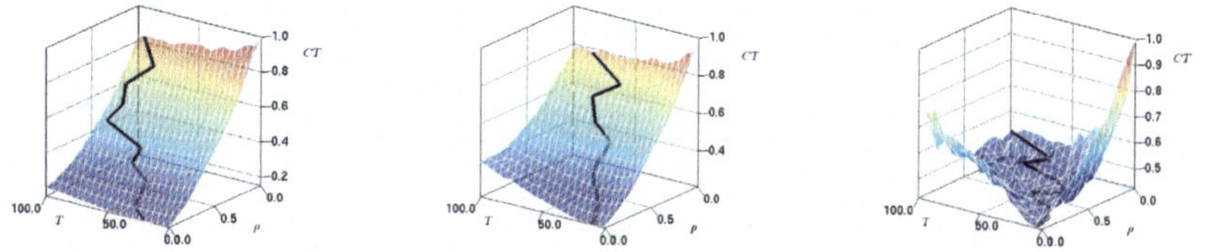

(a) Simulations with $d = 13$ and $N = 17$ corresponding to a packing fraction $\eta = 0.23$

(b) Simulations with $d = 15$ and $N = 20$ corresponding to a packing fraction $\eta = 0.35$.

(c) Simulations with $d = 17$ and $N = 23$ corresponding to a packing fraction $\eta = 0.5$.

Fig. 4 Coverage time as a function of the concentration parameter ρ and memory time T for different choices of packing fraction η averaged over 1000 simulations. The black line in each panel tracks the minimum as a function of T for each different value of ρ. In computing the coverage time a grid of 10^4 rectangles is superimposed on the spatially continuous domain contained in A. Whenever the centroid of a robot is within a distance d from any point of a rectangle, that rectangle is considered covered. Once all rectangles have been covered the simulation runs are halted and the number of time steps are recorded

of robots N as a function of the concentration parameter ρ and the memory time T. We do so in Fig. 4 and for a better appreciation of the robot density we actually use the packing fraction η in place of N. As the robot collision distance is $d/2$ we define $\eta = \pi\, N(d/2)^2/A$, i.e. as the ratio between the maximum total area robots may occupy as they collide with each other, $\pi\, N(d/2)^2$, and the domain size A. For low packing fraction we notice a general decrease of coverage time towards higher persistent walk and lower memory time. We also notice that for a given fixed concentration ρ, the value of memory time T that minimizes the coverage time, drawn as a continuous black curve in all three panels, is neither random nor ballistic, but intermediate between the two, with larger values the smaller the memory time. Eventually for higher packing fraction, the coverage time develops a region in the $T - \rho$ parameter domain where minimization of the CT is possible. The transition to the appearance of a region of global minimization of the coverage time is smooth and is noticeable when $\eta \gtrsim 0.49$.

It is also of interest to know the dependence of the coverage time for a *given* memory time T as a function of ρ and the number of robots, which can also be evinced from Fig. 4. As shown in panel (a) and (b), when packing fraction is sufficiently low, the minimum coverage time is obtained with a ballistic walk. To show clearly this effect we plot in Fig. 5 the coverage time as a function N for different concentration ρ and we compare to the 'perfect' coverage algorithm whereby each location is visited only once by one robot and based on an initial configuration of the robots that gives the lowest possible coverage time. Considering the torus we construct the perfect algorithm by computing $\mathrm{CT} = \sqrt{A + (2d)^2}\sqrt{A}/(2dN) - 2d$ where $\sqrt{A + (2d)^2}$ is the length that a robot needs to travel to wrap around the torus' edge before its original place and $\sqrt{A}/(2d)$ is the number of required trips to go to cover the available area for each robot. The final subtraction is included because robots stop before reaching their original location.

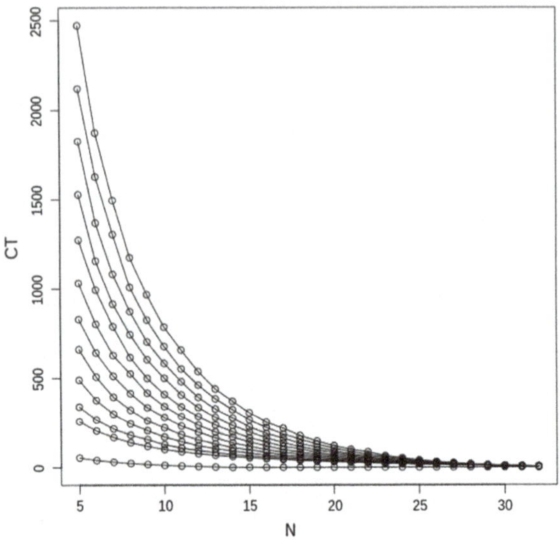

Fig. 5 Coverage time as a function of the number of robots averaged over 1000 simulations. From top to bottom the first ten curves have been drawn with values of ρ increasing by 0.1 starting from the random case at $\rho = 0$ to the ballistic case at $\rho = 1$. The detection distance is $d = 15$ and the memory time is $T = 20$. The bottom curve is the perfect algorithm. Due to optimal initial placement the perfect algorithm possess a zero coverage time beyond 14 robots

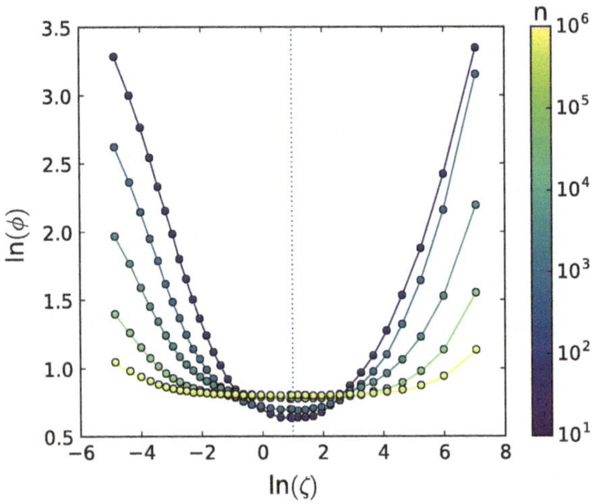

Fig. 6 Dimensionless coverage time as a function of the inverse normalized persistence parameter ζ for a single robot in a circular arena averaged over 1000 simulations following the same movement and interaction (reflection) rules as those in the swarm. Random movement corresponds to $\zeta = 0$ ($\rho = 0$) and ballistic movement corresponds to $\zeta = \infty$ ($\rho = 1$). The circular arena and the speed of the agents were set at 1. From top to bottom the curves correspond to a choice of a square grid superimposed on the arena, respectively, of size L =1000, 316, 100, 31, and 10

To understand better the role of the walk persistence and spatial constraints due to the collisions with other robots, in Fig. 6 we plot CT of a single robot with zero detection distance and zero radius in a circular geometry of radius R. The robot follows the movement rules as in the swarm and reflects the normal component of the velocity vector by colliding with the circular confining wall in the same way robots in the swarm get reflected upon encountering a virtual barrier. We study how the coverage time changes as a function of the robot directional persistence. We do so by plotting CT versus the dimensionless ratio $\zeta = -\lambda/[R\ln(\rho)]$, which represents an effective persistence parameter being the ratio between $-\lambda/\ln(\rho)$, the average distance a correlated random walker would move without turning [26], and the size of the confining domain R.

To measure the coverage time we partition the circular arena with a rectangular grid of $L \times L$ cells, and only the n cells whose centers are inside the arena are taken into account. The coverage time $C(n)$ is then defined as the time that takes a walker to visit all these n sites. We make $C(n)$ dimensionless by considering the ratio $\phi = C(n)/\tau_n$, where τ_n is the time it takes a random walker with no correlation ($\rho = 0$) to cover n distinct sites in open space.

5 Discussions

We have proposed a bio-inspired distributed spatial coverage algorithm that does not require robots to deposit 'marks' on the terrain. Rather than exploiting the stigmergic nature of scent-mediated territorial exclusion, we mimic a form of cognitive territorial behaviour whereby animals remember the locations where they exchange visual or auditory signals with neighbouring individuals to make each others' presence conspicuous. A robotic implementation of this behaviour consists of making each agent consider, for a finite amount of time, the locations of direct collision with or proximity to other robots as virtual barriers that cannot be encroached. This form of interaction generates a dynamical segregation that reduces spatial oversampling between the robots.

Two extreme regimes of spatial heterogeneity of the swarm are present depending on the number and detection distance of the robots. At one extreme, when packing fraction is sufficiently small, robots wander over all space with rare encounter occurrences and the system resembles a fluid-like material with homogeneous mixing. At the other extreme, when packing fraction is relatively high, robots are nearly jammed and the system may be highly heterogeneous. In some areas robots may move very little around their initial placements, while in others robots move very little around their initial placement.s

While the analysis of a single robot moving within a confined arena showed that an intermediate degree of correlation minimizes coverage time, we found no evidence of such minimization in our robotic swarm except by increasing packing fraction (black line Fig. 4). For lower packing coverage minimization was achieved instead when robots moved ballistically with increasingly poorer performance as persistence was reduced (Fig. 5). We ascribe this difference to the fact that virtual barriers, when robot encounters are rare, do provide a degree of confinement but only partially. Although persistence reduces spatial oversampling in a single robot, it also diminishes the chance to revisit or return towards the area where a barrier was created thus preventing the robots to generate collectively long-lasting space partitioning.

As robot confinement is only partial, a natural measure to determine the size of the confining domains is necessary. Home range estimates, being based on an arbitrary integration time, do not offer a proper representation of the spatial confinement. As a consequence it is not evident how to estimate the ζ parameter regime in which robots operate. To explain the shift in optimal coverage between the ballistic and the correlated regime between low and high packing fractions future work should address the lack of a proper tool to estimate the size of the partial confinement.

Promising directions to improve the spatial coverage in the proposed algorithm include the choice of more informed alternatives for the movement paths of the robots, e.g. by avoiding recently visited locations, a variant of the so-called self-avoiding walk [15], or by systematic plowing of the emerging territory of each robot [1].

Finally, we would like to add that although we have left unexplored the impact of robot failures, the presence of obstacles and the confining geometry of a real arena, we do not expect qualitative changes in our findings because the movement response upon encountering an immobile robot, an obstacle or a reflecting wall would follow the same interaction mechanism that a robot undergoes when encountering a virtual mark. On the other hand, the same cannot be said about perception errors as these would affect the degree of confinement of the robots and their effects on spatial coverage would need to be studied. For an empirical test careful considerations should also be made on the robot sensing mechanisms and how malfunctioning of the detection or recording systems would change the efficiency of the algorithm.

Acknowledgements LG thanks the support of the Bar Ilan Robotics Consortium (BIRC) and the office of the vice president during his stay at Bar Ilan University and acknowledges discussions with Adham Sabra and Alan Winfield. The research was supported in part by ISF grant #1511/12 and EPSRC grant EP/I013717/1. As always, thanks to K. Ushi.
Data Access Statement The Java code to run the stochastic simulations is openly available in the data.bris University of Bristol repository under DOI: https://doi.org/10.5523/bris.i1rl4lk2boj410h6ui4cpblfh.

References

1. Agmon, N., Hazon, N., Kaminka, G.A.: The giving tree: constructing trees for efficient offline and online multi-robot coverage. Ann. Math. Artif. Intell. **52**(2–4), 143–168 (2008)
2. Batalin, M.A., Sukhatme, G.S.: Spreading out: a local approach to multi-robot coverage. In: Asama, H., Arai, T., Fukuda, T., Hasegawa, T. (eds.) Distributed Autonomous Robotic Systems 5, pp. 373–382. Springer, Tokyo (2002)
3. Codling, E.A., Planck, M.J., Benhamou, S.: Random walk models in biology. J. R. Soc. Interface **95**(5), 813–834 (2008)

4. Durham, J.W., Carli, R., Frasca, P., Bullo, F.: Discrete partitioning and coverage control for gossiping robots. IEEE Trans. Robot. **28**(2), 364–378 (2012)
5. Galceran, E., Carreras, M.: A survey on coverage path planning for robotics. Robot. Auton. Syst. **61**(12), 1258–1276 (2013). https://doi.org/10.1016/j.robot.2013.09.004
6. Giuggioli, L., Kenkre, V.M.: Consequences of animal interactions on their dynamics: emergence of home ranges and territoriality. Mov. Ecol. **2**, 1–20 (2014)
7. Giuggioli, L., Potts, J.R., Rubenstein, D.I., Levin, S.A.: Stigmergy, collective actions, and animal social spacing. Proc. Natl. Acad. Sci. USA **110**(42), 16904–16909 (2013)
8. Hazon, N., Kaminka, G.: On redundancy, efficiency, and robustness in coverage for multiple robots. Robot. Auton. Syst. **56**(12), 1102–1114 (2008)
9. Huberman, B.A., Glance, N.S.: Evolutionary games and computer simulations. Proc. Natl. Acad. Sci. **90**(16), 7716–7718 (1993)
10. Koenig, S., Liu, Y.: Terrain coverage with ant robots: a simulation study. In: Autonomous Agents, pp. 600–607. ACM (2001). https://doi.org/10.1145/375735.376463
11. Kong, C.S., Peng, N.A., Rekleitis, I.: Distributed coverage with multi-robot system. In: Proceedings of the 2006 IEEE International Conference on Robotics and Automation (2006)
12. Krebs, J.R.: Song and territory in the great tit parus major. In: Stonehouse, B., Perrins, C. (eds.) Evolutionary Ecology, pp. 47–62. Macmillan Education, London (1977)
13. Luo, C., Yang, S.X., Stacey, D.A.: Real-time path planning with deadlock avoidance of multiple cleaning robots. In: Proceedings IEEE International Conference on Robotics and Automation, 2003. ICRA '03, vol. 3, pp. 4080–4085 (2003). https://doi.org/10.1109/ROBOT.2003.1242224
14. Lyons, A.J., Turner, W.C., Getz, W.M.: Home range plus: a space-time characterization of movement over real landscapes. Mov. Ecol. **1**(1), 1 (2013)
15. Madras, N., Slade, G.: The self-avoiding walk. Springer, Berlin (2013)
16. Mardia, K.V., Jupp, P.E.: Directional Statistics. Wiley, Chichester (2000)
17. Menezes, R., Martins, F., Vieira, F.E., Silva, R., Braga, M.: A model for terrain coverage inspired by ant's alarm pheromones. In: Proceedings of the 2007 ACM Symposium on Applied Computing, SAC '07, pp. 728–732 (2007). https://doi.org/10.1145/1244002.1244164
18. Nagavalli, S., Lybarger, A., Luo, L., Chakraborty, N., Sycara, K.: Aligning coordinate frames in multi-robot systems with relative sensing information. In: Proceedings of the IEEE/RSJ International Conference on Intelligent Robots and Systems (IROS), pp. 388–395. IEEE (2014)
19. Okabe, A., Boots, B., Sugihara, K., Chiu, S.N.: Spatial Tessellations: Concepts and Applications of Voronoi Diagrams, 2nd edn. Wiley, Chichester (2000)
20. Potts, J.R., Harris, S., Giuggioli, L.: Quantifying behavioral changes in territorial animals caused by sudden population declines. Am. Nat. **182**(3), E73–82 (2013)
21. Rekleitis, I., New, A.P., Rankin, E.S., Choset, H.: Efficient boustrophedon multi-robot coverage: an algorithmic approach. Ann. Math. Artif. Intell. **52**(2), 109–142 (2008). https://doi.org/10.1007/s10472-009-9120-2
22. Rutishauser, S., Correll, N., Martinoli, A.: Collaborative coverage using a swarm of networked miniature robots. Robot. Auton. Syst. **57**(5), 517–525 (2009). https://doi.org/10.1016/j.robot.2008.10.023
23. Shettleworth, S.J.: Cognition, Evolution, and Behavior, 2nd edn. Oxford University Press, NewYork (2010)
24. Suzuki, I., Yamashita, M.: Agreement on a common x-y coordinate system by a group of mobile robots. In: In proceedings of the 1996 Dagstuhl Workshop on Intelligent Robots: Sensing, Modeling and Planning, pp. 305–321. World Scientific Press (1997)
25. Theraulaz, G., Bonabeau, E.: A brief history of stigmergy. Artif. Life **5**(2), 97–116 (1999)
26. Viswanathan, G.M., Raposo, E.P., Bartumeus, F., Catalan, J., da Luz, M.G.E.: Necessary criterion for distinguishing true superdiffusion from correlated random walk processes. Phys. Rev. E **72**(1), 011–111 (2005)
27. Wagner, I.A., Lindenbaum, M., Bruckstein, A.M.: Distributed covering by ant-robots using evaporating traces. IEEE Trans. Robot. Autom. **15**(5), 918–933 (1999). https://doi.org/10.1109/70.795795
28. White, G.C., Garrott, R.A.: Analysis of Wildlife Radio-Tracking Data. Academic Press, Sand Diego (1990)

Information Based Exploration with Panoramas and Angle Occupancy Grids

Daniel Mox, Anthony Cowley, M. Ani Hsieh and C. J. Taylor

Abstract In this work we present a multi-robot information based exploration strategy with the goal of constructing high resolution 3D maps. We use a Cauchy–Schwarz Quadratic Mutual Information (CSQMI) based objective which operates on a novel angle enhanced occupancy grid to guide robots in the collection of RGBD panoramas, which have been shown to provide memory efficient high quality representations of space. To intelligently collect panoramas, we introduce the angle enhanced occupancy grid which emphasizes perspective in addition to coverage, a characteristic we believe results in the construction of higher quality maps than traditional occupancy grid methods. To show this, we conduct simulations and compare our approach with frontier exploration. Using our angle enhanced occupancy grid, only 11.4% of decimeter wall segments were covered by fewer than 20 pixels as compared with 33.5% for the frontier method.

1 Introduction

A central pillar of robotics research is the development of efficient autonomous mapping and exploration strategies with the attendant development of suitable representations of the environment. Suitable representations serve as a prerequisite for fundamental tasks such as exploration and localization, with a plethora of forms emerging to suit each case. Metric maps that render spatial attributes in terms of their location in a shared coordinate frame have enjoyed immense popularity in the

D. Mox (✉)
Drexel University, Philadelphia, PA, USA
e-mail: dcm64@drexel.edu

A. Cowley · M. A. Hsieh · C. J. Taylor
University of Pennsylvania, Philadelphia, PA, USA
e-mail: acowley@seas.upenn.edu

M. A. Hsieh
e-mail: mhsieh1@drexel.edu

C. J. Taylor
e-mail: cjtaylor@seas.upenn.edu

R. Groß et al. (eds.), *Distributed Autonomous Robotic Systems*,
Springer Proceedings in Advanced Robotics 6,
https://doi.org/10.1007/978-3-319-73008-0_4

literature. Examples of *de facto* standards for metric maps include occupancy grids for exploration [4] and landmark maps for simultaneous localization and mapping (SLAM) [10]. On the other hand, topological maps seek to relate semantic information about the environment gleaned from sensor observations, and have been used to aid in rapid exploration [3] and efficient task allocation [8].

In typical SLAM scenarios, localization and mapping are the motivating factors of the entire system. However, choices made in support of those goals may not be well-suited for other uses of map data. For example, in landmark based maps only a few patches of images representing regions of interest are extracted from source imagery, while in voxel based grids the fine details of a point cloud are lumped into a small number of cells. The effects of filtering out significant portions of sensor data have only been magnified with the emergence of low cost off the shelf RGBD cameras. These sensors combine depth and visual information that can be used to create dense, high fidelity three dimensional maps [11], but the torrent of available data must must be carefully navigated.

New approaches using RGBD cameras include the influential work of Newcombe in which dense mapping and tracking is achieved by fusing depth frames for surface reconstruction [7]. Henry employed surfels, e.g., surface orientation, patch size, and color, gleaned from RGBD frames to build 3D maps of indoor environments [5]. Recent work by Taylor [9] focuses on panoramic views comprising multiple depth frames captured at a location. Panoramic depth images are advantageous for mapping because they provide both spatial information for motion planning and fine grained detail for object recognition. Furthermore, when compared to traditional grid based approaches, panoramic images have the potential to provide significantly more detail while consuming only a modest amount of memory resulting in better scaling characteristics.

In this work, we build upon [9] to develop a novel information maximizing exploration strategy for teams of robots to map an unknown environment. Recent advances in our understanding of information theory and how to apply it to the robot exploration problem have lead to a slew of new algorithms based on a powerful principle: the map provides hints about what observations to expect at different locations and vice versa. This notion is at the heart of mutual information, which seeks to quantify the amount of information one random variable contains about another [2]. Julian demonstrated that mutual information eventually drives the robot towards unknown space and used it as an objective function for autonomous exploration using a range sensor [6]. A similar approach was shown by Charrow who used Cauchy–Schwarz Quadratic Mutual Information (CSQMI), a metric closely related to mutual information, to guide a robot equipped with an RGBD camera to map an unknown environment [1].

Different from existing work, we present a novel spatial grid representation that emphasizes *perspective* in addition to coverage to guide robots in exploring unknown spaces while collecting useful panoramas to create a detailed map of the environment. We leverage existing grid maps and their simplifying power for reasoning about exploration tasks to guide a group of robots to collect panoramas which can then be used to construct a detailed map of the environment. In our strategy, we employ an occupancy grid based representation of the environment to enable each robot to execute a CSQMI based collaborative exploration strategy that uses minimal communication bandwidth. A high fidelity representation of the environment is then reconstructed using the collective sequence of panoramic images.

This paper is organized as follows: A detailed description of our methodology is provided in Sect. 2. Simulation and experimental results are presented in Sect. 3. We conclude with a summary of our work and a discussion on directions for future work in Sect. 4.

2 Methodology

The objective is to efficiently explore an unknown environment using multiple robots to collect high resolution panoramic images that can be used to produce a detailed, memory conscious representation of space.

Using panoramas to represent the environment offer a number of advantages over traditional grid based maps [9]. First, they capture the surface structure of interest and nothing more as free space is implicit in the representation, offering a high level of detail for the memory used as compared to an occupancy grid. While memory usage may not be of significant concern when the workspace is small, it quickly becomes a significant issue when exploring and mapping realistically detailed and expansive spaces. Our work capitalizes on the unique perceptual data provided by RGBD panoramas presented in [9]. By capturing such views of the space surrounding a robot at specific locations in the workspace, a detailed map of the environment can be created. Our work focuses on the development of a suitable exploration strategy and we refer the interested reader to [9] for the details on the synthesis of the panoramas and an overview of comparable state-of-the-art mapping techniques.

From [9], the panoramas are pieced together from a series of color and depth images collected at regular intervals while the robot turns in place. As such, the images can be stitched and refined locally on the robot, alleviating the need for high bandwidth networking. As such, our exploration strategy is decomposed into two main components: (1) collecting and storing panoramas for reconstruction and updating local maps maintained for planning purposes and (2) determining the next best location to collect more panoramas. We briefly describe our approach.

2.1 Angle Enhanced Occupancy Grid

While many exploration techniques focus on coverage of free space, the ideal map should also afford clear and diverse views of surfaces within that space. Consider a scenario where a robot is advancing down a hallway. A panorama is captured, and a sign on the wall just manages to fall within the cameras range. The free space has been identified and, from a traditional free space coverage standpoint, no more value can be gleaned from observing the robot's immediate surroundings. However, from

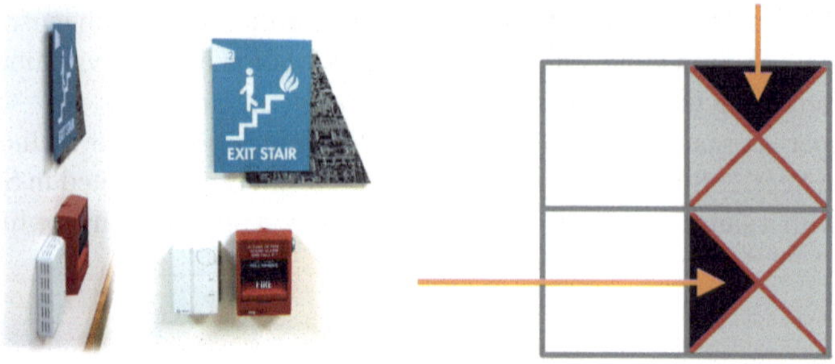

Fig. 1 Objects of interest on display in a hallway captured at different perspectives (left). Illustration of cells in an angle occupancy grid begin intercepted by beams, depicted as orange arrows (right)

a surface reconstruction standpoint, the face of the sign was captured at a shallow angle and may be rendered illegible in the resulting map. Here the utility of capturing space from advantageous perspectives is made plain as can be seen in Fig. 1.

To encourage perspective in addition to coverage, we employ an *angle enhanced* 2D occupancy grid such that each point in space contains four values representing the cardinal directions. As new observations are integrated into the map, the appropriate bin is updated according to the angle at which the cell was observed by the sensor. This creates an incentive for the robot to collect different views of the same physical space. Since free space contains the same information from all perspectives this binning strategy is not applied to unoccupied cells.

The principle behind this choice in map representation is that an occupancy grid naturally breaks environmental surfaces into segments. These segments are, at a fine enough scale, well approximated by star convex functions that we can drastically subsample by only considering a small number of view angles between sensor and surface. The primary tuning parameter then is the occupancy grid resolution, but not, as is typical, to directly capture finer geometry. Instead, occupancy grid resolution must be sufficient for the above convexity assumption to hold. Thus complex surface geometry that may be captured in an image complements the efficient, if coarse, geometric approximation offered by the underlying occupancy grid used for planning.

2.2 *Exploration Planning*

To select useful exploration goals, it is important to estimate how much future observations will tell us about the space they can cover. This notion is captured by computing the mutual information between a predicted sensor reading and the map given by

$$I_{MI}[m; z] = H[m] - H[m \mid z]. \tag{1}$$

Since a future measurement is implicitly conditioned on the robot's position, an objective function naturally arises wherein mutual information is calculated for a collection of candidate poses and used to determine the quality of a candidate panorama capture location.

In this work, we employ the Cauchy–Schwarz Quadratic Mutual Information (CSQMI) between a predicted measurement and the map to determine the next best position for the robot to obtain a panorama during its exploration of the space. In general, mutual information can be computationally expensive since it requires integrating the likelihood over entire space, often represented by a grid. Since a closed form expression for the CSQMI exists [1], it enables the quantification of the value of future measurements in a computationally efficient manner. Furthermore, CSQMI has been shown to produce similar results to mutual information [1]. Thus, when coupled with our angle enhanced 2D occupancy grid representation of the workspace for navigation purposes, the result is a scalable exploration strategy with well bounded computational and memory complexities for any given workspace. We briefly summarize the computation of CSMI using the proposed angle enhanced occupancy grid and refer the interested reader to [1] for the details.

Let z denote the set of random variables representing future sensor measurements that model the distance a beam at image pixel k travels before encountering an obstacle. Since our sensor measurement is a panoramic RGBD image, we evenly distribute k beams on the interval $[0, 2\pi)$ originating at pose x. To model the noise inherent in a depth measurement, we use the piecewise normal distribution given by

$$p(z_k = z \mid d) = \begin{cases} \mathcal{N}(z - 0, \sigma^2) & d \leq z_{min} \\ \mathcal{N}(z - z_{max}, \sigma^2) & d \geq z_{max} \\ \mathcal{N}(z - d, \sigma^2) & \text{otherwise} \end{cases} \tag{2}$$

where z_{min} and z_{max} are the minimum and maximum range of the sensor respectively and $\mathcal{N}(z - \mu, \sigma^2)$ is a Gaussian with mean μ and variance σ^2. While more complex beam based models exist [10], this simple model is sufficient for our purposes.

Using the beam model given by (2), a distribution over possible measurements denoted by $p(z_k)$ can be found by computing the marginal distribution over the list of c cells belonging to the map m intercepted by the beam

$$p(z_k) = \sum_{i=0}^{C} p(c = e_i) p(z_k \mid c = e_i) \tag{3}$$

where for $i > 1$, e_i means that the ith cell is the first occupied cell in c, e_0 means that no cells in c are occupied, and $C = |c|$.

The CSQMI between the map, m, and a predicted observation, z, collected at a location in space, x, is expressed as

$$I_{CS}[m; z \mid x] = \frac{(\sum \int p(m, z \mid x) p(m) p(z \mid x) dz)^2}{\sum \int p^2(m, z \mid x) dz \sum \int p^2(m) p^2(z \mid x) dz} \tag{4}$$

where each sum is over all possible maps and the integrals are over all possible measurements.

Due to the camera's high resolution and the fact that panoramas comprise many images stitched together, the number of beams k in the measurement z_k could be quite large. However, exactly representing each pixel in a panorama with a beam is not necessary. Depending on the resolution of the grid, beams very close together often intercept many of the same cells which leads to the double counting of information gained from each beam. As such, we only consider the subset of beams that can be reasonably considered independent of one another given the range of the camera and the resolution of the grid. Assuming independence between elements of this subset of beams, (4) is computed by summing the individual contribution of each beam as follows

$$\begin{aligned} I_{CS}[c; z_k \mid x] = & \log \sum_{l=0}^{C} w_l \mathcal{N}(0, 2\sigma^2) \\ & + \log \prod_{i=1}^{C} (o_i^2 + (1 - o_i)^2) \sum_{j=0}^{C} \sum_{l=0}^{C} p(e_j) p(e_l) \mathcal{N}(\mu_l - \mu_j, 2\sigma^2) \\ & - 2 \log \sum_{j=0}^{C} \sum_{l=0}^{C} p(e_j) w_l \mathcal{N}(\mu_l - \mu_j, 2\sigma^2). \end{aligned} \tag{5}$$

Here $o_i = p(c_i = 1)$ is the probability that the ith cell in c is occupied and $p(e_j)$ is the probability that the first occupied cell in c is c_j. Additionally, each w_l is calculated as follows

$$w_l = p^2(e_l) \prod_{j=l+1}^{C} (o_j^2 + (1 - o_j)^2). \tag{6}$$

with $0 < l < C$, $w_0 = p(e_0)$, and $w_C = p^2(e_C)$. The final result then becomes

$$I_{CS}[m; z \mid x] = \sum_{i=0}^{k} I_{CS}[c; z_k \mid x]. \tag{7}$$

CSQMI is computed for a list of poses, χ, sampled from known free space. Figure 2 shows the CSQMI reward surface computed for the corresponding workspace using a traditional 2D occupancy grid and using an angle enhanced occupancy grid. By incorporating the notion of perspective into the occupancy grid, the CSQMI reward surface computed using the angle enhanced grid results in better coverage of the all accessible sides of objects and obstacles.

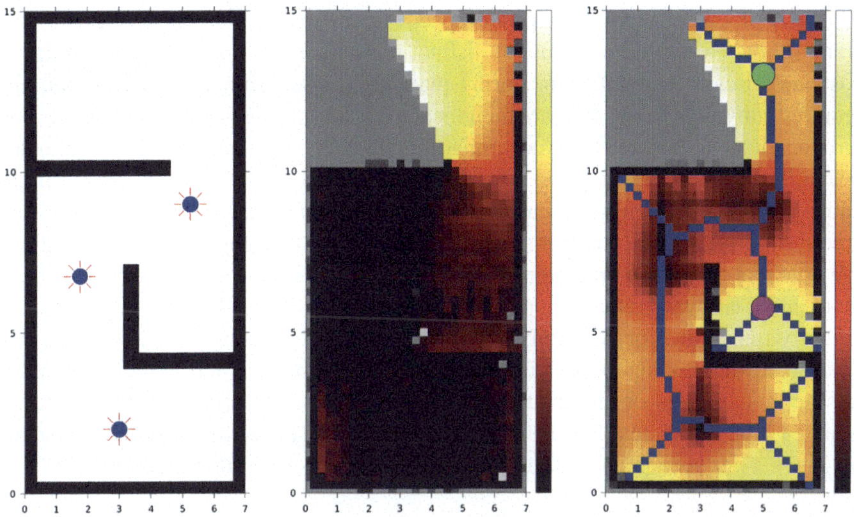

Fig. 2 A sample environment after three panoramas have been captured (left) and the resulting CSQMI reward surface for a traditional 2D occupancy grid (center) and the angle enhanced occupancy grid (right)

To choose between the highest CSQMI poses, travel costs in the form of distance along a path calculated using A* from the robot to the candidate pose are used. Thus, the next best position is given by

$$x^{\star} = \underset{x \in \chi}{\arg\max} \frac{I_{CS}[m; z \mid x]}{Cost(x)} \tag{8}$$

where $Cost(x)$ is the path distance from the robot's current pose to the candidate goal, x. Integrating the travel cost results in emphasizing completion of locally accessible space before advancing towards uncharted territory.

2.3 Multi-robot Strategy

From (4) we see that our approach provides a computationally efficient strategy to determine next best locations for measurements in a scalable way. Accordingly, our strategy is particularly well suited for small, resource constrained platforms since each robot only has to compute the CSQMI objective function once to determine the next best position to capture a panorama. This enables us to develop a simple coordination strategy where the single robot exploration strategy is executed in parallel by a team of robots.

The deployment of multiple robots speeds up the exploration process, especially when robots are tasked to focus on distinct regions of the workspace. Since travel costs are already accounted for during planning, our coordination strategy simply requires individual robots to share their local maps and goals to ensure panorama capture locations do not overlap. Assuming the relative pose of each agent is available, a small grid circumscribing the latest panorama is broadcasted to the other robots and integrated into their local maps. The active goals of other agents are also considered during planning to prevent multiple robots from traveling to the same region. We note that this communication strategy only requires the communication of local maps and goals which are only updated each time a panorama is captured and as such are only transmitted occasionally, reducing the required bandwidth. The result is a distributed strategy for cooperative mapping of an unknown environment. Critically, this approach relies on the exchange of small occupancy grids rather than accumulated imagery or detailed surface reconstructions. This is the difference between megabytes-per-second and kilobytes-per-hour in terms of communication cost.

3 Results

To validate our approach, simulations and experiments were conducted in a variety of indoor environments with teams of one to five robots. An Asus Xtion Pro Live RGBD camera was used to provide observations of the environment. Each panorama comprised of 36 images, one for every 10 degrees of rotation. Throughout our simulations and experiments, we assume the pose of each robot in a global coordinate frame is provided. While robot localization remains a non-trivial problem, our objective is validating the proposed exploration strategy. As such, we assume robot localization can be achieved via existing on-board localization methods.

3.1 Simulations

To evaluate the proposed angle aware exploration strategy, we compare our approach with the well established frontier method [12]. We use two simulated environments shown in Fig. 3a, b to compare the resulting exploration locations. Figure 3c, d show the capture locations of panoramas in a simple indoor environment spanning a space of 8×14 m^2. High level results of each iteration of the simulation are summarized int Table 1.

The benefit of the angle enhanced occupancy grid approach may be seen by a detailed analysis of image coverage of the 2D surfaces in the Hallway environment. As a proxy for data quality, we consider the maximum image size of every 10 cm segment of wall across all simulated panoramas using a linear imaging resolution of 8229 pixels for each 360° panoramic image as used by the experiments in [9]. Visibility is calculated for each decimeter as a whole by ray-casting from the capture

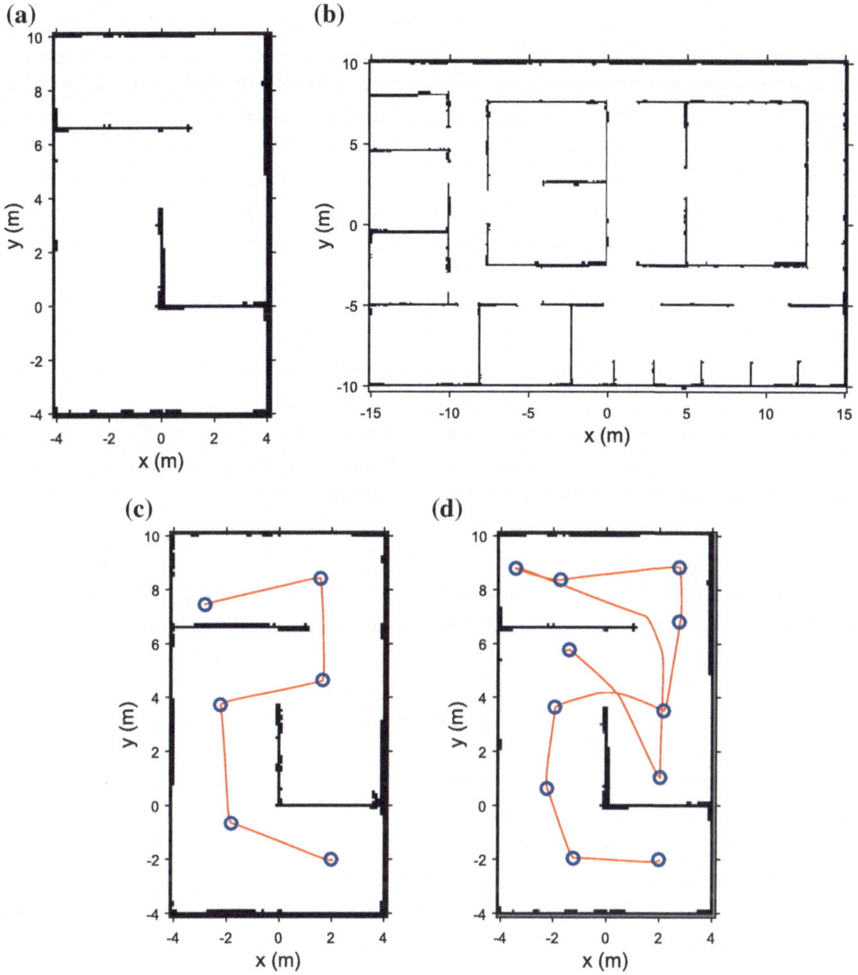

Fig. 3 **a** Curving hallway environment used with 1 robot; **b** larger office environment used with 5 robots; and simulation results for the curved hallway for frontier (**c**) and angle aware (**d**) planning methods. The red line shows the path of the robot and the blue circles represents panorama capture locations

Table 1 Summary of the area covered and panoramas captured by each method in the Hallway (Fig. 3a) and Office (Fig. 3b) environments

Environment	Method	Area covered (m^2)	Panoramas captured
Hallway	Angle aware	110.9	11
Hallway	Frontier	105.9	6
Office	Angle aware	572.0	140
Office	Frontier	571.3	38

location to the center of a wall segment. This is a simplification of the benefits of multiple observations, but is immediately applicable to common tasks such as optical character recognition of writing on walls, or face recognition applied to pictures on walls. Image size relates to our approach in so far as shallow observation angles correspond with lower resolution imaging of a surface. Put plainly, though we are not optimizing specifically for this metric, we expect some correlation. Note that while the numbers used throughout our analysis are arbitrary, they provide a context to compare each method that proves valuable regardless of the exact image resolution chosen.

For the set of simulated experiments shown in Fig. 3c, d, the frontier approach fully explored the Hallway environment after collecting 6 panoramas, while the angle aware approach selected 11 locations to faithfully capture all surfaces. With these panorama collection locations, we can calculate that 33.5% of wall segments observed during the frontier exploration strategy were captured by fewer than 20 pixels in any given panorama. For the angle enhanced occupancy grid approach, only 11.4% of wall segments were observed below the same threshold resolution. If we evenly subdivide the frontier exploration trajectory to result in an equivalent number of panoramas (“augmented frontier”), 17.9% of wall segments are observed below the resolution threshold. This demonstrates that angle sensitivity in the exploration strategy results in *substantially* fewer “blind spots” along surfaces in the environment as it has more information with which to plan. A naïve increase in panorama locations alone, however, *does not* result in the same coverage gains.

The prevalence of low resolution blind spots in the augmented frontier strategy is visible in the overlayed histograms in the lower-right of Fig. 4. In this figure, the green histogram bars of the augmented frontier observations stand out on the left of the chart. Precisely where these blind spots arise may be seen in the top row of Fig. 4 where the light blue wall segments on the right side of the map approximately one third of the way up the image, for example, indicate poor coverage.

Lastly, Fig. 5 shows the panorama capture locations for a team of five robots deployed in the office environment shown in Fig. 3b. Initially each robot balances capturing local space with pushing into uncharted territory. As each robot explores more space, the simple assignment strategy ensures that each robot does not interfere with its neighbor.

3.2 Experiments

We have also conducted live experiments in the space shown in Fig. 6b which covers an area of approximately $5 \times 5\,\text{m}^2$. The two differential drive ground robots shown in Fig. 6a traversed the environment to collect panoramas, sharing maps and goal locations as described in Sect. 2.3. Each robot was equipped with an Asus Xtion Pro Live RGBD camera providing synchronized color and depth frames for the panoramas and spoofing a laser rangefinder for construction of the angle occupancy grid used during planning. All software was written in C++ and executed on an Odroid-XU4

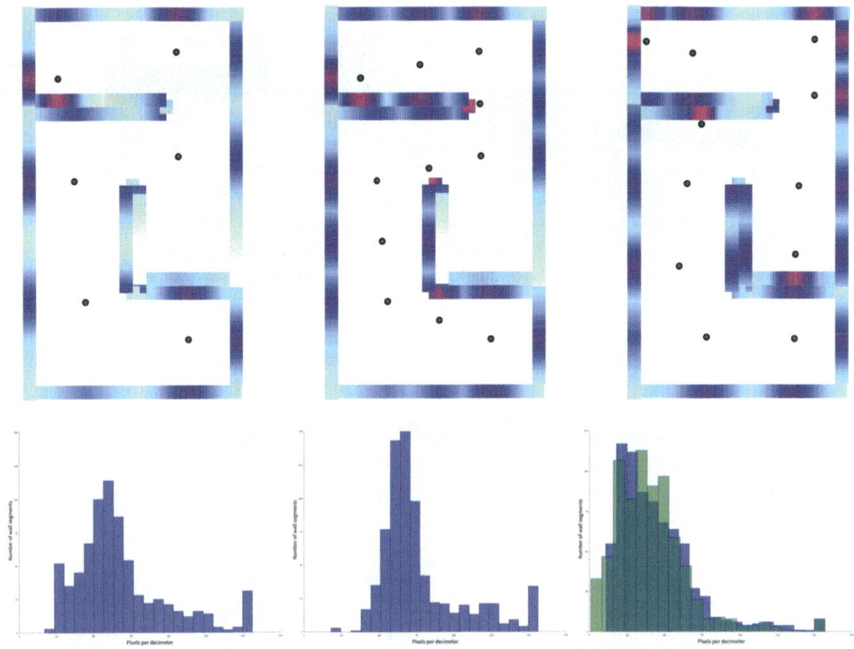

Fig. 4 Top row: map renderings with walls colorized by the resolution at which they were covered for frontier (left), augmented frontier (center), and angle aware (right) approaches. Hue indicates imaging resolution with red indicating high resolution, and blue indicating low resolution. In grayscale, the less saturated wall segments indicate poor coverage. Panorama capture locations are denoted by green circles. Bottom row: histograms of pixels per decimeter for augmented frontier (left), angle aware (center), and augmented frontier overlaid on angle aware (right)

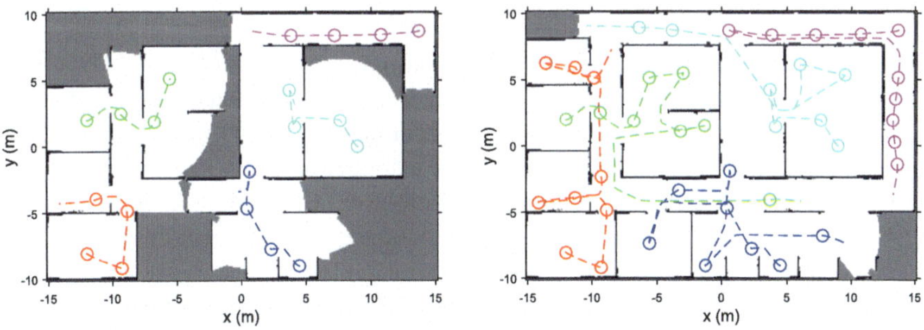

Fig. 5 Five robots collaboratively exploring an office environment after 4 panoramas (left) and after 10 panoramas (right)

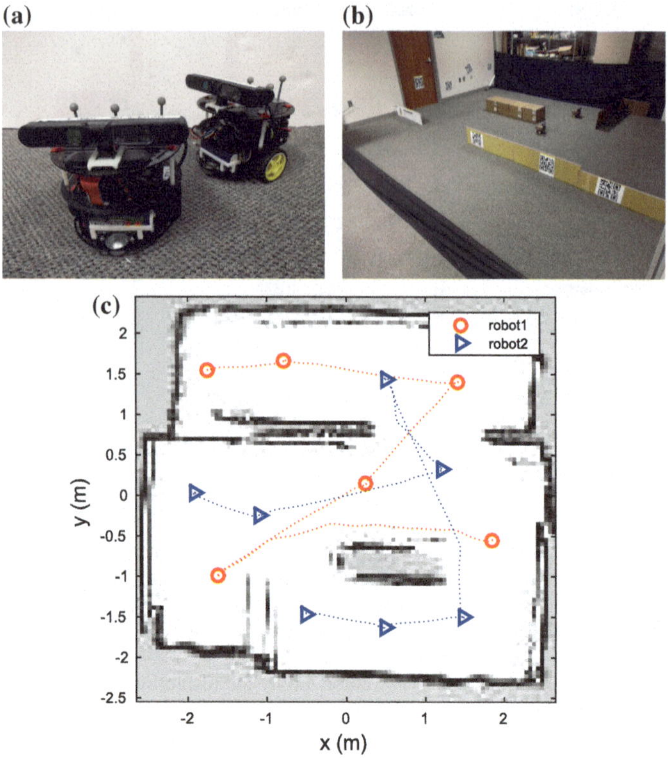

Fig. 6 The ground robots (**a**), lab space used for live experiments (**b**), and experimental results showing the panorama capture locations for the two robot team (**c**)

single board computer on-board the robot running Linux and the Robotic Operating System (ROS). Localization was provided by an external motion capture system.

In order to predict future measurements and compute CSQMI, several parameters were specified. Panoramas were modeled as a collection of 360 beams distributed over the interval $[0, 2\pi)$. Given a map resolution of 0.05 m, increasing the number of beams only resulted in additional dependent beams which intercepted the same cells and would be factored out of the CSQMI calculation. Values of $z_{min} = 0.5$ m and $z_{max} = 4.5$ m were used for the minimum and maximum range of the Xtion, and the noise was modeled with $\sigma = 0.03$.

The results of the experiment can be see in Fig. 6c with the two robots beginning at $\{-2, 0\}$ and $\{-2, 1.5\}$ respectively. Starting at the same time, the team executed our exploration strategy simultaneously and collected a total of 13 panoramas. Operating in proximity, the team successfully avoided capturing panoramas from the same location demonstrating the effectiveness of our proposed multi-robot coordination strategy.

4 Conclusion and Future Work

Panoramas comprised of images collected from an RGBD sensor provide a new way to generate high fidelity representations of environments without demanding significant memory resources. This opens up the need for new exploration strategies that leverages the unique perceptual characteristics provided by the sensor. Our angle aware exploration strategy enables a team of robots to effectively map an unknown environment by searching for locations that yield high information gain while accounting for diverse perspectives when creating a comprehensive 3D map. Since the approach is computationally efficient, it can be used by a wide variety of robotic platforms, even ones with limited computational resources. Furthermore, the approach places minimal demands on inter-agent wireless communication and computation at both the planning and coordination level.

As depth panoramas contain rich information about the environment, an alternative approach could be constructing and using them online directly in the planning phase. This has the added benefit that the final map is included in the planning loop which ensures some level of quality. Additionally, extending this approach to platforms not constrained to the plane such as unmanned aerial vehicles would require modifications to the angle enhanced occupancy grid employed. However, we believe the proposed exploration strategy will enable heterogeneous teams of robots to better leverage the distinct perceptual and mobility capabilities of the various sensing resources within the team. These are all directions we are pursuing for future work.

Acknowledgements The authors gratefully acknowledge the support of ARL grant W911NF-08-2-0004 and NSF grant OISE-1131011.

References

1. Charrow, B., Liu, S., Kumar, V., Michael, N.: Information-theoretic mapping using Cauchy–Schwarz quadratic mutual information. In: 2015 IEEE International Conference on Robotics and Automation (ICRA) pp. 4791–4798 (2015). https://doi.org/10.1109/ICRA.2015.7139865
2. Cover, T.M., Thomas, J.A.: Elements of Information Theory (2005). https://doi.org/10.1002/047174882X
3. Cowley, A., Taylor, C.J., Southall, B.: Rapid multi-robot exploration with topometric maps. In: Proceedings-IEEE International Conference on Robotics and Automation pp. 1044–1049 (2011). https://doi.org/10.1109/ICRA.2011.5980403
4. Elfes, A.: Using occupancy grids for mobile robot perception and navigation. Computer **22**(6), 46–57 (1989). https://doi.org/10.1109/2.30720
5. Henry, P., Krainin, M., Herbst, E., Ren, X., Fox, D.: RGB-D mapping: using depth cameras for dense 3D modeling of indoor environments. Springer Tracts in Adv. Robot. **79**, 477–491 (2014). https://doi.org/10.1007/978-3-642-28572-1_33
6. Julian, B.J., Karaman, S., Rus, D.: On mutual information-based control of range sensing robots for mapping applications. Int. J. Robot. Res. **33**(10), 1375–1392 (2014). https://doi.org/10.1177/0278364914526288
7. Newcombe, R.A., Izadi, S., Hilliges, O., Molyneaux, D., Kim, D., Davison, A.J., Kohli, P., Shotton, J., Hodges, S., Fitzgibbon, A.: KinectFusion: Real-time dense surface mapping and

tracking. In: 10th IEEE International Symposium on Mixed and Augmented Reality, ISMAR 2011 (July 2016), pp. 127–136 (2011). https://doi.org/10.1109/ISMAR.2011.6092378
8. Stachniss, C., Martínez Mozos, Ó., Burgard, W.: Efficient exploration of unknown indoor environments using a team of mobile robots. Ann. Math. Artif. Intell. **52**(2), 205–227 (2008). https://doi.org/10.1007/s10472-009-9123-z
9. Taylor, C.J., Cowley, A., Kettler, R., Ninomiya, K., Gupta, M., Niu, B.: Mapping with depth panoramas. In: 2015 IEEE/RSJ International Conference on Intelligent Robots and Systems (IROS 2015), pp. 6265–6272 (2015)
10. Thrun, S., Burgard, W., Fox, D.: Probabilistic robotics. MIT press, USA (2005)
11. Whelan, T., Kaess, M., Fallon, M., Johannsson, H., Leonard, J., McDonald, J.: Kintinuous: Spatially extended kinectfusion (2012)
12. Yamauchi, B.: Frontier-based exploration using multiple robots. In: Proceedings of the Second International Conference On Autonomous Agents, pp. 47–53. ACM (1998)

Multirobot Persistent Patrolling in Communication-Restricted Environments

Marta Romeo, Jacopo Banfi, Nicola Basilico and Francesco Amigoni

Abstract In multirobot patrolling, a team of robots is deployed in an environment with the aim of keeping under observation a set of locations of interest. In several realistic mission scenarios, only human operators sitting at a base station are able to assess the situation on the basis of data sent by robots. Examples include watching pictures or video streams to detect intruders and correlating measurements to detect leaks of contaminants. We assume that a communication infrastructure is available only in some regions of the environments, from where messages can be exchanged with a sufficient bandwidth between the robots and the base station. In this paper, we first extend the classical multirobot persistent patrolling framework and the related idleness evaluation metrics to such environments with a limited number of "communication zones". Then, we present some centralized and distributed patrolling strategies tailored for this communication-restricted framework. Finally, we evaluate their performance using ROS/Stage simulation.

1 Introduction

In the last few years, a lot of effort has been devoted in the autonomous robotics community to the study of effective patrolling strategies in different problem settings and under different assumptions [1–4, 9, 10, 13, 15]. In the basic formulation of the *multirobot persistent patrolling problem*, a team of robots is deployed in an

M. Romeo · J. Banfi (✉) · F. Amigoni
Politecnico di Milano, Milan, Italy
e-mail: jacopo.banfi@polimi.it

M. Romeo
e-mail: marta.romeo@mail.polimi.it

F. Amigoni
e-mail: francesco.amigoni@polimi.it

N. Basilico
University of Milan, Milan, Italy
e-mail: nicola.basilico@unimi.it

R. Groß et al. (eds.), *Distributed Autonomous Robotic Systems*,
Springer Proceedings in Advanced Robotics 6,
https://doi.org/10.1007/978-3-319-73008-0_5

environment represented as an undirected weighted graph, with the aim of minimizing the *idleness* of each location (vertex) to be patrolled, defined as the time elapsed since that location has received the visit of a robot [10]. Evaluation metrics related to this notion, such as average and worst-case idleness, are introduced and investigated for different patrolling strategies and in a number of different conditions [4, 9, 10, 13, 15].

Some realistic mission scenarios challenge an (often implicit) assumption made in the above formulation, namely that a single robot is able to assess the situation of the visited vertices solely on the basis of the data it collected. In these cases, data must be reported to a base station, acting as a mission central room, where human operators can assess the situation and take appropriate countermeasures. For instance, think about the presence of an intruder or about a leak of contaminants, when the robots might not have the physical capabilities required to detect the problem and intervene to solve it, because operators need to watch pictures or video streams and to correlate measurements. In this paper, we assume that communication between the robots and the base station can only use existing communication infrastructures, such as tactical networks or cellular networks, that are characterized by the fact that only some regions of the environment are covered by a sufficiently strong signal. The presence of such "communication zones" in the environment to patrol was recently investigated in [1], where a centralized inspection tour for a team of robots is calculated with the aim of minimizing the time lag between the visit of a location and its report to the base station.

In this paper, we propose some strategies for multirobot persistent patrolling in communication-restricted environments characterized by the presence of a limited number of communication zones. More precisely, we extend the classical idleness evaluation criteria for the communication-restricted patrolling framework introduced in [1]. Then, we present some patrolling strategies for such a framework. One of these strategies is centralized and optimal, while the other two are distributed. Although non optimal in general, these last strategies are less computationally demanding and are particularly suited for coping with unpredictable events (e.g., temporary unavailability of communication links). The proposed patrolling strategies are implemented within a ROS-based architecture, and ROS/Stage simulations are run to analyze and compare them in a number of different settings. The results show that a centralized planning scheme is not actually needed to obtain good performance and that both the proposed distributed strategies perform well.

2 Related Work

The study of multirobot patrolling strategies is nowadays a rather lively sub-field of autonomous robotics, especially from the theoretical point of view (although some practical systems have been proposed [9]). Approaches can be broadly divided in two categories: those considering an adversarial setting and those considering a non-adversarial setting. In the former category, patrolling strategies are developed considering a specific model of rational intruder [2, 3], while, in the latter category, the patrolling strategies are in general agnostic about the particular events to be monitored, preferring to maximize the frequency of visit of some locations that need to be kept under constant observation [4, 6, 9, 10, 13, 15] (hence the name "persistent patrolling"). In this paper, we concentrate on the last category of works. To the best of our knowledge, there are few works in the literature considering communication constraints in robotic patrolling and in the more general context of robotic information-gathering missions. For example, in [11], the authors consider a generic multirobot routing task, in which robots are required to maintain, by means of local and multi-hop transmissions, a global communication infrastructure while visiting some target areas at minimum traveling cost. More recently, [16] considers the same general setting, but with the additional requirement of planning robots' paths while maintaining a given minimum end-to-end data rate between robots.

Another problem setting is investigated in [7], where robots plan informative paths under the constraint of regaining multi-hop connectivity after a fixed number of time steps. The assumption on the communication infrastructure we make in this paper is rather different from that adopted in these works, because we do not rely on a multi-hop network to deliver data to a base station, but on an existing infrastructure that is available only in some areas. More precisely, motivated by recent military and civil case-studies [12, 17], in [1] we have introduced a patrolling framework modeling the presence of an existing communication infrastructure (for example, a tactical network, a wireless network, or a cellular network) that robots could exploit to reliably communicate with a base station. (This setting is similar to that investigated in [5], but in the context of static target search.)

In this paper, we investigate further along this line through the adoption of the same communication model. However, instead of considering a patrolling mission consisting of a *single* inspection tour computed offline (as in [1]), we focus on *persistent* patrolling, in which vertices need to be visited multiple times in order to keep them under constant observation.

3 Problem Definition

A team $R = \{1, \ldots, m\}$ of robots must persistently patrol some locations of a given environment while sending reports to, and exploiting information provided by, a base station (BS). Robots can communicate with the BS by exchanging messages through a pre-existing communication infrastructure, which is only accessible from some regions of the environment. To formalize this setting, we represent the environment by means of an undirected graph $G = (V, E)$ in which vertices represent all the locations robots can visit, while edges encode physical connections between them (without considering self-loops). Each edge $(v_i, v_j) \in E$ is associated with a positive integer $d(v_i, v_j)$ representing the physical distance between v_i and v_j in the environment. We assume that our graph represents a physical environment. For this reason, we impose that the distance $d(v_i, v_j)$ associated to any physical edge $(v_i, v_j) \in E$ is not greater than that of any other (v_i, v_j)-path in G. Vertices V are used to represent both locations to patrol and communication zones. Accordingly, following the notation introduced in [1], each vertex in V is preassigned to up to two types: m, if the vertex needs to be persistently monitored (patrolled), and c, if the vertex allows to exchange messages with the BS. We denote with $V_m, V_c \subseteq V$ the two subsets of m-type and c-type vertices, respectively. (Notice that V_m and V_c may share a non-empty intersection.) Robots move on the graph G by traversing the physical edges E: once they reach an m-type vertex, they accomplish a (instantaneous) patrolling visit at it; once they reach a c-type vertex, they are allowed to communicate with the BS. At the moment, we do not assume any limit on the communication from a c-type vertex to the BS. In Sect. 5, we consider errors in the transmission of messages.

In order to define the objective of the patrolling, we now generalize the classical evaluation metrics related to the notion of *idleness* [4] to our communication-restricted framework. The persistent patrolling task unfolds across an arbitrary long time horizon T in which time evolves in discrete steps $\{1, \dots, T\}$. The mission undertaken by a robot r is represented by a *walk* on the graph G of the form $w_r = (v_0, v_1, \dots, v_k)$, where v_0 denotes the robot's starting vertex, v_k denotes the last vertex visited before T, and a generic v_i denotes the i-th vertex visited by the robot. (Notice that, being w_r a walk on G, it is not possible to stay still at the same vertex, but this can appear multiple times in w_r.) For each walk w_r, we define a function $t_r : \{1, \dots, k\} \rightarrow \{1, \dots, T\}$ mapping the i-th visited vertex by robot r to the time step in which the visit takes place. In particular, each $t_r(i)$ is determined by the total distance traveled until the i-th vertex visited in w_r ($\sum_{j=0}^{i-1} d(v_j, v_{j+1})$) and by robot r's (possibly varying) speed along that portion of the walk. Now, for any $i \in \{1, \dots, k\}$ such that $v_i \in V_m$, let $v_{\bar{c}}, \bar{c} \geq i$ be the first vertex after the i-th vertex visited s.t. $v_{\bar{c}} \in V_c$, and call $c(i) = \bar{c}$. Following this notation, we define robot r's *reporting time* of visit i to the BS (only for vertices v_i s.t. $v_i \in V_m$) as $t_r(c(i))$, assuming that this quantity will be infinite if the visit is never reported before the patrolling task ends at T. (Notice that $t_r(c(i)) = t_r(i)$ if and only if $v_i \in V_c \cap V_m$.)

Contrarily to many other patrolling settings, it may be possible in our framework that the BS receives reports containing outdated information. This happens when there exist two robots $r, r' \in R$ (not necessarily different), two walks $w_r = (v_0, v_1, \dots, v_k)$, $w_{r'} = (v_0, v_1, \dots, v_{k'})$, and two visits $i \in \{1, \dots, k\}, j \in \{1, \dots, k'\}$ s.t. $v_i = v_j$, $t_r(i) < t_{r'}(j)$, and $t_r(c(i)) \geq t_{r'}(c(j))$. In such a case, we say that the i-th visit of robot r is *outdated*, in the sense that the corresponding report to the BS will contain outdated information. We are now ready for introducing metrics related to *communication idleness*.

We define the *instantaneous communication idleness* of vertex $v \in V_m$ at time $\bar{t} \in \{1, \dots, T\}$ as:

$$IC_v(\bar{t}) = \bar{t} - \bar{t'}, \tag{1}$$

where $\bar{t'} \leq \bar{t}$ denotes the time step of the last not outdated report of vertex v to the BS not after time step $\bar{t}$. We assume that the initial situation of all the vertices to be monitored is known, and hence that $IC_v(\bar{t}) = \bar{t}$ if there exists no visit to v that has been communicated (by any robot) until time $\bar{t}$. Notice that, according to our definition, in case $V_m = V_c = V$ no visit is considered outdated, and the notion of instantaneous communication idleness collapses into the notion of idleness classically adopted in multirobot persistent patrolling (see, e.g., [4]).

The *instantaneous graph communication idleness* at time $\bar{t}$ is defined as the average instantaneous communication idleness among the vertices to be patrolled:

$$IC_G(\bar{t}) = \frac{\sum_{v \in V_m} IC_v(\bar{t})}{|V_m|}. \tag{2}$$

In this paper, we are mainly interested in studying patrolling strategies allowing to minimize the *average graph communication idleness* at the end of the mission, formally defined as:

$$\overline{IC}_G = \frac{\sum_{\bar{t} \in \{1,\ldots,T\}} IC_G(\bar{t})}{T}. \tag{3}$$

Another interesting performance indicator we will investigate is the *worst-case graph communication idless* at time T, WIC_G, which can be informally defined as the largest instantaneous vertex communication idleness encountered through the patrolling task. Average and worst-case idleness for the single vertices can be defined in a similar way.

4 Communication-Aware Patrolling Strategies

In this section, we present three different multirobot patrolling strategies (plus a random baseline one) suitable for the patrolling setting introduced above, aimed at minimizing the average graph communication idleness $\overline{IC}_G$. First, we present a strategy in which the BS performs the planning phase by computing centralized optimal visit plans for all the robots. Then, we present two distributed strategies in which robots query the BS about the current situation of the environment in terms of vertices idlenesses and teammates current commitments (plans) and then calculate their plans.

4.1 Globally Optimal Strategy

The *globally optimal strategy* (G-OPT) is an optimal, centralized, and offline strategy working as follows. At the beginning of the patrolling mission, the BS computes (using brute force) the best set of joint robot walks $\mathbf{w}^* = (w_1^*, w_2^*, \ldots, w_m^*)$ spanning the whole task horizon T and minimizing $\overline{IC}_G$; then, it communicates the plans to the robots (that are assumed to be initially placed in possibly different starting vertices belonging to V_c), which can then start to execute them. The pseudo-code of the planning algorithm of G-OPT is shown in Algorithm 1.

About the computational complexity of such a planning scheme, notice that the evaluation of $\overline{IC}_G$ for a candidate set $\mathbf{w}$ of joint robots' walks may require pseudo-polynomial time $O(mT)$ in general, while the number of (the longest) joint walks of length not exceeding T on G grows not less than exponentially with the size of the input: therefore, this strategy is suitable only for small problem settings. Moreover, it is evident that the walks obtained by adopting this planning approach will be actually optimal w.r.t. the minimization of $\overline{IC}_G$ (if the errors in the predictions of their execution can be assumed to be negligible).

Algorithm 1 G-OPT planning algorithm

```
W ← {all the longest joint walks of m robots on G with maximum length T}
for all w ∈ W do
    simulate the concurrent execution of w and compute the corresponding IC_G (overlined)
end for
w* ← {the obtained joint walk w that minimizes IC_G (overlined)}
return w*
```

We remark that an optimal strategy could also be obtained by formalizing the problem as a Mixed Integer Linear Program, in the spirit of [1]. However, with respect to [1], such an approach would require the additional modeling of the possibility of multiple passages through the same vertices at any time step along the whole mission horizon, making the MILP overly complicated.

4.2 Individually Optimal Strategy

The *individually optimal strategy* (I-OPT) can be thought as the distributed and online version of G-OPT. When adopting this strategy, robots iteratively compute new portions of their own walks as soon as they reach a new c-type vertex, querying the BS about (a) the current m-type vertices instantaneous communication idlenesses and (b) their teammates intentions, expressed in terms of portions of the walks they are currently following. More specifically, let $\bar{t}$ be a generic time step in which a robot r reaches a c-type vertex, and let $\hat{w}_1, \ldots, \hat{w}_{r-1}, \hat{w}_{r+1}, \ldots, \hat{w}_m$ the portions of walks its teammates are going to follow from step $\bar{t}$. (Notice that such portions of walks may be outdated, as they depend on the previous arrivals of the robots to c-type vertices.) Robot r computes a new portion of its walk $\hat{w}_r$ on G of estimated travel time H (the planning horizon) as the walk minimizing the average graph communication idleness between $\bar{t}$ and $\bar{t} + H$, while taking into account the walks followed by its teammates (if a teammate has communicated a plan that ends before $\bar{t} + H$, it is assumed to remain still at the last vertex of the communicated plan). If H is sufficiently small, a complete evaluation of all the possible walks of length

H can be completed in reasonable time; otherwise, a sampling strategy needs to be adopted. The pseudo-code of this planning scheme is reported in Algorithm 2.

Algorithm 2 I-OPT planning algorithm for robot r at time $\bar{t}$

$\hat{w}_1, \ldots, \hat{w}_{r-1}, \hat{w}_{r+1}, \ldots, \hat{w}_m \leftarrow$ {query the BS about the walks followed by other robots}
query the BS about instantaneous communication idlenesses of vertices V_m
$W_r \leftarrow$ {enumerate/sample walks from r's current position along $[\bar{t}, \bar{t} + H]$}
for all $w_r \in W_r$ **do**
  $\hat{\mathbf{w}} \leftarrow (\hat{w}_1, \ldots, \hat{w}_{r-1}, w_r, \hat{w}_{r+1}, \ldots, \hat{w}_m)$
  simulate the execution of $\hat{\mathbf{w}}$ and compute the corresponding $\overline{IC_G}$ in $[\bar{t}, \bar{t} + H]$
end for
$\hat{w}_r \leftarrow$ {the obtained walk w_r that minimizes $\overline{IC_G}$ in $[\bar{t}, \bar{t} + H]$}
return $\hat{w}_r$

4.3 A Simple Reactive Strategy

The *reactive strategy* (RE) is still distributed, but does not involve long-term planning up to horizon H as the previous one; instead, it is inspired by the Cognitive Coordinated strategy studied in [4, 10]. (This strategy empirically showed good performance when considering the classical notion of idleness.) By adopting this strategy, robots iteratively choose a new m-type vertex $\hat{v}$ to reach according to a heuristic in which vertices with the highest instantaneous communication latency have higher priority, and, in case the chosen vertex does not belong also to V_c, subsequently move to the closest c-type vertex in order to communicate back data relative to $\hat{v}$ and to obtain from the BS the data needed for computing a new plan.

More specifically, when a robot r arrives at the c-type vertex selected for communicating the data relative to the previous $\hat{v}' \in V_m$ at a generic time step $\bar{t}$, it queries the BS about (a) the current m-type vertices instantaneous communication idlenesses and (b) their teammates intentions, expressed in terms of portions of walks currently being followed (as in the previous strategy). It then decides to reach the m-type vertex $\hat{v}$ displaying the highest communication idleness that is currently not in the portion of walk followed by any other robot (if each vertex will be visited by at least one robot, it chooses one randomly). The path from the current vertex of robot r to $\hat{v}$ is computed on G, so that, as a side-effect of the choice, additional m-type and c-type vertices can be visited (in the latter case, the data relative to the visited m-type vertices are sent to the BS). In case $\hat{v} \notin V_c$, the path is augmented with the shortest path connecting $\hat{v}$ to the closest c-type vertex. The pseudo-code each robot runs for implementing this strategy is shown in Algorithm 3. Clearly, the algorithm runs in linear time w.r.t. the number of graph vertices (more precisely, in $O(m|V|)$) if the shortest paths composing $\hat{w}_r$ are already available (notice that they can be pre-computed at the beginning of the mission).

Algorithm 3 RE planning algorithm for robot r at time $\bar{t}$

$\hat{w}_1, \ldots, \hat{w}_{r-1}, \hat{w}_{r+1}, \ldots, \hat{w}_m \leftarrow$ {query the BS about the walks followed by others}
query the BS about instantaneous communication idlenesses of vertices V_m
$\hat{v} \leftarrow \arg\max_{v \in V_m, v \notin \hat{w}_i, i \in R \setminus \{r\}} IC_v(\bar{t})$
compute $\hat{w}_r$ as the shortest path on G from r's current position to $\hat{v}$
if $\hat{v} \notin V_c$ **then**
  let $\hat{v}_c$ be the c-type vertex closest to $\hat{v}$
  augment $\hat{w}_r$ with the shortest path on G from $\hat{v}$ to $\hat{v}_c$
end if
return $\hat{w}_r$

In the next section, we will investigate the behavior of these strategies by also comparing them against a baseline *random strategy* (RANDOM) in which each robot chooses randomly a new vertex of G to reach, regardless of its current idleness and membership to V_m and/or V_c.

5 Experimental Evaluation

We validated the proposed patrolling strategies in simulated, yet realistic, scenarios within the ROS/Stage framework [14, 18]. To enrich the Stage simulator with regions providing a communication link with a BS, we implemented an additional ROS node called *Communication Server* (CS) in charge of handling the communication between the robots and the BS. The CS node receives all the messages from the robots (BS) and appropriately forwards them to the BS (robots), considering robots locations in the environment. We selected three representative environments, described by 800×600 pixels bitmap images, upon which we constructed the patrolling graph $G = (V, E)$ by randomly selecting vertices sufficiently far from the obstacles, manually pruning some of them, and by adding edges only between vertices in line-of-sight to obtain a reasonable topological representation of the environment. In all the environments, communication vertices are chosen manually. The three selected environments are (see Fig. 1):

- the Leonardo campus of the Politecnico di Milano (Poli – approx. size 200×150 m), characterized by the presence of several buildings and discretized in 40 m-type vertices, 9 of which also belong to V_c;
- the "acapulco_convention_center" of the Radish dataset repository [8] (Open – approx. size 80×60 m), characterized by a large hall and discretized in 32 vertices, 11 of which only belong to V_c, and the remaining 21 only belong to V_m;
- an imaginary grid-block environment similar to those used in [4, 13] (Grid), discretized in 20 vertices, 4 of which only belong to V_c, 12 only to V_m, and the remaining 4 to both V_m and V_c.

Each simulated robot is equipped with a controller allowing to choose between one of the proposed patrolling strategies, and it is assumed to be able to perfectly

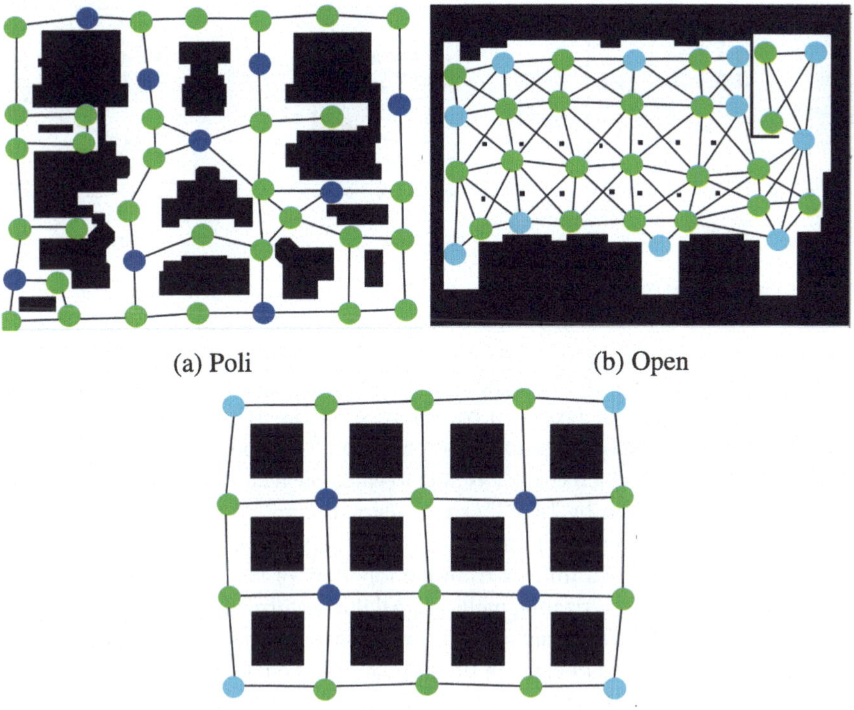

(a) Poli (b) Open (c) Grid

Fig. 1 The three environments. Green, blue, and light-blue vertices define the sets $V_m \setminus V_c$, $V_m \cap V_c$, and $V_c \setminus V_m$, respectively

localize itself in the environment (by using its true position as given by Stage). Given a waypoint to reach, path planning is performed by means of the A* algorithm run on a grid-based discretization of the environment, followed by a simple path-smoothing phase. Robots move at a maximum speed of 10 px/s, avoiding each other by means of a simple behavioral rule according to which the robot with highest ID takes precedence to pass.

Each run is repeated 5 times in order to cope with the non-deterministic outcome of the experiments due to the message exchange protocol adopted by ROS (graphs report averages over the runs and corresponding standard deviation bars). For each team size, we initially place the robots in the corners of the environments, and we keep fixed their initial positions across the different runs. The planning horizon of I-OPT is kept fixed at $H = 150\,\text{s}$, except where indicated otherwise, as this value allows to perform a complete walk enumeration in reasonable time in all the environments. (Notice that this is possible since robots' movements along the edges consume a significant amount of time steps.) All the simulations are performed on a laptop equipped with an Intel P7450 processor and 2 GB of RAM.

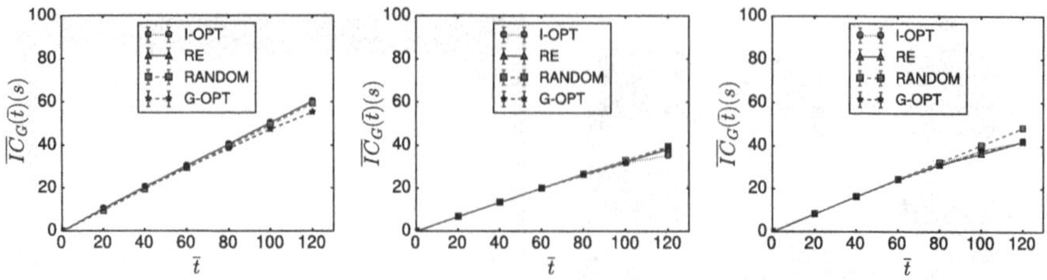

(a) Poli (b) Open (c) Grid

Fig. 2 Comparison of G-OPT, I-OPT, RE, and RANDOM in short mission durations with 2 robots

We start by reporting in Fig. 2 the average idleness values obtained in the three environments by the four proposed patrolling strategies (G-OPT, I-OPT, RE, and RANDOM) through short mission durations T with 2 robots. Such durations are chosen as the ones in which G-OPT is able to calculate a complete plan in the same amount of time required by I-OPT (a few seconds); notice that the durations are not equal for the different environments, as they depend on the complexity of the planning graphs G (which follows from the topology of the environments, as well as from the selected discretization method). All the performance curves remain very close until the end of the mission, where the random baseline starts to behave slightly worse than all the other strategies in the Poli and Grid environments. These results show that I-OPT and RE are able to perform as well as G-OPT on short mission horizons and justify their employment on longer mission durations for which G-OPT is not a viable choice. Notice that, in the Open environment, I-OPT behaves slightly better than G-OPT towards the end. This is due to several factors, such as robots interfering with each other while executing their paths, and a non-perfect simulation of the concurrent walks execution during the planning phase.

We now focus on $T = 30$ min patrolling missions (the average idleness values tend to stabilize after this amount of time), and examine the impact of different team sizes on the average and worst-case idleness values obtained by I-OPT, RE, and RANDOM. Figure 3 shows the average idleness values obtained in the three environments by the three strategies for a number of robots m varying from 1 to 4. In all the environments and for all the values of m, RE outperforms RANDOM and it is in turn outperformed by I-OPT. These results show that the increased planning complexity of the three strategies is immediately reflected in their effectiveness. Increasing the value of m leads to an advantage for all the strategies, but the performance gain becomes smaller. (For larger graphs with similar structures, we argue that the performance should follow the same decreasing trend for a fixed ratio $|V|/m$.) If we now look at the worst-case idleness values (Fig. 4), we can see that I-OPT is outperformed by RE in Poli and Open. This is not surprising, as RE is designed to always lead the robots towards the vertex currently displaying the highest idleness, while I-OPT plans without considering worst-case idleness values. Again, larger team sizes lead to lower worst-case idleness values in all the environments and for all the strategies,

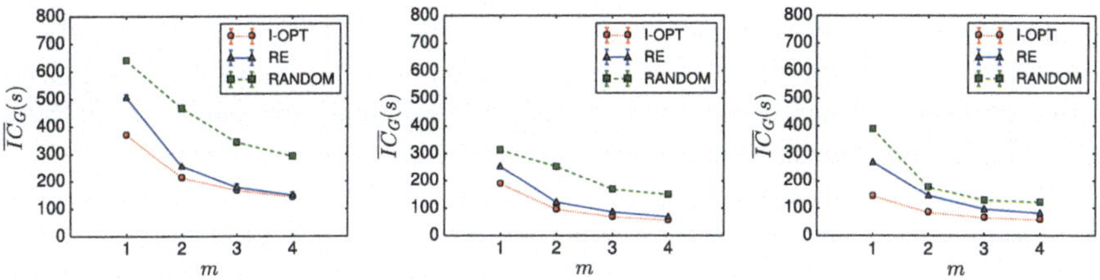

(a) Poli (b) Open (c) Grid

Fig. 3 Average idleness values of I-OPT, RE, and RANDOM for different team sizes for a 30 min mission

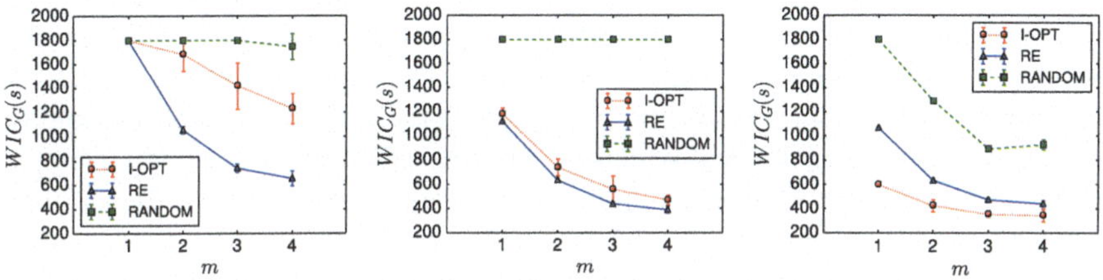

(a) Poli (b) Open (c) Grid

Fig. 4 Worst-case idleness values of I-OPT, RE, and RANDOM for different team sizes for a 30 min mission

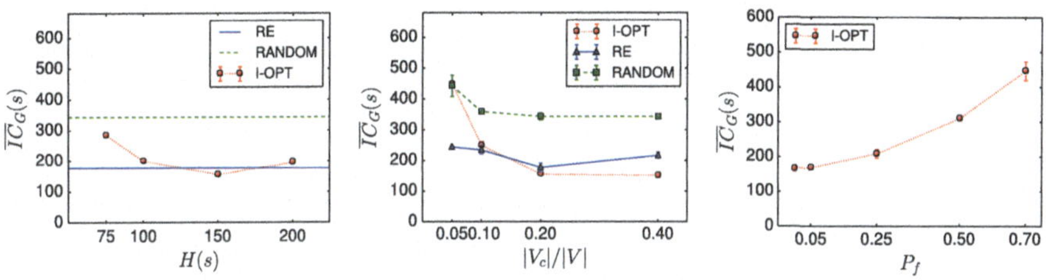

(a) Varying H in I-OPT. (b) Varying the ratio $|V_c|/|V|$. (c) Varying p_f.

Fig. 5 Different parameters in the Poli environment ($m = 3$)

with the only exception of RANDOM in Open, where there is always at least one vertex whose situation is never reported to the BS.

To conclude, we report in Fig. 5 the results obtained with different parameters on the Poli environment for 3 robots. Figure 5a shows the average idleness values obtained for different planning horizons H in I-OPT. As expected, a planning horizon too short or too long leads to worse performance than a balanced one: in the former case, the plan is obtained quickly, but it is not effective; in the latter case, the longer time spent during planning is not compensated by its effectiveness. A way to select a good value for H is to start from a value that allows to reach every node from any node (closely related to the diameter of G) and, then, decrease it if computing strategies takes too long. Figure 5b investigates the impact of varying the ratio $|V_c|/|V|$, while keeping each vertex also in V_m. For $|V_c|/|V| = 0.05$, I-OPT behaves as badly as RANDOM. The reason is that, in this case, there are only two c-type vertices, placed at two opposite corners of the map: therefore, a planning horizon of $H = 150$ s always leaves out from the plans some m-type vertices that are too far. For $|V_c|/|V| = 0.1, 0.2$ the performance is substantially similar for I-OPT and RE, and augmenting the ratio to 0.4 does not lead to substantial improvements for RANDOM and I-OPT, while RE is somehow biased by the presence of several communication vertices since it is more rare that the obtained walk needs to be augmented to reach a c-type vertex, implying that less m-type vertices are visited as a side-effect.

Finally, we focus on a realistic case in which we simulate the temporary unavailability of a communication link. Faults are assumed to happen independently from each other as follows: once a robot reaches a c-type vertex, with probability p_f the expected communication link will not be present (due, e.g., to a temporary network congestion), while with probability $1 - p_f$ the robot and the BS will be able to communicate as before. To deal with these unexpected events, we employ a simple recovery procedure in which the robots decide to move towards the closest c-type vertex. In Fig. 5c, we report results on how performance of I-OPT in the Poli environment degrades with increasing p_f (note that the recovery procedure is independent of the chosen patrolling strategy). Our simple recovery procedure is effectively able to mitigate the impact of communication failures up to 25%, while for higher values of p_f the performance worsens less gracefully, but not dramatically.

6 Conclusions

In this paper, we extended the classical multirobot persistent patrolling framework and the related idleness evaluation metrics in order to account for the presence of a limited number of communication areas and we presented some patrolling strategies tailored for such a setting. Experimental results show the effectiveness of the proposed patrolling strategies in the optimization of our idleness-based evaluation criteria in different environments and settings.

As future works, we intend to adapt the theoretical analysis of cyclic strategies of [4] (relative to the classical worst-case idleness metric) to our framework and to validate our multirobot system also on real robots.

References

1. Banfi, J., Basilico, N., Amigoni, F.: Minimizing communication latency in multirobot situation-aware patrolling. In: Proceedings of the IROS, pp. 616–622 (2015)
2. Basilico, N., Carpin, S.: Online patrolling using hierarchical spatial representations. In: Proceedings of the ICRA, pp. 2163–2169 (2012)
3. Basilico, N., Gatti, N., Amigoni, F.: Leader-follower strategies for robotic patrolling in environments with arbitrary topologies. In: Proceedings of the AAMAS, pp. 57–64 (2009)

4. Chevaleyre, Y.: Theoretical analysis of the multi-agent patrolling problem. In: Proceedings of the IAT, pp. 302–308 (2004)
5. Dames, P., Kumar, V.: Cooperative multi-target localization with noisy sensors. In: Proceedings of the ICRA, pp. 1877–1883 (2013)
6. Elmaliach, Y., Agmon, N., Kaminka, G.: Multi-robot area patrol under frequency constraints. Ann. Math. Artif. Intell. **57**(3–4), 293–320 (2009)
7. Hollinger, G., Singh, S.: Multirobot coordination with periodic connectivity: theory and experiments. IEEE Trans. Robot. **28**(4), 967–973 (2012)
8. Howard, A., Roy, N.: The robotics data set repository (Radish). http://radish.sourceforge.net/ (2003)
9. Iocchi, L., Marchetti, L., Nardi, D.: Multi-robot patrolling with coordinated behaviours in realistic environments. In: Proceedings of the IROS, pp. 2796–2801 (2011)
10. Machado, A., Ramalho, G., Zucker, J.D., Drogoul, A.: Multi-agent patrolling: an empirical analysis of alternative architectures. In: International Workshop on Multi-Agent Systems and Agent-Based Simulation, pp. 155–170 (2002)
11. Mosteo, A., Montano, L., Lagoudakis, M.: Guaranteed-performance multi-robot routing under limited communication range. In: Proceedings of the DARS, vol. 8, pp. 491–502 (2009)
12. Ochoa, S., Santos, R.: Human-centric wireless sensor networks to improve information availability during urban search and rescue activities. Inf. Fus. **22**, 71–84 (2015)
13. Portugal, D., Rocha, R.P.: Multi-robot patrolling algorithms: examining performance and scalability. Adv. Robot. **27**(5), 325–336 (2013)
14. Quigley, M., Gerkey, B., Conley, K., Faust, J., Foote, T., Leibs, J., Berger, E., Wheeler, R., Ng, A.: ROS: an open-source robot operating system. In: ICRA Workshop on Open Source Software (2009)
15. Santana, H., Ramalho, G., Corruble, V., Ratitch, B.: Multi-agent patrolling with reinforcement learning. In: Proceedings of the AAMAS, pp. 1122–1129 (2004)
16. Stephan, J., Fink, J., Kumar, V., Ribeiro, A.: Hybrid architecture for communication-aware multi-robot systems. In: Proceedings of the ICRA, pp. 5269–5276 (2016)
17. Tortonesi, M., Stefanelli, C., Benvegnu, E., Ford, K., Suri, N., Linderman, M.: Multiple-uav coordination and communications in tactical edge networks. IEEE Commun. Mag. **50**(10), 48–55 (2012)
18. Vaughan, R.: Massively multiple robot simulations in stage. Swarm Intell. **2–4**(2), 189–208 (2008)

Part II
Multi-Robot Control

A Comparative Study of Collision Avoidance Algorithms for Unmanned Aerial Vehicles: Performance and Robustness to Noise

Steven Roelofsen, Denis Gillet and Alcherio Martinoli

Abstract Over the past years, the field of small unmanned aerial vehicles has grown significantly and several applications have appeared, requiring always more autonomous flight. An important remaining challenge for fully autonomous unmanned aerial vehicles is collision avoidance between aircraft. In this work, we will compare two collision avoidance algorithms in terms of performance and robustness to sensor noise. We will leverage both experiments with real vehicles and calibrated, realistic simulations to get an insight of the effect of noise on collision avoidance. Our results show that although algorithms that use velocity as input are better in minimizing velocity variation and generally produces more efficient trajectories, they are less robust to perception noise. On the other hand, position-based algorithms that typically generate slower and longer avoidance maneuvers, become competitive at high levels of sensor noise.

Keywords Collision avoidance · Unmanned aerial vehicle · Robustness

1 Introduction

Over the past years, the field of small Unmanned Aerial Vehicles (UAVs) has grown significantly and more applications appear as time goes on, ranging from surveillance and mapping to delivering of goods, and require always more autonomy for the

---

S. Roelofsen (✉) · A. Martinoli
Distributed Intelligent Systems and Algorithms Laboratory (DISAL), School of Architecture, Civil and Environmental Engineering, École Polytechnique Fédérale de Lausanne (EPFL), Lausanne, Switzerland
e-mail: steven.roelofsen@epfl.ch

A. Martinoli
e-mail: alcherio.martinoli@epfl.ch

S. Roelofsen · D. Gillet
Coordination and Interaction System Group (REACT), School of Engineering, École Polytechnique Fédérale de Lausanne (EPFL), Lausanne, Switzerland
e-mail: denis.gillet@epfl.ch

R. Groß et al. (eds.), *Distributed Autonomous Robotic Systems*,
Springer Proceedings in Advanced Robotics 6,
https://doi.org/10.1007/978-3-319-73008-0_6

UAVs. Up to date, research activities have focused mainly on self-localization [12], path planning [6] and navigation [13, 14]. Importantly, ensuring safety is one of the remaining challenges that needs to be overcome in order to achieve fully autonomous UAVs. More specifically, there is yet no Sense And Avoid (SAA) system reliable enough to allow for fully autonomous UAVs.

The robustness of a SAA system is not only determined by the individual robustness of its components (i.e., sensors, estimators or control algorithms) but also the interaction among them. Imperfections in one component of the system (e.g., sensing) can lower the performance and robustness of another one (e.g., control). An example of the effect of sensor imperfection having significant effect on the actuation is presented in [9].

Several collision avoidance algorithms exist in the literature and most of them fall in one of the two following categories. First, the algorithms based on Velocity Obstacle (VO) [5]. VO-based algorithms allow for collision avoidance while minimising the change in velocity and are applicable to a large set of systems [1]. Second, the algorithms derived from Potential Fields (PF) [7]. Those algorithms allow for more diverse behaviors than VO but usually at the expense of optimality in the velocity space (i.e. minimizing changes in velocity). The algorithms are also less generalizable than those of the VO class, but allow for integration of a large range of both actuation [8] and sensing constraints [10].

In this paper we present our effort to assess the robustness of different collision avoidance algorithms in presence of sensor noise by leveraging both real experiments and simulations. We performed real experiments for calibration and validation of simulation tools. We use high-fidelity simulation to go beyond our experimental facility's space limitations and systematically assess the performance avoidance algorithms in more generalizable scenarios.

1.1 Collision Avoidance Algorithms

In this work, two algorithms are investigated as case study: those that only use position information and those that use both position and velocity information of the other aircraft. So far we implemented two collision avoidance algorithms.

ORCA The Optimal Reciprocal Collision Avoidance [2] is based on the concept of Velocity Obstacle (VO) which is the set of all velocities that will lead to a collision. Based on the VO, ORCA builds one half plane of forbidden velocities per obstacle. Putting all the half-planes together results in a convex polygon of allowed velocities from which the optimal velocity (i.e., the closest to the desired velocity) is computed. ORCA also relies on reciprocity where each quadrotor only performs half of the velocity change that would be needed if they would not cooperate. ORCA has already been implemented on real quadrotors in [4].

FOVA The Field Of View Avoidance algorithm was presented in [10]. The algorithm uses a virtual potential field to navigate in its environment and smoothly switches to a turning behavior when another quadrotor approaches. Because the motion of the quadrotor is constrained to always go forward by design, the quadrotors will move away from the collision point as soon as the quadrotor turned enough for the collision point to be out of its field of view (i.e., rear). The turn rate increases as the distance between the quadrotor decreases in order to guarantee avoidance. Once the collision point is out of the field of view of the quadrotor, the quadrotor goes back to its navigation function. A boundary zone makes for a smooth transition. Contrary to ORCA, this algorithm relies only on position information. It has been proven to avoid collision even with constraints on the sensor's field of view. FOVA has already been implemented on real quadrotors in [9].

2 Technical Approach

In this work, the experiments are carried out with two quadrotors, highly agile platforms limiting the impact of the dynamical constraints of the vehicles on the results. To be as generic as possible, the sensory input is emulated, leveraging the millimetric precision of an external Motion Capture System (MCS). Such a solution also allows us to set the level of sensing noise with ease and precision.

To allow for fair comparison between the collision avoidance algorithms, the noise level needs to be equivalent, despite the fact that the two algorithm use different inputs. Both algorithms use position, but only ORCA uses velocity. To remain fair, the errors in position and velocity need to be linked to a single parameter. This is done by considering a sensor that only returns a position measurement that is affected by some Gaussian noise. The velocity is derived from the position using a Kalman filter. This way, both algorithms have access to the same information (filtered position information) but may not use all of the information available (FOVA does not use the velocity information that is contained in the position measurements).

For a meaningful comparison, the algorithms are implemented to take into consideration the effect of sensor uncertainty. All algorithms have been made noise-resistant similarly, using similar techniques to what is presented in [11]. More specifically, in both algorithms, the radius of the other UAV is increased by one standard deviation of the estimated position error given by the Kalman filter's covariance matrix. This increased radius is the only modification needed for the FOVA algorithm. For ORCA, the VO is also shifted by half a standard deviation of the velocity error as described in [11].

Finally, because the FOVA algorithm uses the sensor range as a key parameter, the sensing range is limited in the same way for both algorithms. This also guarantees that the quadrotors do only sense each other in a collision avoidance situation (i.e., not in standby mode). Additionally to a limited range, the field of view of the quadrotors for the FOVA algorithm is 220°, a parameter explained in [10]. This parameter is needed as it is key in the FOVA algorithm. The ORCA algorithm has no such FOV limitation.

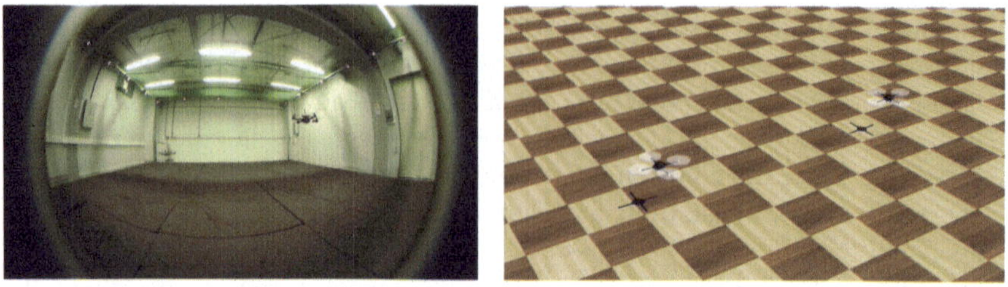

(a) Experimental facility (b) Two quadrotors in Webots

Fig. 1 On the left: photo of the experimental facility with two quadrotors flying (the second quadrotor is in the lower left corner of the flying arena). The bright redish dots all around the flying arena are the cameras of the MCS. On the right: two quadrotors simulated in the high-fidelity robotic simulator

2.1 Experimental Setup: Reality and Simulation

All algorithms are experimentally evaluated with off-the-shelf Hummingbird quadrotors from Ascending Technologies. The quadrotors are equipped with a Gumstix Airstorm computation module, providing a Linux-based operating system and a wireless communication link. The Kalman filter, the collision avoidance algorithms, and most of the control run on the embedded Gumstix. They are implemented using the Robotic Operating System (ROS) framework, leveraging the recording functionality of rosbag. Part of the low-level control (e.g., motor speed control) runs on dedicated, proprietary hardware of the Hummingbird quadrotor and is thus not directly accessible. We used a similar technique for estimating both the parameters of the proprietary, low-level control software as well as the physical parameters of our vehicles (e.g., thrust coefficient of the propellers). The Gumstix and the Hummingbird are interfaced using the *asctec_hl_interface* ROS package. The localization was provided by a MCS, providing millimetric precision pose information. The experiments were performed in a room of size $6 \times 8 \times 3$ m. A picture of the experimental facility with two quadrotors flying can be seen in Fig. 1a.

Due to the limited size of our experimental facility, the number and variety of possible scenarios is limited. For a more thorough study, simulations are leveraged to provide a larger palette of scenarios. Our simulations are performed using Webots, a realistic sub-microscopic simulator. The simulator is able to interface with ROS, allowing it to run the same code as the one implemented on our quadrotors. A screenshot of the simulation environment can be seen in Fig. 1b.

2.2 Implementation

Several software modules are necessary in order to perform the experiments presented in this paper: the MCS management software, a Kalman filter, the avoidance

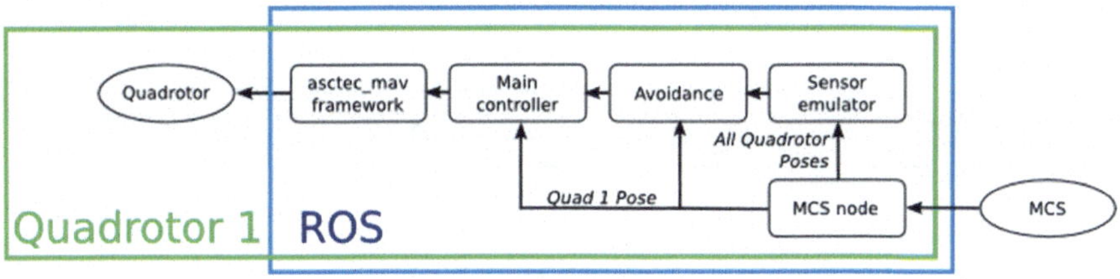

Fig. 2 Interaction between the different rosnodes on one of the two real quadrotors used in the experiments

algorithm, and both the high-level and the low-level control algorithms. The MCS management and the low-level control software are proprietary modules and each one has a dedicated rosnode that allows to communicate with them. The Kalman filter node gets position data from the MCS (which is assumed to be noiseless) at 100 Hz, adds artificial Gaussian noise, and filter the data using a Kalman filter with optimized gain to obtain both a position and velocity estimate. Both position and velocity are forwarded to the collision avoidance algorithm, which computes a desired velocity and heading that both avoids collision and brings the aircraft to the goal. ORCA has been implemented leveraging the existing library [3], as the FOVA algorithm has been implemented by us based on the available literature. The desired velocity and heading are sent to the high-level control node, which translate then to desired thrust and orientation of the quadrotor. The control loop is performed at 20 Hz. Those commands are sent to the low-level control node that is in charge of setting the motors speeds of the quadrotor to ensure convergence to the desired state. The data flow in the quadrotor is shown in Fig. 2.

To automatize the simulations, we added two more types of ROS nodes. Each simulated quadrotor has a Flight Manager node responsible to send commands to other rosnodes (e.g., command to take off the quadrotor). The Flight Manager nodes are supervised by a Scenario Manager common to all quadrotors; the Scenario Manager coordinates the quadrotor on how and when the maneuver should be performed. The MCS is replaced with a Webots supervisor that is able to retrieve the position of the simulated quadrotors and send the data to other rosnodes. The interactions between the rosnodes in simulation is presented in Fig. 3.

2.3 Simulation Calibration

To obtain simulation results comparable to reality, the simulation needs to be calibrated. Physical parameters such as weight are directly measured on the aircraft. Because the internal structure of the low-level control (responsible to control the motor speeds to bring the quadrotor to the correct attitude) is unknown to us, its parameters need also to be estimated. The unknown control scheme is assumed to be a PID controller and to have the structure described by Eqs. 1–3:

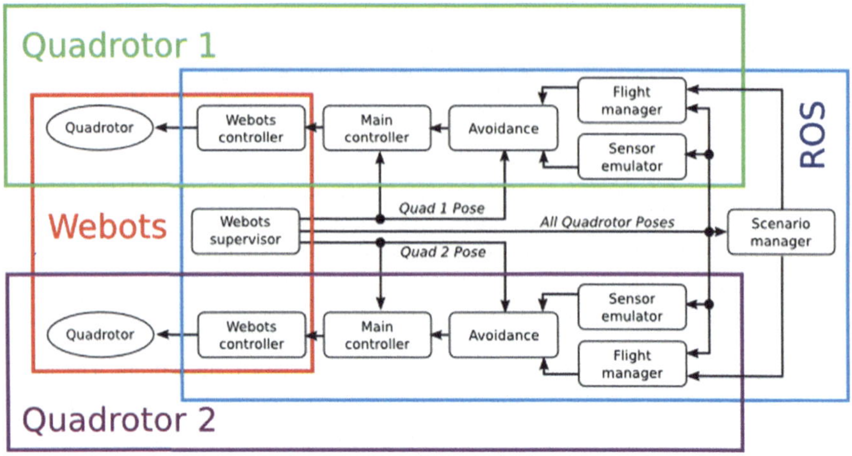

Fig. 3 Interaction between the different rosnodes in the simulation. The simulation framework uses the same avoidance and sensor emulation code that is implemented on the quadrotors

$$M_x = K_a(\phi_t - \phi) - K_{da}\Omega_\phi \tag{1}$$

$$M_y = K_a(\theta_t - \theta) - K_{da}\Omega_\theta \tag{2}$$

$$M_z = K_{dy}(\Omega_{\psi,t} - \Omega_\psi) - K_{ddy}\frac{d\Omega_\psi}{dt} - K_{Iy}\int \Omega_\psi \tag{3}$$

with ϕ, θ and ψ being roll, pitch and yaw respectively, and Ω_ϕ, Ω_θ and Ω_ψ their respective angular rates. ϕ_t and θ_t are target roll and pitch respectively. $\Omega_{\psi,t}$ is a target yaw rate. M_x, M_y and M_z and the desired torques to apply along the x, y and z axis in order to get the quadrotor to the desired state. K_a, K_{da}, K_{dy}, K_{ddy} and K_{Iy} are the control parameters that need to be calibrated.

The five control parameters are optimized to minimize the difference between trajectories obtained from simulation and real experiments. The optimization was carried out using Particle Swarm Optimization (PSO), where each particle is a vector containing the five control parameters. Their fitnesses are defined as the RMS difference between the average trajectory obtained through real experiments and the average trajectory generated in simulation. The real experiment data set is composed of a quadrotor trying to follow a predefined trajectory eleven times. The simulation trajectory is obtained by simulating the quadrotor following the same predefined trajectory. The simulations are performed four times to average out the timing variability of the ROS framework (i.e., messages do not arrive with a deterministic timing).

While performing the experiments, it was observed that the quadrotors were affected by the airflow they were generating, resulting in a not perfectly steady flight. To replicate this disturbance in simulation, we added a first order Gauss–Markov process on the thrust and torques in roll, pitch and yaw. The disturbances for each time step are described as:

$$T_d[n+1] = 0.995T_d[n] + 0.005(W(0, 2)) \tag{4}$$
$$M_{x,d}[n+1] = 0.995M_{x,d}[n] + 0.005(W(0, 0.12)) \tag{5}$$
$$M_{y,d}[n+1] = 0.995M_{y,d}[n] + 0.005(W(0, 0.12)) \tag{6}$$
$$M_{z,d}[n+1] = 0.995M_{z,d}[n] + 0.005(W(0, 0.02)) \tag{7}$$

with $W(\mu, \sigma)$ a Gaussian process of mean μ and standard deviation σ. The time step is 10 ms. The parameters have been set empirically so that the average acceleration between experimental and simulated data is similar. The validation on the simulator's faithfulness to reality is presented in Sect. 3.2.

2.4 Scenarios

The scenario for the real experimental setup is rather simple because of the limited space available; it is a head-on collision between two quadrotors, where each one starts on one side of the room and tries to get to other quadrotor's position. The initial position of the quadrotors are [0, 1.6, 1] m and [0, −1.6, 1] m, the initial inter-vehicle distance is therefore of 3.2 m. The desired speed was set to 0.3 m/s for all algorithms. The experiments have been carried out with Gaussian noise levels of 0.01, 0.03, 0.1, 0.3 and 1.0 m (standard deviation). Table 1 presents the error in position and velocity after the emulated measurements went through the Kalman filter. The values reported in Table 1 are directly used in the avoidance algorithms to make them resistant to noise. For simplicity, it is assumed that both algoritms know the radius, without considering noise, of the other quadrotor to be 0.35 m. For the real experiments, over 20 collision avoidance maneuvers have been carried out for each algorithm and noise level combination, for a total of more than 200 data points.

In simulation, we are not limited by spatial constraints, allowing for more extended scenarios. There are two main differences between simulated and real scenarios. First, the initial distance between the quadrotors is 8 m for a head-on configuration, or put differently, each quadrotor starts 4 m away from the collision point and try to reach a point 8 m in front of them. Second, the angle between the desired directions of the quadrotors is changed. The initial and final positions are modified in order to have the lines defined by the two points to have the desired angle. The angles are 180° (head-on), 150°, 120°, 90°, 60°. For each possible angle, 20 collision avoidance maneuvers are performed. The plots presented in Figs. 6, 7 and 8 report aggregated results for the five angles defined above. As a result, each curve of Figs. 6, 7 and 8 represents

Table 1 Noise levels expressed with their standard deviations, and their corresponding errors in position and velocity obtained after the Kalman filter

Noise level [m]	0.01	0.03	0.1	0.3	1
Position RMSE [m]	0.0071	0.0174	0.0448	0.1044	0.2614
Velocity RMSE [m/s]	0.0770	0.1052	0.1454	0.1936	0.2640

500 avoidance maneuvers for each algorithm (5 noise levels, each with 5 different angles, each angle and noise level combination repeated 20 times). Other parameters (e.g., desired speed of 0.3 m/s) are left the same, unless explicitly mentioned.

2.5 *Metrics*

To compare the algorithms, we use three metrics: first, we compare the proportion of collisions as an indicator of safety; we consider a collision occurs when the horizontal distance between the two quadrotors is below twice their radius, or 0.7 m. Second, we also compare the path length as an approximation of the energy consumption and as a result an indication of the efficiency of the collision avoidance algorithms. Finally, we compare the average acceleration during the avoidance maneuver, as an indicator of overreaction due to noisy sensing. For all plots in Figs. 4, 6, 7 and 8, the thick solid lines represent the median over the multiple runs of the same experiment. For the average acceleration and path length metrics, the color patches represent the interval between the upper and lower quartiles while for the proportion of collision they represent the 95% confidence intervals computed with the Clopper–Pearson method.

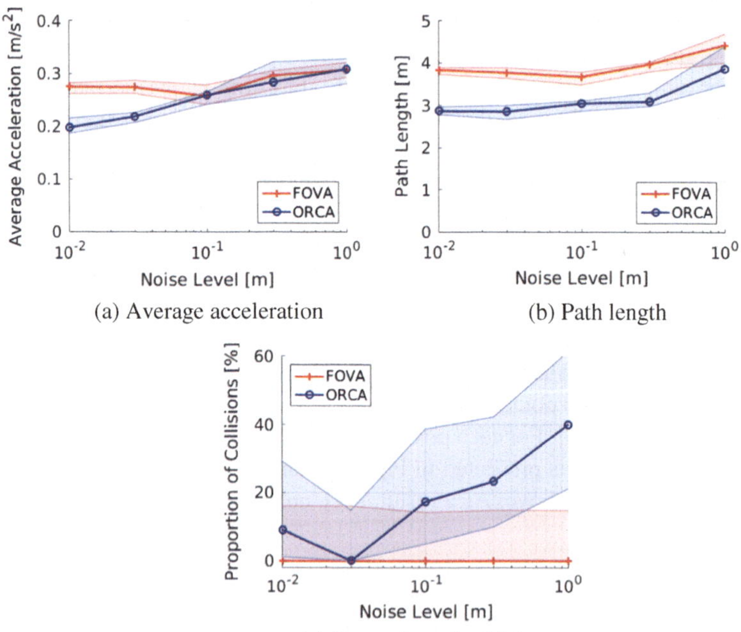

(a) Average acceleration

(b) Path length

(c) Proportion of collisions

Fig. 4 The evolution of different metrics as function of sensor noise. The data were acquired using real quadrotors. **a** The average acceleration for the two algorithms as function of sensor noise. **b** The path length as function of the sensor noise. **c** The proportion of collisions for different noise levels

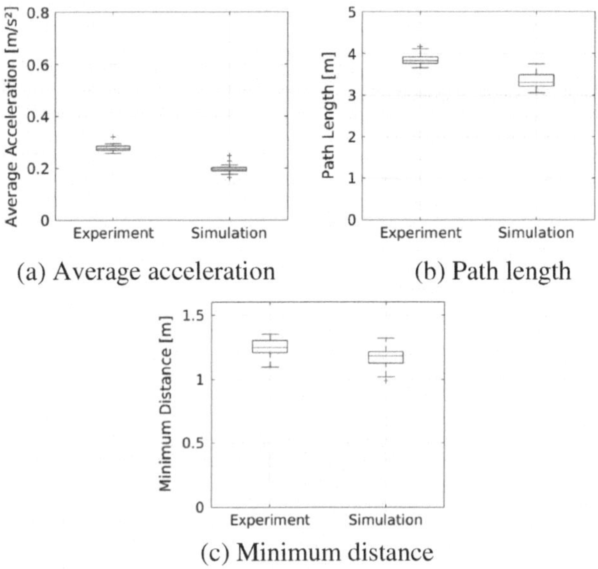

Fig. 5 Comparison between simulation and real experiments

3 Results

In this section, we first present the results of the real experiments followed by simulation and a related discussion on how they compare with reality.

3.1 Real Experiment Results

Figure 4 shows the results obtained for the two algorithms when deployed on real robots. Clearly, the performance of ORCA degrades more as the sensor noise increases. The most significant difference is in the proportion of collisions. Where the FOVA algorithm never had a collision, ORCA has collisions for higher noise levels. The non-zero proportion of collisions for ORCA at the lowest noise level is due to actuation perturbations. In that case, ORCA avoids with such a small distance margin that any actuation perturbation (e.g., turbulent airflow) experienced by the quadrotors can bring them below the collision distance. This is not the case for the FOVA algorithm, which, similarly to most of PF-based approaches, always avoids with a distance larger than the strict minimum. In order for ORCA to become totally noise-resistant, an additional margin on the obstacle radius needs to be considered. However, doing so will lower its performance in the other metrics. For the average acceleration metric, the FOVA algorithm starts with the highest acceleration but remains relatively constant compared to ORCA, the latter increasing as noise increases.

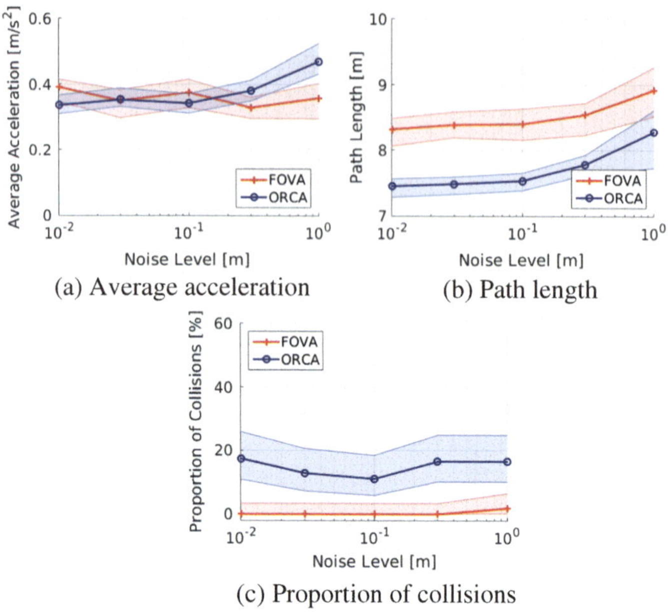

(a) Average acceleration (b) Path length

(c) Proportion of collisions

Fig. 6 The evolution of different metrics as function of sensor noise. The data was obtained in simulation. **a** The average acceleration for the two algorithms as function of sensor noise. **b** The path length as function of the sensor noise. **c** The proportion of collisions for different noise levels

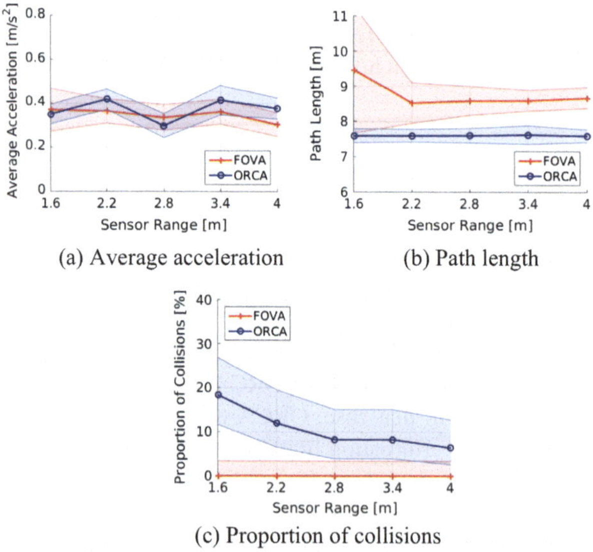

(a) Average acceleration (b) Path length

(c) Proportion of collisions

Fig. 7 The evolution of different metrics as function of sensor range. The data were obtained in simulation. **a** The average acceleration for the two algorithms as function of sensor noise. **b** The path length as function of the sensor noise. **c** The proportion of collisions for different noise levels

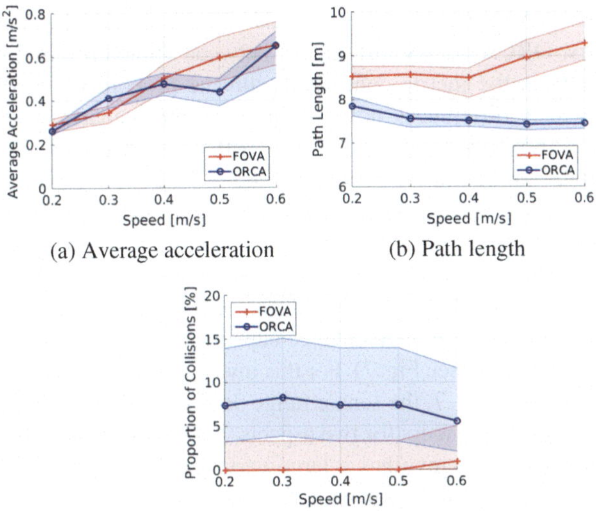

(a) Average acceleration (b) Path length

(c) Proportion of collisions

Fig. 8 The evolution of different metrics as function of vehicles' maximum speed. The data were obtained in simulation. **a** The average acceleration for the two algorithms as function of sensor noise. **b** The path length as function of the sensor noise. **c** The proportion of collisions for different noise levels

Finally, both algorithms show longer path lengths as noise increases, partially due to the increase of the collision radius, correspondingly a priori implemented in the algorithms (see Sect. 2). Again, the difference of performance between position and velocity-based algorithms decreases as sensor noise increases.

3.2 *Simulation Results*

To validate the calibration of our simulator, we compared the performance of the three metrics described above for the FOVA algorithm in simulation and with real data, both with a Gaussian noise level of 0.01 m. The resulting data are shown in Fig. 5. Contrary to all other simulations that use the extended scenario, the simulations carried out for Fig. 5 faithfully reproduced the scenario used in reality (i.e., same initial position and heading as used for the real experiments). We see that for all the metrics considered (acceleration, path lenght, and minimum distance between the quadrotors) the simulated data assume smaller values than those gathered through experiments. This is probably due to the airflow generated by a quadrotor that tends to push away the other one, an effect that is, only roughly approximated in simulation (i.e. the approximation does not simulate that the airflow push the quadrotors away from each other).

Figure 6 presents the evolution of the three metrics for different sensory noise levels. For the FOVA metric, experimental and simulation data show similar trends. On the contrary, ORCA shows differences between simulation and experimental results, especially in the metric concerned with the proportion of collisions. The explanation for such differences is two-fold. First, the scenario is spatially more extended and does not only consists of head-on collisions. Head-on collision is harder to avoid as it is the configuration where the relative speed between the quadrotors is maximal, giving them less time to avoid and therefore explaining why experiments show a higher proportion of collision at high sensor noise levels. Second, the real quadrotors are pushing each other away with their airflow, which is not the case in simulation, explaining the flat curve over different levels of noise of Fig. 6c.

The effect of sensor range on avoidance capability for both algorithms was also investigated in simulation (see Fig. 7). For this investigation, the noise level was kept constant at 0.1 m. From Fig. 7, the sensor range does not appear to have a significant effect on both algorithms, except for two aspects. First, the path length for the FOVA algorithm at small sensor range has a large variance. This is because with such a small sensor range, the FOVA algorithm has to perform very aggressive turns to avoid. Due to its inertia, the quadrotor overshoots its desired yaw angle, doing a full 360° turn. Because the FOVA avoids by only turning in one direction, both quadrotors spin quickly without being able to move away from each-other. When they eventually are able to separate (after some long time), all the spinning sums up to a long distance. The second notable effect of sensor range is that the proportion of collisions decreases as the sensor range increases for the ORCA algorithm. The reason is that the ORCA algorithm does not prefer a specific side on which to avoid; both quadrotors need thus to converge on which side the avoidance will be done. With sensing noise, the ORCA algorithm will oscillate between sides, a phenomenon also known as reciprocal dance. A larger sensor range provides the ORCA algorithm more time to converge to a stable solution.

The effect of the quadrotor's maximal speed was also investigated. As for the sensor range, the simulations were performed with a noise level of 0.1 m. The results of the simulations are shown in Fig. 8. The ORCA algorithm is not significantly affected an increased speed of the vehicles. However, it is notable an increase of the average acceleration for higher speeds due to the corresponding need for more aggressive maneuvers. The FOVA algorithm is more affected by an increase in speed. Besides an increase of the average acceleration for higher speed, the path length is also increased and even suffered from a collision when a maximal speed of 0.6 m/s was allowed. This is the result of the FOVA algorithm relying on turning, which is a less effective type of maneuver on a quadrotor when compared to ORCA's acceleration sideways.

4 Conclusion

In this work, we compare the performance of two collision avoidance algorithms in presence of noisy sensing. In particular, we evaluate the impact of the noise level, the sensing range, and the speed of the vehicles using a combination of calibrated simulation tools and real robot experiments. One algorithm is based on a Potential Field approach and only relies on positional data to perform avoidance, as the second algorithm based on Velocity Obstacles also needs velocity information. For fairness, the sensor only provides position measurements, velocity being acquired through Kalman filtered successive position measurements.

From this work, we draw two main lessons. First, the VO approach, due to its optimal nature, will show better performance in fuel-consumption-related metrics (in this case path length). Second, VO is significantly more collision prone because the algorithm tries to avoid as close as possible the obstacles on the vehicle's path. On the other hand PF-based algorithms stay at a safer distance from encountered obstacles.

Future work will include additional experimental scenarios and a thorough theoretical analysis. It will also consider the performance of the algorithms under specific sensor and actuator constraints (e.g., limited field of view, acceleration limits, different vehicle dynamics).

Acknowledgements This work has been financially supported by Honeywell, and has benefited of the administrative and technical coordination of the EPFL Transportation Center.

References

1. Bareiss, D., van den Berg, J.: Generalized reciprocal collision avoidance. Int. J. Robot. Res. **34**(12), 1501–1514 (2015)
2. van den Berg, J., Guy, S.J., Lin, M., Manocha, D.: Reciprocal n-body collision avoidance. In: Proceedings of International Symposium on Robotics Research 2009, Series Springer Tracts in Advanced Robotics, pp. 3–19. Springer, Berlin (2011)
3. van den Berg, J., Guy, S.J., Snape, J., Lin, M.C., Manocha, D.: RVO2 Library: Reciprocal Collision Avoidance for Real-Time Multi-Agent Simulation (2008–2015). http://gamma.cs.unc.edu/RVO2/
4. Conroy, P., Bareiss, D., Beall, M., van den Berg, J.: 3-D Reciprocal Collision Avoidance on Physical Quadrotor Helicopters with On-board Sensing for Relative Positioning (2014). arXiv:1411.3794
5. Fiorini, P., Shiller, Z.: Motion planning in dynamic environments using velocity obstacles. Int. J. Robot. Res. **17**(7), 760–772 (1998)
6. Forster, C., Faessler, M., Fontana, F., Werlberger, M., Scaramuzza, D.: Continuous on-board monocular-vision-based elevation mapping applied to autonomous landing of micro aerial vehicles. In: IEEE International Conference on Robotics and Automation, pp. 111–118 (2015)
7. Kim, J.O., Khosla, P.K.: Real-time obstacle avoidance using harmonic potential functions. IEEE Trans. Robot. Autom. **8**(3), 338–349 (1992)
8. Panyakeow, P., Mesbahi, M.: Decentralized deconfliction algorithms for unicycle UAVs. In: American Control Conference, pp. 794–799 (2010)

9. Roelofsen, S., Gillet, D., Martinoli, A.: Reciprocal collision avoidance for quadrotors using on-board visual detection. In: IEEE/RSJ International Conference on Intelligent Robots and Systems, pp. 4810–4817 (2015)
10. Roelofsen, S., Martinoli, A., Gillet, D.: Distributed deconfliction algorithm for unmanned aerial vehicles with limited range and field of view sensors. In: American Control Conference, pp. 4356–4361 (2015)
11. Snape, J., van den Berg, J., Guy, S.J., Manocha, D.: The hybrid reciprocal velocity obstacle. IEEE Trans. Robot. **27**(4), 696–706 (2011)
12. Weiss, S., Achtelik, M.W., Lynen, S., Chli, M., Siegwart, R.: Real-time Onboard Visual-Inertial State Estimation and Self-Calibration of MAVs in Unknown Environments. In: IEEE International Conference on Robotics and Automation, pp. 957–964 (2012)
13. Yang, S., Scherer, S.A., Schauwecker, K., Zell, A.: Autonomous landing of MAVs on an arbitrarily textured landing site using onboard monocular vision. J. Intell. Robot. Syst. 27–43 (2013)
14. Zufferey, J., Beyeler, A., Floreano, D.: Autonomous flight at low altitude with vision-based collision avoidance and GPS-based path following. In: IEEE International Conference on Robotics and Automation, pp. 3329–3334 (2010)

A Decentralized Control Strategy for Resilient Connectivity Maintenance in Multi-robot Systems Subject to Failures

Cinara Ghedini, Carlos H. C. Ribeiro and Lorenzo Sabattini

Abstract This paper addresses the problem of topology control for dealing with node failures in networks of multiple robots. While connectivity maintenance has been widely addressed in the literature, issues related to failures are typically not considered in such networks. However, physical robots can fail (i.e. stop working) due to several reasons. It is then mandatory to consider this aspect, as connectivity maintenance is usually critical. In fact, failures of a small fraction of robots — in particular on those that play a crucial role in routing information through the network — can lead to connectivity loss. In this paper, we present a decentralized estimation procedure for letting each robot (a) assess its degree of robustness w.r.t. to connectivity maintenance under the occurrence of failures in its neighborhood, and (b) take actions to improve it when needed. This estimation is combined with a connectivity maintenance control law, thus providing a mechanism that ensures, in the absence of failures, both the network connectivity and an improvement in the overall robustness to failures. In addition, for failures scenarios, the mechanism is able to postpone, or even avoid network fragmentation, as verified through a set of validation experiments.

1 Introduction

Applications based on groups of self-organized mobile robots are becoming pervasive in communication, monitoring, traffic and transportation systems. Groups of

---

C. Ghedini (✉) · C. H. C. Ribeiro
Computer Science Division, Aeronautics Institute of Technology,
São José dos Campos, São Paolo, Brazil
e-mail: cinara@ita.br

C. H. C. Ribeiro
e-mail: carlos@ita.br

L. Sabattini
Department of Sciences and Methods for Engineering (DISMI),
University of Modena and Reggio Emilia, Modena and Reggio Emilia, Italy
e-mail: lorenzo.sabattini@unimore.it

R. Groß et al. (eds.), *Distributed Autonomous Robotic Systems*,
Springer Proceedings in Advanced Robotics 6,
https://doi.org/10.1007/978-3-319-73008-0_7

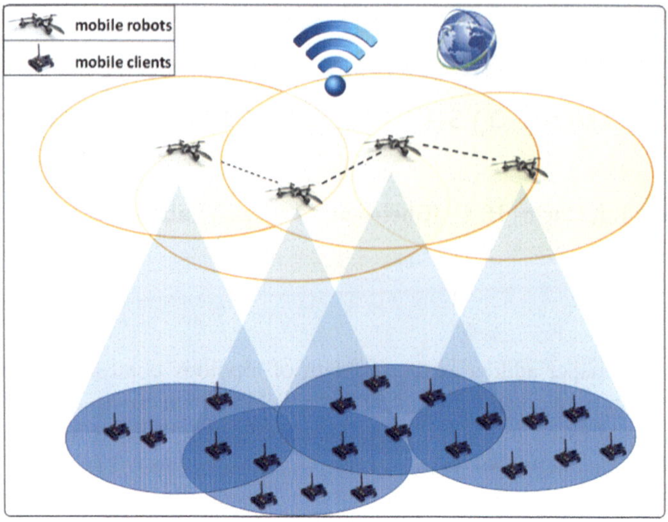

Fig. 1 Example of an application scenario

mobile robots, if equipped with appropriate communication devices, can be exploited to create an infrastructure network to provide communication services. For instance, interconnected mobile robots can provide rescuing, acting as devices in an exploration task, or serving clients (e.g., mobile phones, laptops, tablets, etc.) in a network, with the advantage of doing so without the existence of a previously defined infrastructure and with a high degree of autonomy (Fig. 1). In fact, these features are the key requirements for extending a pressing issue nowadays: how to provide the communication infrastructure that makes it possible for generic clients to access the Internet, cloud technologies, and communication services in several unstructured environments and situations. Thus, the applicability of interconnected mobile robots is expanding from the classical monitoring and exploration tasks to an essential technology that supports several kinds of services to be available everywhere and at any time. In particular, the Internet, considered as an indispensable part of the critical information infrastructure for many personal and business applications, is now expected to be always available, offering uninterrupted service regardless of domain constraints [1, 16, 18].

In this sense, a critical aspect of providing network services for unstructured environments employing interconnected mobile robotic systems is that robots are prone to failures due to hardware or communication issues, and — as it is well known from the literature on Complex Networks — successive or cascading failures, particularly of agents playing a central role in the network topology, may easily result in inoperative or reduced services [3, 5, 8, 11, 15].

Figure 2a illustrates random network topologies highlighting some agents whose failures can strongly affect the network connectivity, characterizing vulnerable topological configurations. In the context of this work, a vulnerable topological configuration means that the network is potentially able to fragment if some nodes fail.

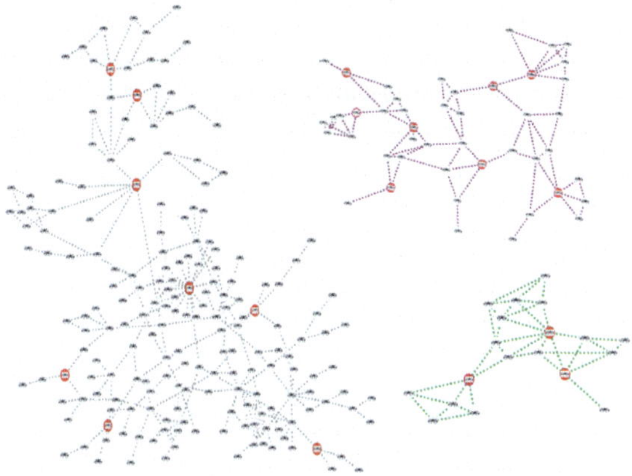

(a) Random network topologies. Failures in highlighted nodes can negatively affect the network connectivity, thus producing inoperative or reduced communication services.

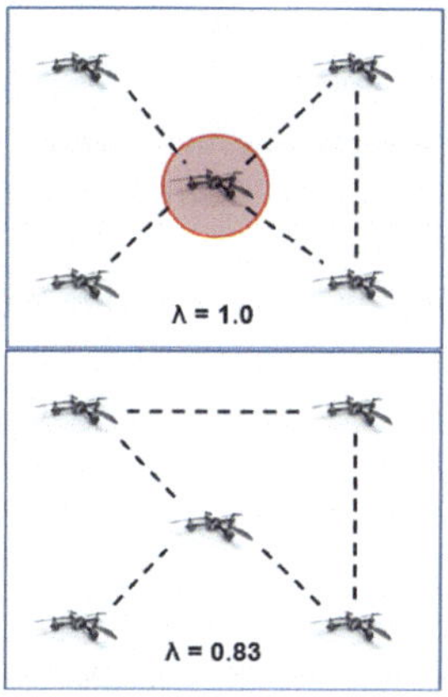

(b) A network with larger algebraic connectivity (top) is more affected w.r.t. central node failure.

Fig. 2 Failures affect the connectivity of the network

Thus, we define the *resilience* of the multi-robot systems regarding its *robustness to failures*, that is the system's capacity to mitigate the effects of node failures through predictive actions that avoid topological configurations vulnerable to such effects.

Detecting such vulnerable configurations is not trivial because the topology emergent from these applications is dynamic; thus, its size and properties are most of the time unknown, costly to estimate and time varying. These features impose constraints on the design of solutions, such as a need for relying mostly on information that is locally available and straightforward to obtain, compute and update.

Being a fundamental issue in multi-robot systems, the connectivity maintenance problem has been widely addressed in the literature. Therefore, several approaches can be found that ensure that, if the communication graph is initially connected, then it will remain connected as the system evolves [2, 4, 6, 12, 13, 17, 19–21]. In particular, the control strategy proposed in [20] ensures the network connectivity maintenance through a decentralized estimation of the algebraic connectivity, using the well-known property that the algebraic connectivity approaches 0 when the network is poorly connected. It is worth noting, however, that the algebraic connectivity *per se* is not effective as an indicator of the overall network fragility. A representative example of this fact is depicted in Fig. 2b: the network on the top has a larger algebraic connectivity (λ) than the network on the bottom, but it is more affected by a central node failure that might disconnect it into three subgraphs.

We, therefore, argue that, despite the algebraic connectivity being an estimator of how well a network is globally connected, it is not a suitable indicator to detect local vulnerable configurations and, thus, does not provide enough information to produce a more resilient network topology. On the other hand, connectivity is a crucial requirement in decentralized multi-robot systems: in order to achieve a common objective robots may need to exchange information. Thus, enhancing the robustness to failures is a fundamental requirement to be addressed in connectivity maintenance approaches. In this sense, a recent work proposes mechanisms, based on locally available information, for detecting and mitigating vulnerable topological configurations, consequently increasing the network resilience [9].

The contribution of this paper is the definition of a novel combined control law that, in the absence of failures, guarantees connectivity maintenance while improving the network robustness to failure and ensuring collision avoidance of robots with obstacles among them. It is important to emphasize that there is no way of guaranteeing that networks will remain connected in case of failures because the failure probability distribution is usually unknown and failures are not controllable. Hence, the central aspect of providing more resilient networks is to improve their capacity to tolerate disturbance without fragmenting. More generally, a resilient network is expected to exhibit the ability to react to undesirable states or unpredictable events through adaptive processes.

The proposed control law is validated comparing its performance regarding different parameterizations of gains in the presence of failures. The results demonstrate that the combined control law is able to increase the resilience of the system by adapting the network topology to accommodate failures, postponing or even avoiding network disconnection.

The rest of this paper is organized as follows. The necessary background on network properties is presented in Sect. 2. Section 3 discusses in details the problem of connectivity maintenance considering failures. Section 4 describes the proposed combined model, and Sect. 5 presents the simulation model and discusses the results. Concluding remarks are given in Sect. 6.

2 Background on Network Properties

We will hereafter define some quantities that can be useful for evaluating node and network connectivity, and robustness to failures.

Consider an undirected graph $\mathscr{G}$, where $\mathscr{V}(\mathscr{G})$ and $\mathscr{E}(\mathscr{G}) \subset \mathscr{V}(\mathscr{G}) \times \mathscr{V}(\mathscr{G})$ are the node set and the edge set, respectively. Moreover, let $W \in \mathbb{R}^{N \times N}$ be the weight matrix: each element w_{ij} is a positive number if an edge exists between nodes i and j, zero otherwise. Since $\mathscr{G}$ is undirected, then $w_{ij} = w_{ji}$.

Now, let $\mathscr{L} \in \mathbb{R}^{N \times N}$ be the Laplacian matrix of graph $\mathscr{G}$ and $D = \text{diag}(\{k_i\})$ be the degree matrix of $\mathscr{G}$, where k_i is the degree of the i-th node of $\mathscr{G}$, i.e. $k_i = \sum_{j=1}^{N} w_{ij}$. The (weighted) Laplacian matrix of $\mathscr{G}$ is then defined as $\mathscr{L} = D - W$.

As is well known from algebraic graph theory, the Laplacian matrix of an undirected graph exhibits some remarkable properties regarding its connectivity [10]. Let $\lambda_i, i = 1, \ldots, N$ be the eigenvalues of the Laplacian matrix, then:

- The eigenvalues are real, and can be ordered such that $0 = \lambda_1 \leq \lambda_2 \leq \ldots \leq \lambda_N$.
- Define now $\lambda = \lambda_2$. Then, $\lambda > 0$ if and only if the graph is connected. Therefore, λ is defined as the **algebraic connectivity** of the graph.
- Considering a weighted graph, λ is a non-decreasing function of each edge weight.

In addition, we want to evaluate the number of node failures a network can stand before disconnecting. It is well-known that failures of nodes playing a central role in the network communication are likely to affect most its connectivity. In particular, referring to connectivity maintenance, we consider the concept of *Betweenness Centrality* (BC) for ranking the network nodes [23]. For a given node i and pair of nodes j, l, the importance of i as a mediator of the communication between j and l can be established as the ratio between the number of shortest paths linking nodes j and l that pass through node i ($g_{jl}(i)$), and the total number of shortest paths connecting nodes j and l (g_{jl}). Then, the BC of a node i is simply the sum of this value over all pairs of nodes, not including i:

$$BC(i) = \sum_{j<l} \frac{g_{jl}(i)}{g_{jl}}. \tag{1}$$

Once the BC has been computed for all the nodes, it is possible to order them from the *most central* (i.e. the node with highest value of BC) to the *less central* (i.e. the node with lowest value of BC). Hence, let $[v_1, \ldots, v_N]$ be the list of nodes ordered by descending value of BC.

We therefore introduce the following definition of *Robustness level.*

Definition 1 (*Robustness level* [9]) Consider a graph $\mathcal{G}$ with N nodes, and let $[v_1, \ldots, v_N]$ be the list of nodes ordered by descending value of BC. Let $\varphi < N$ be the minimum index $i \in [1, \ldots, N]$ such that, removing nodes $[v_1, \ldots, v_i]$ leads to disconnecting the graph, that is, the graph including only nodes $\left[v_{\varphi+1}, \ldots, v_N\right]$ is disconnected. Then, the *robustness level* of $\mathcal{G}$ is defined as:

$$\Theta(\mathcal{G}) = \frac{\varphi}{N}. \tag{2}$$

The robustness level defines the fraction of central nodes that need to be removed from the network to obtain a disconnected network. Small values of $\Theta(\mathcal{G})$ imply that a small fraction of node failures may fragment the network. Therefore, increasing this value means increasing the network robustness to failures. Notice that $\Theta(\mathcal{G})$ is only an estimate of how far the network is from getting disconnected w.r.t. the fraction of nodes removed. In fact, it might be the case that different orderings of nodes with the same BC produce different values of $\Theta(\mathcal{G})$.

From a local perspective, a heuristic for estimating the magnitude of the topological vulnerability of a node by means of information acquired from its 1-hop and 2-hops neighbors is proposed in [9]. Let $d(v, u)$ be the shortest path between nodes v and u, i.e., the minimum number of edges that connect nodes v and u. Subsequently, define $\Pi(v)$ as the set of nodes from which v can acquire information:

$$\Pi(v) = \{u \in V(G) : d(v, u) \leq 2\}.$$

Moreover, let $|\Pi(v)|$ be the number of elements of $\Pi(v)$. In addition, define $\Pi_2(v) \subseteq \Pi(v)$ as the set of the 2-hop neighbors of v, that comprises only nodes whose shortest path from v is exactly equal to 2 hops, namely

$$\Pi_2(v) = \{u \in V(G) : d(v, u) = 2\}.$$

Now define $L(v, u)$ as the *number of paths* between nodes v and u, and let $Path_\beta(v) \subseteq \Pi_2(v)$ be the set of v's 2-hop neighbors that are reachable through at most β paths, namely

$$Path_\beta(v) = \{u \in \Pi_2(v) : L(v, u) \leq \beta\}.$$

Thus, β defines the threshold for the maximal number of paths between a node v and each of its u neighbors that are necessary to include u in $Path_\beta(v)$. Therefore, using a low value for β allows to identify the most weakly connected 2-hop neighbors. Hence, the value of $|Path_\beta(v)|$ is an indicator of the magnitude of node fragility w.r.t. connectivity, and **the vulnerability level of a node regarding failures** is given by $P_\theta(v) \in (0, 1)$:

$$P_\theta(v) = \frac{|Path_\beta(v)|}{|\Pi(v)|}. \tag{3}$$

where $|\Pi(v)|$ is the number of v's 1-hop and 2-hops neighbors, and $|Path_\beta(v)|$ is the number of nodes that are exactly at 2-hops from node v and relying on at most β 2-hops paths to communicate with v.

3 Problem Statement

Consider a multi-robot system composed of N robots that are able to communicate with other robots within the same communication radius R. The resulting communication topology can be represented by an undirected graph $\mathcal{G}$ where each robot is a node of the graph, and each communication link between two robots is an edge of the graph. Let each robot's state be its position $p_i \in \mathbb{R}^m$, and let $p = \left[p_1^T \dots p_N^T\right]^T \in \mathbb{R}^{N \times m}$ be the state vector of the multi-robot system. Let each robot be modeled as a single integrator system, whose velocity can be directly controlled, namely

$$\dot{p}_i = u_i, \tag{4}$$

where $u_i \in \mathbb{R}^m$ is a control input.[1] In order to guarantee the connectivity of $\mathcal{G}$, [20] proposes an approach to solve the connectivity maintenance problem in a decentralized manner, utilizing the algebraic connectivity property. For this purpose, consider a weighted graph, where the edge weights w_{ij} are defined as follows:

$$w_{ij} = \begin{cases} e^{-\left(\|p_i - p_j\|^2\right)/(2\sigma^2)} & \text{if } \|p_i - p_j\| \leq R \\ 0 & otherwise. \end{cases} \quad \text{with} \quad e^{-(R^2)/(2\sigma^2)} = \Delta \tag{5}$$

where Δ is a small predefined threshold.[2]

Define now $\varepsilon > 0$ to be the desired lower-bound for the value of λ. The control strategy is then designed to ensure that the value λ never goes below ε. As in [20], the following *energy function* can then be utilized for generating the decentralized connectivity maintenance control strategy:

$$V(\lambda) = \begin{cases} \coth(\lambda - \varepsilon) & \text{if } \lambda > \varepsilon \\ 0 & \text{otherwise.} \end{cases} \tag{6}$$

The control design drives the robots to perform a gradient descent of $V(\cdot)$, in order to ensure connectivity maintenance. Namely, considering the dynamics of the system introduced in (4), the control law is defined as follows:

$$u_i = u_i^c = -\frac{\partial V(\lambda)}{\partial p_i} = -\frac{\partial V(\lambda)}{\partial \lambda}\frac{\partial \lambda}{\partial p_i}. \tag{7}$$

The connectivity maintenance framework can be enhanced to consider additional objectives. In particular, as shown in [19], the concept of *generalized connectivity* can be utilized for simultaneously guaranteeing **connectivity maintenance and collision avoidance** with environmental obstacles and among the robots. This is achieved considering the following *generalized edge weights*:

$$\omega_{ij} = w_{ij}\gamma_{ij}, \tag{8}$$

$\forall i, j = 1, \ldots, N$. In particular, the edge weights w_{ij} represent the standard connectivity property. The multiplicative coefficients γ_{ij} represent the *collision avoidance edge weights*:

[1] It is worth remarking that, by endowing a robot with a sufficiently good cartesian trajectory tracking controller, it is possible to use this simple model to represent the kinematic behavior of several types of mobile robots, like wheeled mobile robots [22], and UAVs [14].

[2] This definition of the edge–weights introduces a discontinuity in the control action, that can be avoided introducing a smooth bump function, as in [7]. However, from an implementation viewpoint, the effect of the discontinuity can be made negligible by defining the threshold Δ sufficiently small.

Definition 2 The collision avoidance edge weights γ_{ij} exhibits the following properties, $\forall i, j = 1, \ldots, N$:

(P1) $\gamma_{ij} = \gamma_{ij}\left(\left\|p_i - p_j\right\|\right) \geq 0$.
(P2) $\gamma_{ij} = 0$ if $\left\|p_i - p_j\right\| = 0$, and $\gamma_{ij} = 1$ if $\left\|p_i - p_j\right\| \geq d_s$.
(P3) $\gamma_{ij}(d)$ is non-decreasing w.r.t. its argument d.

The parameter $d_s > 0$ represents the *safety distance*: if the distance between two robots is larger than δ_s, then the collision avoidance action is not necessary. As shown in [19], the same formalism can be exploited for avoiding collisions with environmental obstacles as well.

Utilizing the generalized edge weights ω_{ij} defined in (8), we can compute the *generalized Laplacian matrix* $\mathscr{L}^G$, whose second smallest eigenvalue φ represents the *generalized connectivity* of the graph. As shown in [19], guaranteeing positiveness of the generalized connectivity φ simultaneously guarantees maintenance of the algebraic connectivity (i.e. it ensures that λ remains positive) and collision avoidance. This can then be achieved using the control law (7) replacing λ with φ, namely

$$u_i = u_i^c = -\frac{\partial V(\varphi)}{\partial p_i} = -\frac{\partial V(\varphi)}{\partial \varphi}\frac{\partial \varphi}{\partial p_i}. \tag{9}$$

Since φ and its gradient are global quantities, the proposed control law is centralized. Decentralized implementation can be achieved replacing φ and its gradient with their estimates, computed by each robot in a decentralized manner applying the procedure proposed in [20].

This methodology does not consider the fact that robots can unexpectedly fail, thus stopping their activity due to mechanical, electrical or software issues. As described in the introduction, it is necessary to reduce the effects of robot failures on the overall network connectivity, avoiding vulnerable topological configurations. In this paper we address the following problem:

Problem Given a multi-robot system, design a local estimation procedure that allows each robot to assess its vulnerability level, based on locally available information, and subsequently exploit this estimate for controlling the motion.

4 Combined Control Law

This section describes the unified model that aims at improving the robustness of networks to failures while, in the absence of failures, maintaining connectivity and avoiding collisions. In particular, considering the dynamics introduced in (4), we define the following control law:

$$u_i = \sigma u_i^c + \psi u_i^r, \tag{10}$$

where u_i^c is the generalized connectivity maintenance control law introduced in (9), and u_i^r is the additional control law that will be hereafter defined for improving robustness to failures. Moreover, $\sigma, \psi \geq 0$ are design parameters, that represent control gains. Setting either σ or ψ equal to zero leads to removing the effect of one of the control laws. Conversely, if both parameters are greater than zero, both control actions are simultaneously active.

Based on the vulnerability level definition, given in (3), the purpose of the control strategy is to increase the number of links of a potentially vulnerable node i towards its 2-hop neighbors that are in $Path_\beta(i)$, for a given value of β. Hence, define $x_\beta^i \in \mathbb{R}^m$ as the barycenter of the positions of the robots in $Path_\beta(i)$, namely

$$x_\beta^i = \frac{1}{|Path_\beta(i)|} \sum_{j \in Path_\beta(i)} p_j. \tag{11}$$

Considering the dynamics of the system introduced in (4), the control law is defined as follows:

$$u_i^r = \xi_i \frac{x_\beta^i - p_i}{\|x_\beta^i - p_i\|} \alpha(t), \tag{12}$$

where $\alpha(t) \in \mathbb{R}$ is the linear velocity of the robots.[3] The parameter ξ_i is introduced to take into account the vulnerability state of a node i, i.e. $\xi_i = 1$ if node i identify itself as vulnerable or $\xi_i = 0$ otherwise. As we aim at setting as vulnerable those robots i exhibiting high values for $P_\theta(i)$, ξ_i is defined as follows:

$$\xi_i = \begin{cases} 1 \text{ if } P_\theta(i) > r \\ 0 \text{ otherwise} \end{cases} \tag{13}$$

where $r \in (0, 1)$ is a random number drawn from a uniform distribution. Namely, if $P_\theta(i) > r$, then the i-th robot considers itself as vulnerable. It is worth noting that (3) provides a decentralized methodology for each robot to evaluate its vulnerability level.

Summarizing, this control law drives vulnerable robots towards the barycenter of the positions of robots in their $Path_\beta$, thus decreasing their distance to those robots and eventually creating new edges in the communication graph. We will hereafter analyze the performance of the proposed combined control law introduced in (10). In particular, the control law u_i^c was proven in [19, 20] to guarantee positiveness of the generalized connectivity in a disturbance free environment. The following theorem extends these results considering the presence of the additional robustness improvement control law u_i^r.

Theorem 1 *Consider the dynamical system described by* (4), *and the control laws described in* (9), (10) *and* (12). *Then, if the initial value of* $\varphi(t)$, *namely* $\varphi(0)$, *is*

---

[3] We would like to remark that pathological situations exist in which (12) is not well defined, namely when $p_i = x_\beta^i$. However, this corresponds to the case where the i-th robot is in the barycenter of its weakly connected 2-hop neighbors: hence, in practice, this never happens when a robot detects itself as vulnerable.

greater than ε, then the value of $\varphi(t)$ will remain positive, as the system evolves, thus implying algebraic connectivity maintenance and collision avoidance.

Proof It is possible to show that, under the proposed control law, for constant values of P_θ, the energy function $V(\varphi(p))$ in (6) does not increase over time. As a consequence, it is possible to conclude that the generalized connectivity φ remains greater than zero, as the system evolves. Considering the definition of the generalized edge weights ω_{ij} in (8), this implies that the algebraic connectivity λ will remain positive, while avoiding collisions. This result can be extended to the case where P_θ is time-varying, assuming that variations are sufficiently slow. The proof is analogous to that of [20], and is then omitted due to space limitations. ∎

5 Simulations

The proposed control strategy was validated using a simulation model, developed in MATLAB®. The initial network topologies were generated over a bounded area of size $\mathscr{A}$ in $\mathbb{R}^2$, through a random positioning of N robots, connected according to the communication radius R. For this simulation we consider $N = 20$, $\mathscr{A} = 50^2$ and $R = 16$. The experimental setup also considers a set of randomly generated failure times, distributed during the total simulation time of 80 seconds. At every 1 second, the vulnerability level estimation and the network properties were computed. We utilized the following model parametrization: $\varepsilon = 0.25$, $\beta = 1$, and $\sigma, \psi = \{0, 0.1, 0.5, 1\}$. Moreover, we considered a constant linear velocity $\alpha(t) = 0.25R/s$.

For evaluation purposes, at the specific times, a disturbance is introduced into the network by removing its most central node according to the updated BC ranking, as defined in (1). Given an initial network topology and the combined control law, it is expected that the proposed mechanism significantly reduces the impact of failures on the network connectivity, providing robots with means not only for accommodating but also for responding to failures, adapting the interconnection topology.

For demonstrating the fragility of random networks to failures of elements, Fig. 3(a exhibits snapshots of a network during the simulation process without the proposed control law active (i.e. $\psi = 0$ and $\sigma = 0$). These snapshots correspond to simulation time $t = 0$, $t = 20$, $t = 50$ and $t = 80$, depicted by black, blue, green and red nodes, respectively. Notice that at $t = 20$ the network fragmented into two clusters, and as the simulation evolves the number of clusters increases. Of course, different failure distribution might accelerate or delay the network fragmentation. However, cascading failures or a high vulnerable topological configuration are surely harmful to the network connectivity and, as a consequence, to its operation.

Consider now the performance of the combined control law for the same scenario ($\psi = 1$ and $\sigma = 1$), illustrated in Fig. 3b. The resulting network topology is still connected at the end of the simulation despite failures of central elements, emphasizing the effectiveness of the combined control law to produce a more resilient network. As already mentioned, the connectivity can not always be guaranteed, but postponing the

Fig. 3 Snapshots of a network topology at $t = 0$ (black), $t = 20$ (blue), $t = 50$ (green) and $t = 80$ (red) simulation times

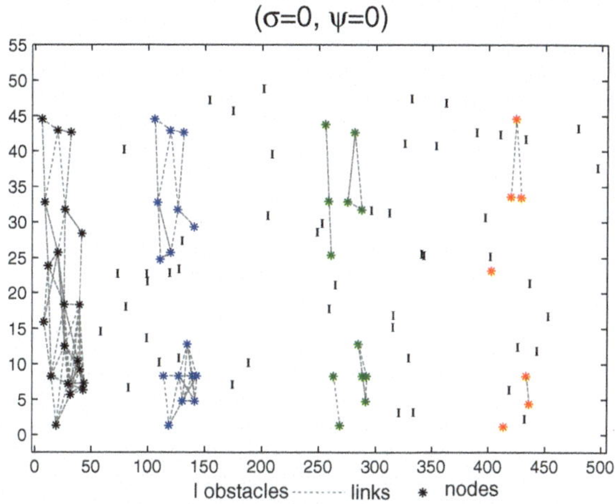

(a) Performance **without** the control law.

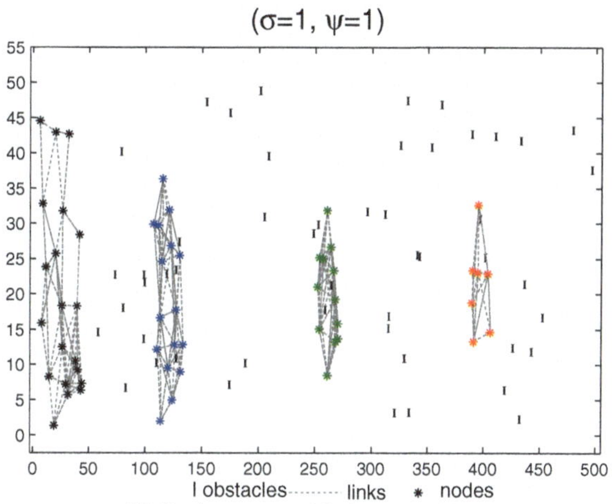

(b) Performance **with** the control law.

network fragmentation or increasing the number of elements in the largest connected component are significant achievements for those applications where connectivity is crucial.

As a proof of concept, the experiments were performed for 50 different network topologies. For each gain combination, the averaged results for the algebraic connectivity and the robustness level are illustrated in Figs. 4 and 5, respectively. The vertical axis represents the initial scores for the respective property and its evolution when exposed to continuous failures through the fraction f of nodes removed from the network (horizontal axis).

As expected, networks were heavily affected by failures in scenarios where the robustness improvement mechanism was not significantly active ($\psi = 0$ and $\psi = 0.1$), as the algebraic connectivity approaches 0 with a few node failures. Notice that

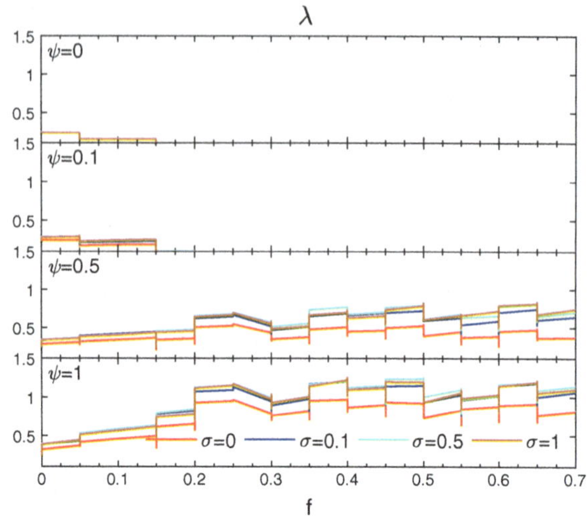

Fig. 4 Control law performance - The algebraic connectivity (λ)

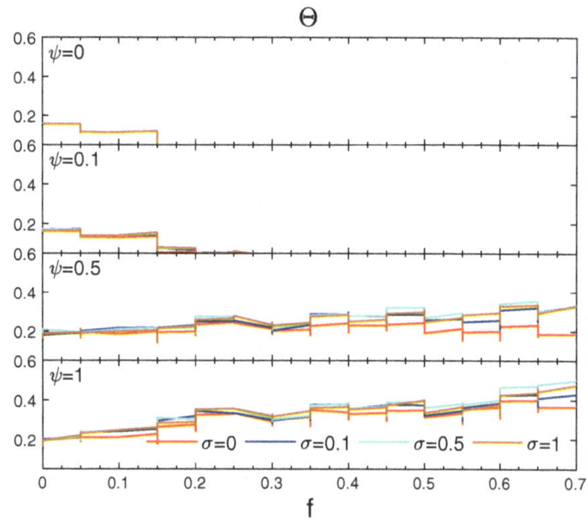

Fig. 5 Control law performance - The robustness level (Θ)

the robustness to failure level increases according to ψ gain setting. On the other hand, it is not significantly affected by the σ gain setting, i.e., improving the weight of the connectivity maintenance control through the algebraic connectivity estimation does not mean improving the robustness to failures. These findings support the claim that the algebraic connectivity is not a suitable property for supporting mechanisms to produce more robust networks.

In contrast to the robustness level evolution, the algebraic connectivity exhibits a slight improvement as the ψ value increases, i.e., improving the robustness to failures implies increasing the overall network connectivity. It is important to highlight that the algebraic connectivity control law plays an important role: when there are no vulnerable nodes in the network the connectivity must be ensured.

In general, the experiments demonstrate the impact of failures on the network connectivity and, mainly, the feasibility of combining connectivity maintenance and robustness to failures improvement mechanisms, even in the presence of obstacles. Besides, the gain modeling allows the design of tools for adaptively setting gains according to the application requirements and the devices/network states (e.g., battery level, sensor feedbacks, failure occurrences), which can result in a desirable control law behavior. Some additional examples can be freely viewed online on https://youtu.be/ueo7nYEAm24.

6 Conclusions

Multi-robot applications that require deploying services in unstructured domains should remain operative, regardless of the possibility of device failures, as service availability is a crucial requirement for most applications. In this paper, we present a model, based on local procedures, that combines control laws for both connectivity maintenance and failure effect mitigation in multi-robot networks, thus allowing a robust operation of the robotic network. The control laws are combined as a weighted sum, and failure mitigation is achieved by a vulnerability assessment that is also performed locally. The results demonstrate the feasibility of the model: the tested networks were able to postpone disconnection or to maintain the network connected even in scenarios of frequently occurring failures. Current work aims at implementing the proposed methodology on real robotic systems, to evaluate its performance in operational scenarios. Moreover, we are investigating the impact of different failure time distributions on the mechanism performance. For future work, we aim at developing methodologies for achieving online adaptation of the gains according to the network configuration and the application requirements, as a means of improving the overall performance. Finally, we aim at considering the presence of additional control objectives, such as formation control or environmental coverage.

References

1. Agarwal, P.K., Efrat, A., Ganjugunte, S., Hay, D., Sankararaman, S., Zussman, G.: The resilience of WDM networks to probabilistic geographical failures. In: INFOCOM, 2011 Proceedings IEEE, pp. 1521–1529 (2011). https://doi.org/10.1109/INFCOM.2011.5934942
2. Ajorlou, A., Momeni, A., Aghdam, A.G.: A class of bounded distributed control strategies for connectivity preservation in multi-agent systems. IEEE Trans. Autom. Control **55**, 2828–2833 (2010)
3. Albert, R., Jeong, H., Barabasi, A.L.: Error and attack tolerance of complex networks. Nature **406**(6794), 378–382 (2000)

4. Cao, Y., Ren, W.: Distributed coordinated tracking via a variable structure approach – part I: consensus tracking. part II: swarm tracking. In: Proceedings of the American Control Conference, pp. 4744–4755 (2010)
5. Dall'Asta, L., Barrat, A., Barthelemy, M., Vespignani, A.: Vulnerability of weighted networks. Theory Exper. **2006**, 04,006 (2006)
6. Dimarogonas, D.V., Johansson, K.H.: Bounded control of network connectivity in multi-agent systems. IET Control Theory Appl. **4**, 1751–8644 (2010)
7. Do, K.D.: Formation tracking control of unicycle-type mobile robots with limited sensing ranges. IEEE Trans. Control Syst. Technol. **16**, 527–538 (2008)
8. Ghedini, C., Ribeiro, C.H.C.: Rethinking failure and attack tolerance assessment in complex networks. Phys. A Stat. Mech. Appl. **390**(23–24), 4684–4691 (2011)
9. Ghedini, C., Secchi, C., Ribeiro, C.H.C., Sabattini, L.: Improving robustness in multi-robot networks. In: Proceedings of the IFAC Symposium on Robot Control (SYROCO). Salvador, Brazil (2015). https://doi.org/10.1016/j.ifacol.2015.12.011
10. Godsil, C., Royle, G.: Algebraic Graph Theory. Springer, Berlin (2001)
11. He, Z., Liu, S., Zhan, M.: Dynamical robustness analysis of weighted complex networks. Phys. A Stat. Mech. Appl. **392**(18), 4181–4191 (2013)
12. Hsieh, M.A., Cowley, A., Kumar, V., Talyor, C.J.: Maintaining network connectivity and performance in robot teams. J. Field Robot. **25**(1), 111–131 (2008)
13. Ji, M., Egerstedt, M.: Distributed coordination control of multiagent systems while preserving connectedness. IEEE Trans. Robot. (2007)
14. Lee, D., Franchi, A., Son, H., Ha, C., Bulthoff, H., Robuffo Giordano, P.: Semiautonomous haptic teleoperation control architecture of multiple unmanned aerial vehicles. IEEE/ASME Trans. Mechatron. **18**(4), 1334–1345 (2013)
15. Manzano, M., Calle, E., Torres-Padrosa, V., Segovia, J., Harle, D.: Endurance: a new robustness measure for complex networks under multiple failure scenarios. Comput. Netw. **57**(17), 3641–3653 (2013)
16. Nelakuditi, S., Lee, S., Yu, Y., Zhang, Z.L., Chuah, C.N.: Fast local rerouting for handling transient link failures. IEEE/ACM Trans. Netw. **15**(2), 359–372 (2007). https://doi.org/10.1109/TNET.2007.892851
17. Notarstefano, G., Savla, K., Bullo, F., Jadbabaie, A.: Maintaining limited–range connectivity among second–order agents. In: Proceedings of the American Control Conference, pp. 2134–2129 (2006)
18. Rak, J.: Resilient Routing in Communication Networks. Computer Communications and Networks, 1st edn. Springer International Publishing, Berlin (2015)
19. Robuffo Giordano, P., Franchi, A., Secchi, C., Bülthoff, H.H.: A passivity-based decentralized strategy for generalized connectivity maintenance. Int. J. Robot. Res. **32**(3), 299–323 (2013)
20. Sabattini, L., Chopra, N., Secchi, C.: Decentralized connectivity maintenance for cooperative control of mobile robotic systems. Int. J. Robot. Res. (SAGE) **32**(12), 1411–1423 (2013)
21. Sabattini, L., Secchi, C., Chopra, N.: Decentralized estimation and control for preserving the strong connectivity of directed graphs. IEEE Trans. Cyber. **45**(10), 2273–2286 (2015)
22. Soukieh, R., Shames, I., Fidan, B.: Obstacle avoidance of non-holonomic unicycle robots based on fluid mechanical modeling. In: Proceedings of the European Control Conference. Budapest, Hungary (2009)
23. Wasserman, S., Faust, K., Iacobucci, D.: Social Network Analysis: Methods and Applications (Structural Analysis in the Social Sciences). Cambridge University Press, Cambridge (1994)

Chase Your Farthest Neighbour

A Simple Gathering Algorithm for Anonymous, Oblivious and Non-communicating Agents

Rotem Manor and Alfred M. Bruckstein

Abstract We consider a group of mobile robotic agents, identical and indistinguishable, having no memory (oblivious) and no common frame of reference (neither absolute location nor a common orientation). Furthermore, these agents are assumed to posses only rudimentary sensing and computational capabilities (limited visibility and basic geometric sorting). We prove that, such robots, implementing a *"Chase the farthest neighbour"* policy, preform the task of gathering to a point within a finite time or a finite expected number of time steps. In continuous time, preforming such a gathering task is rather straightforward, while in the discrete time, we prove that a randomized semi-synchronised timing model leads to gathering within a finite expected number of time-steps.

1 Introduction

Recently there is interest in using swarms of very simple robotic agents in performing various tasks. The simple robots are assumed to have only very basic sensing, motion and computing capabilities. It turns out that oblivious agents with very elementary data processing skills and limited sensing capabilities can be programmed to perform several useful tasks, and the study of what such agents can do is interesting and often quite challenging. A lot of research was already devoted to this topic and led to a wealth of interesting results. Jadbabaie and Moreau and their collaborators, see e.g. [11, 13], dealt with networked agents where the interconnections between them are intermittent, and proved that flocking and gathering can be ensured, provided enough interactions occur over time. Suzuki and Yamashita, [15], suggested several interesting algorithms for gathering agents under limited sensing and computing capabilities, and interestingly, among them also a "chase-the-farthest" neighbor process. Moreau, in [14], proved that moving inside the convex hull of the current

R. Manor (✉) · A. M. Bruckstein
Technion, Haifa, Israel
e-mail: manorrotem@gmail.com

A. M. Bruckstein
e-mail: freddy@cs.technion.ac.il

R. Groß et al. (eds.), *Distributed Autonomous Robotic Systems*,
Springer Proceedings in Advanced Robotics 6,
https://doi.org/10.1007/978-3-319-73008-0_8

agent locations leads, under some mild restrictions, to gathering, while Bruckstein and his co-workers, in a series of papers [3, 8–10], proved that gathering can be ensured in finite time with crude, limited range, and sometimes even bearing only, sensing. Gross, in [6, 7] proposed an aggregation algorithm for groups of robots employing binary sensing, reporting only the existence of other agents in the robots' line of sight.

This paper discusses the multi-agent gathering problem in continuous and discrete time, showing that a swarm of simple agents can gather to a point within a finite expected time by applying a very simple motion law, implementing a *"chase the farthest neighbour"* policy. Suzuki and Yamashita, in an early version of their paper [15], suggested such an algorithm for a discrete-time multi-agent system of oblivious, anonymous and unlimited visibility agents for point convergence in the $\mathbb{R}^2$ plane. In their paper they also proved that preforming this task within a finite number of time-steps is impossible, if the agents cannot agree on a common meeting point.

We here first analyse a similar algorithm in the continuous-time framework for the case of limited-visibility or sensing by the agents. In the discrete case, we suggest a stochastic algorithm using the same, "chase the farthest", idea which provides a solution to the task of gathering within a finite expected number of time-steps.

This paper first discusses some basic concepts, and then proceeds with point gathering of agents having unlimited and limited visibility in a continuous time framework. Then, problems that arise from implementing the *"chase the farthest neighbour"* policy in a discrete time framework are discussed, showing that a system of agents having unlimited visibility clusters to a small region. Gathering to a point under unlimited and limited visibility and randomized semi-synchronized timing of the agents' movements is then proved to occur in finite expected time.

2 Preliminaries

We deal with a system of n identical and anonymous oblivious agents in the $\mathbb{R}^2$ plane specified by their time varying locations $\{p_i(t)\}_{i=1,2,\ldots,n}$. The agents interact with each other in such a way that their position updates are determined by their current location and by interaction with their neighbours. The neighbours of each agent i at time t are defined as the set of agents located within a given visibility range, V, form $p_i(t)$, and this set denoted by $N_i(t)$. The neighbourhood relation between agents is described by a time dependent visibility-graph. Notice that when dealing with unlimited visibility, the set $N_i(t)$ comprises all the agents except i, and the visibility-graph is complete, i.e. all agents "see" each other.

This paper provides proofs of gathering for a swarm of agents modelled as points in the plane, that employ a "chase the farthest neighbour" rule of motion. The robots are assumed to be able to cross each other's paths with no obstructions, in the continuous time there are no decision or computation-induced delays, while in the discrete case we assume either synchronous or semi-synchronous, probabilistic motion schedules.

Note that, we bring only the essence of the proofs of the theorems in this paper. The full proofs can be found in [12].

3 Continuous Time Gathering

3.1 *Unlimited Visibility*

The main idea of the "chase the farthest" gathering law is that each agent continuously moves toward the position of its current farthest neighbour. If more than one neighbour is at the farthest distance from it, the agent arbitrarily selects one of its farthest neighbours, and moves toward its location. This dynamic law is simple, and we shall prove that if all agents of the system act by it, then the system eventually gathers to a point.

In order to define a formal dynamic law, we consider the set of the farthest agents (neighbours). Let $N_i^{Far}(t) \subset N_i(t)$ be the set of the agents located at the maximal distance from agent i at time t.

$$j \in N_i^{Far}(t) \iff \forall q \in N_i(t) \mid \|p_i(t) - p_j(t)\| \geq \|p_i(t) - p_q(t)\|$$

The agents' formal dynamic law is that each agent moves with a constant speed $\sigma > 0$ toward the position of j an arbitrary agent from the set $N_i^{Far}(t)$, unless it is collocated with j (in which case, obliviously the agents are all gathered!).

$$\dot{p}_i(t) = \begin{cases} \sigma \frac{p_j(t)-p_i(t)}{\|p_j(t)-p_i(t)\|} & \|p_j(t) - p_i(t)\| > 0, \quad j \in N_i^{Far}(t) \\ 0 & o.w. \end{cases} \tag{1}$$

Notice that in (1) j is an arbitrary agent from the set $N_i^{Far}(t)$. Hence, a delicate issue here is the continuous need to select the farthest neighbour from a set $N_i(t)$ possibly containing more than one agent (this makes it necessary to assess the well-posedness of the velocity control rule, $\dot{p}_i(t)$, since in principle infinitely many switches in the choice of i are possible. However, one can argue that all choices yield similar effects on $p_i(t)$, and monotonicity argument can be used to prove well-posedness of the evolution with correct bounds on the rates of changes of distances between agents. For a discussion of a similar issue see [4, 5].

Theorem 1 *A system of n agents with dynamic law (1) gathers to a point within a finite time.*

The proof is based on the analysis of the rate of change of $D(P(t))$, the current diameter of the convex-hull of the system's agents, i.e.

$$D(P(t)) = \max_{ij} \|p_i(t) - p_j(t)\|$$

We show that each agent being located at an edge of a diameter necessarily moves in a velocity with a strictly positive projection on the direction to the other edge of the diameter of at least $\sigma \cos(\pi/3)$. As a result we have that the system gathers to a point within a finite time $D(P(0))/\sigma$, proving Theorem 1 (see full proof in [12] Theorem 1).

3.2 Limited Visibility

In this section we apply the concept of "chase the farthest" to agents with limited visibility. By assumption, here each agent senses only the agents located within a visibility range of V. We clearly cannot use the former algorithm, since agents may lose visibility with their neighbours during their movement, and hence cluster into disconnected groups. For example, assume an agent has two neighbours which are both located at a distance V from it and the angle between the vectors pointing to them is larger than $\pi/2$. Then, moving towards one of them will result in losing visibility to the other.

A solution for this problem is given by a slightly more complicated algorithm. We suggest a new dynamic law which addresses the connectivity problem and gathers the system to a point, within a finite time. This algorithm is again based on "chase the farthest" concept, and is similar to the former dynamic law, however in situations where an agent senses more than one farthest neighbour, according to the motion law we consider, it will not move away from any one of them.

The presentation of a formal dynamic law requires us to adjust some of our definitions. Let $N_i^{Far}(t) \subset N_i(t)$ be the subset of the farthest visible neighbours of agent i, i.e. all the agents in $N_i^{Far}(t)$ are equally distanced from i, while the other agents in the set $N_i(t)$ are located closer to i.

Each agent i continuously calculates $\psi_i^{Far}(t)$, the angle of the minimal disk sector anchored at $p_i(t)$ and containing its farthest neighbours $N_i^{Far}(t)$. If $\psi_i^{Far}(t)$ is equal to or greater than π (hence i is "surrounded" by its farthest neighbours), the agent stays put. Otherwise, it moves with speed $\sigma > 0$ in the direction of $U_i^{Far}(t)$ a unit vector in the direction of the bisector of the angle ψ_i^{Far} (see Fig. 1). The motion law of agent i is therefore

$$\dot{p}_i(t) = \begin{cases} \sigma \hat{U}_i^{Far}(t) & \psi_i^{Far}(t) < \pi \\ 0 & \psi_i^{Far}(t) \geq \pi \end{cases} \tag{2}$$

Notice that most of the time an agent has a single farthest neighbour, then $\psi_i^{Far} = 0$, and the agent i moves toward its only farthest neighbour (see Fig. 1a).

Before proving convergence we show that the motion law (2) maintains the connectivity of the system's visibility-graph, i.e. if all the agents of the system obey this law, they maintain visibility with all their current neighbours.

Lemma 1 *The motion law (2) ensures that neighbours in the initial configuration remain neighbours forever.*

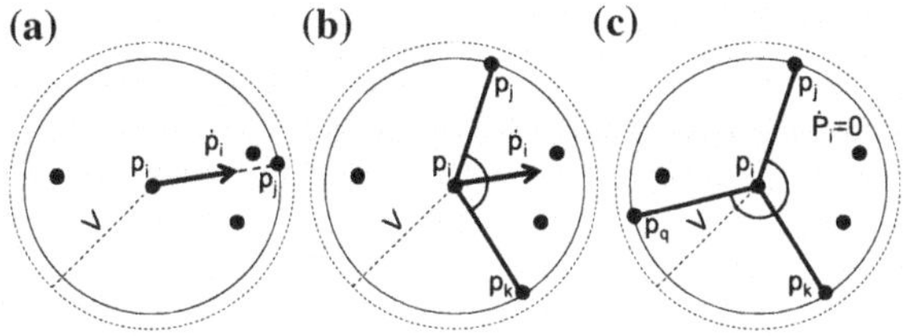

Fig. 1 The motion law (2), in case of limited visibility. **a** $\{\psi_i^{Far}(t) = 0\}$ Agent i moves toward j its only farthest agent within its visibility range. **b** $\{\psi_i^{Far}(t) < \pi\}$ Agent i moves toward the bisector of the minimal sector containing all its farthest neighbours. j and k are the neighbours that define the minimal angular-sector containing all the farthest neighbours. **c** $\{\psi_i^{Far}(t) \geq \pi\}$ Agent i does not move. (Here agents j and q define the sector $S(t)$)

Proof Let $\{i, j\}$ be a pair of neighbours. In order for this pair to disconnect in the visibility-graph, $l_{ij}(t)$, the distance between those agents, must cross V. At that state i and j are necessarily in the set of the farthest neighbours of each other, since none of them may sense agents beyond the range of V, i.e.

$$\|p_i(t) - p_j(t)\| = V \quad \Rightarrow \quad j \in N_i^{Far}(t) \text{ and } i \in N_j^{Far}(t)$$

By the motion rule (2), for the agent i, If $\psi_i^{Far}(t) \geq \pi$, then it stays put. Otherwise (if $\psi_i^{Far}(t) < \pi$), it moves with speed of σ in the direction of $U_i^{Far}(t)$. Hence, if the agent does not stay put, its moves in a direction with an angle smaller than or equal to $\pi/2$ relative to the direction pointing to the agent $j \in N_i^{Far}$. Denote this angle by $\theta_{ij}(t)$. Then considering a pair $\{i, j\}$ where $j \in N_q^{Far}(t)$ and $i \in N_j^{Far}(t)$, we have that

$$\dot{l}_{ij}(t) = \frac{p_i(t) - p_j(t)}{\|p_i(t) - p_j(t)\|}^{\mathsf{T}} (\dot{p}_i(t) - p_j(t)) =$$

$$= -\left(\|\dot{p}_i(t)\| \cos(\theta_{ij}(t)) + \|\dot{p}_j(t)\| \cos(\theta_{ji}(t))\right) \leq 0$$

showing that the distance between i and j can not increase upon reaching V, and hence can not exceed V. This proves the lemma.

Notice that the agents of this system need not be "aware" of their visibility range V in order not to lose neighbours, however we need to have the same visibility range for all the agents (and this is ensured by our assumption that all agents are identical!).

Theorem 2 *A system of n agents moving according to motion rule (2), having a connected initial visibility-graph, will gather to a point within a finite time.*

The proof of this theorem is based on considering the dynamics of s, the agent located at a (currently) sharpest corner of the system's convex-hull, which by Lemma 1, has to be connected to at least one other agent at any time. Let $\varphi_s(t)$ be the inner angle of this corner, which by Proposition 2 in [12] is upper bounded by $\varphi_* = \pi(1 - 2/n)$. Hence, $\psi_s^{Far}(t) \leq \varphi_*$, and we have that agent s necessarily moves inside the systems convex-hull with the speed of $\sigma > 0$, while by the motion law (2), clearly, no other agent may move out of it.

As a consequence, $l_{CH}(t)$, the perimeter of the convex-hull of the system continuously drops with a rate bounded away from zero by a constant as long as its length is not equal to zero, as follows:

$$\dot{l}_{CH}(t) \leq -2\sigma \cos^2(\frac{\varphi_*(t)}{2})$$

Hence, the system gathers to a point within a finite time as claimed in Theorem 2 (see full Proof of Theorem 2 in [12]).

Figure 2 presents simulation result of 5 agents starting in a constellation of a connected visibility net, and gathers to a point. Notice that the discontinuity of the agents' velocity is a consequence of the dynamic law dictating sudden switching events, due to the changes in their farthest neighbours (and/or their selections of the farthest neighbours!).

4 Discrete Time "Chase the Farthest" Gathering

In the sequel we analyse the "chase-the-farthest" concept in the discrete-time framework. We start by showing that a group of agents with unlimited visibility applying a "chase-the-farthest" algorithm with steps of length $\sigma > 0$, gather to a disk of radius σ. We prove this using an unusual geometric Lyapunov function. Then, we further prove that such a group gathers to a point within a finite expected number of steps, in a semi-synchronous model, provided they can also take steps less than or equal to σ in length. We end this section by showing that a simple constraint on the agents' step size, ensures their gathering even if the agents have limited visibility.

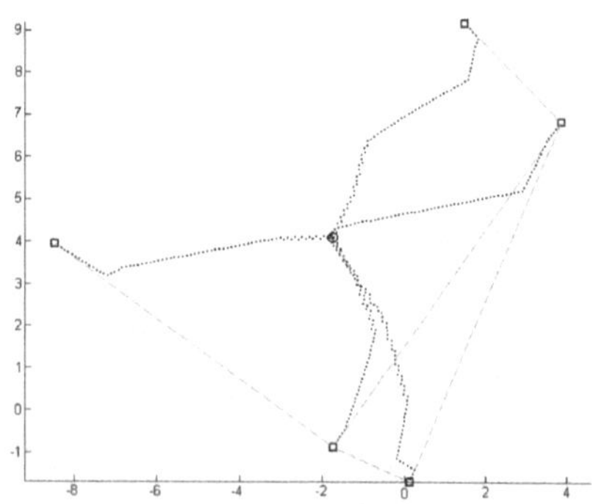

Fig. 2 Simulation result of 5 agents with dynamics (2). The empty big squares are the initial configuration of the agents, the dashed grey lines are the initial connection topology, and the doted lines are the agents' trajectories, which meet in the black circle, i.e. the gathering point

4.1 Unlimited Visibility

We assume next that each agent i jumps a step of size σ towards the position of an arbitrary agent of the set $N_i^{Far}(k)$:

$$p_i(k+1) = p_i(k) + \sigma \frac{p_j(t) - p_i(t)}{\|p_j(t) - p_i(t)\|}$$
$$\text{where } j \text{ is an arbitrary agent in the set } N_i^{Far}(k) \tag{3}$$

The motion rule (3) does not gather the system to a point (except in some special cases, where the initial configuration of the agents leads them to jump exactly to the same point at the same time step), since the agents of the system may jump over each other when they are in close proximity. Nevertheless, we show that the system's dynamics brings all agents together to a small bounded region in $\mathbb{R}^2$ within a finite number of time steps.

Theorem 3 *A system of n agents with dynamic law (3) gathers to a disk of radius σ within a finite number of time steps.*

The proof of this theorem is based on the analysis of the rate of change of the radius of the agents' minimal enclosing circle. By geometry the farthest neighbour of each agent is located in an area beyond $C(k)$, the center of the minimal enclosing circle, as presented in Fig. 3. Therefore, if the distance between $C(k)$ and the current location of an agent is greater than $\sigma/2$, it jumps to a position significantly closer to $C(k)$. Otherwise, it jumps to position with a distance less than or equal to σ from $C(k)$. Hence, if the radius of the enclosing circle at time-step k, is greater then σ, it significantly decreases. As a consequence, we have that all agents of the system gathers to a disk of radius σ within a finite number of time-steps, proving Theorem 3 (see Theorem 3 in [12]).

4.1.1 Gathering to a Point

Suzuki and Yamasita in [15], proved that a multi-agent system of oblivious agents which can not agree on a meeting point are unable to gather to a point within a finite number of time-steps. Agents with the capability to compute their minimal enclosing circle or their convex-hull (calculation that has a complexity of $\mathcal{O}(n \log n)$), can agree on a meeting point, and were used to prove gathering to a point within a finite number of time steps in several previous works, see e.g. [1, 2].

Suzuki and Yamasita in [15], also suggested a "Chase the farthest"algorithm (a calculation that has a complexity of $\mathcal{O}(n)$) which yields asymptotical point convergence. Their algorithm is that each agent jumps towards its farthest neighbour a step with a size equal to the distance to that neighbour multiplied by a positive constant smaller than 1. Hence, at each time-step all agents jump into their convex-hull, and if the initial configuration is not a point, this process will never end, resulting in asymptotic convergence to a point.

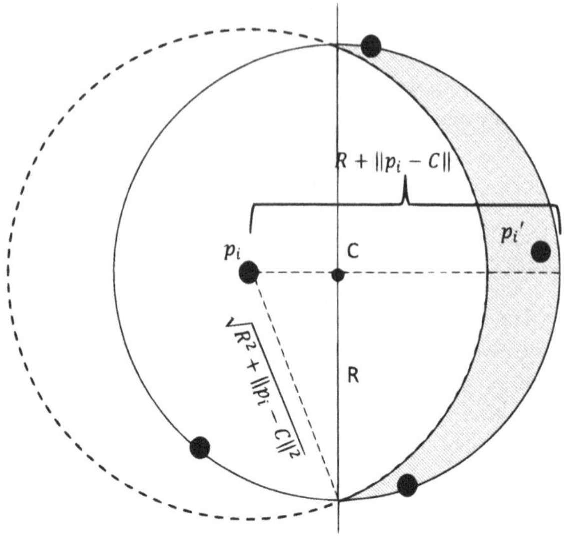

Fig. 3 A valid area for the location of i', the farthest neighbour of agent i, is marked as dashed area. The full-line circle is the minimal enclosing circle of the agents' locations, and the dashed circle is the outcome of the distance between the position of agent i and the closest point where its farthest neighbour may be located at, so that the minimal enclosing circle will be defined properly (Proposition 1 in [12])

We here suggest an alternative simple motion law for the gathering of the oblivious agents, which also requires calculations of $\mathcal{O}(n)$ complexity. The consequence of the simplicity is that the system gathers within a finite *expected* number of time-steps instead of within a finite number of time-steps.

In order to achieve point-gathering, we adjust the motion law (3) as follows. We limit the step-size of each agent to the distance to its current goal, i.e. if the relative position of the goal is within the range of σ, the agent will simply move to the goal agent's position. We therefore define the length of the step of an agent i at time k as:

$$\mu_i(k) \triangleq \min\{\sigma \ , \ \|p_j(k) - p_i(k)\|\}$$

where j is the goal agent.

This restriction is, however, not sufficient, since once all the agents are in close proximity, they may switch locations with each other at every time-step, rather than gather to a point. Therefore, we also adjust the timing of the motion law (3) to a semi-synchronous model, so that at each time-step an agent i may be active with some probability ρ.

$$p_i(k+1) = p_i(k) + \chi_i(k)\mu_i(k)\frac{p_j(t) - p_i(t)}{\|p_j(t) - p_i(t)\|} \tag{4}$$

where j is an arbitrary agent of the set $N_i^{Far}(k)$, and $\chi_i(k)$ equals 1 or 0 with probabilities ρ or $1 - \rho$, respectively.

We next show that a system of n agents with the motion law (4) gathers to a point within a finite expected number of time steps, since it obeys a *"strong asynchronicity assumption"* (as defined in Gordon et al. [9, 10]).

Definition 1 *"Strong asynchronicity assumption"*: There exist a strictly positive constant ε such that for any subset A of the agents, at each time-step k, the probability that A will be the only set of active agents is at least ε.

Theorem 4 *A system of n agents with dynamics law (4) will gather to a point within a finite expected number of time-steps.*

The essence of the proof is that by the motion law (4), clearly, the perimeter of the system's convex-hull cannot increase. Furthermore, if the agents are not confine to a σ-diameter disk, by the "Strong asynchronicity assumption", there is always a strictly positive probability ε, that the agent located at the sharpest corner of the convex-hull, and the agents with close proximities to it will be the only active agents, and therefore will jump into the convex-hull, yielding a bounded away from zero decrease in $l_{CH}(k)$, the perimeter of the system's convex-hull, as follows: (Lemma 3 in [12])

$$l_{CH}(k+1) \leq l_{CH}(k) - \frac{\sigma}{3}\left(1 - \sin\left(\frac{\varphi_*}{2}\right)\right)$$

Consequently, the perimeter decreases until the system gets to a state where all the agents are confined to a σ-diameter disk. At this state, by the motion law (4), an active agent jumps to another agent's current position. Therefore, at each time step there is a strictly positive probability ε that the agents constellation will comprise less and less points in $\mathbb{R}^2$ until it becomes a single point.

As a consequence, the expected number of time-step for a point gathering to occur is upper bounded by

$$\left(\left\lceil \frac{l_{CH}(0) - 2\sigma}{\frac{\sigma}{3}\left(1 - \sin\left(\frac{\varphi_*}{2}\right)\right)} \right\rceil + n - 1\right) \lceil \varepsilon^{-1} \rceil$$

proving Theorem 4 (see Theorem 4 in [12]).

4.2 *Limited Visibility*

In this section we assume that the agents have limited visibility, hence the system's visibility-graph may break into disconnected components while the agents move. The "chase the farthest" motion law (2) maintains the connectivity of the visibility-graph in a continuous-time framework, but a straightforward discretization does not work, since the agents jump steps with significant lengths, and, as a consequence, they may lose connectivity to their neighbours.

We resolve this problem by adding constraint on the step-size of the agents, as was also suggested by Ando et al. in [2].

Let $\theta_{ij}(k)$ be the angle between the two vectors pointing from $p_i(k)$ to $p_j(k)$ and from $p_i(k)$ to the current "goal" of agent i, and let $l_{ij}(k)$ be the current distance between the positions of agents i and j. Then, the maximal step size agent i may take, in order to ensure visibility with j is given by

$$Limit_{ij}(k) = \frac{l_{ij}(k)}{2}cos(\theta_{ij}(k)) + \sqrt{\left(\frac{V}{2}\right)^2 - \left(\frac{l_{ij}(k)}{2}\right)^2 sin^2(\theta_{ij}(k))}$$

and the maximal step size agent i may take, in order to ensure visibility with all its neighbours, is given by

$$Limit_i(k) = \min_{j \in N_i(k)} \{Limit_{ij}(k)\} \tag{5}$$

The meaning of this constraint is that each pair of agents $\{i, j\}$ may not leave a disk of diameter V centered at the average of their locations. Hence, after they both take a step, the distance between them will not exceed V. If an agent has more than one neighbour, it cannot leave the intersection of the disks associated with all its neighbours (see Fig. 4).

The addition of restriction (5) to a "chase the farthest" motion law yields a new law which maintains the connectivity of the system's visibility-graph. In the sequel, we formally present this new motion law, and prove that it gathers the agents of the system to a point.

Prior to presenting the motion law, we also need to adjust the definition of $N_i^{Far}(k)$, the set of the farthest neighbours of an agent i, as follows:

Let $l_i^{Far}(k)$ be the distance between agent i and its farthest neighbour, and let $\delta < 1/2$ be a small but strictly positive constant. Then, agent i's farthest set of neighbours is the subset of its neighbours to which the distance from i is in the range between $l_i^{Far}(k)(1 - \delta)$ and $l_i^{Far}(k)$, i.e.

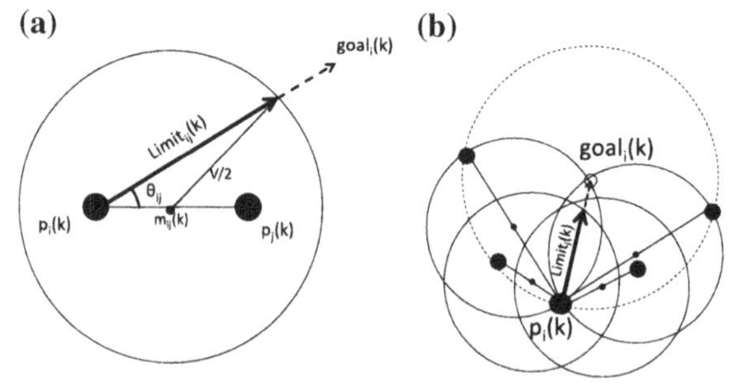

Fig. 4 $Limit_{ij}(k)$ and $Limit_i(k)$: **a** $Limit_{ij}(k)$ - Maximum distance agent i can move towards $c_i(k)$, its "goal" position, without leaving a circle of radius $V/2$ centred at $m_{ij}(k)$, the average position of $p_i(k)$ and $p_j(k)$ at time-step k. **b** $Limit_i(k) = \min_{j \in N_i(k)} \{Limit_{ij}(k)\}$

$$j \in N_i^{Far}(k) \iff l_i^{Far}(k)(1-\delta) \leq \|p_i(k) - p_j(k)\| \leq l_i^{Far}(k)$$

Then, the assumed behaviour of the agents is that at any time step k, each active agent i calculates $\psi_i^{Far}(k)$ the angle of the current minimal angular-sector containing the agents of the set $N_i^{Far}(k)$. If $\psi_i^{Far}(k)$ is greater than or equal to π, agent i stays put. Otherwise, it jumps a step of size $\mu_i(k)$ in the direction of $U_i^{Far}(k)$, the unit vector defining the bisector associated with the angle $\psi_i^{Far}(k)$, where $\mu_i(k)$ is the minimal value from the following quantities:

- $\sigma < V/2$ - maximal step size
- $Limit_i(k)$ - connectivity maintenance restriction
- $l_i^{LR}(k)$ the projection of $U_i^{Far}(k)$ on half the sum of the vectors pointing from $p_i(k)$ to $p_i^{ExtR}(k)$ and to $p_i^{ExtL}(k)$, the extremal right and left agents defining the minimal angular-sector. As a consequence, the current step of agent i cannot cross the line-segment defined by the positions of these two extremal neighbours.

$$p_i(k+1) = p_i(k) + \begin{cases} \chi_i(k)\mu_i(k)U_i^{Far}(k) & \psi_i^{Far}(k) < \pi \\ 0 & \psi_i^{Far}(k) \geq \pi \end{cases} \tag{6}$$

where $\mu_i(k) = \min\{\sigma, Limit_i(k), l_i^{LR}(k)\}$, and $\chi_i(k)$ equals 1 or 0 with probability ρ or $1 - rho$, respectively.

Notice that for a very small δ, except in some rare cases, there is always a single farthest neighbour, and then the minimal sector angle is zero, so that the preformed movement is toward this single farthest neighbour's position.

Lemma 2 *By the dynamic law (6), any pair of current neighbouring agents remain neighbours forever.*

Proof Dynamics (6) yields that if a pair of agents $\{i, j\}$ are neighbours at time-step k, then at time step $k+1$ their positions are limited to a disk of diameter V, due to (5) (one of the step size limits). Therefore, the distance between any pair of current neighbours cannot exceed V at any future time step.

Theorem 5 *A system of n agents with dynamic law (6) and with an initial configuration having a connected visibility-graph, gathers to a point within a finite expected number of time steps.*

In the proof we address two situations in the process of gathering. First, we address a situation where the agents of the system are not contained in a σ-diameter disk. Hence, due to some geometrical consideration and the fact that the constellation of the agents comprises a connected visibility-graph, each sequence of two time-steps, having a specific activity schedule, may result in a significant decrease of the perimeter of the system's convex-hull. By the *"Strong asynchronicity assumption"*, the probability for such a sequence to occur is at least ε^2 (see Lemma 7 in [12]).

Furthermore, clearly, by the motion law (6), the perimeter of the system's convex-hull cannot increase, hence the agents will gather to a disk of diameter σ within finite expected number of time-steps.

Second, we address a situation where the agents of the system are contained in a σ-diameter disk. Hence, any active agent that may move jumps to the position of its farthest neighbour or to a convex combination of the positions of two of its farthest neighbours. Consequently a specific activity schedule of five consequent time-steps gathers the system to a point.

Let agents i and j be the most distanced pair of agents at time-step k, let $m_{ij}(k)$ be the point located at the mean position of point $p_i(k)$ and $p_j(k)$, and let the line crossing point $p_i(k)$ and $p_j(k)$ divide the plane into two half plans, denoted by $Hp1$ and $Hp2$. Then, the mentioned above schedule, given by the active set of agents in each consequent step, starting at time-step k, is as follows:

1. At time-step k, the agents located in $Hp1$ (not include agents i and j)
2. At time-step $k+1$, the agents located in $Hp2$ (not include agents i and j)
3. At time-step $k+2$, the agents located on the segment $(p_i(k), m_{ij}(k))$
4. At time-step $k+3$, the agents located on the segment $(m_{ij}(k), p_j(k)]$
5. At time-step $k+4$, the agents located at point $m_{ij}(k)$

By the "*Strong asynchronicity assumption*" the probability that such a sequence of steps will occur is at least ε^5 (see Lemma 8 in [12]).

Clearly, once all agents gathered to a σ-diameter disk, they remain confined to such a disk, hence the agents will gather to a point within a finite expected number of time-steps, as claimed in Theorem 5 (see full proof in [12] Theorem 5).

Figure 5 presents simulation result of 5 agents starting in a constellation of a connected visibility net, and gather to a point.

5 Discussion

In this work, we analysed several "Chase the farthest" motion laws for multi-agent systems in order to address the problem of gathering identical and oblivious agents. We proved that such simple motion laws provide elegant solutions to the gathering problem in several settings.

First, we showed that in a continuous-time framework, agents that act by a "Chase the farthest" motion law, whether they have unlimited visibility or limited-visibility, gather to a point within a finite time. The suggested motion laws and the proofs of the resulting gathering are quite simple, however even in this case one has to carefully deal with issues of well definedness and non-differentiability.

In the discrete time framework, the situation is different and requires more work in order to define the motion rules and obtain meaningful convergence results. We have proved that agents with unlimited visibility and fixed steps sizes, gather to a disk of diameter twice their step size within a finite number of time-steps. To do so, we used as a Lyapunov function the radius of the minimal enclosing circle of the

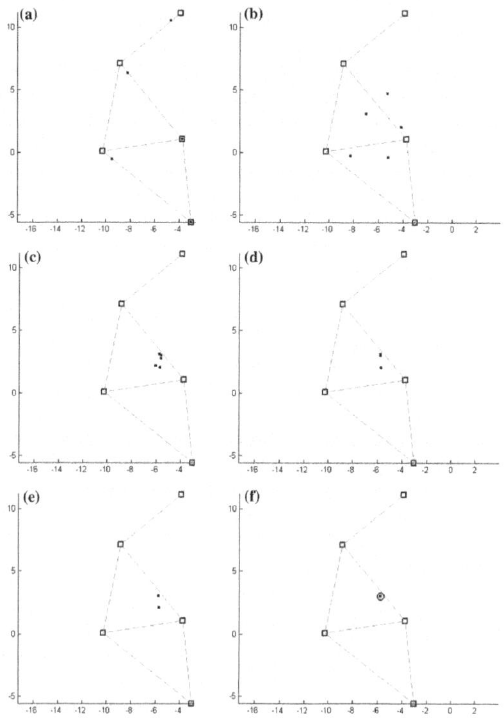

Fig. 5 Simulation results of 5 agents with dynamics (6), where $V = 10$, $\delta = 0.001$ and $\rho = 0.4$ for all the agents. The empty big squares are the initial configuration of the agents, the dashed grey lines are the initial connection topology, and the small squares are the agents. **a** The agent configuration after the first time step; **b** The configuration reacted to a complete visibility graph; **c** Last time step where the agents occupy 5 points in the plane; **d** The agents occupy 3 collinear points in the plane; **e** The agents occupy 2 points in the plane; **f** The agents gathered to a point

agents locations, and showed that it necessarily decreases until it reaches a value below the step size of the agents.

Furthermore, we have suggested an alternative motion law to resolve the problem of gathering within a finite number of time-steps. Solutions of this problem demand the agents to have the ability to agree on a meeting point. This ability requires high computational capabilities, which we try to avoid. Our alternative, allows the agents of the system to gather without agreeing on a meeting point, however as a consequence, the agents gather only within a finite *expected* number of time steps.

We present algorithms and proofs for multi-agent systems in the $\mathbb{R}^2$ plane, however those can simply be adapted to multi dimensional spaces, as well.

References

1. Agmon, N., Peleg, D.: Fault-tolerant gathering algorithms for autonomous mobile robots. SIAM J. Comput. **36**(1), 56–82 (2006)
2. Ando, H., Oasa, Y., Suzuki, Y., Yamashita, Y.: Distributed memory less point convergence algorithm for mobile robots with limited visibility. In: IEEE Transactions on Robotics and Automation, vol. 15(5), pp. 818–828 (1999)

3. Bellaiche, L.I., Bruckstein, A.: Continuous time gathering of agents with limited visibility and bearing-only sensing. Technical report, CIS Technical report, TASP (2015)
4. Bruckstein, A., Zeitouni, O.: A puzzling feedback quantizer. Technical report 879, EE Department Technion IIT (1993)
5. Cortes, J.: Discontinuous dynamical systems. IEEE control Syst. **28**(3), 36–73 (2008)
6. Gauci, M., Chen, J., Dodd, T.J., Gross, R.: Evolving aggregation behaviors in multi-robot systems with binary sensors. Distributed Autonomous Robotic Systems, pp. 355–367. Springer, Berlin (2014)
7. Gauci, M., Chen, J., Li, W., Dodd, T.J., Gross, R.: Self-organized aggregation without computation. Int. J. Robot. Res. (2014). https://doi.org/10.1177/0278364914525244
8. Gordon, N., Wagner, I., Bruckstein, A.: Gathering multiple robotic a(ge)nts with limited sensing capabilities. Ant Colony Optimization and Swarm Intelligence. Lecture Notes in Computer Science, p. 142–153. Springer, Berlin (2004)
9. Gordon, N., Wagner, I., Bruckstein, A.: A randomized gathering algorithm for multiple robots with limited sensing capabilities. In: Proceedings of MARS 2005 workshop at ICINCO Barcelona (2005)
10. Gordon, N., Elor, Y., Bruckstein, A.: Gathering multiple robotic agents with crude distance sensing capabilities. Ant Colony Optimization and Swarm Intelligence. Lecture Notes in Computer Science, pp. 72–83. Springer, Berlin (2008)
11. Jadbabaie, A., Lin, J., Morse, A.S.: Coordination of groups of mobile autonomous agents using nearest neighbor rules. IEEE Trans. Autom. Control **48**(6), 988–1001 (2003)
12. Manor, R., Bruckstein, A.: Chase Your Farthest Neighbour: A simple gathering algorithm for anonymous, oblivious and non-communicating agents. Technical report CIS-2016-01 (2016)
13. Moreau, L.: Stability of continuous-time distributed consensus algorithms. In: Proceedings of the 43rd IEEE Conference on Decision and Control, 2004. CDC. vol. 4, p. 3998–4003 (2004)
14. Moreau, L.: Stability of multi-agent systems with time-dependent communication links. IEEE Trans. Autom. Control **50**, 169–182 (2005)
15. Suzuki, I., Yamashita, M.: A theory of distributed anonymous mobile robots formation and agreement problems. Technical report, DTIC Document (1994)

OuijaBots: Omnidirectional Robots for Cooperative Object Transport with Rotation Control Using No Communication

Zijian Wang, Guang Yang, Xuanshuo Su and Mac Schwager

Abstract We propose a distributed force and torque controller for a group of robots to collectively transport objects with both translation and rotation control. No explicit communication among robots is required. This work goes beyond previous works by including rotation control and experimental demonstrations on a custom built robot platform. We prove that follower robots can synchronize both their forces and torques to a leader (either a robot or human) that guides the group, and thus contribute positively to the transport. We introduce a custom-designed omnidirectional robot platform, called the OuijaBot, with sensing and actuation capabilities for cooperative manipulation. Our approach is verified by experiments with four OuijaBots successfully transporting and rotating a payload through a narrow corridor.

Keywords Multi-robot manipulation · Cooperative mobile manipulation

1 Introduction

Multi-robot cooperative manipulation and object transport is an emerging field [1, 3, 6, 15, 17, 19, 23] that exploits the power of collaboration among a team of robots to move objects that are too large and heavy for any single robot to handle

---

Z. Wang (✉) · M. Schwager
Department of Aeronautics and Astronautics, Stanford University,
Stanford, CA, USA
e-mail: zjwang@stanford.edu

M. Schwager
e-mail: schwager@stanford.edu

G. Yang · X. Su
Department of Mechanical Engineering, Boston University, Boston, MA, USA
e-mail: gyang101@bu.edu

X. Su
e-mail: sxs99@bu.edu

R. Groß et al. (eds.), *Distributed Autonomous Robotic Systems*,
Springer Proceedings in Advanced Robotics 6,
https://doi.org/10.1007/978-3-319-73008-0_9

alone. Most work in this area has focused on controlling the translation of the object, and leaves the rotation of the object uncontrolled. However, rotation control is also an important part of cooperative transport. Rotation control is necessary in reaching an appropriate orientation to navigate through spatially constrained environments, for example, when the robots must transport a long object through a narrow corridor.

In this work, we propose a novel solution for multi-robot cooperative manipulation, with a particular concentration on rotation control. The feature of our approach is that no communication is required between any two robots, yet the robots have to contribute positively to both the object's translation and rotation motion. The robots coordinate themselves by sensing the linear and angular velocity of the object, and then applying a force and torque to reinforce the sensed motion. A leader (either a robot or human), who is the only one in the group that knows the destination or desired trajectory, can steer the entire group by adjusting its input. In our previous work [19–21], we verified that translation can be controlled without communication by properly designing the robots' forces. In this paper, we extend our prior work by incorporating rotation sensing and torque input, so that the orientation of the object can be independently controlled along with the translation.

The main challenge in combining rotation and translation control is due to the complexity of the object dynamics and the kinematics of joint motion of the robots-object assembly. Our method naturally takes into account both the dynamics and kinematics, leading to a provably convergent and physically realizable approach. In order to verify our algorithm, we introduce a new robot platform for cooperative manipulation called the *OuijaBot*. The OuijaBot platform is custom designed to realize the sensing and actuation required for our force and torque controller. Our approach is successfully verified in experiments with four OuijaBot robots carrying a loaded pallet, with rotation being controlled in order to go through a narrow passage.

1.1 Related Work

Our work has been inspired by the multi-disciplinary research on cooperative transport in robotics and in cooperative ant behavior in biology. A multi-robot system was utilized in [17] to move furniture under different sensing and communication configurations. A formation-based approach called caging was studied in [5, 15], where the robots move the object by maintaining a formation to trap the object as if in a cage. An important issue in multi-robot manipulation is impedance control [4, 12], which has been used to regulate internal forces [22] among multiple manipulators. Massive manipulation experiments were done by [16] with up to 100 robots using flocking. The kinematic multi-robot motion controller proposed in [11] is capable of both translating and rotating the object, by specifying the robots' speeds with respect to the centroid of the object. There are also approaches that achieve cooperative manipulation without communication, using vision occlusion [3], passive caster [18], and inter-robot force measurements [8, 9]. In addition to the robotics research, the behavior study of ant colonies also backs our assumption that ants use measured motion information for cooperative object transport [14], and that ants tend to align their manipulation forces during collective transport behaviors [2].

2 Problem Formulation

We consider manipulation in a 2D environment $Q \subset \mathbb{R}^2$, potentially cluttered with obstacles. The object has a mass M and moment of inertia J. The acceleration of the gravity is g, and points downward, perpendicular to the environment plane. The object is subject to two kinds of friction from the ground, kinetic friction and viscous friction, whose coefficients are μ_k and μ_v respectively.

There is a fleet of N robots, indexed as $\{R_1, R_2, \ldots, R_N\}$. We choose R_1 as the leader robot, and others as the followers. Only the leader knows the destination or desired trajectory for the object. The followers do not know where the object needs to be transported, nor the control actions of other robots. The robots are attached to the object, and can apply 2D forces to the object, denoted by $\{\mathbf{F}_1, \mathbf{F}_2, \ldots, \mathbf{F}_N\}$, as well as 1D torques about the z-axis (using the standard right hand rule) represented by $\{T_1, T_2, \ldots, T_N\}$. As the robots cannot communicate, they rely on their sensors to determine what forces and torques to apply. We require the robots to be able to measure the linear and angular velocity of the object, denoted by $\mathbf{v}_c$ and $\boldsymbol{\omega}$. A more detailed sensing model for the robots will be introduced in Sect. 3.1.

Under the manipulation inputs from the robots, the object can translate and rotate in Q. The translation dynamics is governed by Newton's second law,

$$M\dot{\mathbf{v}}_c = \sum_{i=1}^{N} \mathbf{F}_i - \mu_v \mathbf{v}_c - \mu_k M g \frac{\mathbf{v}_c}{\|\mathbf{v}_c\|}, \tag{1}$$

where $\mathbf{v}_c$ denotes the linear velocity of the object at the center of mass. The rotation dynamics, while the object is in motion, can be written as

$$J\dot{\omega} = \sum_{i=1}^{N} T_i + \sum_{i=1}^{N} \mathbf{r}_i \times \mathbf{F}_i - \frac{\mu_v}{M} J\boldsymbol{\omega}, \tag{2}$$

where the position vector $\mathbf{r}_i$ points from the center of mass of the object to the attachment point of robot R_i. The derivation of the rotational dynamics can be found in our previous work [19], where we have shown that the friction model in (2) is linearly damped by $-\mu_v J\boldsymbol{\omega}/M$.

The objective of this paper is to design a communication-free controller for the robots' forces and torques, $\mathbf{F}_i$ and T_i, so that the robots can guide the objects' linear velocity $\mathbf{v}_c$ and angular velocity $\boldsymbol{\omega}$ to navigate the object along a desired trajectory. Furthermore, only the leader (either a robot or a human) knows this desired trajectory.

In order to achieve the coordination with no communication, we need a few assumptions for the robots, which are summarized below.

Assumptions

1. All robots know the value of M, J, μ_k, μ_v, g, N, which correspond to the mass of the object, moment of inertia of the object, coefficients of friction, acceleration of gravity, and the number of robots, respectively;
2. Every robot R_i knows its own position relative to the object's center of mass, $\mathbf{r}_i$;
3. The robots' attachment points are centrosymmetric around the center of mass of the object, meaning that for any robot R_i, there exists another robot $j \neq i$ such that $\mathbf{r}_i = -\mathbf{r}_j$.

Note that Assumption 1 can be achieved by existing distributed parameter estimation approaches [7]. The position information in Assumption 2 can also be estimated online by the robots in practice, for example, by using the *pipelined consensus* algorithm from our previous work [10]. We require Assumption 3 for our convergence analysis, since when force coordination is achieved (i.e., all $\mathbf{F}_i$ are equal), the resultant $\sum_{i=1}^{N} \mathbf{r}_i \times \mathbf{F}_i = 0$ in (2). In practice, our controller is sufficiently robust to tolerate some lack of centrosymmetry.[1]

3 Distributed Force and Torque Controller Design

3.1 Robot Sensing

Our controller requires the follower robots to measure the linear and angular velocity of the object, denoted as $\mathbf{v}_{mi}$ and $\boldsymbol{\omega}_{mi}$, where the subscript "m" refers to "measured" and i specifies the robot R_i. Theoretically, $\boldsymbol{\omega}_{mi}$ should be the same for all the robots since we assume the robots are rigidly attached to the object. However, $\mathbf{v}_{mi}$ will be different for all the robots if the object is rotating, since the robots can only measure the velocities at their local attachment points, which can be characterized by

$$\mathbf{v}_{mi} = \mathbf{v}_c + \boldsymbol{\omega} \times \mathbf{r}_i. \tag{3}$$

In addition to velocity sensing, the robots are also equipped with force and torque sensors in order to perform feedback control, so that the desired force and torque specified by the high-level controller (Sect. 3.3) can be achieved. The measured force and torque of robot R_i are denoted by $\mathbf{f}_{mi}$ and τ_{mi}.

[1] Even though strict centrosymmetry is hard to achieve, when the number of robots is large with respect to the size of the object, it is likely that the robots will spread evenly around the object, so that the centrosymmetry can be nearly satisfied.

3.2 Robot Kinematics

We must deal with both the dynamics and kinematics of our combined robot-object system. On one hand, the movement of the object is governed by the dynamics (1), (2). On the other hand, the robots are physically connected to the object, resulting in complex interactions between the object and the robots. Our approach aims to take into account both kinematics and dynamics involved in the process.

Many previous approaches for multi-robot manipulation use differential-drive robots, which suffer from non-holonomic constraints when moving with the object as a whole system. To effectively reduce the complexity involved in this joint motion, the OuijaBots that we present in this paper feature holonomic dynamics using an omni-directional wheel design. Therefore, for our robots the 2D linear velocity, denoted by $\mathbf{v}_{di}$, and angular velocity, denoted by $\boldsymbol{\omega}_{di}$, can be independently controlled, where the subscript "d" means "desired" and i refers to a specific robot R_i.

Another advantage of the holonomic configuration is that the force and torque generated by the robot can be controlled independently. As was done in [1], we characterize the force/torque generation by the tendency of the robot to move/rotate faster or slower than the object. The larger this tendency is, the greater the resultant force/torque will be and vice versa. Mathematically, we characterize this tendency as the difference between the commanded (or desired) velocity of the robot and the actual velocity of the object. Thus, we use a linear model to describe this phenomenon, as follows,

$$\mathbf{v}_{di} = \mathbf{v}_{mi} + K_f(\mathbf{F}_i - \mathbf{f}_{mi}), \tag{4}$$

$$\boldsymbol{\omega}_{di} = \boldsymbol{\omega}_{mi} + K_\tau(T_i - \tau_{mi}), \tag{5}$$

where $\mathbf{F}_i$ and T_i are the desired force and torque, respectively, of robot R_i, and $\mathbf{f}_{mi}$ and τ_{mi} are the measured force and torque, respectively. The parameters K_f and K_τ are constants that need to be tuned experimentally. Using (4) and (5), the resulting velocity command $\mathbf{v}_{di}$ and $\boldsymbol{\omega}_{di}$ can be implemented by the robot's motion controller (i.e., generate motor voltages to fulfill the desired velocities), so that the desired force and torque can be generated. The physical realization of $\mathbf{v}_{di}$ and $\boldsymbol{\omega}_{di}$, as well as how to measure $\mathbf{v}_{mi}$ and $\boldsymbol{\omega}_{mi}$, will be presented in detail in Sect. 4, where we introduce the hardware design and control system of our OuijaBots.

3.3 Force and Torque Controller

3.3.1 Follower's Controller

Our motivation in designing the follower's controller is that every robot should be equally responsible for $1/N$ share of the effort to overcome the friction or resistance of the motion. Thus, the torque controller for the followers can be written as,

$$\mathbf{F}_i(t) = \frac{1}{N}\left(\mu_v \mathbf{v}_c(t) + \mu_k M g \frac{\mathbf{v}_c(t)}{\|\mathbf{v}_c(t)\|}\right), \; i \in \{2, 3, \ldots, N\}. \tag{6}$$

Since $\mathbf{v}_c$ is not directly measurable, using (3) we can rewrite (6) as

$$\mathbf{F}_i(t) = \frac{1}{N}\left(\mu_v(\mathbf{v}_{mi}(t) - \boldsymbol{\omega}_{mi}(t) \times \mathbf{r}_i) + \mu_k M g \frac{\mathbf{v}_{mi}(t) - \boldsymbol{\omega}_{mi}(t) \times \mathbf{r}_i}{\|\mathbf{v}_{mi}(t) - \boldsymbol{\omega}_{mi}(t) \times \mathbf{r}_i\|}\right), \tag{7}$$

which can be computed by all the robots since they either know or can measure the quantities in (7). Similarly, the torque controller is defined as

$$T_i(t) = \frac{\mu_v J}{NM} \boldsymbol{\omega}_{mi}(t), \; i \in \{2, 3, \ldots, N\}. \tag{8}$$

Note that in (8), the possible resistance from the term $\sum_{i=1}^{N} \mathbf{r}_i \times \mathbf{F}_i$ in (2) is not included. We show in Theorem 1 below that we can safely ignore this term under Assumption 3. Even when this term is not equal to zero, a PI controller on the leader's torque input will effectively reject this disturbance.

A quick observation reveals that the controllers (7), (8) themselves imply no communication and rely on only local measurements. The outcome of (7), (8), $\mathbf{F}_i(t)$ and $T_i(t)$, can be used by (4), (5) to generate appropriate motor commands. Also note that (7), (8) require no global reference frame information. The sensing and force/torque actuation can all be performed in the robots' local reference frames.

3.3.2 Leader's Controller

The goal of the leader robot is to steer the group towards the destination or follow a pre-defined trajectory by adjusting its own force and torque input. Suppose that we already know the desired linear and angular velocities from a higher-level path planning algorithm.[2] Then we define the leader's force controller as follows,

$$\mathbf{F}_1(t) = f_d \frac{\mathbf{v}_d(t)}{\|\mathbf{v}_d(t)\|}, \tag{9}$$

where $\mathbf{v}_d$ is the desired velocity, and

$$\begin{aligned} f_d &= K_1 \max\{\|\mathbf{v}_d(t)\| - \|\mathbf{v}_c(t)\|, 0\} \\ &= K_1 \max\{\|\mathbf{v}_d(t)\| - \|\mathbf{v}_{m1}(t) - \boldsymbol{\omega}_{m1}(t) \times \mathbf{r}_1\|, 0\}, \end{aligned} \tag{10}$$

where K_1 is a proportional gain. The philosophy behind the leader's force controller is as follows. From (9) it can be seen that the leader aligns its force along the direction of $\mathbf{v}_d$, which tends to drag the object's velocity towards $\mathbf{v}_d$. The magnitude of $\mathbf{F}_1$ is

[2]There are many off-the-shelf algorithms we can choose, and this is not the focus of this paper.

determined by a proportional controller (10), aiming to reduce the difference between $\|\mathbf{v}_d\|$ and $\|\mathbf{v}_c\|$. The max function is used to ensure that $\mathbf{F}_1$ does not point opposite to $\mathbf{v}_d$.

The torque controller of the leader is

$$T_1(t) = K_2 e(t) + K_3 \sigma(t), \tag{11}$$

where K_2 is the proportional gain, K_3 is the integral gain, and

$$\dot{\sigma}(t) = e(t) = \boldsymbol{\omega}_d(t) - \boldsymbol{\omega}_{m1}(t). \tag{12}$$

The rotation controller (11) is essentially a PI controller that drives the angular velocity of the object to the desired $\boldsymbol{\omega}_d$.

3.3.3 Convergence Proof

Theorem 1 *Using the followers' force/torque controller (7), (8) and the leader's force/torque controller (9), (11), the object's linear and angular velocity, $\mathbf{v}_c$ and $\boldsymbol{\omega}$ track $\mathbf{v}_d$ and $\boldsymbol{\omega}_d$ as a first order filter. Furthermore, if $\mathbf{v}_d$ and $\boldsymbol{\omega}_d$ are constant, and assuming negligible measurement noise $\boldsymbol{\omega}_{mi} = \boldsymbol{\omega}$, $\forall i$, $\mathbf{v}_c$ and $\boldsymbol{\omega}$ converge asymptotically to $\mathbf{v}_d$ and $\boldsymbol{\omega}_d$.*

Proof For force convergence and the convergence of $\mathbf{v}_c$ to $\mathbf{v}_d$, refer to Lemma 1 and Theorem 1 in our previous work [21], where similar proof procedures can be applied. For rotation, if noise is not present in the sensing, we have $\boldsymbol{\omega}_{mi} = \boldsymbol{\omega}$, $\forall i \in \{1, 2, \ldots, N\}$. Then plugging the followers' torque inputs (8) into the rotation dynamics (2) yields

$$\dot{\boldsymbol{\omega}} = \frac{1}{J} T_1 + \frac{1}{J} \sum_{i=1}^{N} \mathbf{r}_i \times \mathbf{F}_i - \frac{1}{N} \frac{\mu_v}{M} \boldsymbol{\omega}. \tag{13}$$

If we treat $\boldsymbol{\omega}$ as the state and T_1 as the input, (13) is a stable first-order linear system with $\sum_{i=1}^{N} \mathbf{r}_i \times \mathbf{F}_i / J$ as the disturbance, which we know will diminish to zero when the force convergence is achieved under the centrosymmetric assumption. Furthermore, given a constant desired angular velocity ω_d, the leader's torque input (11) implements a PI controller with $\boldsymbol{\omega}_d$ as the set point. Therefore, we know that $\boldsymbol{\omega}$ will converge to $\boldsymbol{\omega}_d$ [13]. □

Note that in (13), the followers' torque inputs help reduce the magnitude of the resistance significantly. Also, in the steady state of (13), the torque inputs from all the robots (including the leader) are equal.

4 OuijaBot Hardware and Control System

We designed and manufactured an omnidirectional robot platform, named OuijaBot, which brings together unique sensing and actuation capabilities to implement the controllers proposed in the previous section, as shown in Fig. 1. The OuijaBot contains four symmetrically-placed omnidirectional wheels, with free rollers along the perimeter that enable sideways movements. The wheels are independently driven by four DC motors, each of which is rated at 5 A stall current under 12 V, generating a maximal torque of 0.78 Nm. Encoders are mounted with the motors to measure the velocity. We designed a custom Printed Circuit Board (PCB) for low-level motor control and processing sensor measurements, including encoders, motor currents, accelerometer, gyroscope, and magnetometer. A Raspberry Pi is used as a more powerful computation hub and for running the Robot Operating System (ROS). OuijaBot weighs 2.7 kg and can move up to 1 m/s.

For cost and complexity considerations, OuijaBot currently does not have a suspension system, which is left for future work. Instead, we test our robots mainly on a resilient surface with plastic mats, to help alleviate the suspension problem.

(a) (b)

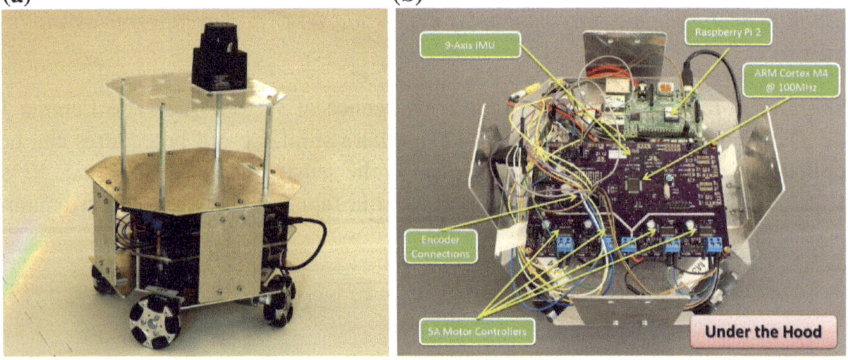

Fig. 1 Our custom-built omnidirectional robot platform called OuijaBot. **a** The overall appearance. Four brushed DC motors are mounted symmetrically underneath the chassis. Battery and all electronics are enclosed compactly inside the aluminum shell. The top layer of the robot can be used to install additional equipment, such as a gripper, robotic arm, camera, or lidar as shown in this picture. **b** Our PCB design. The low-level microcontroller is an ARM Cortex-M4 running at 100 MHz. Four motors are driven by four full H-bridges through PWM signals. The motor currents are measured using four Hall effect current sensors. Other sensors including encoders, 9-axis IMU are also handled by the Cortex-M4. Various interfaces such as Serial, I2C, SPI, and PWM are reserved for connecting future add-on devices

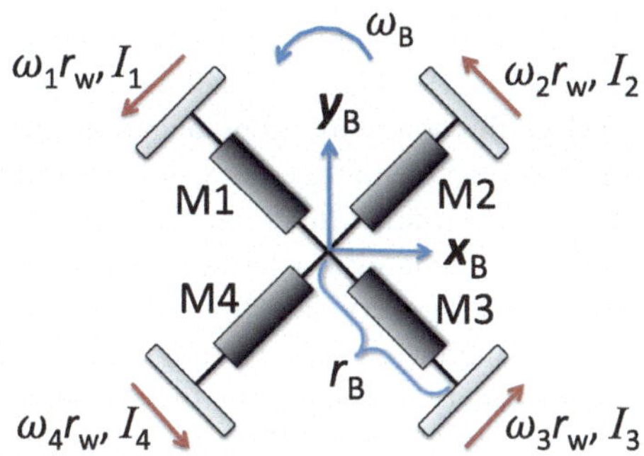

Fig. 2 Configuration of OuijaBot's motors and wheels. M1-4 are the four motors. The positive directions of different quantities are as marked. We choose the body-frame $\mathbf{x}_B$ and $\mathbf{y}_B$ axis as such because these are the two directions in which the robot can generate the maximal amount of force, and also move at the highest speed

4.1 Motion Control

Given a desired linear $\mathbf{v}_{di}$ and angular velocity $\boldsymbol{\omega}_{di}$, as computed by (4), (5), the robot needs to realize the velocity by controlling the speeds of its four wheels. A schematic of the motors and wheels are shown in Fig. 2. We denote the output angular velocity of the wheel as $\boldsymbol{\omega}_i,\ i \in \{1, 2, 3, 4\}$. We assume that each wheel is controlled by the corresponding motor in the nominal direction, while free to slide sideways. Then the mapping from the wheel speeds to the robot speed is

$$\mathbf{v}_{di} = \frac{\sqrt{2}r_w}{4}\left[\omega_3 - \omega_1 + \omega_4 - \omega_2,\ \omega_3 - \omega_1 + \omega_2 - \omega_4\right]^T, \tag{14}$$

$$\omega_{di} = \frac{(\omega_1 + \omega_3)r_w}{2r_B} = \frac{(\omega_2 + \omega_4)r_w}{2r_B}, \tag{15}$$

where r_w is the radius of the wheel and r_B is the radius of the robot. Note that in (15), we impose an additional artificial constraint since four wheels are redundant in generating the 3-DOF velocity. Using (14) and (15), we have four equations to uniquely determine four wheel speeds. Standard PID controllers are implemented on the PCB to handle the motor control and achieve the required wheel speeds $\omega_1, \omega_2, \omega_3, \omega_4$.

4.2 Velocity Measurement and Traction Control

The wheel speeds are measured by four encoders mounted on the back of the motors. In order to measure $\mathbf{v}_{mi}$, as required in (4), (7), (10), we can use the same kinematic equation in (14) and substitute ω_{1-4} with the measured wheel speeds. In order to measure ω_{mi}, as used in (5), (7), (8), (12), the reading from the gyroscope is used rather than solving (15) since this is a more accurate approach.

A possible factor that may degrade the accuracy of measuring the linear velocity is the wheel slipping along the nominal direction. We use a practical approach for detecting wheel slippage and to enhance traction control, similar to what has been done in the automobile industry. In every detection period (200 ms in our case), we calculate the robot velocity using the encoder readings using (14), (15). In parallel, we also estimate the linear velocity by integrating the acceleration, and the angular velocity by reading the gyroscope, using IMU data. If the difference between the two sets of measurements exceeds a pre-defined threshold, then the robot knows that slipping occurred and then tries to reduce speed to regain traction.

4.3 Force and Torque Measurements

The torque generated by each motor can be measured by the current sensor through the linear relationship $T = K_i I$, where K_i is a coefficient. Denote I_1, I_2, I_3, I_4, along directions shown in Fig. 2, as the effective current of each motor corrected with the no-load current, then the force and torque generated by the robot with respect to the center of the mass, as required by (4), (5), can be measured as

$$\mathbf{f}_{mi} = \frac{\sqrt{2}K_i r_w}{2}\left[-I_1 - I_2 + I_3 + I_4,\ -I_1 + I_2 + I_3 - I_4\right]^T, \tag{16}$$

$$t_{mi} = K_i r_w r_b (I_1 + I_2 + I_3 + I_4). \tag{17}$$

5 Simulation

We have successfully verified our distributed force and torque controller first in simulation using Open Dynamic Engine (ODE), a well-known physics engine in the robotics community. Twenty robots are employed to transport a simulated chair which weighs 2 kg. The task setup is visualized in Fig. 3, and the goal is to navigate the chair through the narrow corridor on the right. Since the opening of the corridor is smaller than the length of the chair, the robots have to rotate the chair by 90 degrees. The robots are placed symmetrically around the chair, each of which is able to apply force and torque calculated by (7), (8) or (9), (11). In this simulation we focus on the force and torque rather than the specific mechanics of the robot, so we visualize the robots as spheres. The leader, drawn in blue and located at one corner of the chair, is the only that knows the desired destination and orientation of the object. During the entire process, no communication occurs between any two robots. Additionally,

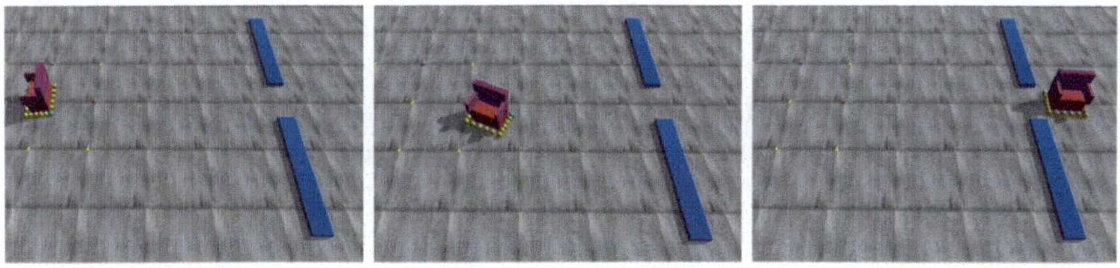

(a) Initial position (b) Intermediate position (c) End position

Fig. 3 Twenty robots successfully transport and rotate a chair to cross a narrow corridor in simulation. The leader robot is drawn in blue while follower robots are in yellow. The robots use the communication-free force and torque controller described in Sect. 3

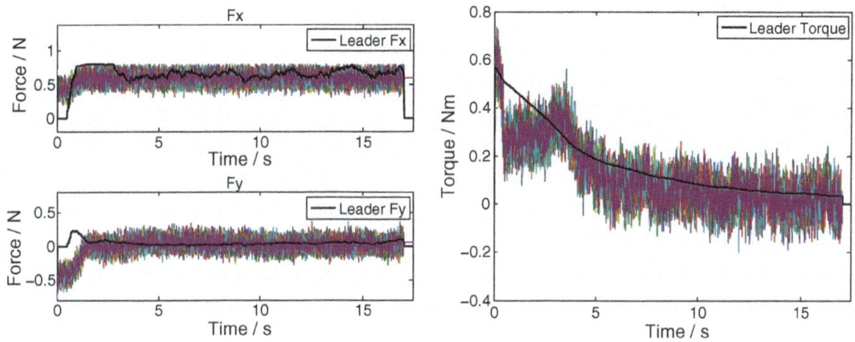

(a) Robots' Forces. (b) Robots' Torques.

Fig. 4 The forces and torques of the robots during the simulation. The leader's force and torque are in bold black lines, while the other lines correspond to the followers. Despite the noises, it is clear that all the followers respond and follow the trend of the leader's input

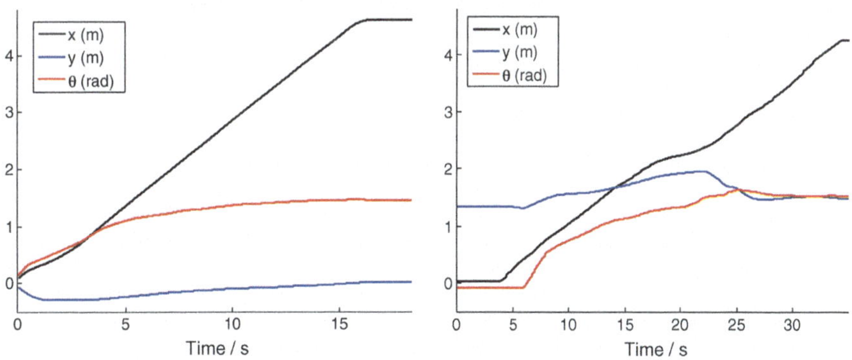

(a) Trajectory in simulation. (b) Trajectory in experiment.

Fig. 5 Trajectory of the object in simulation (left) and experiment (right). In both cases, the robots are successful in transporting the object along x direction, and rotating the object by 90 degrees. The displacements on y axis in both cases are maintained around zero (the offset values are not meaningful), which ensure the desired straight line motion to the right side in order to simulate the noisy sensing and actuation, the force and torque applied by the robots are corrupted by zero-mean Gaussian noise.

The forces and torques of the robots are plotted in Fig. 4, where the leader's steering on the followers' forces and torques is evident. The trajectory of the object is shown in Fig. 5. For the purpose of comparison, we place the trajectory in the simulation next to that of the experiment (Sect. 6), where the task objective is similar. The two trajectories present resemblance to each other in terms of reaching the desired x and θ value, and holding the position in y direction (the offset value does not matter).

6 Experiments

Our approach is validated experimentally using four OuijaBots. The scenario of our experiment is shown in Fig. 6. Four OuijaBots are rigidly attached to the corners of a piece of square wood (0.6 m $\times$ 0.6 m), together serving as a modular pallet system. The imagined payload, three round bars, are fixed onto the pallet. The goal of the experiment is to first track a straight line to move the pallet towards the right of the field, which is about 4.5 m away, and then cross the narrow corridor between the two piles of boxes. As shown in Fig. 6a, the initial orientation of the pallet cannot go through the corridor since the length of the bar is larger than the opening of the

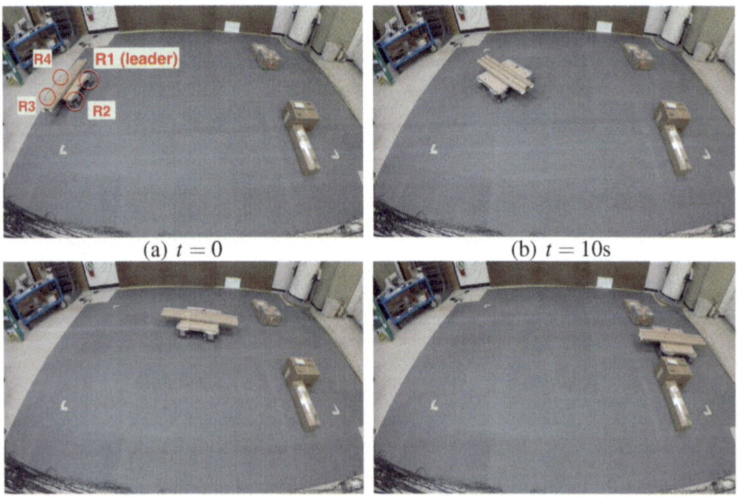

(a) $t = 0$ (b) $t = 10$s (c) $t = 20$s (d) $t = 33$s

Fig. 6 Snapshots of the cooperative transport experiment. The locations of the robots are marked in **a**. The robots are able to move the pallet to the right side and also rotate 90 degrees counterclockwise in order to go through the narrow corridor. The experiment video is available online at https://youtu.be/4nLMYjqUoJ4

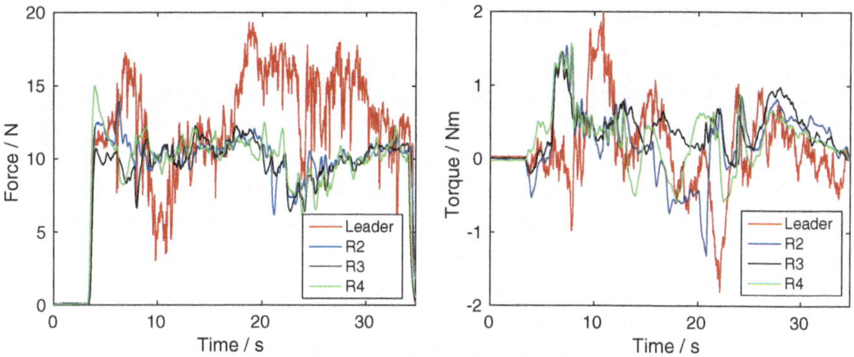

(a) Magnitude of robots' forces

(b) Torques applied by the robots

Fig. 7 Forces and torques applied by the robots during the transport process. The plots confirm that the follower robots contribute positively to the manipulation

corridor. Therefore, the robots must rotate the pallet before reaching the corridor, as shown in the final moment in Fig. 6d.

Upon the start of the experiment, all the robots are issued the same initial speed to establish the initial movement.[3] After that, all the follower robots perform all the sensing and computation as described in Sects. 3 and 4 completely onboard without any inter-robot communication. The leader's controller (9), (11) is implemented offboard by integrating the external positioning information from Vicon,[4] which is consistent with our assumption that only the leader knows the desired trajectory. The actual trajectory of the object is plotted in Fig. 5. The forces and torques applied by all the robots are shown in Fig. 7, which are recorded at 100 Hz. The plots indicate that the follower robots share a large amount of the force in order to help the leader overcome the friction. Also, the torque inputs from the followers tend to follow the changes of the leader's torque, and are mostly positive, verifying the followers' contribution to help rotate the object counterclockwise. Moreover, our approach allows the leader's force and torque to be from a human. We verify this by letting a human operator control the leader robot using a joystick to carry out the same task, while the follower robots run the same program as before. Due to the space limit, this result is shown in the online video: https://youtu.be/4nLMYjqUoJ4.

[3]Initial movement can be also triggered without communication using random trials [19]. For the sake of clear presentation, we skip this phase, which is not the focus of this paper.

[4]http://www.vicon.com/.

7 Conclusion and Future Work

In this paper, we present a decentralized force and torque controller for a group of robots to control both the translation and rotation of an object during cooperative manipulation without communication. We show theoretically and in hardware experiments that the robots can rely on their measurements of the object's motion to coordinate their force and torque inputs, rather than using explicit communication. We designed and built a new omni-directional robot platform, the OuijaBot, and successfully validated our approach in real time experiments with four OuijaBots cooperatively transporting a loaded pallet.

In the future, we plan to leverage machine learning to estimate parameters such as M, μ_k, μ_v, N, $\mathbf{r}_i$ online. Also, we are interested in more advanced filtering on the multi-sensor data, and extending our approach to the 3D using aerial robots.

Acknowledgements This work was supported by NSF grant CNS-1330036, and also by the Toyota-SAIL Center for AI Research. The authors are grateful for this support.

References

1. Alonso-Mora, J., Knepper, R., Siegwart, R., Rus, D.: Local motion planning for collaborative multi-robot manipulation of deformable objects. In: Proceedings of the IEEE International Conference on Robotics and Automation (ICRA), pp. 5495–5502 (2015)
2. Berman, S., Lindsey, Q., Sakar, M.S., Kumar, V., Pratt, S.: Study of group food retrieval by ants as a model for multi-robot collective transport strategies. In: Robotics: Science and Systems (RSS), pp. 259–266 (2010)
3. Chen, J., Gauci, M., Li, W., Kolling, A., Groß, R.: Occlusion-based cooperative transport with a swarm of miniature mobile robots. IEEE Trans. Robot. **31**(2), 307–321 (2015)
4. Erhart, S., Sieber, D., Hirche, S.: An impedance-based control architecture for multi-robot cooperative dual-arm mobile manipulation. In: Proceedings of the International Conference on Intelligent Robots and Systems (IROS), pp. 315–322 (2013)
5. Fink, J., Hsieh, M.A., Kumar, V.: Multi-robot manipulation via caging in environments with obstacles. In: Proceedings of the IEEE International Conference on Robotics and Automation (ICRA), pp. 1471–1476 (2008)
6. Fink, J., Michael, N., Kim, S., Kumar, V.: Planning and control for cooperative manipulation and transportation with aerial robots. Int. J. Robot. Res. **30**(3), 324–334 (2011)
7. Franchi, A., Petitti, A., Rizzo, A.: Decentralized parameter estimation and observation for cooperative mobile manipulation of an unknown load using noisy measurements. In: Proceedings of the IEEE International Conference on Robotics and Automation (ICRA), pp. 5517–5522 (2015)
8. Groß, R., Dorigo, M.: Group transport of an object to a target that only some group members may sense. In: International Conference on Parallel Problem Solving from Nature, pp. 852–861 (2004)
9. Groß, R., Mondada, F., Dorigo, M.: Transport of an object by six pre-attached robots interacting via physical links. In: IEEE International Conference on Robotics and Automation (ICRA), pp. 1317–1323 (2006)
10. Habibi, G., Kingston, Z., Wang, Z., Schwager, M., McLurkin, J.: Pipelined consensus for global state estimation in multi-agent systems. In: Proceedings of the International Conference on Autonomous Agents and Multiagent Systems (AAMAS), pp. 1315–1323 (2015)

11. Habibi, G., Kingston, Z., Xie, W., Jellins, M., McLurkin, J.: Distributed centroid estimation and motion controllers for collective transport by multi-robot systems. In: Proceedings of the IEEE International Conference on Robotics and Automation (ICRA), pp. 1282–1288 (2015)
12. Kennedy, M.D., Guerrero, L., Kumar, V.: Decentralized algorithm for force distribution with applications to cooperative transport. In: ASME International Design Engineering Technical Conferences and Computers and Information in Engineering Conference (2015)
13. Khalil, H.K.: Nonlinear systems, 2nd edn. Prentice Hall, New Jersey (1996)
14. McCreery, H.F., Breed, M.D.: Cooperative transport in ants: a review of proximate mechanisms. Insectes Soc. **61**(2), 99–110 (2014)
15. Pereira, G.A., Campos, M.F., Kumar, V.: Decentralized algorithms for multi-robot manipulation via caging. Int. J. Robot. Res. **23**(7–8), 783–795 (2004)
16. Rubenstein, M., Cabrera, A., Werfel, J., Habibi, G., McLurkin, J., Nagpal, R.: Collective transport of complex objects by simple robots: theory and experiments. In: Proceedings of International Conference on Autonomous Agents and Multi-agent Systems (AAMAS), pp. 47–54 (2013)
17. Rus, D., Donald, B., Jennings, J.: Moving furniture with teams of autonomous robots. In: Proceedings of the International Conference on Intelligent Robots and Systems (IROS), vol. 1, pp. 235–242. (1995)
18. Stilwell, D.J., Bay, J.S.: Toward the development of a material transport system using swarms of ant-like robots. In: Proceedings of the IEEE International Conference on Robotics and Automation (ICRA), pp. 766–771 (1993)
19. Wang, Z., Schwager, M.: Multi-robot manipulation without communication. In: Proceedings of the International Symposium on Distributed Autonomous Robotic Systems (DARS), Daejeon, Korea, 2–5 November, pp. 135–149 (2014)
20. Wang, Z., Schwager, M.: Multi-robot manipulation with no communication using only local measurements. In: Proceedings of the IEEE International Conference on Decision and Control (CDC), pp. 380–385 (2015)
21. Wang, Z., Schwager, M.: Kinematic multi-robot manipulation with no communication using force feedback. In: Proceedings of the IEEE International Conference on Robotics and Automation (ICRA), pp. 427–432 (2016)
22. Williams, D., Khatib, O.: The virtual linkage: A model for internal forces in multi-grasp manipulation. In: Proceedings of the IEEE International Conference on Robotics and Automation (ICRA), pp. 1025–1030 (1993)
23. Wilson, S., Pavlic, T.P., Kumar, G.P., Buffin, A., Pratt, S.C., Berman, S.: Design of ant-inspired stochastic control policies for collective transport by robotic swarms. Swarm Intell. **8**(4), 303–327 (2014)

Persistent Multi-robot Formations with Redundancy

Alyxander Burns, Bernd Schulze and Audrey St. John

Abstract A multi-robot formation composed of autonomous agents may need to maintain an overall rigid shape for tasks such as collective transport of an object. To distribute control, we construct leader-follow formations in the plane that are *persistent*: designated "leader" robots control the movement of the entire formation, while the remaining "follower" robots maintain directed local links sensing data to other robots in such a way that the entire formation retains its overall shape. In this paper, we present an approach based on rigidity theory for constructing persistent leader-follower formations with redundancy; specified robots may experience sensor link failure without losing the persistence of the formation. Within this model, we consider the impact of special positions due to certain geometric conditions and provide simulation results confirming the expected behavior.

1 Introduction

For applications such as collective transport, multi-robot formations need to maintain a global shape. To do so in a distributed fashion, we focus on formations composed of autonomous agents that use local sensing to maintain a global rigid structure. In particular, we consider *persistent leader-follower* formations where designated leader robots control the trajectory of the entire formation; the remaining robots

A. Burns · A. St. John (✉)
Computer Science Department, Mount Holyoke College,
South Hadley, MA, USA
e-mail: astjohn@mtholyoke.edu

A. Burns
e-mail: burns221@mtholyoke.edu

B. Schulze
Department of Mathematics and Statistics, Fylde College,
Lancaster University, Lancaster, UK
e-mail: b.schulze@lancaster.ac.uk

R. Groß et al. (eds.), *Distributed Autonomous Robotic Systems*,
Springer Proceedings in Advanced Robotics 6,
https://doi.org/10.1007/978-3-319-73008-0_10

autonomously sense and adjust their positions locally to follow specified robots in a way that maintains the global structure.

In our model, each autonomous robot is represented as a point in the plane, and we work with range-only measurements, represented as distance constraints between pairs of points. This model is known in an area called "rigidity theory" as the 2D *bar-and-joint framework* (see, e.g., [19]) and is well-understood, with a quadratic algorithm for determining the bar-and-joint rigidity properties [8]. While rigidity theory has been applied to the construction and analysis of formations of autonomous agents [3, 11], the approach assumes undirected constraints, leading to a model where both agents would be responsible for the constraint. Since this may increase sensing and communication costs, a "persistence theory" for directed distance constraints between points was proposed by Hendrickx et al. [5] (see also [4]), effectively cutting costs in half by assigning one of the two agents to be responsible for sensing and maintaining a distance. Unlike decentralized approaches for collective transport where robots maintain constraints to the transported object (e.g., [7, 16]), a persistent formation could be used to carry delicate items, such as a partially constructed vehicle. In particular, we use a leader-follower architecture [2]; as described in [15], local sensing and communication can achieve specific geometric formations, allowing dynamic adaptation based on the surrounding environment.

Contributions. In this paper, we focus on accommodating sensing and communication failures by incorporating redundancy into our model. Redundancy is well-understood in (undirected) rigidity theory, and the associated objects form the foundation for the main contribution of this paper: an approach for constructing (directed) persistent leader-follower formations with redundancy.

We work within the basic model of persistence, following the definitions from [5], and present a class of directed graphs where any edge from a vertex with out-degree 3 is redundant; after removal of such an edge, the resulting formation remains persistent. Algorithms for constructing these graphs, as well as simulation results confirming the expected behavior of acyclic formations, are presented. We also include a discussion of the impact of special geometric conditions that can affect the "generic" behavior of the combinatorial model. To the best of our knowledge, these graphs are the first to incorporate redundancy into persistent formations. While the redundancy is restricted to specified sets of edges, it is a first step towards the stronger notion of *redundantly persistent* formations, defined in Sect. 3, where any edge in the formation could be removed.

Structure. In Sect. 2, we provide an overview of the relevant definitions and results from rigidity and persistence theory. We present an approach for constructing persistent leader-follower formations with redundancy in Sect. 3 before considering special geometric conditions that may affect persistence in Sect. 4. However, restricting to acyclic formations implies that such special conditions do not impact our construction, and we present simulation results (Sect. 5) verifying our approach. We conclude with future directions in Sect. 6.

(a) A flexible framework with another (gray) incongruent embedding satisfying the distance function.

(b) Adding the edge 14 results in a minimally rigid framework.

Fig. 1 Flexible and rigid frameworks in the plane

2 Preliminaries

Let $G = (V, E)$ be an undirected graph with vertex set $V = [1 \ldots n]$ and edge set E of unordered pairs of vertices. An *embedding* of G in the Euclidean plane is an assignment $\mathbf{p} \in (\mathbb{R}^2)^n$ of the vertices to points in the plane; the pair $(G, \mathbf{p})$ is called a *framework*. Another embedding $\mathbf{q}$ is *congruent* to $\mathbf{p}$ if $||\mathbf{p}_i - \mathbf{p}_j|| = ||\mathbf{q}_i - \mathbf{q}_j||$ for every pair of vertices i and j. Given a framework, we can extract a distance function $d : E \to \mathbb{R}$, where $d(ij) = ||\mathbf{p}_i - \mathbf{p}_j||$. If all $\mathbf{q}$ in the neighborhood of $\mathbf{p}$ satisfying the distance function d are congruent to $\mathbf{p}$, the framework is (locally) *rigid* and *flexible* otherwise; refer to Fig. 1. The rigid framework of Fig. 1b is *minimally rigid* as the removal of any edge results in a flexible framework.

For a given graph, almost all associated embeddings, called *generic* embeddings, share the same rigidity properties (see, e.g., [19]). Therefore, we may call a graph *generically* rigid or flexible, referring to the behavior of generic embeddings. The formal definition of genericity is captured by a polynomial whose vanishing indicates a non-generic embedding, or *special position*, and is outside the scope of this paper. The impact of special positions is discussed in Sect. 4.

For multi-robot formations, where minimizing the cost of communication and sensing is desirable, we work with a notion closely related to rigidity called *persistence*. We build upon the foundations of [5] and include here only the relevant definitions and results. Persistence is framed in terms of a directed graph and intuitively defines the directed analog of rigidity. One can interpret each of the vertices as representing an autonomous agent with out-going edges specifying distance constraints to neighbors that it is responsible for satisfying. This eliminates the cost for sensing and communication costs on one endpoint of a constraint edge, which would be required if working with undirected graphs and rigidity. If (1) every agent can find a position to satisfy its distance constraints, and (2) the corresponding framework is rigid, then the formation is considered persistent.

Let $H = (V, E)$ be a directed graph with vertex set $V = [1 \ldots n]$ and edge set E of ordered pairs of vertices. For clarity, we denote a directed edge from the source i to target j with $\overrightarrow{ij}$ to contrast with an undirected edge ij. For a graph H and an embedding $\mathbf{p}$, the pair $(H, \mathbf{p})$ is called a *formation*. Given a formation, we can extract the distance function d as before.

We can now state the technical definitions from [5] for persistence. Given d, let $\mathbf{q} \in (\mathbb{R}^2)^n$ be an embedding of the vertices of H. If $\overrightarrow{ij} \in E$ and $||\mathbf{q}_i - \mathbf{q}_j|| = d(\overrightarrow{ij})$, then the edge $\overrightarrow{ij}$ is *active in* $\mathbf{q}$. The set $A_{\mathbf{q}}(i)$ denotes the set of active edges in $\mathbf{q}$ whose source is i. The position $\mathbf{q}_i$ is *fitting* for vertex i if there does not exist another embedding $\mathbf{q}^*$ such that (1) $\mathbf{q}_i^* \neq \mathbf{q}_i$, (2) $\mathbf{q}_j^* = \mathbf{q}_j$ for all $j \neq i$, and (3) $A_{\mathbf{q}}(i) \subsetneq A_{\mathbf{q}^*}(i)$. Intuitively, a position is fitting for a vertex if it cannot be moved to satisfy additional constraints. If the positions for all vertices are fitting, then $\mathbf{q}$ is a *fitting embedding*. With this notion of fitting embeddings formalized, we can define persistence.

Definition 1 Let $(H, \mathbf{p})$ be a formation. If all fitting embeddings $\mathbf{q}$ in the neighborhood of $\mathbf{p}$ are congruent to $\mathbf{p}$, then the formation is *persistent*.

See Fig. 2 for examples of persistent and non-persistent formations. While the formations in Fig. 2a, b have the same underlying (rigid) undirected graph, only one is persistent. The embedding shown in Fig. 2c is fitting for the formation in Fig. 2d as each vertex is in a position that maximizes the number of its out-going constraints; vertex 4 cannot satisfy the dashed constraint to 3 without violating at least one of its other constraints to 1 or 2. Since it is not congruent, this certifies that the formation in Fig. 2b is not persistent.

As with rigidity, almost all embeddings of a directed graph exhibit the same persistence properties, so we may refer to a directed graph as *generically* persistent. We rely on the following result of [5]:

Theorem 1 (Theorem 3 of [5]) *A graph is generically persistent if and only if the underlying undirected graph of every subgraph obtained by removing out-edges from vertices with out-degree >2 until all vertices have out-degree ≤2 is generically rigid.*

Leader-follower formations. In this paper, we consider a specific type of persistent formations called *leader-follower* formations. In the persistent formations of Fig. 2a, d: only vertex 1 has out-degree 0 and is called the *leader*, only vertex 2 incident to it has out-degree 1 and is called the *co-leader*, and all other vertices (3 and 4) have out-degree ≥2 and are called *followers*. Since a point in 2D has two degrees of freedom (translation along the x- and y-axes), but an entire formation has three (translation

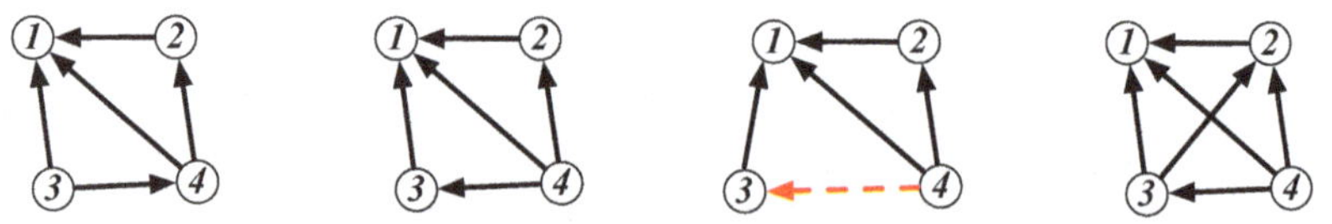

(a) A persistent formation with leader 1, co-leader 2 and followers 3 and 4.

(b) Reversing the edge from 3 to 4 results in a non-persistent formation.

(c) A fitting embedding for the formation in (b) that is not congruent.

(d) Adding an edge from 3 to 2 results in a persistent formation.

Fig. 2 Persistent and non-persistent frameworks in the plane along the x- and y-axes along with rotation about the origin), the simplest leader-follower formation must have both a leader and a co-leader; the entire formation cannot be controlled by a single lead point agent.

3 Redundancy for Persistence Theory

Communication links and sensors can fail, motivating the need for redundancy in a multi-robot system. In this section, we present an approach for constructing persistent leader-follower formations with redundancy.

We begin by reviewing redundancy in rigidity; a graph is generically *redundantly rigid* if removing any edge results in a rigid graph. Minimality is defined as follows: an undirected graph is a generic *rigidity circuit* if removing any edge results in a generically minimally rigid graph. Circuits are standard in matroid theory (see, e.g., [12]) and the *2D bar-and-joint rigidity matroid* captures the behavior of the distance constraints described in Sect. 2 [18]. In this section, we only consider the generic behavior of graphs; for brevity, we omit the word "generically" for the remainder. The smallest example of a rigidity circuit can be seen in Fig. 2d; removing any edge from the (undirected) K_4 graph gives a minimally rigid graph.

We analogously define a directed graph to be *redundantly persistent* if the removal of any edge results in a persistent graph. However, the behavior of redundant rigidity does not easily extend to the persistence model. Refer back to the formation in Fig. 2d. While its underlying undirected graph is redundantly rigid, the formation is not redundantly persistent; without the edge $\overrightarrow{32}$, the resulting formation (of Fig. 2b) is no longer persistent. We leave the question of redundantly persistent graphs open; it is challenging to even come up with a simple formation that satisfies the definition.

In the remainder of this section, we present a class of graphs with a more restricted notion of redundancy. These arise from considering leader-follower formations whose underlying undirected graphs are rigidity circuits, beginning with the following result.

Proposition 1 *Let G be a rigidity circuit. Then there exists a leader-follower orientation of G that is persistent.*

Proof Let $G = (V, E)$ be a rigidity circuit and let $e = ij \in E$ be any edge. Then $G' = (V, E \setminus \{e\})$ is a minimally rigid graph. By using an algorithm called the (2, 3)-pebble game [8, 10], there exists an orientation of G' where every vertex has out-degree at most 2. Furthermore, we can use "pebble collection" moves in the pebble game to find an orientation H where exactly one vertex v_L has out-degree 0, another vertex v_C incident to v_L, has out-degree 1 and all other vertices have out-degree 2. By Theorem 1, this formation is persistent; there are no vertices with out-degree > 2, so we only need to consider the underlying undirected graph G' of H, which is (minimally) rigid.

We now add the edge ij back to the formation, orienting it as $\overrightarrow{ij}$ if $i \neq v_C, v_L$ and $\overrightarrow{ji}$ otherwise. By this construction, the source of the edge e now has out-degree 3. Consider the undirected graphs underlying the three subgraphs obtained by dropping each out-edge from the source of e; since G is a rigidity circuit, each is (minimally) rigid. Therefore, by Theorem 1, this formation is persistent. □

Note that the proof is constructive, as captured by Algorithm 1.

Algorithm 1 Constructing a persistent leader-follower formation from a rigidity circuit.

Input: a rigidity circuit $G = (V, E)$, a desired leader vertex v_L and a desired co-leader vertex v_C incident to v_L.
Output: a persistent leader-follower formation.

1. Remove any edge $e = ij \neq v_C v_L \in E$.
2. Play the (2, 3)-pebble game on $G' = (V, E \setminus \{e\})$ to obtain a directed graph H.
3. Use pebble collection moves on H to collect 2 pebbles on v_L and one pebble on v_C.
4. Output the resulting directed graph with the additional edge $\overrightarrow{ij}$, if $i \neq v_C, v_L$, or $\overrightarrow{ji}$ otherwise.

Algorithm 1 runs in $O(n^2)$ time; Steps 1 and 4 are constant, and Steps 2 and 3 take $O(n^2)$ and $O(n)$ time, respectively [10]. The produced persistent formation has the following properties: (1) there is a single leader vertex v_L; (2) there is a single co-leader vertex v_C incident to the leader; (3) there is exactly one follower vertex (the source of e) that has out-degree 3 and removal of any of its out-edges will maintain persistency; and (4) every other follower vertex has out-degree 2.

As an example, refer again to Fig. 2d. It was constructed by first dropping the edge 24. Then the pebble game algorithm was executed, giving the orientation without $\overrightarrow{42}$, with vertex 1 as the leader and 2 the co-leader. Adding the edge back in with direction $\overrightarrow{42}$ (since 2 is the co-leader) gives a persistent leader-follower formation, where vertex 4 is the one follower vertex with out-degree 3.

The formations produced by Algorithm 1 contain redundancy via the out-going edge set of the vertex with out-degree 3. For the K_4 example, this implies that, if a sensor link were to fail, there is a 50% chance of that failure not impacting the persistence of the formation; exactly 3 of the 6 edges are in the redundant set of edges leaving vertex 4. While this does not give a formation where the redundancy is uniformly distributed through the formation, it is a step towards a more robust theoretical model.

Recursively constructing persistent formations. By using the formations from Algorithm 1 as *seed* formations, we can recursively construct larger formations with more vertices permitting sensor failures. We rely on the following result from [6], specialized to our setting. See Fig. 3a for a visual depiction of the construction.

Proposition 2 (Proposition 3 of [6]) *Let $H = (U, E)$ and $I = (V, F)$ be persistent leader-follower graphs with leaders $u_L \in U$, $v_L \in V$ and co-leaders $u_C \in U$ and*

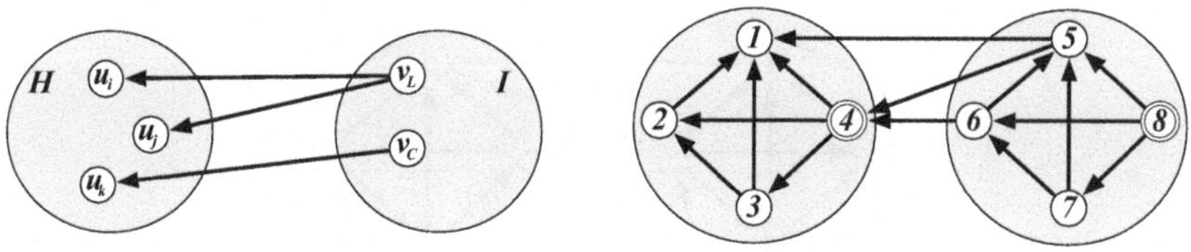

(a) The construction adds 3 out-edges from the leader and co-leader of one formation to the other.

(b) Applying the construction to two persistent leader-follower "seeds" formations.

Fig. 3 Proposition 2's recursive construction produces additional (double-outlined) vertices with out-edge sets containing redundancy

$v_C \in V$. Then the graph $(U \cup V, E \cup F \cup \{e = \overrightarrow{v_L u_i},\ f = \overrightarrow{v_L u_j},\ g = \overrightarrow{v_C u_k}\}$ is a persistent leader-follower graph, where $u_i, u_j, u_k \in U$ and $|\{u_i, u_j, u_k\}| > 1$.

Proposition 2 gives a recursive approach for constructing persistent leader-follower formations with any desired number of vertices whose out-edge sets each contain redundancy. We refer to the leaves of the recursion as *seeds* in the construction.

Lemma 1 *Let $H = (V, E)$ be a graph resulting from any number of applications of the construction step described in Proposition 2 using formations produced by Algorithm 1 as seeds. Let $R_1, \ldots, R_k$ denote the out-edge sets of the vertices $v_1, \ldots, v_k$ with out-degree 3 and let $r_1 \in R_1, \ldots, r_k \in R_k$. Then $H' = (V, E' = E - \{r_1, \ldots, r_k\})$ is a persistent leader-follower formation.*

Proof By (strong) induction on k. For the base case of $k = 1$, no applications of the construction step have occurred and H' is persistent from Algorithm 1. For the inductive step, assume the statement holds for graphs with less than K vertices of out-degree 3 and suppose H has $k = K > 1$. Then at least one construction step has occurred. Let $I = (V_I, E_I)$ and $J = (V_J, E_J)$ be the two input graphs with $V = V_I \cup V_J$ and $E \setminus (E_I \cup E_J) = \{e_1, e_2, e_3\}$, so that e_1, e_2, e_3 are the edges added by the construction. Since the number of vertices of out-degree 3 in I and J must both be at least 1 and thus less than K, $I' = (V_I, E'_I = E_I \cap E')$ and $J' = (V_J, E'_J = E_J \cap E')$ are persistent leader-follower graphs by induction. Since e_1, e_2, e_3 have sources in H with out-degree exactly 2, H' is precisely $(V_I \cup V_J, E'_I \cup E'_J \cup \{e_1, e_2, e_3\})$, a persistent leader-follower formation constructed using Proposition 2. □

Figures 3b and 4 depict examples of this recursive construction, using the persistent K_4 formation of Fig. 2d as the seeds. An out-edge from each of the double-outlined vertices with out-degree 3 may be dropped without losing persistence.

Acyclic persistent formations for simulation. Persistent formations without cycles are of particular interest when considering autonomous agents. We consider the situation where the leader and co-leader may move first (e.g., via tele-operation); the followers will then move to satisfy their constraints. If a directed cycle is present in

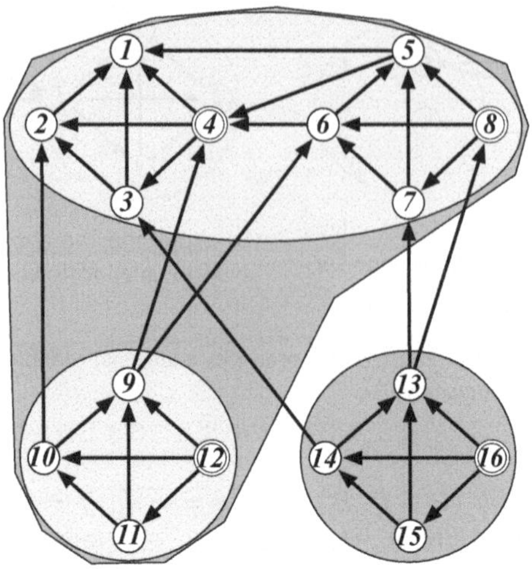

Fig. 4 The recursive construction can be applied several times to obtain additional vertices whose out-edge sets have redundancy. In this case, an out-edge from each of the 4 double-outlined vertices (4, 8, 12, 16) can be dropped without losing persistence

the graph, the formation may not be able to converge to an embedding satisfying the constraints, even if it is persistent.

However, if a persistent graph is acyclic, then there exists an ordering of the vertices such that (1) the first vertex has out-degree 0, (2) the second out-degree 1, and (3) every other vertex has ≥ 2 out-edges to vertices earlier in the ordering [5]. Graphs of this type are a generalization of *Henneberg I graphs*, given the use of the "vertex addition" step for the followers that was first described by Henneberg for minimally rigid graphs [18, 19]. The formations in Figs. 2d, 3b and 4 are acyclic, certified by the orderings indicated via the vertex labels. Such an ordering permits a simple algorithm for the formation to satisfy all constraints within $O(n)$ movements, one per robot, as described in Sect. 5.

We conclude this section by observing that, if the seeds of the recursive construction described in Proposition 2 are acyclic, the resulting formation is also acyclic. This allows us to construct persistent acyclic formations with vertices whose out-edge sets contain redundancy. Section 5 provides simulation results that verify the ability of such formations to remain persistent even when sensing links fail.

4 Special Geometric Conditions

In Sect. 3, we presented approaches for constructing generically persistent graphs, applying to situations where agents are positioned with a generic embedding in the plane. In practice, however, agents are often placed in a systematic way so that the formation takes on a specific shape or pattern. Using a symmetric or repetitive configuration to execute certain tasks, or simply for aesthetic reasons, may lead to a non-generic embedding. In this section we discuss how such special geometry in the formation of the agents can impact the formation's redundancy and persistence.

As noted in Sect. 2, for particular geometric configurations, a generically rigid graph may become flexible. Such configurations, however, are in general difficult to detect; the problem of determining whether a framework is rigid is coNP-hard [1]. A highly active research area in geometric rigidity theory is to study under what conditions non-trivial symmetries in a framework lead to flexibility in a generically rigid graph. We refer the reader to [9, 13, 14] for some key results in this area.

Given a formation of autonomous agents, it follows that special geometric conditions (as induced by symmetry in the formation, for example) can affect redundancy. For the situation we study, where the underlying undirected graph is a generic rigidity circuit, a special geometric embedding may: (1) cease to be redundant, with the removal of some edge yielding a flexible graph; or (2) cease to be rigid. We illustrate these two types of special positions with some simple examples.

Figures 5 and 6 depict two examples for type 1. Each of the graphs is a generic rigidity circuit, but are in special positions that cause the dashed edges to cease to be redundant; their removal yields flexible frameworks. For the triangular prism "Desargues" graph obtained by removing the edge 24 from the graph in Fig. 5a, it is well-known (and easy to verify) that if 1346 and 2345 are parallelograms then

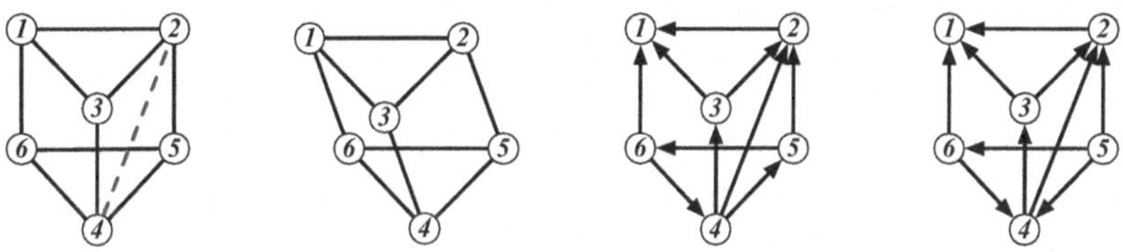

(a) The edge 24 is no longer redundant. (b) Without the edge 24, the framework is flexible. (c) A non-persistent formation contains the cycle 456. (d) Reversing edge 45 gives a persistent, acyclic formation.

Fig. 5 A generic rigidity circuit in a special geometric position

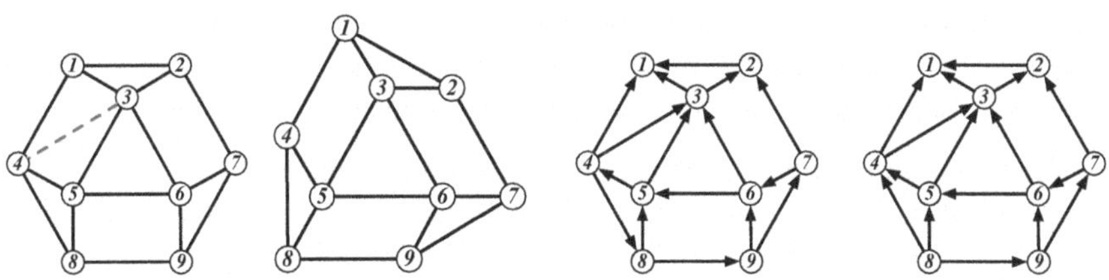

(a) The edge 34 is no longer redundant. (b) Without the edge 34, the framework is flexible. (c) A non-persistent formation contains the cycle 485. (d) Reversing edge 48 gives a persistent, acyclic formation.

Fig. 6 A generic rigidity circuit in a special geometric position

(a) A special geometric embedding with points on the x-axis and y-axis.
(b) In this position, the framework is flexible.

Fig. 7 The bipartite graph $K_{3,4}$, a generic rigidity circuit, is flexible in this special position

the framework becomes flexible, as in Fig. 5b. A geometric analysis verifies that an embedding of the graph obtained by removing the edge 34 from the graph in Fig. 6a becomes flexible if the three congruent faces 1354, 2367 and 5698 are parallelograms, as shown in Fig. 6b.

As observed in [5], the formation shown in Fig. 5c is not persistent. This follows from the result:

Theorem 2 (Remark 2(b) of [5]) *A formation with no vertex having a position collinear with two or more of its neighbours is persistent if and only if the framework of every subgraph obtained by removing out-edges from vertices with out-degree* >2 *until all vertices have out-degree* ≤ 2 *is rigid.*

In Fig. 5c, only vertex 4 has out-degree 3, but removing the edge $\overrightarrow{42}$ leads to a flexible framework. However, the formation in Fig. 5d is persistent; only vertex 5 has out-degree 3 and removal of any of its edges maintains rigidity. Similarly, Fig. 6c depicts a formation that is not persistent (the only vertex of out-degree 3 is 4, and removal of the edge $\overrightarrow{43}$ gives a flexible framework), while Fig. 6d depicts a formation that is persistent (the only out-degree 3 vertex is 8).

Figure 7 depicts an example for type 2: the complete bipartite graph $K_{3,4}$, which is again a generic rigidity circuit. It can be shown that if the vertices of each partite set are collinear and the two lines are perpendicular, then the framework becomes flexible. We note that there exists a range of further examples of generically rigid (and redundant) graphs which become flexible due to special geometric configurations. See [9, 14], for example, for situations where continuous flexibility is induced by symmetry in the framework. It follows from Theorem 2 that any formation with an underlying framework of this type will not be persistent.

Acyclic formations with underlying special positions. Observe that the persistent formations of Figs. 5 and 6 are acyclic, while the non-persistent formations are not. This is not a coincidence; removing edge 24, respectively 34, is the only way to obtain a flexible framework. Therefore, a generically persistent graph would become a non-persistent formation with these embeddings exactly when the vertex of out-degree 3 is incident to the "essential" edge. Without the essential edge, though, one can check that any generically persistent orientation contains a cycle. The same is true if we drop *any* edge in the graph in Fig. 7. However, if we restrict ourselves to acyclic leader-follower formations on generic rigidity circuits as described in Sect. 3, the following result shows that special positions cannot destroy persistence.

Lemma 2 *If H is an acyclic generically persistent graph with: (1) exactly one vertex of out-degree 0, (2) exactly one vertex of out-degree 1, (3) exactly one vertex of out-degree 3 and (4) an underlying undirected generic rigidity circuit, then any formation of H (having no vertex collinear with two neighbors) is persistent.*

Proof Remove an edge from the vertex with out-degree 3 in H. Then, by Propositions 1 and 3 in [5], we obtain an acyclic generically minimally persistent graph G. As shown in [5], such graphs can be constructed from a single edge using only vertex additions, and hence any embedding of G is rigid. Moreover, every vertex of G has an out-degree of at most 2. The result now follows from Theorem 2. □

5 Simulation

To verify the expected behavior of the approach in Sect. 3, we simulated three formations using Webots [17]: one produced by Algorithm 1 and depicted in Fig. 2d, one using a single construction step of Proposition 2 as per Fig. 3b, and one using 3 construction steps as depicted in Fig. 4.

Setup. Each vertex in the graph is implemented by a robot that is equipped with a generic emitter, and each directed edge by equipping the source robot with a receiver that listens to the target robot's emitter. From the strength and direction of the signal, the source robot computes the relative position of the target; these measurements are without noise. Every follower computes and moves to a goal position based on its assigned distance constraints: with two constraints, the closest point of the (generally) 2 intersection points of 2 circles, and, with three constraints, the average of 3 points computed for each pair of constraints. To simplify control of the formation, the leader and co-leader vertices are attached to a single leader robot with emitters at each location.

The implementation assumes an acyclic persistent formation. While the associated ordering is not explicitly used, each robot waits for a "go" signal from the 2 or 3 robots it is following. We know that each follower robot must adjust its position only once by using this approach, but we do not know a priori the time required for actuation to its goal position. Once it has moved within a specified *constraint accuracy threshold*, it emits a "go" signal that can be received by any subsequent followers. We accommodate for the unknown actuation time by moving the leader robot in steps with a delay between steps that will allow for the formation's movements; after each such *simulation step*, we assume the formation has converged to an embedding that satisfies the constraints within the constraint accuracy threshold.

Results. As a control, one simulation was executed for each formation with all edges present. To confirm expected redundancy, an edge was randomly chosen and dropped from each robot with out-degree 3; the leader robot follows the same path as the control. Random simulations were repeated 20 times.

The expected position of the follower robots is computed using the position of the leader and co-leader positions; since the co-leader is on the same robot as the leader, its distance constraint is always satisfied. The results of the simulations are summarized in Table 1, with leader and co-leader vertices omitted from the calculations; the expected and actual paths for Formation 1 are shown in Fig. 8. For each simulation step t, the difference δ_{ti} between the expected and actual positions of follower i was computed to obtain $m_t = \min_i \delta_{ti}$, $\mu_t = \text{mean}_i \delta_{ti}$ and $M_t = \max_i \delta_{ti}$ values. Data for each simulation was then computed as the mean of the m_t, μ_t and M_t across all simulation steps. The mean distance constraint length was 1.66325 ± 0.577504 meters across all formations; simulations were performed with a 0.02 m constraint accuracy threshold. We expect the accuracy threshold to accumulate error for robots later in the sequence, as reflected in the maximum data values. The results confirm the expected behavior of the formations; the simulations testing redundancy performed comparably to the control with all edges present.

Table 1 Mean with standard deviation data for simulations (constraint accuracy threshold 0.02); for random simulations, mean values across the 20 experiments are provided

	Mean of minimums	Mean of means	Mean of maximums
Formation 1 control	0.0132058 ± 0.004	0.0173324 ± 0.006	0.021459 ± 0.010
Formation 1 random	0.0131554 ± 0.004	0.0179166 ± 0.006	0.0235468 ± 0.010
Formation 2 control	0.00826846 ± 0.004	0.0810221 ± 0.011	0.191657 ± 0.034
Formation 2 random	0.00769588 ± 0.004	0.0872681 ± 0.012	0.261483 ± 0.104
Formation 3 control	0.0087836 ± 0.004	0.0766168 ± 0.051	0.362899 ± 0.361
Formation 3 random	0.00835823 ± 0.004	0.0608315 ± 0.045	0.197745 ± 0.162

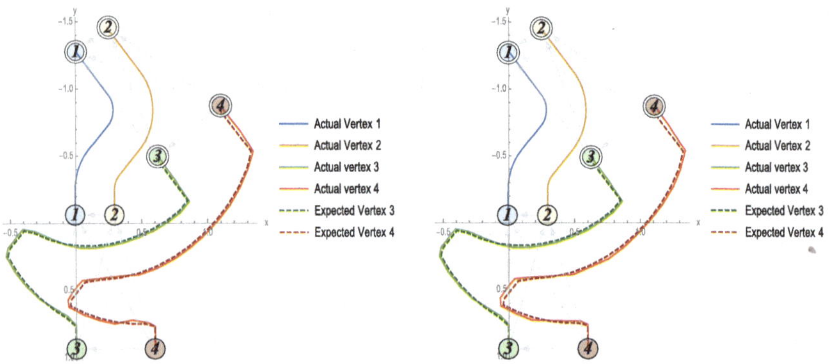

(a) Control simulation, with all edges included. (b) Random simulation, with 1 redundant edge randomly dropped (from one of 20 simulations).

Fig. 8 Simulation paths for Formation 1, with starting (single-outlined) and ending (double-outlined) positions highlighted. Coordinates are in meters

6 Conclusions and Future Work

This paper introduces a new class of generically persistent leader-follower graphs with redundancy for formations in the plane along with a constructive approach for generating them. The approach can be restricted to generate acyclic formations, for which a simple approach can be used to satisfy all constraints within $O(n)$ movements. Simulations verified the expected behavior of the formations when redundant constraints were dropped.

Under special geometric conditions, certain non-generic embeddings may no longer maintain the persistence and redundancy properties; while acyclic formations are not susceptible, it remains open to characterize the behavior of persistent formations with cycles. We recognize that the redundancy property exhibited by the class of graphs presented here is quite restrictive and wish to address the strong notion of *redundantly persistent* formations defined in Sect. 2. Finally, our model does not incorporate noise, as it is based on equality constraints stemming from classical rigidity theory. We hope to further develop and evaluate a theoretical framework that would be robust to noise and disturbances.

Acknowledgements We are grateful for discussions at the 2014 AIM Workshop on Configuration spaces of linkages, 2015 BIRS Workshop on Advances in Combinatorial and Geometric Rigidity and the 2016 ICMS Workshop on Geometric Rigidity Theory and Applications. We would also like to thank Joydeep Biswas for insightful conversations that stimulated the results in this paper and reviewers for their helpful feedback. A. Burns and A. St. John were partially supported by NSF IIS-1253146 and the Clare Boothe Luce Foundation. B. Schulze was supported by EPSRC grant EP/M013642/1.

References

1. Abbott, T.G.: Generalizations of Kempe's universality theorem. Master's thesis, Massachusetts Institute of Technology (2008). http://web.mit.edu/tabbott/www/papers/mthesis.pdf
2. Das, A.K., Fierro, R., Kumar, V., Ostrowski, J.P., Spletzer, J., Taylor, C.J.: A vision-based formation control framework. IEEE Trans. Robot. Autom. **18**(5), 813–825 (2002)
3. Eren, T., Anderson, B.D.O., Morse, A.S., Whiteley, W., Belhumeur, P.N.: Operations on rigid formations of autonomous agents. Commun. Inf. Syst. **3**(4), 223–258 (2004)
4. Eren, T., Whiteley, W., Anderson, B.D.O., Morse, A.S., Belhumeur, P.N.: Information structures to secure control of rigid formations with leader-follower structure. In: Proceedings of the American Control Conference, pp. 2966–2971 (2005)
5. Hendrickx, J.M., Anderson, B.D.O., Delvenne, J.C., Blondel, V.D.: Directed graphs for the analysis of rigidity and persistence in autonomous agent systems. Int. J. Robust Nonlinear Control **17**(10–11), 960–981 (2007)
6. Hendrickx, J.M., Yu, C., Fidan, B., Anderson, B.D.O.: Rigidity and persistence for ensuring shape maintenance in multiagent meta-formations. Asian J. Control (special issue on Collective Behavior and Control of Multi-Agent Systems) **10**(2), 131–143 (2008). arXiv:0710.2659
7. Hichri, B., Adouane, L., Fauroux, J.C., Mezouar, Y., Doroftei, I.: Cooperative mobile robot control architecture for lifting and transportation of any shape payload. In: The 12th International Symposium on Distributed Autonomous Robotic Systems, pp. 177–191 (2016)
8. Jacobs, D., Hendrickson, B.: An algorithm for two-dimensional rigidity percolation: the pebble game. J. Comput. Phys. **137**(CP975809), 346–365 (1997)

9. Jordán, T., Kaszanitzky, V., Tanigawa, S.: Gain-sparsity and symmetry-forced rigidity in the plane. Discret. Comput. Geom. **55**(3), 314–372 (2016)
10. Lee, A., Streinu, I.: Pebble game algorithms and sparse graphs. Discret. Math. **308**(8), 1425–1437 (2008)
11. Olfati-Saber, R., Murray, R.M.: Graph rigidity and distributed formation stabilization of multi-vehicle systems. In: Proceedings of the IEEE International Conference on Decision and Control, pp. 2965–2971 (2002)
12. Oxley, J.G.: Matroid Theory. Oxford University Press, Oxford (1992)
13. Schulze, B.: Symmetry as a sufficient condition for a finite flex. SIAM J. Discret. Math. **24**(4), 1291–1312 (2010)
14. Schulze, B., Whiteley, W.: The orbit rigidity matrix of a symmetric framework. Discret. Comput. Geom. **46**(3), 561–598 (2011)
15. Vilca, J., Adouane, L., Mezouar, Y.: Adaptive leader-follower formation in cluttered environment using dynamic target reconfiguration. In: The 12th International Symposium on Distributed Autonomous Robotic Systems, pp. 237–254 (2016)
16. Wang, Z., Schwager, M.: Multi-robot manipulation without communication. In: The 12th International Symposium on Distributed Autonomous Robotic Systems, pp. 135–149 (2016)
17. Webots (2016). http://www.cyberbotics.com. Commercial Mobile Robot Simulation Software
18. Whiteley, W.: Some matroids from discrete applied geometry. In: Bonin, J., Oxley, J.G., Servatius, B. (eds.) Matroid Theory. Contemporary Mathematics, vol. 197, pp. 171–311. American Mathematical Society, Providence (1996)
19. Whiteley, W.: Rigidity and scene analysis. In: Goodman, J., O'Rourke, J. (eds.) Handbook of Discrete and Computational Geometry, 2nd edn, pp. 1254–1315. Chapman & Hall/CRC, Boca Raton (2004)

Triangular Networks for Resilient Formations

David Saldaña, Amanda Prorok, Mario F. M. Campos and Vijay Kumar

Abstract Consensus algorithms allow multiple robots to achieve agreement on estimates of variables in a distributed manner, hereby coordinating the robots as a team, and enabling applications such as formation control and cooperative area coverage. These algorithms achieve agreement by relying only on local, nearest-neighbor communication. The problem with distributed consensus, however, is that a single malicious or faulty robot can control and manipulate the whole network. The objective of this paper is to propose a formation topology that is resilient to one malicious node, and that satisfies two important properties for distributed systems: *(i)* it can be constructed incrementally by adding one node at a time in such a way that the conditions for attachment can be computed locally, and *(ii)* its robustness can be verified through a distributed method by using only neighborhood-based information. Our topology is characterized by triangular robust graphs, consists of a modular structure, is fully scalable, and is well suited for applications of large-scale networks. We describe how our proposed topology can be used to deploy networks of robots. Results show how triangular robust networks guarantee asymptotic consensus in the face of a malicious agent.

D. Saldaña (✉) · A. Prorok · V. Kumar
University of Pennsylvania, Philadelphia, PA, USA
e-mail: dsaldana@seas.upenn.edu

A. Prorok
e-mail: prorok@seas.upenn.edu

V. Kumar
e-mail: kumar@seas.upenn.edu

M. F. M. Campos
Universidade Federal de Minas Gerais, Belo Horizonte, Brazil
e-mail: mario@dcc.ufmg.br

R. Groß et al. (eds.), *Distributed Autonomous Robotic Systems*,
Springer Proceedings in Advanced Robotics 6,
https://doi.org/10.1007/978-3-319-73008-0_11

1 Introduction

Coordination of distributed autonomous systems with nearest neighbor communication has been widely used in swarms and multi-robot systems [5, 9, 15]. A majority of these applications build on information diffusion (or consensus) algorithms to synchronize the network with respect to a specific variable. However, such systems rely on the fact that all robots in the network are reliable, and contribute only legitimate information. As a consequence, the systems are susceptible to failure when one or several robots are non-cooperative and share wrong information. This situation can be due to malicious attacks (e.g., a malicious outsider trying to manipulate the whole network) or due to platform-level faults (e.g., a robot sharing an incorrect location due to a defective GPS sensor). Hence, the question of network resilience is of utmost importance. In this work, we focus on *topological properties* that guarantee resilience to faults and attacks on individual nodes. We build on a distributed consensus algorithm, termed Weighted-Mean Subsequence-Reduced (W-MSR) algorithm [20], which guarantees resilience if certain topological requirements are met. Throughout this paper, we use the terms node and agent to refer to a robot in a network.

1.1 Related work

The topic of robustness has received considerable attention, in particular in the domain of complex networks [4, 8]. A main result of this body of work states that robustness can be achieved through sufficiently high connectivity: if the connectivity of the network is $2F$ or less, then F malicious nodes can prevent some of the nodes from receiving legitimate information from other nodes in the network. Conversely, when the network connectivity is $2F + 1$ or higher, there are various algorithms that enable a reliable diffusion of information [13, 16]. However, these algorithms not only depend on high connectivity, but also require non-local information in order to compute updates. As a consequence, Zhang et al., and later LeBlanc et al. [7, 20]introduced an alternative definition of network robustness that can deal with purely local update rules, and that provides resilient asymptotic consensus. The strength of the proposed approach is that it can deal with an arbitrary number F of malicious agents, the identities of which are unknown to the normal nodes in the network. However, in order to build the required topologies, the method assumes that any node can be connected to any other node in the network (i.e., in absence of sophisticated node placement algorithms, the networks must provide full connectivity). Hence, it remains unclear how the approach is practically implemented in networks where nodes have constrained communication radii. Also, it is noteworthy that the prior approach mainly deals with the problem of distributed estimation, and it remains to be explored how it applies to problems that require robust control of shapes and distributions, such as for cooperative exploration and coordination tasks [18, 19]. Finally, the problem of robustness has also been considered in mobile robot systems [2, 11, 12], with a particular focus on rendez-vous (i.e., the application of consensus algorithms to induce a gathering of robots in n-dimensional space). Most similarly to our work, the work in [12] considers the presence of non-compliant robots, and develops a solution that controls a team of robots so that rendez-vous can be achieved robustly. However, since their systems are mobile, they assume that communication radii can be adjusted over time (to ensure connectivity). As a consequence, their method requires a relatively large local neighborhood size: for a single malicious robot, the neighborhood size must be at least 5 (cf. assumption 5.1 in their paper).

1.2 Contribution

The main contribution of our work is the definition of a network topology that can be constructed in a distributed and fully scalable manner, and that guarantees resilient asymptotic consensus in the face of malicious agents. In addition, our method is also able to deal with networked robots that have fixed communication radii. We consider the special case when a network is susceptible to a single fault or malicious node, and we identify a particular class of networks (which we term *triangular robust graphs*) for this case. We derive proofs of all our claims.

2 Preliminaries

Consider a network composed of a set of nodes $\mathcal{V} = \{1, 2, ..., n\}$ that are represented by points in a planar space $v_i \in \mathbb{R}^2$, for all $i \in \mathcal{V}$. Every node is equipped with a communication module that allows it to communicate with other nodes. The ability to communicate with adjacent nodes defines the set of connections $\mathcal{E} \subseteq \mathcal{V} \times \mathcal{V}$. Therefore, we model the network as an undirected graph $\mathcal{G} = (\mathcal{V}, \mathcal{E})$. The neighbors of node i are $\mathcal{N}_i = \{j|(i, j) \in \mathcal{E}\}$. For a node subset $S \subset \mathcal{V}$, we denote its complement by $\bar{S} = \{i|\, i \notin S, i \in \mathcal{V}\}$.

2.1 Consensus

In distributed networked systems, each node is an autonomous entity that can adapt to changing conditions based on incoming data streams originating from neighboring nodes. As there is no central master, the nodes need to find an agreement with respect to the shared information in order to make unified decisions. The question of how to do this is solved by *consensus algorithms* [5, 9, 10, 14]. When performing a consensus algorithm, each node i has a variable of interest x_i, e.g., that describes the

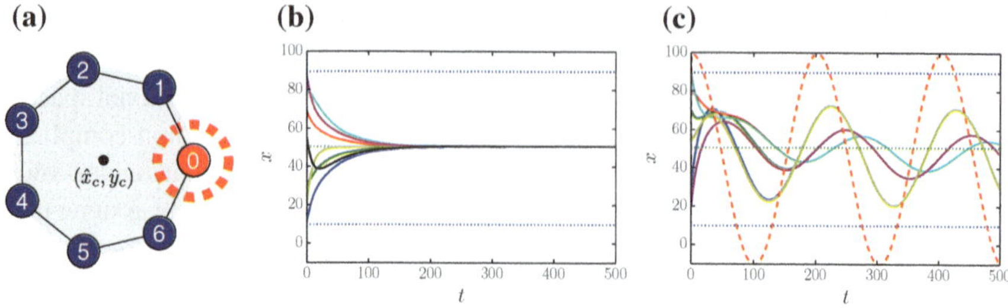

Fig. 1 Consensus in a circular formation with seven robots. Panel (**a**) shows the network topology. Panel (**b**) shows a successful consensus for the variable x_c, where all nodes perform the same averaging update rule. Panel (**c**) shows an unsuccessful consensus that is manipulated by node 0, who shares an oscillatory value to avoid convergence for the non-malicious robots

locations of the nodes, or that measures local temperatures. Subsequently, the whole network may want to estimate a global variable, such as the centroid of the network, or the average temperature of the environment, respectively, based on the distributed information available to the network as a whole. This goal can be achieved by local interaction, where each node i updates its own value at time-step t based on an update function:

$$x_i[t+1] = f(x_i[t], \{x_j[t] | j \in \mathcal{N}_i\}). \tag{1}$$

In [5], the authors show that, given a connected and balanced graph $\mathcal{G}$, every node $i \in \mathcal{V}$ reaches consensus on the average of the initial values, $x_i[t] \rightarrow \bar{x}[0] = \frac{1}{n}\sum_{i\in\mathcal{V}} x_i[0]$, when $t \rightarrow \infty$ by exchanging messages with the local neighborhood and applying an averaging function $x_i[t+1] = \frac{1}{|\mathcal{N}_i|+1}(x_i[t] + \sum_{j\in\mathcal{N}_i} x_j[t])$. We illustrate an example of consensus in Fig. 1 for a circular formation in a *biconnected* ring topology (Fig. 1a). In this context, each robot i moves to the location (x_i, y_i), given by $x_i = \hat{x}_c + \rho\cos(i\ 2\pi/n)$ and $y_i = \hat{y}_c + \rho\sin(i\ 2\pi/n)$, where $(\hat{x}_c, \hat{y}_c)$ is the center of the circle, estimated by consensus, and $\rho > 0$ is the known radius of the circle. Figure 1b shows how the robots achieve consensus on the average of the variable $\hat{x}_c$ (consensus for the variable $\hat{y}_c$ is analogous). This kind of robotic system works scales well when every node is functional and trustworthy. However, when a malicious node[1] stops adhering to the consensus update rule due to a hardware failure or a malicious attack, that robot can affect the behavior of all other robots in the network. As we observe in Fig. 1c, when robot 0 starts to share an arbitrary oscillatory value, the values of all robots in the network are affected and consensus is not achieved. We see that a single malicious agent can hinder convergence and manipulate the whole network. For this reason, it is desirable to devise a resilient strategy. The problem of consensus in the presence of malicious nodes (such as robot 0 seen above) can be solved by deploying a *resilient consensuss* algorithm on all nodes. The following section introduces this method.

[1]We consider an attack or a failure as the same case, where a node shares a value that does not adhere to the consensus update rule. We call this kind of node a *malicious node*.

2.2 Resilient Consensus

A robust network is defined as a network that can reach consensus, even in the presence of F malicious nodes. Neither the identity nor the strategy of the malicious nodes is known. A known method that achieves consensus by converging to a weighted average is the *Weighted Mean-Subsequence-Reduced (W-MSR)* algorithm [7, 20]. Yet, for this method to work in the presence of malicious nodes, the network must satisfy certain topological conditions, which we detail below.

The W-MSR algorithm consists of three steps, executed at time t. First, node i creates a sorted list, from smallest to largest, with the received values from its neighbors $\mathcal{N}_i$. Second, the list is compared to $x_i[t]$, and if there are more than F values that are larger than $x_i[t]$, the F largest values are removed. The same removal process is applied to the smaller values. The remaining values in the list are denoted by $\mathcal{R}_i[t]$. Third, node i updates its value with the following rule:

$$x_i[t+1] = w_{ii}[t]x_i[t] + \sum_{j \in \mathcal{R}_i[t]} w_{ij}[t]x_j[t], \tag{2}$$

where $w_{ij} > 0$, and $\sum_j w_{ij}[t] = 1$. In the remainder of this paper, we consider all weights $w_{ij} = 1/(|\mathcal{R}_i[t]| + 1)$. An extended explanation of this algorithm is given in [7]. Using this algorithm, it is possible to achieve asymptotic consensus in a network with at most F malicious nodes, if the communication graph $\mathcal{G}$ is $(F+1, F+1)$-*robust.*

Definition 1 A graph $\mathcal{G} = (\mathcal{V}, \mathcal{E})$ is (r,s)-*robust*, with constants $r \in \mathbb{Z}_{\geq 0}$, and $0 \leq s \leq |\mathcal{V}|$, if for any pair of disjoint subsets $S_1, S_2 \subset \mathcal{V}$, at least one of the following conditions is satisfied:

1. $|\mathcal{X}^r_{S_1}| = |S_1|$;
2. $|\mathcal{X}^r_{S_2}| = |S_2|$;
3. $|\mathcal{X}^r_{S_1}| + |\mathcal{X}^r_{S_2}| \geq s$,

where the r-reachable set $\mathcal{X}^r_{S_k} = \{i \in S_k \mid \mathcal{N}_i \backslash S_k \geq r\}, k \in \{1, 2\}$ is composed of the nodes in S_k with at least r neighbors outside S_k.

Based on this definition of the communication topology, LeBlanc et al. [7] stated the following theorem, which specifies the conditions for asymptotic consensus in presence of F malicious agents.

Theorem 1 (Theorem 1, [7]) *Consider a network modeled by a graph* $\mathcal{G} = (\mathcal{V}, \mathcal{E})$ *where each normal node updates its value based on the W-MSR algorithm with an upper bound of F malicious agents. Then, resilient asymptotic consensus is achieved if the* $\mathcal{G}$ *is* $(F+1, F+1)$*-robust.*

Although these recent works provide a rigorous study of the topological characteristics that are necessary to provide resilience against a number of malicious agents [7, 12], they do not consider the physical constraints that real-world systems often have, such as limited or non-adjustable communication radii. Hence, it is not clear how the methods are applicable in real settings, and it is still an open question if their implementations are suitable to distributed actuator/sensor networks.

2.3 Biconnected Graphs

In the following, we describe a concept that is relevant to the derivations in the subsequent sections.

Definition 2 A graph $\mathcal{G} = (\mathcal{V}, \mathcal{E})$ is *biconnected* if it stays connected after removing any vertex $i \in \mathcal{V}$.

An example of a biconected graph in shown in Fig. 1a. Biconnected graphs have two special properties. First, they have two disjoint paths (no common vertices) between every pair of vertices (Theorem 2), and second, they can be extended/grown by adding nodes iteratively, as described by the expansion lemma below (Lemma 1).

Theorem 2 (Theorem 4.2.1, [17]) *If a graph $\mathcal{G} = (\mathcal{V}, \mathcal{E})$, with at least three nodes, is biconnected, then there exist at least two disjoint paths between any pair of nodes $i, j \in \mathcal{V}$.*

Lemma 1 (Lemma 4.2.2, [17]) *If $\mathcal{G}$ is a biconnected graph, and $\mathcal{G}'$ is obtained from $\mathcal{G}$ by adding a new vertex k adjacent to at least two vertices of $\mathcal{G}$, then $\mathcal{G}'$ is biconnected.*

3 Triangular Robust Graphs

The creation of robust and (r, s)-robust networks is challenged by three main items: *(i)* high connectivity is required; *(ii)* verifying if a graph is (r, s)-robust is an NP-hard problem [6]; *(iii)* default algorithms for creating robust networks are not designed for physically embedded systems, which potentially have hard constraints on edge lengths. In this section, we propose a particular network topology, termed *triangular robust graph*, that is resilient to one malicious node (i.e., $F = 1$) whose identity is unknown to the rest of the nodes. Our proposed network topology has the following convenient properties: it is (2, 2)-robust; it is incrementally expandable; and it can be verified in polynomial time and in a distributed manner. Furthermore, due to the inherent geometric properties of triangles, the topology is well suited to networked robotic systems with agents that have homogeneous, constrained communication radii. As a consequence, triangular robust graphs are well-suited to distributed robotic systems.

Definition 3 A graph $\mathcal{G} = (\mathcal{V}, \mathcal{E})$ is *triangular robust* if it has at least three nodes and satisfies the following conditions:

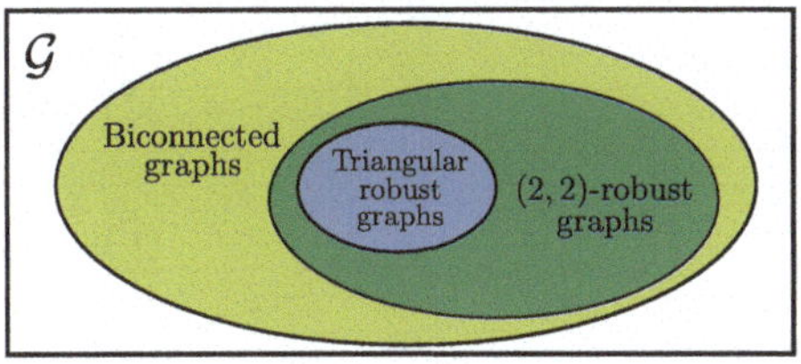

Fig. 2 In the whole set of connected graphs, this Venn diagram represents the set of the triangular robust graphs and its relationship with biconnected graphs and (2, 2)-robust graphs

1. The graph $\mathcal{G}$ is *biconnected*.
2. The neighbors of node i form a connected sub-graph, $\mathcal{G}_i = (\mathcal{N}_i, \mathcal{E})$, for all $i \in v$.

Property 1 *Based on Condition 1, the minimum degree of any node is two, $Deg(i) \geq 2$ for all $i \in \mathcal{V}$.*

In Fig. 2, we show a Venn diagram that represents in the whole set of connected graphs, our proposed topology is a subset of the Biconnected graphs and also a subset of the (2, 2)-robust graphs. We highlight that not all the (2, 2)-robust graphs have the properties that we describe along this section.

Theorem 3 *Consider a network modeled by the graph $\mathcal{G} = (\mathcal{V}, \mathcal{E})$. If $\mathcal{G}$ is triangular robust, then it is (2,2)-robust.*

Proof We show that the triangular robust graph $\mathcal{G}$ is *(2,2)-robust*, as it satisfies the conditions of Definition 1. Given any pair of disjoint subsets $S_1, S_2 \subset \mathcal{V}$, by Definition 3, and Theorem 2, the graph $\mathcal{G}$ is *biconnected* and there exist two disjoint paths between any source node $s \in S_1$ and any target node $g \in S_2$. It implies that there are also two edges, $e_1 = (i, j)$ and $e_2 = (o, p)$, for each path respectively, such that $i, o \in S_1$ and $j, p \in \bar{S}_1$. Figure 3 illustrates both paths and their intermediate edges e_1 and e_2. The conditions of robustness are satisfied by checking the neighbors of j. There are only the following two cases:

1. If the node j does not have any neighbors outside S_1, *i.e.* $|\mathcal{N}_j \cap \bar{S}_1| = 0$. It implies that the node j is the target node, $j = p = g$, and $S_2 = \{j\}$. Then, the condition of robustness $|\mathcal{X}^r_{S_2}| = |S_2| = 1$ is satisfied for any S_1 (Definition 1, Condition 2). Figure 4a illustrates this case.
2. Otherwise, the node j has at least one neighbor outside S_1, i.e., $|\mathcal{N}_j \cap \bar{S}_1| \geq 1$. By Definition 3, Condition 2, since the neighbors of j are connected, there exists an edge $(k, l) \in \mathcal{E}$ such that $k \in \mathcal{N}_j \cap S_1$, and $l \in \mathcal{N}_j \cap \bar{S}_1$. Then the node k has two neighbors (j and l) outside S_1, and the 2-reachable set of S_1 contains at least one element $|\mathcal{X}^2_{S_1}| \geq 1$. Figure 4b illustrates this case.
   Applying the same sequence of statements for S_2, there exist at least one node with two neighbors outside S_2, then $|\mathcal{X}^2_{S_2}| \geq 1$. In this way, we show that the condition of robustness $|\mathcal{X}^2_{S_1}| + |\mathcal{X}^2_{S_2}| \geq 2$ is satisfied.

With the two cases above, we show that the conditions of robustness are satisfied for any pair of sets in a triangular robust network. □

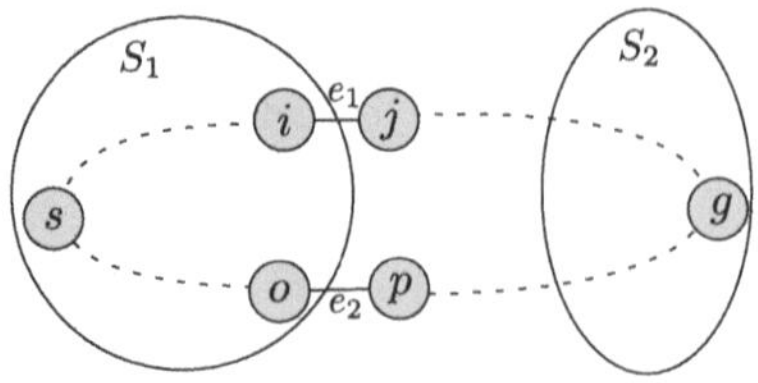

Fig. 3 A graph with two partitions S_1 and S_2 and two disjoint paths between the nodes $s \in S_1$ and $g \in S_2$

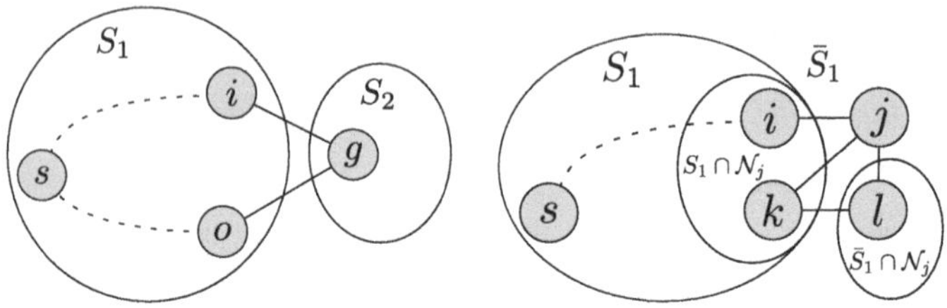

(a) $|\mathcal{N}_j \cap S_1| = 0$. Node j is g. (b) $|\mathcal{N}_j \cap \bar{S}_1| \geq 1$.

Fig. 4 Possible cases based on the neighbors of node j

Remark 1 The converse of the Theorem 3 does not hold, i.e., not all (2, 2)-robust graphs are triangular robust (see Fig. 6 in [7]).

3.1 Determining Triangular Robustness

Checking if a graph is *(r,s)-robust* is an NP-hard problem [6], but we can check if a graph is triangular robust in polynomial time, based on the two conditions of Definition 3.

Condition 1: A simple centralized algorithm to check if $\mathcal{G}$ is *biconnected* can also be done in polynomial time. For each node $i \in \mathcal{N}$, we check if the graph maintains connectivity after removing i. It is possible to check connectivity based on *breadth-first-search* (BFS) with time complexity $O(|\mathcal{V}| + |\mathcal{E}|)$. Then, the result of checking connectivity $|\mathcal{V}|$ times is $O(|\mathcal{V}|^2 + |\mathcal{V}||\mathcal{E}|)$. It is also possible to check it in distributed way by using the algorithm presented in [1].

Condition 2: Checking if the neighbors are connected can be compute in linear time and in a distributed way. First, every node $i \in \mathcal{V}$ only needs to know the neighbors of its neighbors to create the sub-graph $\mathcal{G}_i = (\mathcal{N}_i, \mathcal{E})$. The worse case is the complete graph, for which the time complexity to check connectivity for the neighbors of node i is $O(|\mathcal{V}| + |\mathcal{E}|)$. For the complete network it is $O(|\mathcal{V}|^2 + |\mathcal{V}||\mathcal{E}|)$.

Since Conditions 1 and 2 are checked independently, we conclude that it is possible to check triangular robustness in polinomial time $O(|\mathcal{V}|^2 + |\mathcal{V}||\mathcal{E}|)$.

3.2 *Inductive Construction*

In the following, we show a simple method that expands a $(2, 2)$-robust network, starting from an initial network configuration. Our starting point is a strongly connected graph with three nodes, which is the most basic form of a triangular robust graph. This configuration is then extended by adding one node at a time, while maintaining the $(2, 2)$-robust property at all times, by ensuring that each new node is connected to two or more nodes (that are also connected among themselves). The following theorem shows that a triangular robust graph is expandable.

Theorem 4 *If $\mathcal{G}$ is a triangular robust graph, and $\mathcal{G}'$ is obtained from $\mathcal{G}$ by adding a node k, which is connected to a subset of nodes $S \subset \mathcal{V}$, $|S| \geq 2$ that are connected among themselves. Then $\mathcal{G}'$ is also a triangular robust graph.*

Proof We check that the conditions of Definition 3 are satisfied by $\mathcal{G}'$.

Condition 1: By the Expansion Lemma 1, $\mathcal{G}'$ is *biconnected* as it is an expansion of $\mathcal{G}$ by adding the node k, which is adjacent to two nodes.

Condition 2: As node k is connected to a set of connected nodes, its neighbors are also connected.

It follows that $\mathcal{G}'$ satisfies the conditions of a triangular robust graph. □

Property 2 *By constructing a triangular graph iteratively, we have that the minimum number of edges of a triangular robust graph is $2n - 3$, $n \geq 3$. This is the same minimum number of edges for a $(2, 2)$-robust graph [6].*

Using our inductive construction method, we guarantee that the resulting graph has the minimum number of edges, which is in constrast to the construction method proposed in [7].[2] Also, Theorem 4 leads to the particularly practical property that triangular robust graphs can be constructed incrementally, in a fully distributed manner. In fact, triangles are geometric configurations that can be easily formed during the deployment of networked robot teams, where each agent has a communication module that is constrained by a fixed radius $R > 0$. These robotic networks determine the set of connections based on the Euclidean distance $\mathcal{E} = \{(i, j) | \, \|v_j - v_i\| \leq R, \forall i, j \in \mathcal{V}\}$. For example, a simple formation with three robots maximizes its connectivity and its covered area by positioning the robots at the extremes of a regular triangle. Figure 5 shows a basic example of how an initial formation can be incrementally extended in order to cover a certain area with a triangular robust network.

[2] In the specific case of $(2, 2)$-robustness, the algorithm of LeBlanc et al. requires three new edges for every new node (cf. Th. 5), whereas our algorithm requires only two new edges.

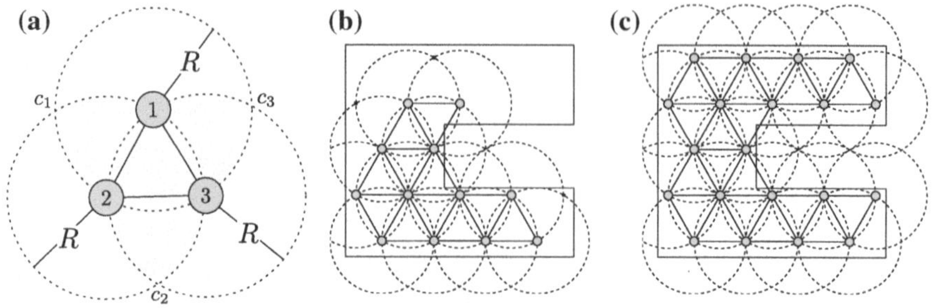

Fig. 5 Incremental expansion of a triangular robust network. Panel (**a**) shows the initial triangular robust network. By placing a subsequent node in the shaded area, the conditions for inductive construction are satisfied. Panels (**b**) and (**c**) depict how the network is grown to cover a given area iteratively, for $n = 12$ and $n = 20$, respectively

4 Consensus in Triangular Robust Networks

In this section, we compare three different formations, and show how consensus is affected by their differing topologies. The formations consist of seven nodes. The nodes' initial values are $x[0] = [60, 75, 2, 85, 66, 83, 20]$. We consider a malicious node that shares an arbitrary (oscillatory) value, attempting to hinder convergence. Figure 6 shows the three different topologies, and demonstrates how the values of the nodes behave after applying the W-MSR algorithm (see Eq. (2)). Panel 6a shows a *biconnected*, *rigid* and *pseudo-triangular* graph, which is known as Laman graph [3]. Although the Laman graph has similar properties, as well as the same number of edges $2n - 3$, it is neither $(2, 2)$-robust,[3] nor triangular robust (the neighbors of node 3 are not connected among themselves). We can see in Fig. 6d that the node values do not reach consensus, with two different values among the normal nodes that remain unchanged as time progresses. A triangular robust network is shown in Fig. 6b. It is a simple variation of the previous Laman graph, where the edge $(0, 6)$ is replaced by the edge $(2, 4)$. As we stated earlier in Section 3, a successful convergence will reach consensus to a value within the range of initial values, i.e., a weighted average. This is demonstrated in Fig. 6e. Finally, in Fig. 6c, we show a triangular robust graph that is not planar (with intersecting edges), and that has two fully connected nodes. This topology represents the worst possible case, since one of the fully connected nodes is the malicious node (node 6 in the graph), with maximal influence over the other nodes in the graph. Also in this case, we see that the nodes reach consensus to a weighted average, as shown in Fig. 6f. This particular example shows how the maximum value of the normal nodes does not change when the malicious node's value is the greatest value; instead, we see that the minimum value of the normal nodes increases. An analogous behavior is observed when themalicious

[3]Consider the sets $S_1 = \{0, 1, 2\}$ and $S_2 = \{4, 5, 6\}$ since there are not 2-reachable nodes, the conditions for $(2, 2)$-robustness are not satisfied.

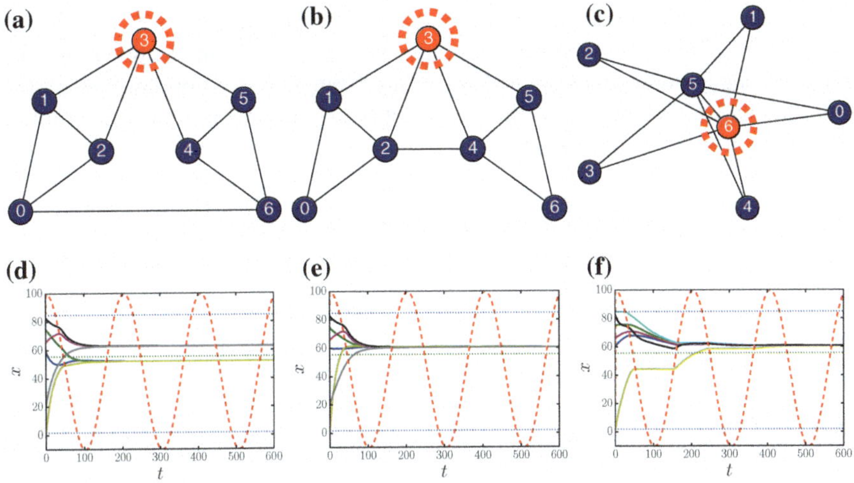

Fig. 6 Results of the resilient convergence method for different network topologies in the presence of a malicious agent. The network topologies are shown in (**a**)–(**c**), with the malicious agent shown in red, encircled by the dashed line. Panel (**a**) shows a Laman graph, panel (**b**) shows a planar triangular robust graph, and panel (**c**) shows a triangular robust graph with two fully-connected nodes. All normal nodes in the network perform the W-MSR update rule. The corresponding convergence behavior is shown in graphs (**d**)–(**f**), respectively. The average of the initial values is represented by the green dotted line, and the minimum and maximum of the initial values are depicted by the blue dotted lines. The malicious agent does not follow the update policy by sharing an arbitrary oscillatory value, denoted by the red dashed line

node's value the smallest value. As a consequence, the difference between the maximum and the minimum values of the normal nodes is continuously reduced, and the network achieves asymptotic convergence.

5 Conclusions and Future Work

In this paper, we proposed a particular network topology that is resilient to a malicious member node. We showed that this topology provides robustness, and guarantees asymptotic consensus in the presence of illegitimate information originating from one of the nodes. Especially, we showed that it satisfies two important properties for distributed systems: *(i)* it can be constructed incrementally by adding one node at a time such that the conditions for attachment can be computed locally, and *(ii)* its robustness can be verified through a distributed method by using only neighborhood-based information. The topology is fully scalable, and its robustness property can be validated in polynomial time. Also, its geometric properties lend themselves elegantly to applications of coverage and formation control for networked robot teams. In future work, we intend to study how our current method can be extended to withstand more than one malicious agent, i.e., $(F+1, F+1)$-robust graphs for $F > 1$. This is challenging because prior approaches assume unbounded communication radii (i.e., unlimited edge lengths between nodes that are being deployed in physical space). Furthermore, we will investigate the applicability of our methods to mobile problem settings, such as robust formation control.

Acknowledgements We gratefully acknowledge the support of the Colombian Innovation Agency (COLCIENCIAS), and the Brazilian agencies CAPES, CNPq, FAPEMIG. We also acknowledge the support of ONR grants N00014-15-1-2115 and N00014-14-1-0510, ARL grant W911NF-08-2-0004, NSF grant IIS-1426840, and TerraSwarm, one of six centers of STARnet, a Semiconductor Research Corporation program sponsored by MARCO and DARPA.

References

1. Ahmadi, M., Stone, P.: A distributed biconnectivity check. In: Distributed Autonomous Robotic Systems 7, pp. 1–10. Springer, Berlin (2006)
2. Cortés, J., Martínez, S., Bullo, F.: Robust rendezvous for mobile autonomous agents via proximity graphs in arbitrary dimensions. IEEE Trans. Autom. Control **51**(8), 1289–1298 (2006)
3. Haas, R., Orden, D., Rote, G., Santos, F., Servatius, B., Servatius, H., Souvaine, D., Streinu, I., Whiteley, W.: Planar minimally rigid graphs and pseudo-triangulations. In: Proceedings of the Nineteenth Annual Symposium On Computational Geometry, pp. 154–163. ACM (2003)
4. Hromkovič, J.: Dissemination Of Information In Communication Networks: Broadcasting, Gossiping, Leader Election, And Fault-tolerance. Springer Science & Business Media (2005)
5. Jadbabaie, A., Lin, J., Morse, A.S.: Coordination of groups of mobile autonomous agents using nearest neighbor rules. In: Proceedings of the 41st IEEE Conference on Decision and Control, 2002, vol. 3, pp. 2953–2958 (2002). https://doi.org/10.1109/CDC.2002.1184304
6. LeBlanc, H.J., Koutsoukos, X.D.: Algorithms for determining network robustness. In: Proceedings of the 2nd ACM International Conference On High Confidence Networked Systems, pp. 57–64. ACM (2013)
7. LeBlanc, H.J., Zhang, H., Koutsoukos, X., Sundaram, S.: Resilient asymptotic consensus in robust networks. IEEE J. Sel. Areas Commun. **31**(4), 766–781 (2013). https://doi.org/10.1109/JSAC.2013.130413
8. Lynch, N.A.: Distributed Algorithms. Morgan Kaufmann (1996)
9. Murray, R.M, Olfati Saber, R.: Consensus protocols for networks of dynamic agents. In: Proceedings of the 2003 American Controls Conference (2003)
10. Olfati-Saber, R., Fax, J.A., Murray, R.M.: Consensus and cooperation in networked multi-agent systems. Proc. IEEE **95**(1), 215–233 (2007)
11. Park, H., Hutchinson, S.: A distributed robust convergence algorithm for multi-robot systems in the presence of faulty robots. In: 2015 IEEE/RSJ International Conference on Intelligent Robots and Systems (IROS), pp. 2980–2985. IEEE (2015)
12. Park, H., Hutchinson, S.: An efficient algorithm for fault-tolerant rendezvous of multi-robot systems with controllable sensing range. In: 2016 IEEE International Conference on Robotics and Automation (ICRA), pp. 358–365. IEEE (2016)
13. Pasqualetti, F., Bicchi, A., Bullo, F.: Consensus computation in unreliable networks: a system theoretic approach. IEEE Trans. Autom. Control **57**(1), 90–104 (2012)
14. Ren, W., Beard, R.W., Atkins, E.M.: A survey of consensus problems in multi-agent coordination. In: American Control Conference (ACC), pp. 1859–1864. IEEE (2005)
15. Ren, W., Beard, R.W., Atkins, E.M.: Information consensus in multivehicle cooperative control. IEEE Control Syst. **27**(2), 71–82 (2007). https://doi.org/10.1109/MCS.2007.338264

16. Sundaram, S., Hadjicostis, C.N.: Distributed function calculation via linear iterative strategies in the presence of malicious agents. IEEE Trans. Autom. Control **56**(7), 1495–1508 (2011)
17. West, D.B., et al.: Introduction To Graph Theory. Prentice hall Upper Saddle River (2001)
18. Yang, P., Freeman, R.A., Lynch, K.M.: Multi-agent coordination by decentralized estimation and control. IEEE Trans. Autom. Control **53**(11), 2480–2496 (2008)
19. Zhang, F., Leonard, N.E.: Cooperative filters and control for cooperative exploration. IEEE Trans. Autom. Control **55**(3), 650–663 (2010)
20. Zhang, H., Sundaram, S.: Robustness of information diffusion algorithms to locally bounded adversaries. In: American Control Conference (ACC), pp. 5855–5861 (2012). https://doi.org/10.1109/ACC.2012.6315661

Part III
Multi-Robot Estimation

Construction of Optimal Control Graphs in Multi-robot Systems

Gal A. Kaminka, Ilan Lupu and Noa Agmon

Abstract Control graphs are used in multi-robot systems to maintain information about which robot senses another robot, and at what position. Control graphs allow robots to localize relative to others, and maintain stable formations. Previous work makes two critical assumptions. First, it assumes edge weights of control graphs are deterministic scalars, while in reality they represent complex stochastic factors. Second, it assumes that a single robot is pre-determined to serve as the global anchor for the robots' relative estimates. However, optimal selection of this robot is an open problem. In this work, we address these two issues. We show that existing work may be recast as graph-theoretic algorithms inducing control graphs for more general representation of the sensing capabilities of robots. We then formulate the problem of optimal selection of an anchor, and present a centralized algorithm for solving it. We evaluate use of these algorithm on physical and simulated robots and show they very significantly improve on existing work.

1 Introduction

Control graphs are used in multi-robot systems to maintain information about which robot senses another robot, and at what position. In such control graphs, nodes represent robots in given positions. Weighted edges represent sensing capabilities; an edge from node A to node B, with weight w, represents the fact that robot A can sense robot B, with preference w (typically, smaller weight indicates stronger prefer-

G. A. Kaminka
Computer Science Department and Gonda Brain Research Center, Bar Ilan University, Ramat Gan, Israel
e-mail: galk@cs.biu.ac.il

I. Lupu · N. Agmon (✉)
Computer Science Department, Bar Ilan University, Ramat Gan, Israel
e-mail: agmon@cs.biu.ac.il

I. Lupu
e-mail: lupu.ilan@gmail.com

R. Groß et al. (eds.), *Distributed Autonomous Robotic Systems*,
Springer Proceedings in Advanced Robotics 6,
https://doi.org/10.1007/978-3-319-73008-0_12

ence). On the basis of such graphs, it is possible to build a shared coordinate system (e.g., [10]), compute message passing paths in ad-hoc networks, and maintain stable formations (e.g., [2, 6]).

Existing work utilizing control graphs raises several open challenges. First, it offers no systematic treatment of the edge weights, how they are determined, and how they should be utilized in the computation of optimal control graphs. Different tasks (e.g., building a coordinate system versus formation maintenance) utilizes the edge weights differently. Second, it makes the assumption that a single robot is given, chosen to serve as the *global anchor* of the shared coordinate system, *leader* of the formation, or origin of a message whose position is taken as the basis for the robots' relative positioning and location estimates. Third, it ignores uncertainty in the weights of edges, such that, for instance, if the edge weight denotes a distance, it assumes the distance is known with certainty, despite the inherent uncertainty that exists in real-world sensing. In this work, we tackle these open challenges.

First, we synthesize from existing work, and then generalize the notion of control graphs and their uses. We begin by refining the definition of *monitoring multi-graphs* [6], which distinguish between different sensing configurations of robots. We show how existing techniques (e.g., for computing shared coordinate systems) can be optimized by re-casting them in terms of graph-theoretic algorithms for inducing directed trees from the multi-graphs, such that the trees optimize for a given criteria (e.g., team costs, individual position error). Each such tree is an *optimal control graph* for a given task (e.g., message passing, formation maintenance).

Second, on the basis of this more general understanding of how control graphs are generated from monitoring multi-graphs, we formulate the problem of optimal selection of *leader* or *global anchor* in a given monitoring multi-graph. A leader robot serves as the root of the control graph (tree) generated from it. We present a centralized algorithm that efficiently determines the optimal leader for a given task, as well as the resulting control graph.

We evaluate use of the novel algorithms on physical and simulated robots equipped with depth and image sensors (RGB-D cameras), and contrast them with results obtained from existing work. The results show very significant improvements from using these algorithms for coordinate frame alignment, in both simulated and real robots, in static and dynamic settings.

2 Related Work

The use of graph theory for reasoning about roles of robots in cooperative multi-robot tasks has a long history. We survey below only the most related, recent work.

Formation maintenance. Here the robots move while maintaining a shape, dictated by their relative positions. Desai et al. [2] defined a *control graph* as an unweighted directed graph (digraph) whose vertices are the robots in the formation. An edge from A to B represents that robot A monitors robot B's position. They show that a formation can be stably maintained if the control graph implies each robot (except a single *leader*) maintains its bearing (angle) and separation (distance) with respect to one other robot (*target*). This type of formation control is known as SBC (Separation-Bearing Control). Without referring to control graphs, Fredslund and Matarić [3] propose a distributed algorithm for generating SBC monitoring rules (i.e., which robot monitors whom) given a target placement of the robots and the leader. In contrast, we consider weighted edges in our control graphs, and show how to induce *optimal* control graphs for different tasks (not just formations). We also address the question of leader selection. However, our algorithms here are centralized.

Kaminka et al. [6] generalized on these works. They defined a weighted monitoring multi-graph, which compactly represents all possible SBC control graphs for a given placement of robots. Each edge represents a possible configuration of the follower robot by which it can sense a target robot, and its weight represents its cost. They present a centralized algorithm for inducing a specific control graph, which optimizes the selection of targets, assuming a pre-determined leader. We show that their representation and algorithm is in a special case of a broader definition of monitoring multi-graphs, and we address the question of leader selection, which they leave open.

Lemay et al. [7] present a distributed method of assigning robots to formation positions. The computation relies on a cost function that considers distances and angles to the teammates; it outputs the lowest-cost assignment of robots to positions, and a leader that minimizes costs over all possible assignments. In contrast, we begin with robots already assigned to positions, and only then select a leader and SBC targets. However, we explicitly consider sensor capabilities, including errors.

Shared Coordinate Systems (Coordinate Frame Alignment). Another common task is that of multiple robots agreeing on a common coordinate system (axes and origin), e.g., as the basis for multi-robot mapping. There are several studies regarding the construction and alignment of coordinate systems (e.g., [4, 9, 11, 13]). Briefly, the task here is for robots to identify their alignment (translation and rotation) with respect to each other (typically one of the robots serves as a global anchor). As not all robots can sense the global anchors, they may instead localize via *anchor chains*, i.e., localize with respect to local anchors, who sense other anchors, etc. This is also referred to as coordinate frame alignment.

Most such work focuses on the filtering mechanisms able to cope with the uncertainty inherent to this process, and with various types of errors (e.g., receiving only range information). However, recently, Nagavalli et al. [10] presented a distributed method for improving the accuracy of such alignments, by utilizing a breadth-first search (BFS) to minimize the number of anchors in anchor chains, all beginning with a selected global anchor. In this paper we present a centralized algorithm for selecting an optimal global anchor in this task, and show that this further improves (significantly) the position estimates of the robots. Moreover, this works with anchors that are not part of the team, such as objects in the team surroundings that the robots can identify.

3 Optimal Construction of Control Graphs

We begin with robots placed in fixed relative positions, and no leader assigned. In Sect. 3.1 we show how to compactly represent all the different possibilities for robots to sense each other in their positions, using a refined definition of *monitoring multi-graphs*, originally presented in [6]. Then, in Sect. 3.1 we show how existing work can be re-cast in terms of graph-theoretical algorithms, properly extended to run on monitoring multi-graphs. Existing work leaves open the question of optimal leader selection, which we address in Sect. 3.2.

3.1 Monitoring Multi-graphs

A monitoring multigraph captures all the potential control graphs for a group of robots in fixed positions. As defined in [6], it is a directed, weighted multigraph $G = \langle V, E \rangle$, where V is a set of vertices representing robots, and E is *a bag* (multi-set) of weighted edges between vertices.

Each $v_i \in V$ represents a unique robot i, identified by its index, and having a specific pose in space. The function $pos : V \mapsto \Re^n$ identifies the unique pose of each robot $v \in V$ (typically, $n = 3$, with the pose determined by the position and orientation of the robot v).

Let $v_i, v_j \in V$ be two robots. Suppose v_i can use a specific configuration of its sensors to sense v_j, i.e., v_i computes an estimate of $pos(v_j)$, denoted by $\widehat{pos}(v_j)$. Denote the specific configuration by x. For instance, it may refer to a specific pan of a camera or Lidar, combined with a specific sensor processing algorithms (e.g., visual marking recognition, depth perception), or a specific choice of resolution or focus. Reference [6] propose using a single scalar value c^x_{ij} as the edge weight, indicating preferences for using the sensor in this configuration, e.g.. based on reliability. We depart from this definition in two ways. First, we distinguish between *directly measurable resource costs* (such as expenditure of power, computation time, or sensor processing latency), and errors in the estimate $\widehat{pos}(v_j)$, which are given in terms of deviations from the ground truth. Second, we accept that realistically, costs and errors can only be estimated with uncertainty. Thus we model them as random variables, with a known probability distribution function.

More precisely, with each measurable cost factor k in the operation of the sensor, and each component of error m resulting from it in $\widehat{pos}(v_j)$, we associate a known probability distribution $C^{x,k}_{ij}$ ($R^{x,m}_{ij}$, respectively), explicitly or parametrically represented. For instance, if the perception latency l is known to be *uniformly* distributed in the range 20–30 ms, this may be explicitly represented by setting $C^{x,l}_{ij} \equiv \mathcal{U}(20, 30)$. If the distance from v_i to v_j is d, measured by a Lidar with a 3% error, we may set $R^{x,d}_{ij} \equiv \mathcal{U}(-0.015d, +0.015d)$. As v_i only approximates the true position of v_j with $\widehat{pos}(v_j)$), we use an approximate distance measure d, and update it as additional measurements are made. The overall costs associated with the edge e_{ij} are then drawn from the joint distribution of all $C_{ij}^{x,k}$, denote $\overline{C_{ij}^{x}}$. Likewise, we denote the errors by $\overline{R_{ij}^{x}}$.

Given these definitions, we define the edges in E as follows. An edge $e_{ij}^{x} \in E$ is a tuple $e_{ij}^{x} = \langle v_i, v_j, \overline{C_{ij}^{x}}, \overline{R_{ij}^{x}} \rangle$. When clear from the context, we omit the superscript ^{x}. This definition departs from [6] in that we add the representation of errors, and distinguish multiple components in costs and errors. We also depart from [6] in that we assume that the sensing robot can identify the sensed robot id and contrast the graph with the existing edges without assuming all possible edges can exist and eliminating edges that are occluded by other robots. Alternative configurations may result in improved costs or lower errors; often a robot may trade these off, e.g., by spending more computation time or more energy to improve its position estimate of the other robot. Given $|X|$ configurations for robot v_i to monitor v_j (which are usually determined by the number of different sensors the robot has), there exist edges $e_{ij}^{1}, e_{ij}^{2}, \ldots, e_{ij}^{|X|} \in E$.

Inducing Control Graphs with Uncertainty: Managing Risk. Following [8], we refer to a multigraph with random-variable weights as a *stochastic multigraph*. Different tasks, such as *formation maintenance*, may reduce to selecting paths in the multigraph. The length of a path in a stochastic graph is a function of random events characterized by the probability distributions associated with the cost along the path. We therefore have to decide how we would like to deal with the uncertainty. The common approach to dealing with uncertainty is by considering the risk involved in the decision. Standard policies include *risk-aversion* (hoping to reduce risk, even at higher cost, i.e., minimize the expected maximal cost/error); *risk-seeking* (inversely); and risk-neutrality (perfectly balancing risk and costs). Different decision strategies can lead to different shortest path selections.

Several such algorithms appear elsewhere [5, 8], and are outside the scope of this paper. However, it has been shown that risk-neutral selection both works correctly [8], and is safe, in the sense that it minimizes notions of regret [12]. For the remainder of the work, and in the experiments, we therefore used the risk-neutral policy, by using the expected (mean) value of the distributions $E[\overline{C_{ij}^{x}}]$ (or, as needed, $E[\overline{R_{ij}^{x}}]$) as the edge weights. Here $E[P]$ is the expected (mean) value of the probability distribution P.

Inducing Control Graphs (for a Given Robot). Monitoring multigraphs compactly represent all potential ways in which robots could monitor each other in their positions. Given a task which requires robots to monitor each other's positions (e.g., formation maintenance), we want to induce a *control graph*: a subset of the monitoring graph, which specifies for each robot which sensor configuration to use, and what other robot(s) to monitor, in order to improve task performance.

Table 1 summarizes the progression in previous work. In the column marked "Arbitrary leader, arbitrary control graph" we list previous works which utilize heuristic algorithms for constructing control graphs which are not guaranteed to be optimal (in the sense of reducing accumulating errors or costs). In the next column, marked "Arbitrary leader, Optimal control graph", we list investigations which, for a pre-

Table 1 Related work utilizing accumulating factors, re-cast by type of algorithm and problem settings. Reference [6] uses costs to represent errors. Reference [10] assumes uniform errors, allowing use of BFS instead of Dijkstra's algorithm

	Arbitrary leader, arbitrary control graph	Arbitrary leader, optimal control graph	Optimal leader, optimal control graph
Algorithm type	Heuristic	Dijkstra's	All pairs shortest path
Formation maintenance	[3]	[6]	This work
Relative localization	[4, 13]	[10]	

determined leader, generate an optimal control graphs minimizing accumulating errors or costs (assuming scalar edge weights). A variant of Dijkstra's single-source shortest path (S3P), described in [6] is optimal for such cases.

3.2 *Inducing Control Graphs with Optimal Global Anchor*

Thus the challenge remains of determining the optimal leader (i.e., one whose associated control graph is superior to those of other leaders). Our task here is to select a single robot which will serve as a leader of a formation, or the origin point (global anchor) for an agreed-upon shared coordinate system. We will therefore optimize the leader selection and associated control graph to reduce the errors $\overline{R_{ij}^x}$.

3.2.1 Problem Formulation

K robots are positioned in space. Each robot is equipped with sensors, allowing it to identify (some) other robots in its vicinity, and to estimate their position with respect to itself (i.e., their position in its own ego-centric coordinate frames). Furthermore, we assume robots are able to communicate with their peers, at least with those they are able to observe. The settings are captured by a monitoring multi-graph G_K. The task is to extract a control graph where the coordinate frame of a single robot (global anchor) is used as the origin, and all robots align their coordinate frames to it. Because not all robots can directly sense the global anchor, each robot can decide to align its coordinate system with respect to one other robot (called local anchor), who aligns itself to the global anchor, or to another local anchor. Thus a coordinate frame alignment control graph has the following properties:

- The vertex representing the global anchor has an out-degree of 0.
- All other vertices (robots) have an out-degree of 1.
- There exist a path from every vertex (robot) to the vertex representing the global anchor.

A coordinate frame alignment control graph is optimal with respect to the selected global anchor v_A if it minimizes the errors in position estimates of the robots. Suppose we have a robot v_0. Its position estimate in the shared coordinate system accumulates errors with every local anchor it uses on a path from itself to the global anchor in the control graph. It thus seeks to minimize the sum of expected errors $\sum_{e_{ij}} E[\overline{R_{ij}}]$ where e_{ij} is an edge on the path from v_0 to v_A. The question is how to choose v_A.

3.2.2 Optimal Global Anchor Selection

A global anchor v_A is called optimal, if its associated control graph is superior to the control graphs associated with any other potential global anchor. We consider two different ways a control graph may be superior to another: It may reduce the average position error for the group (a societal view of errors), or it may reduce the maximal position error (an individual view of errors). Our task here is to determine the optimal global anchor for both definitions. The process includes the following steps (see details next).

1. Transform the stochastic monitoring multigraph G_K into an intermediate representation, G'_K, which is a deterministically-weighted regular digraph (embedding errors, and reversing direction of edges). This step is carried out in time $\mathcal{O}(|E|)$, where E is the bag of edges in G_K.
2. Apply an All Pairs Shortest Path (APSP) algorithm to the graph G'_K. The time needed depends on the algorithm chosen, but is generally $\mathcal{O}(|V|^3)$, where V is the set of vertices in G'_K (normally, $|V| = K$).
3. Determine for each robot $v \in V$ the set of shortest paths leading from it P_v. For each such set P_v, determine the sum of the path lengths S_v, or the maximal path length M_v, depending on the global anchor selection criteria. This is carried out in time $\mathcal{O}(|V|^2)$.
4. The global anchor v_A is one that minimizes S_{v_A} or M_{v_A}. This is determined in time $\mathcal{O}(|V|)$.

Transformation of G_K into G'_K. This step is carried out to transform the stochastic directed monitoring multigraph into a deterministic graph, which embeds the necessary information, yet amenable to the execution of familiar graph-theoretic algorithm. The graph $G'_K = \langle V', E' \rangle$ is built as follows (see example in Fig. 1).

First, we set $V' \leftarrow V$. Then, for each pair of vertices $v_i, v_j \in V$, we do the following: (1) If an edge e^x_{ij} exists, with error distribution $\overline{R^x_{ij}}$, then create a temporary reversed edge, e'^x_{ji}, with scalar weight $r^x_{ji} = E[\overline{R^x_{ij}}]$. (2) Among all edges e'^x_{ji}, select the one with minimum r^x_{ji}, i.e., $e_{ji} = \arg\min_{e'^x_{ji}}(r^x_{ji})$. Finally, (3) add e_{ji} to E'. The result is a directed graph, with scalar deterministic edge weights, in which all errors have been folded into the edge weights using the risk-neutral policy, redundant edges in the multigraph removed, and edge direction reversed.[1]

[1]Note that one can decide at this step to use any function combining C and R.

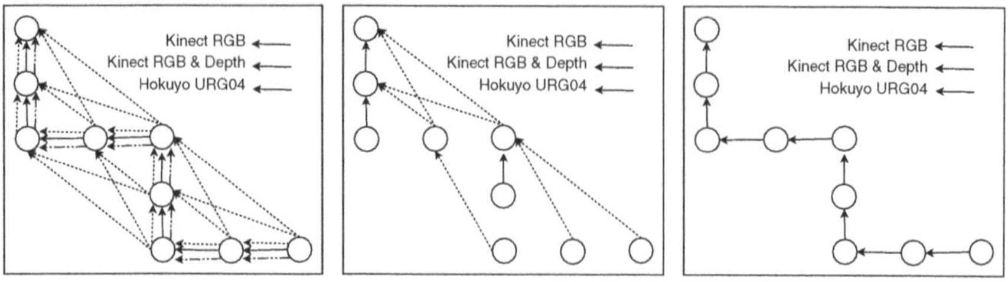

Fig. 1 An example for a monitoring multigraph (left), and two resulting monitoring graphs: one that minimizes the maximal path length (middle), and one that minimizes the sum of path lengths (right)

All Pairs Shortest Paths. We now run an algorithm for determining the shortest paths for all pairs of vertices. In our implementation we utilized Johnson's algorithm [1]. Given the size of V' is the number of robots K, the algorithm runs in $\mathcal{O}(K^2 \log K + K|E|)$. The result is often represented in a matrix L, such that matrix cell l_{ji} contains the length of the shortest path from vertex j to vertex i (or ∞ if none exists). As edges are reversed in direction compared to the sensing direction, l_{ji} is the accumulating error in position estimates, from robot v_i to robot v_j, where $v_i, v_j \in V$.

Determine S_v **and/or** M_v. We propose two different criteria for selecting a global anchor that, if used as the origin for a shared coordinate system, would result in smaller position estimate errors for the team of K robots. One possible criterion is to minimize the mean position error of all K robots. This is a societal criterion, as it balances the errors across all robots. An alternative criterion is to minimize the worst-case error of any single robot, possibly resulting in some robots accepting a larger error than individually needed, in order to reduce the error of the other robots.

We examine the matrix L. Let S, M be vectors of dimension K. We denote S_v the component of S associated with a given v (and similarly, M_v). For all $v \in V$, $S_v = \frac{1}{K}\sum\limits_{i=1}^{K} l_{vi}$, i.e., the sum of all cells in row v divided by K, or more intuitively, the mean length of shortest paths from all robots i to robot v. As these shortest path represent smallest errors, this is the mean smallest error in position estimates, if v is selected as global anchor. Similarly, for all $v \in V$, $M_v = \max\limits_{i=1}^{K} l_{vi}$, i.e., the maximal smallest error in position estimate for any robot i, if v is the global anchor.

Determine global anchor v_A. Finally, a new global anchor can be chosen, by setting $v_A = \arg\min_{v \in V'} S_v$, if we prefer a global anchor that minimizes the average position error, or $v_A = \arg\min_{v \in V'} M_v$, if we prefer to minimize the maximal error instead. If there are ties, they can be broken by preferring according to the other criterion, or arbitrarily.

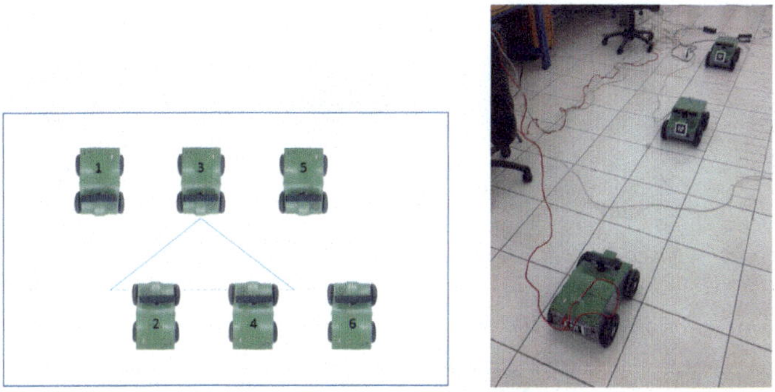

(a) Six simulated robots.

(b) Three real robots. Simulated robots placed like wise.

Fig. 2 Formation in static experiments

4 Evaluation

To evaluate the effects of using the techniques presented in this work, we implemented the algorithms for optimal global-anchor selection and coordinate frame alignment in ROS (Robot Operating System), to be used on Gazebo-simulated and real RoboTICan Lizi robots (shown in Fig. 2b). All robots in the team were marked with unique visual markers identifying each robot. Using image and depth data from an RGB-D sensor, each robot identified its neighbors and measured their relative position in its reference frame. A calibrated sensor model was used to estimate the error measurements R_{ij}.

We compared the global position errors resulting from using the optimal v_A algorithm above, to the errors resulting from using an arbitrary robot [10]. Specifically, we contrast the robots' estimates with the ground truth measured externally. This was done by carrying out five repeated trials in each setting, each lasting two minutes, resulting in thousands of data points, *for each robot*.

We have carried out experiments in three types of settings: robots standing still, robots moving while maintaining a static formation, and robots moving while changing formation. In the first two settings, the relative positions of the robots are maintained: by definition in the first setting, and using feedback control in the second. In the third setting, moving robots changed their initial formation, requiring them to select a new global anchor.

Our first experiment recreates an experiment in [10]. Six Lizi robots are placed as shown in Fig. 2a. All robots are static, and align their coordinate system with respect to the selected global anchor. Similar experiments involve placing three robots as shown in Fig. 2b. These were conducted both in simulation, as well as in real robots. Robot 1 (bottom of the image) could monitor robot 2 (center) and vice versa; robot 3 could see robot 2.

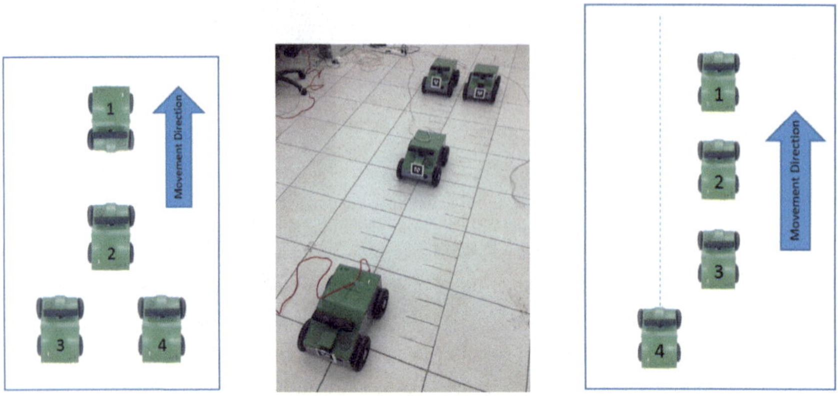

(a) Static formation placement.

(b) Static real robots.

(c) Dynamic formation. Robot #4 overtaking others.

Fig. 3 Formations maintained while moving

We then turned to experiments where robots moved while continually estimating their position based on a shared coordinate system, with the origin at the selected global anchor. We placed four robots in the formation shown in Fig. 3a, again both in simulation as well as in the lab. Robot 1 (front of the formation) could monitor robot 2 (center) and vice versa, robots 3 and 4 (side by side, bottom) could monitor robot 2. Figure 3b shows the real robots in one of the trials. In the arbitrary ID settings, robot 1 was selected as the global anchor. In the optimal settings, our algorithm chose robot 2 as the global anchor.

As a final experiment, we tested the ability of the algorithm to adjust the global anchor while moving, when the relative position of robots is changed. Four simulated robots were placed as shown in Fig. 3c. All robots moved forward; robots 1–3 at constant speed, and robot 4 three time faster, along the dotted path shown in the figure, and until it pulled ahead of everyone else. While moving, the robots continually checked and recomputed the global anchor appropriate to their current settings. At the beginning of each run, robot 1 was chosen as global anchor v_A, and the algorithm chose local anchors for all other robots: robot 4 monitored 3, which monitored 2, which monitored 1. However, as robot 4 begins to overtake it peers, its local anchor changes from 3 to 2, then to 1, until finally it overtakes robot 1, at which point it becomes the global anchor, and root 1 switches to monitor it.

Figure 4 shows the mean error (error bars indicate standard deviation) of robot 4 during the experiment. It shows that between 0.1 and 0.5 min into a trial, when robot 4's local anchor is robot 3, the error in position (in the shared coordinate system where robot 1 is the origin) is around 40 cm. After passing robot 3, robot 4 changes local anchor based on the optimal selection, first to robot 2 and then to robot 1. Approximately 0.95 min into the run, and until 1.15 min in it, robot 4's local anchor is robot 1 which is still the global anchor v_A. We see a corresponding decrease in robot 4's position error as it now monitors the global anchor directly. After 1.15 min,

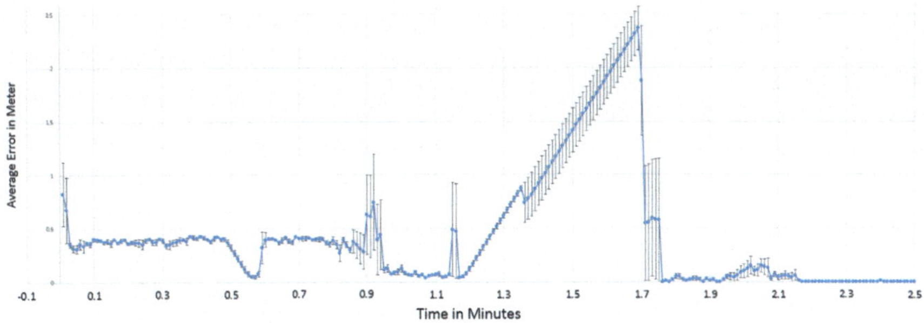

Fig. 4 Changing control graph in real time

robot 4 cannot see any other robot and its error increases due to moving and assuming location in its last position. With real robot it is possible to change the localization method to less accurate one such as GPS in this situation. After robot 4 enters robot 1's field of view, the algorithm sets robot 4 to serve as v_A.

Results. The results, summarized in Table 2, show the use of the leader-selection algorithm leads to very significant improvements in the position estimates of the robots in the shared coordinate system. In many cases, the mean error is reduced by 50% or more. For example, in the experiment with six standing robots, when using the minimal robot ID as a global anchor the farthest robot (#6) was located five hops away, and accumulated approximately 13 cm in error. However, using the global anchor selected by our algorithm, the average error for the same robot, now located 3 hops away, drops to 6 cm. This improvement is statistically significant (one tailed t-test, $p < 7.49 \times 10^{-16}$). Similar improvements can be seen in the other experiments, both in simulated and real robots. Over all trials, these results are over approximately 5000 measurements in each settings, for each robot.

5 Conclusions and Future Work

Control graphs are used in multi-robot systems to maintain information about which robot senses another robot, and at what position. On the basis of such graphs, it is possible to compute a shared coordinate system, localize relative to others, and maintain stable formations. In this work, we demonstrated that previous work assumes that a robot is pre-determined, to serve as global anchor (origin point) for coordinate frame alignment. We extended previous notions of *monitoring multigraphs*, a construct intended to compactly represent all possible control graphs. We focused on risk-neutral decision policy, which allows us to replace stochastic edge weights with the deterministic expected value of the distributions. Second, we demonstrated

Table 2 All experiment results, including mean errors in meters (standard deviations), and t-test significance testing. Robot ID is shown for robots not acting as global anchor v_A in either setting. The optimal global anchor column shows significant improvement in *all experiments*

Type	Experiment	Robot ID	Arbitrary v_A error in meters	**Optimal** v_A error in meters	Significance p value (one-tailed t-test)
Standing	3-line (simulation)	3	0.058 (0.102)	**0.036 (0.009)**	7.12×10^{-15}
	3-line (real robots)	3	0.107 (0.019)	**0.049 (0.001)**	0 (below excel limit)
	6 zigzag (simulation)	2	0.031 (0.021)	**0.014 (0.006)**	2.62×10^{-78}
		4	0.073 (0.142)	**0.030 (0.005)**	4.12×10^{-15}
		5	0.086 (0.181)	**0.032 (0.005)**	4.90×10^{-15}
		6	0.134 (0.239)	**0.061 (0.033)**	7.49×10^{-16}
Moving	Simulation 4 center	3	0.036 (0.014)	**0.019 (0.018)**	$1.72 \times 10-98$
		4	0.032 (0.017)	**0.013 (0.012)**	5.92×10^{-143}
	Real moving 4 center	3	0.155 (0.076)	**0.095 (0.009)**	8.31×10^{-6}
		4	0.140 (0.105)	**0.084 (0.038)**	0.00056

that an All Pairs Shortest Path algorithm can be utilized, on the extended monitoring multi-graph, through some transformations. This facilitates the automatic determination of an optimal robot to lead a formation or serve as a global anchor. We conducted extensive experiments in real and simulated robots; these show very significant improvement to the robots' position estimates. In future work, we hope to examine alternative methods for dealing with decision policies that are risk-averse, or risk-seeking.

The algorithms presented herein assume that all information about the sensing capabilities and location of the robots is known - either to a centralized unit, or to one of the robots. Using this information, the optimal local and global anchors are determined. It would be interesting to extend these results to a decentralized setting. In this case, choosing a local anchor may be straightforward, yet choosing an optimal global anchor would require using innovative methods.

Acknowledgements We gratefully acknowledge support by ISF grants #1511/12 and #1337/15. As always, thanks to K. Ushi.

References

1. Cormen, T.T., Leiserson, C.E., Rivest, R.L.: Introduction to Algorithms. MIT Press, USA (1990)
2. Desai, J.P.: Modeling multiple teams of mobile robots: a graph-theoretic approach. IROS **1**, 381–386 (2001)
3. Fredslund, J., Mataric, M.J.: A general algorithm for robot formations using local sensing and minimal communications. IEEE Trans. Robot. Autom. **18**(5), 837–846 (2002)
4. Howard, A., Matarić, M.J., Sukhatme, G.S.: Putting the 'i' in 'team': an ego-centric approach to cooperative localization. In: ICRA, pp. 868–892 (2003)
5. Hwang, L.K.: Stochastic shortest path algorithm based on lagrangian relaxation. Master's thesis, University of Illinois at Urbana-Champaign (2010)
6. Kaminka, G.A., Schechter-Glick, R., Sadov, V.: Using sensor morphology for multi-robot formations. IEEE Trans. Robot. **24**, 271–282 (2008)
7. Lemay, M., Michaud, F., Létourneau, D., Valin, J.M.: Autonomous initialization of robot formations. In: ICRA-04 (2004)
8. Loui, R.P.: Optimal paths in graphs with stochastic or multidimensional weights. Technical report TR115, Computer Science Department, University of Rochester (1982)
9. Martinelli, A., Pont, F., Siegwart, R.: Multi-robot localization using relative observations. In: ICRA-05, pp. 2797–2802. IEEE (2005)
10. Nagavalli, S., Lybarger, A., Luo, L., Chakraborty, N., Sycara, K.: Aligning coordinate frames in multi-robot systems with relative sensing information. In: IROS-14, pp. 388–395. IEEE (2014)
11. Piovan, G., Shames, I., Fidan, B., Bullo, F., Anderson, B.D.O.: On frame and orientation localization for relative sensing networks. Automatica **49**(1), 206–213 (2013)
12. Traub, M., Kaminka, G.A., Agmon, N.: Who goes *there*? using social regret to select a robot to reach a goal. In: AAMAS-11 (2011)
13. Trawny, N., Zhou, X.S., Zhou, K., Roumeliotis, S.I.: Interrobot transformations in 3-d. IEEE Trans. Robot. **26**(2), 226–243 (2010). https://doi.org/10.1109/TRO.2010.2042539

Decision-Making Accuracy for Sensor Networks with Inhomogeneous Poisson Observations

Chetan D. Pahlajani, Indrajeet Yadav, Herbert G. Tanner and Ioannis Poulakakis

Abstract The paper considers a network of sensors which observes a time-inhomogeneous Poisson signal and has to decide, within a fixed time interval, between two hypotheses concerning the intensity of the observed signal. The focus is on the impact of information sharing among individual sensors on the accuracy of a decision. Each sensor computes locally a likelihood ratio based on its own observations, and, at the end of the decision interval, shares this information with its neighbors according to a communication graph, transforming each sensor to a decision-making unit. Using analytically derived upper bounds on the decision error probabilities, the capacity of each sensor as a decision maker is evaluated, and consequences of ranking are explored. Example communication topologies are studied to highlight the interplay between a sensor's location in the underlying communication graph (quantity of information) and the strength of the signal it observes (quality of information). The results are illustrated through application to the problem of deciding whether or not a moving target carries a radioactive source.

1 Introduction

We currently rely on networks of distributed sensors for triggering a timely response to emergency situations, including natural disasters such as hurricanes [8], earthquakes [10] and tsunamis [6]. Existing work on decision-making over networks of

C. D. Pahlajani
Indian Institute of Technology, Gandhinagar, India
e-mail: cdpahlajani@iitgn.ac.in

I. Yadav · H. G. Tanner · I. Poulakakis (✉)
University of Delaware, Newark, DE, USA
e-mail: poulakas@udel.edu

I. Yadav
e-mail: indragt@udel.edu

H. G. Tanner
e-mail: btanner@udel.edu

R. Groß et al. (eds.), *Distributed Autonomous Robotic Systems*,
Springer Proceedings in Advanced Robotics 6,
https://doi.org/10.1007/978-3-319-73008-0_13

observers that monitor a physical process and have to decide on its state, suggests that the performance of decision-making is affected by the structure of the network. For example, in the context of a network of stochastic evidence accumulators, each represented by a drift-diffusion process accruing evidence in continuous time by observing a noisy signal, information exchange among individual nodes affects the certainty of each node in a way that is governed by information centrality [14, 15]. Information centrality is a structural property of the underlying interconnection graph that depends on the totality of paths connecting that node with the rest of the network [17]. Heterogeneity in the network has been introduced in [5] by allowing only a limited number of units—termed leaders—to observe the external signal directly. Although these results refer to continuous-time implementations of sequential probability ratio tests inspired by human decision-making models [2], they still highlight the effect of general network topologies on decision-making performance.

Only specific communication topologies for the network have been explored, due to the explosive combinatorial complexity of decision making in distributed sensor systems [20]. There is emphasis on particular directed rooted trees [19, 21–23], where the root is a designated fusion center node that makes the final decision. In this context, and for this limited range of topologies where information from all observers eventually trickles down in some compressed form to the designated fusion center, results show that decision performance is affected by the structure of the network. In addition, to avoid the complexity associated with distributed decision making, the results for more general tree topologies are restricted to an asymptotic analysis, where the order of the graph grows unbounded. For some applications, however—such as nuclear detection—practical and economic considerations preclude the deployment of large-scale networks [18]. In that smaller-scale regime, and particularly for the case where decision makers do not have access to (compressed) information from all other observers, there is not enough knowledge and insight to determine the specific effect of general network topologies on decision-making performance.

The present paper focuses on the case of a sensor network which observes a time-inhomogeneous Poisson process, and has to decide, within a fixed time interval, between two hypotheses concerning the intensity of the observed process. Armed with analytical expressions for bounds on error probabilities for this test in the classical networking case where all sensors, after observing this inhomogeneous process for a time period, submit a statistic to a central decision maker (fusion center) [11], this paper explores alternative cases where the topology of the network can vary, and any sensor can potentially become the decision maker. The goal here is to explore the effect of information sharing on the capacity of individual sensors in the network to make accurate decisions.

The topology of the sensor network is modeled by a directed graph; each sensor is represented by a node, and a directed edge from node i to node j signifies that there is directed flow of information from sensor i to sensor j. Sensors do not need to communicate until time instant T, at which a decision needs to be made by the network. At that time instant each sensor sends its local likelihood ratio $L_T(i)$ along outgoing edges in the communication graph. Next, each sensor—now referred to as a Decision Maker (DM)—implements a Likelihood Ratio Test (LRT) where the product of the available $L_T(i)$'s is compared against a threshold to make a (local) decision. The problem is now to assess the relative accuracy of the DMs.

Intuitively, the factors improving decision accuracy are large numbers of incoming edges in the information-sharing graph, and availability of $L_T(i)$'s collected from sensors which observe strong signals. The principal challenge now lies in formalizing *exactly* how these factors mathematically influence the decision error probabilities, given that these probabilities are, in all but perhaps the simplest cases, intractable to analytic computation. The paper circumvents this difficulty by working with Chernoff upper bounds on the error probabilities for each DM, borrowing the analytic expressions of which from existing work [12]. Treating these explicitly computable bounds as proxies for the true (unknown) error probabilities, an index for decision-making accuracy can be formulated and subsequently used to rank the DMs. The advantage of this approach is that it recasts the problem of ranking incomputable probabilities for each of the DMs in terms of ranking natural computable surrogates, at the expense of some sharpness.

These results find natural application in the problem of detecting, using a distributed network of radiation sensors, illicit nuclear material in transit. The severity of the threat associated with radioactive materials falling into the hands of potential terrorists has been recognized [4]. Possible mitigation strategies include networks of detectors deployed along roads and highways, tasked with detecting illicit material that has slipped through border checks or portal alarms [4].

There are at least two reasons that make nuclear detection—irrespective of whether the sensors are static or mobile—extremely challenging. The first is that radiation detectors pick up signals emitted not just by the illicit radioactive material to be detected, but also from ubiquitous, naturally occurring, background radiation. From a counter's perspective, the two signals are of identical nature and indistinguishable once superimposed. The second reason relates to attenuation: although a kilogram of Highly Enriched Uranium (HEU) can emit as many as 4×10^7 gamma rays per second [4], shielding and attenuation can limit the effective detection range to a few feet, and require detection times that can range from several minutes to hours. To put these in perspective, the gamma-ray emission of nuclear missiles containing HEU becomes comparable to background just 25 cm away from the warhead [16]. The problem is exacerbated by the motion of the signal source. Not only does the mathematical model of the physical phenomenon change (becoming time-inhomogeneous), but now detectors have limited time to decide before the target disappears from sight: the sensors are faced with a problem of detecting in a matter of seconds, a weak time-varying signal, buried inside another signal of same nature.

It is important to note that while the technical approach in this paper is particularized in the context of nuclear detection, the impact of the methods proposed here can also reach other application domains which involve networked decision-making with Poisson process observations.

2 Background

As noted above, our analysis focuses on the effect of information sharing on decision-making by a network of sensors observing a time-inhomogeneous Poisson process, and as such, is applicable in a variety of domains. That said, it will be convenient from this point on to frame the discussion largely in terms of applications to nuclear detection. It goes without saying that the treatment can be carried over to other applied problems after suitably reinterpreting various mathematical quantities.

With the aforementioned application domain in mind, consider a network of k radiation sensors that is deployed over a spatial region of interest for the purpose of detecting illicit nuclear materials in transit; see Fig. 1. The typical detection scenario involves a vehicle (target) suspected of being a carrier of nuclear material (source) moving through the sensing field of this network. The objective is to decide, at the end of a fixed time interval $[0, T]$, whether the counts recorded at the sensors can be attributed solely to ubiquitous background radiation (hypothesis H_0) or whether they contain, in addition, radiation from a source carried by the moving target (hypothesis H_1). Local decision making can be enhanced through allowing, at the terminal time T, limited communication among sensors according to a suitable communication topology, following which each sensor is enabled to act as a DM operating on the information available to it. The primary goal of the present work is to formulate a metric for ranking the decision-making accuracy of the individual DMs. To wit, it is desirable to develop an index that mathematically captures the interplay between centrality of a DM in the network (quantity of information) and proximity to the suspected source (quality of information) in a way that naturally mimics the true, but incomputable, error probabilities.

A review of some existing notation and results [11] helps set the stage for the decision-making rules considered in this paper. The probabilistic setup is as follows.

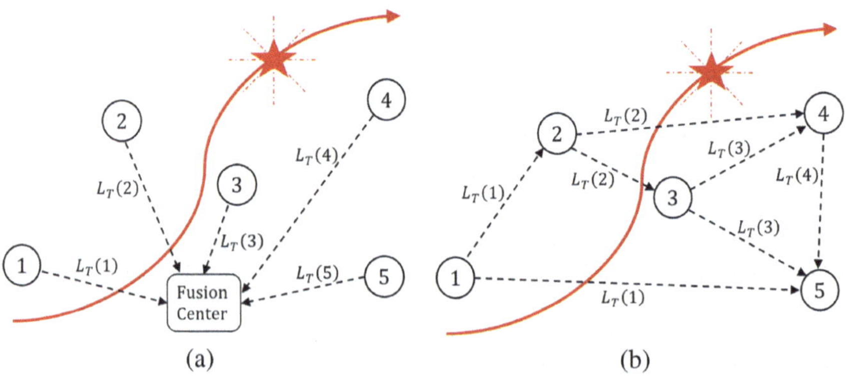

Fig. 1 A source moving (red line) in the sensing field of a network of sensors. Different information sharing scenarios are depicted. **a** All sensors send their L_T's to a fusion center. **b** The sensors share their L_T's according to a directed graph. Communication is allowed only at time T

On the measurable space $(\Omega, \mathscr{F})$, there is a k-dimensional vector of counting processes $\boldsymbol{N}_t \triangleq (N_t(1), \ldots, N_t(k))$, $t \in [0, T]$. Here, $N_t(i)$ represents the number of counts recorded at sensor $i \in \{1, 2, \ldots, k\}$ up to (and including) time $t \in [0, T]$. The two hypotheses, H_0 and H_1, regarding the state of the environment correspond, respectively, to two distinct probability measures $\mathbb{P}_0$ and $\mathbb{P}_1$ on $(\Omega, \mathscr{F})$. With respect to $\mathbb{P}_0$, the $N_t(i)$'s, $1 \leq i \leq k$, are assumed to be independent Poisson processes with $N_t(i)$ possessing intensity $\beta_i(t)$, while with respect to $\mathbb{P}_1$, the $N_t(i)$'s, $1 \leq i \leq k$, are assumed to be independent Poisson processes with intensities $\beta_i(t) + \nu_i(t)$, respectively. The functions $\beta_i(t)$, $\nu_i(t)$, $1 \leq i \leq k$, defined for $t \in [0, T]$ are assumed to be bounded, continuous and strictly positive as in [11]. Here, $\beta_i(t)$ is the (possibly time-varying) intensity at time t due to background radiation at the spatial location of sensor i, while $\nu_i(t)$ represents the intensity due to the source (if present) as perceived by sensor i at time t. Note that the time-dependence of $\nu_i(t)$ arises from relative motion between the source and the sensor; indeed, in the context of radiation measurement it is generally accepted that $\nu_i(t)$ is proportional to the inverse square of the distance $r_i(t)$ between the source and sensor i [9].

A test for deciding between H_0 and H_1 can be thought of as an event B_1, whose occurrence or non-occurrence can be ascertained on the basis of sensor observations over $[0, T]$, and has the following significance: If the outcome $\omega \in B_1$, decide H_1; if the outcome $\omega \in B_0 \triangleq \Omega \setminus B_1$, decide H_0. For such a test, two types of errors can occur. A false alarm occurs if $\omega \in B_1$ with H_0 being the correct hypothesis; this occurs with probability $\mathbb{P}_0(B_1)$. A miss occurs if the outcome $\omega \in B_0$ while H_1 is the true hypothesis; this occurs with probability $\mathbb{P}_1(\Omega \setminus B_1)$.

In this setting, the optimal test for deciding between H_0 and H_1 is an LRT obtained as follows [3, 11]. For $i \in \{1, 2, \ldots, k\}$, let $\tau_n(i)$ for $n \geq 1$ denote the nth jump time, i.e., time at which the sensor generates a count, of $N_t(i)$, and let

$$L_T(i) \triangleq \exp\left(-\int_0^T \nu_i(s)\mathrm{d}s\right) \prod_{n=1}^{N_T(i)} \left(1 + \frac{\nu_i(\tau_n(i))}{\beta_i(\tau_n(i))}\right) \tag{1}$$

with the convention that $\prod_{n=1}^{0}(\ldots) = 1$. Assuming that $\mathbb{P}_1$ is absolutely continuous with respect to $\mathbb{P}_0$, the test

$$\{L_T \geq \gamma\} \quad \text{where} \quad L_T \triangleq \prod_{i=1}^{k} L_T(i) \text{ and } \gamma > 0 \tag{2}$$

is optimal in the (Neyman–Pearson) sense that if B is any test whose probability of false alarm $\mathbb{P}_0(B) \leq \mathbb{P}_0(L_T \geq \gamma)$, then the probability of miss for the test (2) is at least as low as that for B, i.e., we have $\mathbb{P}_1(L_T < \gamma) \leq \mathbb{P}_1(\Omega \setminus B)$.

In the context of the test (2), the decision is made at a single network node that receives all sensory information from the network, and processes it by computing the product L_T to issue the decision. This node is called the *fusion center*; see Fig. 1a. Before proceeding further, note that the only information needed from sensor

$i \in \{1, 2, \ldots, k\}$ includes the functions $\beta_i(\cdot)$ and $\nu_i(\cdot)$ together with the single real number $L_T(i)$. Put another way, once the problem parameters $\beta_i(\cdot)$, $\nu_i(\cdot)$ and the local likelihood ratios $L_T(i)$ are known, there is no increase in accuracy that can be obtained through knowledge of the sample path $t \mapsto N_t(i)$.

3 Deciding Without a Fusion Center

One of the drawbacks of the setup of Fig. 1a described above is the vulnerability of the system to targeted attacks or failures: if the fusion center is lost, the entire detection system collapses. This motivates one to study the following problem: Suppose that at the terminal time T (or just after), there is some sharing of the $L_T(i)$'s (and the problem parameters $\beta_i(\cdot)$, $\nu_i(\cdot)$) among the sensors according to a directed graph, as depicted in Fig. 1b. Assume that no single sensor has access to all local likelihood ratios, i.e., there is no obvious choice of fusion center. If each sensor is now enabled to act as a DM operating on the information available to it, one can ask which DM is the most reliable in terms of decision-making accuracy.

To formulate this problem precisely, a directed graph is first specified. This graph encodes the allowed communication among the sensors. Recall that a directed graph is an ordered pair $G = (V, A)$ where V is a set of vertices or nodes, and A is a set of ordered pairs of nodes, referred to as arcs or directed edges. For the problem at hand, the vertex set $V = \{1, 2, \ldots, k\}$ indexes the set of observers (sensors). The arcs in A correspond to inter-sensor communication in the sense that $(i, j) \in A$ if, and only if, there is directional flow of information *from* sensor i *to* sensor j. Thus, if there is two-way communication between sensors i and j, both (i, j) and (j, i) are included in A. Self-loops are meaningless in this context and therefore excluded. Finally, it will be convenient to assume that information travels exactly one directed edge, and no further. In other words, if there are directed edges (i, j) and (j, ℓ) in A, it is assumed that $L_T(i)$ is sent from sensor i to sensor j and no further, while $L_T(j)$ is sent from sensor j to sensor ℓ and no further.[1]

The probabilistic setup and notation are exactly as in Sect. 2. The quantities $L_T(i)$ and L_T are defined as in (1), (2); and each sensor $i \in \{1, 2, \ldots, k\}$ observes $N_t(i)$ over time interval $t \in [0, T]$ and computes the quantity $L_T(i)$ at time T. Now that each sensor is armed with its own $L_T(\cdot)$, inter-sensor communication takes place. For each sensor i, let $\mathscr{S}_i$ comprise the set of sensors whose information is made available to sensor i just after time T. Thus, for $1 \leq i \leq k$,

$$\mathscr{S}_i \triangleq \{j \in \{1, 2, \ldots, k\} : \text{Sensor } i \text{ knows } L_T(j) \text{ and } \beta_j(\cdot), \nu_j(\cdot) \text{ just after time } T\} \ .$$

Since a sensor always has access to its own information, we have $i \in \mathscr{S}_i$ for all $1 \leq i \leq k$. Thus, $\mathscr{S}_i$ consists of the index i, together with the indices corresponding to *incoming* edges. Once inter-sensor communication has taken place, each sensor

[1] This entails no loss of generality; indeed, if information is to be sent from sensor i to sensor ℓ, this can be accommodated at the expense of introducing the additional directed edge (i, ℓ).

can be considered a DM. Thus, for $1 \leq i \leq k$, DM(i) refers to sensor i once it has access to the quantities $\{L_T(j), \beta_j(\cdot), \nu_j(\cdot) : j \in \mathscr{S}_i\}$. Letting DM($i$) use the test

$$\{\mathscr{L}_T(i) \geq \gamma_i\} \quad \text{where} \quad \mathscr{L}_T(i) \triangleq \prod_{j \in \mathscr{S}_i} L_T(j) \quad \text{and } \gamma_i > 0$$

the probabilities of false alarm and miss for DM(i) are given by

$$P_{F,i} \triangleq \mathbb{P}_0 \{\mathscr{L}_T(i) \geq \gamma_i\} \ , \qquad P_{M,i} \triangleq \mathbb{P}_1 \{\mathscr{L}_T(i) < \gamma_i\} \ ,$$

respectively. If function $\Lambda_i : \mathbb{R} \to \mathbb{R}$ is defined by $\Lambda_i(p) \triangleq \log \mathbb{E}_0 \left[(\mathscr{L}_T(i))^p\right]$ for $p \in \mathbb{R}$, then it follows [12, Theorem 8] that

$$P_{F,i} \leq \exp\left(\inf_{p>0}[\Lambda_i(p) - p\eta_i]\right) \ , \quad P_{M,i} \leq \exp\left(\inf_{p<1}[\Lambda_i(p) + (1-p)\eta_i]\right) \ , \tag{3}$$

where $\eta_i \triangleq \log \gamma_i \in \mathbb{R}$, and $\Lambda_i(p)$ is explicitly computable via

$$\Lambda_i(p) = \sum_{j \in \mathscr{S}_i} \int_0^T [\mu_j(s)^p - p\mu_j(s) + p - 1]\beta_j(s)\mathrm{d}s \ . \tag{4}$$

Note that the bounds (3), (4) hold for *any* $\eta_i = \log \gamma_i \in \mathbb{R}$. The reader is referred to [12] for a detailed proof of the bounds in (3).

In order to effectively use these bounds, one would need to know where, if at all, the infima in (3) are realized, and further, whether the infima are negative—to ensure that the bounds are non-trivial. One proceeds here by making repeated use of convexity of the functions $p \mapsto \Lambda_i(p)$. Indeed, it follows [12, Lemmas 16–19] that the function $p \mapsto \Lambda_i(p)$ is C^2 with derivatives given by differentiating under the integral sign; further, $\Lambda_i''(p) > 0$ for all $p \in \mathbb{R}$, implying that the function $p \mapsto \Lambda_i(p)$ is strictly convex; and finally that $\Lambda_i'(0) < 0$, $\Lambda_i'(1) > 0$. It now follows [12, Proposition 13] that if η_i is chosen to lie in $(\Lambda_i'(0), \Lambda_i'(1))$, then the infima in (3) are attained at a unique $p_i^* \in (0, 1)$ and the infima are negative. More precisely, there exists a unique $p_i^* \in (0, 1)$ given by $\Lambda_i'(p_i^*) = \eta_i$ such that

$$\inf_{p>0}[\Lambda_i(p) - p\eta_i] = \mathscr{E}_{F,i}(p_i^*) < 0 \ , \qquad \inf_{p<1}[\Lambda_i(p) + (1-p)\eta_i] = \mathscr{E}_{M,i}(p_i^*) < 0 \ ,$$

where the *error exponents* $\mathscr{E}_{F,i}(p)$, $\mathscr{E}_{M,i}(p)$ mapping $(0, 1)$ to $\mathbb{R}$ are

$$\mathscr{E}_{F,i}(p) \triangleq \Lambda_i(p) - p\Lambda_i'(p) \quad \text{and} \quad \mathscr{E}_{M,i}(p) \triangleq \Lambda_i(p) + (1-p)\Lambda_i'(p) \ . \tag{5}$$

Thus, if $\eta_i \in (\Lambda_i'(0), \Lambda_i'(1))$, then the tightest error probability bounds for DM(i) are given by

$$P_{F,i} \leq \exp[\mathscr{E}_{F,i}(p_i^*)] < 1 \ , \qquad P_{M,i} \leq \exp[\mathscr{E}_{M,i}(p_i^*)] < 1 \ . \tag{6}$$

To rank the DMs in terms of their capacity to make accurate decisions, one *ideally* solves the following problem: Let $\alpha \in (0, 1)$ (acceptable upper bound on the probability of false alarm) be given. Allowing each DM to choose its own threshold to comply with the constraints that $P_{F,i} \leq \alpha$ and $P_{M,i}$ is minimized, rank the nodes in increasing order of $P_{M,i}$. The node with the smallest $P_{M,i}$ is the best DM, at least for the particular α. The challenge, as noted earlier, is that the error probabilities $P_{F,i}$ and $P_{M,i}$ are not amenable to analytic computation. We therefore work with the corresponding Chernoff bounds (6) and study the problem stated above with $P_{F,i}$ and $P_{M,i}$ replaced by the corresponding tightest upper bounds.

To solve this problem, the threshold selection algorithm [12, Proposition 14] is employed. The algorithm implies that if, for some $1 \leq i \leq k$, we have $\log \alpha > -\Lambda_i'(1)$, then there exists a unique $p_i^\dagger \in (0, 1)$ which solves the equation

$$\mathscr{E}_{F,i}(p_i^\dagger) = \log \alpha \ . \tag{7}$$

Moreover, $p_i^\dagger$ minimizes $\mathscr{E}_{M,i}(p)$ over all $p \in (0, 1)$ which satisfy $\mathscr{E}_{F,i}(p) \leq \log \alpha$. This minimum value of $\mathscr{E}_{M,i}(p)$ is given by

$$\mathscr{E}_{M,i}(p_i^\dagger) = \log \alpha + \Lambda_i'(p_i^\dagger) \ .$$

Thus, choosing $\eta_i = \Lambda_i'(p_i^\dagger)$, i.e. letting DM($i$) select the threshold $\gamma_i = \exp(\Lambda_i'(p_i^\dagger))$, yields

$$P_{F,i} \leq \alpha \ , \qquad P_{M,i} \leq \alpha \exp(\Lambda_i'(p_i^\dagger)) = \alpha \gamma_i \ .$$

To provide some insight on the foregoing expressions—see [12] for details—one can use convexity to show that on (0, 1), the functions $\mathscr{E}_{F,i}$, $\mathscr{E}_{M,i}$ are differentiable and negative, with $\mathscr{E}_{F,i}$ being strictly decreasing while $\mathscr{E}_{M,i}$ is strictly increasing. The threshold selection algorithm can be summarized as follows: Take the threshold $\gamma_i = \exp(\Lambda_i'(p))$ and select $p \in (0, 1)$ as small as possible to make $\mathscr{E}_{M,i}$ as small as possible, while ensuring that the false alarm constraint is met, i.e., $\mathscr{E}_{F,i}(p) \leq \log \alpha$. The latter is possible only if $\log \alpha$ is strictly greater than the infimum of $\mathscr{E}_{F,i}$ on (0, 1), which is easily computed to be $-\Lambda_i'(1)$. Of course, one should now find $p_i^\dagger$ by solving (7).

One can now rank the DMs from most reliable to least reliable by ranking the quantities $\mathscr{E}_{M,i}(p_i^\dagger)$, $1 \leq i \leq k$ (or, equivalently, $\exp(\mathscr{E}_{M,i}(p_i^\dagger)) = \alpha\gamma_i$) from smallest to largest. These findings are summarized as follows.

Theorem 1 *For* $\alpha \in (0, 1)$ *with* $\log \alpha > \max_{1 \leq i \leq k}(-\Lambda_i'(1))$, *define the functions* $\mathsf{M}_i(\alpha)$, $1 \leq i \leq k$, *by*

$$\mathsf{M}_i(\alpha) \triangleq \exp(\Lambda_i'(p_i^\dagger)) \ , \tag{8}$$

where $p_i^\dagger \in (0, 1)$ *solves* $\mathscr{E}_{F,i}(p_i^\dagger) = \log \alpha$*; thus,* $\mathsf{M}_i(\alpha)$ *is the threshold at DM(*i*). Then,* $\mathsf{M}_i(\alpha)$ *can be used as an index for decision-making accuracy in the following sense: If* $(i_1, i_2, \ldots, i_k)$ *is an ordering of* $\{1, 2, \ldots, k\}$ *such that*

$$\mathsf{M}_{i_1}(\alpha) \le \mathsf{M}_{i_2}(\alpha) \le \cdots \le \mathsf{M}_{i_k}(\alpha) \ ,$$

then $(i_1, i_2, \ldots, i_k)$ *provides a ranking of the DM's from most accurate to least accurate, as measured by the Chernoff bounds.*

Remark 1 Note that the accuracy index $\mathsf{M}_i(\alpha)$ is tied to the *specific* $\alpha \in (0, 1)$. This naturally prompts the question: for a given network and set of problem parameters, does the sensor which has the smallest $\mathsf{M}_i(\alpha)$ vary with α. It is also natural to ask, for a given network and set of problem parameters, whether the ranking based on $\mathsf{M}_i(\alpha)$ coincides with the ranking based on minimizing $P_{M,i}$ subject to $P_{F,i} \le \alpha$, i.e., do the Chernoff bounds *faithfully* capture the relative decision-making accuracies of various sensors. These questions are the subject of ongoing work.

4 Examples

This section presents two examples of networks of nuclear detectors deciding about the observed process. In the first case, it is assumed that the process at each sensor has identical characteristics, and heterogeneity is introduced only through the network topology. In the second example, in addition to the differences among sensors due to their location in the underlying interconnection graph, heterogeneity is introduced in the quality of each sensor's observations. The objective is to highlight the interplay between the network topology and the quality of individual observations, and its impact on the ability of individual sensors to make decisions.

4.1 Example 1: Identical Measurement Characteristics

This example addresses the following question. If the measurement quality characteristics at all sensors are the same, is it true that the DM with the most $L_T(\cdot)$'s is the most accurate? The answer, as will be seen below, is affirmative, implying that the node with the largest in-degree is better equipped to make a decision.

Indeed, suppose that $\beta_i \equiv \beta$, $\nu_i \equiv \nu$, $\mu_i \equiv \mu = 1 + \nu/\beta$. We will use $|\mathscr{S}_i|$ to denote the cardinality of $\mathscr{S}_i$, $1 \le i \le k$, that is, the in-degree of node i that corresponds to the number of incoming edges to i. It is easily seen from (4), (5), that $\mathscr{E}_{F,i}(p) = -g(p) \cdot |\mathscr{S}_i|$ where $g(p) \triangleq -\beta T[\mu^p - p\mu^p \log \mu - 1]$ and $\mathscr{E}_{M,i}(p) = -h(p) \cdot |\mathscr{S}_i|$ where $h(p) \triangleq -\beta T[\mu^p + (1-p)\mu^p \log \mu - \mu]$. Since $\mathscr{E}_{F,i}(p)$, $\mathscr{E}_{M,i}(p)$ are negative for $p \in (0, 1)$ with the former being strictly decreasing and the latter strictly increasing (see [12, Lemma 19]), it follows that $g(p)$ is positive and strictly increasing for $p \in (0, 1)$, while $h(p)$ is positive and strictly decreasing for $p \in (0, 1)$.

Fix α as in Theorem 1. Let $i, j \in \{1, 2, \ldots, k\}$ with $i \neq j$. The claim is that

1. If $|\mathscr{S}_i| > |\mathscr{S}_j|$, then $\mathsf{M}_i(\alpha) < \mathsf{M}_j(\alpha)$,
2. If $|\mathscr{S}_i| = |\mathscr{S}_j|$, then $\mathsf{M}_i(\alpha) = \mathsf{M}_j(\alpha)$,
3. If $|\mathscr{S}_i| < |\mathscr{S}_j|$, then $\mathsf{M}_i(\alpha) > \mathsf{M}_j(\alpha)$.

The first of these three claims will only be proven here; the arguments can be easily modified to prove the other two. Suppose $|\mathscr{S}_i| > |\mathscr{S}_j|$. By (7), one has $\mathscr{E}_{F,i}(p_i^\dagger) = \log\alpha$ and $\mathscr{E}_{F,j}(p_j^\dagger) = \log\alpha$. Recalling the notation above,

$$g(p_i^\dagger) = -\frac{\log\alpha}{|\mathscr{S}_i|} \quad \text{and} \quad g(p_j^\dagger) = -\frac{\log\alpha}{|\mathscr{S}_j|} \,.$$

Noting that $-\log\alpha > 0$ and $|\mathscr{S}_i| > |\mathscr{S}_j|$, one writes $g(p_i^\dagger) < g(p_j^\dagger)$. Since g is strictly increasing, it now follows that $p_i^\dagger < p_j^\dagger$. If it can be shown that $\mathscr{E}_{M,i}(p_i^\dagger) < \mathscr{E}_{M,j}(p_j^\dagger)$, then it follows (see Theorem 1) that $\mathsf{M}_i(\alpha) < \mathsf{M}_j(\alpha)$. The argument for establishing the former statement is as follows. Since the function h is strictly decreasing, it must be the case that $h(p_i^\dagger) > h(p_j^\dagger)$. Since h is positive and $|\mathscr{S}_i| > |\mathscr{S}_j|$ (by assumption), $|\mathscr{S}_i| \cdot h(p_i^\dagger) > |\mathscr{S}_j| \cdot h(p_j^\dagger)$. Taking negatives, yields $\mathscr{E}_{M,i}(p_i^\dagger) < \mathscr{E}_{M,j}(p_j^\dagger)$ as required.

This example shows that, under the assumption that the measurement process has identical statistical characteristics at each node, the nodes with the largest in-degree are the most accurate DMs. It is interesting to note that this observation is in contrast to the diffusive models studied in [14, 15], in which local centrality measures such as those based on nodal degrees cannot capture the certainty of each unit in terms of the collected evidence.

4.2 Example 2: Non-identical Measurement Characteristics

Particular examples of directed graphs in the setting of Fig. 1b help examine the effect of heterogeneity not only in the underlying communication topology, but also in the characteristics of the measurements process at each sensor. Similarly to [9], five sensors are arranged along a straight line at fixed locations with 0.5 m distance from each other and with the first sensor positioned at 0.5 m. A radioactive source moves parallel to this array of sensors with a constant speed of 0.03 m/s. As a result of the source's motion, the measurement characteristics among different sensors are not the same, since certain sensors may spend longer intervals in close proximity to the source than others. The source's straight line path is at a distance of 0.5 m above the array of the sensors, and the initial source position is 0.5 m behind the sensor array. The activity of the source and background radiation are measured in gamma rays emitted per second, i.e., counts per second (cps). For the source, it is taken at $a = 3$ cps, while for the background we assume 0.167 cps. The numerical simulation of the

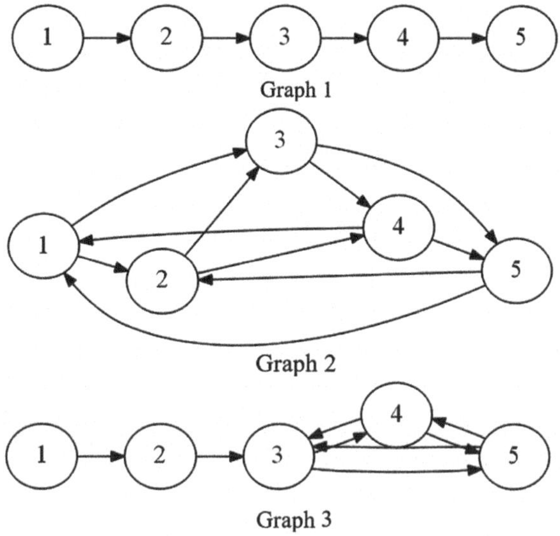

Fig. 2 Different communication topologies studied in Example 2. A directed edge from node i to node j indicates that at time T, the locally computed L_T and μ at node i is transmitted to node j

corresponding arrival process with the aforementioned intensities has been generated using a thinning algorithm [7, 13]. The maximum acceptable probability of false alarm is taken as $\alpha = 10^{-3}$ and for numerical purposes the sensor cross-section is assumed to be 1 m^2. Note that these figures are too big for practical applications, and are used here only for reasons of numerical convenience to emphasize the differences in sensor performance; see [12] for details.

Figure 2 depicts three different interconnection topologies with which locally processed information can be disseminated among sensors. Directed edges mark unidirectional flow of likelihood ratios L_T and the histories of the intensities β, ν between (network) adjacent sensors at time T. These three communication topologies (graphs) result in different sensor network performance in terms of accuracy of decision making, at least to the degree that the latter can be reflected on the computed bounds on the probability of missed detection. Various detection scenarios are depicted in Figs. 3 and 4, showing that for a constant background activity, the performance of each sensor as a decision-making unit depends on the strength of the signals perceived by the sensors and the time available to make the decision. In these figures, individual sensors are denoted by S1 to S5, and the motion of the source is indicated by the blue line at the top of each figure. Different decision times T are examined, resulting to different time intervals over which the source remains in the close vicinity of each sensor, thereby introducing heterogeneity in the measurement characteristics of each sensor. In each of the Figs. 3 and 4, the vertical axis corresponds to the logarithm of the bound of the probability of missed detection computed by (6) and the horizontal axis shows the distance between sensors.

Figure 3 shows that the performance of the sensors under more information sharing (graph 2 in Fig. 2) is better than their performance under less information sharing (graph 1 in Fig. 2). A major advantage of exchanging information among individual

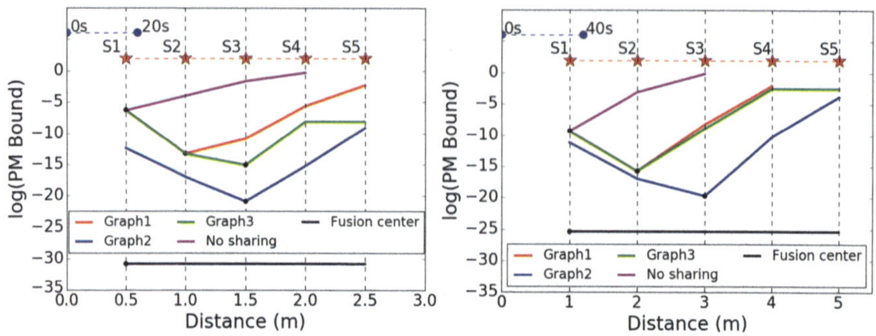

Fig. 3 The communication topologies of Fig. 2 result in different bounds for the probability of missed detection for each sensor S1 to S5. These are bounded by the two extreme cases: no information is shared (magenta curve), and fusion center having data from all sensors (black curve). Missing line segments indicate that the corresponding sensors do not satisfy the constraint on the probability of false alarm

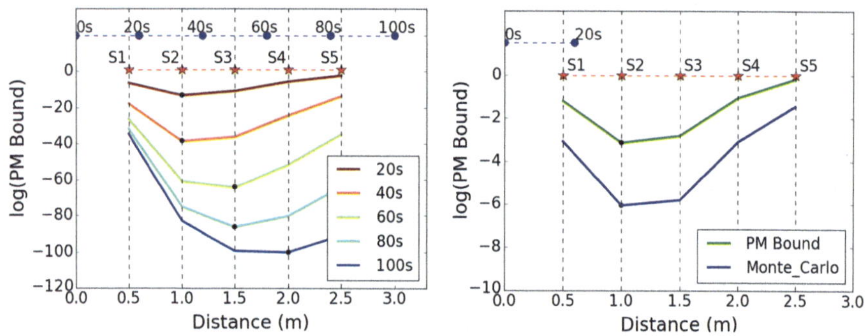

Fig. 4 The left plot shows the logarithm of the bound on the probability on missed detection with decision time T using graph 1 of Fig. 2. The right plot shows Monte-Carlo estimates of the probability of missed detection using graph 1 of Fig. 2

sensors is evident from the performance of node 3 in Fig. 3 (left), which corresponds to graph 3 of Fig. 2. Indeed, due to the larger distance from the source, sensor 3 neither makes good quality measurements nor it receives signals from sensor 1; yet it performs better than 1 or 2 because of its location in the graph. However, more incoming edges to a sensor does not always result into better performance. To see this, Fig. 3 (right) describes network performance when the spatial arrangement is scaled up by 2 such that all the distances are doubled (1 m instead of 0.5 m). Doubling the distances between sensors, results in weakening the observed signal, in the sense that the intensity of the arrival process perceived by each sensor decreases. In this case of larger inter-sensor distance, more sensors do not satisfy the constraint on the probability of false alarm at the end of the decision interval $T = 40\,\text{s}$, and the additional information available to sensor 3 due to its location in graph 3 does not compensate for the overall weak quality of the signals received by all the sensors.

The dependency of node ranking on the available decision time is illustrated in Fig. 4 (left) for the case of graph 1. It can be seen that as more time is allowed to make the decision, the location of the best performing sensor shifts in the direction of the motion of the source. This observation reflects the intuitive fact that the sensor that remains closer than the rest to the source during the detection time window performs better. As a final remark it should be mentioned that the aforementioned observations on node ranking in terms of their decision-making capability are based on the Chernoff bounds (6). To compare the prediction of the bounds with that obtained by the actual probabilities, Fig. 4 (right) depicts an estimate for the logarithm of the probability of missed detection obtained using Monte-Carlo simulations. In this particular example, the source intensity is assumed to be 1.75 cps and the network topology corresponds to graph 1 of Fig. 2. As can be seen, the prediction of the bounds is the same with that of the estimated probabilities of missed detection. It is important to note, however, that although the general trend of actual error probabilities is captured by the bounds [12], one must be aware that the loss of sharpness associated with the use of bounds may result in node rankings that do not faithfully reflect the actual ranking; for example, this is the case when neighboring sensors exhibit comparatively similar decision-making performance in a way that cannot be differentiated by the bounds.

5 Conclusions

The paper argues that the decision-making accuracy of individual nodes in a network of radiation detectors can be quantified using explicitly computed Chernoff upper bounds as proxies for true error probabilities. The resulting ranking of nodes can be used to choose which node should make the final decision in a distributed setup where each individual node can act as a potential decision maker. Examples show how the capability of each sensor to make a decision about the observed process depends on the interplay between the location of the sensor in the underlying network architecture (reflecting the quantity of information available to the sensor) and the spatial location of the sensors with respect to the source (reflecting the quality of individual sensor measurements). Performance comparable to that of decision making with a fusion center can be achieved by allowing partial information sharing between the nodes over some fixed, directional communication topology. This paper lays the foundation for future work in which the network performance is optimized by reconfiguring the underlying communication topology (cf. [1]) given the number of sensors, the strength of the signal, and the time available to make the decision.

Acknowledgements This work is supported in part by DTRA under award #HDTRA1-16-1-0039.

References

1. Bhambhani, V., Valbuena, L., Tanner, H.G.: Spatially distributed cellular neural networks. Int. J. Intell. Comput. Cybern. **4**(4), 465–486 (2011)
2. Bogacz, R., Brown, E., Moehlis, J., Holmes, P., Cohen, J.D.: The physics of optimal decision making: a formal analysis of models of performance in two-alternative forced-choice tasks. Psychol. Rev. **113**(4), 700–765 (2006)
3. Brémaud, P.: Point Processes and Queues. Martingale Dynamics. Springer, Berlin (1981)
4. Byrd, R., Moss, J., Priedhorsky, W., Pura, C., Richter, G.W., Saeger, K., Scarlett, W., Scott Jr., S., Wagner, R.: Nuclear detection to prevent or defeat clandestine nuclear attack. IEEE Sens. J. **5**(4), 593–609 (2005)
5. Fitch, K., Leonard, N.E.: Information centrality and optimal leader selection in noisy networks. In: IEEE Intenrational Conference on Decision and Control, pp. 7510–7515 (2013)
6. González, F.I., Bernard, E.N., Meinig, C., Eble, M.C., Mofjeld, H.O., Stalin, S.: The NTHMP tsunameter network. Nat. Hazards **35**, 25–39 (2005)
7. Lewis, P.A.W., Shedler, G.S.: Simulation of nonhomogeneous Poisson processes by thinning. Nav. Res. Logist. Q. **26**(3), 403–413 (1979)
8. Morreale, P., Qi, F., Croft, P.: A green wireless sensor network for environmental monitoring and risk identification. Int. J. Sens. Netw. **10**(1–2), 73–82 (2011)
9. Nemzek, R.J., Dreicer, J.S., Torney, D.C., Warnock, T.T.: Distributed sensor networks for detection of mobile radioactive sources. IEEE Trans. Nucl. Sci. **51**(4), 1693–1700 (2004)
10. Ogata, Y.: Seismicity analysis through point-process modeling: a review. Pure Appl. Geophys. **155**, 471–507 (1999)
11. Pahlajani, C.D., Poulakakis, I., Tanner, H.G.: Networked decision making for Poisson processes with applications to nuclear detection. IEEE Trans. Autom. Control **59**(1), 193–198 (2014)
12. Pahlajani, C.D., Sun, J., Poulakakis, I., Tanner, H.G.: Error probability bounds for nuclear detection: improving accuracy through controlled mobility. Automatica **50**(10), 2470–2481 (2014)
13. Pasupathy, R.: Generating nonhomogeneous Poisson processes. Wiley Encyclopedia of Operations Research and Management Science. Wiley, New York (2009)
14. Poulakakis, I., Scardovi, L., Leonard, N.E.: Node classification in collective evidence accumulation toward a decision. In: Proceedings of the IEEE International Conference on Decision and Control (2012)
15. Poulakakis, I., Young, G.F., Scardovi, L., Leonard, N.E.: Information centrality and ordering of nodes for accuracy in noisy decision-making networks. IEEE Trans. Autom. Control **61**(4), 1040–1046 (2016)
16. Srikrishna, D., Chari, A.N., Tisch, T.: Deterance of nuclear terrorism with mobile radiation detectors. Nonproliferation Rev. **12**(3), 573–614 (2005)
17. Stephenson, K., Zelen, M.: Rethinking centrality: methods and examples. Soc. Netw. **11**(1), 1–37 (1989)
18. Sundaresan, A., Varshney, P.K., Rao, N.S.V.: Distributed detection of a nuclear radioactive source using fusion of correlated decisions. In: Proceedings of the International Conference on Information Fusion, pp. 1–7. IEEE (2007)
19. Tenney, R.R., Sandell Jr., N.R.: Detection with distributed sensors. IEEE Trans. Aerosp. Electron. Syst. **17**(4), 501–510 (1981)
20. Tsitsiklis, J., Athans, M.: On the complexity of distributed decision problems. IEEE Trans. Autom. Control **AC-30**, 440–446 (1985)
21. Tsitsiklis, J.N.: Decentralized detection. Adv. Stat. Signal Process. **2**, 297–344 (1993)
22. Varshney, P.K.: Distributed Detections and Data Fusion. Springer, New York (1997)
23. Viswanathan, R., Varshney, P.K.: Distributed detection with multiple sensors: part I - fundamentals. Proc. IEEE **85**(1), 54–63 (1997)

Distributed Laplacian Eigenvalue and Eigenvector Estimation in Multi-robot Systems

Mehran Zareh, Lorenzo Sabattini and Cristian Secchi

Abstract In many multi-robot systems applications, obtaining the spectrum and the eigenvectors of the Laplacian matrix provides very useful information. For example, the second smallest eigenvalue, and the corresponding eigenvector, can be used for connectivity maintenance (see for example Freeman et al., Stability and convergence properties of dynamic average consensus estimators, 2006, [5]). Moreover, as shown in Zareh et al. (Decentralized biconnectivity conditions in multi-robot systems, 2016, [22], Enforcing biconnectivity in multi-robot systems, 2016, [23]), the third smallest eigenvalue provides a metric for ensuring robust connectivity in the presence of single robot failures. In this paper, we introduce a novel decentralized gradient based protocol to estimate the eigenvalues and the corresponding eigenvectors of the Laplacian matrix. The most significant advantage of this method is that there is no limit on the multiplicity of the eigenvalues. Simulations show the effectiveness of the theoretical findings.

1 Introduction

The last decade has witnessed a dramatically growing interest in multi-robot systems. The application of such systems can be found in environment monitoring, submarine exploration, and distributed sensing [11, 12, 17–19]. A multi-robot system is composed of distributed nodes, and each of them has the ability of sensing, processing, communicating, and moving. Recent technological advances have led to the emergence of pervasive networks of small, low-power devices that integrate sensors and

M. Zareh · L. Sabattini (✉) · C. Secchi
Department of Sciences and Methods for Engineering (DISMI),
University of Modena and Reggio Emilia, Modena and Reggio Emilia, Italy
e-mail: lorenzo.sabattini@unimore.it

M. Zareh
e-mail: mehran.zareh@unimore.it

C. Secchi
e-mail: cristian.secchi@unimore.it

R. Groß et al. (eds.), *Distributed Autonomous Robotic Systems*,
Springer Proceedings in Advanced Robotics 6,
https://doi.org/10.1007/978-3-319-73008-0_14

actuators with limited on-board processing and wireless communication capabilities. Several researches in the control system community are dedicated to provide algorithms for coordination and estimation in networks of multi-robot systems [4–6, 20, 21].

The network topology of multi-robot systems can be described by a graph, in which nodes represent robots and edges represent links between the nodes. The behavior of such systems depend both on the interactions between the nodes and on the network topology. Some useful characteristics of a graph can be provided using algebraic graph theory tools. For instance, the spectrum of the Laplacian matrix associated to a graph can be used to estimate topological properties such as connectivity, average degree, diameter, spectral gap, biconnectivity, etc [2]. In the context of multi-robot systems, these quantities provide relevant information regarding the performance of the overall system. As stated in [13], the algebraic connectivity, i.e., the second smallest eigenvalue, is a fundamental parameter to estimate the worst case convergence rate of consensus algorithms. The algebraic connectivity can also be exploited, as shown in [14], for implementing connectivity maintenance control strategies. Conditions for enforcing biconnectivity, based on the third smallest eigenvalue of the Laplacian matrix, are introduced in [22, 23]. Algebraic conditions for ensuring k-connectivity are discussed in [10].

While several centralized methods exist in the literature that allow to computing the spectrum of the Laplacian matrix [1, 3], there are only a few articles that present decentralized approaches. In [16], the authors proposed a distributed estimator for the algebraic connectivity of a weighted Laplacian matrix based on the power iteration algorithm, that exploits the estimation of the corresponding eigenvector. Since this approach is based on power iteration and matrix deflation, its convergence is slow and requires a large computational effort. A decentralized algorithm to estimate the eigenvalues of the Laplacian matrix of undirected graphs is proposed in [4], based on local interaction rules among the nodes defined in such a way that their state trajectory is a linear combination of sinusoids oscillating only at frequencies function of the eigenvalues of the Laplacian matrix. Eigenvalues are then found by using standard signal processing techniques. Using this method, only the value of the eigenvalues can be determined and the multiplicity of each eigenvalue is not given. Reference [15] presents an algorithm for estimating eigenvalues of the Laplacian matrix associated with the graph describing the network topology a multi-agent system, while [8] studies a constrained consensus optimization problem for estimating the Laplacian eigenvalues of the network graph. None of these works discuss eigenvector estimation problem.

Also the eigenvectors of the Laplacian matrix provide some information about the graph. For example, [9] studies the relation between the eigenvector associated to the second smallest eigenvalue, or the Fiedler vector, and the Perron components of a graph. In [10], k-connectivity based on the elements of the Fiedler vector is studied. Reference [16] proposes a gradient controller to maximize the value of the second smallest eigenvalue, in which the controller is formed from the Fiedler vector. Very recently, in [23], we presented a controller, which is a function of the eigenvector associated to the third smallest eigenvalue of the Laplacian matrix, to enforce biconnectivity in the network.

In this paper we develop a novel distributed estimator to find all the eigenvalues and eigenvectors of the Laplacian matrix simultaneously. Each node runs a local consensus protocol to estimate an eigenvalue, and the associated eigenvector, of the Laplacian matrix. Once the estimation error has become sufficiently small, the nodes renovate this protocol to obtain another eigenvalue-eigenvector pair. This continues until each node estimates all the eigenvalues and the associated eigenvectors. A major novelty of this paper, with respect to the existing literature, is that the eigenvectors and the eigenvalues are computed all at the same time. Moreover, the simulations show that the estimation procedure can converge very rapidly if proper gains are selected and the computational loads are not constrained. The convergence rate and the computational constraints are left for the future studies. In addition, this method has the benefit of giving the algebraic multiplicity for each eigenvalue (for example, the method introduced in [4] does not determine the algebraic multiplicity).

The outline of the paper is as follows. In Sect. 2 we introduce notations and some basic notions on graph theory, which will be used in this work. The problem statement is introduced in Sect. 3. Section 4 provides the main contribution of this paper. We provide some theorems on decentralized eigenvector and eigenvalue estimation. In Sect. 5, the simulation results are given to verify the theoretical findings. Finally, in Sect. 6, we conclude the paper and describe the open problems.

2 Preliminaries and Notations

The topology of bidirectional communication channels among the robots is represented by an undirected graph $\mathcal{G} = (\mathcal{V}, \mathcal{E})$ where $\mathcal{V} = \{1, \dots, n\}$ is the set of nodes (robots) and $\mathcal{E} \subset \mathcal{V} \times \mathcal{V}$ is the set of edges. An edge $(i, j) \in \mathcal{E}$ exists if there is a communication channel between robots i and j. Self loops (i, i) are not considered. The set of robot i's neighbors is denoted by $\mathcal{N}_i = \{j \ : \ (j, i) \in \mathcal{E}; j = 1, \dots, n\}$. The network graph $\mathcal{G}$ is encoded by the so-called *adjacency matrix*, an $n \times n$ matrix A whose (i, j)th entry a_{ij} is greater than 0 if $(i, j) \in \mathcal{E}$, 0 otherwise. Obviously, in an undirected graph matrix A is symmetric. The degree matrix is defined as $D = \text{diag}(d_1, d_2, \dots, d_n)$ where $d_i = \sum_{j=1}^{n} a_{ij}$ is the degree of node i. The Laplacian matrix of a graph is defined as $\mathcal{L} = D - A$. The Laplacian matrix of a graph has several structural properties. It has non-negative real eigenvalues for any undirected graph $\mathcal{G}$. Furthermore, let $\mathbf{1}$ and $\mathbf{0}$ be, respectively, the vectors of ones and zeros with proper dimensions, then $\mathcal{L}\mathbf{1} = \mathbf{0}$ and $\mathbf{1}^T\mathcal{L} = \mathbf{0}^T$. If the weights a_{ij} are not equal to 1, the graph and the associated adjacency and Laplacian matrices are called weighted. Denote by $\lambda_i(\cdot)$ the ith leftmost eigenvalue, and by $v_i(\cdot)$ and $w_i(\cdot)$ the right and left eigenvectors associated with $\lambda_i(\cdot)$. In this way, the eigenvalues of the Laplacian matrix can be ordered as $0 = \lambda_1(\mathcal{L}) \le \lambda_2(\mathcal{L}) \le \cdots \le \lambda_n(\mathcal{L})$. In $\mathcal{G}$ a node i is reachable from a node j if there exists an undirected path from j to i. If $\mathcal{G}$ is connected then $\mathcal{L}$ is a symmetric positive semi-definite irreducible matrix. Moreover, the algebraic multiplicity of the null eigenvalue of $\mathcal{L}$ is one.

3 Problem Statement

We study the problem of distributed estimation of the eigenvalues and eigenvectors of the Laplacian matrix in multi-robot systems. Communications are assumed to be between each robot and its 1-hop neighbors. The connectivity of the initial network is also assumed. Then, we raise the following problem.

Problem 1 For a multi-robot system with a connected undirected interaction graph $\mathscr{G}$, find a decentralized protocol to estimate the eigenvalues and eigenvectors of the Laplacian matrix.

4 Proposed Algorithm

In this section we introduce an algorithm for estimating, in a decentralized manner, the eigenvalues of the Laplacian matrix and the corresponding eigenvectors. The proposed method can be summarized as follows:

- For estimating n eigenvalue-eigenvector pairs, n steps are necessary. Each step includes a run of a distributed estimator.
- At the cth step, the ith robot computes α_i as the estimate of the eigenvalue λ_c, and z_i as the estimate of the corresponding eigenvector v_c. Note that the eigenvalue-eigenvector pairs are computed without following any specific order.
- The cth step concludes when the estimator reaches a steady state. The estimator is then re-initialized, and a new eigenvalue-eigenvector pair is estimated.

We will now describe in details the proposed methodology.

Assume that, in a multi-robot system, the network graph $\mathscr{G}$ is connected and undirected. Define by α_i the estimation of the eigenvalue λ_c of $\mathscr{L}$ performed by the robot i at step c. Let $\tilde{\mathscr{L}}_i(\alpha_i) = \mathscr{L} - \alpha_i I$. From the eigenvector properties [7], we know that the eigenvectors of $\mathscr{L}$ and $\tilde{\mathscr{L}}_i(\alpha_i)$ are identical. Denote by $\tilde{l}^i \in \mathbb{R}^n$ the ith column of $\tilde{\mathscr{L}}_i(\alpha_i)$. Let $P_i(\alpha_i) = \tilde{l}^i \tilde{l}^{i^T}$, $i = 1, \ldots, n$, and define a block-diagonal matrix $P(\alpha) = \text{diag}(P_1(\alpha_1), \ldots, P_n(\alpha_n))$.

Now, consider the following decentralized estimator

$$\dot{z}_i(t) = k_z \left[\sum_{j=1}^{n} a_{ij}(z_j(t) - z_i(t)) - P_i z_i(t) \right], \quad i = 1, \ldots, n, \tag{1}$$

in which z_i, $i = 1, \ldots, n$ is the ith agent's normalized estimation for the eigenvector v_c of $\mathscr{L}$, at step c. $k_z \in \mathbb{R}^+$ is the estimator gain. The estimate of the eigenvalue-eigenvector pair is achieved by means of the following protocol:

$$\dot{\alpha}_i(t) = k_\alpha \left[\sum_{j=1}^{n} a_{ij}(\alpha_j(t) - \alpha_i(t)) - 2z_{ii}\tilde{l}^{i^T} z_i(t) \right], \tag{2}$$

where z_{ii} is the ith element of z_i, and $k_\alpha \in \mathbb{R}^+$. We can rewrite the above equations in state-space form as

$$\begin{cases} \dot{z}(t) = -k_z M(\alpha)z(t), \\ \dot{\alpha} = -k_\alpha(\mathscr{L}\alpha(t) - G(\alpha)z(t)). \end{cases} \tag{3}$$

in which $z = [z_1^T, \ldots, z_n^T]^T$, $M(\alpha) = \mathscr{L} \otimes I + P(\alpha)$, $\alpha = [\alpha_1, \ldots, \alpha_n]^T$, and

$$G(\alpha) = 2\text{diag}(z_{11}\tilde{l}^{1^T}, z_{22}\tilde{l}^{2^T}, \ldots, z_{nn}\tilde{l}^{n^T}).$$

We need the following lemmas to explain some properties of $M(\alpha)$.

Lemma 1 *$P(\alpha)$ is a positive semi-definite matrix.*

Proof According to its definition, $P(\alpha)$ is a symmetric matrix, and we can rewrite it as

$$P(\alpha) = P_0^T(\alpha)P_0(\alpha),$$

where $P_0(\alpha) = \text{diag}(\tilde{l}^{1^T}, \ldots, \tilde{l}^{n^T})$. Consequently

$$x^T P(\alpha)x = x^T P_0^T(\alpha)P_0(\alpha)x = (xP_0(\alpha))^T(xP_0(\alpha)) \geq 0.$$

This completes the proof. □

In the next lemma we use the fact that any two eigenvectors of any real symmetric matrix are perpendicular.

Lemma 2 *The eigenvalues of $\mathscr{L} \otimes I$ are achieved by n-times repeating each eigenvalue of $\mathscr{L}$, i.e.,*

$$\lambda_{kn+1}(\mathscr{L} \otimes I) = \cdots = \lambda_{(k+1)n}(\mathscr{L} \otimes I) = \lambda_k(\mathscr{L}), \ k = 0, \ldots, n-1.$$

The set $\{v_k(\mathscr{L}) \otimes v_l(\mathscr{L}), \ l = 1, \ldots, n\}$ forms an orthogonal basis for the eigenspace of $\mathscr{L} \otimes I$ associated with $\lambda_k(\mathscr{L})$, $k = 1, \ldots, n$.

Proof See the proof of Lemma 2 in [23]. □

Corollary 1 *For a connected undirected graph $\mathscr{G}$, the set $\{\mathbf{1} \otimes v_1(\mathscr{L}), \ldots, \mathbf{1} \otimes v_n(\mathscr{L})\}$ forms an orthogonal basis for the kernel of $\mathscr{L} \otimes I$.*

Proof Since G is a connected undirected graph, $\mathscr{L}$ has a unique null eigenvalue and, based on Lemma 2, $\mathscr{L} \otimes I$ has n null eigenvalues. Let $v_1(\mathscr{L}) = \mathbf{1}$ be the first eigenvector. The result then follows from straightforward computations. □

Lemma 3 *For a connected graph $\mathscr{G}$, the kernels of $\mathscr{L} \otimes I$ and $P(\alpha)$ intersect if and only if $\alpha = \lambda_k(\mathscr{L})\mathbf{1},\ k = 1, \ldots, n$. This unique intersection lies in span$(\mathbf{1} \otimes v_k(\mathscr{L}))$.*

Proof By direct calculation, we can show that, for $\alpha = \lambda_k\mathbf{1}$, span$(\mathbf{1} \otimes v_k(\mathscr{L}))$ is in the kernel of $P(\alpha)$ and, based on Corollary 1, it is in the intersection of the kernels of $\mathscr{L} \otimes I$ and P.

Now, using contradiction, we show that this intersection is unique. Suppose that the intersection of the kernels is not unique. Suppose $\alpha_i \neq \lambda_m(\mathscr{L})\mathbf{1}$ and let span$\{\mathbf{1} \otimes v_m(\mathscr{L})\},\ m \in \{i = 1, \ldots, n, m \neq k\}$, be another intersection for the kernels of $\mathscr{L} \otimes I$ and P. Then $P(\mathbf{1} \otimes v_m(\mathscr{L})) = 0$. From the definition of P we get that

$$\tilde{l}^i\tilde{l}^{i^T} v_m(\mathscr{L}) = \tilde{l}^i \left(\tilde{l}^{i^T} v_m(\mathscr{L})\right) = 0,\ i = 1, \ldots, n,$$

which is the product of a vector and a scalar. Since the graph $\mathscr{G}$ is connected, $\tilde{l}^i \neq 0$, and hence

$$\tilde{l}^{i^T} v_m(\mathscr{L}) = 0,\ \Rightarrow\ l^{i^T} v_m(\mathscr{L}) = \alpha_i v_{m_i}(\mathscr{L}),\ i = 1 \ldots, n.$$

which gives

$$\lambda_m(\mathscr{L})v_{m_i}(\mathscr{L}) = \alpha_i v_{m_i}(\mathscr{L}),\ i = 1, \ldots, n,$$

or

$$\lambda_m(\mathscr{L})v_m(\mathscr{L}) = \text{diag}(\alpha)v_m(\mathscr{L}). \tag{4}$$

If $\alpha_i \neq \lambda_m(\mathscr{L}),\ i = 1, \ldots, n$, then the unique solution for the above equation is $v_m(\mathscr{L}) = \mathbf{0}$, which is not acceptable. As a consequence, the intersection of the kernels of $\mathscr{L}$ and $P(\alpha)$ exists, if and only if $\alpha = \lambda_m(\mathscr{L})\mathbf{1},\ m = 1, \ldots, n$, and this intersection that lies in span$(\mathbf{1} \otimes v_m(\mathscr{L}))$. □

Lemma 4 *For a connected undirected graph $\mathscr{G}$, the matrix $M(\alpha)$ in (3) is positive semi-definite, and has, at most, one null eigenvalue, and z lies in the kernel of $M(\alpha)$ if and only if $z \in$ span$\{\mathbf{1} \otimes v_m(\mathscr{L})\}$ and $\alpha = \lambda_m(\mathscr{L})\mathbf{1}$, $m = 1, \ldots, n$.*

Proof For any non-zero normalized vector $x \in \mathbb{R}^{n^2},\ x^T x = 1$, the Rayleigh quotient [7] of M is defined as

$$R\left(M(\alpha), x\right) = x^T M(\alpha)x. \tag{5}$$

Let $\gamma_1 \leq \cdots \leq \gamma_{n^2}$ be eigenvalues of M. Since $M(\alpha)$, $\mathscr{L} \otimes I$, and P are symmetric matrices, and hence Hermitian, from the min-max theorem [7] we get

$$\begin{aligned}\gamma_1 &= \min\{R(M(\alpha), x) : x \neq 0\} = \min\{R(\mathscr{L} \otimes I, x) + R(P(\alpha), x) : x \neq 0\}\\ &= \min\{R(\mathscr{L} \otimes I, x) : x \neq 0\} + \min\{R(P(\alpha), x) : x \neq 0\}\\ &= \lambda_{min}(\mathscr{L}) + \lambda_{min}(P) = 0.\end{aligned}$$

From Lemma 2, we know that $\mathbf{1} \otimes v_j, j = 1, \ldots, n$, forms an orthogonal basis for the kernel of $\mathscr{L} \otimes I$. From Lemma 3, we know that the kernels of $\mathscr{L} \otimes I$ and P intersect if and only if $\alpha = \lambda_m(\mathscr{L})\mathbf{1},\ m = 1, \ldots, n$, and otherwise there is no intersection. This means that

$$\gamma_1 = \min\{R(M(\alpha), x)\} = 0$$

if and only if x is in the intersection of the kernels of P and $\mathscr{L} \otimes I$ which, from Lemma 3, is $x \in \text{span}\{\mathbf{1} \otimes v_m(\mathscr{L})\}$. □

In the next step, by adding an additional term to the estimator in (3), we propose a protocol that leads the system to estimate the eigenvalues and eigenvectors of the Laplacian matrix. Consider

$$\dot{z}_i(t) = k_z \left(\sum_{j=1}^{n} a_{ij}(z_j(t) - z_i(t)) - k_z \left[P_i + \sum_{k \in \mathscr{S}} v_k(\mathscr{L}) v_k^T(\mathscr{L}) \right] z_i(t) \right), \quad i = 1, \ldots, n, \tag{6}$$

where $\mathscr{S} \subset \{1, \ldots, n\}$. $\alpha_i(t)$ is obtained from (2). In the state-space form we get

$$\begin{cases} \dot{z}(t) = -k_z M_1(\alpha) z(t), \\ \dot{\alpha} = -k_\alpha(\mathscr{L}\alpha(t) - G(\alpha) z(t)), \end{cases} \tag{7}$$

where

$$M_1(\alpha) = M(\alpha) + M_2,$$

with

$$M_2 = I \otimes \left[\sum_{k \in \mathscr{S}} v_k(\mathscr{L}) v_k^T(\mathscr{L}) \right].$$

Notice that the protocol (6) uses only locally known information, and hence, it is a decentralized estimator. The next lemmas describe some properties of $M_1(\alpha)$ and M_2.

Lemma 5 *M_2 is a positive semi-definite matrix and any $x = \mathbf{1} \otimes v_m$, where $m \notin \mathscr{S}$ lies in the kernel of M_2.*

Proof Note that M_2 is a block diagonal matrix with all blocks $\sum_{k \in \mathscr{S}} v_k(\mathscr{L}) v_k^T(\mathscr{L})$ positive semi-definite. Hence, M_2 is a positive semi-definite matrix. The second part can be shown by direct calculations. □

Lemma 6 *For a connected graph $\mathscr{G}$, the matrix $M_1(\alpha)$ in (3) is positive semi-definite, and has, at most, one null eigenvalue. Moreover, z lies in the kernel of $M(\alpha)$ if and only if $z \in$ span$\{\mathbf{1} \otimes v_m(\mathscr{L})\}$ and $\alpha = \lambda_m(\mathscr{L})\mathbf{1}$, $m \notin \mathscr{S}$.*

Proof Based on Lemmas 4 and 5, $M_1(\alpha)$ is a sum of two positive semi-definite matrices, hence is a positive semi-definite matrix. Similar to what we showed in the proof of Lemma 4, for any non-zero normalized vector $x \in \mathbb{R}^{n^2}$, $x^T x = 1$, define the Rayleigh quotient [7] of M_1 as

$$R(M_1(\alpha), x) = x^T M_1(\alpha)x = x^T M(\alpha)x + x^T M_2 x. \tag{8}$$

From Lemma 4, $M(\alpha)$ has at most one null eigenvalue, and therefore, $R(M_1(\alpha), x)$ may be equal to zero if $x \in \ker(M(\alpha))$. Hence, $M_1(\alpha)$ can get at most one null eigenvalue where, due to Lemma 4, it occurs for $x = 1 \otimes v_m$ and $\alpha = \lambda_m(\mathscr{L})\mathbf{1}$. Now from Lemma 5, we can conclude that the intersection of the kernels of $M(\alpha)$ and M_2 is

$$x = \mathbf{1} \otimes v_m(\mathscr{L}),\ m \in \{1, \ldots, n\} - \mathscr{S}.$$

Therefore, $M_1(\alpha)$ is a positive semi-definite matrix with at most one null eigenvalue and the kernel which is given by the above equation. □

Now, we are ready to show the stability of the estimators in (3) and (7).

Theorem 1 *If the graph $\mathscr{G}$ is undirected and connected, the system in* (3) *is asymptotically stable, and the estimator reaches*

$$\lim_{t\to\infty} z_i = \mathbf{1} \otimes v_m(\mathscr{L}),\ and\ \lim_{t\to\infty} \alpha_i = \lambda_m(\mathscr{L}),\ m \in \{1, \ldots, n\} - \mathscr{S}$$

where $\lambda_m(\mathscr{L})$ and $v_m(\mathscr{L})$ are a pair of associated eigenvalue and eigenvector of the $\mathscr{L}$.

Proof From (7), we know that the equilibrium points must be located in the kernel of $M_1(\alpha)$. Again, due to the fact that the vectors z_i are normalized, we get

$$z^T z = z_1^T z_1 + \cdots + z_n^T z_n = n,$$

So the point $z = 0$ cannot be an equilibrium point. Then, according to Lemma 6, the equilibrium points of the system in (7) are $z = \text{span}(\mathbf{1} \otimes v_m(\mathscr{L}))$ and $\alpha = \lambda_m(\mathscr{L})\mathbf{1}$ where $m \in \{1, \ldots, n\} - \mathscr{S}$.

Consider the following candidate potential function

$$V = k_z z^T(t) M_1(\alpha) z(t) + k_\alpha \alpha^T(t) \mathscr{L} \alpha(t). \tag{9}$$

Notice that

$$\frac{\partial V}{\partial z} = k_z M_1(\alpha) z,$$

and

$$\frac{\partial V}{\partial \alpha} = k_\alpha \mathscr{L}\alpha + k_\alpha \frac{\partial}{\partial \alpha}(z^T M(\alpha) z) = k_\alpha \mathscr{L}\alpha + k_\alpha \frac{\partial}{\partial \alpha}(z^T(\mathscr{L} \otimes I)z) + \frac{\partial}{\partial \alpha}(z^T P(\alpha) z). \tag{10}$$

From the definition of $P(\alpha)$ we get

$$\frac{\partial}{\partial \alpha}(z^T P(\alpha)z) = \frac{\partial}{\partial \alpha}(z^T P(\alpha)z) = \frac{\partial}{\partial \alpha}\sum_{i=1}^{n}(z_i^T P_i(\alpha_i)z_i) = \sum_{i=1}^{n}\left(z_i^T \frac{\partial P_i(\alpha_i)}{\partial \alpha_i} z_i\right) \tag{11}$$

Let l^i be the ith column of $\mathscr{L}$ and μ_i be a vector with all elements equal to zero, except for the ith element which is equal to 1. From the definition of l_i we have $\tilde{l}^i = l^i - \alpha_i \mu_i$

$$P_i(\alpha_i) = l^i l^{i^T} - \alpha_i(l^i \mu_i^T + \mu_i l^{i^T}) + \alpha_i^2 \mu_i \mu_i^T.$$

Derivation with respect to α_i gives

$$\frac{\partial P_i(\alpha_i)}{\partial \alpha_i} = -(l^i \mu_i^T + \mu_i l^{i^T}) + 2\alpha_i \mu_i \mu_i^T.$$

Hence

$$z_i^T \frac{\partial P_i(\alpha_i)}{\partial \alpha_i} z_i = -2z_{ii} l^{i^T} z_i + 2\alpha_i z_{ii}^2 = -2z_{ii}\tilde{l}^{i^T} z_i. \tag{12}$$

By replacing from (12) in (10), we get

$$\frac{\partial V}{\partial \alpha} = k_\alpha \mathscr{L}\alpha - G(\alpha)z. \tag{13}$$

In order to guarantee convergence to a minimum of the potential function V in (9), we need to show its time derivative $\dot{V}$ to be non-positive. It is worth noting that the proposed estimator in (7) implements a gradient descent of the potential function V. In fact:

$$\begin{cases} \dot{z}(t) = -k_z M_1 z = -\dfrac{\partial V}{\partial z} \\ \dot{\alpha}(t) = -k_\alpha(\mathscr{L}\alpha - G(\alpha)z) = -\dfrac{\partial V}{\partial \alpha}. \end{cases}$$

We can then ensure that

$$\dot{V} = \frac{\partial V}{\partial z}\dot{z} + \frac{\partial V}{\partial \alpha}\dot{\alpha} = -\dot{z}^2(t) - \dot{\alpha}^2(t) \le 0.$$

In addition, notice that $\dot{V}$ is zero if and only if $\dot{z} = 0$ and $\dot{\alpha} = 0$, i.e., when the system reaches the equilibrium point. This proves that the estimator (7) asymptotically converges to the equilibrium point $z = \mathbf{1} \otimes v_m(\mathscr{L})$ and $\alpha = \lambda_m(\mathscr{L})\mathbf{1}$, $m \in \{1, \ldots, n\} - \mathscr{S}$. □

Now, we introduce our main algorithm to estimate the eigenvalues and eigenvectors of the Laplacian matrix.

As the diagram in Fig. 1 shows, each node i first runs the protocol (7) with $M_1 = M$ until the distance from the equilibrium point (here we measured the distance

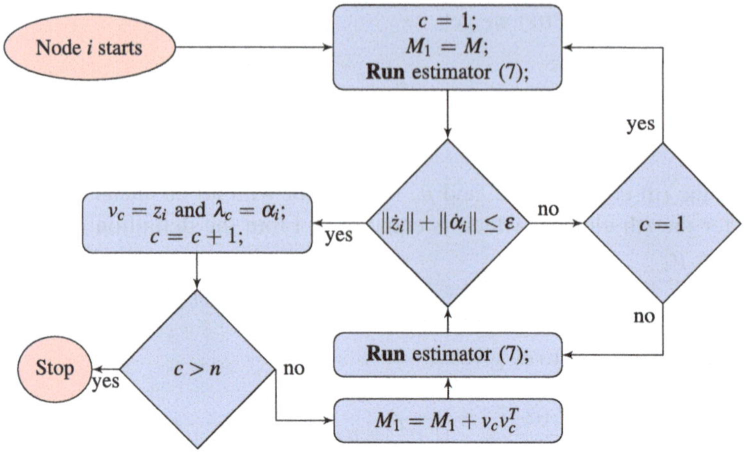

Fig. 1 Distributed eigenvalue and eigenvector estimation algorithm

by $\|\dot{z}_i\| + \|\dot{\alpha}_i\|$) gets a sufficiently small value ε. Since $\mathscr{S}$ is an empty set, from Theorem 1, this leads to an estimate of a random eigenvalue and its associated eigenvector. Parameter c is a counter and counts the number of eigenvalues already calculated. Based on Theorem 1, α_i converges to an eigenvalue of $\mathscr{L}$ and z_i converges to the corresponding eigenvector. Without loosing generality, we call the estimated eigenvalue λ_c and the eigenvector v_c at each step. Notice that this does not indicate the order of the eigenvalues from the minimum to the maximum. In order to find another eigenvalue and eigenvector, node i implements the protocol (7) by updating $M_1 = M_1 + v_c v_c^T$. As shown in Theorem 1, the estimator (7) is asymptotically stable and α_i converges to an eigenvalue of $\mathscr{L}$, which is different from the previously estimated $\lambda_1, \ldots, \lambda_c$. Accordingly, z_i converges to the corresponding eigenvector. Again, when the distance from the equilibrium point gets close enough to zero, the counter is increased by 1. This procedure is continued until the counter is equal to n, that is when each agent has estimated all the eigenvalues and eigenvectors of $\mathscr{L}$.

Remark 1 In this algorithm, we do not have any assumption on the algebraic multiplicity. Notice that each eigenvalue can be repeated several times. Since $\mathscr{L}$ is a real symmetric matrices, is has orthogonal eigenvectors, and the conditions in Lemmas 5, 6, and Theorem 1 hold. This means that we can achieve orthogonal (orthonormal in our case) eigenvectors associated with the equal eigenvalues.

Remark 2 Based on the connectivity assumption, $\lambda_1(\mathscr{L}) = 0$ is always a simple eigenvalue of $\mathscr{L}$, and $v_1 = \mathbf{1}/\sqrt{n}$ is the corresponding normalized eigenvector. Hence, we exclude this eigenvalue by initially updating $M_1 = M + \mathbf{1}\mathbf{1}^T/n$.

Remark 3 At each step, the eigenvector obtained from the previous step is used. Therefore, an accumulative error can be expected. Moreover, from the asymptotic stability, one cannot estimate any bounded convergence time. We leave these open problems, as well as the effect of k_z and k_α, for the future studies.

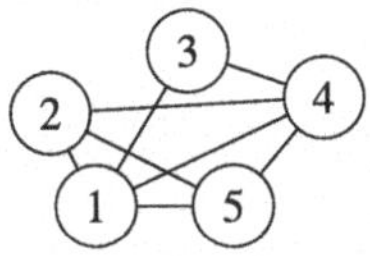

(a) Unweighted graph in Example 1

$$\mathscr{L} = \begin{bmatrix} 1.0288 & 0 & -0.0055 & -0.0148 & -0.4501 & -0.0174 & -0.5334 & -0.0076 \\ 0 & 0.8480 & -0.4320 & -0.2045 & 0 & -0.2046 & -0.0025 & -0.0044 \\ -0.0055 & -0.4320 & 2.1480 & -0.7064 & -0.0097 & -0.8531 & -0.0868 & -0.0545 \\ -0.0148 & -0.2045 & -0.7064 & 1.6826 & -0.0092 & -0.5354 & -0.2004 & -0.0119 \\ -0.4501 & 0 & -0.0097 & -0.0092 & 0.8623 & -0.0410 & -0.2329 & -0.1194 \\ -0.0174 & -0.2046 & -0.8531 & -0.5354 & -0.0410 & 1.9691 & -0.1525 & -0.1651 \\ -0.5334 & -0.0025 & -0.0868 & -0.2004 & -0.2329 & -0.1525 & 1.2260 & -0.0175 \\ -0.0076 & -0.0044 & -0.0545 & -0.0119 & -0.1194 & -0.1651 & -0.0175 & 0.3804 \end{bmatrix}$$

(b) Weighted graph in Example 2

Fig. 2 Examples

5 Simulations

In this section we provide simulation results. Two different cases are demonstrated. In the first one, a small unweighted graph with some non-simple eigenvalues is given. Conversely, the second one presents a larger weighted graph.

Example 1 Consider the network graph given in Fig. 2a. We suppose that the edge weights are all equal to one. The eigenvalues of the Laplacian matrix are 0, 3, 3, 5, 5. Select $k_z = 1$ and $k_\alpha = 10$. Figure 3 demonstrates the estimated eigenvalues and eigenvectors in the order in which the algorithm calculated them, which is random and varies from one simulation to another. In our future work, we want to study the relation between the initial values and the estimated eigenvectors and eigenvalues.

Example 2 Now, we use the proposed algorithm to estimate the eigenvalues and eigenvectors of a weighted graph. Consider the Laplacian matrix in Fig. 2b. The estimator parameters are selected $k_z = 20$ and $k_\alpha = 50$. Due to space limitations, only the first two estimated eigenvalues and eigenvectors are shown in Fig. 4a–d, and the estimation error for the other eigenvalues and eigenvectors are given in the Table in Fig. 4e. The order of the estimated eigenvalues and eigenvectors is kept as obtained from the algorithm.

The results show that the estimation errors are sufficiently small. This errors depend on the initial values and the estimator gains. We leave finding the relation between the errors and the mentioned parameters for the future works.

6 Conclusions

In this paper we developed a novel distributed algorithm to estimate all the eigenvalues and their corresponding eigenvectors of the Laplacian matrix. The major advantage of the proposed algorithm with respect to the main approaches in the literature is that it is possible to estimate simultaneously both eigenvalues and the eigenvectors. In addition, there is no limit on the multiplicity of the eigenvalues. The algorithm was tested for weighted and unweighted graphs. In each of the cases the

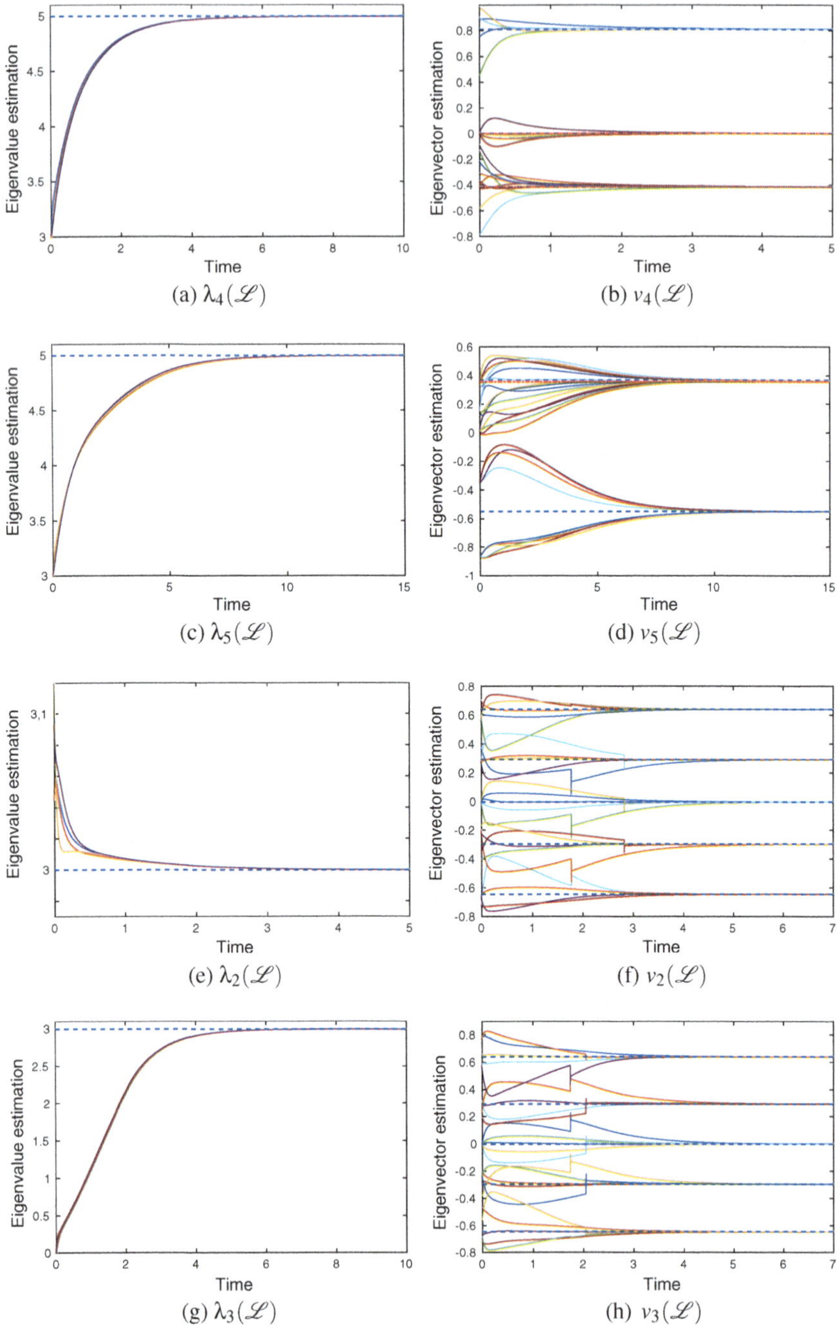

Fig. 3 Laplacian eigenvalue and eigenvector in Example 1 (estimated (solid) and exact value (dashed))

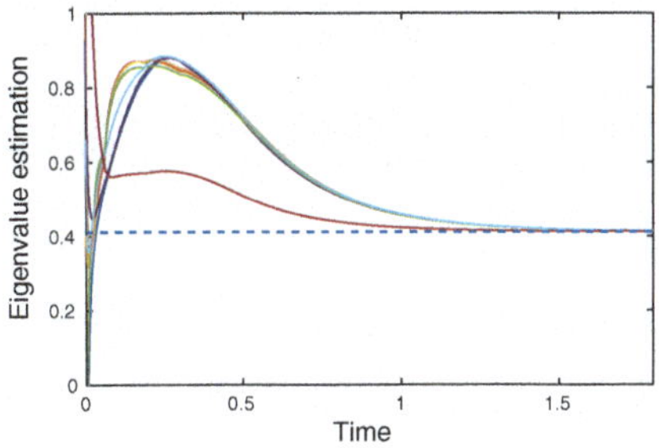

(a) $\lambda_3(\mathscr{L})$

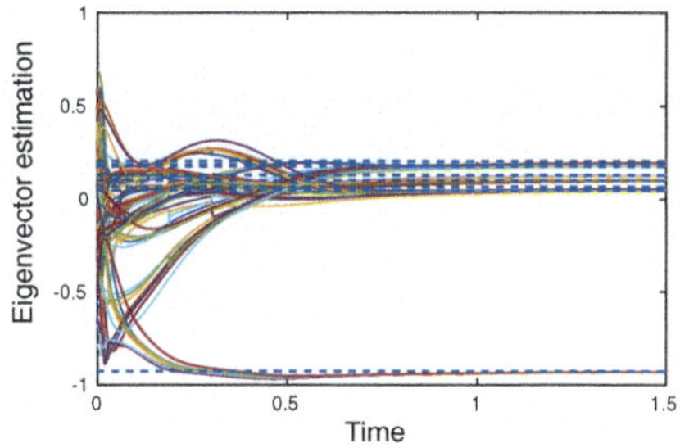

(b) $v_3(\mathscr{L})$

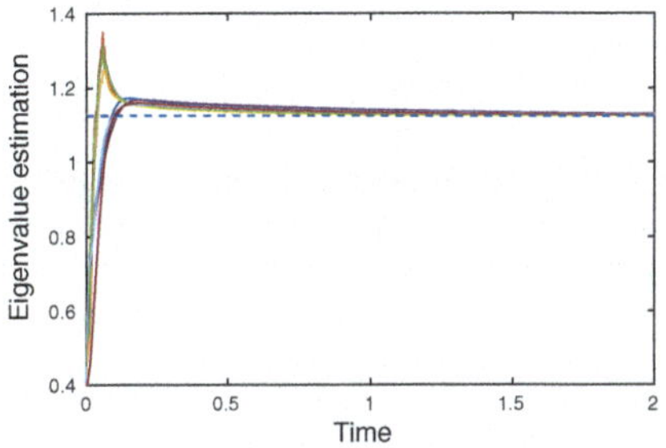

(c) $\lambda_4(\mathscr{L})$

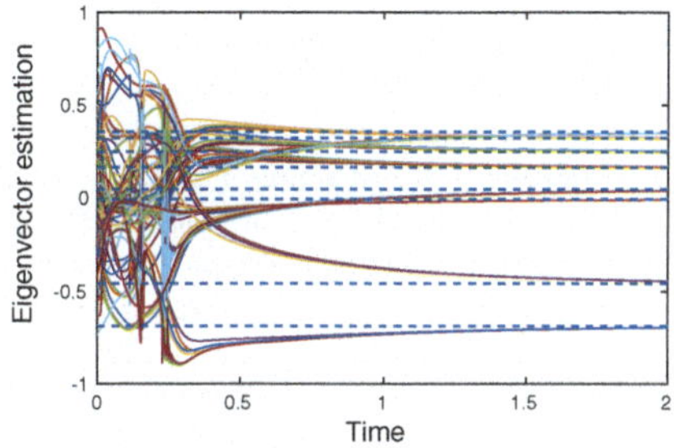

(d) $v_4(\mathscr{L})$

	Estimation error		Estimation error
$\lambda_5(\mathscr{L})$	4.04e-6	$v_5(\mathscr{L})$	3.03e-6
$\lambda_2(\mathscr{L})$	1.65e-6	$v_2(\mathscr{L})$	1.10e-6
$\lambda_7(\mathscr{L})$	4.71e-7	$v_7(\mathscr{L})$	1.98e-7
$\lambda_6(\mathscr{L})$	1.24e-6	$v_7(\mathscr{L})$	3.52 e-7
$\lambda_8(\mathscr{L})$	7.54e-7	$v_8(\mathscr{L})$	3.97e-7

(e)

Fig. 4 Laplacian eigenvalue and eigenvector in Example 2 (estimated (solid) and exact value (dashed))

estimation errors were very small. The relation between the errors and the design parameters and the initial values for the estimator, as well as the convergence rate, are left for the future studies.

References

1. Bapat, R.B., Pati, S.: Algebraic connectivity and the characteristic set of a graph. Linear Multilinear Algebra **45**(2–3), 247–273 (1998)
2. Bapat, R.B., Lal, A.K., Pati, S.: On algebraic connectivity of graphs with at most two points of articulation in each block. Linear Multilinear Algebra **60**(4), 415–432 (2012)
3. Fiedler, M.: Algebraic connectivity of graphs. Czechoslov. Math. J. **23**(2), 298–305 (1973)

4. Franceschelli, M., Gasparri, A., Giua, A., Seatzu, C.: Decentralized estimation of Laplacian eigenvalues in multi-agent systems. Automatica **49**(4), 1031–1036 (2013)
5. Freeman, R., Yang, P., Lynch, K.: Stability and convergence properties of dynamic average consensus estimators. In: 2006 45th IEEE Conference on Decision and Control, pp. 338–343 (2006). https://doi.org/10.1109/CDC.2006.377078
6. Guo, J.M.: The Laplacian spectral radius of a graph under perturbation. Comput. Math. Appl. **54**(5), 709–720 (2007)
7. Horn, R.A., Johnson, C.R.: Matrix Analysis. Cambridge University Press, Cambridge (2012)
8. Kibangou, A.Y., et al.: Distributed estimation of Laplacian eigenvalues via constrained consensus optimization problems. Syst. Control Lett. **80**, 56–62 (2015)
9. Kirkland, S., Fallat, S.: Perron components and algebraic connectivity for weighted graphs. Linear Multilinear Algebra **44**(2), 131–148 (1998)
10. Kirkland, S., Rocha, I., Trevisan, V.: Algebraic connectivity of k-connected graphs. Czechoslov. Math. J. **65**(1), 219–236 (2015)
11. Olfati-Saber, R.: Flocking for multi-agent dynamic systems: algorithms and theory. IEEE Trans. Autom. Control **51**(3), 401–420 (2006)
12. Olfati-Saber, R., Shamma, J.S.: Consensus filters for sensor networks and distributed sensor fusion. In: 44th IEEE Conference on Decision and Control, 2005 and 2005 European Control Conference. CDC-ECC'05, pp. 6698–6703. IEEE (2005)
13. Olfati-Saber, R., Fax, A., Murray, R.M.: Consensus and cooperation in networked multi-agent systems. Proc. IEEE **95**(1), 215–233 (2007)
14. Sabattini, L., Chopra, N., Secchi, C.: Decentralized connectivity maintenance for cooperative control of mobile robotic systems. Int. J. Robot. Res. **32**(12), 1411–1423 (2013)
15. Tran, T.M.D., Kibangou, A.Y.: Consensus-based distributed estimation of Laplacian eigenvalues of undirected graphs. In: 12th Biannual European Control Conference (ECC 2013), pp. 227–232 (2013)
16. Yang, P., Freeman, R.A., Gordon, G.J., Lynch, K.M., Srinivasa, S.S., Sukthankar, R.: Decentralized estimation and control of graph connectivity for mobile sensor networks. Automatica **46**(2), 390–396 (2010)
17. Zareh, M.: Consensus in multi-agent systems with time-delays. Ph.D. thesis, University of Cagliari (2015)
18. Zareh, M., Seatzu, C., Franceschelli, M.: Consensus of second-order multi-agent systems with time delays and slow switching topology. In: 2013 10th IEEE International Conference on Networking, Sensing and Control (ICNSC), pp. 269–275 (2013). https://doi.org/10.1109/ICNSC.2013.6548749
19. Zareh, M., Seatzu, C., Franceschelli, M.: Consensus on the average in arbitrary directed network topologies with time-delays. In: 4th IFAC Workshop on Distributed Estimation and Control in Networked Systems, pp. 342–347 (2013). https://doi.org/10.3182/20130925-2-DE-4044.00022
20. Zareh, M., Dimarogonas, D.V., Franceschelli, M., Johansson, K.H., Seatzu, C.: Consensus in multi-agent systems with non-periodic sampled-data exchange and uncertain network topology. In: 2014 International Conference on Control, Decision and Information Technologies (CoDIT), pp. 411–416. IEEE (2014)
21. Zareh, M., Dimarogonas, D.V., Franceschelli, M., Johansson, K.H., Seatzu, C.: Consensus in multi-agent systems with second-order dynamics and non-periodic sampled-data exchange. In: Proceedings of the 2014 IEEE Emerging Technology and Factory Automation (ETFA), pp. 1–8. IEEE (2014)
22. Zareh, M., Secchi, C., Sabattini, L.: Decentralized biconnectivity conditions in multi-robot systems. In: IEEE International Conference on Decision and Control (CDC). IEEE (2016)
23. Zareh, M., Secchi, C., Sabattini, L.: Enforcing biconnectivity in multi-robot systems. In: IEEE International Conference on Decision and Control (CDC). IEEE (2016)

Distributed Object Characterization with Local Sensing by a Multi-robot System

Golnaz Habibi, Sándor P. Fekete, Zachary Kingston and James McLurkin

Abstract This paper presents two distributed algorithms for enabling a swarm of robots with local sensing and local coordinates to estimate the dimensions and orientation of an unknown complex polygonal object, i.e., its minimum and maximum width and its main axis. Our first approach is based on a robust heuristic of distributed Principal Component Analysis (DPCA), while the second is based on turning the idea of Rotating Calipers into a distributed algorithm (DRC). We simulate DRC and DPCA methods and test DPCA on real robots. The result show our algorithms successfully estimate the dimension and orientation of convex or concave objects with a reasonable error in the presence of noisy data.

1 Introduction

Computing the geometric features of an object has many applications in robotics and autonomous manufacturing. Collective transport, imaging, fitting the bounding box, assembly and manipulation are some examples that involve object characterization. In collective transport, agents require the dimensions of the object and the orientation to transport the object safely through narrow corridors or environments with obstacles [6]. Another application is to use object shape and orientation to manipulate an object by a robot or by an industrial arm. If we can measure the object's main axes vectors (which implies the object orientation) as well as the object's main dimensions, fitting

G. Habibi (✉) · Z. Kingston · J. McLurkin
Rice University, Houston, TX 77005, USA
e-mail: golnaz.habibi@gmail.com

Z. Kingston
e-mail: zk11@rice.edu

J. McLurkin
e-mail: mclurkin@rice.edu

S. P. Fekete
TU Braunschweig, Braunschweig, Germany
e-mail: s.fekete@tu-bs.de

R. Groß et al. (eds.), *Distributed Autonomous Robotic Systems*,
Springer Proceedings in Advanced Robotics 6,
https://doi.org/10.1007/978-3-319-73008-0_15

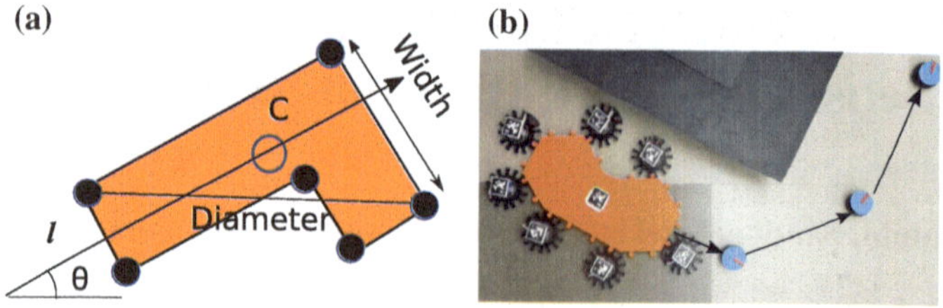

Fig. 1 **a** A polygonal object. The three key dimensions we need to measure are object minimum width (*width*), object maximum width (*diameter*), and orientation(θ). The centroid is marked with a blue circle. Major axis is vector l. **b** Seven r-one manipulator robots with grippers (black circles) [10] are moving the orange object. Our goal is to transport the object along the path marked by the guide robots (blue circles). In our previous work [6], we described algorithm that enable each guide robot to compute a collision-free pose for the object at the robots location. This path planning algorithm requires the *object dimension* to generate a safe path through narrow corridors. Previously, we presented distributed controllers that use the guide robots navigation information to control a group of manipulator robots [7]. In order to navigate the object and avoid obstacles, manipulator robots measure the *object orientation* and use it as a feedback to rotate the object during the transport. This paper presents distributed algorithms to estimate object dimensions and orientation

the bounding rectangle or bounding box is easy; the latter has its own applications in manufacturing and pattern recognition.

A common approach is to use Principal Component Analysis for finding the major axes of the object by finding the eigenvector of the points on the boundary [3] or inside the object [4, 8]. However, most of the previous work requires global sensing and central processing [3]. See Fekete et al. [4, 8] for the use of Distributed Principal Component Analysis in the context of a sensor grid of smart pixels with only local information and communication, based on very limited computation; however, [4, 8] makes use of the discretized grid geometry of the underlying sensor field, while this paper considers robot positions at a limited number of continuous locations.

We aim to address situations in which global sensing, communication, and geometry are not readily available or too costly to implement. Multi-robot systems offer the potential to estimate a global state by using distributed algorithms. In situations without GPS or global positioning, inferring global information from local information is essential. We use multi-hop communications [1] to exchange local information *and* local geometry in order to cooperate with other robots, avoiding the need for a shared global coordinate frame [9].

Using distributed algorithms, we can break a complex task into simple subproblems. In the same way, multi-robot systems are usually simple at the individual level, but they collaboratively accomplish a task that cannot be achieved by a single robot.

We consider a scenario in which there are only sensors or robots around the boundary of an object. We consider three key parameters of the object including (1) centroid or center of geometry, (2) object minimum and maximum dimension, and (3) object orientation (see Fig. 1a). These agents use local information and may not be fully connected. These include line-of-sight communication, so robots on different sides of the object cannot communicate directly. Calculating these features helps

Fig. 2 A potential application of object characterization in collective transport: by computing the centroid and orientation of the object, a swarm of manipulator robots can maneuver the object to valid free configurations and safely transport the object

in estimating the shape of a closed region by using sensors on the corners of the region with limited sensing. These features are also useful for manipulating a large polygonal object by robots without prior knowledge of the object.

Our objective is to estimate object geometric features (centroid, object dimension and orientation) in order to use them in collective transport problem, assuming a subgroup of robots attach themselves to the object [2, 12]. These robots, called *manipulator robots*, are equipped with grippers [7]. These are responsible for estimating the object dimensions, orientation and centroid. This information is provided to the rest of the robots that cover the environment [6]; we call these *guide robots*. Given the dimension of the object, guide robots generate a safe path from the object to the goal. The last step is to transport the object through the path. By computing the centroid and orientation of the object, the manipulator robots can maneuver and safely transport the object (see Fig. 2). Figure 1b shows a result of a collective transport. The path is generated based on the dimension of the object. During the transport, manipulator robots rotate the object to avoid the collision. The details of collective transport are out of the scope of this paper and are discussed in future work. We developed two distributed algorithms for estimating the centroid of the object in previous work [5, 7]. Here we present two algorithms for *object characterization*, which compute three key geometric properties of our polygonal object: its minimum width, diameter, and orientation. The minimum width determines the narrowest corridor the object can negotiate, assuming it is in the proper orientation. We define the orientation as the direction that is perpendicular to the object's minimum width. The object diameter determines the minimum distance from an obstacle at which it is safe to rotate the object.

Our first approach is Distributed Principal Component Analysis (DPCA). This is an approximation, but it is easy to implement on real hardware and produces good results for symmetric objects. Our second approach is a distributed version (called DRC) of a Rotating Calipers algorithm [13] that computes exact geometry if the manipulator robots are positioned at object vertices. We test our algorithms in simulation and on hardware. Our results are promising, the algorithm successfully estimates the dimension of most convex or concave objects.

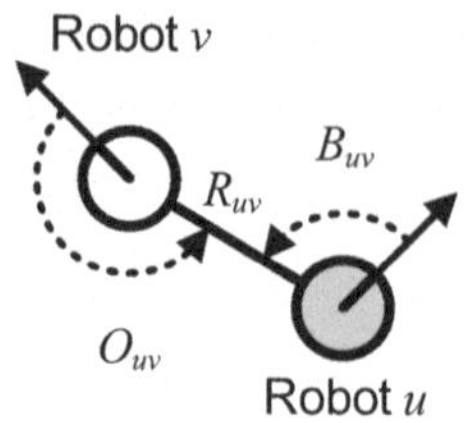

Fig. 3 Local network geometry of robot v measured from robot u. B_{uv} is the bearing, the relative angle of robot's heading u from robot v. O_{uv} is the orientation, the relative heading of robot v from robot u, and R_{uv} is the distance between two robots

The rest of the paper is organized as follows. The model and assumptions are explained in Sect. 2. We briefly describe our previous algorithm of average consensus in Sect. 3. Our object characterization algorithms are described in Sect. 4. Our results are discussed in Sect. 5, and we conclude the paper in Sect. 6.

2 Model and Assumptions

Object characterization can be divided into two steps. (1) Detect the boundary of the object. (2) Estimate the object parameter based on the boundary points. We assume the first is done by an existing algorithm [2, 12], and robots have already attached to the object, ideally at the vertices of the object, which we assume to be simply connected, so that there are no interior boundaries. The robots have no prior knowledge of the shape or size of the object. A communication network is built by the robots using inter-robot communications between nearby robots within a fixed distance d, where d is much smaller than the size of the environment. We can model the robot's communications network, $G = (V, E)$, as an undirected unit disk graph, obstructed by the geometry of the object. Each robot constitutes a node $u \in V$, where V is the set of all robots and E is the set of all robot-to-robot communication links. The set V_m is the set of all manipulator robots. The neighbors of each node u are the set of robots with unobstructed line of sight and communication range d of robot u, denoted $N(u) = \{v \in V \mid \{u, v\} \in E\}$. We assume that G is connected and that it contains a cycle that surrounds the object, so that a message sent by a robot to one of its neighbors can be passed all the way around until it reaches the other neighbor.

Our robots are homogeneous and are modeled as disks, moving with the help of a non-holonomic differential drive. Each robot u has a unique id, $u.id$, and is situated at the origin of its own local coordinate frame with the $\hat{x}$-axis aligned with its current heading. Robots can measure the relative pose of their neighbors (See Fig. 3). Note that the message-passing protocol used for Distributed Rotating Calipers works even in an asynchronous manner; for easier description, we still describe the algorithm execution as proceeding in a series of discrete *rounds*. While the robots actual operation is asynchronous, implementing a synchronizer simplifies analysis greatly and is easy to implement [9].

3 Pipelined Consensus

In some parts of this paper, we need to estimate the average of values, including the centroid. By definition, the centroid of a polygon with m vertices is the average position of its vertices:

$$(x_c, y_c) = \frac{1}{m}\sum_{i=1}^{m}(x_i, y_i) \tag{1}$$

Thanks to consensus-based algorithms, one can find the global estimate of variables such as max/min/average by continuously finding a local estimate of that value [11]. In our previous work, we presented a pipelined consensus algorithm, an extension of pairwise gossip-based consensus for multi-agent systems [5] to estimate global values including the object's centroid as the average of the object's vertices. As a result, each robot estimates the average value (i.e., its centroid) in its local coordinate; see [5] for more detail. In this approach, each agent starts a new consensus in each round of gossiping, and stores the intermediate results for the previous k consensus in a pipeline message with size k. After k rounds of gossiping, the results of the first consensus are ready. In this paper, pipelined consensus is used to estimate the object's centroid as well as estimating the other parameters that are essential for computing the object orientation. We will explain these key parameters in the next section.

4 Object Characterization

In this section we present two distributed algorithms for extracting key geometric features of the object including: object dimension (width and diameter) and object orientation (see Fig. 1a).

4.1 Distributed Principal Component Analysis (DPCA)

Principal Components Analysis allows us to compute the orientation of the main axis of the object using the vertices around the boundary [3]. We assume that the robots are in heading consensus [11], i.e., they have agreed to a common heading. This condition does not necessarily mean that the robots have the same orientation. Instead, the robots can reach a common *virtual* heading alignment. In the beginning, each robot estimates the position of the objects centroid at its reference frame by using the pipelined consensus algorithm [5]. In the next part, we show that this is sufficient information to estimate the object's dimension and orientation.

4.1.1 Computing the Object Orientation

Given the position of centroid and robots, Chaudhuri et al. [3] presented an algorithm to compute the object orientation θ, as follows.

$$\tan 2\theta = 2\frac{\sum_{i=1}^{m}(x_i - x_c)(y_i - y_c)}{\sum_{i=1}^{m}[(x_i - x_c)^2 - (y_i - y_c)^2]} \tag{2}$$

In this equation, $(x_i, y_i), i = 1, \ldots, m$ is the global position of vertices on the boundary of the object, and (x_c, y_c) is the centroid position in a global reference.

Lemma 1 *Given the object's centroid and a common heading for all robots (either virtual or real), the object orientation θ is calculated by Eq. (2) in the local coordinate frame of each robot. The communication complexity is $O(1)$ per robot, i.e., each robot a constant number of messages of constant size.*

Proof We have to show that the components of Eq. (2), i.e.,$(x_i - x_c)$ and $(y_i - y_c)$, are invariant with respect to local frames if the local frames reach the common heading. In other words, a vector in a coordinate frame does not change when it is transformed to another coordinate frame if the axes of two frames are parallel. To show this, we use homogeneous coordinates. Assume there are two different coordinate frames i and j with the same orientation. If the origin of i with respect to j is $(x_t, y_t, 1)$ and these frames are parallel, the vector **AB** in the i-coordinate, i.e. ${}^{i}\mathbf{AB} = (x_A - x_B, y_A - y_B, 1)$, is transformed to j-coordinates as follows. ${}^{j}_{i}T$ is the transformation matrix in homogeneous coordinates.

$${}^{j}_{i}T = \begin{bmatrix} 1 & 0 & x_{ij} \\ 0 & 1 & y_{ij} \\ 0 & 0 & 1 \end{bmatrix}, ({}^{j}\mathbf{AB}) = ({}^{j}_{i}T)({}^{i}\mathbf{AB}) = \begin{bmatrix} 1 & 0 & x_{ij} \\ 0 & 1 & y_{ij} \\ 0 & 0 & 1 \end{bmatrix}\left(\begin{bmatrix} x_A \\ y_A \\ 1 \end{bmatrix} - \begin{bmatrix} x_B \\ y_B \\ 1 \end{bmatrix}\right) \tag{3}$$

$$({}^{j}\mathbf{AB}) = \begin{bmatrix} x_A + x_{ij} \\ y_A + y_{ij} \\ 1 \end{bmatrix} - \begin{bmatrix} x_B + x_{ij} \\ y_B + y_{ij} \\ 1 \end{bmatrix} = \begin{bmatrix} x_A - x_B \\ y_B - y_B \\ 1 \end{bmatrix} = ({}^{i}\mathbf{AB}) \tag{4}$$

□

This gives us exactly the same vector in coordinate frame i. This holds for arbitrary points in space, as well as for centroid positions. Therefore, we can show the following.

$$(x_i - x_{c_i}, y_i - y_{c_i}, 1) = (x'_{c_i}, y'_{ci}, 1), i = 1, \ldots, m. \tag{5}$$

Here, (x_i, y_i) is the position of the robot i in the global reference, (x_{c_i}, y_{c_i}) is the estimated position of the centroid by robot i in global coordinate. (x'_{c_i}, y'_{c_i}) is the estimated position of the centroid in local coordinate i. This proves that vectors are invariant under translation transformation of endpoints of the vector. The second step is to show that Eq. (2) can be computed in a distributed fashion. By applying the invariance feature of centroid vector, Eq. (2) can be rewritten in local coordinates, as follows.

$$\tan(2\theta) = 2\frac{\frac{1}{m}\sum_{i=1}^{m}(x'_{c_i})(y'_{c_i})}{\frac{1}{m}\sum_{i=1}^{m}[(x'_{c_i})^2 - (y'_{c_i})^2]} \tag{6}$$

We have inserted $\frac{1}{m}$ in both enumerator and denominator to simplify this equation. By using the definition of the average $\overline{S} = \frac{1}{m}\sum_{i=1}^{m} s_i$, we get

$$\overline{[(x'_c)(y'_c)]} = \frac{1}{m}\sum_{i=1}^{m}[x'_{c_i} y'_{c_i}] \quad \text{and} \quad \overline{(x'^2_c - y'^2_c)} = \frac{1}{m}\sum_{i=1}^{m}[x'^{\,2}_{c_i} - y'^{\,2}_{c_i}]. \tag{7}$$

Equation (2) is simplified to

$$\tan(2\theta) = 2\overline{x'_c y'_c}/\overline{x'^2_c - y'^2_c}. \tag{8}$$

Equation (8) requires consensus algorithms to estimate the averages $\overline{x'_c y'_c}$ and $\overline{x'^2_c - y'^2_c}$. We use our pipelined consensus (see Sect. 3) to compute these averages, as well as the centroid vector on each robot $(\overline{x}, \overline{y})$ and η as the heading consensus. This is the total of five averages, requiring a message of a constant size $5W$ for each robot per round, where W is constant. Therefore, the message complexity is $O(1)$ per robot. □

4.1.2 Approximation of Diameter and Minimum Width by DPCA

Once we have estimated the orientation of the object, we calculate the minimum width and the diameter with two leader election algorithms [9]. Computing the diameter is straightforward: the robots run a leader election algorithm to find the largest centroid distance. Using the triangle inequality, we see that the diameter is bounded by twice this distance. Computing the minimum width requires a bit more work. Figure 4a shows how each robot u can compute its distance to the principal axis, $d_u = R_u \sin(\alpha_u)$. We define vector l as a vector passing through the centroid (C) and its orientation is θ. We also define $d_{max} = \max_{j=1}^{m} d_u$. If the polygon is symmetric across its principal axis and the vertices are balanced around the boundary, the

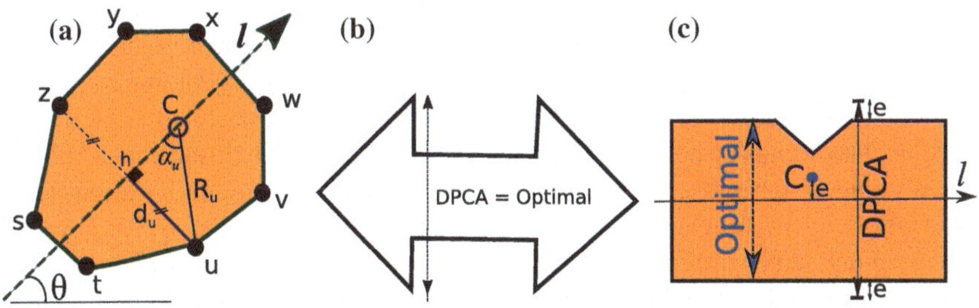

Fig. 4 **a** Geometric computation of distance from a vertex (robot) to the main axis l. **b** DPCA illustration on minimum-width estimation. The DPCA error estimate is zero and **c** the DPCA estimation error is bounded by $2e$ (top), where e is the offset of object centroid (letter C) from the principal axis l

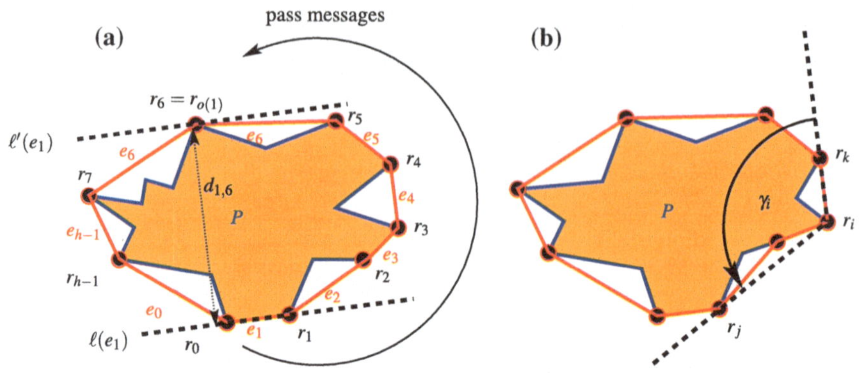

Fig. 5 **a** The basic setup for the algorithm DRC. The object contour is blue, convex hull edges are shown in red. Messages are passed along the chain of robots in order to compute and compare geometric information. **b** The aperture angle γ_i at a robot r_i. r_i is on the (strict) convex hull if and only if γ_i is at most (strictly smaller than) 180°

object's centroid lies on the principal axis (l). In that case, the minimum width of the polygon is exactly $2d_{max}$ for all robots u (Fig. 4b). Otherwise, the minimum width of the polygon is no greater than $2d_{max}$, which gives us an upper bound estimate of min-width $\leq 2max(d_u)$ (Fig. 4c). DPCA has a good performance for most objects and the straightforward implementation on physical systems.

4.2 Distributed Rotating Calipers

Tightening the bounds of our estimates requires more algorithmic machinery. We must determine the best of a quadratic number of pairs of object vertices, or pairs of vertices and object edges. In a centralized setting, reducing the computation time can be achieved with the method of *Rotating Calipers*. The idea was first conceived by Shamos [13], the name coined by Toussaint [14]; the key is to keep track of a pair of opposite tangents enclosing the object; updating the contact points during a full rotation gives rise to a (centralized) $\Theta(n)$ algorithm for computing minimum and maximum width, i.e., the diameter. In our distributed setting, a straightforward implementation ends up being quadratic, as a single update of opposite contact points requires long-distance communication, which may take $\Omega(n)$ communication steps. In the following, we develop a distributed variant with overall time $O(n)$. The key idea is to use pipelined communication along the perimeter of the object, with geometric updates performed on the fly, such that only the minimum and maximum width for each object vertex and each object edge are tracked. See Fig. 5a for the basic idea. This method yields exact results when we have accurate coordinate measurements, but requires a more sophisticated overall protocol. This model also uses the assumption that the robots can perceive any part of the object or any other robot that can be reached by an unobstructed line of sight.

4.2.1 Convex Hull

First, each robot determines whether it lies on the convex hull by checking the angle under which it sees the object P, based on the following lemma; see Fig. 5b.

Lemma 2 *A robot r_i on the perimeter of P is on the (strict) convex hull, if and only if it sees P within an aperture angle γ_i at most (strictly smaller than) 180°.*

This yields the set of h corner robots that lie on the boundary of the convex hull. In the following, we focus on communication between corner robots; implicitly, this may use non-hull robots as relays. We assume that, adjacent hull robots are connected, and non-hull robots lie between precisely between two hull robots, with direct access to other hull vertices blocked by the geometry of the object.

4.2.2 Computing Minimum Width

The minimum width of P is the the width of a narrowest corridor that can be passed by the object. This can be evaluated as follows.

Lemma 3 *Let P be a convex polygon with h vertices. Then for polygon vertices $r_0, \ldots, r_{h-1}$ and polygon edges $e_0 = (r_{h-1}, r_0), \ldots, e_{h-1} = (r_{h-2}, r_{h-1})$, the minimum width of P is $\min_{i=0}^{h-1} \max_{j=0}^{h-1} d_{i,j}$, where $d_{i,j} := d(\ell(e_i), r_j)$ is the Euclidean distance between the line $\ell(e_i)$ through edge e_i and r_j.*

The following observation is the basis for the idea of rotating calipers, i.e., parallel tangents at opposite sides of the polygon: a pair of opposite sides that attains minimum distance induces a pair of parallel tangents, i.e., a minimum-width corridor. For any edge e_i, we denote by $o(i)$ the corresponding "opposite" index, such that vertex $r_{o(i)}$ is the first one after r_i (in counterclockwise order) that attains $\max_{j=0}^{h-1} d_{i,j}$.

Lemma 4 *Let P be a convex polygon with h vertices. For i^* and j^* with $\min_{i=0}^{h-1} \max_{j=0}^{h-1} d_{i,j} = d(\ell(e_{i^*}), r_{j^*})$, there is a tangent, $\ell'((e)_{i^*})$, to P through $r_{j^*} = r_{o(i^*)}$ that is parallel to $\ell(e_{i^*})$, such that P lies between $\ell((e)_{i^*})$ and $\ell'((e)_{i^*})$.*

Based on this lemma, we describe a distributed algorithm. In the following, the *vertex description* D_i for a corner robot r_i consists of its own coordinates (x_i, y_i), along with the coordinates of both of its neighbors, (x_{i-1}, y_{i-1}) and (x_{i+1}, y_{i+1}). The angle α_i of (r_i, r_{i+1}) with the x-axis and the angle β_i of (r_i, r_{i-1}) with the x-axis can be deduced from this information; they describe the visibility cone in which robot r_i sees P. (In a practical setting, it is easiest to simply measure these angles, rather than computing them by means of trigonometry.) Originally, D_i is *unappended*, if it contains only the vertices; it is *appended* and denoted by D_i^*, if it also contains an "enclosure bit", i.e., the information by robot $r_{o(i)})$ opposite to edge e_i that the parallel tangents $\ell(e_i)$ to P through (r_{i-1}, r_i) and $\ell'(e_i)$ through $r_{o(i)}$ enclose P, along with distance $d_{i,o(i)}$ between those tangents. Overall, the smallest of these distances is computed as follows.

Distributed Rotating Calipers (DRC)

---

(1) Any robot checks whether it is a corner robot by considering its visibility cone.
(2) Elect a leader corner robot, r_0, as the one with the smallest ID.
(3) By passing a message from r_0 around the hull, establish the (counterclockwise) cyclic order of corner robots along the hull; let this be $r_0, \ldots, r_{h-1}, r_0$, such that each corner robot knows its predecessor and successor. This also determines the hull edges, $e_0, \ldots, e_{h-1}$.
(4) Pass around vertex descriptions, as follows.

  (4.1) All robots start in "unappended" mode.
  (4.2) Robot r_0 begins with sending its own (unappended) D_0 to robot r_1.
  (4.3) While in "unappended" mode, a robot r_j:
  - based on angle information, checks for any incoming unappended D_i (originating from some robot $r_i \neq r_j$) whether the line parallel to $\ell(e_i)$ through r_j separates r_{j-1} from r_{j+1}, i.e., whether the angle of (r_i, r_{i-1}) with the x-axis lies between α_j and β_j;
  - if not, then r_j is a robot furthest from the line $\ell(e_i)$, i.e., $j = o(i)$, and D_i is appended with $d_{i,j}$, turning D_i into D_i^*;
  - passes on D_i or D_i^* (whether appended or not) to its successor;
  - upon receiving D_{j-1} from its predecessor r_{j-1}, passes it on to its successor r_{j+1}, followed by its own (newly minted) D_j in the next round;
  - upon receiving its own D_j^*, switches into "appended" mode.

  (4.4) While in "appended" mode, a robot r_j:
  - keeps track of the smallest encountered $d_{i,o(i)}$ for a D_i^*;
  - when receiving D_0^* for the second time, passes on D_0^*, then STOPs;
  - passes on any received D_i^*.

We claim the following; note that bookkeeping applies to the convex hull vertices, with possible relays counted implicitly.

Theorem 1 *After the preprocessing steps (1)–(3), the algorithm DRC stops after time* $3h$*, with at most* $2h+1$ *messages passed on by any robot, with all robots knowing the minimum* $d_{i^*o(i^*)}$*, the indices* i^* *and* $o(i^*)$ *at which it is attained, and the orientation of the corresponding tangents. Thus, the total number of messages is* $O(h^2)$*, with* $O(h)$ *per robot. Each message size depends only on the encoding size of coordinate information.*

Proof Full details are omitted due to limited space. To see that the algorithms stops with the required information as claimed, note that any message D_j must have come through a robot $r_{o(j)}$ opposite to r_j after being passed around P once, so any robot r_j receives its own annotated D_j^* in h communication rounds after sending out the unannotated D_j. When receiving D_j^* for the second time, all D_i^* must have been encountered, so the current minimum $d((r_{i^*-1}, r_i^*), r_{o(i^*)})$ is the global minimum. Even if non-hull relay robots are used, the number of messages per robot remains $O(h)$; the total number of messages becomes $O(ah)$ for a total of a active robots.

4.2.3 Computing Diameter

The diameter of a polygon is attained between two vertices of the convex hull. We augment the above algorithm to compute the maximum distance between hull vertices simultaneously by keeping track of the maximum encountered distance in the vertex descriptions.

5 Results

5.1 Simulation Results

In this section we analyze our algorithms in simulation and compare their performance. In the first experiment, we assume there is no error in measurement and we have enough robots to be placed at the vertices of the object (Fig. 6). As expected, DRC estimates the exact dimension and orientation for the objects, while the DPCA estimate is quite good for most of the objects.

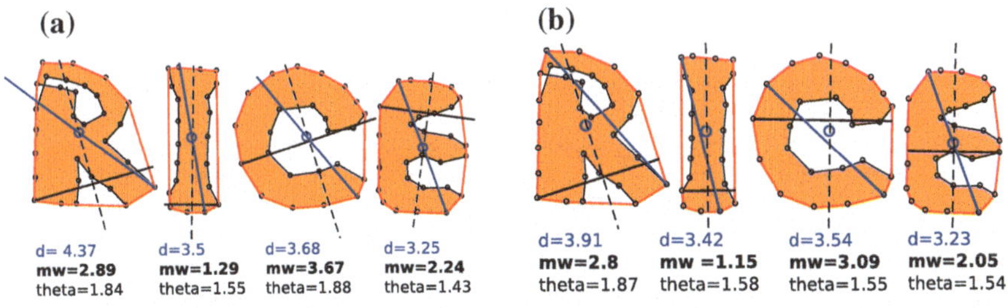

Fig. 6 Comparison of DPCA (**a**) and DRC (**b**) for four different objects. Shown are object minimum width (solid black), diameter (blue line), orientation (dashed black) and centroid (blue circle). The convex hull of the objects is shown in red color. Robots (blue circles) are placed at the vertices of orange objects. The estimates are also compared quantitatively. The letter d stands for diameter, mw is the minimum width, $theta$ is the orientation estimate in radians

We also analyzed DPCA and DRC performances when the number of robots around the object varies from 4 robots to 45 for the R-shaped object in Fig. 6. Robots are assumed to be randomly placed at the vertices of the object. The sensors are assumed with perfect measurement. Figure 7(top) shows the improvement of DRC and DPCA for estimating the object dimension by increasing the number of robots. For small numbers of robots, the polygon that is induced by the robots is very different from the original object, causing a large error. The mean of the object orientation error is small even for a small number of robots, because of the symmetric nature of the orientation value. By picking vertices randomly, the average for orientation estimation gets close to the actual value. By increasing the number of robots, the orientation errors of DPCA and DRC tend to zero. Increasing the number of robots may cause an imbalanced distribution of robots around the object, which affects the centroid estimation and thus the object dimension estimates. While DRC tends to picks the closest value to the optimum, which causes it to underestimate dimensions, DPCA always picks the maximum distance to the centroid, causing it to overestimate.

Lastly, we consider the setting in which robots are placed at all vertices of an almond-shaped object, as shown in Fig. 7 (bottom), and measurement errors exist. As shown, DRC usually underestimates the minimum width, while DPCA overestimates it, for the same reason described above. However, the error of estimating diameter by DRC has a normal distribution around zero, while DPCA overestimates the polygon diameter.

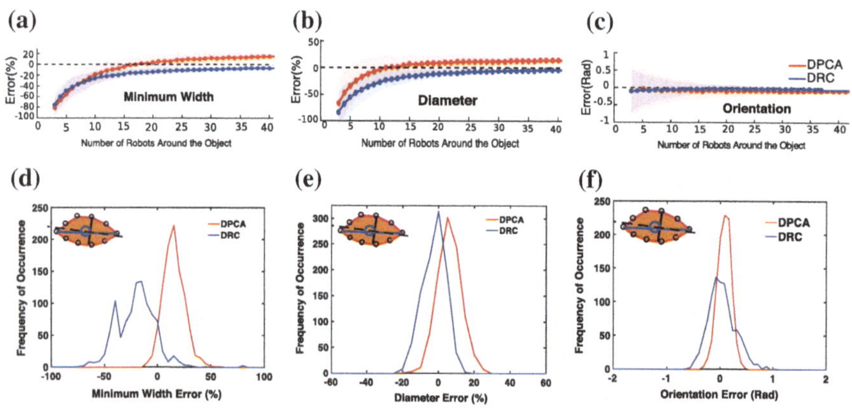

Fig. 7 (Top) The mean and standard deviation of the estimation error for **a** minimum width, **b** diameter, **c** orientation, by DRC (blue) and DPCA (red) for 1000 trials, when the sensors are ideal, but the number of robots varies from $m = 4$ to $m = 40$. The R-shaped object has 30 vertices, with 12 convex vertices. (Bottom) The almond-shaped object and estimation error distribution of DPCA (red) and DRC (blue), when sensor errors exist: **d** minimum width **e** diameter **f** object orientation

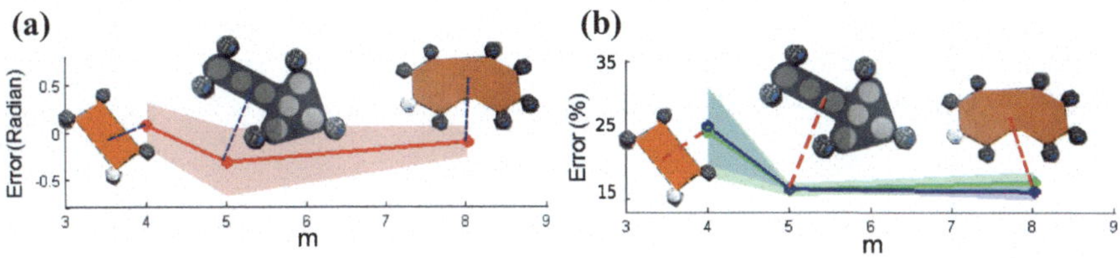

Fig. 8 Experimental result of object characterization by DPCA for three different objects. 12 trials for each experiment are shown with standard deviation (shadows) and mean error (solid lines): **a** orientation estimation error (radians). **b** object diameter (blue) and object minimum width (green) estimation error

5.2 Experimental Results

We used th r-one robot platform [10] to implement DPCA on a real robotic system. DPCA is used to estimate dimensions and orientation of three different symmetric, concave and convex objects (Fig. 8). We used 4, 5 and 8 robots to estimate the dimension and orientation of rectangle, arrow, and bean-shaped objects, respectively. In this setup, robots are placed on the vertices of the convex hull of the object. Robots use our pipelined consensus algorithm to reach the heading consensus and estimate the object dimension and orientation simultaneously. As shown, DPCA successfully estimates the object orientation and dimensions with a reasonable error.

6 Conclusion

We have presented two distributed algorithms for estimating the dimension and orientation of polygonal complex objects. Our algorithms are useful in different applications in robotics when global sensing is not available. We have tested our algorithms in simulations and experiment. Our algorithms successfully estimate the dimension and orientation of convex and concave objects. We compared our algorithm in different experiments. While DRC estimates the optimal values when there are enough robots to sample the object boundary, DPCA is more sensitive to the distribution of the robots around the object. In the presence of small errors, the more accurate DRC yields the better results; in the presence of larger errors, DPCA is to be preferred, as its inherent tendency to overestimate object width may be safer for avoiding tight corridors in applications of collective object transport. Our algorithms are self-stabilizing and robust to dynamic network topology and population changes. We will show these features in our future work. One of the future applications of these algorithms is in collective transport. Manipulator robots estimate the orientation of the object and adjust the object orientation during motion. Another potentially interesting direction for future work is to intelligently select subsets of vertices that lead to accurate estimates of the shapes.

Acknowledgements We thank several anonymous reviewers for helpful input that improved the presentation of this paper. We also thank Madeleine Nikirk, James Gringe, Sam Caroll, and Randy Zhang for helping us in data collection.

References

1. Akkaya, K., Younis, M.: A survey on routing protocols for wireless sensor networks. Ad Hoc Netw. **3**(3), 325–349 (2005)
2. Bicchi, A., Kumar, V.: Robotic grasping and contact: a review. In: ICRA, pp. 348–353. Citeseer (2000)
3. Chaudhuri, D., Samal, A.: A simple method for fitting of bounding rectangle to closed regions. Pattern Recognit. **40**(7), 1981–1989 (2007)
4. Fekete, S.P., Fey, D., Komann, M., Kröller, A., Reichenbach, M., Schmidt, C.: Distributed vision with smart pixels. In: Proceedings of the 25th ACM Symposium Computional Geometery, pp. 257–266. ACM (2009)
5. Habibi, G., Kingston, Z., Wang, Z., Schwager, M., McLurkin, J.: Pipelined consensus for global state estimation in multi-agent systems. In: Proceedings of the 2015 International Conference on Autonomous Agents and Multi-agent Systems, AAMAS '15. International Foundation for Autonomous Agents and Multiagent Systems (2015)
6. Habibi, G., Xie, W., Jellins, M., McLurkin, J.: Distributed path Planning for collective transport using homogeneous multi-robot systems. In: Proceedings of the International Symposium on Distributed Autonomous Robotics Systems (2014)
7. Habibi, G., Zachary, K., Xie, W., Jellins, M., McLurkin, J.: Distributed centroid estimation and motion controllers for collective transport by multi-robot systems. In: International Conference on Robotics and Automation (ICRA). IEEE (2015)
8. Komann, M., Kröller, A., Schmidt, C., Fey, D., Fekete, S.P.: Emergent algorithms for centroid and orientation detection in high-performance embedded cameras. In: Proceedings 5th Conference on Computing Frontiers, pp. 221–230. ACM (2008)
9. McLurkin, J.: Analysis and implementation of distributed algorithms for multi-robot systems. Ph.D. thesis, MIT, USA (2008)
10. McLurkin, J., McMullen, A., Robbins, N., Habibi, G., Becker, A., Chou, A., Li, H., John, M., Okeke, N., Rykowski, J., Kim, S., Xie, W., Vaughn, T., Zhou, Y., Shen, J., Chen, N., Kaseman, Q., Langford, L., Hunt, J., Boone, A., Koch, K.: A robot system design for low-cost multi-robot manipulation. In: 2014 IEEE/RSJ International Conference on Intelligent Robots and Systems, Chicago, IL, USA, 14–18 September 2014, pp. 912–918. IEEE (2014)
11. Olfati-Saber, R., Fax, J.A., Murray, R.M.: Consensus and cooperation in networked multi-agent systems. Proc. IEEE **95**(1), 215–233 (2007)
12. Ota, J., Miyata, N., Arai, T., Yoshida, E., Kurabatashi, D., Sasaki, J.: Transferring and regrasping a large object by cooperation of multiple mobile robots. In: Intelligent Robots and Systems 95. Human Robot Interaction and Cooperative Robots', Proceedings. 1995 IEEE/RSJ International Conference on, vol. 3, pp. 543–548. IEEE (1995)
13. Shamos, M.I.: Computational geometry. Ph.D. thesis, Yale University (1978)
14. Toussaint, G.T.: Solving geometric problems with the rotating calipers. In: Proceedings IEEE Melecon, vol. 83, p. A10 (1983)

Optical Wireless Communications for Heterogeneous DARS

Patricio J. Cruz, Christoph Hintz, Jonathan West and Rafael Fierro

Abstract Distributed autonomous robotic systems (DARS) can provide essential support to human task forces in a variety of missions. Maintaining reliable communications on these mobile networks is critically important for cooperative autonomy. Radio-frequency is the common wireless communication technology employed for task coordination. However, it has some limitations that can be mitigated by complementing radio-frequency systems with optical wireless communications. In this chapter, we present our efforts for the development and implementation of an optical wireless communication link in DARS, more specifically between aerial (e.g., quadrotors) and ground robots. We describe a line-of-sight directed optical link between these two platforms. In addition, we detail our strategy to maintain an adequate transmitter-receiver relative position to enhance the optical signal strength. Furthermore, we provide some details about our optical transceiver prototype that can be mounted on small aerial and ground autonomous robotic systems.

1 Introduction

Surveillance of complex environments, search and rescue and localization of targets are just few mission examples that place special requirements on robotic systems. Man-portable autonomous robots with lengths between 0.1–0.5 m and weights ranging from hundreds of grams to a few kilograms can be deployed and coordinated

P. J. Cruz · C. Hintz · J. West · R. Fierro (✉)
Department of Electrical & Computer Engineering, University of New Mexico, Albuquerque, NM, USA
e-mail: rfierro@unm.edu

P. J. Cruz
e-mail: pcruzec@unm.edu

C. Hintz
e-mail: chintz@unm.edu

J. West
e-mail: jmwest@unm.edu

R. Groß et al. (eds.), *Distributed Autonomous Robotic Systems*,
Springer Proceedings in Advanced Robotics 6,
https://doi.org/10.1007/978-3-319-73008-0_16

to successfully perform these tasks. Furthermore, the use of multiple small robotic platforms with different dynamics and/or capabilities can overcome some of the limitations of using a homogeneous robotic network. For example, aerial autonomous vehicles have a large sensor footprint but with a low resolution. On the other hand, ground mobile robots can sense a limited area but they do so with much more accuracy. Therefore, these autonomous robotic systems are complementary and by coordinating them, it is possible to mitigate their weaknesses.

For this type of distributed autonomous system, radio-frequency (RF) communication is the common method that allows the robotic network to operate wirelessly. However, RF has some limitations such as security issues, a congested spectrum, a limited available data rate and hostile jamming [3, 10]. One possible alternative to overcome these limitations is to complement RF with optical wireless (OW) communications [3, 9, 10, 14]. In fact, free-space optical (FSO) links have augmented the transmission capacity of RF networks for large-scale applications [2, 12, 15]. These hybrid OW/RF systems offer temporary point-to-point links within the network and provide wireless access when RF can be jammed or can create undesired interference. In addition, indoor OW systems are currently being revisited as part of the visible light communication (VLC) framework that aims to combine lighting and communications employing commercially visible light emitting diodes [5, 16].

OW communications have been explored for the case of distributed sensor networks [13] where this technology can provide coherent connectivity to large numbers of compact sensor nodes. For the case of robotic platforms, OW systems have been principally applied to underwater vehicles. RF is not effective in the aquatic domain due to the high absorption of electromagnetic radiation by water, so acoustic modems are commonly used for underwater operations. However, they are extremely slow with a high latency. Thus, OW communication has become an alternative technology for wireless transmission in underwater settings [8, 18, 20]. For example, [8] presents a real-time video streaming solution for autonomous underwater vehicles based on a VLC system. For the case of land applications, different tracking and pointing mechanisms for establishing an OW link between mobile agents have been proposed in the literature [1, 7, 17]. However, the size, weight and power requirements for these pointing systems generally limits their application to robots with large payload and energy capacity and not micro/mini autonomous robotic platforms.

To the best of our knowledge, OW communications has not been proposed for the case of a robotic team of small aerial and ground vehicles. Possibly the major shortcoming of OW technology is the line-of-sight (LOS) pointing and tracking requirement. This challenge has to be addressed to fully exploit the benefits of using an OW link. In our previous work [6], we introduced a model for a directed line-of-sight optical link between an unmanned aerial robot and a ground mobile vehicle. Based on the link model, we defined a *connectivity cone* over the receiver where a minimum transmission rate is guaranteed. This chapter is an extension of this preliminary work. Here, we present our framework to track a ground mobile receiver by an aerial transmitter to establish a point-to-point optical link. Also, we detail our efforts for the development and implementation of a prototype of an OW transceiver that can be mounted and used by small aerial and ground robotic platforms.

2 OW Communications

Compared to RF, OW systems replace the radio waves with light and the antennas with free-space optical transceivers. Indeed, OW has unique advantages respect to RF [3, 10, 14]. For example, OW systems use a wide unlicensed spectrum, i.e., the infrared and visible spectrum, offering a cost-effective link with large modulation bandwidths for communication. Optical components are also smaller, lighter and cheaper than high-speed RF components. Moreover, OW links have a high level of security against jamming and eavesdropping due to the high directionality of the optical beam. Even though OW links can support high data rate, their reliability is severely affected by atmospheric conditions like snow or fog [3, 10]. Besides, the main disadvantage for optical-based communications is the necessity of alignment and pointing between the transmitter and receiver. Thus, directed LOS systems require tracking solutions to allow user mobility.

A simple block diagram of an indoor OW communication system is shown in the inset of Fig. 1a. A light-emitting diode (LED) or a laser diode (LD) is used as optical transmitter and a PIN photodiode or an avalanche photodiode (APD) is used as optical receiver. Due to its simplicity and reduced cost, intensity modulation with direct detection (IM/DD) is de-facto method of implementing OW systems. A modulated signal $m(t)$ drives the current of the optical source varying its intensity $I(t)$. The optical receiver integrates the incident optical signal generating a photocurrent $i(t)$ which is directly proportional to the instantaneous optical power incident on it.

In our previous work [6], we introduced a model of a directional LOS optical link between an aerial and a ground robot. Next, we summarize this model. A diagram showing the main parameters is illustrated in Fig. 1a. Let d be the distance between the two optical transceivers, ϕ be the pointing angle relative to the optical transmitter axis, and ψ be the incidence angle relative to the optical receiver axis. Notice that the transmitter is perfectly pointed at the receiver when $\phi = 0°$, so ϕ is also known as the *pointing error*. Similarly, the receiver is perfectly pointed at the transmitter

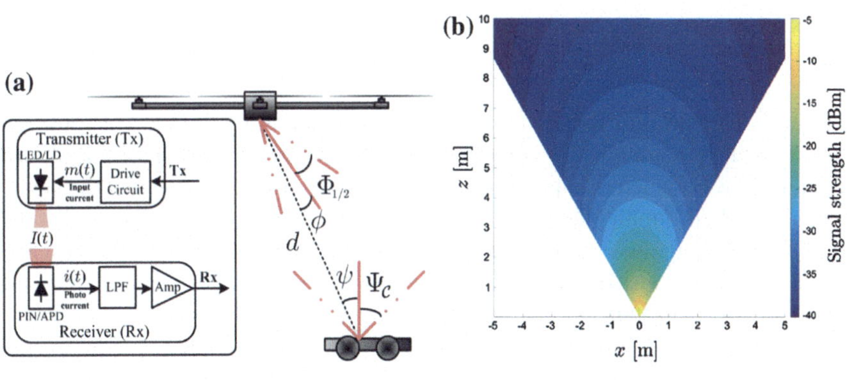

Fig. 1 **a** Diagram of the LOS optical link with its main parameters. The inset presents a block diagram of the OW communication system. **b** Signal strength contour map for the OW link

when $\psi = 0°$, so ψ is sometimes referred to as the *receiver pointing error*. Even though all the angles in Fig. 1a are in the same plane, the link model presented next is generally valid with circularly symmetric beam and FOV.

The radiant intensity of the optical source can be described by

$$I_s = P_{\mathrm{Tx}} \frac{m+1}{2\pi d^2} \cos^m \phi, \tag{1}$$

where P_{Tx} is the average transmitted optical power, and m is the Lambert's mode number expressing directivity of the source beam. This number is defined as $m = \frac{-\ln 2}{\ln(\cos \Phi_{1/2})}$.

Here, $\Phi_{1/2}$ is the half-angle at half-power which describes the transmitter beam width.

The optical receiver can be modeled as an effective area A_{eff} collecting the incident radiation. This area is given by

$$A_{\mathrm{eff}}(\psi) = g(\psi) A \cos \psi, \tag{2}$$

where A is the receiver active area, and $g(\psi)$ is the light-concentrator gain which for an ideal case is given by

$$g(\psi) = \begin{cases} \frac{n^2}{\sin^2 \Psi_C} & \text{if } |\psi| \leq \Psi_C, \\ 0 & \text{otherwise.} \end{cases} \tag{3}$$

Here, n is the refractive index and Ψ_C is the half-angle field-of-view (FOV) of the optical detector.

Using (1) and (2), the received signal power is given by

$$P_{\mathrm{Rx}} = I_s A_{\mathrm{eff}}. \tag{4}$$

2.1 Connectivity Cone

Using (4), we can plot the signal strength as a function of the receiver position relative to the transmitter. The contour map of the signal strength in dBm is shown in Fig. 1b. The parameters to plot this map are adopted from [19]. We plot the signal strength for the azimuth plane (x-y plane) assuming that the receiver is fixed at the origin. For plotting this contour map, we assume that the pointing error is zero, i.e., $\phi = 0°$. Also, we consider that the receiver is always pointing up. The white regions in the contour map indicate that no optical signal can be detected in these areas. These regions appear because the concentrator gain $g(\psi)$ becomes zero, according to (3), if the receiver pointing error is greater than the half-angle FOV, i.e., $\psi = \Psi_C$.

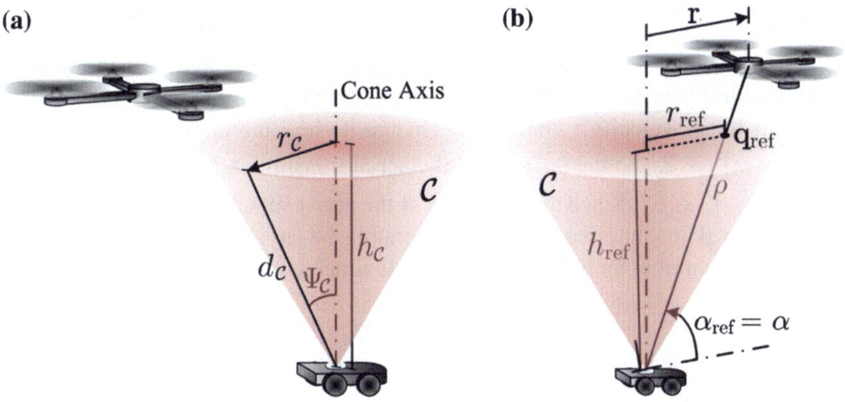

Fig. 2 **a** Connectivity cone C and its parameters. **b** Sketch of the required parameters to find the reference position by using Algorithm 1

Therefore, the receiver does not detect any optical signal when the transmitter is within these areas.

Using (1) to (4), it is possible to find the range ρ to achieve a desired received signal power P_{Rx} assuming a maximum pointing error $\phi = \phi_{\text{max}}$

$$d = \sqrt{\frac{A_{\text{eff}}(m+1)(\cos^m \phi_{\text{max}}) P_{\text{Tx}}}{2\pi P_{\text{Rx}}}}. \tag{5}$$

Combining this range with the maximum receiver pointing error that happens when $\psi = \Psi_C$, it is possible to define a right circular cone as shown in Fig. 2a. Hence, a received signal strength greater or equal to the desired one would be sufficiently guaranteed if the transmitter lies in this cone. We call this the connectivity cone and denote it as C. The apex or vertex of C is given by the position of the ground receiver. Furthermore, C can be defined by its apex angle Ψ_C and by its slant height d_c which can be found using Eq. (5). Employing these parameters, it is easy to find the base radius of the cone r_C and the height of the cone h_C, see Fig. 2a. Also, notice that the normal vector of the cone axis is given by $\mathbf{e}_3 = [0\ 0\ 1]^T$.

3 Target Tracking to Maintain OW Communications

From our discussion in Sect. 2, the aerial transmitter sis required to track the ground mobile receiver in order to approach and stay inside of the connectivity cone C. For chapter completeness, we summarize next our target tracking controller presented in [6].

Let $\left[x_{\text{Rx}}\ y_{\text{Rx}}\ \theta\right]^T \in \text{SE}(2)$ be the state of the mobile ground receiver, where $\mathbf{q}_{\text{Rx}} = \left[x_{\text{Rx}}\ y_{\text{Rx}}\ 0\right]^T$ is its 3-D position, θ is its heading angle and SE(2) is the Special Euclidean Group in two dimensions. We assume the ground receiver moves with a constant linear velocity v, but it can change its heading angle. Now let $\mathbf{q}_{\text{Tx}} = \left[x_{\text{Tx}}\ y_{\text{Tx}}\ z_{\text{Tx}}\right]^T \in \mathbb{R}^3$ be the position of the aerial transmitter. We define its state as $\mathbf{x} = \left[\mathbf{q}_{\text{Tx}}\ \dot{\mathbf{q}}_{\text{Tx}}\right]^T \in \mathbb{R}^6$. Since the dynamics of the aerial vehicle can be approximated as those of a point mass capable of accelerating in any direction [11], the aerial transmitter dynamics can be written in discrete time as

$$\mathbf{x}_{t+1} = \mathbf{A}\mathbf{x}_t + \mathbf{B}\mathbf{u}_t, \tag{6}$$

where

$$\mathbf{A} = \begin{bmatrix} 1\,0\,0\,\tau\,0\,0 \\ 0\,1\,0\,0\,\tau\,0 \\ 0\,0\,1\,0\,0\,\tau \\ 0\,0\,0\,1\,0\,0 \\ 0\,0\,0\,0\,1\,0 \\ 0\,0\,0\,0\,0\,1 \end{bmatrix}, \quad \mathbf{B} = \begin{bmatrix} 0\,0\,0 \\ 0\,0\,0 \\ 0\,0\,0 \\ \tau\,0\,0 \\ 0\,\tau\,0 \\ 0\,0\,\tau \end{bmatrix},$$

$\mathbf{u} \in \mathbb{R}^3$ is the acceleration control input, t is the time index, and τ is the time step.

We assume that the aerial transmitter knows the state of the ground vehicle and its linear velocity at all times. Then our goal is to design a control input $\mathbf{u}$ such that the flying transmitter gets first inside the connectivity cone C and stays within it. As control input, we consider a proportional-derivative controller given by

$$\mathbf{u} = \text{K}_\text{p}\left(\mathbf{q}_{\text{ref}} - \mathbf{q}_{\text{Tx}}\right) + \text{K}_\text{v}\left(\mathbf{v}_{\text{ref}} - \dot{\mathbf{q}}_{\text{Tx}}\right). \tag{7}$$

Here, K_p and K_v are control gains, $\mathbf{q}_{\text{ref}}$ is a reference position to be determined, and $\mathbf{v}_{\text{ref}} = \left[\text{v}\cos\theta\ \text{v}\sin\theta\ 0\right]^T$ is the reference velocity.

In order to find $\mathbf{q}_{\text{ref}}$, we need to compute ρ the distance between transmitter and receiver, and α the elevation angle of the transmitter respect to the receiver. These parameters are illustrated in Fig. 2b. Notice that if ρ is less than or equal to d_C and α is greater than or equal to $(\pi/2 - \Psi_C)$ then the aerial transmitter is inside of C. Algorithm 1 details the method to find the reference position $\mathbf{q}_{\text{ref}}$. Notice that this algorithm gives as result $\mathbf{q}_{\text{ref}} = \mathbf{q}_{\text{Tx}}$ for the case that the flying transmitter is inside the connectivity cone.

3.1 Results

The proposed target tracking controller is validated by running numerical simulations. The 3-D environment developed for visualization purposes is shown in Fig. 3a. The aerial transmitter is represented as a thin cross with four circles at each side,

Algorithm 1 Reference position $\mathbf{q}_{\text{ref}}$

Require: d_C slant height of C, Ψ_C apex angle of C, $\mathbf{q}_{\text{Rx}}$ receiver position, $\mathbf{q}_{\text{Tx}}$ transmitter position

1: $\rho \leftarrow \|\mathbf{q}_{\text{Tx}} - \mathbf{q}_{\text{Rx}}\|$ ▷ transmitter-receiver distance
2: **if** $\rho > d_C$ **then**
3: $\quad h_{\text{ref}} \leftarrow d_C \cos \Psi_C$ ▷ transmitter outside of C
4: **else**
5: $\quad h_{\text{ref}} \leftarrow z_{\text{Tx}}$
6: **end if**
7: $\mathbf{r} \leftarrow \left[x_{\text{Tx}} - x_{\text{Rx}} \;\; y_{\text{Tx}} - y_{\text{Rx}} \;\; 0\right]^T$ ▷ transmitter distance with respect to the cone axis
8: $\alpha \leftarrow \text{atan}\,(z_{\text{Tx}}/\|\mathbf{r}\|)$ ▷ elevation angle
9: **if** $\alpha < (\pi/2 - \Psi_C)$ **then**
10: $\quad \alpha_{\text{ref}} \leftarrow \pi/2 - \Psi_C$ ▷ elevation angle outside of cone region
11: **else**
12: $\quad \alpha_{\text{ref}} \leftarrow \alpha$
13: **end if**
14: $r_{\text{ref}} \leftarrow h_{\text{ref}}/\tan \alpha_{\text{ref}}$ ▷ reference point at the cone base

Output: $\mathbf{q}_{\text{ref}} \leftarrow \mathbf{q}_{\text{Rx}} + r_{\text{ref}} \frac{\mathbf{r}}{\|\mathbf{r}\|} + h_{\text{ref}} \mathbf{e}_3$ ▷ reference point inside C

while the mobile ground receiver is depicted as a cuboid with a sensing field-of-view in its front. We have drawn in red the connectivity cone C on top of the receiver, the reference position $\mathbf{q}_{\text{ref}}$ found using Algorithm 1 by an asterisk marker "*", and the distance between the aerial transmitter and $\mathbf{p}_{\text{ref}}$ by a dash-dot line "–·", see Fig. 3a.

For the simulation, the transmitter-receiver distance at the beginning is approximately 9.18 m. The parameters for the optical link used for simulation are the same ones we employed to plot the contour maps in Fig. 1b. The control constants are set up to $K_p = 8$ and $K_v = 2.5$. The constant linear velocity of the mobile ground robot is assumed equal to 12 m/s. The time step τ is assumed equal to 0.01 s and we ran the simulations for 10 s.

The ground receiver follows the trajectory shown in Fig. 3b by the black dotted line. The empty square and circle denotes the initial xy position of receiver (Rx) and transmitter (Tx), respectively. The filled square and circle indicates the position of Rx and Tx after 10 seconds, respectively. The transmitter-receiver distance d is shown in Fig. 3c. The slant height of the connectivity cone d_C is drawn by the dash-dot line in this figure. Notice that after around 0.92 s, d is less than d_C and it remains less than this value during the rest of the simulation. Furthermore, the aerial transmitter remains within the connectivity cone from this time on. This time is outlined in the plots by a dashed black line. The transmitter remains inside the cone after 0.92 s, the received signal power exceeds the desired P_{Rx} which is set to −34 dBm. The received signal power is shown in Fig. 3d where the desired limit of −34 dBm is represented by the dash-dot line. The minimum bit rate after the 0.92 s is −34.42 dBm with an average of −31.28 dBm. Consequently, our goal of establishing and maintaining an optical link of at least −34 dBm is accomplished after 0.92 s.

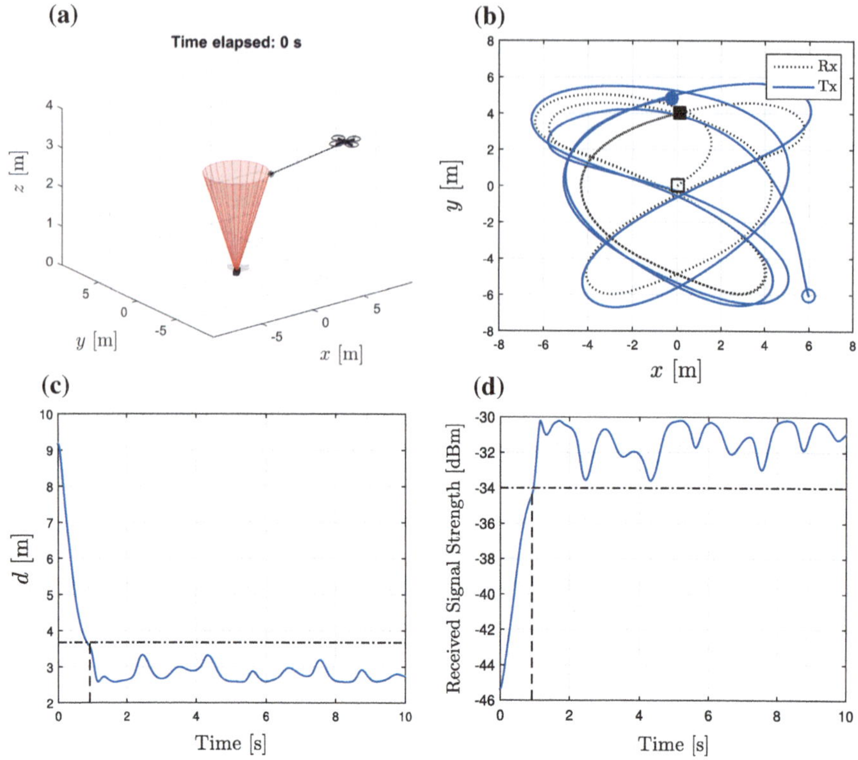

Fig. 3 **a** 3-D environment for numerical simulations. **b** xy trajectories described by the mobile ground receiver (Rx) and the aerial transmitter (Tx). **c** The distance d between transmitter and receiver. The dash-dot line is the value of d_C. **d** The received signal strength (RSS) P_{Rx} calculated applying (4). The variations in the RSS is due to the changing distance between transceivers

4 Experimental System

4.1 OW Communication System

We have designed and prototyped a bidirectional OW communication system so that each unit can both send and receive data. It consists of two main components: (1) a transducer comprising the physical receiver and transmitter units, and (2) a tracking mechanism based on a vision sensor, color LEDs, and a pan/tilt (PT) unit. The transducer, the vision sensor and the color LED array are all housed on one board which is then mounted to the PT mechanism. In particular, the board has the option of mounting the Pixy in two configurations: right-side up and up-side down. These options are shown in Fig. 4a. Having two units of our OW communication system, one for each configuration facing each other, the transmitter will line up with the receiver and the color LED array will be in the center of the field of view of the

Fig. 4 **a** OW transceiver prototypes. OW transceiver mounted in **b** a Swarmie mobile ground robot, and **c** an AscTec hummingbird quadrotor

Table 1 Specifications of the receiver PIN photodiodes

Part #	Active area A [mm^2]	Half-angle FOV Ψ_C [°]	Responsivity R [A/W]	Response time t_r [ns]	Wavelength λ_{max} [nm]
SFH 230 P	1	±75	0.62	5	850
PDB-C160SM	8	±60	0.62	20	850

vision sensor on each board, see Fig. 4a. Therefore, it is possible to establish and maintain the LOS between these two units by learning the color LEDs and using the vision sensor to track this feature. In this fashion, the modules will lock and the data transmission can occur. To transmit and receive data to/from our OW communication system, we have used in our tests a serial UART interface supporting a data transfer rate up to 1 Mbps.

4.1.1 Optical Transmitter and Receiver

The transmitter consists of a high-speed infrared (IR) emitting diode produced by Vishay Semiconductors (part number VSLY-5850). This IR diode has a peak wavelength $\lambda_p = 850$ nm, a radiant power $P = 55$ mW, an angle of half intensity $\Phi_{1/2} = \pm 3°$, and a response time $t_r = 10$ ns. The narrow transmitting beam determines the necessary accuracy of the tracking system.

Two different types of photodiodes are used for the receiver as shown in Table 1. Since the receiver has a wide viewing angle, the tracking of the receiver is not as critical as that of the transmitter.

The data transmission uses amplitude modulation. The communication protocol is a standard UART at 1Mbps which is converted to USB using an FTDI TTL-232R-5V-PCB converter for communication with the computer.

Custom printed circuit boards were designed to support the Pixy, LED tracking target, and the data transmitter and receiver. The Pixy is located in the center of the board with the LED target above it. The board has the transmitter and receiver on the opposite sides of the Pixy so that they are properly aligned when one board is rotated 180° relative to the other. The entire circuit assembly is mounted on a servo driven pan-tilt head which is directly controlled by the Pixy. The boards are powered by 12VDC. The assembled and mounted boards are shown in Fig. 4a.

4.1.2 Tracking System

The key to maintaining an optimal wireless communication is to have the receiver and transmitter properly aligned.

As an initial proof of concept, the tracking function is achieved using the Pixy CMUcam5 [4]. The Pixy is chosen because it is light, low-power, and relatively small. It is readily available and is functional off-the-shelf, which saves a great deal of development time. It comes with servo connections for controlling a pan-tilt mount and can be programmed to track objects through the free software PixyMon.

The tracking update rate is 50 Hz which allows tracking at reasonable speeds. The Pixy is a color based tracker which requires a distinctly colored object as a target. In our application an LED strip is used as a tracking target to improve the robustness of the system under varying light conditions.

4.2 Ground Mobile Robot

As ground robotic agent, we use a platform known as a Swarmie which is available as part of the NASA Swarmathon competition. The Swarmie is controlled by an Intel NUC computer running the Robot Operating System (ROS) which allowed us to easily integrate our serial communication software to the existing system. It communicates with a server using WiFi so it is ideally suited to demonstrate a Hybrid OW/RF system. The OW system can be mounted to the top of the Swarmie as shown in Fig. 4b.

4.3 Quadrotor Unmanned Aerial Vehicle

In order to incorporate the OW system into an unmanned aerial vehicle (UAV), an Asctec Hummingbird, shown in Fig. 4c, is being developed to carry the optical wireless link. This quadrotor is a widely used research platform, which is well documented and supported and is equipped with an internal flight control processor. For our experiment we plan to add an Odroid XU4 micro-controller which runs ROS to provide the control and data processing on the quadcopter. The Odroid will receive position data from the Vicon motion capture system and send it to the Humming-

Fig. 5 Snapshot of the experimental evaluation

bird's controller. The controller will use this feedback for position control loop that is running at 1 kHz which allows us to control the hummingbird precisely. A 12V regulator board will be added to supply power to the optical communication system.

Supporting hardware elements are designed and printed to mount the hardware and protect the communication system during landing. Legs are added to raise the hummingbird by 5.5 inches to allow clearance for the optical link. The mount for the optical communication system is printed to mount the board at the center of gravity of the platform to minimize its effects on the flight dynamics. A mount is also printed to secure the Odroid XU4 to the top of the body. The 3D prints utilize an internal honeycomb structure to reduce weight.

4.4 Experimental Evaluation

We carry out preliminary experimental evaluations of the optical communication prototype. The initial experiment consists of two Swarmies with OW systems communicating with each other by having one stream video to the other. Two ground robots are used in order to simplify the dynamics of the tracking system and to validate the performance of the OW system under stable conditions which are difficult to achieve with UAVs. The OW system is able to maintain communication as the two Swarmies moved throughout the area. The video from one Swarmie is transmitted optically to the other and then relayed to the server where it is displayed. Figure 5 shows a snapshot of the experiment, while Fig. 6 shows the result in the video transmission when the LOS between modules is blocked or they are not aligned. A video of this experiment can be found at https://youtu.be/gCyFTMlMi9o.

Fig. 6 Faulty video transmission due when the LOS between transceivers is blocked

5 ROS/Gazebo Simulation Environment

To facilitate the development and validation of the algorithms to maintain connectivity when an optical link is used between mobile platforms, a detailed simulation environment is required since hardware testing is time consuming and expensive. The ROS system includes the Gazebo simulator which allows highly accurate, physics based modeling of complex robotic systems. The ROS software is tightly integrated with the Gazebo simulator allowing the same code that will run on the hardware platforms to run in a simulated environment. The Gazebo system receives the control outputs from the rover ROS software, models the physical response in the simulated world and then produces simulated sensor feedback to the control software.

The simulations take place in a Gazebo world which is customizable to reproduce almost any terrain or situation. For our simulation, the MARHES indoor test bed as seen in Fig. 7a was used as a model. The test bed includes two ground robots and two aerial vehicles. One ground robot and one aerial vehicle are equipped with the OW transceiver. A Vicon Motion capture system helps to coordinate the robots. The corresponding Gazebo environment is shown in Fig. 7b. This room includes a flat floor surface and several obstacles as would be present during the indoor testing at our lab.

In this environment we place 3 ground robots using existing models of the Swarmathon rovers. On top of each of these rovers we place a cone which visually represents the area in which the optical receiver can maintain communication as shown in Fig. 7c.

A quadcopter model is then added using the existing ROS "hector" model. This is a generic quadcopter model but it incorporates all of the necessary simulation elements and will be refined in the future to closely match the actual flight characteristics of our UAVs.

For this simulation, the ground rovers move according to a random walk/collision avoidance pattern and collect images of the environment. The UAV is currently operator controlled using a joystick.

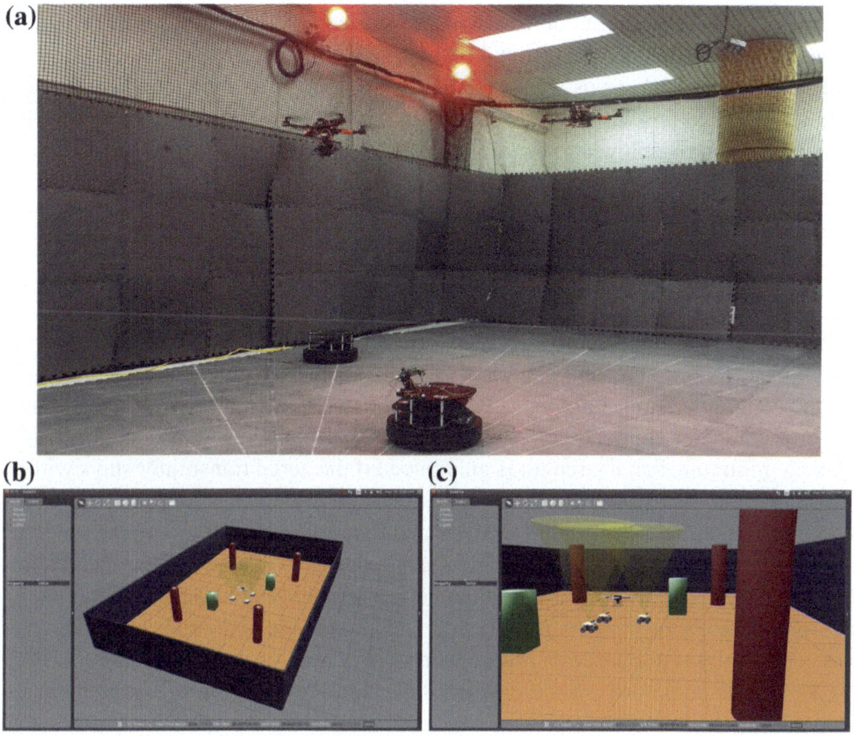

Fig. 7 **a** Test bed at MARHES Lab with 2 aerial robots and 2 ground robots. 2 Robots are equipped with OW transceivers. **b** Gazebo simulation of our lab testing area. **c** Gazebo simulation showing connectivity cones

We plan to use this simulation environment to develop new algorithms under the assumption of a hybrid RF/OW communication system which can be used for accessing to cloud-based services. As the ground rovers move about, the UAV will go among them and as it enters each connectivity cone, it will establish communication, upload the images and other data, transmit this to the server and then proceed to the other rovers.

A simulation of the RF communication will provide low-bandwidth command and control information which will be used to give the ground rovers their route information and to guide the UAV to each ground rover. As the UAV enters the connectivity cone of a ground rovers, the quality of the communication channel will be simulated by calculating the line of sight distance and incident angle between the UAV and the ground rover. These distances are readily available from the Gazebo system and will be used in conjunction with the formulas given in Sect. 2 to compute a maximum usable bitrate and applicable bit error rates which will then be used to simulate the data transfer. The UAV will then transfer the collected data to a simulated server which will process the data, develop the maps and then send instructions to the rovers. The UAV flight controls will use the same ROS code as will be run on the actual hardware UAV and ground rovers.

The simulation allows us to run the hardware flight code on a simulated system to validate algorithms without lengthy real-time tests. Once validated in Gazebo, the same code is then deployed to the actual hardware and the results can be validated.

6 Conclusions

In this chapter, we studied an optical wireless link between aerial and ground autonomous vehicles. The combination of OW and RF has the potential to dramatically expand communications rates in DARS. Based on a model for a directed LOS optical link, we defined a connectivity cone on top of the mobile ground optical receiver where a minimum signal strength is guaranteed if the aerial transmitter stays within. The numerical simulation results validates our control strategy for the aerial transmitter so that it tracks the ground receiver and remains inside this connectivity cone. In addition, we presented the details and the initial tests of our optical transceiver prototype to implement a point-to-point link between small ground and aerial robots. As part of our assumptions, we considered perfect knowledge of the state of the receiver which generally is not the case. Future work will consist on modifying our approach to take into account noisy measurements of the receiver. Also, we will implement the proposed methodology through hardware experiments by using the designed optical transceiver. We will also validate the performance of a larger network of robots by expanding our simulation and through real world experiments.

Acknowledgements This work was supported in part by the Army Research Lab Micro Autonomous Systems and Technology Collaborative Alliance ARL MAST-CTA #W911NF-08-2-0004. We would like to thank the Ecuadorian scholarship program administrated by the *Secretaría de Educación Superior, Ciencia, Tecnología e Innovación* (SENESCYT) for providing part of the financial support for Patricio J. Cruz.

References

1. Ahmad, M.H., Kerr, D., Bouazza-Marouf, K.: Fast pointing and tracking system for mobile robot short range control via free space optical laser line of sight communication link. In: Tokhi, M.O., Virk, G.S., Hossain, M.A. (eds.) Climbing and Walking Robots, pp. 147–154. Springer, Berlin (2006). https://doi.org/10.1007/3-540-26415-9_17
2. Bagley, Z.C., Hughes, D.H., Juarez, J.C., Kolodzy, P., Martin, T., Northcott, M., Pike, H.A., Plasson, N.D., Stadler, B., Stotts, L.B., Young, D.W.: Hybrid optical radio frequency airborne communications. Opt. Eng. **51**(5), 055,006–1–055,006–25 (2012). https://doi.org/10.1117/1.OE.51.5.055006
3. Borah, D., Boucouvalas, A., Davis, C., Hranilovic, S., Yiannopoulos, K.: A review of communication-oriented optical wireless systems. EURASIP J. Wirel. Commun. Netw. **2012**(1), 91 (2012). https://doi.org/10.1186/1687-1499-2012-91

4. Charmed Labs: CMUcam5 Pixy - Overview (2016). http://charmedlabs.com/default/pixy-cmucam5/ (Pixy is a collaboration between Charmed Labs and Carnegie Mellon University. Retrieved April 2016)
5. Chowdhury, H., Katz, M.: Data download on move in indoor hybrid (radio-optical) WLAN-VLC hotspot coverages. In: Proceedings of the IEEE Vehicular Technology Conference (VTC Spring), pp. 1–5 (2013). https://doi.org/10.1109/VTCSpring.2013.6692499
6. Cruz, P.J., Fierro, R.: Towards optical wireless communications between micro unmanned aerial and ground systems. In: Proceedings of the International Conference on Unmanned Aircraft Systems (ICUAS), pp. 669–676 (2015). https://doi.org/10.1109/ICUAS.2015.7152349
7. Derenick, J., Thorne, C., Spletzer, J.: On the deployment of a hybrid free-space optic/radio frequency (FSO/RF) mobile ad-hoc network. In: Proceedings of the IEEE/RSJ International Conference on Intelligent Robots and Systems (IROS), pp. 3990–3996 (2005). https://doi.org/10.1109/IROS.2005.1545193
8. Doniec, M., Xu, A., Rus, D.: Robust real-time underwater digital video streaming using optical communication. In: Proceeding of the IEEE International Conference on Robotics and Automation (ICRA), pp. 5117–5124 (2013). https://doi.org/10.1109/ICRA.2013.6631308
9. Elgala, H., Mesleh, R., Haas, H.: Indoor optical wireless communication: potential and state-of-the-art. IEEE Commun. Mag. **49**(9), 56–62 (2011). https://doi.org/10.1109/MCOM.2011.6011734
10. Ghassemlooy, Z., Popoola, W., Rajbhandari, S.: Optical Wireless Communications: System and Channel Modelling with MATLAB®. Taylor & Francis, Boca Raton (2012)
11. Hoffmann, G.M., Tomlin, C.J.: Mobile sensor network control using mutual information methods and particle filters. IEEE Trans. Autom. Control **55**(1), 32–47 (2010). https://doi.org/10.1109/TAC.2009.2034206
12. Izadpanah, H., ElBatt, T., Kukshya, V., Dolezal, F., Ryu, B.: High-availability free space optical and RF hybrid wireless networks. IEEE Wirel. Commun. **10**(2), 45–53 (2003). https://doi.org/10.1109/MWC.2003.1196402
13. Kahn, J.M., Katz, R.H., Pister, K.: Emerging challenges: mobile networking for “smart dust”. J. Commun. Netw. **2**(3), 188–196 (2000). https://doi.org/10.1109/JCN.2000.6596708
14. Majumdar, A., Ricklin, J.: Free-Space Laser Communications: Principles and Advances. Optical and Fiber Communications Reports. Springer, Berlin (2010)
15. Milner, S.D., Davis, C.C.: Hybrid free space optical/RF networks for tactical operations. In: Proceedings of the IEEE Military Communications Conference (MILCOM), vol. 1, pp. 409–415 (2004). https://doi.org/10.1109/MILCOM.2004.1493303
16. O’brien, D., Zeng, L., Le-Minh, H., Faulkner, G., Walewski, J., Randel, S.: Visible light communications: challenges and possibilities. In: IEEE 19th International Symposium on Personal, Indoor and Mobile Radio Communications (PIMRC), pp. 1–5 (2008). https://doi.org/10.1109/PIMRC.2008.4699964
17. Rabinovich, W.S., Murphy, J.L., Suite, M., Ferraro, M., Mahon, R., Goetz, P., Hacker, K., Freeman, W., Saint Georges, E., Uecke, S., Sender, J.: Free-space optical data link to a small robot using modulating retroreflectors. In: Proceedings of the SPIE, vol. 7464, pp. 746,408–746,408–9 (2009). https://doi.org/10.1117/12.828869
18. Rust, I.C., Asada, H.H.: A dual-use visible light approach to integrated communication and localization of underwater robots with application to non-destructive nuclear reactor inspection. In: Proceedings of the IEEE International Conference on Robotics and Automation (ICRA), pp. 2445–2450 (2012). https://doi.org/10.1109/ICRA.2012.6224718
19. Shen, T.C., Drost, R.J., Davis, C.C., Sadler, B.M.: Design of dual-link (wide- and narrow-beam) LED communication systems. Opt. Express **22**(9), 11107–11118 (2014)
20. Tian, B., Zhang, F., Tan, X.: Design and development of an LED-based optical communication system for autonomous underwater robots. In: Proceedings of the IEEE/ASME International Conference on Advanced Intelligent Mechatronics (AIM), pp. 1558–1563 (2013). https://doi.org/10.1109/AIM.2013.6584317

Part IV
Multi-Robot Planning

Bundling Policies for Sequential Stochastic Tasks in Multi-robot Systems

Changjoo Nam and Dylan A. Shell

Abstract This paper studies multi-robot task allocation in settings where tasks are revealed sequentially for an infinite or indefinite time horizon, and where robots may execute bundles of tasks. The tasks are assumed to be synergistic so efficiency gains accrue from performing more tasks together. Since there is a tension between the performance cost (e.g., fuel per task) and the task completion time, a robot needs to decide when to stop collecting tasks and to begin executing its whole bundle. This paper explores the problem of optimizing bundle size with respect to the two objectives and their trade-off. Based on qualitative properties of any multi-robot system that bundles sequential stochastic tasks, we introduce and explore an assortment of simple bundling policies. Our experiments examine how these policies perform in a warehouse automation scenario, showing that they are efficient compared to baseline policies where robots do not bundle tasks strategically.

1 Introduction

Multi-robot task allocation (MRTA) considers optimizing collective performance of a team of robots that execute a set of tasks [5]. In the canonical formulation, the sets of robots and tasks are fixed, and a decision-maker has full access to all information about the tasks. In practice, knowledge of the complete set of tasks may unavailable beforehand. In many applications, tasks are only revealed sequentially in an online fashion, e.g., dial-a-ride, e-commerce orders, etc. Compared to the case where the tasks are known a priori, work examining online instances of MRTA is scant (cf. discussion at length in [6]).

C. Nam (✉)
Robotics Institute, Carnegie Mellon University, Pittsburgh, PA, USA
e-mail: cjnam@cmu.edu

D. A. Shell
Department of Computer Science and Engineering, Texas A&M University,
College Station, TX, USA
e-mail: dshell@cs.tamu.edu

R. Groß et al. (eds.), *Distributed Autonomous Robotic Systems*,
Springer Proceedings in Advanced Robotics 6,
https://doi.org/10.1007/978-3-319-73008-0_17

Fig. 1 A simple example of online synergistic tasks: cherries must be transported from the conveyor to the storage facility. The robot may reduce traveling costs (fruitility = Joules per fruit) by waiting for multiple items, aggregating and transporting them together at once (i.e., filling a bundle). But as the robot forms larger sets of items, the items wait longer on average

In addition to considering sequential revelation of information, here we are concerned with *synergistic tasks*. By this we mean that work performed toward one task may be useful for others too, and planning with larger sets of tasks is beneficial. Also, in addition to conventional cumulative cost measures (e.g., fuel, time), we consider the timespan of tasks, viz. the elapsed time from creation until completion. There can be a tension between these two objectives. If robots wait and execute multiple tasks together (i.e., as a *bundle*), the costs per task may be less than for independent executions. But one incurs a delay in waiting for tasks to arrive in order to fill a bundle, so bundling drives up the timespan. Figure 1 shows an example of a setting where the robot fills its bundle by waiting for more items. Over and above the standard question of 'how should the tasks be allocated among robots?' one asks 'how many tasks should the robots bundle?'

This paper explores the structure of the bundling question at a high-level, abstracting away details of task performance itself, so that the findings can apply to a variety of settings. We begin with a qualitative study of the most basic instances in which tasks are revealed deterministically at a fixed frequency and where the task cost is a function of the task's location, and the location is independently and identically drawn from a known probability distribution. Also, robots are initially assumed to have no interactions. Based on the models of task arrival and execution, we compute the bundle sizes that minimize each objective. The simplifying assumptions that give this result are often violated in reality; to consider more realistic and complex settings, we explore a set of policies which efficiently adapt in order to improve performance.

This paper formulates of the problem of optimal task bundling for MRTA for sequentially revealed, synergistic tasks (Sect. 3). After analyzing the most basic scenario (Sect. 4), we introduce models that describe performance objectives as a function of bundle size. Using these models, our study of iterated bundle execution leads us to propose simple and efficient bundling policies suitable for variations of the problem which generalize the basic instance (Sect. 5). Evaluation of our policies is carried out quantitatively with extensive experiments (Sect. 6).

2 Related Work

Most previous work in online MRTA focuses on the question of *how to allocate tasks*. Early work on auction and market mechanisms [3, 4] studied the allocation of tasks when no model of their arrival was known. Some recent work [1, 10] considers online tasks with unpredictable arrivals, but the option of bundling is not considered. Online bipartite matching solves the underlying optimization problem of the online MRTA, but does not consider the bundling as a first-class question.

Bundling has been the focus of some prior attention. Koenig et al. [8] proposed Sequential Single-Item (SSI) auctions with bundles, wherein robots submit bids for subsets of tasks from a known set of tasks. The bidding phase and the winner determination phase iterate sequentially until all tasks have been assigned. Compared to standard parallel auctions, SSI reduces the team's cost by exploiting synergies among tasks. It also reduces the time spent bidding compared to the standard combinatorial auctions since not all permutations of assignments are considered. Zheng et al. [13] propose SSI with roll-outs where the cost of a task is evaluated together with the previously allocated tasks in order to exploit synergies. The aspect we study, which is absent for these prior works, is the meta-reckoning that weighs the consequences of choosing to delay decision-making since doing so may improve per-task system costs but worsen completion times.

Bullo et al. [2] show a variety of routing policies for multi-vehicle servicing demands in various scenarios, applying spatial-queuing theoretic models to vehicle routing where visiting locations arrive through a stochastic process. They provide theoretical analyses showing the stability (i.e., the number of waiting demands is bounded at all times) and the service quality of each routing policy. Their work provides thorough analyses and extensive comparisons of routing policies, demonstrating deep understanding in the routing domain. The techniques have yet to be shown to be directly applicable to more generic synergistic tasks. This paper adopts the hypothesis that the fundamentals of bundling and the principle trade-offs are not tied to the particular task-type or setting. It is possible that our study concerning the optimal bundle size would improve some of the routing policies of [2]. For example, the unbiased TSP policy has a design parameter n determining the size of sets to plan a TSP tour for all n demands. The set is analogous to our task bundle so n could be determined based on our result.

3 Problem Description

This section formulates the problem, expresses constraints on the problem, and describes the objectives. Warehouse automation is used as an example.

3.1 Problem Formulation

Given n robots, a task arrives and is inserted to a structure $\mathbf{T}$ every α seconds ($\alpha > 0$). The total number of tasks is unknown and the sequence may be unbounded. Here α could be deterministic or a random variable—in the latter case, we abuse notation slightly by using α to denote the mean of a distribution. The robots are assumed to share all available task information (i.e., $\mathbf{T}$) through a communication network or some other similar mechanism. The cost of performing a task is a function of both the robot and the task locations; we assume the locations of tasks are drawn from a probability distribution. Performing multiple tasks together enables some common work to be lumped together so that the cost, as a function of number of tasks performing in a bundle, is sub-linear.

Definition 3.1 (*Synergistic task*) Let $c(S)$ be the cost of performing a set of tasks S. Tasks are *synergistic* if $c(S_1) + c(S_2) > c(S_1 \cup S_2)$ where $S_1 \neq S_2$.

Definition 3.2 (*Task bundle*) Each robot owns a task bundle which is a group of $x \in \mathbb{Z}^+$ tasks extracted from $\mathbf{T}$. The *bundling time* is the average time that a task waits until the bundle of itself is completed.

We assume that the tasks in $\mathbf{T}$ are taken by order of arrival. Once a task is assigned to a bundle, it is no longer available to other robots. Once a robot has its bundle, it must finish the tasks without further modifications of that bundle. Tasks continue to arrive concurrently while the robots perform the work assigned to them. The robots iterate bundling and executing tasks in turn. A robot may be idle while waiting to fill its bundle and so idleness will depend on x. The simplifications in the preceding (e.g., no task reordering, no out-of-order execution, no pre-emption) ensure the operation will be starvation-free; judicious relaxation of these constraints may improve overall performance, but remains for future work.

Strategies for assigning tasks to robots make use of flexibility in (*i*) making the choice of whom to assign to a certain task, and (*ii*) when to assign the task. In general, waiting increases the available opportunities to optimize performance but waiting induces delays. Since (*ii*) is a central consideration in the present work, it is important to delineate the requirements of the strategies for assigning tasks. We do this by noting two necessities for the performance of online tasks:

- **Unconditional task acceptance:** any task that arrives, must be inserted to $\mathbf{T}$.
- **Non-starvation:** no task may be abandoned to remain in $\mathbf{T}$ indefinitely.

We consider two objectives to minimize, subject to fulfillment of both these requirements. Since there is no fixed set of tasks, the conventional sum-of-cost measure is no longer directly applicable, though average values of the following are:

Definition 3.3 (*System cost or execution cost*) The system cost per task $\bar{c} \in \mathbb{R}^{\geq 0}$ is the average cost spent to finish a task by the system. The average is taken over all robots and task pairs.

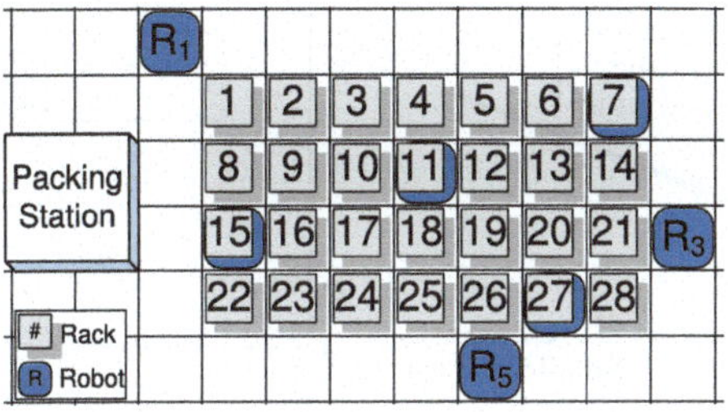

Fig. 2 A warehouse scenario where delivery tasks arrive sequentially. Delivering multiple items in one tour reduces the traveling cost but would delay the shipment of some items. Bundling strategically can manage this efficiency trade-off

Definition 3.4 (*Timespan or end-to-end time*) The timespan of a task $\bar{\tau} \in \mathbb{R}^{\geq 0}$ is the average time elapsed for a task from its insertion into **T** until its completion. The average is taken over all tasks.

3.2 An Example: Warehouse Automation

We use a warehouse automation problem (Fig. 2) as a running example to explore the properties of strategic bundling. Our study is not tied to this particular example as we have studied the optimization of bundle size for the vehicle routing problem too, see [11] for details. The reader interested in the generality of the model may find those additional examples beneficial.

In a warehouse, n robots are tasked with transporting items from racks to a packing station using baskets. Each rack has its own fixed location and contains identical items. Each rack has a space below permitting robots to navigate under it. The floor is discretized into grid cells where one cell contains one robot or one rack. Robots must navigate to their destinations while avoiding collisions. Once a robot is underneath its assigned rack, the rack dispenses its items in the quantity requested. The robot can visit multiple racks until its basket is full.

A job managing server (JM) receives orders that arrive sequentially and indefinitely according to a stochastic process. The JM splits each order into atomic transportation tasks: robots must deliver items from racks to the packing station. The JM inserts these into **T**. A robot bundles k tasks from **T** after it has completed its previous tasks. The extraction of tasks is done from the head of **T**. Also, robots execute their bundles from the head.[1]

[1] Again, robots may employ more advanced methods rather than this in-order task bundling and execution. The focus of this paper is not on elaborate methods for these aspects, so we leave them as modules that can be replaced by well-designed allocation and planning methods.

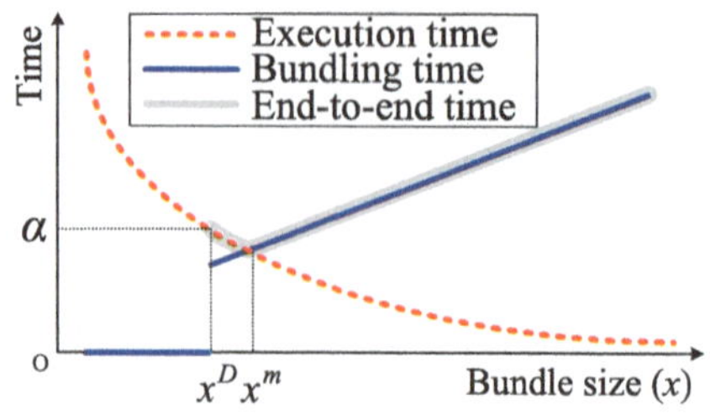

(a) The case where $x_g^* = x^m$.

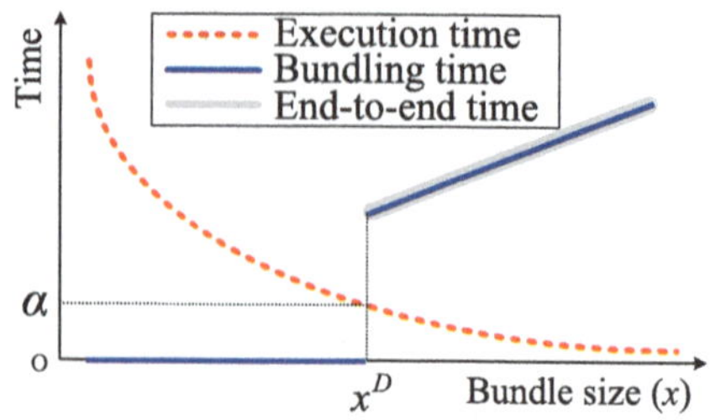

(b) The case where $x_g^* = x^D$.

Fig. 3 Illustrative functions that describe the system cost (red) and timespan (gray) per task. The time in **T**, $s(\cdot)$, is infinite for $x < x^D$ and the same with $f(\cdot)$ otherwise. There exists a finite bundle size x that makes $g(\cdot)$ minimum, and $g(\cdot)$ for $x < x^D$ is not shown since the value is infinite

4 An Analysis of Bundle Size

This section develops models for the optimization objectives. We begin with a simplified setting in which enables introduction of basic execution cost and task arrival models. Next, some complexity is added to help improve realism and applicability.

4.1 *The Basic Case: Independent Robots and Regular Task Arrivals*

4.1.1 The General Model

Consider a multi-robot system with negligible physical interference among robots and tasks in **T** that are sufficiently abundant so no contention occurs. The degree to which this is an over-simplification depends on practical circumstances, but this model results in objective values possessing invariant, steady-state properties. We assume stochastic tasks with locations independently and identically distributed from a uniform distribution over a rectangular grid with area S. A new task is revealed every α seconds. In what follows, refer to Fig. 3 as it shows models of both the system cost $\bar{c}$ per task and the average timespan $\bar{\tau}$ of a task.

A model for $\bar{c}$ is given by $f(x|S, v)$ where x is the bundle size and v is the task performance rate (e.g., velocity) of a robot (red in Fig. 3). Task synergies imply that $f(\cdot)$ is decreasing and, hence, the bundle size that minimizes the system cost is infinite ($x_f^* = \infty$). A model for bundling time is given by a functional $h(x|\alpha, f)$. Note that $h(\cdot)$ is discontinuous since $h(x|\alpha, f) = 0$ if $x < x^D$ (blue in Fig. 3). The quantity x^D denotes the bundle size when $f(x|S, v) = \alpha$, that is, the point of balance between the rate of task arrivals and (average) executions. Below equilibrium x^D, a task arrives before an existing task is completed. Thus, tasks accumulate ($|\mathbf{T}|$ is unbounded) and robots remove tasks from **T** without waiting, i.e., the bundling time is zero. For

$x \geq x^D$, tasks do not accumulate in **T**, and a robot must wait for tasks in order to fill its bundle and so the bundling time is nonzero. So, $h(x|\alpha, f) = h'(x|\alpha)$ for $x \geq x^D$, where $h'(\cdot)$ represents the bundling time without considering the potential unbounded accumulation of tasks in **T**. But there is also another component $s(x|\alpha, f)$, the average time a task resides in **T** before it is taken by a robot. We have $s(x|\alpha, f) = \infty$ for $x < x^D$ because **T** keeps accumulating tasks so it has a significant number of tasks that wait indefinitely to be bundled. Otherwise, $s(x|\alpha, f) = f(\cdot)$ since tasks only stay in **T** while robots are executing their bundles.

A model of timespan per task $\bar{\tau}$ is given by $g(x|S, v, \alpha, f)$, which describes the component that dominates the time:

$$g(x|S, v, \alpha, f) := \max(f(x|S, v), h(x|\alpha, f), s(x|\alpha, f)), \tag{1}$$

illustrated with the thick gray curves in Fig. 3. For $x < x^D$, $g(\cdot)$ the value is infinite. The x value that minimizes $g(\cdot)$ is the optimal bundle size x_g^*. The value x_g^* can be determined via two cases, shown in Fig. 3a, b respectively. In Fig. 3a, x^m is the equilibrium between the task bundling time and the execution time. At x^m, the robot finishes executing a bundle when the next bundle has just been filled, thus $g(\cdot)$ takes the minimum at this point, so $x_g^* = x^m$. It is worth noting that there is no waiting time between iterations. In Fig. 3b, x^m does not exist because the number of tasks in **T** are unbounded. The execution time dominates the (zero) bundling time, and $f(x^D|\cdot)$ takes the minimum at x^D, so $x_g^* = x^D$.

4.1.2 The Warehouse Example

We derive analytic models of $\bar{c}$ and $\bar{\tau}$ for the warehouse example. We assume that robots move one grid cell (with no diagonal) at each step. Robots bundle k tasks and execute them sequentially using a path planner (e.g., A*). Collisions are avoided by dynamic replanning. In an $a \times b$ rectangular space, racks are located within an $a' \times b'$ rectangle. The packing station is located at (x_0, y_0). Figure 4a shows an example setting where $a = b = a' = 20$, $b' = 5$, and $(x_0, y_0) = (10, 1)$. Tasks arrive every α steps and their locations are uniformly distributed within the $a' \times b'$ rectangle.

At steady-state, executing a bundle consists of three trips: (*i*) the trip from the station to a rack, (*ii*) the trip among the k racks, and (*iii*) the trip to the station. Let d be the expected distance from the station to a random rack and E_d be the expected distance between two random racks. We do not write out the expressions for d and E_d owing to limited space, but their computation requires nothing more than basic calculus. From (*i*) and (*iii*) we have $2d$. From (*ii*), we have $(k-1)$ trips among k racks. Thus, it takes $2d + (k-1)E_d$ time steps for k tasks. The system cost (time traveled) per task is

$$f(k|S, v) = \left(\frac{2d + (k-1)E_d + (k+1)}{k}\right). \tag{2}$$

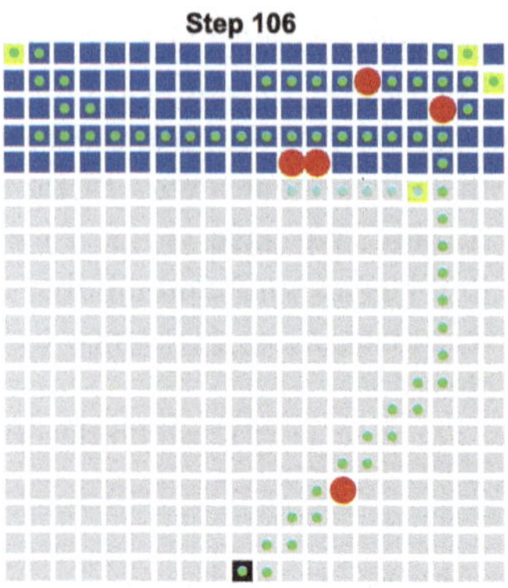

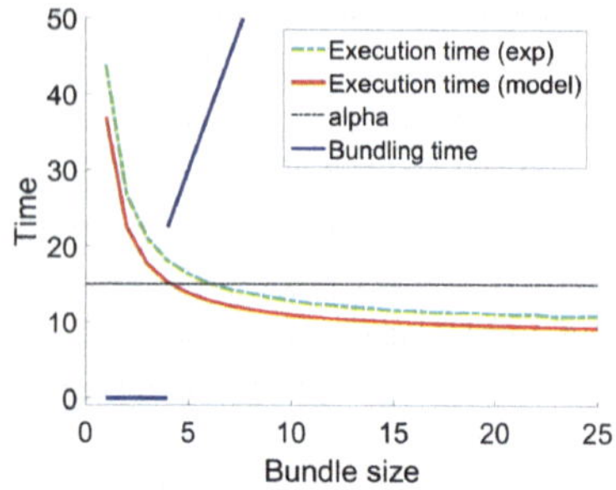

(a) A simulation of the warehouse problem. Red circles and yellow squares represent robots and tasks. The blue and black squares represent racks and the station, respectively. The small dots show paths where the cyan dots are replanned paths to avoid collisions.

(b) The models (2) in red dotted and (3) in blue solid. The horizontal line represents α. The green curve shows the experimental result from Fig. 4a.

Fig. 4 A snapshot of the warehouse simulator and the validation of the models from simulations

The term $(k+1)$ is added because we assume that planning a path between two points takes one time step. There are total $k+1$ runs of the A* planner. Notice that all environmental variables are represented by S.

A task waits in a bundle for $k\alpha - j\alpha$ steps where j is the step when the task is inserted. Then, the sum of the bundling time for all tasks is $\sum_{j=1}^{k} k\alpha - j\alpha = k^2\alpha - \frac{k(k+1)}{2}\alpha = \frac{\alpha}{2}k(k-1)$. The function describing the bundling time per task is

$$h(k|\alpha, f) = \begin{cases} 0 & \text{if } k < x^D, \\ h'(k|\alpha) = \frac{\alpha}{2}(k-1) & \text{otherwise.} \end{cases} \tag{3}$$

Interestingly, $h'(\cdot)$ is a special case of the mean residual life of a customer in a renewal process presented in [7]. The residual life is the amount of time that the customer must wait until being served. The general form of (3) when task arrivals follow a Poisson process is

$$h'(k|\alpha, \lambda) = \frac{\alpha}{2}\left(1 + \frac{\lambda}{\alpha^2}\right)(k-1), \tag{4}$$

where λ is the variance of the arrival interval. If $\lambda = 0$, then (4) reduces to (3).

Figure 4b shows (2) as a function of bundle size (red) along with the values from experiments (green) where $\alpha = 15$, and $a = b = 20$. The blue line represents (3). One may derive models analytically as above or make use of a body of research that provides such the models (e.g., the optimal length of a TSP tour, which is a function of the number of visit locations [9, 12]).

4.2 *Adding Realism: Robot Interactions and Stochastic Task Arrivals*

Next, we consider settings where tasks involve uncertainty via a stochastic arrival process. In addition, non-negligible robot interactions are taken into account. These generalizations introduce a gap between the basic models and the performance observed in the system (different from Fig. 3). Factors responsible for the gap include:

1. *Task location:* Owing to the stochasticity of task locations, the system cost described above should really describe the mean of a random variable. Therefore $f(\cdot)$ is no longer deterministic, but has some variance.
2. *The task arrival process:* Stochastic arrivals mean that $h(\cdot)$ is no longer deterministic, making $h(\cdot)$ a random function with some variability.
3. *Physical interference:* Robots may interfere with one another, increasing the system cost so that $f(\cdot)$ in the basic model underestimates the actual cost.
4. *Task contention:* Robots may experience contention for tasks, increasing the bundling time while waiting to fill bundles. Thus, the $h(\cdot)$ in the basic model underestimates the actual bundling time.
5. *Robot coordination:* Robots that coordinate to optimize performance will reduce the system cost per task; thus, the basic model $f(\cdot)$ may overestimate actual costs.

In sum, the optimal bundle size x_g^* in the basic case is likely to differ from the optimal value for realistic settings. But the modifications needed depend on aspects of the domain and many details of the particular instance. This fact motivates our exploration of adaptive *bundling policies* to adjust to circumstances, improving performance of idealized treatments which ignore complicated aspects of the system. In Sect. 5, we propose model-free policies with a dynamic bundle sizes.

4.2.1 Remarks on Synergies in the Realistic Case

The factors given above may have ramifications, not only for the parameters (e.g., slope, y-intercept) of $f(\cdot)$, but also its monotonicity. We assume that the tasks remain synergistic as negative robot interactions caused by resource contention can often be mitigated by (far-sighted or globally aware) planning algorithms—even if some optimality must be sacrificed, approximations should suffice in this regard. Thus, we believe that $f(\cdot)$ usually continues to decrease even if synergism diminishes.

Nevertheless, some extreme configurations (e.g., a very narrow warehouse causing heavy congestion on every passage) could make $f(\cdot)$ non-synergistic when no algorithm can reduce these negative effects. The analysis remains valid even though it becomes more involved when handling non-idealized cases: the question of what bundle size, x, makes $g(\cdot)$ a minimum still persists. If $g(\cdot)$ is always infinite or constant, the fact that the optimal set is empty or all of $\mathbb{Z}^+$ is informative. And, if there are multiple bundle sizes which minimize $g(\cdot)$, any of them can be chosen. And, of course, when $f(\cdot)$ is no longer synergistic, the optimal bundle size is 1, as there is no benefit from executing tasks in concert.

5 Bundling Policies

Before considering adaptive policies, we describe static policies where bundle sizes remain constant. Later, these static policies will serve as a performance baseline. Next, we propose simple policies that are flexible with agreeable behavior across a range of circumstances. We provide a condition for a stable policy (i.e., ensuring $|\mathbf{T}|$ is bounded) and a bound that describing the best possible performance, which no bundling policy can exceed. We consider only policies that minimize the timespan $\bar{\tau}$, since minimizing $\bar{c}$ has limited practical value, leading to infinite bundles.

- **Baseline static policies**:
  - **The ideal policy**: If the basic model $g(\cdot)$ is available, finding a k that minimizes (1), the timespan, gives x_g^*. In this ideal static policy, each robot keeps $k = x_g^*$. When a robot finishes its current bundle and tries to execute the next bundle, there may be insufficient tasks in $\mathbf{T}$ to form that next bundle, causing the robot to wait, idly, for new tasks. If $|\mathbf{T}| \geq k$, the robot takes k tasks and executes them immediately. Since this policy does not handle uncertainties in the task profile (discussed in Sect. 4.2) it is possible that the timespan can diverge if $\mathbf{T}$ is unbounded.
  - **The min- and max-load policies**: Another reasonable policy is to execute, instantaneously, tasks one by one (i.e., $k = 1$). This min-load policy does not exploit synergies but would yield a small timespan since robots never bundle multiple tasks. We also consider the max-load policy where bundles take up to some given capacity (e.g., the capacity of baskets of robots, the memory size of robots where task information is stored).
- **Model-free policies**:
  - **The sweeping policy**: This policy takes all tasks $k = |\mathbf{T}|$ if $|\mathbf{T}| \geq 1$, never incurring any bundling time as long as at least one task is available. The policy exploits synergies among available tasks. Figure 5a shows the bundle size *vs.* time (30,000 steps). The black dotted line shows the ideal bundle size reflecting the equilibrium in which the execution and bundling times are equal. The average bundle size (blue dotted) would reflect the equilibrium when the task uncertainties and robot interactions are accounted for.
  - **The averaging policy**: The sweeping policy's equilibrium is constant unless the stochastic parameters describing the task location and arrival process change. The sweeping policy does not make explicit use of any representation of the equilibrium. Instead, it tracks the equilibrium via history: the averaging policy begins with $k = 1$ and averages the previous bundle sizes saved in a history window W. The smaller the window size, the more sensitive the policy to variability.

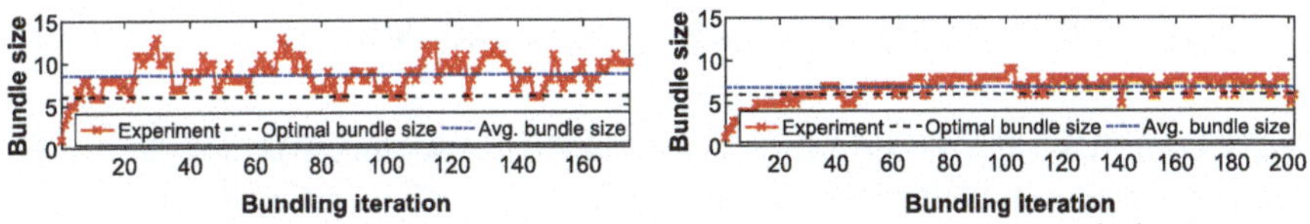

(a) The sweeping policy **(b)** The averaging policy ($|W| = 20$)

Fig. 5 Plots showing bundle size versus time steps. Bundle sizes (red) stay around the blue value, which is the optimal bundle size for the given task setting

5.1 An Analysis of the Bundling Policies

We say a policy is *stable* if the number of waiting tasks in **T** is bounded.

Proposition 5.5 *Let $x^{D'}$ be the value of x^D, incorporating the various non-idealizations for more realistic settings. A bundling policy is stable if $k \geq x^{D'}$.*

Proof Even in circumstances outside of the basic case, the shape of the objectives (Fig. 3) is invariant: non-idealized circumstances elevate or lower the lines, but do not affect their shape. Thus, the analysis of the models in Sect. 4.1.1 may still holds. Task queue **T** does not accumulate tasks if the bundle size is greater than the equilibrium between the task execution rate and the arrival rate (i.e., $x \geq x^{D'}$). In other words, a task is completed before or when a new task arrives and, thus, a bundling policy with $k \geq x^{D'}$ is must be stable. □

The ideal policy should be stable in the basic case but is cannot be guaranteed to be stable in more realistic settings, since $x^{D'}$ is not known exactly. Also, it is not obvious whether the sweeping and the average polices are stable. In practice, however, they appear to be stable as Fig. 5 shows. Specifically, $x_g^* \geq x^{D'}$ in any case as shown in Sect. 4.1.1. In Fig. 5, the average bundle size (blue) is larger than x_g^* (black) so larger than $x^{D'}$. The stability of the two is shown empirically in the following experiments.

Next, we show the lower bound in which any bundling policy cannot exceed.

Proposition 5.6 *The lower bound of $\bar{\tau}$ is $\lim_{k\to\infty} f(k|\cdot)$ for any policy.*

Proof From (1), the lower bound of $\bar{\tau}$ is $g(\cdot) = \max(\min f(\cdot), \min h(\cdot), \min s(\cdot))$. The minimum of $f(\cdot)$ is at $k \to \infty$. The minimum of others are zero (i.e., no bundling at $k = 1$ and no residing time for $k < x^D$). Then, the maximum is $\min f(k|\cdot) = \lim_{k\to\infty} f(k|\cdot)$. No policy can exceed this bound. □

For example, in the warehouse problem, $\lim_{k\to\infty} f(k|\cdot) = E_d + 1$.

6 Quantitative Study: Comparisons of the Policies

This section describes experiments in the warehouse setting. We provide experimental settings and analyze the results where task locations are i.i.d. with a fixed task interval. Then, we show results when tasks arrive according to a Poisson process whose arrival intervals and locations are non-i.i.d. Both cases involve physical interference and task contention among robots.

6.1 *Experimental Settings and Results*

The size of the warehouse is $a = b = 30$ where the number of racks is 300 so $a' = 30$, $b' = 10$, $E_d = 11.11$. The packing station is at $(15, 1)$ so $d = 25.50$. For a fixed number of robots ($n = 5$), we assume that all robots move at the same velocity, one grid cell per one time step. As discussed above, using x_f^* is unrealistic so we minimize the timespan only and scrutinize how the system cost changes.

We set $\alpha = 2$ for the regular *intense* task arrival.[2] The parameter for the Poisson arrival process is $\lambda = \frac{1}{\alpha}$, where the mean arrival interval is $\lambda^{-1} = \alpha$. Those two arrival processes have the same mean interval. In addition, we ran experiments for $\alpha = 30$ which represent *intermittent* arrivals (i.e., $E_d + 1 < \alpha$). For the ideal policy, $k = 10$ for $\alpha = 2$ and $k = 2$ for $\alpha = 30$. The max-load and the min-load have $k = 30$ and $k = 1$, respectively. The size of window is $|W| = 20$ for the averaging policy. We measure the two objective values over 10,000 steps and run 10 repetitions. Table 1 summarizes the results. Figure 6 shows the size of $\mathbf{T}$ over time.

6.2 *Analysis of Results*

6.2.1 Regular Task Arrivals and an i.i.d. Spatial Distribution

The results of the intense arrivals (Table 1a) show that any bundling policy outperforms non-bundling (the min-load). The min-load policy does not make use of the synergies of tasks so its task execution time (system cost) is larger than other policies. Thus, tasks accumulate rapidly in $\mathbf{T}$ (Fig. 6a) which increases the timespan significantly. The max-load policy yields the smallest system cost since it exploits synergies maximally. Bundling that many tasks increases the bundling time which contributes to the total timespan. Thus, tasks are accumulated moderately in $\mathbf{T}$ (Fig. 6b) while robots spend time bundling. The ideal policy is outperformed by the max-load policy. It is possible that $k = 30$ is closer to the actual optimal bundle size than $k = 10$ when robot interactions are taken into account. Task accumulation (Fig. 6c) is faster than

[2]The lower bound of the execution time per task is greater than the arrival interval of a task, i.e., $E_d + 1 = 26.50 > \alpha = 2$.

Table 1 Comparisons of policies. The values are the mean and standard deviation (10 repetitions)

(a) Regular task arrivals with an i.i.d. task distributions

Bundling policy	Intense ($\alpha = 2$)		Intermittent ($\alpha = 30$)	
	System cost	Timespan	System cost	Timespan
Min-load	64.64 (0.2614)	4231 (4.557)	67.54 (5.599)	65.74 (5.532)
Max-load	16.47 (0.0906)	2142 (30.41)	24.83 (14.49)	1131 (246.7)
Ideal	22.89 (0.1766)	2845 (23.22)	51.51 (7.291)	116.7 (14.43)
Sweeping	20.73 (1.784)	1488 (49.41)	71.98 (9.627)	69.87 (9.200)
Averaging	21.43 (0.3958)	1441 (25.59)	70.13 (9.627)	74.34 (0.7071)

(b) A Poisson task arrival process with a non-i.i.d. task distribution

Bundling policy	Non-i.i.d. interval	
	System cost	Timespan
Min-load	65.28 (0.5991)	4046 (38.98)
Max-load	11.38 (0.0675)	396.9 (12.79)
Sweeping	12.57 (0.4441)	301.3 (65.04)
Averaging	18.69 (2.234)	1921 (98.11)

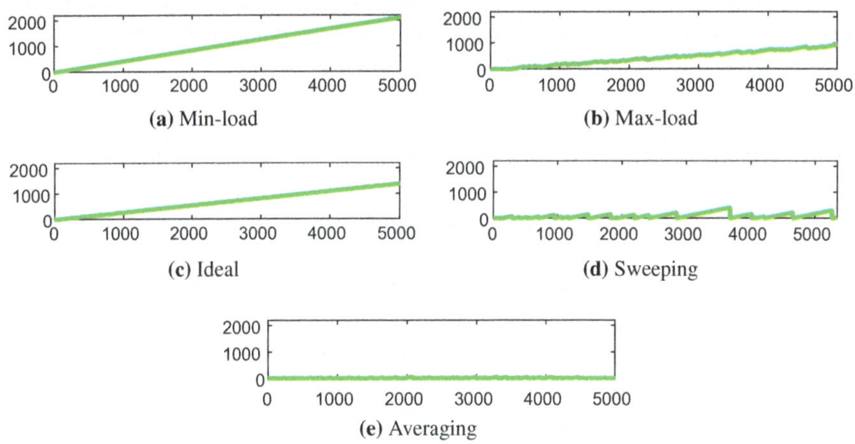

Fig. 6 Plots showing bundle size (y-axis) over time (x-axis). We show the case of regular and intense intervals. The sweeping and averaging policies are able to bound **T**

the max-load policy but slower than the min-load policy. The sweeping and averaging policies show the similar timespan, which outperform all the baseline polices because they can ensure **T** remains bounded as shown in (Fig. 6d, e). Their system costs are not the most efficient found, but are competitive.

The most important aspect of the performance is how a policy ensures **T** is bounded because it relates directly to the timespan at steady-state. If **T** is not bounded, the timespan, which includes the time a task resides in **T** will itself diverge. The max-load policy seems to have a moderate timespan, but must diverge if experiments are run longer. On the other hand, the proposed model-free policies are able to keep **T** short.

To determine the appropriate policy, one's overall purpose must be borne in mind. To help in choosing a policy, we show the results in the objective space in Fig. 7. For intense arrivals, we would use one of the model-free policies for the timespan. In considering system cost, one would choose the max-load policy.

The intermittent case shows a slightly different result. The min-load policy has the smallest timespan. For the system cost, the max-load policy still has the smallest value. Since tasks arrive slowly, only few robots (most times only one robot) executes tasks, resulting in little physical interference, and, thus, $f(\cdot)$ in Fig. 3 is the same as the basic case. On the other hand, severe task contention means that $h(\cdot)$ is steeper than the basic model. This causes the optimal bundle size to shrink. The min-load policy has the closest bundle size to this small value so its timespan is the smallest. All policies ensure that **T** is bounded since tasks arrive slowly. It would be beneficial for the robots to move to the centroid of the area of racks when they are idle since the expected distance to a random rack is the shortest at the point. This waiting location changes depending on the spatial distribution of tasks.

6.2.2 A Poisson Arrival of Tasks and a Non-i.i.d. Spatial Distribution

We also ran experiments to explore how the policies work in more complex situations with task locations chosen so as *not* to be independently and identically distributed. Specifically, a task is drawn from a uniform distribution within the area that robots work with a probability of 0.5. With a probability of 0.5, a task is drawn from a normal distribution that has the location of the previous task as the mean. The task arrival process also has a mean interval between tasks that is non-i.i.d. With a probability of 0.5, the arrival process follows the Poisson process with λ which is a sinusoidal function. With a probability of 0.5, the interval is drawn from a uniform distribution where the upper bound is related to the previous value of λ.

In Table 1b and Fig. 7c, we show the results (no ideal policy is reported as we have no appropriate model). Except for the averaging policy, the policies exhibit a tendency similar to that of intense and regular arrivals before. As noted, the max-load policy accumulates tasks over time, so the timespan will increase with longer experiments. In the regular arrival case, the averaging policy shows performance that is competitive with the sweeping policy, but is now far worse. This is because of the frequent switching of the task profile (both arrivals and locations) disrupts the averaging policy's local estimates of the rate and various system costs.

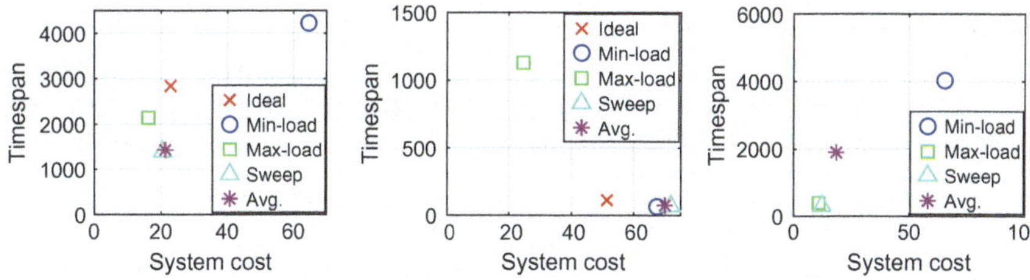

(a) Regular intense (b) Regular intermittent (c) Poisson non-i.i.d.

Fig. 7 The objective space showing a Pareto frontier

This problem is resolved by having a short history: setting $|W| = 1$, the result is $\bar{c} = 11.98$ and $\bar{\tau} = 427.9$, which is now comparable to sweeping. The size of $\mathbf{T}$ in this experiment is essentially the same as with Fig. 6 and is therefore omitted.

7 Conclusion and Future Work

This paper treats a variant of the MRTA problem where stochastic tasks arrive continuously, and the system must determine how to bundle tasks in order to make best use of synergies between tasks. First, we proposed idealized models to understand the foundations of the bundling question. Then we explored how the models change for more realistic cases where task uncertainties and robot interactions are involved. We proposed adaptive bundling policies in order to deal with uncertainties, comparing the results to baseline policies which do not have a particular strategy for bundling. Evidence is provided to show that the proposed policies outperform the baseline. More importantly, the proposed policies can bound the number of waiting tasks at all time for intense task arrivals whereas the baseline policies cannot. Also, the results show that the policies are able to deal with other sources of complexity and uncertainty, such as probabilistic task arrivals or non-i.i.d. task locations and arrival intervals—which can express aspects of spatio and temporal locality.

Further study of related strategies can improve the performance of bundling. For example, preemption of bundle executions may be useful so that some robots can stop working to perform other tasks. Or robots may swap the tasks in their bundles to reduce costs during execution. Out-of-order task insertion to $\mathbf{T}$ or bundles is also interesting. Several applications naturally impose temporal constraints between tasks which are worth considering. Lastly, it would be desirable to give the analytic bound of the team performance using the policies to compute the minimum number of robots to bound $\mathbf{T}$ for a given frequency of task arrivals.

Acknowledgements This work was supported in part by NSF awards IIS-1302393 and IIS-1453652.

References

1. Amador, S., Okamoto, S., Zivan, R.: Dynamic multi-agent task allocation with spatial and temporal constraints. In: International Conference on Autonomous Agents and Multi-agent Systems, pp. 1495–1496 (2014)
2. Bullo, F., Frazzoli, E., Pavone, M., Savla, K., Smith, S.: Dynamic vehicle routing for robotic systems. Proc. IEEE **99**, 1482–1504 (2011)
3. Dias, M., Stentz, A:. A market approach to multirobot coordination. Technical report, Carnegie Mellon University, 2000
4. Gerkey, B., Matarić, M.: Sold!: auction methods for multi-robot coordination. IEEE Trans. Robot. **18**, 758–768 (2002)
5. Gerkey, B., Matarić, M.: A formal analysis and taxonomy of task allocation in multi-robot systems. Int. J. Robot. Res. **23**, 939–954 (2004)
6. Heap, B.: Sequential single-cluster auctions for multi-robot task allocation. Ph.D. thesis, The University of New South Wales, 2013
7. Kleinrock, L.: Queuing Systems. Wiley, New York (1975)
8. Koenig, S., Tovey, C., Zheng, X., Sungur, I.: Sequential bundle-bid single-sale auction algorithms for decentralized control. In: Proceedings of International Joint Conference on Artificial intelligence, pp. 1359–1365 (2007)
9. Lee, J., Choi, M.: Optimization by multicanonical annealing and the traveling salesman problem. Phys. Rev. E **50**, R651 (1994)
10. Meir, R., Chen, Y., Feldman, M.: Efficient parking allocation as online bipartite matching with posted prices. In: International Conference on Autonomous Agents and Multi-agent Systems, pp. 303–310 (2013)
11. Nam, C., Shell, D.: An empirical study of task bundling for sequential stochastic tasks in multi-robot task allocation. Technical Report TAMU-CSE-16-7-1, CSE Department, Texas A&M University, 2016
12. Stein, D.: An asymptotic, probabilistic analysis of a routing problem. Math. Oper. Res. **3**, 89–101 (1978)
13. Zheng, X., Koenig, S., Tovey, C.: Improving sequential single-item auctions. In: Proceedings of IEEE/RSJ International Conference on Intelligent Robots and System, pp. 2238–2244 (2006)

Decomposition of Finite LTL Specifications for Efficient Multi-agent Planning

Philipp Schillinger, Mathias Bürger and Dimos V. Dimarogonas

Abstract Generating verifiably correct execution strategies from *Linear Temporal Logic* (LTL) mission specifications avoids the need for manually designed robot behaviors. However, when incorporating a team of robot agents, the additional model complexity becomes a critical issue. Given a single finite LTL mission and a team of robots, we propose an automata-based approach to automatically identify possible decompositions of the LTL specification into sets of independently executable task specifications. Our approach leads directly to the construction of a team model with significantly lower complexity than other representations constructed with conventional methods. Thus, it enables efficient search for an optimal decomposition and allocation of tasks to the robot agents.

1 Introduction

High-level planning based on Linear Temporal Logic (LTL) specifications creates the opportunity to deploy robots in increasingly sophisticated scenarios while being able to provide guarantees regarding correctness and optimality, e.g. [14, 16, 17]. In such scenarios, systems often benefit from utilizing multiple agents in order to flexibly distribute workload. Instead of executing tasks sequentially, they can be allocated to

Dimos V. Dimarogonas was supported by the H2020 ERC Starting Grant BUCOPHSYS and the Swedish Research Council (VR).

P. Schillinger (✉) · M. Bürger
Corporate Research - Cognitive Systems (CR/AEY2), Robert Bosch GmbH Renningen, Stuttgart, Germany
e-mail: philipp.schillinger@de.bosch.com; schillin@kth.se

M. Bürger
e-mail: mathias.buerger@de.bosch.com

P. Schillinger · D. V. Dimarogonas
EES, KTH Centre for Autonomous Systems and ACCESS Linnaeus Center, KTH Royal Institute of Technology, Stockholm, Sweden
e-mail: dimos@kth.se

R. Groß et al. (eds.), *Distributed Autonomous Robotic Systems*, Springer Proceedings in Advanced Robotics 6,
https://doi.org/10.1007/978-3-319-73008-0_18

different agents and carried out in parallel. Nonetheless, considering multiple agents for a set of tasks introduces significant additional planning complexity.

Consider as a motivating example a hospital station. In addition to caring for patients, nurses are required to refill medical supplies, guide visitors, deliver meals, or clean equipment. To support them, a multi-robot system can automate required transportation. Similarly, workflows in an intelligent factory can be improved using a multi-robot system. Machines for assembly can monitor required components and request supplies if they run low. In such scenarios, missions are typically given as one specification, implicitly based on a set of independent finite tasks, e.g., deliver meals to all rooms or supply a machine with several components. In the following, we formally define problems of this type, based on an LTL mission specification and a system model, with the goal to obtain execution strategies for all robots such that the mission is guaranteed to be fulfilled with minimal costs.

Approaches for multi-agent task planning include Mixed-Integer Linear Program (MILP) formulations [9, 21], numerical vector-based allocation [1], or market-based contracting [23]. However, these approaches involve forming a team product or require separate cost calculation for all combinations of allocation options, given that these options are explicitly known. An automata-based approach is proposed in [10] and assumes synchronous team motions to plan motion sequences from an LTL specification. This assumption is relaxed in [19] by a two-phase model reduction approach, and in [3, 20] by employing trace-closed languages [18] to abstract over asynchronous motions of the agents. However, these approaches only decompose the LTL specification into independent tasks in the special case of disjoint agent alphabets. Their focus is rather to coordinate execution between a known team of agents given explicit allocation options.

If task allocation to the agents is explicitly provided, approaches like [8, 15] take communication and motion coordination between the agents into account; [11, 13] choose a Petri-Net based approach for explicit coordination whereas [6] uses a game theoretic approach for negotiation, able to consider adversarial agents. In contrast, our approach does not assume known allocation and instead decomposes the mission such that execution does not require coordination between the agents, enabling them to operate independently based on their assigned task.

The contributions of this paper are as follows: (i) A formal definition of the decomposition of finite LTL specifications into tasks is introduced and several relevant properties of such decompositions are discussed. (ii) It is shown how such decompositions can be efficiently identified in an automaton representation of the LTL specification. (iii) Based on these results, a method is presented for constructing a team model of tractable complexity with respect to the team size that can be used for efficient multi-agent planning. The proposed approach and its computational advantages are illustrated on an example system setup motivated by the scenarios mentioned above. Specifically, we evaluate state space complexity of the team model and planning time based on a ROS implementation of our approach.

2 Preliminaries

2.1 LTL Semantics

An LTL specification ϕ over a set of atomic propositions Π identifies a set of temporal sequences $\sigma = \sigma(1)\sigma(2)\ldots$ which fulfill this specification, written as $\sigma \models \phi$. At each discrete time $t \in \mathbb{T}$ with $\mathbb{T} \subseteq \mathbb{N}$, a set of propositions $\sigma(t) \subseteq \Pi$ is true, i.e., a sequence is defined as $\sigma : \mathbb{T} \to 2^{\Pi}$. A sequence is called finite if it is bounded by a maximum time T and then we say it has length T. Finite LTL is a variant of LTL that can be interpreted over finite sequences. Classes of finite LTL are formulas which are insensitive to infiniteness [5], for example co-safe LTL [12] or LTL_f [4].

The semantics of LTL over a finite sequence $\sigma(1)\ldots\sigma(T)$ are defined as follows: **(1)** $\sigma(t) \models \pi$ iff $\pi \in \sigma(t)$; **(2)** $\sigma(t) \models \neg\varphi$ iff $\sigma(t) \not\models \varphi$; **(3)** $\sigma(t) \models \varphi_1 \wedge \varphi_2$ iff $\sigma(t) \models \varphi_1$ and $\sigma(t) \models \varphi_2$; **(4)** $\sigma(t) \models \varphi_1 \vee \varphi_2$ iff $\sigma(t) \models \varphi_1$ or $\sigma(t) \models \varphi_2$; **(5)** $\sigma(t) \models \circ\varphi$ iff $\sigma(t+1) \models \varphi$; **(6)** $\sigma(t) \models \varphi_1 \mathcal{U} \varphi_2$ iff there exists $t_2 \in [t, T]$ such that $\sigma(t_2) \models \varphi_2$ and $\sigma(t_1) \models \varphi_1$ for all $t_1 \in [t, t_2 - 1]$; **(7)** $\sigma(t) \models \varphi_1 \mathcal{R} \varphi_2$ iff for all $t_2 \in [t, T]$ either $\sigma(t_2) \models \varphi_2$ or there exists a $t_1 \in [t, t_2 - 1]$ such that $\sigma(t_1) \models \varphi_1$.

Furthermore, we derive the operators *eventually* $\diamond\varphi = \top \mathcal{U} \varphi$ and *always* $\Box\varphi = \Omega \mathcal{R} \varphi$ with $\Omega = \bot$ for all $t \le T$, where $\top$ denotes *true* and $\bot$ denotes *false*. Note that, as usual for finite LTL [5], the scope of *always* is limited to the range $t \in \{1, \ldots, T\}$, not considering an infinite future.

In the case of finite LTL specifications, a finite automaton can be constructed from the given LTL formula ϕ [2].

Definition 1 *(NFA)* A nondeterministic finite automaton (NFA) is given as the tuple $\mathcal{F} = (Q, Q_0, \alpha, \delta, F)$ consisting of **(1)** a set of states Q, **(2)** a set of initial states $Q_0 \subseteq Q$, **(3)** an alphabet α of Boolean formulas over $\pi \in \Pi$, **(4)** a set of transition conditions $\delta\colon Q \times Q \to \alpha$, **(5)** a set of accepting (final) states $F \subseteq Q$.

Note that we define the set of transitions as Boolean conditions which need to be fulfilled for taking a transition. Especially, the absence of a transition is denoted by the formula $\delta(q_1, q_2) = \bot$, which can never be fulfilled. A finite sequence of states $q \in Q$ is called a run $\rho\colon \{0, 1, \ldots, T\} \to Q$ with $\delta(\rho(t-1), \rho(t)) \neq \bot$ for all $t \in \{1, \ldots, T\}$. We say that a sequence σ describes ρ if $\sigma(t) \models \delta(\rho(t-1), \rho(t))$ for all $t \in \{1, \ldots, T\}$. In the case that a run ρ leads from an initial state $\rho(0) \in Q_0$ to an accepting state $\rho(T) \in F$, this run is called accepting. If $\mathcal{F}$ is constructed appropriately [22], a finite sequence σ fulfills a finite LTL formula ϕ if and only if σ describes an accepting run ρ in the NFA $\mathcal{F}$ constructed from ϕ.

In general, constructing an NFA from an LTL formula has worst-case time and space complexity exponential in the length of the formula $|\phi|$ [2]. The actual complexity might be lower depending on the specific formula. However, this construction complexity is usually still a limiting factor.

2.2 Closure Labeling

In order to decide if a sequence σ fulfills an LTL specification ϕ, [22] defines a closure labeling $\tau : \mathbb{N} \rightarrow 2^{cl(\phi)}$ corresponding to σ and constructed as given below. The closure of ϕ is the set of its subformulas, given by $cl(\phi)$ with $\phi \in cl(\phi)$ and recursively for all operations $\circ\varphi_1 \implies \varphi_1 \in cl(\phi)$ and $(\varphi_1 \wedge \varphi_2), (\varphi_1 \vee \varphi_2), (\varphi_1 \mathscr{U} \varphi_2), (\varphi_1 \mathscr{R} \varphi_2) \in cl(\phi) \implies \varphi_1, \varphi_2 \in cl(\phi)$.

The closure labeling enables to formally reason about requirements imposed by the part of a sequence σ from 1 to t, in the following denoted by $\sigma(1, \ldots, t)$. Following the intuition of [7], we divide the construction of τ into two parts, one defining *expectations* from the previous step τ^e and one consequently required *observations* τ^o. Then, τ is given by $\tau(t) = \tau^e(t) \cup \tau^o(t)$.

Definition 2 *(Expectations)* Expectations $\tau^e(t+1)$ on the next time step are constructed such that:

- if $\circ\varphi_1 \in \tau(t)$ then $\varphi_1 \in \tau^e(t+1)$
- if $\varphi_1 \mathscr{U} \varphi_2 \in \tau(t)$ and $\varphi_2 \notin \tau(t)$, then $\varphi_1 \mathscr{U} \varphi_2 \in \tau^e(t+1)$
- if $\varphi_1 \mathscr{R} \varphi_2 \in \tau(t)$ and $\varphi_1 \notin \tau(t)$, then $\varphi_1 \mathscr{R} \varphi_2 \in \tau^e(t+1)$.

Definition 3 *(Observations)* Observations $\tau^o(t)$ on the current time step are constructed such that:

- $\perp \notin \tau^o(t)$
- if $\varphi_1 \wedge \varphi_2 \in \tau(t)$ then $\varphi_1 \in \tau^o(t)$ and $\varphi_2 \in \tau^o(t)$
- if $\varphi_1 \vee \varphi_2 \in \tau(t)$ then $\varphi_1 \in \tau^o(t)$ or $\varphi_2 \in \tau^o(t)$
- if $\varphi_1 \mathscr{U} \varphi_2 \in \tau(t)$ then either $\varphi_2 \in \tau^o(t)$, or $\varphi_1 \in \tau^o(t)$ and $\varphi_1 \mathscr{U} \varphi_2 \in \tau^e(t+1)$
- if $\varphi_1 \mathscr{R} \varphi_2 \in \tau(t)$ then $\varphi_2 \in \tau^o(t)$, and either $\varphi_1 \in \tau^o(t)$ or $\varphi_1 \mathscr{R} \varphi_2 \in \tau^e(t+1)$.

Requirements on observations $\tau^o(t)$ at time t initially come from the expectations $\tau^e(t)$ on this time step. Thus, starting from $t = 1$, the closure labeling τ can be constructed consecutively. Finally, the following rules regarding propositions need to hold true for all $\pi \in \Pi$ with respect to σ. If $\pi \in \tau(t)$ then $\pi \in \sigma(t)$ and if $\neg\pi \in \tau(t)$ then $\pi \notin \sigma(t)$. Otherwise, we say that σ violates the requirements imposed by the LTL specification ϕ.

By this construction of τ, $\sigma(t, \ldots, T)$ fulfills all $\varphi \in \tau(t)$ if and only if $\tau^e(T+1) \cap cl(\varphi) = \emptyset$. In particular, we can check if $\sigma \models \phi$ by constructing τ from the expectation $\tau^e(1) = \{\phi\}$ and then check if $\tau^e(T+1) = \emptyset$. Note that [22] additionally requires that there exists a $t' \geq t$ such that $\varphi_2 \in \tau(t')$ for the until operation $\varphi_1 \mathscr{U} \varphi_2 \in \tau(t)$. However, this requirement is already covered in our extended acceptance condition $\tau^e(T+1) = \emptyset$ and does not need to be explicitly required here.

Illustratively speaking, observations τ^o formally describe what needs to be observed at a given time, while expectations τ^e specify expected future observations. Intuitively, if not all observations are fulfilled at a certain time, the corresponding LTL formula is violated. If this is not the case and at some point, there are no implied expectations anymore, the LTL formula is satisfied.

2.3 System Model

Every agent is represented by a transition system to model its available actions and define which propositions are true consequently. It usually combines a topological map of the environment with discrete actions, which can be executed at certain locations, and is formally defined as follows.

Definition 4 *(Agent Model)* An agent model is given as the transition system $\mathscr{A} = (S_{\mathscr{A}}, s_{0,\mathscr{A}}, A_{\mathscr{A}}, \Pi, \lambda, C_{\mathscr{A}})$ consisting of **(1)** a set of states $S_{\mathscr{A}}$, **(2)** an initial state $s_{0,\mathscr{A}} \in S_{\mathscr{A}}$, **(3)** a set of actions $A_{\mathscr{A}} \subseteq S_{\mathscr{A}} \times S_{\mathscr{A}}$, **(4)** a set of propositions Π, **(5)** a labeling function $\lambda \colon S_{\mathscr{A}} \to 2^{\Pi}$, **(6)** action costs $C_{\mathscr{A}} : A_{\mathscr{A}} \to \mathbb{R}$.

Forming a product between the agent model $\mathscr{A}$ and the NFA $\mathscr{F}$ constructed from the LTL specification ϕ creates an automaton that combines properties of $\mathscr{A}$ and $\mathscr{F}$.

Definition 5 *(Product Automaton)* A product automaton is a tuple $\mathscr{P} = \mathscr{F} \otimes \mathscr{A} = (S_{\mathscr{P}}, S_{0,\mathscr{P}}, A_{\mathscr{P}}, C_{\mathscr{P}})$ consisting of **(1)** a set of states $S_{\mathscr{P}} = Q \times S_{\mathscr{A}}$, **(2)** a set of initial states $S_{0,\mathscr{P}} = \{(q, s_{0,\mathscr{A}}) \in S_{\mathscr{P}} : q \in Q_0\}$, **(3)** a set of actions $A_{\mathscr{P}} = \{((q_s, s_s), (q_t, s_t)) \in S_{\mathscr{P}} \times S_{\mathscr{P}} : (s_s, s_t) \in A_{\mathscr{A}} \wedge \lambda(s_s) \models \delta(q_s, q_t)\}$, **(4)** action costs $C_{\mathscr{P}} : A_{\mathscr{P}} \to \mathbb{R}$ with $C_{\mathscr{P}}(a_{\mathscr{P}}) = C_{\mathscr{A}}(a_{\mathscr{A}})$.

Consequently, only actions that do not violate ϕ are contained in the model. A run ending in a state $s_n = (q, s_{\mathscr{A}})$ with $q \in F$ being an accepting state in $\mathscr{F}$ gives an action sequence which fulfills ϕ.

Definition 6 *(Action Sequence)* An action sequence β is given by $\beta = s_0 a_1 s_1 \ldots a_n s_n$ with $s_i \in S_{\mathscr{P}}$, $s_0 \in S_{0,\mathscr{P}}$, and $a_j = (s_{j-1}, s_j) \in A_{\mathscr{P}}$.

3 LTL Decomposition

Considering a multi-agent system with N agents, we can represent each robotic agent $r \in \{1, \ldots, N\}$ according to the above definitions by an individual product automaton $\mathscr{P}^{(r)}$ created from its agent model $\mathscr{A}^{(r)}$ and the NFA $\mathscr{F}$, obtained from the complete LTL mission specification denoted by $\mathscr{M}$.

Problem 1. Given a team of agents $r \in \{1, \ldots, N\}$, each modeled as $\mathscr{A}^{(r)}$, and the finite LTL mission specification $\mathscr{M}$, provide independent action sequences $\beta^{(r)}$ for all agents such that $\mathscr{M}$ is fulfilled in an optimal way for the specified team cost.

In order to utilize the team of agents, it is desirable to decompose the mission $\mathscr{M}$ such that parts of it can be allocated to different agents, i.e., automatically identify independently executable task specifications if there exist some. This decomposition will allow us to distribute $\mathscr{M}$ and solve Problem 1 as if the mission has been specified by one LTL formula for each agent. Based on the above LTL semantics, we define what we accept as a semantically valid decomposition of a finite LTL mission $\mathscr{M}$ following the motivation of specifying $\mathscr{M}$ as a set of independent tasks.

Definition 7 *(Finite Decomposition)* Let $\mathcal{T}_i$ with $i \in \{1, \ldots, n\}$ be a set of finite LTL task specifications and σ_i denote any sequence such that $\sigma_i \models \mathcal{T}_i$. These tasks are called a decomposition of the finite LTL mission specification $\mathcal{M}$ if and only if:

$$\sigma_{j_1} \ldots \sigma_{j_i} \ldots \sigma_{j_n} \models \mathcal{M} \quad (1)$$

for all permutations of $j_i \in \{1, \ldots, n\}$ and all respective sequences σ_i.

Note that this decomposition condition (1) includes that each $\mathcal{T}_i$ is a safe, i.e., non-violating, prefix of $\mathcal{M}$. Furthermore, by requiring all permutations of sequences to be feasible, we make sure that no σ_i implies expectations to be respected by other sequences of the decomposition, and that the set of tasks completely covers $\mathcal{M}$. In the following, we discuss how to decompose the mission $\mathcal{M}$ into independently executable tasks $\mathcal{T}_i$ such that $\mathcal{M}$ is fulfilled if the set of tasks is fulfilled.

Example. Consider the LTL mission specification $\mathcal{M} = \diamond a \wedge \diamond b \wedge \Box(b \rightarrow c)$. In this simple case, as will be more clear at the end of this section, a possible decomposition of $\mathcal{M}$ is given by $\mathcal{T}_1 = \diamond a \wedge \Box(b \rightarrow c)$ and $\mathcal{T}_2 = \diamond b \wedge \Box(b \rightarrow c)$. For example, the sequence $\sigma = \sigma_1\sigma_2$ with $\sigma_1 = \{c\}\{a\}$, $\sigma_2 = \{a\}\{b, c\}$ would fulfill $\mathcal{M}$, and also the permutation $\sigma_2\sigma_1 = \{a\}\{b, c\}\{c\}\{a\}$ would be valid. However, note that the modification of the above solution such that $\mathcal{T}_1 = \diamond a$ would not constitute a valid decomposition. In this case, for example, $\sigma_1' = \{b\}\{a\} \models \mathcal{T}_1$ and still, $\sigma_2 \models \mathcal{T}_2$, but $\sigma_1'\sigma_2 \not\models \mathcal{M}$.

For more complex LTL formulas, the explicit LTL formulation of a decomposition can be significantly different from simply splitting the mission specification or replicating some parts. However, a boolean conjunction of all tasks $\mathcal{T}_i$ always gives the complete specification $\mathcal{M}$.

Notation Remark. The following notation conventions are used throughout the rest of this paper. $\mathcal{M}$ is the finite LTL mission specification, σ a sequence such that $\sigma \models \mathcal{M}$ and τ the closure labeling of σ. A different σ will have a different τ. Furthermore, $\mathcal{T}_i$ is a finite LTL task specification, i.e., a subformula of $\mathcal{M}$, σ_i a sequence such that $\sigma_i \models \mathcal{T}_i$ and τ_i the closure labeling of σ_i. Note that τ_i is defined over the closure $cl(\mathcal{T}_i) \subset cl(\mathcal{M})$ while τ is defined over $cl(\mathcal{M})$. Accordingly for τ and τ_i, T and T_i denote the ending times, τ^e and τ_i^e the expectations, and so on.

In order to efficiently determine a valid LTL decomposition as discussed above, note the following observation.

Theorem 1 (Transitivity) *If $\mathcal{T}_1$, $\mathcal{T}_2$ is a decomposition of $\mathcal{M}$ and $\mathcal{T}_3$, $\mathcal{T}_4$ is a decomposition of $\mathcal{T}_2$, then $\mathcal{T}_1$, $\mathcal{T}_3$, $\mathcal{T}_4$ is a decomposition of $\mathcal{M}$.*

Proof We need to show that all six permutations of the sequences $\sigma_1, \sigma_3, \sigma_4$ fulfill $\mathcal{M}$. Four of them are rather trivial by substitution of σ_2 with $\sigma_3\sigma_4$ or $\sigma_4\sigma_3$. However, for proving $\sigma_3\sigma_1\sigma_4 \models \mathcal{M}$ and $\sigma_4\sigma_1\sigma_3 \models \mathcal{M}$, the closure labeling is used. Specifically, we need to show that we can construct a closure labeling τ of $\mathcal{M}$ from the closure labelings τ_i of the tasks $\mathcal{T}_i, i \in \{1, 3, 4\}$, such that $\tau^e(1) = \{\mathcal{M}\}$ and $\tau^e(T + 1) = \emptyset$ for completion time $T = T_1 + T_3 + T_4$.

For this purpose, we construct a candidate τ with $\tau^e(1) = \{\mathcal{M}\}$ from the given set of τ_i, and then show that this always leads to $\tau^e(T+1) = \emptyset$. Since $\mathcal{T}_1, \mathcal{T}_2$ are a decomposition of $\mathcal{M}$, following decomposition condition (1), $\tau^e(1) = \{\mathcal{M}\}$ is equivalent to $\tau^e(1) = \{\mathcal{T}_1, \mathcal{T}_2\}$, which itself is equivalent to $\tau^e(1) = \{\mathcal{T}_1, \mathcal{T}_3, \mathcal{T}_4\}$ for $\mathcal{T}_3, \mathcal{T}_4$ being a decomposition of $\mathcal{T}_2$. Consequently, we have $\tau(1) = \tau^e(1) \cup \tau^o(1)$ with $\tau^o(1)$ following Definition 3.

We start by considering the permutation $\sigma_3\sigma_1\sigma_4$. For the first part, $t \in [2, T_3]$, we construct the candidate τ from τ_3 such that $\tau(t) = \tau_3(t) \cup (\tau(1) \setminus cl(\mathcal{T}_3))$. From using τ_3 we get that this part fulfills $\mathcal{T}_3$, and extend τ_3 by all requirements which are not covered by $\mathcal{T}_3$, i.e., which are not in $cl(\mathcal{T}_3)$. τ is still valid because the tasks are decomposition pairs as stated in the theorem. Specifically, $\sigma_{2,1} = \sigma_2\sigma_1 \models \mathcal{M}$ implies $\mathcal{T}_1 \in \tau_{2,1}(t)$ with $t \in [1, T_2 + 1]$ for the closure labeling $\tau_{2,1}$ of the permutation $\sigma_2\sigma_1$. This means that the requirement $\mathcal{T}_1$ cannot be violated at any time during execution of $\sigma_2 = \sigma_3\sigma_4$, and thus, also during σ_3.

We can repeat this construction for the remaining two parts, continuing with $\tau^e(T_3 + 1) = \{\mathcal{T}_1, \mathcal{T}_4\}$. Finally, this leads to $\tau^e(T_3 + T_1 + T_4 + 1) = \emptyset$, meaning that $\mathcal{M}$ is fulfilled. Thus, we see that the constructed candidate is a valid closure labeling respecting all requirements and consequently, $\sigma_3\sigma_1\sigma_4 \models \mathcal{M}$. The proof for the last permutation $\sigma_4\sigma_1\sigma_3 \models \mathcal{M}$ follows accordingly. □

Theorem 1 has especially two consequences. First, only n specific permutations instead of all $n!$ permutations need to be checked in order to decide if a set of n tasks $\mathcal{T}_i$ is a valid decomposition of $\mathcal{M}$. This is obtained by forming pairs, each of one task $\mathcal{T}_i$ and combination of the other $n-1$ tasks, for example by a conjunction. Then, it is sufficient to check the decomposition condition (1) only for these n permutations in order to decide if these tasks form a decomposition of $\mathcal{M}$. Illustratively, the specific n permutations individually separate tasks $\mathcal{T}_i$ from the rest to decide whether $\mathcal{T}_i$ is independent.

Second, it is not required to find a complete set of tasks decomposing $\mathcal{M}$ at once. Instead, it is possible to step-wise identify individual parts to be isolated into a separate task $\mathcal{T}_i$ of the final decomposition and continue with the rest of $\mathcal{M}$. This progress can be repeated until no further task is found to be isolated and especially enables automata-based approaches for finding possible decompositions.

3.1 Decomposition Set

In general, different decompositions of $\mathcal{M}$ can exist and a task $\mathcal{T}_i$ does not need to be minimal in the sense that it cannot be further decomposed. Thus, we propose an efficient automata-based approach to identify all possible choices of decomposition as shown in the remainder of this section. First, note the following relation between states $q \in Q$ of the NFA $\mathcal{F}$ constructed from the LTL mission specification $\mathcal{M}$ and the closure $cl(\mathcal{M})$.

Lemma 1 (Subformula Labeling) *Each state $q \in Q$ of the NFA $\mathscr{F}$ constructed from $\mathscr{M}$ can be labeled with subformulas $\Phi_q \in 2^{cl(\mathscr{M})}$ which are required to be true at this particular state.*

We refer the interested reader to [22], Sect. 4.4, for a detailed proof, covering the more general case of infinite sequences. In summary, $\mathscr{F}$ is explicitly constructed from $\mathscr{M}$ such that its state space Q is given by $2^{cl(\mathscr{M})}$ as discussed in Sect. 2.1, and there is a transition if and only if the successor state fulfills the requirements of the closure labeling of its predecessor. In particular, note that $\mathscr{M} \in \Phi_q$ for all $q \in Q_0$ and $\Phi_q = \emptyset$ for all $q \in F$. Lemma 1 shows the connection between the NFA $\mathscr{F}$ and the closure labeling τ, since τ is as well defined over the subformulas $2^{cl(\mathscr{M})}$ and construction of $\mathscr{F}$ respects the requirements imposed by τ.

Furthermore, we introduce the following notion of *essential sequences* to generalize over sequences by associating them with runs ρ in the NFA.

Definition 8 *(Essential Sequence)* A sequence σ is called essential for an NFA $\mathscr{F}$ if and only if it describes a run ρ in $\mathscr{F}$ and $\sigma(t) \setminus \{\pi\} \nvDash \delta(\rho(t-1), \rho(t))$ for all t and propositions $\pi \in \sigma(t)$, i.e., σ only contains required propositions.

This notation is motivated by the closure labeling τ of σ. By restricting σ to satisfy only the conditions explicitly required by τ, we get the following property.

Lemma 2 (Closure Coverage) *Let τ denote the closure labeling of a sequence σ. If σ is an essential sequence, any other τ' satisfied by σ is at most as restrictive as τ, in the sense that $\tau'(t) \cap \Pi \subseteq \tau(t) \cap \Pi$ for every t and the set of propositions Π.*

Proof Assume there would be a $\pi \in \Pi$ such that $\pi \in \tau'(t)$ and $\pi \notin \tau(t)$. τ' would then require that $\pi \in \sigma(t)$. However, this cannot be the case since σ is essential. □

This property ensures that if an essential sequence describes a run in one part of the NFA corresponding to τ as well as one corresponding to τ', any other non-essential sequence conforming with τ will not violate τ' neither. This can be used to generalize over sequences without explicitly constructing the closure labeling, but instead finding an essential sequence.

Finally, we can associate a pair of tasks $\mathscr{T}_1^q$, $\mathscr{T}_2^q$ with a state $q \in Q$ of $\mathscr{F}$. Every sequence σ_1 describing a run ρ_1 from an initial state $q_0 \in Q_0$ to q satisfies $\mathscr{T}_1^q$, specified by the set of fulfilled subformulas $\Phi_{q_0} \setminus \Phi_q$. $\mathscr{T}_2^q$ is given accordingly by $\Phi_q \setminus \Phi_{q_F} = \Phi_q$ with $q_F \in F$ and represents the rest of $\mathscr{M}$ not fulfilled by $\mathscr{T}_1^q$.

It remains to decide if the pair $\mathscr{T}_1^q$, $\mathscr{T}_2^q$ resulting from a split forms a valid decomposition of $\mathscr{M}$, and we define the *decomposition set* of $\mathscr{F}$ as follows.

Definition 9 *(Decomposition Set)* The decomposition set $D \subseteq Q$ of the NFA $\mathscr{F}$ constructed from $\mathscr{M}$ contains all states q for which the pair of tasks $\mathscr{T}_1^q$, $\mathscr{T}_2^q$ defines a valid decomposition of $\mathscr{M}$ according to Definition 7.

This decomposition set can then be constructed as follows, giving all possible decomposition choices of the finite LTL mission specification $\mathscr{M}$.

Theorem 2 (Decomposability) *Let $q \in Q$ be a state in the NFA $\mathcal{F}$ constructed from $\mathcal{M}$, and $\sigma = \sigma_1\sigma_2$ be an essential sequence such that σ_1 describes a run from an initial state to q and σ_2 describes a run from q to an accepting state of $\mathcal{F}$. Then, $q \in D$ if and only if $\widehat{\sigma} = \sigma_2\sigma_1$ describes an accepting run in $\mathcal{F}$.*

Proof The "only if"-part follows directly from the decomposition condition in Definition 7. For the "if"-part, it remains to show that σ generalizes over all possible $\sigma_1' \models \mathcal{T}_1^{\mathrm{q}}$ and $\sigma_2' \models \mathcal{T}_2^{\mathrm{q}}$, i.e., all pairs of sequences describing a run through q. Note that, given that $\sigma = \sigma_1\sigma_2$ is an essential sequence, also σ_1 and σ_2 are essential.

First, we show that the essential sequence σ_1, generalizes over $\sigma_1' \models \mathcal{T}_1^{\mathrm{q}}$. This means, if $\widehat{\sigma} = \sigma_2\sigma_1$ describes an accepting run, then also any other $\sigma_2\sigma_1'$ describes an accepting run. According to Lemma 2, the closure labeling $\widehat{\tau}(t)$, $t \in [T_2 + 1, T_2 + T_1]$ of $\widehat{\sigma}$ is at most as restrictive as $\tau(t)$, $t \in [1, T_1]$ of σ. This means that no sequence can violate $\widehat{\tau}$ if it conforms with τ.

Next, following Lemma 1, we can retrieve the closure labeling τ' of a sequence σ_1' from a run ρ' described by σ_1', given by $\tau'(t) = \Phi_q$ for $q = \rho'(t)$. By construction of the NFA and $\mathcal{T}_1^{\mathrm{q}}$, all sequences leading to the respective last state $\rho'(T)$ fulfill all requirements imposed by $\rho'(0)$. Although these sequences may have a different closure labeling τ', this always satisfies the same requirements as τ, given by $\sigma_1' \models \mathcal{T}_1^{\mathrm{q}} = \Phi_{\mathrm{q}_0} \setminus \Phi_{\mathrm{q}}$ with $q_0 = \rho'(0) = \rho(0)$ and $q = \rho'(T_1) = \rho(T_1)$ where ρ is described by σ_1. Consequently, σ_1' cannot violate $\widehat{\tau}$ as shown by Lemma 2.

Finally for the permutation $\sigma_2\sigma_1'$, this gives that any $\sigma_1' \models \mathcal{T}_1^{\mathrm{q}}$ applied to the same state as σ_1 leads to an accepting state and thus, $\sigma_2\sigma_1'$ describes an accepting run if $\widehat{\sigma} = \sigma_2\sigma_1$ does, i.e., the essential sequence σ_1 indeed generalizes over possible different realizations of $\mathcal{T}_1^{\mathrm{q}}$. The same then holds true accordingly for σ_2 and thus, we get $\sigma_2'\sigma_1' \models \mathcal{M}$ if and only if $\sigma_2\sigma_1 \models \mathcal{M}$, given that σ_1 and σ_2 are essential. □

Note that Theorem 2 only requires to check one essential sequence, which is much more efficient than the requirement to check every single possible sequence. Furthermore, an essential sequence σ to a specific state q can be easily constructed from an NFA $\mathcal{F}$, for example by representing the set of transition conditions α of $\mathcal{F}$ in disjunctive normal form (DNF). Then, the essential sequence to q is given by the propositions which are true in one of the conjunctive clauses along the path to q. By step-wise constructing these sequences σ for all states first, all essential sequences can be found in linear time with respect to $|Q|$, which is non-critical compared to constructing $\mathcal{F}$ as discussed earlier in Sect. 2.1.

4 Team Model Construction

Based on the results of the previous section, a team model can be constructed as follows in order to solve Problem 1. First, the mission specification $\mathcal{M}$ is translated to an equivalent NFA $\mathcal{F}$. Next, we form a local product automaton $\mathcal{P}^{(r)} = \mathcal{F} \otimes \mathcal{A}^{(r)}$ for each agent $r \in \{1, \ldots, N\}$. Unlike previous task allocation approaches, we do not explicitly calculate the costs for each subset of tasks resulting from a possible decomposition choice in each of the local $\mathscr{P}^{(r)}$. Instead, we combine these local product automata into a team model of tractable complexity in which the optimal task allocation can be calculated much more efficiently.

The basis for this team model is given by a union of all $\mathscr{P}^{(r)}$, resulting in N unconnected partitions. Afterwards, additional *switch transitions* connect these partitions. They represent an option in the planning process to consider a different agent for allocation of the part of the mission which is not yet assigned. Such a switch transition is only present if both parts of the mission form a valid decomposition. Formally, the team automaton is defined as follows.

Definition 10 *(Team Automaton)* The team automaton $\mathscr{G}$ is a union of N local product automata $\mathscr{P}^{(r)}$ with $r \in \{1, \ldots, N\}$ given by $\mathscr{G} = (S_{\mathscr{G}}, S_{0,\mathscr{G}}, A_{\mathscr{G}}, C_{\mathscr{G}})$ consisting of **(1)** a set of states $S_{\mathscr{G}} = \{(r, q, s) : r \in \{1, \ldots, N\}, (q, s) \in S^{(r)}_{\mathscr{P}}\}$, **(2)** a set of initial states $S_{0,\mathscr{G}} = \{(r, q, s) \in S_{\mathscr{G}} : r = 1\}$ equivalent to the initial states of one arbitrary agent, **(3)** a set of actions $A_{\mathscr{G}} = \bigcup_r A^{(r)}_{\mathscr{P}} \cup \zeta$ where the individual agent actions $A^{(r)}_{\mathscr{P}}$ are extended by the set of switch transitions ζ as defined below, **(4)** action costs $C_{\mathscr{G}} : A_{\mathscr{G}} \to \mathbb{R}$ with $C_{\mathscr{G}}(a_{\mathscr{G}}) = C^{(r)}_{\mathscr{P}}(a^{(r)}_{\mathscr{P}})$ and $C_{\mathscr{G}}(\varsigma) = 0$ for all $\varsigma \in \zeta$.

Core part of this composition is constructing the set of switch transitions ζ connecting states in the partitions of two different agents and preserving mission progress, restricted to states corresponding to a valid mission decomposition.

Definition 11 *(Switch Transition)* The set $\zeta \subset S_{\mathscr{G}} \times S_{\mathscr{G}}$ denotes switch transitions in the team automaton $\mathscr{G}$ and $\varsigma \in \zeta$ for $\varsigma = ((i, q_{\mathrm{s}}, s_{\mathrm{s}}), (j, q_{\mathrm{t}}, s_{\mathrm{t}}))$ if and only if it **(i)** connects different agents: $i \neq j$, **(ii)** preserves the NFA progress: $q_{\mathrm{s}} = q_{\mathrm{t}}$, **(iii)** is directed: $r_i \prec r_j$ for an arbitrary ordering of agents $r_1 \prec r_2 \prec \cdots \prec r_N$, **(iv)** points to an initial agent state: $s_{\mathrm{t}} = s^{(j)}_{0,\mathscr{A}}$, **(v)** implies a valid decomposition of $\mathscr{F}$: $q_{\mathrm{s}} \in D$.

While condition (i) is trivial, (ii) characterizes the main purpose of a switch transition, which is transferring the mission progress to another agent. Condition (iv) in combination with (iii) requires to account for the initial state of each agent. Specifically, (iii) ensures that each agent is considered exactly once for participating in solving the mission. Finally, (v) guarantees that any possible decomposition resulting from switch transitions is valid.

The model $\mathscr{G}$ has a much lower state space complexity than the complete product $\mathscr{C}_{\mathrm{Prod}} = \mathscr{P}^{(1)} \otimes \cdots \otimes \mathscr{P}^{(N)}$ of all local automata, which would be required if we did not decompose the LTL mission into independent tasks. Specifically, the number of states of $\mathscr{G}$ is linear in the number of agents N and given by $O(N \cdot |Q| \cdot |S_{\mathscr{A}}|)$. $|S_{\mathscr{A}}|$ denotes the number of states of the agent model and $|Q|$ the number of states of the NFA $\mathscr{F}$. In contrast, the state space complexity of $\mathscr{C}_{\mathrm{Prod}}$ would be exponential in the number of agents with $O(|Q| \cdot |S_{\mathscr{A}}|^N)$.

A team model $\mathscr{G}$ constructed as defined above enables to employ conventional graph-search algorithms for obtaining optimal action sequences $\beta^{(r)}$ for all agents such that the LTL mission specification $\mathscr{M}$ is fulfilled. Consequently, this solves Problem 1 and is summarized by the following properties.

Correctness. $\mathscr{P}^{(r)} = \mathscr{F} \otimes \mathscr{A}$ preserves the acceptance criterion of $\mathscr{F}$. A union of the state space when constructing $\mathscr{G}$ out of all $\mathscr{P}^{(r)}$ does not add any new transitions except ζ and because condition (ii) requires all $\varsigma \in \zeta$ to preserve the NFA component, any accepting run ρ in $\mathscr{G}$ satisfies $\mathscr{M}$.

Independence. Given by switch condition (v) and the construction of the decomposition set D as discussed above, parts of ρ referring to different agents r solve independent tasks $\mathscr{T}_r$ and thus, do not affect each other. Especially, each $\mathscr{T}_r$ is a safe prefix of all other tasks, i.e., all constraints of $\mathscr{M}$ are covered by $\mathscr{T}_r$.

Completeness. Any set of individual agent action sequences $\beta^{(r)}$ resulting from a run in the complete product automaton $\mathscr{C}_{\text{Prod}}$ is also present in the reduced team model $\mathscr{G}$, since the parts referring to different agents are independent. Especially, also the optimal solution in $\mathscr{C}_{\text{Prod}}$ is as well contained in $\mathscr{G}$.

5 Evaluation

The presented approach has been implemented in ROS and evaluated both in simulation and on a real system. In the following, we discuss our performance evaluation results for a set of simulated scenarios and compare them to the conventional product model approach. For planning the optimal action sequence based on the constructed team model, we used a conventional Bellman–Ford graph-search and minimize the largest individual agent costs, i.e., aim to distribute the mission equally. Note that, although action costs are usually positive, we cannot use a greedy graph-search such as Dijkstra or A* because we aim to distribute the mission equally and not to minimize the sum of all action costs.

To evaluate applicability in scenarios as motivated in Sect. 1, we assume a hospital environment, depicted left in Fig. 1, and form the agent model $\mathscr{A}$ as the product between this topological map (left) and a transition model of robot actions (right). The set of states $S_{\mathscr{A}}$ is given as the product between the map locations and the robot states. Propositions Π according to the state labels describe specific properties of locations and robot states, e.g., p for pick-up locations, s for station rooms, c for carrying an object. The actions $A_{\mathscr{A}}$ consist of navigation actions according to the undirected edges in the map and further robot actions according to the robot

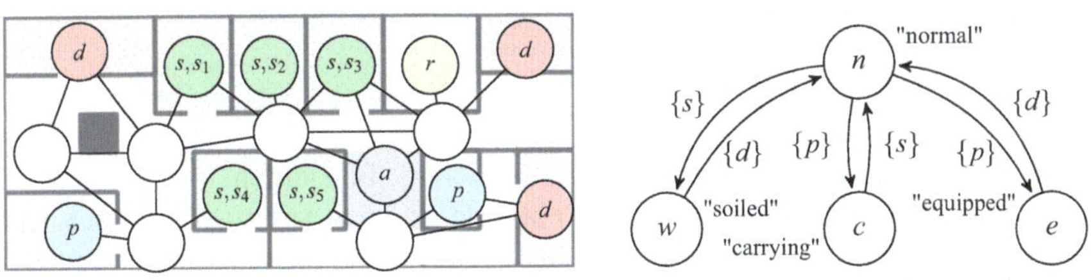

Fig. 1 Map (left) and robot capabilities (right) used in the scenarios. State labels denote propositions which are true at this state, transition labels denote requirements for performing an action

model. These robot actions are limited to certain locations such that an action is only feasible if the respective state contains the propositions listed by the transition label, e.g., the transition from "normal" to "carrying" is only possible at pick-up locations p. Finally, action costs $C_{\mathscr{A}}$ are chosen to approximately represent the execution times of actions.

5.1 Scenarios

Before we present illustrative scenarios in the presented environment, we discuss two mission specifications representing corner cases with respect to the decomposition of the mission, specifically for the size of the decomposition set D. First, requiring the team to visit the five station rooms in any order is essentially a multi-agent traveling salesman problem (TSP) and given by the mission $\mathscr{M}_{TSP} = \diamond s_1 \wedge \diamond s_2 \wedge \diamond s_3 \wedge \diamond s_4 \wedge \diamond s_5$. Consequently, all 32 states of the NFA $\mathscr{F}$ are in the decomposition set: $D = Q$. In this case, construction of $\mathscr{F}$ took approximately 56 ms and determination of D around 2.2 ms. Note that, although the decomposition set contains the full state space of $\mathscr{F}$, the proposed team automaton $\mathscr{G}$ still has a significantly lower state space complexity than the complete product $\mathscr{C}_{\text{Prod}}$. In fact, the state space complexity is independent of the size of the decomposition set and $|D|$ only determines the density of switch transitions.

In contrast, requiring visits to occur in a specific order is given by the mission $\mathscr{M}_{Seq} = \diamond(s_3 \wedge \diamond(s_4 \wedge \diamond(s_2 \wedge \diamond(s_5 \wedge \diamond s_1))))$. Different robots cannot execute parts of $\mathscr{M}_{Seq}$ independently since the correct order could not be guaranteed. Thus, the decomposition set only contains trivially the initial and accepting states: $D = Q_0 \cup F$, totaling to 2 of 6 states and reducing the mission to an allocation problem of choosing the single best robot to execute the mission alone. Construction of $\mathscr{F}$ took approximately 50 ms and determination of D around 0.1 ms. For both cases, specialized solutions exist to solve problems of this type. However, most missions in the motivated scenarios usually combine characteristics of both cases.

In the following, we consider three scenarios with different characteristics in the presented hospital environment to represent the most common use cases. Teams consist of three robots, although also the performance for varying team sizes is investigated at the end of this section as well.

Scenario 1 (Station Tour)

$$\mathscr{M}_1 = \diamond s_1 \wedge \diamond s_2 \wedge \diamond s_3 \wedge \diamond s_4 \wedge \diamond s_5 \wedge \square(s \rightarrow e) \wedge \square(e \rightarrow \neg a)$$

The robots are required to visit all five station rooms. In addition, a robot needs to carry medical equipment (proposition e) in order to be at any station room s and should avoid the public area a while being equipped. As presented before, robots know from the agent model $\mathscr{A}$ that equipment can only be picked up at pick-up locations p. This mission is similar to a constrained TSP, but requires robots to perform additional actions before visiting a room, regardless of which room they would choose to visit first.

Scenario 2 (Room Cleaning)

$$\mathcal{M}_2 = \Diamond(s_3 \wedge \circ\delta) \wedge \Diamond(s_4 \wedge \circ\delta) \wedge \Diamond(s_5 \wedge \circ\delta) \wedge \Box((\neg s \wedge \circ s) \rightarrow n)$$

with $\delta := w\ \mathcal{U}\ (d \wedge \circ(d\ \mathcal{U}\ \neg w))$. The robots need to pick up waste w at three of the rooms. In this scenario, robots are only required to be in "normal" state n in order to enter a state room s. But in addition, they are required to visit a dispose location d as consequence of visiting a room, given by δ. Again, this combines goal allocation with sequential action planning as consequence of servicing one of the goals.

Scenario 3 (Medication Delivery)

$$\mathcal{M}_3 = \Diamond(s_1 \wedge n) \wedge \Diamond(s_2 \wedge n) \wedge \Diamond(s_3 \wedge n) \wedge \Diamond(s_4 \wedge n) \wedge \Diamond(s_5 \wedge n) \wedge \Box((\neg s \wedge \circ s) \rightarrow c)$$

The robots need to deliver medication to all station rooms. They can only enter a room s when carrying medication c and need to deliver it by switching back to their "normal" state n. Consequently, robots need to repeatedly visit pick-up locations.

Figure 2 summarizes our performance results for the three scenarios, each randomly initialized. t is the average planning time in seconds, including model construction, calculation of the decomposition set and planning, and $|S|$ the total number of states in the model. We compare the team model $\mathcal{G}$ of our presented approach with the conventional complete product model $\mathcal{C}_{\text{Prod}}$. Already for the small team of three robots, our approach is much more efficient.

	$t_{\mathcal{G}}$	$\lvert S_{\mathcal{G}}\rvert$	$t_{\mathcal{C}_{\text{Prod}}}$	$\lvert S_{\mathcal{C}_{\text{Prod}}}\rvert$
$\mathcal{M}_1$	0.965	6,912	1.71×10^4	1.19×10^7
$\mathcal{M}_2$	0.946	9,504	1.52×10^4	1.64×10^7
$\mathcal{M}_3$	2.908	13,824	$>4.32 \times 10^4$	2.39×10^7

Fig. 2 Analysis of the evaluation scenarios for teams of three agents, time given in seconds

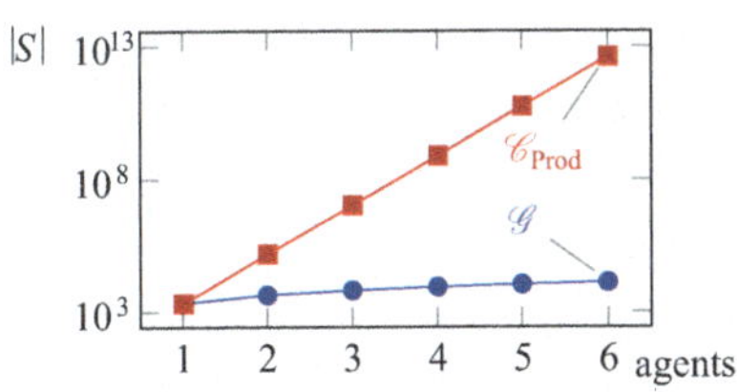

agents	$t_{\mathcal{G}}$	$\lvert S_{\mathcal{G}}\rvert$	$t_{\mathcal{C}_{\text{Prod}}}$	$\lvert S_{\mathcal{C}_{\text{Prod}}}\rvert$
1	0.324	2.3×10^3	0.342	2.3×10^3
2	0.525	4.6×10^3	94.75	1.6×10^5
3	0.965	6.9×10^3	1.7×10^4	1.2×10^7
6	2.324	1.4×10^4	n/a	4.5×10^{12}
10	3.896	2.3×10^4	n/a	1.2×10^{19}
100	92.46	2.3×10^5	n/a	1.7×10^{187}

Fig. 3 Complexity analysis with respect to the team size, performed for scenario $\mathcal{M}_1$. Planning time in seconds, missing entries exceeded the maximum time of 8 h

Furthermore, Fig. 3 provides an analysis of how both approaches scale with an increasing number of agents, evaluated for scenario $\mathscr{M}_1$. The significantly increasing planning times on the product model $\mathscr{C}_{\text{Prod}}$ reflect the exponential growth of its state space. In contrast, our team model $\mathscr{G}$ scales well with increasing team size and produces reasonable results even for large teams.

6 Conclusion

Motivated by the need for efficient methods for multi-robot team planning, we presented an approach for decomposition of finite LTL mission specifications into independent tasks, resulting in a team model of tractable complexity for increasing team sizes. On this model, graph-search algorithms can efficiently distribute action sequences for the available robots, such that the LTL mission is completed by the team best suitable, which can dynamically change between missions. We illustrated the computational advantages of our approach over the conventional product model in example scenarios, resulting in significantly lower planning times.

References

1. Agarwal, M., Kumar, N., Vig, L.: Non-additive multi-objective robot coalition formation. Expert Syst. Appl. **41**(8), 3736–3747 (2014)
2. Baier, C., Katoen, J.P.: Principles of Model Checking. MIT Press, Cambridge (2008)
3. Chen, Y., Ding, X.C., Stefanescu, A., Belta, C.: Formal approach to the deployment of distributed robotic teams. IEEE Trans. Robot. **28**(1), 158–171 (2012)
4. De Giacomo, G., Vardi, M.: Linear temporal logic and linear dynamic logic on finite traces. In: International Joint Conference on Artificial Intelligence (IJCAI), pp. 854–860. Association for Computing Machinery (2013)
5. De Giacomo, G., De Masellis, R., Montali, M.: Reasoning on LTL on finite traces: insensitivity to infiniteness. In: AAAI, pp. 1027–1033. Citeseer (2014)
6. Fu, J., Tanner, H., Heinz, J.: Concurrent multi-agent systems with temporal logic objectives: game theoretic analysis and planning through negotiation. IET Control Theory Appl. **9**(3), 465–474 (2015)
7. Gerth, R., Peled, D., Vardi, M., Wolper, P.: Simple on-the-fly automatic verification of linear temporal logic. In: International Symposium on Protocol Specification, Testing and Verification. IFIP (1995)
8. Guo, M., Dimarogonas, D.V.: Bottom-up motion and task coordination for loosely-coupled multi-agent systems with dependent local tasks. In: CASE, pp. 348–355. IEEE (2015)
9. Karaman, S., Frazzoli, E.: Vehicle routing problem with metric temporal logic specifications. In: Conference on Decision and Control (CDC), pp. 3953–3958. IEEE (2008)
10. Kloetzer, M., Belta, C.: Automatic deployment of distributed teams of robots from temporal logic motion specifications. IEEE Trans. Robot. **26**(1), 48–61 (2010)
11. Kloetzer, M., Mahulea, C.: Accomplish multi-robot tasks via Petri net models. In: International Conference on Automation Science and Engineering (CASE), pp. 304–309. IEEE (2015)
12. Kupferman, O., Vardi, M.: Model checking of safety properties. Form. Methods Syst. Des. **19**(3), 291–314 (2001)

13. Lacerda, B., Lima, P.: LTL-based decentralized supervisory control of multi-robot tasks modelled as Petri nets. In: International Conference on Intelligent Robots and Systems (IROS), pp. 3081–3086. IEEE (2011)
14. Luna, R., Lahijanian, M., Moll, M., Kavraki, L.: Asymptotically optimal stochastic motion planning with temporal goals. In: Algorithmic Foundations of Robotics XI, pp. 335–352. Springer, Berlin (2015)
15. Raman, V., Kress-Gazit, H.: Synthesis for multi-robot controllers with interleaved motion. In: International Conference on Robotics and Automation (ICRA), pp. 4316–4321. IEEE (2014)
16. Raman, V., Finucane, C., Kress-Gazit, H.: Temporal logic robot mission planning for slow and fast actions. In: International Robots and Systems (IROS), pp. 251–256. IEEE (2012)
17. Smith, S., Tumova, J., Belta, C., Rus, D.: Optimal path planning for surveillance with temporal logic constraints. Int. J. Robot. Res. **30**, 1695–1708 (2011)
18. Stefanescu, A.: Automatic synthesis of distributed transition systems. Ph.D. thesis, University of Stuttgart (2006)
19. Tumova, J., Dimarogonas, D.V.: Decomposition of multi-agent planning under distributed motion and task LTL specifications. In: CDC, pp. 1775–1780. IEEE (2015)
20. Ulusoy, A., Smith, S., Ding, X.C., Belta, C.: Robust multi-robot optimal path planning with temporal logic constraints. In: International Conference on Robotics and Automation (ICRA), pp. 4693–4698. IEEE (2012)
21. Wolff, E., Topcu, U., Murray, R.: Optimization-based trajectory generation with linear temporal logic specifications. In: International Conference on Robotics and Automation (ICRA), pp. 5319–5325. IEEE (2014)
22. Wolper, P.: Constructing automata from temporal logic formulas: a tutorial. In: Lectures on Formal Methods and Performance Analysis, pp. 261–277. Springer, Berlin (2001)
23. Zlot, R., Stentz, A.: Complex task allocation for multiple robots. In: International Conference on Robotics and Automation (ICRA), pp. 1515–1522. IEEE (2005)

Informative Path Planning and Mapping with Multiple UAVs in Wind Fields

Doo-Hyun Cho, Jung-Su Ha, Sujin Lee, Sunghyun Moon and Han-Lim Choi

Abstract Informative path planning (IPP) is used to design paths for robotic sensor platforms to extract the best/maximum possible information about a quantity of interest while operating under a set of constraints, such as dynamic feasibility of vehicles. The key challenges of IPP are the strong coupling in multiple layers of decisions: the selection of locations to visit, the allocation of sensor platforms to those locations; and the processing of the gathered information along the paths. This paper presents an systematic procedure for IPP and environmental mapping using multiple UAV sensor platforms. It (a) selects the best locations to observe, (b) calculates the cost and finds the best paths for each UAV, and (c) estimates the measurement value within a given region using the Gaussian process (GP) regression framework. An illustrative example of RF intensity field mapping is presented to demonstrate the validity and applicability of the proposed approach.

1 Introduction

Unmanned autonomous vehicles (UAVs) have been used as mobile sensor platforms for data collection in a variety of fields, including intelligence, surveillance, and reconnaissance missions, where the vehicles' mobility enables a large-scale sensing. Nonetheless, it can be difficult with just a single UAV to cover a huge area in a

D.-H. Cho (✉) · J.-S. Ha · S. Moon · H.-L. Choi
Department of Aerospace Engineering, KAIST, Daejeon, Republic of Korea
e-mail: dhcho@lics.kaist.ac.kr

J.-S. Ha
e-mail: jsha@lics.kaist.ac.kr

S. Moon
e-mail: shmoon@lics.kaist.ac.kr

H.-L. Choi
e-mail: hanlimc@kaist.ac.kr

S. Lee
Agency for Defense Development, Daejeon, Republic of Korea
e-mail: sjlee@lics.kaist.ac.kr

R. Groß et al. (eds.), *Distributed Autonomous Robotic Systems*,
Springer Proceedings in Advanced Robotics 6,
https://doi.org/10.1007/978-3-319-73008-0_19

short time, particularly when the environment is changing rapidly. Also, from an economical perspective, data collection using multiple small UAVs with inexpensive onboard sensors is relatively cheaper and safer than carrying out the mission with a single UAV.

Path planning for UAV missions is one of the key methods of ensuring efficient information acquisition, measurement, and mapping of a region of interest. Path planning problem have been studied for a long time, and most of that research has focused on minimizing the path length or the moving cost of the robot. However, in most cases, the highest priority of UAV path design should be to make the best use of the sensor and maximize the amount of information acquired in a given mission. Especially considering that the value or quantity of information is not equally distributed at every point in most cases, the planned path should prioritize areas that are information-rich.

At the same time, one of the main constraints in path generation when operating a UAV system is the limited power source. In multiple studies, it has been shown that exploiting wind-energy is one of the most effective ways of decreasing UAV energy consumption [1, 11, 20]. Especially in mountainous regions, the wind blows strongly and can have a strong influence on the system dynamics of UAVs. This raises the question of whether a wind field can be exploited to operate the UAV system in a more efficient way.

Several information measurement criteria have previously been used to formulate informative path planning (IPP) problems, such as the Fisher information [12], the Kullback–Leibler divergence (relative entropy) [14], the Gaussian process uncertainty [2], and the mutual information [3]. Of these, we borrowed the meaning of information from [17], and the concept of 'mutual information' with entropy is used in the following sections. In papers dealing with IPP problem information is generally defined as a signal obtained by missions, such as reconnaissance about a region or targets of interest. Also, there exist some studies [13, 18] which use an adaptive path planning in IPP problem since the non-adaptive setting of the problem is known as NP-hard [6, 19], the non-linear solver was used in this study to avoid the local optimal solution.

In this study, we focus on a mission which involves mapping the magnitude of measured data (e.g. an RF signal) in a specific region using a set of fully connected UAVs which are affected by wind. It is assumed that the mathematical model of the collected data is unknown, and therefore a probabilistic model, Gaussian Process, is applied to calculate an estimate [16]. We develop a procedure for online mapping using path planning for multiple UAVs which maximizes the acquisition of information. Although there exist many studies involving the informative path planning problem, all of them were focusing on the partial side of the problem and there does not exist any studies which presented the whole procedure of the problem as mentioned above.

Most of the studies about IPP have formulated target visitation to acquire information as a constraint [10] or a reward [2]. Depending on the original intention, the obtained information can be used for classification of the target or for a decision making process. But that is beyond the scope of this paper, and here we only focus on the acquisition of the information.

2 Problem Formulation

The problem to be handled in this paper is as follows: given a set of initial locations of UAVs in a bounded non-convex region with a wind field exerting an influence on the UAV dynamics, generate a path set which traverses every sensing points with minimum cost, and finally have all of the UAVs return to their initial locations. After the path set is generated, the UAV creates a map of the information to be measured along the path. Since the scope of the problem is wide and can be considered obscure, we subdivide the problem into 4 subproblems:

1. Given the number of tasks, optimize the set of task locations (representative spots to visit) for the maximum acquisition of information within the region of interest. (Sect. 3.1, Eq. (2).)
2. Calculate the paths and moving costs of a UAV between every pair of task locations obtained from subproblem #1. Because of the wind field dynamics, the paths and costs for both directions should be calculated separately. (Sect. 3.2, Eq. (4).)
3. Using the output of the subproblem #2, determine the minimum cost route which traverses all of the task locations and finally returns to the starting point. Since the situation is a multiple-UAV case, this subproblem can be modeled as a Multiple-Depot Multiple Traveling Salesmen Problem (MDMTSP). (Sect. 3.3, Eqs. (5), (6))
4. Each UAV obtains an information along the generated path. During travel, a map of the information to be measured is created for the region of interest. (Sect. 3.4, (15))

Below are the assumptions that outline the constraints used in the problem.

- The energy consumption of each UAV is affected by the wind field dynamics.
- The UAVs are fixed wing and homogeneous; all of them share the same dynamics.
- The wind field dynamics and amount of information are stationary within a given region.
- The wind field dynamics is known in advance.
- Sensing is carried out by an onboard omni-directional sensor (e.g., radar, sonar) mounted on the UAVs, discretely along the paths (discrete time measurement).
- After sensing is finished, the UAVs return to their initial location.

3 Procedure

3.1 *Task Location Optimization Based on Entropy and Mutual Information with Gaussian Random Variables*

3.1.1 Entropy and Mutual Information

Entropy, which measures the amount of uncertainty, in a random variable (R.V.) X with probability mass function $p_X(x)$ is $H(X) = -\mathbb{E}\left[\log(p(x))\right]$ [5]. It can be also said that $H(X)$ is approximately equal to how much information the R.V. X has on average. For multiple R.V.s, the joint entropy can be derived from the above definition and is $H(X_1, X_2, \cdots, X_n) = \sum_{i=1}^{n} H(X_i|X_{i-1}, \cdots, X_1)$. Specifically, if a set of R.V.s follows the multivariate Gaussian distribution with mean μ and covariance matrix K, $\mathbf{X} \sim \mathcal{N}(\mu, K)$, then $H(\mathbf{X}) = \frac{1}{2}\log(2\pi e)^n|K|$ where $|K|$ denotes the determinant of K and n for the dimension of $\mathbf{X}$.

The mutual information is the relative entropy (of KL-divergence) between the joint distribution and the product distribution between R.V. X_1 and R.V. X_2,

$$I(X_1; X_2) = \mathbb{E}\left[\log \frac{p(X_1, X_2)}{p(X_1)p(X_2)}\right] = H(X_1) - H(X_2).$$

It is also said that $I(X_1; X_2)$ is a measure of the amount of information that one R.V. X_1 contains about another R.V. X_2, and can be interpreted as the reduction of entropy X_1 by conditioning on X_2. The mutual information for two sets of R.V.s which follow the multivariate Gaussian distribution, $\mathbf{X}_1 \sim \mathcal{N}(\mu_1, K_1)$, $\mathbf{X}_2 \sim \mathcal{N}(\mu_2, K_2)$, and $(\mathbf{X}_1, \mathbf{X}_2) \sim \mathcal{N}(\mu, K)$, can be written as

$$I(\mathbf{X}_1; \mathbf{X}_2) = \frac{1}{2}\log \frac{\det(K_1)\det(K_2)}{\det(K)}. \tag{1}$$

3.1.2 Task Location Optimization

To solve the first subproblem mentioned in Sect. 2, assume the interesting region $A \subset \mathbb{R}^2$ and the number of task locations n are given. The optimal set of locations $P_T^* \subset A$ for a task set $T = \{1, 2, \cdots, n\}$ which maximizes the mutual information, $I(\mathbf{X}_T(P_T); \mathbf{X}_o(P_o))$, is [3, 4]

$$P_T^* = \underset{P_T \in A}{\operatorname{argmax}}\, I(\mathbf{X}_T(P_T); \mathbf{X}_o(P_o)). \tag{2}$$

Here, $\mathbf{X}_T(P_T)$, the function of P_T, is termed the *verification variable* which represents the variables (or data set) obtained from the set of task locations T, where $\mathbf{p}_t =$

$(x_t, y_t) \in P_T$ is the location of tth task, and $t \in T$. $\mathbf{X}_o$ is a set of variables from the points called *test points*. To calculate the entropy and the mutual information for the whole region of interest, locations for the *test points* are equally spaced inside the region A, and denoted as P_o. The verification variable $\mathbf{X}_T$ is obtained from the set of locations P_T, and $\mathbf{X}_o$ is obtained from P_o. In this study, Eqs. (1) and (2) were taken into the nonlinear programming solver[1] to obtain the set of optimal points P_T^*. The number of task locations n depends on the effective range and the noise level of the sensor.

3.1.3 Meaning of Assigning Task Locations

The set of optimal task locations, P_T^*, is considered to be the set of representative spots, where the maximum amount of information can be obtained when the UAV tours around those locations. Suppose a set of task locations are given, and a UAV moves along the path which traverses all of the given locations. Since it is assumed that the UAV is gathering information discretely during a tour even when it is not near one of the task locations, it is important to set the number of task locations properly for the path to be set well. In Fig. 1, the tasks with the Hamiltonian path in the region $A \in [0, 100] \times [0, 100]$ are drawn for each case - the number of tasks $N = 200, 40$, and 5. Here the figure shows three cases of the generated path, *overfitted, normal,* and *underfitted.* If the number of tasks is set too high, then a region of overlapped sensing exists which is superfluous. Costs of moving increase unnecessarily, and we define this situation as *overfitted.* Or if the number of tasks is too small to cover all of the region, the information obtained will be insufficient to estimate the data of interest after finishing the mission. We define this situation as *underfitted.*

The noise level and the length scale of a system model are important factors when setting the proper number of tasks for the mission. Described below is the relationship between the number of task locations and the number of sensing points. Assume that if the effective range of the sensor is very wide compared to the region of interest, then only few sensing points are enough to cover the whole region and there would be no problem to consider the task locations as the sensing points. However, in most of the cases, the inverse situation is given, and a lot of sensing points are needed to obtain the information of the whole region of interest. In this situation, if the number of task locations is given as the number of sensing points as mentioned above, and each UAV has to visit every task locations, then the *overfitting* situation happens naturally. To avoid the *overfitting* situation, the number of task locations should be less than the number of sensing points, and the proper number depends on the measuring frequency and the effective range of the sensor.

To reduce the gap between the amount of mutual information obtained only on the task locations and the sensing points along the generated path for each UAV, it is necessary to assume that the noise level is less and the length scale is longer than the effective sensor range when optimizing the locations of tasks. Let MI_1 be the expected

[1] MATLAB *fminsearch* function was used in this study.

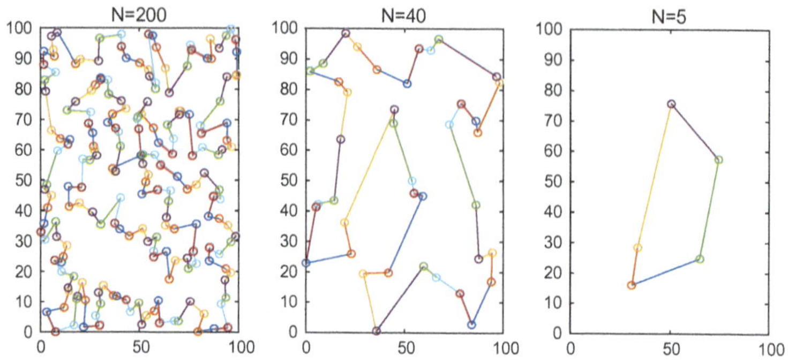

Fig. 1 Randomly generated task locations with Hamiltonian path. Left: #Tasks: 200, Mid: #Tasks: 40, Right: #Tasks: 5. Left plot shows the path is overfitted, and right is underfitted

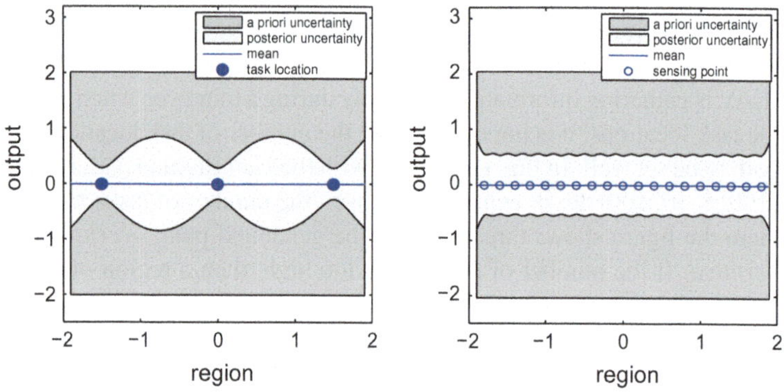

Fig. 2 Uncertainty difference between before and after sensing. Left plot shows the concept of a role of task locations, and right plot shows the uncertainty level decreases with sensing

amount of mutual information gathered only from optimal task locations with low noise level and wide sensor range, and MI_2 be the expected amount of a mutual information gathered along the path by sensors mounted on a UAV. Using the above procedure mentioned in Sect. 3.1.2 with proper leveling, the difference between the amount of MI_1 and MI_2 converges to 0. The amount of mutual information can be interpreted as the difference between *a priori uncertainty* and *posterior uncertainty*, and this is indicated by the gray field in Fig. 2. In the plots, the blue dots represent the verification variables with equally distributed locations in the 1-dimensional region, and the output was set to be 0. The output can be regarded as the measurement value of the sensor. We assume that the variables follow a Gaussian distribution with a linear mean function and a non-isotropic squared exponential covariance function. Details are shown in Sect. 3.4.

3.2 Calculation of Moving Cost with FMT* Algorithm

In order to compute the distances between every pair of task locations while considering the wind field, we adopted the Fast Marching Tree star (FMT*) algorithm which was recently proposed by Janson et al. [8]. The FMT* algorithm is a batch processing sampling-based path planning algorithm and it performs a direct dynamic programming process and lazy collision checking which dramatically accelerates the speed of the algorithm.

The algorithm was modified to be suitable for the problem stated in this study. First, while assuming that the UAVs follow the planned path, $\sigma : [0, \tau] \rightarrow \mathbb{R}^2$ (where τ denotes the duration of the path), with constant speed, v_0, the effect of the wind field, $\mathbf{w} \in \mathbb{R}^2$, was incorporated in the cost of the path planning problem. For the path direction, suppose that the UAV has a simple 2nd order dynamics with external force and drag as:

$$\dot{v}(t) = \frac{F(t)}{m} - k(v(t) - w_\sigma),$$

where F, m and k denote the force, mass and drag coefficients of the UAV, respectively; $w_\sigma = \mathbf{w}(\sigma(t)) \cdot \frac{\dot{\sigma}(t)}{||\dot{\sigma}(t)||}$ represents wind speed along the path direction. In order to maintain the constant speed, v_0, the required force is given as $F_0(t) = mk(v_0 - w_\sigma)$. Then, we obtained the total energy consumption along the trajectory as:

$$\begin{aligned} E(\sigma) &= \int_0^\tau F_0(t) d||\sigma(t)|| \\ &= \int_0^\tau (mkv_0 - mkw_\sigma) d||\sigma(t)|| \\ &= mkv_0^2\tau - mk \int_0^\tau \mathbf{w}(\sigma(t)) \cdot d\sigma(t), \end{aligned} \tag{3}$$

where we assume $v_0 > w_\sigma$ so that $F_0(t) > 0$ for simplicity. The cost function of the path planning problem is obtained by simplifying the energy equation $C(\sigma) = E(\sigma)/mkv_0^2$:

$$C(\sigma) = \tau - \frac{1}{v_0^2} \int_0^\tau \mathbf{w}(\sigma(t)) \cdot d\sigma(t). \tag{4}$$

Note that there is a trade-off between path length and path direction: the cost penalizes the longer trajectory for the high desired speed v_0, while the path direction is encouraged to be aligned with the wind direction for lower v_0. Second, rather than solving the planning problem for every pair of task locations individually, we modified the algorithm into a multi-query version: with a fixed starting location, one planning problem finds all the paths to the other locations. As a result, if there are n task locations, only n (rather than n^2) planning problems need to be solved. Finally, we made all of the n planning problems share the edge information. This was done

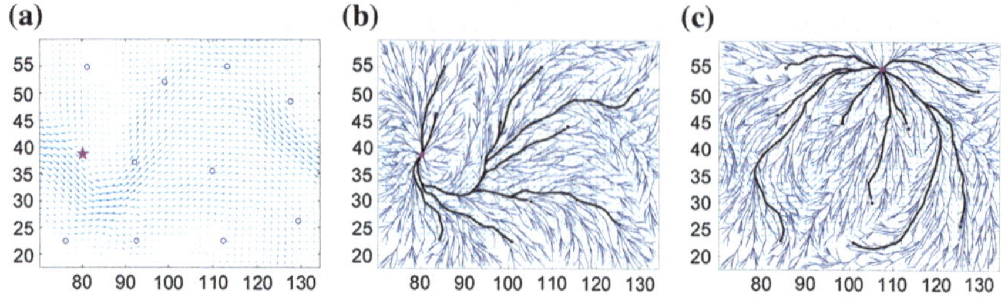

Fig. 3 **a** Wind field and task locations. **b–c** Some resulting paths. Magenta stars and black lines represent starting locations and resulting paths, respectively

because, in general, cost evaluation and the collision checking of edges are computational bottlenecks for the planning algorithm. Sharing edge information through the problems significantly improves the scalability of the algorithm. Figure 3 shows some of the resulting paths in an environment with an arbitrary wind field. It can be seen that the overall directions of the resulting paths are aligned with those of the wind field.

3.3 Mathematical Formulation of the Min-Max Multiple Depots Multiple Traveling Salesmen Problem (MMMDMTSP)

The MMMDMTSP [9] is defined with a set of task locations $T = \{1, 2, \cdots, n\}$ and a set of depot locations (the initial location of the UAVs) $D = \{n+1, n+2, \cdots, n+m\}$. The cardinality of T and D is denoted as $|T| = n$ and $|D| = m$, and it is assumed that $n, m \geq 1$. In this study, we let each UAV have a separate location, so if there are multiple vehicles in a depot, then the location of the depot will be repeated with how ever many vehicles exist in there. Therefore, m also represents the number of UAVs.

Let $G = (V, E)$ be a directed graph where $V = T \cup D = \{1, 2, \cdots, n, n+1, \cdots, n+m\}$ is the union of sets of the task locations and depot locations, and $E = \{(i, j) | \forall (i, j) \in (V \times V) \backslash (D \times D)\}$ is the edge set (note that E doesn't include any edge between depots) with the cost denoted by c_{ij}. The reason that G is not an undirected graph is that the cost from i to j, c_{ij} and the cost from j to i, c_{ji} are different because of the wind field dynamics model.

For each edge, $\theta_{ijk} \in \{0, 1\}$, is defined to be equal to one if UAV k takes the edge $e = (i, j)$ as a part of its route and zero otherwise, where $i, j \in V$, $i \neq j$, and $k \in D$. To prevent a UAV taking a loop for a certain task location, $\theta_{iik} = 0$ should be added as a constraint.

From the above definitions, the cost sum of the kth UAV route can be formulated as:

$$C_k = \sum_{j=1}^{n} \theta_{kjk} c_{kj} + \sum_{i=1}^{n} \sum_{j=1}^{n} \theta_{ijk} c_{ij} + \sum_{i=1}^{n} \theta_{ikk} c_{ik}. \tag{5}$$

The first term represents the cost from the kth UAV's depot to the first task location. The second term is a sum of the costs along the route for set of task locations, and the final term is the cost from the last task location to the UAVs' depot.

To minimize the longest tour for every UAV, $C_{max} = \max_{k \in D} C_k$ is defined as a continuous decision variable, and the following is the formulation of the MMMDMTSP:

$$\text{Minimize} \quad C_{max} \tag{6}$$

such that

$$\sum_{k=1}^{m} \theta_{kik} + \sum_{j=1}^{n} \sum_{k=1}^{m} \theta_{jik} = 1 \qquad \forall i \in N \tag{7}$$

$$\theta_{kik} + \sum_{j=1}^{n} \theta_{jik} - \left(\theta_{ikk} + \sum_{j=1}^{n} \theta_{ijk} \right) = 0 \qquad \forall i, j \in T, \forall k \in D \tag{8}$$

$$\sum_{i=1}^{n} \theta_{kik} \leq 1 \qquad \forall k \in D \tag{9}$$

$$\sum_{i=1}^{n} \theta_{kik} - \sum_{i=1}^{n} \theta_{ikk} = 0 \qquad \forall k \in D \tag{10}$$

$$C_k \leq C_{max} \qquad \forall k \in D \tag{11}$$

$$\sum_{i \in S} \sum_{j \in S} \theta_{ijk} \leq |S| - 1 \qquad i \neq j, \forall k \in D, \; |S| = 2, 3, \cdots, n-1 \tag{12}$$

The constraint (7) ensure that all the task locations are visited exactly once by the set of routes, and (8) guarantees that each task is not the final destination of the routes. Equations (9) and (10) are the constraints that respectively ensure that each UAV has at most one route, and if it does have more, then the UAV should return to its initial depot (this makes the problem a TSP(traveling salesman problem), not a VRP(vehicle routing problem)). Equation (11) balances the cost of each route for every UAV. Finally, (12) is a constraint for subtour elimination. Here, a subtour is a route which only has vertices $i \in T$ and is not from the set of depots, D [9].

Since it is intractable to set up the subtour elimination constraint equation for every cases, the number of this constraint equation is a determining factor for solving this MILP formulation. The GA (Genetic algorithm) was adopted in this work, and we followed the details shown in [15].

3.4 Gaussian Process Regression

Gaussian Process Regression [16] (GPR) is one of the supervised modeling scheme that approximates the interesting target points (output points) using the function of training points (input points). This scheme regards the relationship between input and output points as one of the examples of Gaussian process, and thus the output points can be approximated as a kind of Bayesian inference which computes the posterior distribution of output points of interest conditioned on the experimentally obtained input points. In particular, it is assumed that all of the relevant probability distributions of input and output points follow a joint Gaussian distribution. With input points, the GPR procedure consists of defining a mean function and a covariance function and learning their hyperparameters, to maximize the probability of the GPR generating interesting output points. Here the hyperparameters determine the shape and characteristics of the GPR.

The most used mean function and covariance in GPR are described below; for the mean function, a constant function is enough to estimate the model in many cases:

$$\mathbb{E}[f(\mathbf{x})] = m(\mathbf{x}). \tag{13}$$

For the covariance function, the non-isotropic squared exponential covariance function is used:

$$\text{cov}\left(f(\mathbf{x}_p), f(\mathbf{x}_q)\right) = k(\mathbf{x}_p, \mathbf{x}_q) = \sigma_f^2 \exp\left[-\frac{1}{2}(\mathbf{x}_p - \mathbf{x}_q)^T \Sigma_l^{-2} (\mathbf{x}_p - \mathbf{x}_q)\right] \tag{14}$$

where σ_f is the signal's standard deviation, and $\Sigma_l = \text{diag}\left(\sigma_{l_1}, \cdots, \sigma_{l_n}\right)$, which represents the characteristic length-scale for each input dimension. These parameters, σ_f and Σ_l, are called hyperparameters.

In Gaussian Process, for an arbitrary input point set $\{\mathbf{x}_1, \cdots, \mathbf{x}_n\} \in \mathbf{X}$, the output point set $\{f(\mathbf{x}_1), \cdots, f(\mathbf{x}_n)\}$ follows the following Gaussian distribution.

$$\begin{bmatrix} f(\mathbf{x}_1) \\ \vdots \\ f(\mathbf{x}_n) \end{bmatrix} \sim \mathcal{N}\left(\begin{bmatrix} m(\mathbf{x}_1) \\ \vdots \\ m(\mathbf{x}_n) \end{bmatrix}, \begin{bmatrix} k(\mathbf{x}_1, \mathbf{x}_1) & \cdots & k(\mathbf{x}_1, \mathbf{x}_n) \\ \vdots & \ddots & \vdots \\ k(\mathbf{x}_n, \mathbf{x}_1) & \cdots & k(\mathbf{x}_n, \mathbf{x}_n) \end{bmatrix}\right)$$

In the case of predicting with noisy observations, it is known that the predictive equations for GPR are

$$\mathbf{f}_* | X, \mathbf{y}, X_* \sim \mathcal{N}\left(\bar{\mathbf{f}}_*, \text{cov}(\mathbf{f}_*)\right) \tag{15}$$

where $\bar{\mathbf{f}}_* = K_{OI}\left[K_{II} + \sigma_n^2 I\right]^{-1} \mathbf{y}$ and $\text{cov}(\mathbf{f}_*) = K_{OO} - K_{OI}\left[K_{II} + \sigma_n^2 I\right]^{-1} K_{IO}$. Here $\mathbf{y}$ is the observation vector with additive Gaussian noise. The subscript I indicates the input data of size I, and O is the interesting point to be verified. K_{II} denotes the $I \times I$ matrix of the covariances evaluated at all pairs of input and input points, and similarly for the other matrices K_{OI} and K_{OO}.

As mentioned above, the hyperparameters determine the characteristics of the GPR structure; so an optimization process to obtain the proper hyperparameters is needed for an accurate approximation. The most commonly used method of optimization is to choose the hyperparameters which maximize the log likelihood of the given input points, in other words:

$$(\sigma_f, \Sigma_l) = \underset{\sigma_f, \Sigma_l}{\text{argmax}} \log p(\mathbf{y}|X) \tag{16}$$

In subproblem #4, an information map of an interesting region A can be constructed with GPR, (15) and (16). From Sect. 3.1.2, the training points are P_T with the values $X_T(P_T)$, and the target points are the set of locations P_o to obtain $X_o(P_o)$.

4 Numerical Example

This section shows the detailed results of the simulation for the problem given in Sect. 2, using the suggested procedure (Sect. 3). The information to be measured is assumed to be the intensity of the RF signal, and a signal intensity map is generated with the obtained information. The parameters and conditions used in the simulation are as follows.

4.1 Simulation Parameters and Conditions

4.1.1 Wind-Field

There are several kinds of wind field models that can be used for generating wind field data. Each model depends on the degree of simplification of the *Navier–Stokes* equation, and this point determines the spatial resolution of the output data. It is necessary to choose the wind field model with the appropriate spatial resolution when the size of the region is given. In this simulation the size was fixed as 20 km × 20 km. The wind field data was generated with software called *WindSim*, which is based on the *Computational Fluid Dynamics* model, with a spatial resolution of 20 m, and this resolution size fits the assumed size of the interesting region. If the temperature is assumed to be constant, the streamline of the wind differs with the following factors: variation in the altitude of the surface, and land uses (forest, farm, downtown, mountain, etc.). To obtain wind field data, a digital elevation model (DEM), land uses, and the boundary conditions of the wind field are needed as inputs of the program. In the simulation the region of interest was assumed to be a mountainous area. The boundary condition of the wind was set to be 10 m/s from the northeast. The 3D wind field data was obtained with these inputs and conditions, and the data at specific altitude, which is the 2D data along the xy plane, was chosen to be used in the simulation below. In Fig. 5, the generated wind field is shown as a vector plot (blue arrow).

4.1.2 RF Signal Intensity Data

The basic RF propagation model can be expressed using the *free space path loss* model, $P_L(dB) = \log_{10} \frac{P_t}{P_r} = -10 \log_{10} \frac{G\lambda^2}{(4\pi d)^2}$ where λ is the signal wavelength, d is the distance from the transmitter, and G is the gain value. But a change in intensity always happens because of geographical features, and this effect is called *shadowing*. It is usually assumed that shadowing follows lognormal distribution, and an RF signal intensity map can be generated for a region with complex geographical features using the *combined path loss and shadowing* model [7]. The program called *Radio mobile* was used to obtain an RF signal intensity map of an interesting region at the frequency band of 146 MHz. The type of transmitter antenna was assumed to be omni-directional, and the gains of the transmitter/receiver antenna were 6.0/2.0 dBi respectively. It was assumed that one transmitter was in the center of the region. A signal intensity map of the region is shown in Fig. 4. The intensity plot is drawn for an altitude of 3,000 m.

4.1.3 Information Acquisition with Onboard Sensor

It was assumed that a total of 3 UAVs were used to obtain information in this simulation, and that each of them moved along the generated path inside the given region. The initial locations of each UAV can be arbitrarily chosen, but it was assumed that the locations were given far away moderately from each other. The altitude of the UAVs was fixed at 3,000 m. The UAV dynamics followed the simple Dubins path model, where $\dot{x} = v\cos(\theta)$, $\dot{y} = v\sin(\theta)$, and $\dot{\theta} = u$ where u is bounded. The speed and the minimum turning radius were $v = 100$ m/s, $r_{min} = 50$ m each. To generate an

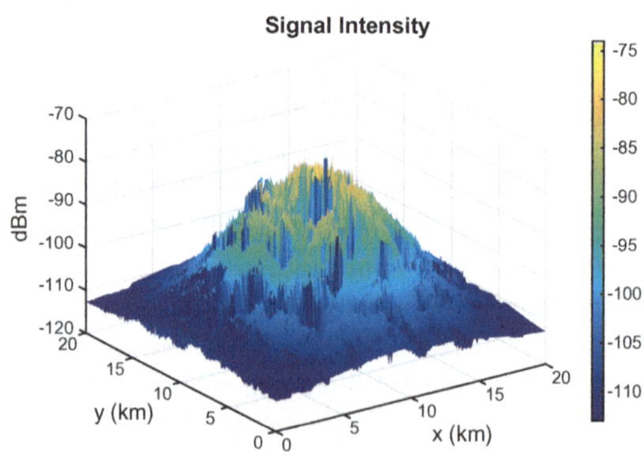

Fig. 4 RF Signal Intensity with altitude 3 km. Unit: dBm

RF signal intensity map using GPR, Eqs. (13), (14) were used as the mean and covariance function. We set the hyperparameters as follows: $\sigma_f = 30\,\text{dBm}$, $\sigma_n = 1\,\text{dBm}$, and $l_x, l_y = 4\,\text{km}$. Each sensor measured the data for every 10 s, and during the data acquisition the optimization of these hyperparameters were carried out with Eq. (16) to maximize the amount of mutual information. σ_n is a sensor dependent value which is not the subject of a hyperparameter optimization.

4.2 *Simulation Result*

The results are shown in Figs. 5 and 6a, b. These were obtained with the parameters and conditions in Sect. 4.1. Figure 5 shows the results of the subproblems #1 to #3. Since it is assumed that the amount of information is equally distributed within the region, the task locations are spread apart from each other. The moving costs for every pair of tasks were obtained using the FMT* algorithm to construct a distance matrix. This matrix becomes an input of the MDMTSP problem, and the output of this problem (reference paths for each UAV) is drawn as a black line. The depot locations are marked as red stars; one is located in the top center, and the others are on the left and right bottom in the given region. Red lines are the paths generated by the UAV dynamics for the given reference paths.

Figure 6a, b show the mapping results for RF signal intensity and the level of uncertainty within the region. The bottom contour shows how far the UAVs have gone along the paths. The 3D plot in the middle is the estimated RF map using GPR. Each of the blue dots are the points where sensing is performed, and the mapping is based on this obtained data. The level of uncertainty is shown in the upper contour; a domain with bright color indicates that the uncertainty level is low (or the information has been obtained), otherwise if the color is dark, the uncertainty level is high. Figure 6a shows the early stage of the mission, and the final result is shown in Fig. 6b. As the UAVs move and take measurements, the mapping becomes similar to the original data and the level of uncertainty gets lower.

5 Conclusion

An IPP procedure to measure and map a region of interest using multiple UAVs has been proposed, in the framework of mutual information and Gaussian process regression. The validity of the procedure was demonstrated by simulation using a realistic wind field and RF signal intensity data, and the target region was assumed to be a mountainous area.

In future work, a more realistic MDMTSP with Dubins path concept will be considered as well as various amounts and distribution of information inside the region. Also, the problem can be extended to time-varying situations, which makes the problem harder to analyze and solve.

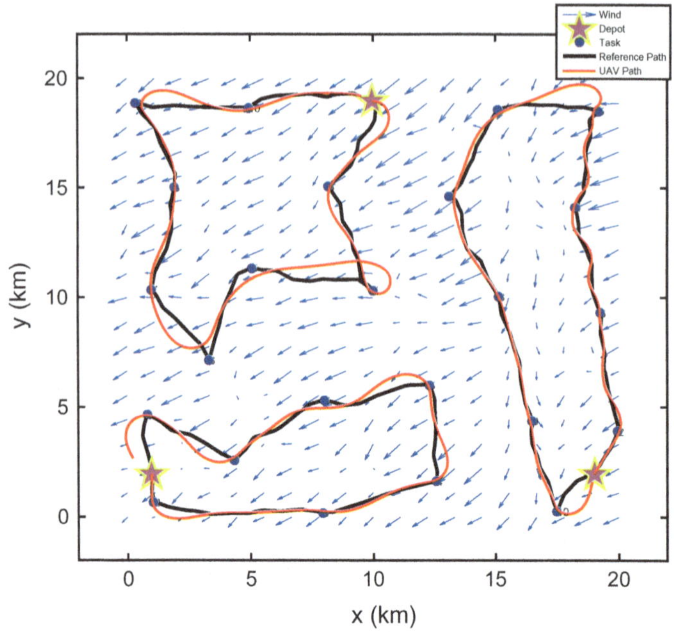

Fig. 5 Locations of depots (red stars) and tasks (blue dots), reference paths (black line), and UAV paths (red line)

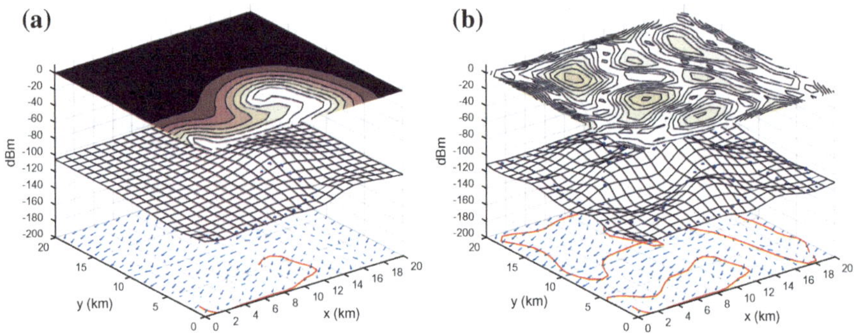

Fig. 6 Simulation Results. **a** Initial stage - High uncertainty level, low information gained. **b** Final stage - Low uncertainty level, high information gained. Top: The uncertainty level of the region of interest. Middle: The estimated RF map. Bottom: The portion of the paths executed by the UAVs

Acknowledgements This work was supported by Agency for Defense Development (contract #UD140053JD).

References

1. Al-Sabban, W.H., Gonzalez, L.F., Smith, R.N.: Wind-energy based path planning for unmanned aerial vehicles using markov decision processes. In: IEEE International Conference on Robotics and Automation (ICRA), pp. 784–789. IEEE (2013)
2. Binney, J., Sukhatme, G.S.: Branch and bound for informative path planning. In: ICRA, pp. 2147–2154. Citeseer (2012)
3. Choi, H.L., How, J.P.: Continuous trajectory planning of mobile sensors for informative forecasting. Automatica **46**(8), 1266–1275 (2010)
4. Choi, H.L., Ahn, J., Cho, D.H.: Information-maximizing adaptive design of experiments for wind tunnel testing. Eng. Optim. **2014**, 329 (2014)
5. Cover, T.M., Thomas, J.A.: Elements of Information Theory. Wiley, New York (2012)
6. Guestrin, C., Krause, A., Singh, A.P.: Near-optimal sensor placements in gaussian processes. In: Proceedings of the 22nd International Conference on Machine learning, pp. 265–272. ACM (2005)
7. Hufford, G.A., Longley, A.G., Kissick, W.A., et al.: A guide to the use of the ITS irregular terrain model in the area prediction mode. US Department of Commerce, National Telecommunications and Information Administration (1982)
8. Janson, L., Schmerling, E., Clark, A., Pavone, M.: Fast marching tree: a fast marching sampling-based method for optimal motion planning in many dimensions. Int. J. Robot. Res. **34**, 883–921 (2015)
9. Kivelevitch, E., Cohen, K., Kumar, M.: A binary programming solution to the multiple-depot, multiple traveling salesman problem with constant profits. In: Proceedings of 2012 AIAA Infotech Aerospace Conference (2012)
10. Krause, A., Singh, A., Guestrin, C.: Near-optimal sensor placements in gaussian processes: theory, efficient algorithms and empirical studies. J. Mach. Learn. Res. **9**, 235–284 (2008)
11. Langelaan, J.W., Alley, N., Neidhoefer, J.: Wind field estimation for small unmanned aerial vehicles. J. Guid. Control. Dyn. **34**(4), 1016–1030 (2011)
12. Levine, D.S.: Information-rich path planning under general constraints using rapidly-exploring random trees. Ph.D. thesis, Citeseer (2010)
13. Lim, Z.W., Hsu, D., Lee, W.S.: Adaptive informative path planning in metric spaces. Int. J. Robot. Res. **35**(5), 585–598 (2016)
14. Lu, W.: Autonomous sensor path planning and control for active information gathering. Ph.D. thesis, Duke University (2014)
15. Ombuki-Berman, B., Hanshar, F.T.: Using genetic algorithms for multi-depot vehicle routing. Bio-inspired Algorithms for the Vehicle Routing Problem, pp. 77–99. Springer, Berlin (2009)
16. Rasmussen, C.E.: Gaussian Processes for Machine Learning. Citeseer (2006)
17. Shannon, C.E.: A mathematical theory of communication. ACM SIGMOBILE Mob. Comput. Commun. Rev. **5**(1), 3–55 (2001)
18. Singh, A.: Nonmyopic adaptive informative path planning for multiple robots. Center for Embedded Network Sensing (2009)
19. Singh, A., Krause, A., Guestrin, C., Kaiser, W.J., Batalin, M.A.: Efficient planning of informative paths for multiple robots. IJCAI **7**, 2204–2211 (2007)
20. Ware, J., Roy, N.: An analysis of wind field estimation and exploitation for quadrotor flight in the urban canopy layer. In: 2016 IEEE International Conference on Robotics and Automation (ICRA), pp. 1507–1514. IEEE (2016)

Multi-robot Informative and Adaptive Planning for Persistent Environmental Monitoring

Kai-Chieh Ma, Zhibei Ma, Lantao Liu and Gaurav S. Sukhatme

Abstract To gain a better understanding of environmental processes we are interested in the problem of deploying multi-robot systems for efficient collection of environmental data. For long-term autonomy, enabling persistent monitoring, it is important to consider the spatio-temporal variations of environmental phenomena. We develop a multi-robot persistent path planning method that reduces uncertainty in the environmental model. Our framework contains two components: the first component computes potential observation points that minimize model prediction uncertainty, and the second component uses this for online planning of multi-robot paths, while also taking into account the efficiency of information collection. We validated our method via simulations, and the results show that it produces multi-robot routing paths that are conflict-free, informative, and adaptive to the environmental dynamics.

1 Introduction

We are interested in the problem of deploying multiple robots for efficient collection of environmental data, to gain a greater understanding of environmental processes. In particular, we are interested in reconstruction of physical, chemical or biological scalar fields. One example is the use of autonomous underwater vehicles (AUVs) for ocean monitoring, to map physical or biological properties of the ocean, such as temperature, salinity, and chlorophyll contents. Environmental monitoring is inher-

K.-C. Ma · Z. Ma · L. Liu (✉) · G. S. Sukhatme
Department of Computer Science, University of Southern California,
Los Angeles, CA, USA
e-mail: lantao.liu@usc.edu

K.-C. Ma
e-mail: kaichiem@usc.edu

Z. Ma
e-mail: zhibeima@usc.edu

G. S. Sukhatme
e-mail: gaurav@usc.edu

R. Groß et al. (eds.), *Distributed Autonomous Robotic Systems*,
Springer Proceedings in Advanced Robotics 6,
https://doi.org/10.1007/978-3-319-73008-0_20

ently a continuous and persistent task, because many of the underlying environmental processes vary both spatially and temporally. Therefore, in order to obtain a good estimate of the state of the environment at any time, robots need to perform persistent monitoring [15, 16].

One aspect that sets apart persistent monitoring from conventional path planning methods, is that travel costs (e.g. travel time and distance) are not the only concern, because the robots are performing the task in a continuous, lasting manner. Instead, the objectives of a planning framework for multi-robot, long-term autonomy missions, are:

- Maximization of information gain: At any time, the robots' observations along their paths can not cover the entire environmental space. We will need to model and predict continuous environmental phenomena with these limited observations, which inevitably causes uncertainty. Any planning approach should thus minimize model uncertainty, or equivalently, maximize information gain.
- Multi-robot coordination: Any paths planned for all robots should resolve potential conflicts. For example, two paths should avoid cross or transit the same location. Furthermore, each robot's path should collaboratively optimize for the global objective, namely the collective informativeness of the model.
- Adaptive and online routing: The robots should be capable of adapting to the collected data. Given the spatio-temporal variability of the environmental fields, it is crucial that the paths are adapted as the robots progress. This requires online routing of the vehicles; dynamic goals and re-planning of paths.

We use Gaussian Process (GP) regression to model the phenomenon of interest [17]. To characterize the amount of information collected, we utilize the mutual information between visited locations and the remainder of the space [19]. This allows us to obtain a set of "most informative" future observation points. However, these observation points do not yet form a path (or multiple paths), because no routing information is provided. Many traditional path planning methods require all routing goals to be determined in advance. However, such goals are unrealistic for long-term autonomy path planning, because vehicles need to continuously visit infinite number of goals. Therefore, we extend an existing matching graph-based routing method [12], such that the routing destinations can be dynamically determined, and conflict-free paths can be adaptively computed, while taking into account the information gain.

2 Related Works

Planning methodologies designed for the spatio-temporal environmental monitoring are often called *informative path planning*, because the objective is to maximize the collected information (informativeness) [1]. Representative informative path planning approaches include approaches based on recursive-greedy path planning using mutual information on top of Gaussian Process regression [2, 15, 19], where the informativeness is generalized as submodular functions built on which a sequential-allocation mechanism is designed in order to obtain subsequent waypoints. Recently, a differential entropy based method was proposed, in which a batch of waypoints can be obtained through solving a dynamic program [3, 13]. However, the framework is formulated with an assumption that the underlying map is transected (sliced) column-wise, so that each algorithmic iteration computes waypoints within a separate column and the navigation paths are obtained by connecting those waypoints among the pairwise adjacent columns. In recent works, we have extended such a framework by allowing the path to be searched and computed across the entire space at any stage [14]. In this work, we further extend our approach to persistent monitoring tasks for multi-robot systems.

Recent works that investigate informative path planning approaches for persistent ocean monitoring include [9, 16]. In [16], an active sensing based method was proposed, which uses a criterion that trades off between gathering the most informative observations for estimating unknown local regions, and predicting the phenomenon given the current estimates of those regions. To capture and adapt to the environment's model dynamics, we plan paths within short time horizons. In [9], paths were also planned over short time horizons, using receding horizon planning. However, they used a different metric, and they did not consider multi-robot coordination.

Other related works include the *Orienteering Problem* (OP). The OP is a routing problem in which the goal is to determine a subset of nodes to visit, and in which order, such that the total collected score is maximized, and the given time budget is not exceeded [7, 8]. Heuristics have been designed to approximate this NP-hard problem in an efficient way [5, 6, 11]. However, one drawback of approaching this problem as an OP lies in that the time limit or cost can be hard to determine for long-term autonomy scenarios. In this paper, we extend an efficient matching graph based planning method by strategically integrating metrics of information gain and travel cost. We compare our method to a popular heuristic for the OP, and our results show that our method performs better for persistent multi-robot environmental monitoring.

3 Informative and Adaptive Planning Framework

Environmental phenomena vary not only spatially but also temporally. We regard the temporal process as a sequence of short horizons of equal length, and assume that within each short horizon the latent environmental phenomena are time-invariant. This allows us to eliminate the temporal parameter and focus on constructing the environment's spatial properties, using existing methods from spatial statistics, such as Gaussian Processes (GPs). In this section, we explain how a set of potential informative observation points can be obtained from the GP. Following that, we construct routes over observation points using conflict-free paths. The observations along the paths are then used as a prior for generating a new set of potential observation points for the next time horizon. Note that these priors are time-varying, which means that the entropy (model uncertainty) of earlier observed points grows again gradually after the last observation. Therefore we need to be able to update the routing solution for each new horizon. Our observation point selection procedures thus repeat, and routes are updated, such that we carry out environmental monitoring persistently.

3.1 Gaussian Process Regression and Information Gain

We model the environment using Gaussian Process (GP) regression [17], similar to previous works [13, 19]. A GP's behavior is specified by its prior covariance function (also known as *kernel*), which describes the relation between two independent data points. The GP is further defined by its hyperparameters, which can be estimated using training data, typically through maximum likelihood estimation [17]. In our implementation, we use the *squared exponential automatic relevance determination* kernel function. The mean and variance of each sample location can be predicted via the GP. The variance represents the uncertainty of the predicted data value, which can be used to find future observation points.

To assess prediction uncertainty, we use mutual information as a metric. In information theory, the mutual information is used to describe the mutual dependence between two variables. It is derived from the concept of entropy which is defined to quantify the uncertainty of random variables. For two arbitrary vectors of sampling points A, B, the mutual information between A and B can be expressed in terms of (conditional) entropy

$$I(Z_A; Z_B) = I(Z_B; Z_A) = H(Z_A) - H(Z_A|Z_B) \tag{1}$$

where Z represent random variables, $H(Z_A)$ is the entropy of Z_A, and the conditional entropy $H(Z_A|Z_B)$ can be calculated via

$$H(Z_A|Z_B) = \frac{1}{2}\log\left((2\pi e)^k|\Sigma_{A|B}|\right). \tag{2}$$

The conditional covariance matrix $\Sigma_{A|B}$ can be calculated from the GP's posterior covariance matrix.

3.2 Generating Informative Observation Points

Let W denote the sampling set of the grid map, and let n be the desired number of observation points. The objective is to find a subset of sampling points, $P \subset W$ with a size $|P| = n$, which gives us the most information for our model. This is equivalent to the problem of finding observation points that maximize the mutual information between observed and unobserved locations of the map. The optimal subset of sampling points, P^*, with maximal mutual information is

$$P^* = \arg\max_{P \in \mathscr{X}} I(Z_P; Z_{W \setminus P}) \tag{3}$$

where $\mathscr{X}$ represents all possible combinatorial sets, each of which is of size n. P^* can be computed efficiently using a *dynamic programming* approach [14].

The dynamic programming approach is as follows: Formally, let $w_i \in W$ denote an arbitrary sampling point at stage i and $w_{a:b}$ represent a sequence of sampling points from stage a to stage b. Following Eq. (9), the mutual information between P and the unobserved part at the final stage n can then be written as $I(Z_{w_{1:n}}; Z_{W \setminus \{w_{1:n}\}})$. This mutual information can be expanded using the chain rule:

$$I(Z_{w_{1:n}}; Z_{W \setminus \{w_{1:n}\}}) = I(Z_{w_1}; Z_{W \setminus \{w_{1:n}\}}) + \sum_{i=2}^{n} I(Z_{w_i}; Z_{W \setminus \{w_{1:n}\}} | Z_{w_{1:i-1}}). \tag{4}$$

One can utilize this form of mutual information to calculate w_i step by step. However, at every stage i before the final stage, the entire unobserved set $W \setminus \{w_{1:n}\}$ is not known in advance, therefore we make an approximation:

$$I(Z_{w_{1:n}}; Z_{W \setminus \{w_{1:n}\}}) \approx I(Z_{w_1}; Z_{W \setminus \{w_1\}}) + \sum_{i=2}^{n} I(Z_{w_i}; Z_{W \setminus \{w_1, \ldots, w_i\}} | Z_{w_{1:i-1}}), \tag{5}$$

which can be formulated in a recursive form, i.e. for stages $i = 2, \ldots, n$, the value $V_i(w_i)$ of w_i is:

$$V_i(w_i) = \max_{w_i \in W \setminus \{w_1, \ldots, w_{i-1}\}} I(Z_{w_i}; Z_{W \setminus \{w_1, \ldots, w_i\}} | Z_{w_{1:i-1}}) + V_{i-1}(w_{i-1}), \tag{6}$$

with the base case for this recursion: $V_1(w_1) = I(Z_{w_1}; Z_{W \setminus \{w_1\}})$. Note that the optimal waypoint in the last stage n is

$$w_n^* = \arg\max_{w_n \in W} V_n(w_n). \tag{7}$$

With the optimal solution in the last stage, w_n^*, we can backtrace all optimal sampling points (optimal with respect to the approximation made in Eq. (9)) until the first stage w_1^*, and get the whole set of observation points $w^* = \{w_1^*, w_2^*, \ldots, w_n^*\}$.

3.3 *Planning Multi-robot Paths Among Observation Points*

Given the most informative observation points, we can then plan the paths for each robot. Our path planning framework differs from traditional path planning methods in three ways: First, we need to plan paths for multiple robots, where each path starts from the robot's current location and ends at a unique destination, and these paths should not interfere (e.g. no intersection). Second, the metric for path quality is not only the travel distance/time (a minimization problem), but combined with information gain (a maximization problem). In our work, we evaluate the path quality via the *information gain in unit time*, i.e. the relative information gain given the time needed to collect such information. Third, the planning needs to adapt to the spatio-temporal dynamics. Observation points with time-varying priors are generated online, and the routing of paths needs to be able to adapt to such variations. We address these problems as follows.

Multi-robot Conflict-Free Path Planning: We use a graph $G = (V, E)$ to describe the possible paths between observation points V. Each point $v_i \in V$ is weighted by its information gain $\tau(v_i)$. Each edge $e_{ij} = (v_i, v_j) \in E$ is weighted by $w(v_i, v_j)$, the travel time between the two ending vertices. Motivated by the embedding of both vertex weight and edge weight as well as the capacity for describing multi-agent assignment, we opt to extend our routing method based on bipartite graphs (also called matching graphs) [12], to plan the multi-robot informative and conflict-free paths. In essence, the bipartite graph $\tilde{G} = (V, V', \tilde{E})$ is an augmented version of the standard graph $G = (V, E)$, if we regard it in the way that each vertex weight in G is uniquely transformed to some edge weight in $\tilde{G}$ (such that all vertex weights are eliminated). Such a bipartite graph can well represent the matching (assignment) problem, and the optimal matching solution to it can be converted and interpreted as a routing path on the standard graph. We briefly describe the idea as follows, more details can be found in [12].

A bipartite graph $\tilde{G}$ has two sets of nodes, V and V', where V' is simply a copy of $V \in G$ such that $|V| = |V'|$, and an edge $\tilde{e}_{ij} = (v_i, v'_j) \in \tilde{E}$ connects the vertices $v_i \in V$ and $v'_j \in V'$ if there is an edge $e_{ij} = (v_i, v_j) \in E \in G$. Edge $\tilde{e}_{ij} = (v_i, v'_j)$ is weighted the same as the counterpart edge $e_{ij} = (v_i, v_j)$, i.e., $\tilde{w}(v_i, v'_j) = \tilde{w}(v'_i, v_j) = w(v_i, v_j)$.

Figure 1 shows an example of bipartite graph. If we insert some starting vertices V_s and some goal/ending vertices V_g, a new matching problem is formed and we can

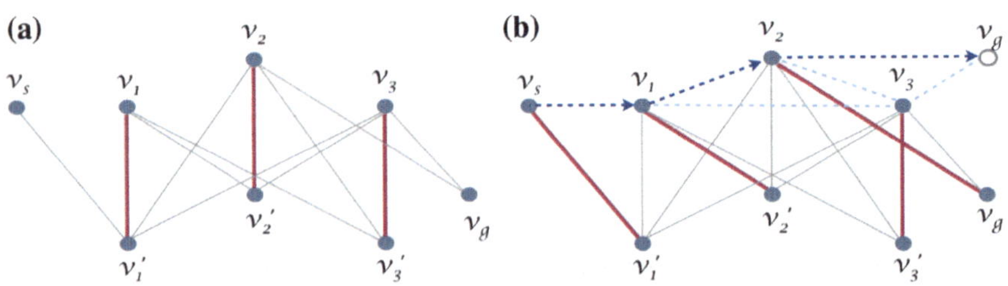

Fig. 1 Bipartite graph in the form of a 3D mesh, where $V = \{v_1, v_2, v_3\}$, $V' = \{v'_1, v'_2, v'_3\}$. The starting nodes (i.e. robots' current locations) are put in a set $V_s = \{v_s\}$; similarly, the goal nodes for each robot are in set $V_g = \{v_g\}$. In this example, we have only one start node and one goal node for each robot, and these are mutually exclusive. **a** Matched edges are in red bold, others are unmatched edges; **b** Optimal matching solution after running the Hungarian Method. The projected routing path is v_s—v_1—v_2—v_g, the vertices of which are only in routing graph G. The path is illustrated by dashed arrows in the top layer

employ the Hungarian Method [10] (with time complexity $O(n^3)$) to solve it. The output is a mapping that matches each $v_i \in V \cup V_s$ to a unique $v'_j \in V' \cup V_g$. The matched pair (v_i, v'_j) form a matched edge in $\tilde{G}$. Matched edges in Fig. 1 are colored in red. To retrieve a routing path, all vertices on the matched edges except those in the set V', form the path waypoints. For instance, in Fig. 1b, the path starting from v_s and ending at v_g is: v_s—v_1—v_2—v_g. Note that, from multiple starting vertices V_s to multiple pre-specified goal nodes V_g, multiple paths can be obtained. Because each vertex can not simultaneously be on more than one matched edge, the retrieved routing paths are conflict-free with no shared vertices. The paths do not cross or overlap due to the matching optimization mechanism [20].

Incorporating Informativeness: A useful property of the bipartite graph method is that paths can be tuned. This can be achieved by manipulating the weights $\tilde{w}(v_i, v'_i)$ of the corresponding *vertical* edges in Fig. 1. The path tuning feature allows us to incorporate the information gain metric. Specifically, we set each weight:

$$\begin{aligned} \tilde{w}(v_i, v'_i) &= \lambda_i \tilde{w_0}(v_i, v'_i) \\ \lambda_i &= f(\tau(v_i, t)), \end{aligned} \tag{8}$$

where $\tilde{w_0}(v_i, v'_i)$ is initialized to be the minimum weight among all the outgoing edges, and $0 \leq \lambda_i \leq 1$ is a parameter for scaling the importance of the information gain versus the travel cost. λ_i is a function of $\tau(v_i, t)$, which is the information gain for vertex $v_i \in G$ at time t. Function f is empirically pre-defined to express how the raw information gain should be transformed to reflect the importance. For example, f can be a linearly increasing function. Intuitively, as λ increases, the paths become more winding and include more nearby informative observation points.

Adaptive Routing for Spatio-Temporal Dynamics: We want our path planning approach to be able to adapt to the spatio-temporal dynamics, and to handle online routing. Pre-defining routing goals for all future horizons is impractical, because the persistent monitoring task can be infinitely long. Instead, we want the planner to determine the goals online. We achieve this by further extending the above routing mechanism to address the online goal selection and path optimization problem. Specifically, we start by setting V as an empty set, V_s as the current locations of all robots, and V_g as all potential observation points, within the current time horizon. Then we solve the matching problem, which matches V_s to a set, say, $V'_g \subset V_g$. Note that this step is optimal only with respect to the one step planning horizon since it does not account for the future observation points. Therefore, we manipulate the sets by letting $V = V \cup V'_g$, $V_g = V_g \setminus V'_g$ and solve the new matching again. By repeating this process, vertices are incrementally moved from V_g to V, and the sequentially obtained vertices in V'_g form the routing paths, with the last waypoint at the end of each path as its routing destination. Algorithm 1 shows this incremental adaptive planning in pseudo-code.

Algorithm 1: Incremental Adaptive Path Planner

Given the starting locations of k robots $s = \{s_1, \dots, s_k\}$, the sampling waypoints $w = \{w_1, \dots, w_n\}$, and the planning horizon, h.

Initialize, $V = \emptyset$, $V_s = s$, $V_g = w$, $V'_g = \emptyset$

for $t = 1$ `to` h **do**

  $V = V \cup V'_g$, $V_g = V_g \setminus V'_g$

  Build a bipartite graph $\tilde{G} = (V, V', E)$, where $\tilde{e}_{ij} = (v_i, v'_j) \in E$, $\tilde{w}(v_i, v'_j) = d(v_i, v'_j)$, and the Euclidean distance $d(v_i, v'_j)$ represents travel cost.

  Parameterize V: $\tilde{w}(v_i, v'_i) = \lambda_i \min_{\forall v'_j} \tilde{w}(v_i, v'_j)$

  Insert V_s, V_g to the above graph as starting/ending vertices, then solve the matching problem using the Hungarian Method.

  Let V'_g be the resultant matchings.

Transform V'_g to k routing paths as described in Fig. 1b

4 Experimental Results

Experimental Set-Up: We validate our method through simulations, using the scenario of ocean environmental monitoring. The simulation environment is constructed as a two-dimensional ocean surface which is tessellated into a grid map. We use salinity and ocean currents data, observed in the Southern California Bight region, obtained via ROMS [18]. Figure 2a, b show visualizations of these data. The grid map resolution, as well as the hyperparameters of GP, are manually tuned and pre-set, such that approximately 30 observation points from the entire space can be generated and they can well cover the space. The robot used in simulation is an underwater AUV (marine glider).

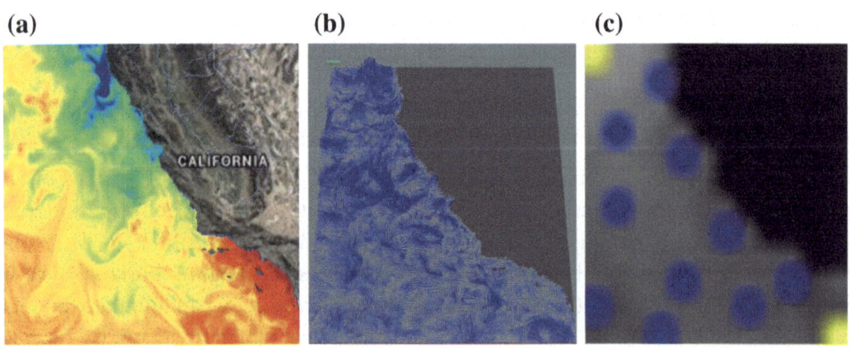

Fig. 2 **a** Ocean temperature field near southern California generated by ROMS. **b** Ocean currents predicted by ROMS. **c** Potential observation points (blue) with priors in the corners (yellow)

We use a hierarchical model for motion planning, with two levels. At the higher level, the robots follow the planned paths presented in Sect. 3. At the lower level, the robots follow disturbance-aware motion policies, built on Markov Decision Process (MDP) (formulation details of the low level planner can be found in [14]). These motion policies let us integrate external disturbances (such as ocean currents) into the stochastic transition model. Therefore we take into account the robots' motion uncertainty caused by the ocean currents. By setting succeeding path waypoints as the short-horizon goal states, the low-level motion planner generates policies for local guidance.

Figure 2c demonstrates a set of 10 observation points that maximize mutual information in the map. In the figure, the black region represents land and the gray area represents the ocean. The yellow dots in the corners represent prior observation points, and the blue blobs are the resultant observation points. With the observation points obtained, we run the path planner described in Algorithm 1 to generate the informative and adaptive paths for the multi-robot system. As shown in Fig. 3, the routing paths are incrementally and adaptively augmented in each time step. Figure 3d, e show that the paths are adapted to avoid conflicts.

Results: Figure 4a–d shows the simulation results for two robots. Each robot follows its local decision-policy computed from an MDP model, combining both the ocean current disturbances and the reward information for the next waypoint. The colormap in these figures denotes the significance of uncertainty (red = low uncertainty, green = high uncertainty), from which we can see that the proposed method produces informative paths that explore and cover the regions with high uncertainty. As noted previously, the information gain is time-varying, i.e. the uncertainty of an observed point starts increasing again after a robot finishes its observation at this location and moves to somewhere else. Therefore, we incorporate the fact that the uncertainty of predictions increases as time elapses. Figure 4c, d show us that the earlier explored regions become uncertain again as time elapses, and that the robots always explore the most uncertain parts of the environment, as the environment changes.

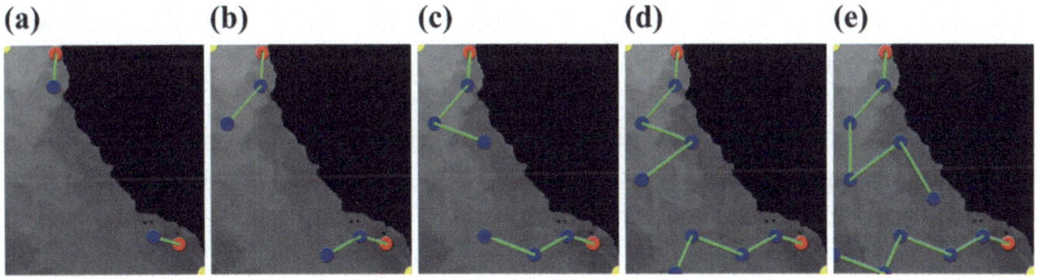

Fig. 3 Intermediate results of our multi-robot routing process. The routing points are incrementally and adaptively chosen during the process, as illustrated from **d** and **e**.

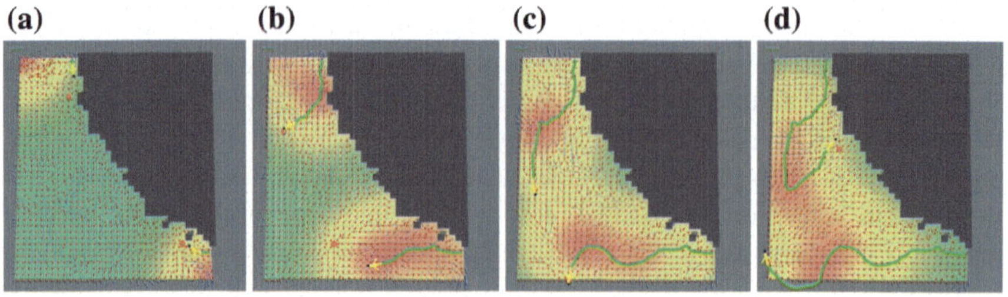

Fig. 4 Demonstration of environmental monitoring with 2 AUVs. The regions with warmer colors indicate less uncertainty (high confidence), whereas regions with colder colors indicate high uncertainty (low confidence). For the purpose of clarity, only one robot's MDP policy map (small arrows) is shown

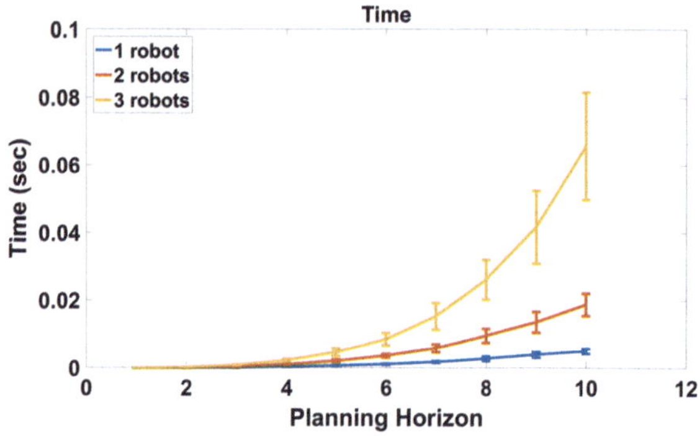

Fig. 5 Average computation time for different numbers of deployed robots and planning horizons

Computational Performance: We also evaluated the computational performance of our approach. Figure 5 shows the computation times given different numbers of deployed robots and planning horizons, as run on a computer with an 8-core 2.6GHz CPU and 12GB DDR3 memory. All statistics are mean values of 20 trials for each setting. In every simulation, 30 observation points are generated, and the prior data are randomly selected points. We can see that the computation time generally grows polynomially with the planning horizon increases, for each fixed number of robots. This can be justified by inspecting Algorithm 1. We can see that the bottleneck step is the Hungarian Method, whose time complexity is $O(n^3)$ and therefore the overall complexity is polynomial. The growth in computation time is mostly due to the generation of observation points. Table 1 shows the comparison of computation time between the generation of observation points, and the multi-robot path planning. Three robots are deployed and the planning horizon for path planning is set to cover as many observations points as possible. The informative observation point generation part is more costly due to the large search space; the observation point algorithm needs to evaluate all grid points in the grid map, whereas the path planning method only needs to compute routing solutions from the subset of obtained observation points.

Table 1 Computation time (sec) for two components of our informative path planning approach: finding informative observation points, and calculating a path between these

	Stages = 5	10	15	20	25	30
Observation points	5.5752	12.8784	20.7266	28.6449	36.9960	47.2544
Path planning	0.0001	0.0007	0.0038	0.0070	0.0216	0.0659

To assess the quality of planned paths, we compare our method with an algorithm that solves the Orienteering Problem (OP). The OP solver aims at maximizing the collected score along the paths within some given time limits, thus it also considers two competing metrics (score collection and travel time). We implemented a well-known heuristic called the *centre-of-gravity heuristic* [7], which combines other local refining heuristics such as the well-known 2-Opt heuristic [4]. One drawback of such OP solution lies in that both the ending vertices and the time limits must be specified. In contrast, our method plans the paths with their goals adaptively, in order to achieve better scores. To address the goal specification requirement for the OP, we first run our method and obtain the goals, then we assign these obtained goals to the centre-of-gravity OP. The time limits fed to the OP are the recorded time of each path computed from our method.

Figure 6 provides two sets of results for our proposed planning method (green paths) and the orienteering algorithm (red paths). The score for each vertex is $score_i \in [0, 100]$ and λ_i is set to be

$$\lambda_i = \begin{cases} score_i/100, & \text{if } score_i > 20 \\ 0, & \text{otherwise,} \end{cases} \tag{9}$$

Specifically, Fig. 6a, b are planned paths from the two methods on the same set of artificially created observation points. The physical size of a node in the environment represents the significance of information gain (or score). We can observe that the paths produced from our method transit many high-score waypoints, whereas the *centre-of-gravity heuristic* transit fewer. Then, we manipulated the scores so that the score distribution is imbalanced, see Fig. 6c, d. We can see that our method can skip those low-score regions and transit only those high score nodes. Similar behavior can also be observed from the orienteering algorithm.

Next we compare these two approaches on their scoring performance. Figure 7 shows the detailed numerical results for 50 trials with randomly generated locations and scores. The y-axis is the average score collected, corrected for the path length. Figure 7b shows the average statistics of the two scenarios depicted in Fig. 6, from which we can also conclude that our method is superior in terms of scoring performance.

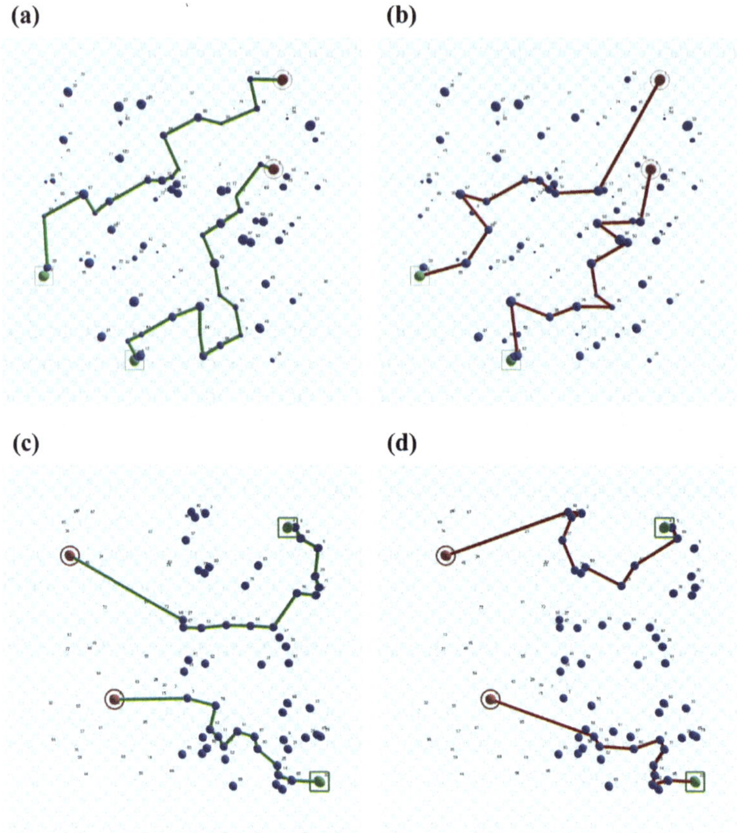

Fig. 6 Two different scenarios used to compare results between our method (green **a**, **c**) and the OP algorithm (red, **b**, **d**). Two robots are deployed. The circled nodes indicate their starting locations, and the squared nodes the ending locations. The size of each node reflects their significance of score. **a** and **b** compare performance on artificially created observation points, **c** and **d** compare performance on a skewed score distribution

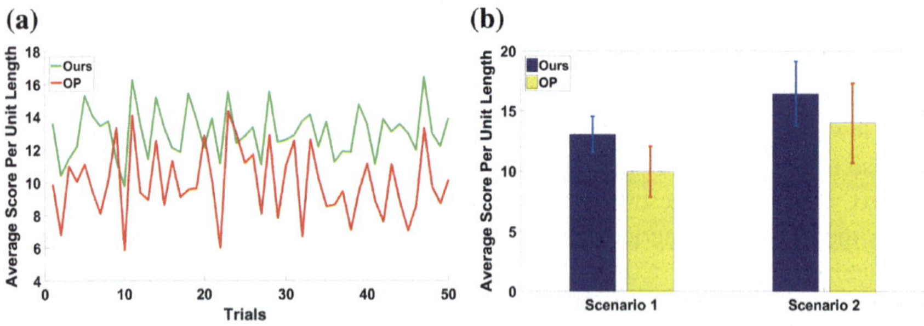

Fig. 7 The scoring performance comparison between our method and the OP solver: **a** average scores for 50 trials with uniformly distributed scores, **b** average scores for the scenarios shown in Fig. 6

5 Conclusions

In this paper, we presented an informative path planning approach for persistent multi-robot environmental monitoring. Taking into account the spatio-temporal variations of ocean phenomena, we first developed an information-driven component that computes the observation points, by minimizing the environmental model's prediction uncertainty. Multi-robot paths are then obtained by extending a matching graph based routing method, which allows the vehicles to transit the obtained informative observation points in an efficient manner. We validated our method through simulations with real ocean data. The results show that our method generates informative paths, which are conflict-free for multiple robots, and adaptive to the dynamics of the environment. Our approach is polynomial in the planning horizon, and linear in the number of robots. Furthermore, we have shown that our approach outperforms a well-known orienteering problem solver. Thereby we have developed an approach well suited for persistent monitoring with a multi-robot system.

Acknowledgements The authors would like to thank Stephanie Kemna and Hordur Heidarsson for their valuable inputs on this paper.

References

1. Binney, J., Krause, A., Sukhatme, G.S.: Informative path planning for an autonomous underwater vehicle. In: International Conference on Robotics and Automation, pp. 4791–4796 (2010). http://robotics.usc.edu/publications/642/
2. Binney, J., Krause, A., Sukhatme, G.S.: Optimizing waypoints for monitoring spatiotemporal phenomena. Int. J. Robot. Res. (IJRR) **32**(8), 873–888 (2013)
3. Cao, N., Low, K.H., Dolan, J.M.: Multi-robot informative path planning for active sensing of environmental phenomena: a tale of two algorithms. In: Proceedings of the 2013 International Conference on Autonomous Agents and Multi-agent Systems, pp. 7–14 (2013)
4. Croes, A.: A method for solving traveling salesman problems. Oper. Res. **5**, 791–812 (1958)
5. Fomin, F.V., Lilngas, A.: Approximation algorithms for time-dependent orienteering. Inf. Process. Lett. **83**(2), 57–62 (2002)
6. Gendreau, M., Laporte, G., Semet, F.: A tabu search heuristic for the undirected selective travelling salesman problem. Euro. J. Oper. Res. **106**(2–3), 539–545 (1998)
7. Golden, B.L., Levy, L., Vohra, R.: The orienteering problem. Nav. Res. Logist. (NRL) **34**(3), 307–318 (1987)
8. Gunawan, A., Lau, H.C., Vansteenwegen, P.: Orienteering problem: a survey of recent variants, solution approaches and applications. Euro. J. Oper. Res. **255**(2), 315–332 (2016)
9. Hitz, G., Gotovos, A., Garneau, M.É., Pradalier, C., Krause, A., Siegwart, R.Y., et al.: Fully autonomous focused exploration for robotic environmental monitoring. In: 2014 IEEE International Conference on Robotics and Automation (ICRA), pp. 2658–2664. IEEE(2014)
10. Kuhn, H.W.: The Hungarian method for the assignment problem. Nav. Res. Logist. Q. **2**, 83–97 (1955)
11. Laporte, G., Martello, S.: The selective travelling salesman problem. Discret. Appl. Math. **26**(2), 193–207 (1990)
12. Liu, L., Shell, D.A.: Physically routing robots in a multi-robot network: flexibility through a three dimensional matching graph. IJRR **32**(12), 1475–1494 (2013)

13. Low, K.H., Dolan, J.M., Khosla, P.: Active markov information-theoretic path planning for robotic environmental sensing. In: Proceedings of the 10th International Conference on Autonomous Agents and MultiAgent Systems (AAMAS-11), pp. 753–760 (2011)
14. Ma, K.C., Liu, L., Sukhatme, G.S.: An information-driven and disturbance-aware planning method for long-term ocean monitoring. In: Proceedings of IEEE/RSJ International Conference on Intelligent Robots and Systems (2016)
15. Meliou, A., Krause, A., Guestrin, C., Hellerstein, J.M.: Nonmyopic informative path planning in spatio-temporal models. In: Proceedings of National Conference on Artificial Intelligence (AAAI), pp. 602–607 (2007)
16. Ouyang, R., Low, K.H., Chen, J., Jaillet, P.: Multi-robot active sensing of non-stationary gaussian process-based environmental phenomena. In: Proceedings of the 2014 International Conference on Autonomous Agents and Multi-agent Systems, pp. 573–580 (2014)
17. Rasmussen, C.E., Williams, C.K.I.: Gaussian Processes for Machine Learning (Adaptive Computation and Machine Learning). The MIT Press, Cambridge (2005)
18. Shchepetkin, A.F., McWilliams, J.C.: The regional oceanic modeling system (ROMS): a split-explicit, free-surface, topography-following-coordinate oceanic model. Ocean. Model. **9**(4), 347–404 (2005)
19. Singh, A., Krause, A., Guestrin, C., Kaiser, W., Batalin, M.: Efficient planning of informative paths for multiple robots. In: Proceedings of the 20th International Joint Conference on Artifical Intelligence, IJCAI'07, pp. 2204–2211 (2007)
20. Turpin, M., Michael, N., Kumar, V.: Capt: Concurrent assignment and planning of trajectories for multiple robots. Int. J. Robot. Res. **33**(1), 98–112 (2014)

The Effectiveness Index Intrinsic Reward for Coordinating Service Robots

Yinon Douchan and Gal A. Kaminka

Abstract Modern multi-robot service robotics applications often rely on coordination capabilities at multiple levels, from global (system-wide) task allocation and selection, to local (nearby) spatial coordination to avoid collisions. Often, the global methods are considered to be the heart of the multi-robot system, while local methods are tacked on to overcome intermittent, spatially-limited hindrances. We tackle this general assumption. Utilizing the *alphabet soup* simulator (simulating *order picking*, made famous by Kiva Systems), we experiment with a set of myopic, local methods for obstacle avoidance. We report on a series of experiments with a reinforcement-learning approach, using the *Effectiveness-Index* intrinsic reward, to allow robots to learn to select between methods to use when avoiding collisions. We show that allowing the learner to explore the space of parameterized methods results in significant improvements, even compared to the original methods provided by the simulator.

1 Introduction

Modern service robotics applications often rely today on multiple robots which coordinate tasks between them, and the user, in order to fulfill their design goals. Such multi-robot systems require coordination capabilities at multiple levels, from global (system-wide) task allocation and task selection, to local (nearby) spatial coordination to avoid collisions.

Often, the global methods are considered to be the heart of the multi-robot system, as they are responsible for the scheduling and allocation of tasks to the robots. Local

Y. Douchan
School of Mechanical Engineering, Faculty of Engineering, Tel Aviv University, Tel Aviv, Israel
e-mail: yinondouchan@mail.tau.ac.il

G. A. Kaminka (✉)
Computer Science Department and Gonda Brain Research Center, Bar Ilan University, Ramat Gan, Israel
e-mail: galk@cs.biu.ac.il

R. Groß et al. (eds.), *Distributed Autonomous Robotic Systems*,
Springer Proceedings in Advanced Robotics 6,
https://doi.org/10.1007/978-3-319-73008-0_21

methods, on the other hand, are often added on, to overcome intermittent, spatially-limited hindrances to the tasks assigned to individual robots.

The Kiva Systems (Amazon Robotics) pick ordering and material handling system is a highly successful example of this design [13]. Here, hundreds of individual robots are sent around a warehouse to fetch and place back shelves containing products to be shipped to online customers. A centralized market allocation algorithm is used to optimize decisions as to which robot should move which shelf, and where. Each robot is then responsible to plan and execute a path from its current location to the target location, in order to maximize its own individual performance. Potential collisions between robots are resolved dynamically, ad-hoc.[1]

This paper examines the role of such local, dynamic collision-avoidance methods within multi-robot systems. We utilize the *alphabet soup* simulator [7], written by the founders of Kiva Systems to encourage research into their application domain. Here, as described in [13], a centralized optimization algorithm globally assigns tasks to robots. But planning paths and moving, to execute these tasks, is left to the robots.

We first experiment with a set of myopic, local methods for obstacle avoidance, which we demonstrate to work much worse than the methods provided with the simulator. Indeed, it turns out that it is quite difficult to get a local method to work well: the simulator's best method is a non-trivial stochastic combination of a myopic random direction selection, and predictive collision avoidance, a simple variant of the *dynamic window* algorithm [5].

Then, we report on a series of experiments with a reinforcement-learning approach using an *intrinsic* reward, to allow robots to learn which method to use when handling collisions. The reward function, called *effectiveness index* (EI), first proposed in [8], seeks to minimize resource usage when avoiding obstacles, and does not rely on communications or any information about the decisions or state of other robots. It is thus easily implementable in existing service robots, as it only relies on their own estimates of their own resource use.

While a first attempt at such learning generates mediocre results, we show that allowing the learner to explore the space of parameterized methods, results in significant improvements, even compared to the original methods provided by the simulator. The results demonstrate clearly that using the effectiveness-index reward, a reinforcement-learning technique can outperform manually-designed coordination methods, in a non-trivial real-world task. This, despite the simplicity of the learning mechanism itself.

A second key lesson touches on the significant impact of local collision avoidance within a sophisticated multi-robot system utilizing global coordination methods. The extreme differences in system performance when using different local methods—but all using the same centralized task allocation method—shows that good coordination cannot be carried out only at the global task allocation and team-plan level, but is instead an important factor at all levels of decision-making.

[1]This is actually not stated explicitly in [13], but is implied by the design, which explicitly leaves path-planning and motion-planning to each robot's individual controlling agent.

2 Related Work

Early work on local decision-making investigated reactive methods that proved useful in canonical multi-robot tasks, such as foraging, formations, or exploration. Balch and Arkin [1] describe a method, which we term *noise* in this paper, where once a robot is about to collide with another robot it moves backwards with some random directional noise. Vaughan et al. [12] describe *aggression*, a method in which when two robots collide with each other the robot with the highest aggression factor goes forward while the other backs away. Vaughan et al. describe three ways to determine the aggression factor for each of the two colliding robots: randomly, fixed, or based on each robot's free personal space behind them. We use the random version in the experiments reported below. Rosenfeld et al. [9] discuss the *repel* method, where by a collision is avoided by the robots moving backwards for a fixed amount of time.

Reactive methods are myopic, in the sense that they do not engage in, or rely on, any prediction of the future position or velocity of colliding robots. At a cost of more computation, a variety of sophisticated methods utilize predictions of others motions. *Dynamic window* [5] is one such method, which takes the motion constraints of the robot into account. It works by generating a search space for a navigation solution, and then selecting a heading and velocity within the space, that maximizes clearance from obstacles and other robots. The Reciprocal Velocity Obstacles (RVO) family of predictive methods takes into account not only the motions allowed for the robot, but also the responses of other robots to the these motions [11]. PassPMP [3] and related algorithms focus on safety in collision avoidance, at a cost of significant computation. In this paper, we utilize the *best-evade* and *original* methods supplied with the simulation (and seem to be variants of the dynamic window algorithm), and add the reactive methods discussed above.

Regardless of the method used—whether myopic or predictive—no method is always best. Both Rosenfeld et al. [9] and Rybski et al. [10] demonstrated that no local method is always better than others, but instead its performance depends on the density of robots. To address this, learning approaches have been utilized to try to dynamically adjust the collision-avoidance method to the density and other dynamic settings.

Rosenfeld et al. [9] measured a *Combined Coordination Cost* (CCC), the sum total of costs (time, energy, communications) that rise from a robot's attempt to coordinate away from collisions. The CCC was shown to be negatively correlated with performance in multi-robot foraging, and then used to guide dynamic selection of a reactive method for each robot. Each robot estimated its own CCC—relying on the estimation of the local density—and switched method if thresholds were passed. The thresholds were determined offline (before deployment), via hill-climbing. They do not change with robot deaths, or with other dynamic changes to the task. Recent work by Godoy et al. [6] uses reinforcement learning to tackle the problem of multi-agent navigation. Their ALAN reward function controls the robot's velocity according to a weighted sum of two factors: a goal-oriented component factors the robot's distance to the goal; a second component considers how the robot's velocity vector affects others' velocities—using a variant of RVO to carry out this computation. A key difficulty with this reward function is that both of its components are *extrinsic*, relying on information external to the robot (e.g., the distance to the goal), which can be challenging to sense, and requires more sophisticated—and costly—capabilities.

To address this challenge, Erusalimchik et al. [8] proposed an *intrinsic* reward function, called the *effectiveness index* (EI). EI is a measure of the robot's own resource usage dealing with collisions. The robot only evaluates its own resources, and does not need any extrinsic information. However, EI has only been demonstrated to use in foraging, and its application in more structured settings is not straightforward (as we show) in particular where the navigation goals are given by a top-level decision process. We address this in this paper.

3 Reinforcement Learning Using Effectiveness Index

We briefly remind the reader of the EI reward function and its rationale. Please see [8] for details. The execution time of each robot's task can be divided into intervals where the robot is carrying out its task, uninterrupted, and these are interleaved with intervals where the robot is avoiding or otherwise actively handling an impending collision. This division of the task's time into interleaved execution and interruption intervals is characteristic of many service robotics tasks.

From the perspective of collision-handling there is a repeating pair of intervals: an active time interval, denoted I_a, where the robot is busy coordinating to avoid a collision, followed by a passive time interval I_p where the robot performs its task. Given these, the EI for a given conflict, where a coordination method α was used, is

$$EI(\alpha) = \frac{I_a^\alpha}{I_a^\alpha + I_p^\alpha}. \tag{1}$$

This measures how costly a coordination method was (time spent I_a) against how effective it was (large I_p). The smaller EI is, the more effective the method. For example, assume it took for a robot 1 millisecond to coordinate but had a passive time of 1 millisecond afterwards EI will be 0.5. It took it a short time to coordinate but it also entered another conflict almost immediately. If it took a robot 5 seconds to coordinate but it had a passive time of 20 seconds then EI will be 0.2. It coordinated for quite a long time, but focused on the task afterwards for a much longer time and therefore, it is more effective.

Each robot independently uses the Q-learning algorithm with EI as a reward. We defined the action space to be the coordination methods. In particular, we use a single-state version of Q-learning [4], $Q_t(\alpha) = Q_{t-1}(\alpha) + \rho(R_t(\alpha) - Q_{t-1}(\alpha))$, with $R_t(\alpha) = EI$. When a collision occurs, the robot calculates EI using the active and passive times before this collision and updates the Q-value of the method used in the previous collision. It then proceeds to making a selection as to the next method to be used.

4 Experiments in Learning Order Picking

We introduce the experiment test-bed and setup in Sect. 4.1. The following sections then systematically explore the use of learning in this environment.

4.1 Experiment Setup: Robotic Order Picking

Order picking is the task of bringing products from various locations in a warehouse, to a centralized handling station where they can be packed together to fulfill an incoming order. It is a task carried out by people all around the world, and is considered to be particularly arduous: In some warehouses, pickers walk up to 15 miles a day in the warehouse.

In the mid 2000s, Kiva Systems began developing and selling robotic order picking systems, where robots would carry entire shelves and bring them to a human employee, who would be responsible for grasping the relevant products off the shelf, and packing them together. Kiva was sold in 2012 to Amazon, becoming *Amazon Robotics*, for a reported amount of $775 million.

4.1.1 The Alphabet Soup Simulator

Alphabet soup is a simulator used to simulate a multi robot system in a warehouse [7]. It was developed by one of the Kiva Systems co-founders, in hope of encouraging research into algorithms that can be effective in this task. Multiple simulated robots deliver buckets containing letters to "word stations". Each word station contains words to be completed and robots must deliver letters for these stations in order to complete the demanded words. In order to place a letter in a word station, a robot must take the bucket, move it to the word station and take it back. There also exist letter stations in order to replenish a bucket's letter stash.

The simulator comes with a default set of settings (which we use, unless otherwise noted), including two local coordination methods: a *best-evade* method which makes a prediction as to a good obstacle-free direction, and a default *original* method, which is a stochastic combination of best-evade and random motions. The simulator also contains a task-allocation procedure, which makes allocations to the robots. We used it as is. A screen shot appears in Fig. 1.

4.1.2 Coordination Methods

Learning occurs over a set of base coordination methods, which can be applied to handle impending collisions. In addition to the best-evade and original methods, we implemented three additional coordination methods: *repel*, *noise*, and *aggression*

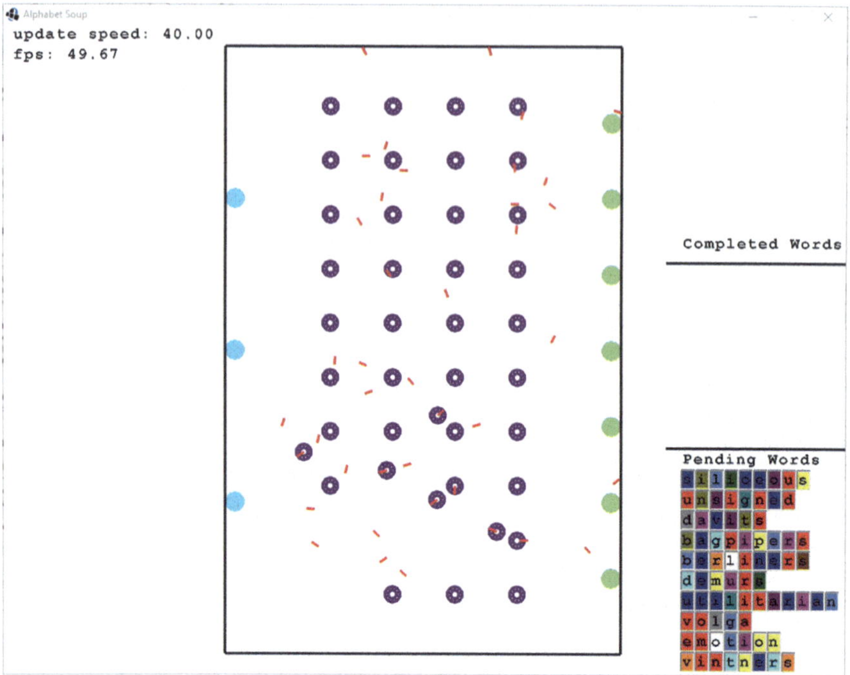

Fig. 1 The alphabet soup simulator. Circles on the right side are the word stations, on the left side are letter stations, in the center are the buckets and the small lines are the robots

(as discussed in Sect. 2). All methods, except *original* which cannot be adjusted, are time-based in the manner that they act for a limited time interval. For example, Repel with a parameter 200 milliseconds, given a conflict, will go back for 200 milliseconds.

4.1.3 Initial Results Using EI

Despite its success in *foraging* [8], EI is not a magic bullet that works any time. Figure 2 shows the *baseline* results from applying EI to the five base methods, without any tweaking, and with arbitrary parameters for the different methods' timing set at 20ms. Each data point is the mean of 60 runs, each 10 min with measurement taking place only in the latter 5 min, to allow training to take place. The learning rate parameter ρ was set to 0.5. Exploration rate was set at 0.1. In all the figures below, beginning with Fig. 2, the X axis measures the number of robots within the fixed default working area. The Y axis measures the total number of letters (products) collected by all robots. Error bars mark a single standard deviation.

The results are shown to highlight that applying learning blindly does not work here, unlike in foraging [8]. The figure shows that the three reactive methods (*repel*,

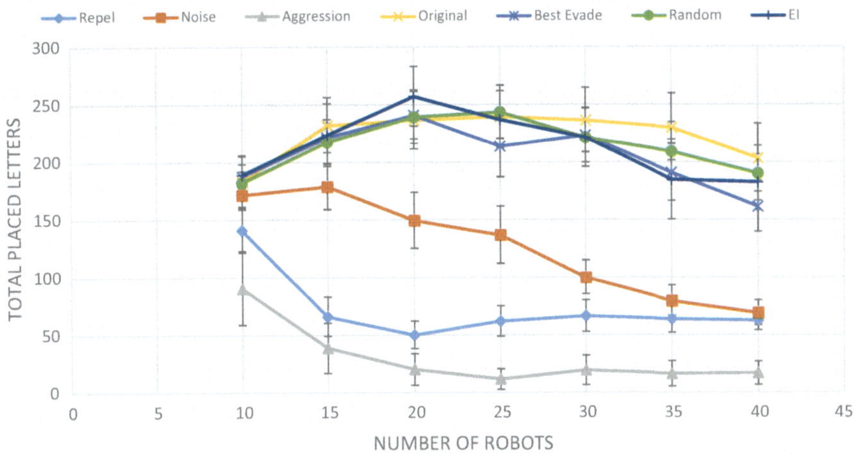

Fig. 2 Initial results using *EI*

noise, *aggression*) are significantly and clearly inferior to the *original* and *best-evade* methods. Indeed, they do worse than random selection of coordination methods (line marked *random*). EI-based learning improves on the base methods by combining them effectively (as each robot makes its own selections), but is only on par with *random* selection, the *Original* method and the *Best Evade* method.

4.2 *Stateless EI Q-Learning with Parameters*

We have several directions to try to improve these results. One is to improve the base methods used in the learning, by allowing modification of the parameters used with each method. We demonstrate that a hierarchical learning of these parameters works much better than any manual tweaking. This is the approach we take in this section. In Sect. 4.3 we instead focus on enriching the state description at the basis of learning.

4.2.1 Choosing Parameters According to Area Under Curve

We first begin by attempting to choose a single *best* parameter for each coordination method. To do this, we first conducted simulations for each of the method, for a variety of timing parameter values: 20, 100, 200, 500, 1000 and 2000. We then computed the area under curve for the resulting curves, with the rationale that the parameter with the largest area under curve would be the most reliable, across changes in density. The resulting parameters are: repel(200ms), noise(500ms), aggression(2000ms), and best evade(200ms).

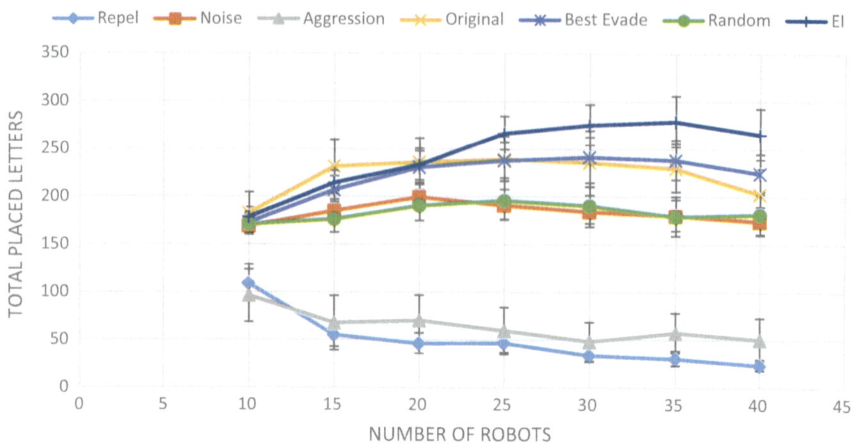

Fig. 3 Applying stateless Q-learning [4] with EI reward, with reactive *best*-parameter methods

Armed with this improved set of actions, we ran the EI learning. The results are shown in Fig. 3. They show *repel* and *aggression* were the worst with at most around 100 placed letters, only slightly improving on their earlier versions (different parameters). *Noise* improves clearly using the best parameter chosen for it. *Random* selection and the *EI*-based learned selection are now clearly distinguished: a random mix of these improved fixed-parameter methods does worse than earlier, while the learned selection improves on the *original* and *best evade* methods.

Choosing parameters for coordination methods based on area under curve is a long and dull process, and offers no guarantees to produce good results, despite its relative success in this case. This is likely because the result of the process is a parameter that works well *on average*, and remains fixed through out the task. Indeed, we show that combinations of *worse* parameterized methods (in terms of their area-under-curve) can do *better*.

Figure 4 shows the results from using EI to learn when the underlying methods have been changed. Instead of using best-evade with a parameter of 200ms, we instead switch to using best-evade at 2000ms (which is worse, as the figure shows), and keeping all other methods as above. The combination of the other methods with best-evade 2000ms (worse by itself), does much better than with best-evade 200ms.

4.2.2 Stateless EI with Parameter Sweep

We now move away from fixed parameters, and use EI to not only choose the best coordination method, but also in order to adjust the parameters of the coordination methods. This was done by learning using a hierarchical version of EI and Q-learning. Q-learning with EI was used to select a general class coordination method. A second Q-learning with EI was used to learn the parameter that works best with this method.

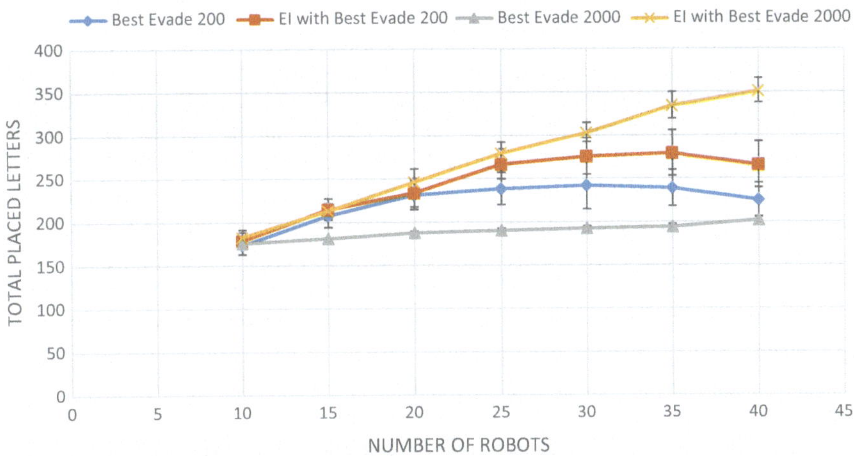

Fig. 4 A comparison between Best Evade 200 and 2000 and stateless EI with best-evade 200 and best-evade 2000. X-Axis is the number of robots and Y-Axis is the total amount of letters placed

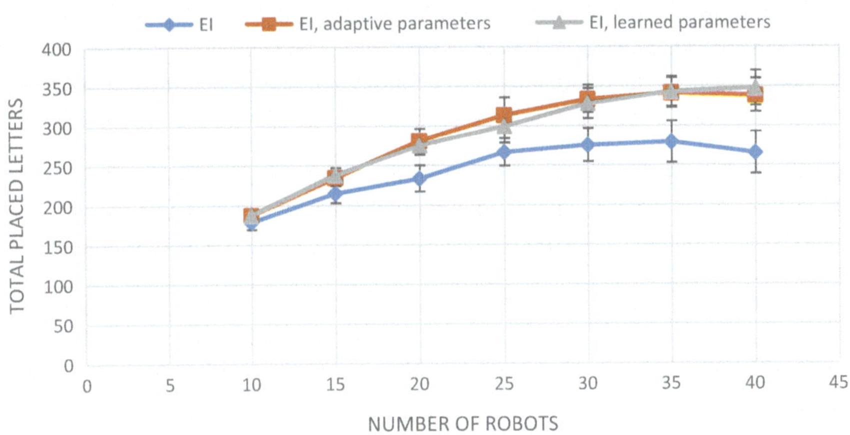

Fig. 5 Stateless EI with fixed parameters learned from adaptation against EI with highest area under curve parameters. X-Axis is the number of robots and Y-Axis is the total amount of letters placed

The two learning mechanisms run simultaneously, both using the same learning rate as before. The results, in Fig. 5, show that learning also what parameter to select improves results significantly. To convince ourselves of the benefit of the learned parameters, we fixed the parameters for each method based on the results of the learning, and went back to trying Q-learning over the (small) set of methods, with the fixed parameters: Repel 700ms, Noise 540ms, Aggression 500ms and Best Evade 600ms. The results are shown in Fig. 5, and match the simultaneous learning results.

4.3 Stateful (Multi-state) EI

We had also experimented with a different approach to improving the performance of the team using reinforcement-learning and EI. In particular, in this section we report on experiments where the set of coordination methods was once again restricted to atomic, non-parameterize simple actions, while the Q-learning method was used to maintain multiple states. In this section we use the original Q-learning formula for multiple states $Q_t(s, \alpha) = Q_{t-1}(s, \alpha) + \rho(R_t(\alpha) + \gamma \cdot max_{a'} Q(s', a') - Q_{t-1}(s, \alpha))$ with $\gamma = 0.7$ and $\rho = 0.5$. Each data point is the mean of 10 runs, each 10 min with measurement taking place only in the latter 5 min, to allow training to take place.

We had explored two different ways of specifying state. The first state factor maintained position of the robot. This was done by a discretization of the work area into a grid (not necessarily regular), and keeping track of the grid cell the robot was in. A second factor consisted of the heading of the robot. A final factor consisted of the task state of the robot (e.g., "taking a bucket to be processed" vs. "taking a bucket back to home position").

Using stateful (multi-state) Q-learning with these state spaces and the five aforementioned coordination methods as the action space yielded results that were, at best, like stateless EI. We therefore moved to defining new sets of coordination methods and tested stateful EI with them. We first used motion in absolute directions—West, East, North and South. Given a conflict the robot goes to this direction for a fixed amount of time. We arbitrarily determined the parameters of these methods to be 500 milliseconds. We ran the same simulations with the above set of coordination methods and compared group performance of random method selection, stateless EI and stateful EI with different state spaces.

Figure 6 shows that with absolute directions to resolve collisions, stateless EI improves performance significantly in comparison to random selection. However, stateful EI with direction as a state (40 Directions) does not improve results in relation to stateless EI. Using position on a 10×4 grid (40 positions) as the basis for the learning improved results significantly over the original method. However, it still is not as good as simply using stateless EI with parameter sweeping, as introduced in the previous section.

We have also experimented with trying relative motions, instead of absolute motions. Just like we can define absolute directions as coordination methods, we can define relative directions: Left, Right, Forward, or Back. Figure 7 shows the results with these relative motions. Clearly, they are quite disappointing: None of the state feature combinations tested lead to results that improve over the *original* method used in the simulation.

It is shown that with relative motions stateless EI does not improve performance over random selection. Stateful EI, using various combinations of state features, incrementally improves over stateless learning. First by incorporating the position on a 10×4 grid as the state, then using the heading (distinguishing 40 angles), and then finally in combination of position and heading. We have also attempted to

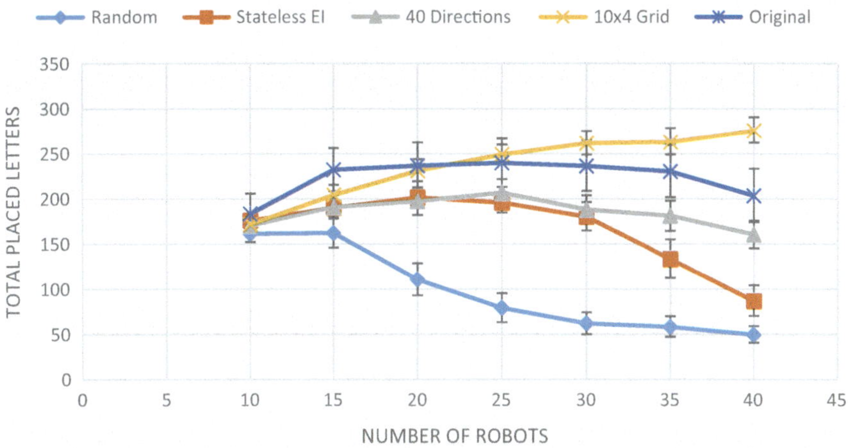

Fig. 6 Learning over absolute actions: West, East North and South

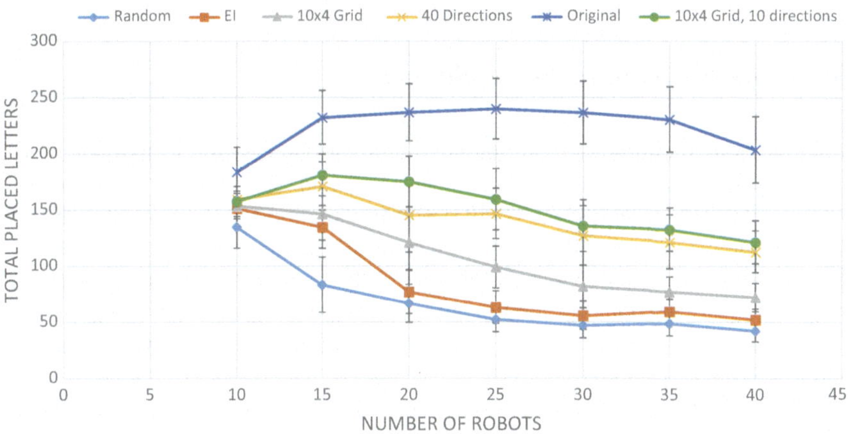

Fig. 7 Relative directions Left, Right and Back. X-Axis is the number of robots and Y-Axis is the total number of letters placed

incorporate the task state ("bucket to station", "bucket to storage", etc.) into the state, but this did not change the results.

What can be concluded from here is that the problem is not only choosing what state space, but what state space we should choose given an action space or what action space should we choose given a state space. This leaves an opening for a more formal investigation as to when a state-action space works better than another state-action space for a specific problem, given a learning method.

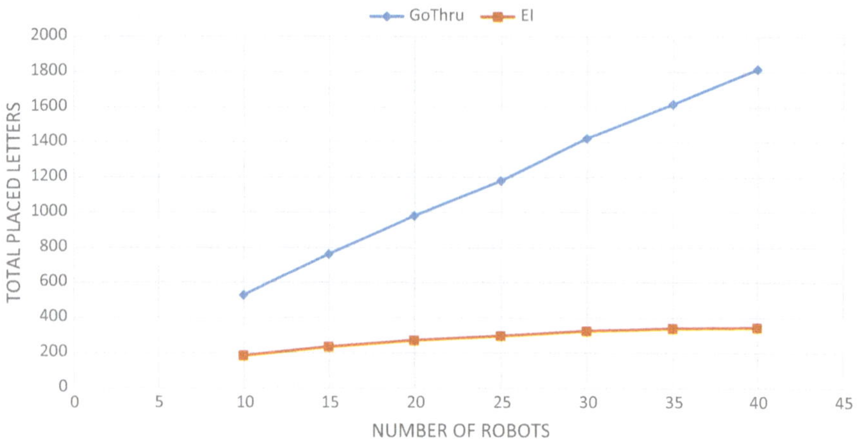

Fig. 8 EI with learned parameters (best EI variant) compared to GoThru. X-Axis is the number of robots and Y-Axis is the total amount of letters placed

4.4 Comparison to Ideal Conditions

In the previous sections we experimented with various learning techniques which improve performance to a certain extent. We would like to know how an ideal system—one with no collisions—will perform comparing to the best we achieved with learning. Therefore, following [9] we modified the simulator so that robots were able to go through other robots and measured the system's performance with this modification. We will call this modification *GoThru*. We ran simulations for GoThru in the same manner we did for EI. For GoThru averaging was done over 10 runs per sample and for EI over 60.

Figure 8 shows that GoThru performs at least almost 3 times better than EI and at most more than 5 times better than EI. This shows that not only there is a large room for improvement, but also shows the significance of local collision avoidance even in lower densities.

5 Summary and Future Work

This paper explores the use of an intrinsic reward function, called *effectiveness index*, applied to local multi-robot coordination, in particular in addressing impending collisions. The function is used with simple, plain Q-learning [4]. The results demonstrate clearly that with an appropriate set of base actions, a reinforcement-learning technique using the effectiveness index reward can outperform complex, manually-designed coordination methods, in non-trivial tasks for service robots. Indeed, the improved learning system utilizes simultaneous learning in two layers: the robots learned which method to select, while also learning which parameter to use for the selected method. We look forward to experimenting with additional advanced method for parameter learning (e.g., [2]). In addition, we plan to apply this learning approach to settings requiring more complex, structured task execution.

Acknowledgements We gratefully acknowledge support by ISF grants #1511/12, and #1865/16, and good advice from Avi Seifert. As always, thanks to K. Ushi.

References

1. Balch, T., Arkin, R.C.: Behavior-based formation control for multirobot teams. IEEE Trans. Robot. Autom. **14**(6), 926–939 (1998)
2. Bergstra, J., Bengio, Y.: Random search for hyper-parameter optimization. J. Mach. Learn. Res. **12**, 281–305 (2012)
3. Bouraine, S., Fraichard, T., Azouaoui, O., Salhi, H.: Passively safe partial motion planning for mobile robots with limited field-of-views in unknown dynamic environments. In: Proceedings of the IEEE International Conference on Robotics and Automation (2014)
4. Claus, C., Boutilier, C.: The dynamics of reinforcement learning in cooperative multiagent systems. In: Proceedings of the Fifteenth National Conference on Artificial Intelligence (AAAI-98), pp. 746–752 (1998)
5. Fox, D., Burgard, W., Thrun, S.: The dynamic window approach to collision avoidance. IEEE Robot. Autom. Mag. **4**(1), 23–33 (1997)
6. Godoy, J.E., Karamouzas, I., Guy, S.J., Gini, M.: Adaptive learning for multi-agent navigation. In: Proceedings of the Fourteenth International Joint Conference on Autonomous Agents and Multi-Agent Systems (AAMAS-15), pp. 1577–1585 (2015)
7. Hazard, C.J., Wurman, P.R.: Alphabet soup: a testbed for studying resource allocation in multi-vehicle systems. In: Proceedings of the 2006 AAAI Workshop on Auction Mechanisms for Robot Coordination, pp. 23–30 (2006)
8. Kaminka, G.A., Erusalimchik, D., Kraus, S.: Adaptive multi-robot coordination: a game-theoretic perspective. In: Proceedings of IEEE International Conference on Robotics and Automation (ICRA-10) (2010)
9. Rosenfeld, A., Kaminka, G.A., Kraus, S., Shehory, O.: A study of mechanisms for improving robotic group performance. Artif. Intell. **172**(6), 633–655 (2008)
10. Rybski, P., Larson, A., Lindahl, M., Gini, M.: Performance evaluation of multiple robots in a search and retrieval task. In: Proceedings of the Workshop on Artificial Intelligence and Manufacturing, pp. 153–160. Albuquerque, NM (1998)
11. van den Berg, J., Guy, S., Lin, M., Manocha, D.: Reciprocal n-body collision avoidance. In: Robotics Research pp. 3–19 (2011)
12. Vaughan, R., Støy, K., Sukhatme, G., Matarić, M.: Go ahead, make my day: robot conflict resolution by aggressive competition. In: Proceedings of the 6th International Conference on the Simulation of Adaptive Behavior. Paris, France (2000)
13. Wurman, P.R., DÁndrea, R., Mountz, M.: Coordinating hundreds of cooperative, autonomous vehicles in warehouses. AI Mag. (2008)

United We Move: Decentralized Segregated Robotic Swarm Navigation

Fabrício R. Inácio, Douglas G. Macharet and Luiz Chaimowicz

Abstract A robotic swarm is a particular type of multi-robot system that employs a large number of simple agents in order to cooperatively perform different types of tasks. In this context, a topic that has received much attention in recent years is the concept of segregation. This concept is important, for example, in tasks that require maintaining robots with similar features or objectives arranged in cohesive groups, while robots with different characteristics remain separated on their own groups. In this paper we propose a decentralized methodology to navigate heterogeneous groups of robots whilst maintaining segregation among different groups. Our approach consists of extending the ORCA algorithm with a modified version of the classical flocking behaviors to keep robots segregated. A series of simulations and real experiments show that the groups were able to navigate in a cohesive fashion in all evaluated scenarios. Furthermore, the methodology allowed for a faster convergence of the group to the goal when compared to state-of-the-art algorithms.

Keywords Swarm robotics · Segregative navigation · Flocking · ORCA

1 Introduction

The use of multi-robot systems in different contexts can bring several advantages over single-robot systems. In this sense, we consider the specific scenario of robotic

---

This work was developed with the support of CNPq, CAPES, FAPEMIG, and ITV - Instituto Tecnologico Vale.

F. R. Inácio (✉) · D. G. Macharet · L. Chaimowicz
Computer Vision and Robotics Laboratory (VeRLab), Department of Computer Science, Universidade Federal de Minas Gerais, Belo Horizonte, Brazil
e-mail: fabricio.rod@dcc.ufmg.br

D. G. Macharet
e-mail: doug@dcc.ufmg.br

L. Chaimowicz
e-mail: chaimo@dcc.ufmg.br

R. Groß et al. (eds.), *Distributed Autonomous Robotic Systems*, Springer Proceedings in Advanced Robotics 6,
https://doi.org/10.1007/978-3-319-73008-0_22

swarms, which are systems composed of a large number of agents seeking to cooperatively accomplish a particular task. Inspired by colonies of social insects that cooperate with each other to carry out tasks of common interest, robotic swarms emerged as an alternative to solve complex problems.

In general, swarms usually consist of simple agents with little processing power and able to perform a limited set of actions. Another common feature in swarms is the limitation of perception and communication between agents. Normally, each agent is capable of sensing a small portion of the environment and communicating locally, i.e., the communication takes place only between agents who are close to each other.

Similarly to different swarms found in nature (e.g. some insects colonies, bird flocks or fish schools), an important aspect in robotic swarms is the absence of a central entity responsible for coordinating the entire swarm. Hence, each individual behavior contributes to emergent group behaviors that are able to guide the swarm to its objectives [9].

One of these group behaviors is segregation, a natural phenomenon that is commonly used as a sorting mechanism by several biological systems and can be useful in many different tasks and scenarios. This behavior can be important for maintaining robots with similar features or objectives arranged in cohesive groups, while the groups with different characteristics remain separated [11].

In this paper we present a methodology capable of navigating heterogeneous groups of robots whilst maintaining segregation among distinct groups. Our approach consists of extending the Optimal Reciprocal Collision Avoidance(ORCA) algorithm [14] with a modified version of the classical flocking behaviors [8]. The set of possible velocities is informed with the flocking demands, which enables different robot groups to avoid collisions and navigate in a cohesive manner. This methodology improves our previous approach, which was based on velocity obstacles [10], presenting a much better performance. Moreover, differently from other segregation works found in the literature [7, 11], the proposed technique is fully decentralized and assumes local perception only.

2 Related Work

2.1 Navigation and Collision Avoidance

Finding feasible paths for mobile agents that are either length or time optimized is fundamental in any task that requires the navigation from an initial to a goal position in the environment. However, in general, path planners consider single-agent systems, a static and known environment with a given map, and may not be able to adapt to dynamic issues that may arise during the execution of the path.

An approach that has received much attention in recent years is called Velocity Obstacles (VO). Proposed in [4], this technique uses the agents' velocities and obstacles' positions to calculate the VO-set of all velocities that, if applied to the agents, will result in a collision and then must be considered obstacles. Next, linear and angular speeds that do not belong to the VO-set must be chosen, which ensures a safe navigation to the agents. Several extensions over the VO algorithm have been proposed in the literature [6, 12, 13, 15], which aim to improve its performance and increase the possibilities for use.

The concept of using the agents velocity space in order to choose acceptable velocities, i.e., the ones that avoid collisions, was further extended resulting in the ORCA algorithm [14]. This technique computes for each other robot a half-plane of allowed velocities. The robot must select its optimal velocity from the intersection of all permitted half-planes, which can be done efficiently by solving a low-dimensional linear system. Moreover, differently from the VO, the ORCA algorithm guarantee local collision-free navigation [14].

2.2 Group Segregation in Robotic Swarms

A characteristic found in social insect communities is the existence of specific roles that segregate the individuals into distinct groups (e.g. bees and ants). Each group of individuals has a distinct set of activities and responsibilities that promote the perpetuation of the colony.

In robotic swarms, as well, it might also be interesting to keep different groups of agents segregated, allowing for example to assign specific tasks according to certain characteristics of each group.

One of the first works to deal with this problem was [7]. It presented an algorithm capable of segregating distinct types of agents by applying different potential functions on the robot according to the type of the agents which are in its neighborhood. The technique produced satisfactory results, however, it can only be applied in scenarios where there are only two different types of agents. In applications that use a very large number of robots, there may be the need to segregate the swarm in more than two groups.

The aforementioned technique was later extended in [11] in order to deal with more than two types of robots. In order to calculate the potential to be used, each agent evaluates whether the neighbors are of their same type or of a different type. Thus, it is possible to obtain the artificial force which must be applied to each agent in order to make the robots with the same type come closer, while robots of different types remain spaced apart. The main restrictions of this technique are the need for global sensing and a balanced number of agents in all groups. More recently, an approach based on abstractions and using an artificial potential function to segregate the groups was proposed [3]. Differently from other works on swarm segregation, it is mathematically guaranteed that the system will always converge to a state where multiple dissimilar groups are segregated. A different segregation algorithm, based on the *Brazil Nut Effect*, is discussed in [5]. A distributed controller considers robots as having distinct virtual sizes, and local interactions make "larger" robots move outwards, segregating the robots in annular structures.

A segregative navigation strategy is proposed in [10], combining the concepts of flocking [8] and the use of abstractions to represent the groups. Built upon the classical VO algorithm [4], it introduces a novel concept called Virtual Group Velocity Obstacles (VGVO). In this algorithm, a robot i senses the relative position and velocity of every robot j within its neighborhood n_i and builds shapes containing other teams of robots, with the exception of its own. These shapes can be considered as the smallest enclosing disc, the convex hull, or the more general class of α-shapes [2]. In the workspace of robot i, these shapes are considered as virtual obstacles moving at the average velocity of their respective underlying robots. Thus, robot i can build a virtual velocity obstacle specifying all velocities that will lead to a collision with these shapes, assuming that they maintain their current average velocities. However, the cost of computing the virtual shape for each group can be prohibitive, slowing down the navigation.

To cope with this problem, in this paper we propose a decentralized methodology in order to navigate heterogeneous groups of robots whilst maintaining segregation among different groups. We propose a modification over the classical flocking model established by Reynolds [8] and combine it with the ORCA algorithm [14] in order to avoid collisions and keep the group cohesion.

3 Methodology

As mentioned, our methodology is based on a combination of flocking behaviors with the ORCA algorithm. Using local sensors, each robot is able to sense its neighborhood and compute the velocity that will guide it to the goal, while keeping it close to its group. This velocity is used as the preferred velocity by the ORCA algorithm, which is responsible for navigating the robot while avoiding other agents. As in the original algorithms, we assume that an agent has access to the position and velocity of all other agents present in its neighborhood, or can infer these values based on its observations. They can also detect if its neighbors belong to its group and infer their state. Finally, we consider that the robots start in a segregated state and know the direction of a specific goal in the environment. The next sections detail our methodology.

3.1 Robot and Group Modeling

We consider a scenario in which a swarm $\mathscr{R} = \{\mathscr{R}_1, \ldots, \mathscr{R}_\eta\}$ of η robots must navigate in a static 2D environment. Each robot i is represented by its pose $q_i = \langle x_i, y_i, \theta_i \rangle$, with kinematic model given by $\dot{q}_i = u_i$.

Robots are holonomic and can move in any direction given by the velocity vector u_i. Thus, θ_i is the direction of the movement of robot i in a global frame.

The entire swarm is formed by distinct types (groups) of robots, which we represent by the partition $\Gamma = \{\Gamma_1, \ldots, \Gamma_m\}$, where each Γ_k contains all agents of type k. We assume that $\forall j, k : j \neq k \rightarrow \Gamma_j \cap \Gamma_k = \emptyset$, i.e., each robot is uniquely assigned to a single type.

3.2 Flocking

In order to maintain a cohesive behavior during navigation, we use a slightly modified version of the classical flocking behaviors presented in [8]. These behaviors are obtained by applying three simple rules, which are described next.

Initially, we define a robot's neighborhood based on its sensing range. During navigation, robot i maintains a neighborhood $\mathcal{N}_i$ around its current position. A neighborhood consists of a circular region of radius λ around the current position of a robot. Within the neighborhood, we may have robots from the same group and robots from different groups. Therefore, we define $\mathcal{N}_i^+ = \{\mathcal{R}_j : \|p_j - p_i\| \leq \lambda \wedge \mathcal{R}_j, \mathcal{R}_i \in \Gamma_k\}$ as the set of all robots of the same group currently located in the neighborhood. Similarly, we define $\mathcal{N}_i^- = \{\mathcal{R}_j : \|p_j - p_i\| \leq \lambda \wedge (\mathcal{R}_j \in \Gamma_u \wedge \mathcal{R}_i \in \Gamma_k, u \neq k)\}$ the set of all robots of other groups in its neighborhood.

The first flocking rule is related to cohesion. This rule is used to keep the agents close to each other. Therefore, agent i must compute the midpoint of the positions of all agents in $\mathcal{N}_i^+$. Next, a velocity vector that moves the agent to the computed point is calculated. Formally:

$$v_{\text{cohesion}} = \left(\frac{1}{|\mathcal{N}_i^+|} \sum_{j \in \mathcal{N}_i^+} p_j \right) - p_i. \tag{1}$$

The second rule was originally proposed to keep a safe distance among agents, with the objective of avoiding collisions between them. However, as will be further detailed, we use the ORCA algorithm in order to ensure obstacle avoidance during navigation. Therefore, we use the separation rule in order to keep distinct groups of agents apart from each other. The agent should move to a position that respects a minimum distance from agents of other groups in its neighborhood, given by:

$$v_{\text{separation}} = \sum_{j \in \mathcal{N}_i^-} (p_i - p_j). \tag{2}$$

We emphasize that the safety distance among robots of the same group are guaranteed by the ORCA algorithm. Therefore, the proposed algorithm does not use the separation rule for agents that belong to the same type.

Finally, the third rule evaluates the alignment of agents of the same group. In order to respect this rule, each agent calculates the average of the spatial orientation of all agents in its group inside its neighborhood. After that, the agent is guided by the calculated average alignment, i.e.,

$$v_{\text{alignment}} = \frac{1}{|\mathcal{N}_i^+|} \sum_{j \in \mathcal{N}_i^+} \theta_j. \tag{3}$$

The final resulting vector v_{flock} that will be used in the control phase is given by:

$$v_{\text{flock}} = k_c \cdot v_{\text{cohesion}} + k_s \cdot v_{\text{separation}} + k_a \cdot v_{\text{alignment}}, \tag{4}$$

which is composed by the weighted sum of the three vectors. Constants k_c, k_s, and k_a are determined empirically, and normally k_c receives a larger value.

3.3 Setting Robot Velocities

The robot velocities can be adjusted according to the different situations an agent may face during its navigation. For example, it can be surrounded by agents from its own group only, or it may be in a position where there are no other agents between itself and the goal or may be close to agents who belong to another group. Thus the control input is composed by a combination of different velocities:

$$u = \alpha.v_{goal} + \beta.v_{flock} + \gamma.v_{aux}. \tag{5}$$

v_{goal} is an attractive velocity that drives the agent towards its goal, v_{flock} is a flocking velocity set according to Eq. 4 and v_{aux} is an auxiliary velocity. Constants α, β and γ, as well as the auxiliary velocity, are set according to the scenario faced by the robot. This is controlled by a finite state machine depicted in Fig. 1 and explained next.

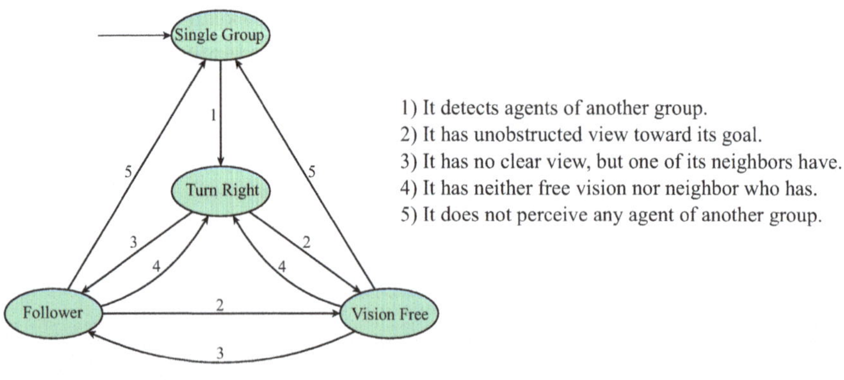

Fig. 1 Finite state machine describing the possible situations faced by each robot

Single Group: This first state is enabled whenever a robot does not perceive any agent of a different group within its neighborhood. In this case, its action is to move straight to the goal while also keeping cohesion with its group. For this behavior, the constants are set in such way that $\alpha \gg \beta$ and $\gamma = 0$.

Vision Free: The state *Vision Free* is enabled if the agent perceives the presence of one or more robots of different groups, but still has a free line of sight to its goal. More specifically, in a sector defined from the current position of the agent towards its goal and within his field of view there is no agent of a different group. In this case, the agent still tends to move to the goal, but also gives more importance to the flocking factor. Thus, constants are set so that $\alpha > \beta$ and $\gamma = 0$.

Follower: If the robot does not have a clear view towards the goal but detects a neighbor of its own group that is in one of the previous states, it enters in a follower mode, following the agent that has a clear path to the goal, but also keeping the flocking behavior. In this case, the v_{aux} component is set to move the robot towards this neighbor, with $\alpha = 0$ and $\beta < \gamma$.

Turn right: Finally, if the agent does not fit into any of the situations described above, it probably has a congested situation in front of it and should move to avoid this. So, a "traffic rule" is imposed that makes the robot move perpendicularly to its goal direction, i.e., v_{aux} is set in a direction that is perpendicular to v_{goal}. With this behavior, the agents try to get around this congested area rather than trying to cross it. Constants are set to balance this velocity and the flocking component: $\alpha = 0$ and $\beta \simeq \gamma$.

3.4 ORCA

The ORCA algorithm [14] is a velocity-based navigation strategy based on the concept of velocity obstacles [4].

Consider two robots $\mathscr{R}_A$ and $\mathscr{R}_B$ with radii r_A and r_B, positions p_A and p_B and velocities v_A and v_B, respectively. Robot $\mathscr{R}_A$ tries to reach an assigned goal point g_A by selecting a *preferred velocity* v_A^{pref}. The objective is to choose an optimal v_A^*, which lies as close as possible to v_A^{pref}, such that collisions among the robots are avoided for at least a time horizon τ.

The velocity obstacle $VO_{A|B}^{\tau}$ for $\mathscr{R}_A$ induced by $\mathscr{R}_B$ in the local time interval $[0, \tau]$ is the set of velocities of $\mathscr{R}_A$ relative to $\mathscr{R}_B$ that will cause a collision between $\mathscr{R}_A$ and $\mathscr{R}_B$ at some moment before time τ has elapsed. It is assumed that both robots maintain a constant trajectory within that time interval. Formally:

$$VO_{A|B}^{\tau} = \{v | \exists t \in [0, \tau] :: t(v - v_B) \in D(p_B - p_A, r_A + r_B)\} \tag{6}$$

where $D(p_B - p_A, r_A + r_B)$ denote an open disc of radius $(r_A + r_B)$ centered at position $(p_B - p_A)$. If $\mathscr{R}_A$ and $\mathscr{R}_B$ each choose a velocity outside $VO_{A|B}^{\tau}$ and $VO_{B|A}^{\tau}$, respectively, then they will be collision-free for at least the period of time τ.

The half-plane of velocities $ORCA^{\tau}_{A|B}$ can be constructed geometrically as follows. Let us assume that $\mathscr{R}_A$ and $\mathscr{R}_B$ adopt velocities v_A and v_B, respectively, and that these velocities causes $\mathscr{R}_A$ and $\mathscr{R}_B$ to be on collision course, i.e. $v_A - v_B \in VO^{\tau}_{A|B}$. Let $\mathbf{w}$ be the vector from $v_A - v_B$ to the closest point on the boundary of the velocity obstacle:

$$\mathbf{w} = (\text{argmin}_{v \in \partial VO^{\tau}_{A|B}} ||v - (v_A - v_B)||) - (v_A - v_B). \tag{7}$$

Then, $\mathbf{w}$ is the smallest change required to the relative velocity of $\mathscr{R}_A$ and $\mathscr{R}_B$ to avert collision within τ time. To "share the responsibility" of avoiding collisions among the robots, $\mathscr{R}_A$ adapts its velocity by (at least) $\frac{1}{2}w$ and assumes that B takes care of the other half. Hence, the set $ORCA^{\tau}_{A|B}$ of permitted velocities for $\mathscr{R}_A$ is the half-plane pointing in the direction of $\mathbf{n}$ starting at the point $v_A + \frac{1}{2}w$, where $\mathbf{n}$ is the outward normal of $VO^{\tau}_{A|B}$ at $v_A - v_B + w$ [14]:

$$ORCA^{\tau}_{A|B} = \{v|(v - (v_A + \frac{1}{2}\mathbf{w})).\mathbf{n} \geq 0\}. \tag{8}$$

In our methodology, the finite state machine described in the previous section is used to calculate the preferred velocity. This velocity is passed along to the classic ORCA algorithm that sets the new velocity to be applied to the agent. In other words, the velocity computed by Eq. 5 is fed to ORCA as the *preferred velocity*, which is a value used as a reference to determine the actual velocity that will be assigned to the robot during navigation. Consequently, the value belonging to that half-plane that most closely matches the preferred velocity is calculated using linear programming and passed as the new velocity that the robot should take.

4 Experiments

To evaluate the proposed methodology with regards to its performance and feasibility, we executed a series of simulations and compared it with the classical versions of the ORCA and VGVO algorithms. We also performed proof-of-concept experiments with a group of e-puck robots to show its applicability with real robots.

The parameters used in these experiments are shown in Table 1. Our main objective when selecting the values was to keep the robots segregated, even if this requires spending more time for the groups to achieve their goals.

In Fig. 2, we show snapshots of the execution of the three algorithms. In this first example, we have 4 distinct groups, represented by different colors, each one formed by 40 agents. Each group must navigate to its opposite direction, changing sides with another group. As can be seen in Fig. 2a, as expected, ORCA does not keep robots segregated when groups meet in the center of the scenario. Figures 2b, c show, respectively, the results obtained by the VGVO algorithm and the proposed methodology for the same scenario. As shown, with both algorithms, robots belonging to

Table 1 Parameters used in the experiments

Agent status	k_c	k_s	k_a	α	β	γ
Single group	10	0	10	10	1	0
Vision free	15	5	5	3	1	0
Follower	15	10	0	0	20	50
Turn right	5	2.5	1	0	20	30

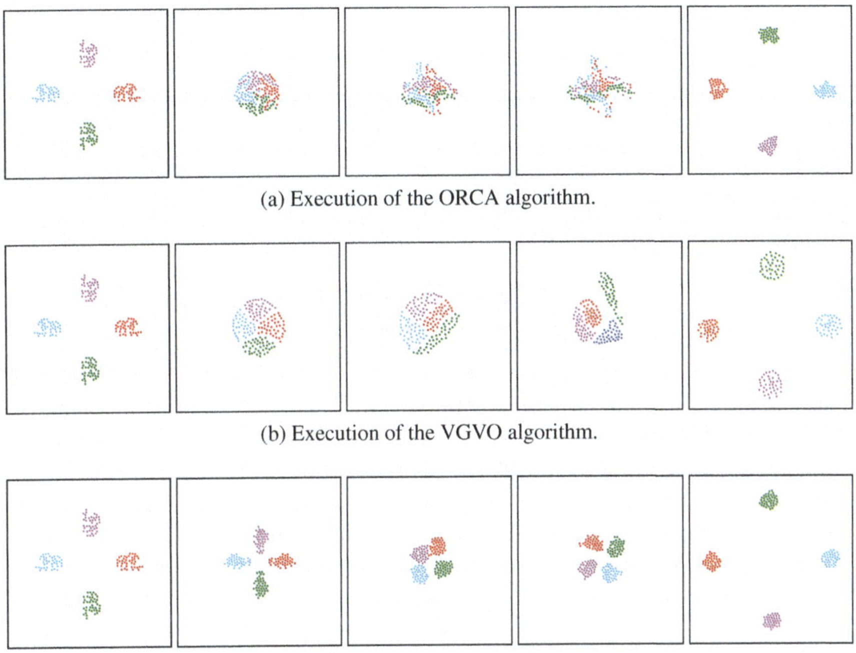

(a) Execution of the ORCA algorithm.

(b) Execution of the VGVO algorithm.

(c) Execution of the proposed algorithm.

Fig. 2 Execution of the evaluated algorithms in a scenario composed of 160 agents evenly distributed into 4 groups

the same group remain united during the entire navigation, while seeking to move away from other groups.

To better evaluate the segregation, we use a metric proposed in [7], which consists of calculating the average distances between agents of the same group and agents from different groups. According to this metric, two different groups of agents, e.g. A and B, are said to be segregated if the average distance between the agents of the alike types (type A or type B) is less than the average distance between the agents of the unlike types (i.e., between the agents of type A and type B). More formally, we should have:

$$d_{AA} < d_{AB} \quad \text{and} \quad d_{BB} < d_{AB}, \tag{9}$$

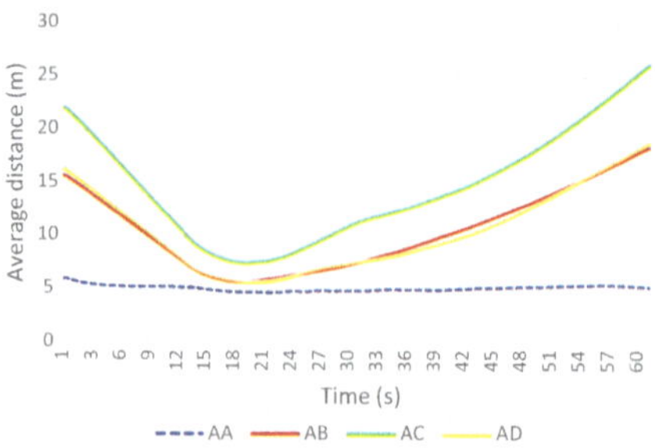

(a) 4 distinct groups.

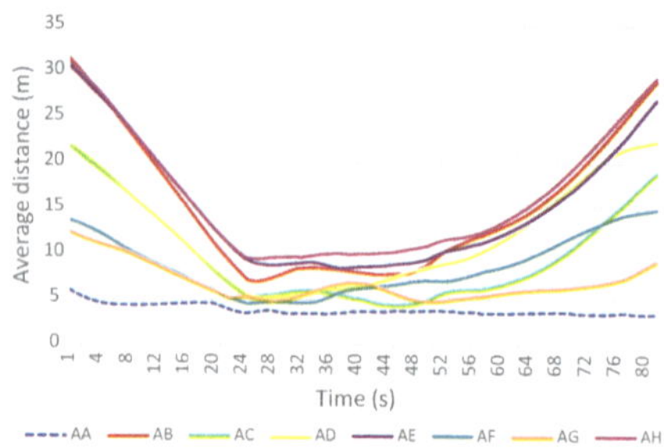

(b) 8 distinct groups.

Fig. 3 Average distance among agents during the execution of the proposed algorithm. Two scenarios were considered: **a** 4 distinct groups and **b** 8 distinct groups

where d_{XY} is the average distance between the agents of types X and Y. Formally:

$$d_{XY} = \frac{1}{|\Gamma_X|} \sum_{i \in \Gamma_X} \left(\frac{1}{|\Gamma_Y|} \sum_{j \in \Gamma_Y} (p_i - p_j) \right). \tag{10}$$

The segregative behavior of the methodology may be confirmed by using the aforementioned metric pairwise. As can be seen in Fig. 3, the average distance between agents in the group A are smaller than the average distance between agents from group A relative to agents of all other groups, i.e., $d_{XX} < d_{XY}\ \forall\ \Gamma_X, \Gamma_Y \in \mathscr{R}$.

In the following experiments, we varied the number of robots and groups. Initially, the algorithm was evaluated considering a fixed number of groups and an increasing number of agents in each group. Next, we have fixed the number of agents and varied the quantity of distinct groups in the environment. Finally, we have evaluated the methodology in a scenario where each group has a different number of agents.

In the first experiment we investigated the performance of the approach relative to the variation in the number of agents on each group. The simulations were performed with 4 groups consisting of 10, 20, 30 and 40 agents. We emphasize that on these first set of simulations all the groups have the same number of agents.

The methodology was evaluated considering the average time the groups take to converge to their goals and the results were compared to the ones obtained by the ORCA and the VGVO algorithms. We recall that ORCA does not sort (segregate) agents into distinct groups, i.e., each agent is considered as a single entity and no group information is used during its execution. The main objective of this evaluation is to verify if the restriction to keep the groups segregated would affect the navigation time and if our approach is able to increase the performance of the VGVO algorithm.

Since the proposed technique (as well as the VGVO algorithm) may produce different results during executions (due to a non-deterministic factor on the velocity selection), we performed 100 simulations for each experiment and the average time

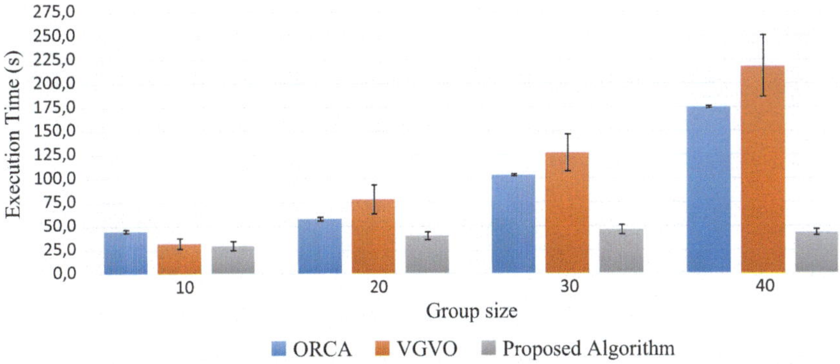

Fig. 4 Average execution time of the algorithms varying group size

was used for comparison. Figure 4 presents the results for each technique for this scenario.

As might be expected, the time required for all the agents to reach their goals increases with the number of agents. Also, we can observe that the proposed algorithm has a much better performance than ORCA and VGVO when the number of robots increase. This may be explained by the fact that in ORCA, robots have to deal with a very congested situation when all groups meet, which makes it difficult for the algorithm to find feasible velocities. Regarding VGVO, as explained in [10], the algorithm prioritizes slower speeds in order to maintain stable relative distances and velocities among robots. We can also note that, due to its characteristics, VGVO has a large variation in its running time within the 100 runs, while in ORCA and our algorithm the standard deviation is relatively small.

Next, the second experiment aims to evaluate our approach in scenarios with a fixed number of agents that were distributed in a different number of groups in each set of runs. The simulations consist of scenarios composed by a total of 160 agents evenly distributed into 2, 4, 6, and 8 groups. As shown in Fig. 5, the increase in the number of groups causes the time required for the agents to reach their targets to also increase. However, we observe that the proposed approach improves considerably the average navigation time. Another point to note is the difference between the times obtained using VGVO algorithm and the proposed algorithm. The VGVO algorithm makes the groups to be positioned uniformly in the conflict area, causing the agents to take a long time to find a new path that would allow the group to contour the obstacles that hinder its movement. On the other hand, our approach allows the agents to start diversion maneuvers when a new group is detected, causing a reduction on the time needed to solve the problem.

Finally, the last experiment was performed in order to investigate the methodology behavior in scenarios with groups of different sizes. One of the experiments evaluated a scenario with 220 agents divided into 8 groups with sizes ranging from 10 to 40 agents. Figure 6 shows snapshots of the navigation over time.

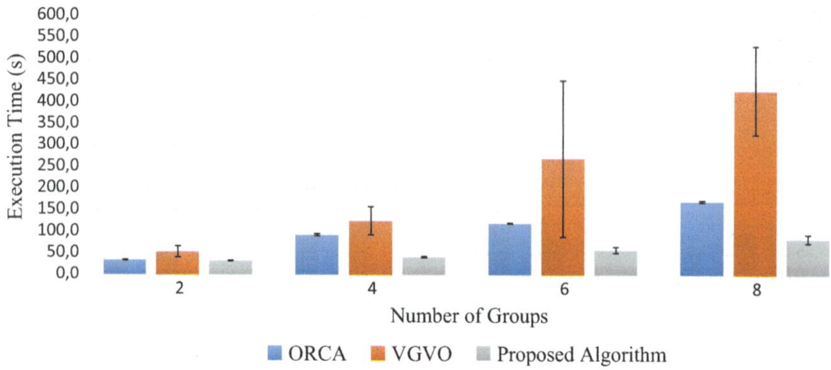

Fig. 5 Average execution time of the algorithm with different number of groups

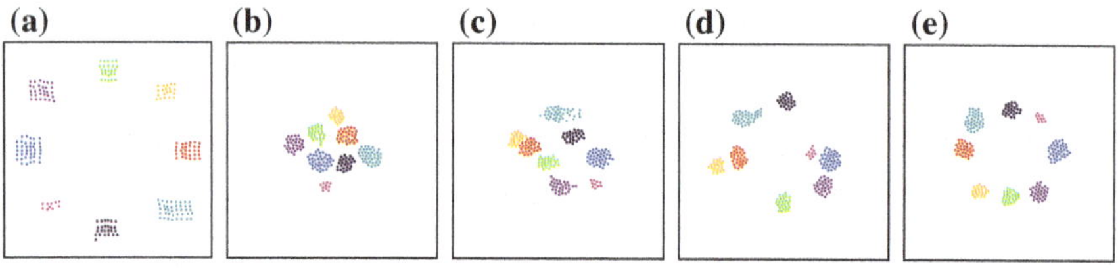

Fig. 6 Execution of the proposed algorithm in a scenario with 220 agents distributed in 8 groups of different sizes

It is possible to observe that the proposed algorithm has the same segregating behavior when used in environments having different group sizes. We call attention to the trajectory performed by the pink group during this simulation. As shown in Fig. 6c–e, despite the definition of an explicit rule that requires the agents to move to the right when another group is blocking its way, the pink group managed to find a better path than the path resulting from applying the rule. This occurs because at least one of the agents on the pink group could directly see the goal and, consequently, the other agents began to follow it.

Finally, real experiments were conducted indoors using six e-puck robots. These proof-of-concept experiments are important in order to show the feasibility of the algorithm in real scenarios, where uncertainties caused by sensing and actuation errors may have a great impact on the results.

In these experiments, we used a swarm localization framework based on an overhead camera and fiduciary markers for estimating robot's pose, orientation. Also, as the e-puck's IR sensors have a very small range, we implemented a virtual sensor based on the localization system to detect neighboring agents. To account for non-holonomic constraints, input velocities were transformed following the approach presented in [1].

The Figs. 7 and 8 show snapshots from executions of the proposed algorithm with two and three groups of robots (we overlay colored circles to highlight the

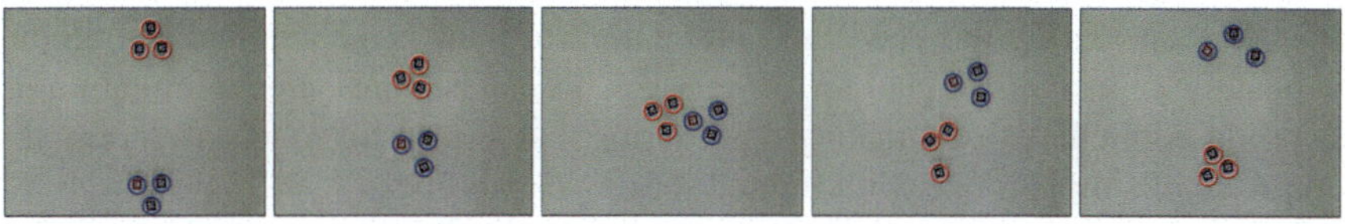

Fig. 7 Execution of the proposed algorithm in a scenario with 6 e-puck robots distributed in 2 groups

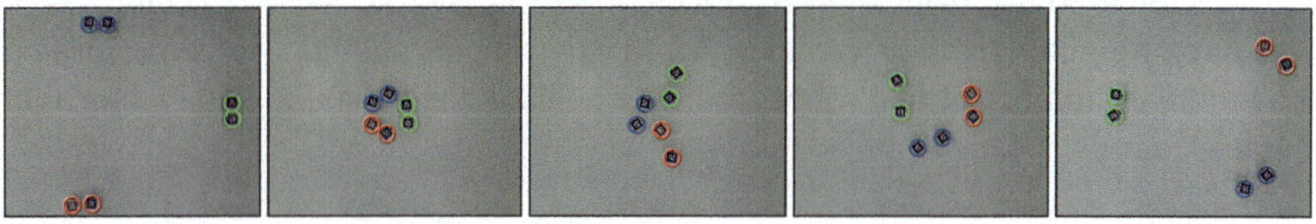

Fig. 8 Execution of the proposed algorithm in a scenario with 6 e-puck robots distributed in 3 groups

different groups). We can visually inspect that the behaviors obtained with the real robots is pretty similar to the simulation results, i.e., robots maintain cohesion and segregation during navigation. Despite not showing the graphs, we observed that average distances follow the trend shown in Fig. 3: the average distance between robots in the same group is always less than the average distance among robots in different groups. These proof of concept experiments indicate that the algorithm can work well to coordinate groups of real robots, allowing them to navigate while maintaining a segregative behavior in an efficient way.

5 Conclusion

In this paper we proposed a decentralized methodology to navigate heterogeneous groups of robots maintaining segregation among different groups. Our approach consists of extending the ORCA algorithm with a modified version of the classical flocking behaviors: using local sensors, each robot is able to sense its neighborhood and, using a variation of the flocking rules, compute the velocity that will guide it to the goal, while keeping it close to its group. This velocity is used as the preferred velocity by the ORCA algorithm, which is responsible for navigating the robot while avoiding other agents.

Several experiments were performed to evaluate the proposed methodology and the results showed that our algorithm is an effective alternative to maintain different robots groups segregated whilst they navigate on a shared environment. By choosing better trajectories and avoiding congested areas, it achieved a better performance when compared to other state-of-the-art algorithms.

As in ORCA and VGVO, in our methodology the robots must be able to infer the position and velocities of the neighboring robots. Moreover, robots should infer the state (considering the FSM) and orientation of their neighbors and also know the position of a common goal for the group. This can be considered a limitation of the proposed methodology, mainly considering that swarms of robots are normally comprised of simple robots with limited sensing capabilities.

Future research directions include the evaluation of different segregation metrics, such as the intersection area of the convex hull formed by the agents and a possible network connectivity inside the group. We also intend to consider standard metrics of separation used in cluster analysis, for example, the distance between centroids of the groups weighted by their variance. More experiments should also be performed using a larger number of real robots to better analyze the behavior of the algorithm in real scenarios.

References

1. De Luca, A., Oriolo, G., Vendittelli, M.: Stabilization of the unicycle via dynamic feedback linearization. In: 6th IFAC Symp. on Robot Control, pp. 397–402 (2000)
2. Egerstedt, M., Hu, X.: Formation constrained multi-agent control. In: ICRA'01. IEEE International Conference on Robotics and Automation, 2001, vol. 4, pp. 3961–3966 (2001)
3. Ferreira Filho, E.B., Pimenta, L.C.A.: Segregating multiple groups of heterogeneous units in robot swarms using abstractions. In: Proceedings of the IEEE International Conference on Intelligent Robots and Systems (IROS), pp. 401–406 (2015)
4. Fiorini, P., Shiller, Z.: Motion planning in dynamic environments using velocity obstacles. Int. J. Robot. Res. **17**(7), 760–772 (1998)
5. Groß, R., Magnenat, S., Mondada, F.: Segregation in swarms of mobile robots based on the brazil nut effect. In: Proceedings of the IEEE International Conference on Intelligent Robots and Systems (IROS), pp. 4349–4356 (2009)
6. He, L., van den Berg, J.: Meso-scale planning for multi-agent navigation. In: Proceedings of the IEEE International Conference on Robotics and Automation (ICRA) (2013)
7. Kumar, M., Garg, D.P., Kumar, V.: Segregation of heterogeneous units in a swarm of robotic agents. IEEE Trans. Autom. Control **55**(3), 743–748 (2010)
8. Reynolds, C.W.: Flocks, herds and schools: a distributed behavioral model. In: ACM Siggraph Computer Graphics, vol. 21, pp. 25–34. ACM (1987)
9. Şahin, E.: Swarm robotics: from sources of inspiration to domains of application. In: Swarm robotics, pp. 10–20. Springer (2005)
10. Santos, V.G., Campos, M.F., Chaimowicz, L.: On segregative behaviors using flocking and velocity obstacles. In: Distributed Autonomous Robotic Systems, pp. 121–133. Springer (2014)
11. Santos, V.G., Pimenta, L.C., Chaimowicz, L., et al.: Segregation of multiple heterogeneous units in a robotic swarm. In: 2014 IEEE International Conference on Robotics and Automation (ICRA), pp. 1112–1117. IEEE (2014)
12. Snape, J., Guy, S.J., Vembar, D., Lake, A., Lin, M.C., Manochal l, D.: Reciprocal collision avoidance and navigation for video games. In: Game Developers Conference, San Francisco (2012)
13. Van Den Berg, J., Guy, S.J., Lin, M., Manocha, D.: Optimal reciprocal collision avoidance for multi-agent navigation. In: Proceedings of the IEEE International Conference on Robotics and Automation, Anchorage (AK), USA (2010)
14. Van Den Berg, J., Guy, S.J., Lin, M., Manocha, D.: Reciprocal n-body collision avoidance. In: Robotics research, pp. 3–19. Springer (2011)
15. Wilkie, D., Van den Berg, J., Manocha, D.: Generalized velocity obstacles. In: IEEE/RSJ International Conference on Intelligent Robots and Systems, 2009. IROS 2009, pp. 5573–5578. IEEE (2009)

Part V
Modular Robots and Smart Materials

A Rule Synthesis Algorithm for Programmable Stochastic Self-assembly of Robotic Modules

Bahar Haghighat and Alcherio Martinoli

Abstract Programmable self-assembly of modular robots offers promising means for structure formation at different scales. Rule-based approaches have been previously employed for distributed control of stochastic self-assembly processes. The assembly rate in the process directly depends on the concurrency level induced by the employed ruleset, i.e. the number of concurrent steps necessary to build one instance of the target structure. Our aim here is to design a formal synthesis algorithm to automatically derive rulesets of high concurrency for a given target structure composed of robotic modules. In the literature, self-assembly of (simulated or real) robotic modules has been realized through manually designed rulesets or manually adjusted rulesets generated by employing graph-grammar formalisms or metaheuristic methods. In this work, we employ an extended graph-grammar formalism, adapted for self-assembly of robotic modules, and propose a novel formal synthesis algorithm capable of generating rulesets for robotic modules by natively considering the morphology of their connectors. The synthesized rulesets induce a high level of concurrency in the self-assembly scheme by exploiting controlled information propagation, using solely local communication. Simulation results of microscopic (non-spatial) and submicroscopic (spatial) models of our robotic platform confirm higher performance of rulesets synthesized by our algorithm compared to related work in the literature.

B. Haghighat (✉) · A. Martinoli
École Polytechnique Fédérale de Lausanne (EPFL), Distributed Intelligent Systems and Algorithms Laboratory (DISAL), Lausanne, Switzerland
e-mail: bahar.haghighat@epfl.ch

A. Martinoli
e-mail: alcherio.martinoli@epfl.ch

R. Groß et al. (eds.), *Distributed Autonomous Robotic Systems*,
Springer Proceedings in Advanced Robotics 6,
https://doi.org/10.1007/978-3-319-73008-0_23

1 Introduction

Self-assembly (SA) is defined as the reversible and spontaneous phenomenon of an ordered spatial structure emerging from the aggregate behavior of simpler preexisting entities, through inherently local and random interactions in the system. In recent years, SA has been extensively studied both as an enabling technique for micro/nano-fabrication, and as a coordination mechanism for distributed robotic systems of miniaturized modules with limited capabilities, where highly stochastic sensing, actuation, and interactions are inevitable [1, 4, 13]. Various implementations of SA have been demonstrated in engineered systems. For instance, a deterministic and quasi-serial approach to shape formation through programmable SA has been implemented in a large swarm of miniaturized Kilobot robots [13]. A completely different approach is to achieve SA by taking advantage of the stochastic ambient dynamics for module transportation. Such approach typically enables simpler internal design of the programmable robotic modules as well as their on-board algorithms. The robotic modules in [9] stochastically self-assemble on an air table based on their behavioral ruleset which is first derived for SA of a similar abstract target graph by a synthesis algorithm, then tuned to suit the specific modules' morphology.

The problem of ruleset synthesis for programmable SA of graphs was first addressed in [8]. In principle, all proposed formal synthesis algorithms in this context generate rules by iteratively browsing and parsing a description of the target structure. In [10], the formalism of graph-grammar is applied to the SA of graphs and two rule synthesis algorithms are proposed for acyclic and then cyclic target graphs. Both algorithms result in rulesets which build the target graph serially, i.e. by adding one atomic agent (graph vertex) at a time. The same work also discusses the deadlock situation, where the number of copies of the target being built in parallel is higher than the maximally feasible number, considering the total number of available agents. In order to avoid deadlocks the authors then propose a disassociating rule which requires implementation of a consensus algorithm among the agents. In [3, 12], SA of graphs is achieved while avoiding deadlocks by employing probabilistic dissociating rules. In order to accelerate the SA process induced by an inherently serial ruleset, the authors in [12] employ a broadcast radio communication scheme to decompose less advanced sub-assemblies in favor of the others. The formalism of graph-grammar is employed in [3] for modeling and control of SA of abstract graphs. Two formal synthesis algorithms, Singleton and Linchpin, are then introduced. While Singleton generates utterly serial rulesets, Linchpin induces a more parallel scheme and allows for sub-assemblies of bigger sizes to merge and form the target graph. Two main factors limit the concurrency induced by the Linchpin rulesets. First, in an attempt to fully avoid information propagation for scalability reasons, all sub-assemblies including the ones with identical structure are labeled distinctly, determining their final placement in the target structure. As a consequence, several rules are required to form the target, meaning more assembly time. Second, at each branching point in the target, i.e. at any vertex with two or more neighbors, the algorithm generates rules to build the branches in parallel and eventually join them, regardless of the size

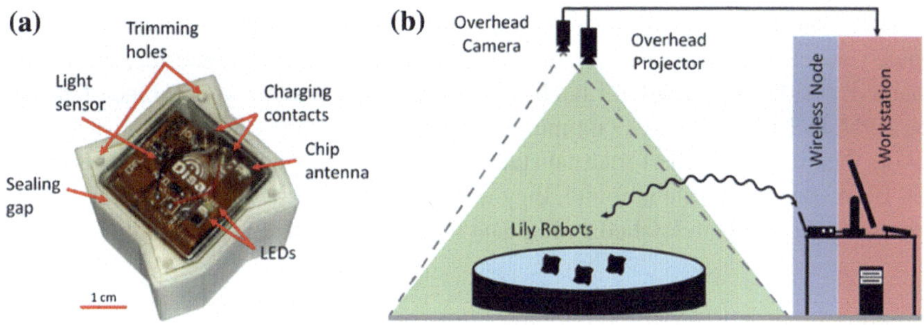

Fig. 1 **a** The Lily robot [5]. **b** Sketch of the full experimental setup [6]

of the branches. This imbalance in the size of the concurrently built branches can result in unnecessary delay, as the longer branch may need more rules to form. The slow nature of stochastic SA highlights the importance of inducing a high level of concurrency in the process in order to reduce the total assembly time.

In a previous contribution, we presented an extended graph-grammar formalism adapted for formulating the problem of SA of robotic modules. The formalism is based on the notion of extended vertices [7]. In addition, we showed that our extended formalism allows for a ruleset of complexity $O(N)$ compared to the $O(N^2)$ in the previous literature. In this work, we employ the extended graph-grammar formalism presented in [7], and propose a new synthesis algorithm with two key features: first, our algorithm automatically derives rules for robotic modules considering their latching connectors morphology, and second, it achieves a parallel assembly scheme by exploiting update rules (i.e. information propagation) of controllable depth between neighboring robots. In order to avoid deadlocks, our algorithm incorporates reversible rules. We consider two target structures composed of our floating Lily robots (see Fig. 1) and compare the performance of rulesets derived by our algorithm and those of the extended Linchpin, adapted to generate rules for robotic modules [7], in simulation. To this end, we leverage non-spatial microscopic simulations in Matlab and high-fidelity spatial submicroscopic simulations in Webots [11] (see Fig. 2).

2 Fluidic Self-assembly of Lily Robots

In this paper, we study the SA of our robotic modules in simulation. For the sake of thoroughness, here we give a brief summary of the SA process in our real world experimental system [6]. Our system consists of two main components: (1) the Lily robots, originally presented in [5], which serve as the building blocks of the SA process, and (2) the experimental setup built around them. Lilies are not self-locomoted, instead, they float on water and are stirred by the flow field produced within a tank by several peripheral pumps, colliding randomly with other robots. The robots are endowed with four custom-designed Electro-Permanent Magnets (EPM) to latch and also to communicate locally with their neighbors. Given a target structure, an appropriate ruleset is programmed on all robots. The robots' EPM latches are by default enabled, resulting in a default latching upon meeting another robot. Once latched, the EPM-to-EPM inductive communication channel is physically established. The robots then exchange their internal states and look for applicable rules in their ruleset. If no applicable rule is found, they unlatch by switching off their EPM latches; otherwise they remain latched and update their internal states accordingly.

3 Graph Grammars for Self-assembly of Robotic Modules

In this section, we summarize the graph-grammar formalism adapted for formulating SA of robotic modules [7].

Definition: An extended labeled graph is a quadruple $G = (V, E, S, \ell)$ where $V = \{1, \dots, N\}$ is the set of extended vertices, $E \subset V \times V$ is the set of edges, $S : E \rightarrow K \times K$, with $K = \{1, \dots, N_s\}$ and N_s the total number of link-slots associated with an extended vertex, defines which slots are involved in a link between two vertices. $\ell : V \rightarrow \Sigma$ is a labeling function, with Σ being a set of extended labels. A pair of vertices $\{x, y\} \in E$ is represented by xy. The $n_E(k)$ represents the neighbors of vertex k relative to the edge set E. Two extended graphs are considered to be isomorphic when there exists a bijection $h : V_{G1} \rightarrow V_{G2}$ such that $\forall ij \in E_{G1} \Leftrightarrow h(i)h(j) \in E_{G2}$. The function h is called a witness. A label-preserving isomorphism has the additional property that $\ell_{G1}(x) = \ell_{G2}(h(x)), \forall x \in V_{G1}$.

Definition: An extended vertex has ordered link-slots which correspond to the latching connectors of a robotic module. The numbering on the slots is assumed to match the one of the physical module, following a counter-clockwise (CCW) rotation convention. We assume that the robotic modules have a rotational symmetry. As a result, for an isolated module the connectors are anonymous.

Definition: An extended label is a pair $l = (l_a, l_n)$ encoding the internal state of a module. l_a represents the control state of the robotic module and l_n represents the index of the most recently engaged connector.

Once two modules are latched, each communicates its internal state in the form of a relative extended label of $l = (l_a, l_h)$ with l_a being the module's control state and l_h being a relative hop number representing the relative orientation of the currently engaged connector with respect to its predecessor, assuming a CCW hop convention. For a vertex with an extended label of (l_a, l_n) on a robotic module with N connectors, the communicated l_h is $[(l_n - l_c) mod N] + 1$, where l_n and l_c are the indexes of the most recently and the currently engaged connectors, respectively. For further details on application of extended labels see [7].

Definition: An extended rule is an ordered pair of extended graphs $r = (L, R)$. An extended binary rule can be depicted as $l_1 \quad l_2 \rightharpoonup l_3 - l_4$, with the $l_i = (l_{ia}, l_{ih})$ values being the relative extended label of the engaged vertex i.

Definition: A rule $r = (L, R)$ is applicable to a graph G if there exists $I \subset V_G$ such that the subgraph $G \cap I$ has a label-preserving isomorphism $h : I \rightarrow V_L$.

Definition: The triple (r, I, h) is called an action. Application of an action with $r = (L, R)$ to G gives a new graph $G' = (V_G, E_{G'}, l_{G'})$ defined by

$$E_{G'} = (E_G - xy : xy \in E_G \cap I \times I) \cup (xy : h(x)h(y) \in E_R)$$

$$\ell_{G'}(x) = \begin{cases} \ell_G(x), & \text{if} x \in V_G - I \\ \ell_R(h(x)), & \text{otherwise} \end{cases}$$

Definition: The complement or reverse of a rule $r = (L, R)$, is $\bar{r} = (R, L)$, such that $G \xrightarrow{r,I,h} G' \xrightarrow{\bar{r},I,h} G'' = G$.

Note: When forming an edge, the rule is denoted as a "link rule". When severing an edge, the rule is denoted as an "unlink rule".

Definition: An update rule can be depicted as $l_1 - l_2 \rightharpoonup l_3 - l_4$; it does not create or sever an edge between two vertexes, instead it modifies their labels.

Definition: A trajectory of a system (G_0, ϕ), where G_0 is the initial graph of the system and ϕ is a ruleset, is a sequence of $G_0 \xrightarrow{r_1,I,h} G_1 \xrightarrow{r_2,I,h} G_2 \xrightarrow{r_3,I,h} \ldots$

Given a set of rules ϕ, we can study the sequences of graphs obtained from successive application of the rules contained in ϕ. For a probabilistic ruleset, a probability may be associated with each rule by the mapping $P : \phi \rightarrow (0, 1]$, indicating the tendency for the corresponding event to take place provided that the conditions under which the rule is applicable are met.

4 Proposed Synthesis Algorithm

Given an acyclic target structure composed of rotationally symmetrical robotic modules with any number of connectors, our proposed synthesis algorithm derives rulesets based on two principles: (1) limiting the size of the concurrently built sub-assemblies to a user-defined value, and (2) unifying the rules which give rise to sub-assemblies with similar structures. The algorithm comprises two stages, each realizing one of the two principles. The first stage parses a graphical description of the target, and derives a ruleset which builds the target by merging sub-assemblies with sizes no more than the user-defined value and with distinct labelings, as a result of employing distinct rules. The second stage then processes this ruleset to identify the rules producing structures with identical morphology; such rules are then merged in a single one. As a result of the second principle, i.e. unifying the rules and consequently the labelings, the rulesets need to include update rules. Consider the case where the maximum user-defined size is 2. With the rules unified properly, all dimers (sub-assemblies composed of 2 modules) are labeled similarly. As the dimers join to build the target, the labelings of both consisting modules need to be updated to reflect their placement in the forming target structure to allow for proper further reactions, completing the target eventually. In other words, the use of update rules is an alternative to building the target out of distinctly labeled sub-assemblies, as a result of being formed through distinct link rules, according to their intended placement in the target structure. Thus in general, introducing update rules into the rulesets can reduce the number of link rules necessary to build the target, at the expense of possibly increasing the total ruleset size to include several update rules. However, this can offer a significant advantage in terms of assembly time. The occurrence of the link rules is probabilistic and is determined by the stochastic nature and dynamics of the system which is relied upon to provide proper interactions in order for the SA process to progress. The occurrence rate for link rules is usually in the order of once in tens of seconds. On the other hand, update rules are purely communicational rules and do not depend on the system dynamics. Once a proper interaction has happened and two modules have bonded successfully, the occurrence of a proper update rule is solely determined by the modules communication rate, usually in the order of once in less than a second. Fewer link rules can thus significantly decrease the total assembly time. The first principle addresses the propagation delay concerns which can cause scalability issues. Limiting the size of the concurrently built sub-assemblies allows for restricting the extent by which the update rules need to propagate. Therefore, in a robotic system with measurable propagation delay and interaction rate, the update propagation depth can be set accordingly to allow for a parallel assembly scheme while minimizing possible propagation delay faults. In order to avoid deadlocks, we employ probabilistic dissociating rules. Appropriate reverse rules are generated at the end of Stage II. It should be noted that the algorithm only generates the ruleset. Appropriate probabilities should be assigned to the rules to reliably build the target while avoiding deadlocks.

4.1 Stage I: Grow Subtrees (GS)

Stage I allows for creating concurrently built sub-assemblies similar to the concurrency created by the Linchpin algorithm [3], with the additional capability to control the maximum permitted size of such sub-assemblies. The GS algorithm employed in Stage I tries to build a given target structure using as many sub-assemblies of a defined size as possible built in parallel, before trying to join them to make a bigger sub-assembly and eventually form the target. In principle, the algorithm addresses the second issue with the Linchpin algorithm (see Sect. 1). Linchpin generates rules to build parallel substructures for every branch split in the target recursively in order to build the target using one final finishing rule. With one finishing rule in the ruleset, it is shown in [3] that the target can be built reliably while avoiding deadlocks by having probabilisitic dissociating rules for all rules except for the finishing rule. The GS algorithm on the other hand, permits a maximum size for the concurrently built sub-assemblies and as a result may end up building the target using several concurrent finishing rules. Stage II then processes this ruleset and results in one finishing rule. Algorithm 1 shows the pseudo code of the GS algorithm. The first call to GS is by $Size = 0$. The algorithm then recursively proceeds to create extended labels and corresponding rules, moving outwards from a starting vertex k. The ruleset returned by GS for a chain structure of size 6 and maximum sub-assembly size of 2, is depicted below along with the link rules generated by Linchpin. In order to simplify the comparison, the rulesets have been designed for an abstract graph.

$$\phi_{GS} = \begin{cases} 0\ 0 & \rightarrow \quad 1-2 \quad (r_1) \\ 0\ 0 & \rightarrow \quad 3-4 \quad (r_2) \\ 0\ 0 & \rightarrow \quad 5-6 \quad (r_3) \\ 4\ 5 & \rightarrow \quad 7-8 \quad (r_4) \\ 2\ 3 & \rightarrow \quad 9-10 \quad (r_5) \end{cases}$$

$$\phi_{Linchpin} = \begin{cases} 0\ 0 & \rightarrow \quad 1-2 \quad (r1) \\ 0\ 0 & \rightarrow \quad 3-4 \quad (r2) \\ 0\ 2 & \rightarrow \quad 5-6 \quad (r3) \\ 0\ 4 & \rightarrow \quad 7-8 \quad (r4) \\ 5\ 7 & \rightarrow \quad 9-10 \quad (r5) \end{cases}$$

Consider a set of initially isolated atomic agents, all labeled 0. The rules generated by GS allow for the target to be built in two concurrent steps, i.e. first (r_1, r_2, r_3) and then (r_4, r_5), while the rules synthesized by Linchpin require three concurrent steps, i.e. first (r_1, r_2), then (r_3, r_4), and eventually (r_5).

4.2 Stage II: ReGroup Subtrees (RGS)

Stage II processes the ruleset generated by Stage I to unify the link rules which create up to the maximum size sub-assemblies and add proper update rules. The key idea of processing is to apply the rules synthesized by GS to two graphs with initially fully isolated vertices. The two graphs evolve identically in structure but differ in labeling, one graph is labeled according to the original ruleset, while in the other graph the forming sub-assemblies are processed to identify the rules with products of identical shapes. In order to identify structures with identical shapes the shape recognition algorithm explained in Sect. 5.1.2 is utilized. We omit the pseudo code for the RGS algorithm employed in this stage for brevity. The RGS algorithm can be explained in four phases:

```
C : (V, E, S, L, k, l, Size, S_max)
procedure GS(C)
    φ ← ∅
    φ̄ ← ∅
    Size ← Size + 1
    if |n_E(k)| = 0 then
        return (l, φ)
    else
        {v_j : j = 1, 2, ..., |n_E(k)|} ← n_E(k)
        for j = 1 to |n_E(k)| do
            s_k ← S(v_k, v_j)
            s_j ← S(v_j, v_k)
            l_k ← GVL(L, s_k, v_k)
            if Size < S_max then
                l_j ← GVL(L, s_j, v_j)
                l̄ ← INCREMENTSTATE(l, 1)
                l ← INCREMENTSTATE(l, 2)
                φ̄ ← φ ∪ {l_k l_j ⇌ l̄ − l}
                SVL(L, v_k, s_k, l̄)
                SVL(L, v_j, s_j, l)
                Size_j ← Size
            else
                φ̄ ← ∅
                Size_j ← 0
            end if
            Let (V^j, E^j, S^j) be the
        component of (V, E − {kv_j}) containing v_j
            C : (V^j, E^j, S, L, v_j, l, Size_j, S_max)
            (l, φ_j, Size_j) ← GS(C)
            φ ← φ ∪ φ̄ ∪ φ_j
            if Size == S_max then
                l_j ← GVL(L, s_j, v_j)
                l̄ ← INCREMENTSTATE(l, 1)
                l ← INCREMENTSTATE(l, 2)
                φ ← φ ∪ {l_k l_j ⇌ l̄ − l}
            else
                Size ← Size_j
            end if
        end for
    end if
    return (l, φ, Size)
end procedure

procedure GVL(L, s, v)
    (l_a, l_n) ← L(v)
    l_h ← (l_n − s + 1) (mod N)
    return (l_a, l_h)
end procedure

procedure SVL(L, v, s, l)
    (l_a, l_h) ← l(1 : 2)
    l_n ← s
    L(v) ← (l_a, l_n)
end procedure

procedure INCREMENTSTATE(l, k)
    return (l_a + k, l_n)
end procedure
```

Algorithm 1 Pseudo code of the GS algorithm employed in Stage I

Forming dimers Unify the rules (of Stage I) which form dimers.

Forming larger sub-assemblies Grow on the dimers. Recognize the shape of the resulting sub-assemblies. Unify the rules producing identical structures.

Relabeling max-size sub-assemblies Create update rules for relabeling all the modules in sub-assemblies of up to the max-size (i.e. user-defined value) size.

Growing on max-size sub-assemblies Create necessary rules, both link and update, to form the target assembly out of the max-size sub-assemblies.

Considering the target shape of a chain of size 6 for an abstract graph, the rules generated by RGS are as below:

$$\phi_{RGS} = \begin{cases} 0 \;\; 0 & \rightarrow \quad 1-2 \qquad (r_1) \\ 1 \;\; 2 & \rightarrow \quad 4-3 \qquad (r_2) \\ 1 \;\; 8 & \rightarrow \quad 10-9 \qquad (r_3) \\ 1-3 & \rightarrow \quad 6-5 \qquad (r_4^u) \\ 2-4 & \rightarrow \quad 8-7 \qquad (r_5^u) \\ 7-9 & \rightarrow \quad 12-11 \qquad (r_6^u) \\ 2-10 & \rightarrow \quad 14-13 \qquad (r_7^u) \end{cases}$$

Two points are noteworthy here. First, the existence of the update rules in the ruleset (r_4^u to r_7^u). And second, the number of concurrent steps necessary for forming the target being equal to 3. Assuming that the update events are instantaneous and that the number of available modules is limited, RGS can on average build the target faster than Linchpin with the same number of necessary concurrent steps (see Sect. 6). This is due to the fact that RGS makes a better use of the available modules by limiting the number of distinctly labeled sub-assemblies with identical shapes. At the end of Stage II, the ruleset is augmented with proper reverse rules accounting for both the link and the update rules. The application of a reverse rule essentially takes the SA process back in time by reversing the labeling and/or the bonding.

Proposition 1 *The complete ruleset ϕ_{full} generated by our proposed method for assembling a target structure described as an extended graph $G = (V, E, S, l)$ will eventually achieve the maximum possible number of copies of the target structure (i.e. maximum yield) provided that the available assembly modules executing the ruleset interact often enough and that the corresponding execution probability is set to $p = 1$ for link and update rules and to $p < 1$ for reverse rules.*

Proof The ruleset ϕ_{full} contains an unlink rule for each link and update rule. Only the last link rule has no corresponding reversal rule. Therefore, while all partially formed structures dis-assemble with a non-zero probability, the finishing rule is reversed with zero probability, therefore leading to a stable target structure. □

Proposition 2 *The complete ruleset ϕ_{full} generated by our proposed method for assembling a target structure described as an extended graph $G = (V, E, S, l)$ will achieve an assembly rate at least as fast as that of a ruleset derived by the Linchpin algorithm, assuming that the update rules are applied instantaneously.*

Proof Rulesets generated by Stage I have as many link rules as the ones generated by the Linchpin algorithm, i.e. the number of edges in the target graph, as a result of forming the target out of uniquely labeled sub-assemblies. As a result of merging the link rules in Stage II, ϕ_{full} contains at most as many link rules as the ruleset ϕ_{GS} created in Stage I. Therefore, ϕ_{full} achieves the target structure in the same or fewer concurrent steps than rulesets derived by Linchpin. □

4.3 Synthesized Rulesets for Lily Robots

We consider two targets, a chain and a cross structure, each composed of six Lily robots (see Fig. 3). The rulesets returned by our algorithm (with the maximum user-defined size set to 2) for the chain structure ϕ_{-}, and for the cross structure ϕ_{+}, are reported below. The (l_a, l_h) notation is used for the relative extended labels and the reverse rules are separated. Note that the reverse rules do not correspond to a single link or update rule, but rather have a time reversal effect, taking the labeling back in time. For the sake of brevity we skip functional explanation of these rulesets, further details can be found in [7].

$$\phi_- = \begin{cases} (0,0) \quad (0,0) & \xrightarrow{r_1} \quad (1,1)-(2,1) \\ (1,3) \quad (2,3) & \xrightarrow{r_2} \quad (4,1)-(3,1) \\ (1,3) \quad (8,3) & \xrightarrow{r_3} \quad (10,1)-(9,1) \\ (1,1)-(3,3) & \xrightarrow{r_4^u} \quad (6,1)-(5,1) \\ (2,1)-(4,3) & \xrightarrow{r_5^u} \quad (8,1)-(7,1) \\ (7,1)-(9,3) & \xrightarrow{r_6^u} \quad (12,1)-(11,1) \\ (2,1)-(10,3) & \xrightarrow{r_7^u} \quad (14,1)-(13,1) \\ (1,1)-(2,1) & \xrightarrow{\bar{r}_1} \quad (0,0) \quad (0,0) \\ (5,3)-(7,3) & \xrightarrow{\bar{r}_{2/4/5}} \quad (3,1) \quad (4,1) \\ (3,3)-(6,1) & \xrightarrow{\bar{r}_{2/4}} \quad (2,1)-(1,1) \\ (4,3)-(8,1) & \xrightarrow{\bar{r}_{2/5}} \quad (1,1)-(2,1) \end{cases} \qquad \phi_+ = \begin{cases} (0,0) \quad (0,0) & \xrightarrow{r_1} \quad (1,1)-(2,1) \\ (0,0) \quad (2,4) & \xrightarrow{r_2} \quad (4,1)-(3,1) \\ (0,0) \quad (5,2) & \xrightarrow{r_3} \quad (8,1)-(7,1) \\ (2,3) \quad (7,2) & \xrightarrow{r_4} \quad (10,1)-(9,1) \\ (1,1)-(3,2) & \xrightarrow{r_5^u} \quad (6,1)-(5,1) \\ (1,1)-(10,3) & \xrightarrow{r_6^u} \quad (12,1)-(11,1) \\ (1,1)-(2,1) & \xrightarrow{\bar{r}_1} \quad (0,0) \quad (0,0) \\ (4,1)-(5,4) & \xrightarrow{\bar{r}_2} \quad (0,0) \quad (3,1) \\ (6,1)-(3,2) & \xrightarrow{\bar{r}_{2/5}} \quad (1,1)-(2,1) \\ (7,1)-(8,1) & \xrightarrow{\bar{r}_3} \quad (5,3) \quad (0,0) \end{cases}$$

5 Modeling Levels and Simulation Frameworks

In order to compare the performance of our rulesets for SA of robots and to study the transient behavior, we conduct simulated studies using two different simulation frameworks. In the non-spatial microscopic framework implemented in Matlab, the system is modeled as an extended graph (see Sect. 3), with each extended vertex in the extended graph corresponding to a robotic module. In the high-fidelity sub-microscopic simulation framework realized in Webots [11], the physics of the system including the embodiment of the robotic modules, the hydrodynamic forces acting on the robots, and the software running on them are recreated.

5.1 Microscopic Model and Simulation Framework

In this framework, the physical dynamics of our robotic system is abstracted into randomized interactions between atomic units in favor of gaining simulation speed. The system is represented as an extended graph which evolves over time. Each robotic modules corresponds to an extended vertex in the system graph. In order to model interactions between the robotic modules a randomized scheme along with appropriate geometrical constraints is utilized. In order to track the progress of the SA process in the system, we employ a shape recognition algorithm which is an extension over a graph isomorphism check.

5.1.1 Random Pairwise Interactions

In our extended formalism, a random pairwise interaction dynamics is defined as a quadruple (G, F, ϕ, P). Rule probabilities are assigned by $P : \phi \rightarrow (0, 1]$. The set of pairs of disjoint vertices is defined as $PW(G) = \{(x, y) : \nexists I \subset G|(x, y) \in V_I, x \neq y\}$, where I is a connected subgraph of G. The set $PW(G)$ defines modules among which an interaction is feasible as they are not on the same sub-assembly. $F(G)$ maps an extended graph G to probabilities of pairwise vertex selections from V_G. A random trajectory of the system, is generated by sampling $F(G_t)$ at each time instant to obtain a pair (x, y) and then executing an appropriate action on the selected pair. For two selected vertices to interact, the link-slots are chosen randomly from the available slots. Sampling from $F(G_t)$ introduces an inherent stochasticity to the trajectories even if the ruleset contains only deterministic rules. The interaction probabilities, defined by $F(G_t)$, depend on the current graph G_t and can be calibrated.

5.1.2 Shape Recognition

Tracking the progress of the SA process of the simulated system requires a mapping between the connected components of the graph of the system and the shape of the corresponding sub-assemblies. For the case of SA of graphs where the system is represented by an abstract graph at each time instant, this describes a problem of graph isomorphism. However, for the case of our extended graphs, the relative position of the engaged slots need to be taken into account to recognize the shapes. We employ a simple method for recognizing the shapes based on traversing the connected components of the extended graph and constructing a series of locations of the Center Of Mass (COM) of the robotic modules. The relative ordering of the slots of neighboring modules determines the orientation of each traverse. The series of locations are then rotated and translated such that all coordinates are positive. The resulting ordered set is used as the identifier of the structure. This method can be applied to modules with a variety of shapes forming structures in 2D.

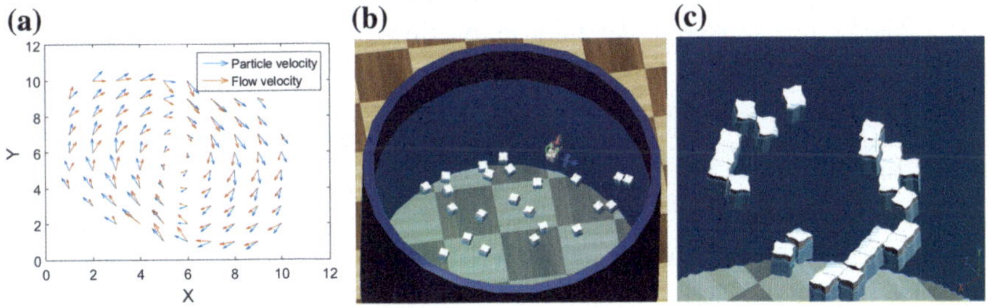

Fig. 2 **a** Velocity field of the tracked globe and the computed fluid flow. **b** and **c** Simulated world of Lily robots in Webots

5.2 Sub-Microscopic Model and Simulation Framework

In this context, submicroscopic means that it provides a higher level of detail than a canonical microscopic model, faithfully reproducing intra-robot details (e.g., body shape, individual sensors and actuators). In order to realistically recreate our self-assembling system in simulation, we use Webots [11], a physics-based robotics simulator. Webots uses the Open Dynamics Engine (ODE) for simulating rigid body dynamics. Additionally, in order to simulate specific not natively supported physics, it is possible to employ custom designed physics plugins. The Lily modules' CAD design as well as the robots' controller software is imported into the Webots simulated world. The rulesets programmed on the simulated robots are also identical to the case of the microscopic simulation in the previous section. The latest version of Webots supports a basic fluid node which allows for a simple uniform stream velocity, but is not capable of simulating a complex fluidic field. We used a similar approach as [2] to reproduce the complex flow field and the hydrodynamic forces.

We record the trajectory of a single floating globe (diameter of 3 cm), roughly the same size of a Lily robot, for 3 experiments with random starting positions and duration of 10 min each. The globe's weight is tuned such that the submersion level is similar to that of a Lily robot (25 mm below water level). The captured velocity fields are then augmented and discretized on a regular grid of 50 cells on each side, for our water tank of 120 cm in diameter. For each cell of the grid, the observed velocity vectors are averaged and assigned as the velocity of that cell. The fluid velocity field can be computed considering the drag force. The value of the Reynolds number Re determines the flow regime and the form of the drag force:

$$Re = \frac{\rho V L}{\mu} \simeq 2000 \qquad |\overrightarrow{F}_{drag}| = \frac{1}{2}\rho A C |\overrightarrow{v}_{block} - \overrightarrow{v}_{flow}|^2 \tag{1}$$

where $\rho = 10^3\,\text{kg/m}^3$ is the density of water, $V \simeq 20\,\text{cm/s}$ the experimentally-measured mean velocity of the globe, L = 3 cm the characteristic dimension, and $\mu = 8.90\ .\ 10^{-4}$ Pa.s the dynamic viscosity of water. The submerged area of the globe is, $A = 7\,\text{cm}^2$ and the drag coefficient constant in all directions $C = 0.47$. The velocity and acceleration of the globe are computed using the captured trajectory data. Considering the mass of the globe m, the flow velocity is then computed as:

$$\overrightarrow{v}_{flow} = \overrightarrow{v}_{block} + \frac{m\overrightarrow{a}}{\sqrt{\frac{1}{2}\rho A C m \sqrt{a_x^2 + a_y^2}}} \tag{2}$$

We developed a physics plugin for Webots that applies the drag force to the simulated Lily module based on the velocity of the robot and the flow velocity at its location at each time instant. In order to account for rotational effects, the drag force is integrated over each face of the Robot. Each face is divided into $N = 10$ sections, and the drag force is computed for each section using Eq. 1 with C being the estimated

Lily robot's drag coefficient C_{Lily}. The physics plugin also adds a stochastic force $F_s \sim \mathcal{N}(0, \sigma_f^2)$ to the center of mass of each robotic module in order to account for non-modeled effects (e.g., physical irregularities, turbulences). The values of $C_{Lily} = 0.7$ and $\sigma_f = 50$ mN were empirically set based on our previous findings in [2]. Two independent Kolmogrov-Smirnov (KS) tests showed that the simulated and real distributions of the step lengths and step angles extracted from the trajectories have a KS distance of less than 0.2.

6 Experiments and Results

We evaluate our algorithm leveraging the two modeling levels and corresponding simulation frameworks of Sect. 5, studying the SA process in a swarm of 24 initially isolated Lily robots. Two target shapes of a chain and a cross shape, each composed of 6 robotic modules, are considered. A maximum of 4 copies of each target can be assembled thus. The microscopic simulation framework (see Sect. 5.1) employs a random pairwise interaction dynamics. All interactions among microscopic nodes are set to be equiprobable, i.e. we make the assumption that the system is perfectly mixed. We employ rulesets synthesized by our algorithm (see Sect. 4.3) and the extended Linchpin algorithm [7], for our simulated Lily robots. For forward and update rules $P(.) = 1$ and for reverse rules $P(.) = 0.01$ is set. The finishing rule is set to be irreversible in all the rulesets, giving rise to stable target assemblies once they are formed. Figure 3 depicts the performance of the rulesets derived by our algorithm along with the ones of the extended Linchpin for the two target shapes. While for the submicroscopic simulations the results are reported as a function of the experimental time (emulating the real time progress in a real experiment), the results of the microscopic simulations are reported as a function of steps, each step representing a formation event in the system. While such choice makes the results of the two modeling levels not directly comparable, the adopted progress unit is well suited for measuring the concurrency of the rulesets. It can be seen that our proposed algorithm achieves higher assembly rates in all cases. Interestingly, the maximum yield of 4 is not obtained in the case of the submicroscopic simulations within the 1 hour simulated time. This can be ascribed to several reasons. First, it is observed in the submicroscopic simulation that larger structures with several corners trap other sub-assemblies and stall the SA process. In addition, as the structures grow the conditions diverge from perfect mixing since the shape of the sub-assemblies affects their orientation in the fluidic field and certain interactions tend to be less probable.

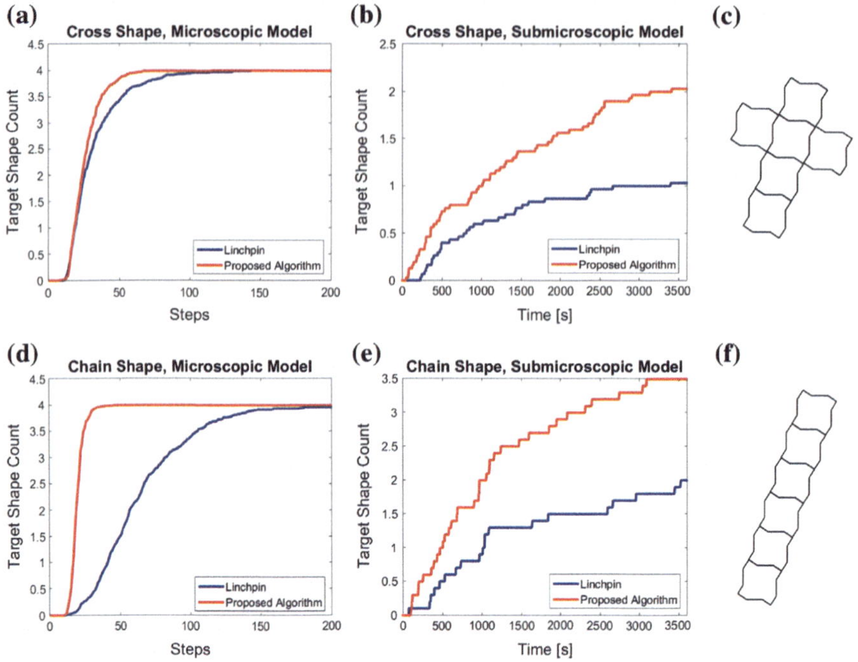

Fig. 3 Results of microscopic (**a** and **d**, 100 runs averaged) and submicroscopic models (**b** and **e**, 30 runs averaged) for two target shapes, cross **c** and chain **f**

7 Conclusion

In this paper, we addressed the problem of synthesizing parallel rulesets for programmable SA of robotic modules. We employed an extended graph-grammar formalism to account for the morphology of the modules and proposed a formal synthesis algorithm to automatically synthesize rules. Using our new algorithm, we synthesized rulesets for two target structures. Studies on the synthesized rulesets in simulation, using both non-spatial microscopic and spatial submicroscopic models, demonstrated the superior performance of our algorithm compared to the Linchpin algorithm [3], appropriately extended to be deployed on our robotic modules [7]. This evidenced the functionality of update rules in increasing the concurrency in the SA process resulting in higher assembly rates. In the future, we plan to conduct studies in simulation and on real robots to investigate the effects and mitigation strategies of propagation delays on the performance of the rulesets generated by our algorithm. Additionally, we plan to fully utilize our real experimental setup to conduct systematic real experiments involving up to 50 Lily robots.

Acknowledgements We gratefully acknowledge the contributions of Loic Waegeli and Brice Platerrier to the microscopic simulation framework. This work has been sponsored by the Swiss National Science Foundation under the grant numbers 200021_137838/1 and 200020_157191/1.

References

1. Cademartiri, L., Bishop, K.J.: Programmable self-assembly. Nat. mater. **14**(1), 2–9 (2015)
2. Di Mario, E., Mermoud, G., Mastrangeli, M., Martinoli, A.: A trajectory-based calibration method for stochastic motion models. In: IEEE/RSJ International Conference on Intelligent Robots and Systems, pp. 4341–4347 (2011)
3. Fox, M., Shamma, J.: Probabilistic performance guarantees for distributed self-assembly. IEEE Trans. Autom. Control **60**(12), 3180–3194 (2015)
4. Gilpin, K., Rus, D.: Modular robot systems. IEEE Robot. Autom. Mag. **17**(3), 38–55 (2010)
5. Haghighat, B., Droz, E., Martinoli, A.: Lily: A miniature floating robotic platform for programmable stochastic self-assembly. In: IEEE International Conference on Robotics and Automation, pp. 1941–1948 (2015)
6. Haghighat, B., Martinoli, A.: Characterization and validation of a novel robotic system for fluid-mediated programmable stochastic self-assembly. To appear in IEEE/RSJ International Conference on Intelligent Robots and Systems (2016)
7. Haghighat, B., Platerrier, B., Waegeli, L., Martinoli, A.: Synthesizing rulesets for programmable robotic self-assembly: A case study using floating miniaturized robots. In: International Conference on Swarm Intelligence (ANTS), pp. 197–209 (2016)
8. Klavins, E.: Automatic synthesis of controllers for distributed assembly and formation forming. In: IEEE International Conference on Robotics and Automation, pp. 3296–3302 (2002)
9. Klavins, E.: Programmable self-assembly. IEEE Control Syst. **27**(4), 43–56 (2007)
10. Klavins, E., Ghrist, R., Lipsky, D.: A grammatical approach to self-organizing robotic systems. IEEE Trans. Autom. Control **51**(6), 949–962 (2006)
11. Michel, O.: Webots: professional mobile robot simulation. Adv. Robot. Syst. **1**(1), 39–42 (2004)
12. Rai, V., Van Rossum, A., Correll, N.: Self-assembly of modular robots from finite number of modules using graph grammars. In: 2011 IEEE/RSJ International Conference on Intelligent Robots and Systems, pp. 4783–4789. IEEE (2011)
13. Rubenstein, M., Cornejo, A., Nagpal, R.: Programmable self-assembly in a thousand-robot swarm. Science **345**(6198), 795–799 (2014)

Distributed Adaptive Locomotion Learning in ModRED Modular Self-reconfigurable Robot

Ayan Dutta, Prithviraj Dasgupta and Carl Nelson

Abstract We study the problem of adaptive locomotion learning for modular self-reconfigurable robots (MSRs). MSRs are mostly used in unknown and difficult-to-navigate environments where they can take a completely new shape to accomplish the current task at hand. Therefore it is almost impossible to develop the control sequences for all possible configurations with varying shape and size. The modules have to learn and adapt their locomotion in dynamic time to be more robust in nature. In this paper, we propose a Q-learning based locomotion adaptation strategy which balances exploration versus exploitation in a more sophisticated fashion. We have applied our proposed strategy mainly on the ModRED modular robot within the Webots simulator environment. To show the generalizability of our approach, we have also applied it on a Yamor modular robot. Experimental results show that our proposed technique outperforms a random locomotion strategy and it is able to adapt to module failures.

1 Introduction

Modular self-reconfigurable robots (MSRs) are composed of multiple autonomous modules which can change their connections with each other to form different configurations [20]. This enables the robot to navigate through different types of terrains and environments where it is usually difficult to do so. Therefore MSRs can be

A. Dutta (✉) · P. Dasgupta
Computer Science Department, University of Nebraska Omaha, Omaha, NE, USA
e-mail: adutta@unomaha.edu

P. Dasgupta
e-mail: pdasgupta@unomaha.edu

C. Nelson
Mechanical and Materials Engineering Department, University of Nebraska Lincoln, Lincoln, NE, USA
e-mail: cnelson5@unl.edu

R. Groß et al. (eds.), *Distributed Autonomous Robotic Systems*,
Springer Proceedings in Advanced Robotics 6,
https://doi.org/10.1007/978-3-319-73008-0_24

used for unmanned extra-terrestrial exploration, information collection in nuclear radiation plants or for search-and-rescue tasks [20].

In our recent work [9], we have proposed a solution for configuration formation from singleton modules. After the configuration is formed, the newly built configuration needs to move to different locations to complete the tasks at hand, such as exploration or information collection. If both the size (number of modules) and the topology of the configuration are known, then coordinated locomotion for all the member modules can be planned *a priori* in a deterministic manner [15]. But the difficulty arises when either the size or the shape, or both, is unknown. In that case, determining the control sequences *a priori* for locomotion is impossible. Finding the best set of actions for a particular configuration has exponential complexity on the number of modules in the configuration $O(|A|^{|M|})$ where A and M are the set of actions available to each module and the set of modules present in the configuration [5]. Moreover, if one or some of the modules in the configuration become faulty during the operation, then finding the best set of actions for the other operational modules is an important task.

To solve this problem, we propose a Q-learning-based adaptive locomotion strategy which learns the best action for each module. We have observed that each module's locomotion performance (e.g., distance traveled) is highly dependent not only on the action that particular module is taking, but also on the actions taken by its neighboring modules (i.e., the modules directly connected to it). This observation led us to build our learning strategy in such a way that it does not only learn from its own previous actions, but it also learns from the correlation of that past action with the neighboring modules' actions at that particular time-step.

2 Related Work

Although locomotion planning in modular robotic systems has been studied in literature extensively, it has still remained a big challenge for MSR researchers to find a solution which quickly adapts to the shape of the current configuration of the MSR and consequently, the configuration learns to move within a short amount of time [1]. Most of the research in MSR locomotion tried to develop pre-defined locomotion plans for any particular configuration, one of the first of which is due to [19]. Many of these works on locomotion planning in MSRs have proposed solutions based on gait tables. Gaits are synchronized patterns of locomotion used by animals, humans and machines such as robots [14]. Hand-coded locomotion patterns using gait control tables for the ModRED MSR have been proposed in [6]. Though popular, this approach can be mostly useful for chain type modular robotic systems [5]. In [5], the authors have proposed a novel reinforcement learning based technique for gait adaptation for locomotion learning in MSRs. Central pattern generators (CPGs) have also been studied for locomotion planning in MSRs [11]. The goal of the CPG is to generate synchronized oscillations (locomotion patterns/rhythms) in connected oscillators (modules) [17]. Locomotion in MSRs using CPGs has been first studied by Kamimura et al. for their M-TRAN MSR [12]. In [17], the authors have proposed a CPG-based approach for locomotion learning in Yamor, a chain-type modular robotic system.

Synchronization plays a very important role in locomotion planning. If the movements of multiple modules are not synchronized, then the locomotion is not synchronized and therefore the configuration will not move towards the correct direction and/or the speed of the configuration will be very slow [15]. Several different approaches for maintaining synchronization among modules in a configuration have been proposed. A biology-inspired hormone based approach has been proposed in [15] where modules pass hormones (information-coded) among themselves in the configuration and can detect any change in the topology of the current configuration; this is also used for synchronization of their actions in a distributed manner. In a leader-follower approach, every module detects its local leader module and coordinates its actions with that leader module only, and thus the synchronized behavior flows through the configuration [16]. On the other hand, in a master-slave approach, a central leader is elected for the whole configuration (called the master) and control is passed to all other slave modules [4].

Our proposed approach for locomotion learning is inspired by the online locomotion adaptation works [5, 7, 10] in which the modules learn to adapt the locomotion for the current configuration on the fly. We also employ a leader-follower-like distributed synchronization method which is more robust against failures than the master-slave approach.

3 Problem Setup

Let $M = \{m_1, m_2, \ldots, m_N\}$ denote the set of N modules connected together forming a certain configuration. The configuration is connected, i.e., any two modules in the configuration are connected either physically or through other modules. $neigh(m_i)$ denote the set of neighboring modules, i.e., the modules which are physically connected to m_i. Each module has a unique identification (ID). Module m_i's position and orientation are denoted by (x_i, y_i, θ_i). We assume that each module knows the topology of the configuration [3]. We also assume that each module is able to calculate its own position using a GPS or an overhead tracking system.

Each module performs an action by actuating its motors. An action a_j is a vector which specifies the actuation provided to each of the K motors present in the module, i.e., $a_j = \{ac_1, ac_2, \ldots, ac_K\}$. Each module is provided a library of actions, A, from which it can choose its action at any given time-step. The actions present in the library are known beforehand and given as an input by the user. Each module, m_i, receives a reward by performing an action a_j in a specific time-step t, denoted by $R_i(a_j, t)$. Reward can be calculated as the Euclidean distance traveled by a module since its last time step, given by:

$$R_i(a_j, t) = ||p_i(t) - p_i(t-1)|| \tag{1}$$

where $p_i(t)$ is the position of the module m_i at time-step t. Let a_i^{best} denote the best (highest reward earning) action. We have modeled the learning strategy as a stateless Q-learning approach [5, 13, 18]. Let $Q(a_j)$ denote the Q-value of an action a_j. The Q-value provides an estimate of the usefulness of executing any action in the next iteration, and this value is updated after each learning cycle according to the reward received for the action. Also, let ε denote the ratio between exploration of new actions and exploitation of past high reward-earning actions.

4 Q-Learning Based Approach for Distributed Locomotion Learning

In our proposed approach for locomotion learning, we try to capture the following idea: each module not only learns from its own current and past actions, but also from the relationship among the actions performed by its neighboring modules at any particular time-step. The pseudo-code of our proposed approach is shown in Algorithm 1.

Algorithm 1: Q-Learning Based Distributed Locomotion Learning Algorithm

Perform every action $a_j \in A$ in sequential order.
Receive corresponding rewards $R_i(a_j, t)$.
$Q(a_j) \leftarrow R_i(a_j, t), \forall a_j \in A$.
$A_{DS} \leftarrow$ Data structure for storing best action-pair.
Loop
  $a_i^{best} \leftarrow \arg\max_{a_k \in A} Q(a_k)$.
  With ε probability, $a_{nxt} \leftarrow a_i^{best}$.
  With $(1-\varepsilon)$ probability, $a_{nxt} \leftarrow a_{rand}$.
  $a_{pres} \leftarrow$ Prescribed best action for a_{nxt} to be send to neighbor modules.
  Send $\{a_{nxt}, a_{pres}, R_i(a_{nxt}, t)\}$ to the neighboring modules.
  Receive similar message from the neighbor modules: $\{a'_{nxt}, a'_{pres}, R_j(a_{nxt}, t)'\}$.
  Let a''_{pres} be the already prescribed action stored with m_i in A_{DS} for a'_{nxt} with corresponding reward $R_i(a'_{nxt}, t)''$.
  if $R_j(a_{nxt}, t)' > R_i(a_{nxt}, t)''$ **then**
    With τ probability, $a_{nxt} \leftarrow a'_{pres}$./*switch to prescribed action*/
  Perform the action a_{nxt} and receive reward $R_i(a_{nxt}, t)$.
  Update Q-value: $Q(a_{nxt}) = Q(a_{nxt}) + \alpha \cdot (R_i(a_{nxt}, t) - Q(a_{nxt}))$.
  Update a_i^{best} and $\{a'_{nxt}, a''_{pres}, R_i(a'_{nxt}, t)''\}$ if necessary.

First each module, $m_i \in M$, performs every action, a_j, available in the action library, A, in a sequential order and calculates the rewards, $R_i(a_j, t)$, for all the actions. Q-values of all the actions are also initialized to these reward amounts, i.e., $Q(a_j) = R_i(a_j, t)$. Next, the modules follow the modified Q-learning strategy as shown in Algorithm 1. In the beginning of every learning iteration, each module finds out which action has maximum Q-value associated with it so far, i.e., the best action. This information is local to every module, i.e., each module has its local best action, a_i^{best}, calculated by the following equation: $a_i^{best} \leftarrow \arg\max_{a_k \in A} Q(a_k)$. We have observed that the behavior of the modules are highly coupled and if the actions taken by them are not synchronized in any way, then the modules take a longer time to reach the final learned behavior [1]. To alleviate this problem, each module communicates its next action information with its neighboring modules before performing any action. Each module chooses its best action seen so far a_i^{best}, to execute in the next learning iteration with ε probability or a random action a_{rand} with $(1-\varepsilon)$ probability. Each module sends this calculated next action for the next iteration, a_{nxt}, to the neighboring modules. Similarly, it receives the next actions of its neighboring modules for the next iteration.

Each module m_i maintains an action-pair data-structure A_{DS} which can be imagined as a $|A| \times 3$ matrix. Each row in A_{DS} contains an action $a_k \in A$ (first column), a prescribed best (highest reward earning) action $a_{pres} \in A$ for a_k (second column) and corresponding reward received by m_i by performing a_{pres}: $R_i(a_{pres}, t)$ (third column). $a_k \in A_{DS}$ represents the action performed by a neighboring module, $m_j \in neigh(m_i)$, at any particular time-step t. a_{pres} represents the corresponding action from the action set performed by m_i at that time-step t and which earned m_i a reward of $R_i(a_{pres}, t)$. For any module m_i, A_{DS} contains only one copy of each action $a_k \in A$ in its first column, even if multiple neighboring modules might have performed that action. Only the corresponding best action performed by m_i at those time-steps and the reward earned by it (column 2 and 3) change over time. Initially all the prescribed actions in A_{DS} and the corresponding rewards earned for those actions are unknown and therefore initialized to a null value. Over time, when the modules start communicating their current actions, this data structure gets populated accordingly. While communicating, each module not only sends its next action, a_{nxt}, but also the information about the corresponding prescribed best action available with it, a_{pres}, for its neighboring modules and the reward received for that action pair, $R_i(a_{nxt}, t)$.

Once the module receives this information from its neighboring modules, it checks whether any of the neighboring modules has any prescribed best action for that current learning iteration or not. Only if the prescribed action received from the neighboring module, a_{pres}, has been shown to earn higher reward than the prescribed action already available with the module, then the module selects a_{pres} as its next action with τ probability or keeps a_{nxt} to be its next action with $(1-\tau)$ probability. After the next action is decided, the module performs that action and receives reward for that action. Following [13, 18], each module then updates the Q-value of the action performed following the update equation:

$$Q(a_{nxt}) = Q(a_{nxt}) + \alpha \cdot (R_i(a_{nxt}, t) - Q(a_{nxt})) \tag{2}$$

where $\alpha \in [0, 1]$ is the learning rate. Also, it updates the best action and action-pair data structure A_{DS} as necessary.

Table 1 **a** Simulated ModRED Modules within Webots Robot Simulator and **b** Hardware of ModRED

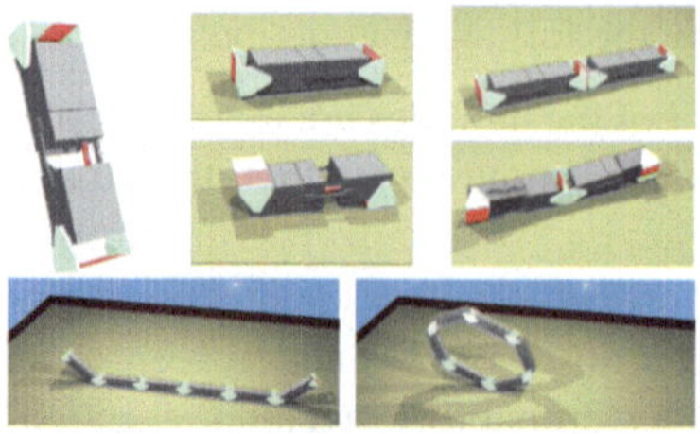

(a)

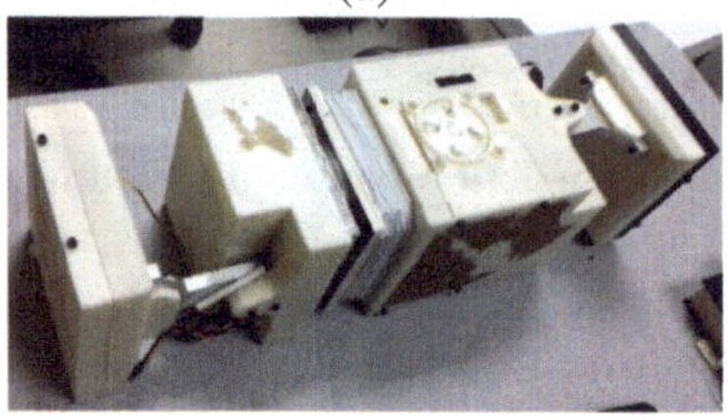

(b)

5 Experimental Evaluation

5.1 Settings

We have mainly implemented our proposed adaptive locomotion learning strategy on simulated ModRED modules within the Webots robot simulator. Each ModRED module is a 4-DOF robot with connectors in both ends. For more details on ModRED hardware architecture and features, readers are referred to [2]. Simulated ModRED modules within the Webots simulator and actual ModRED hardware are shown in Table 1. We have tested our approach on ModRED chain configurations having 2, 3, 4, 5 modules (denoted by M2, M3, M4, and M5 respectively).[1]

ModRED actions: Each ModRED module can have 54 unique actions [6]. In this paper, we have shortlisted 10 of them for inchworm locomotion and 5 actions for rolling motion which have been used for our testing purpose. These actions have also been used in [6] for creating hand-coded gaits. More actions can always be used for more robust locomotion behavior, but at the same time it will slow down the learning process exponentially.

[1]Because each module has 4 DOF, testing with ModRED modules becomes computationally intensive with more than 5 modules. Testing with larger configurations is reported for a 1-DOF robot called Yamor.

Table 2 Actions available to single ModRED module for inchworm locomotion

	FC	T	R	RC
Action 1	0	−1	0	0
Action 2	−1	−1	0	1
Action 3	−1	1	0	1
Action 4	1	−1	0	−1
Action 5	−1	−1	0	0
Action 6	0	1	0	−1
Action 7	1	1	0	1
Action 8	−1	−1	0	−1
Action 9	0	−1	0	−1
Action 10	−1	1	0	0

Table 3 Actions available to single ModRED module for rolling locomotion

	FC	T	R	RC
Action 1	0	−1	0	0
Action 2	−1	−1	0	1
Action 3	−1	−1	1	1
Action 4	1	−1	0	−1
Action 5	1	−1	1	−1

In [6], the authors have described hand-coded gait tables for ModRED's locomotion (inchworm and rolling). As described earlier, ModRED has 4 degrees of freedom and consequently 4 motors. Tables 2 and 3 summarize the action set used for ModRED modules for inchworm and rolling locomotion. Two end connectors (front (FC) and rear (RC) connector) can go up (0.7 rad), down (−0.7 rad) or stay in neutral (0 rad) positions represented as +1, −1 and 0 respectively in the action table. Similarly, the translational motor (T) of ModRED helps to extend or contract the body of the module and is represented by +1 and −1 in the gait tables. The rotational degree of freedom (R) of the module either rotates the module in clockwise (+1: 6.28 rad) or anti-clockwise (−1: −6.28 rad) directions or just stays neutral (0: 0 rad). For inchworm locomotion, the rotational motor is always inactive (0 throughout the column). After one action is executed, the module calculates its local reward for that particular performed action using Eq. 1.

We have also tested our approach on Yamor modular robots, which is a 1-DOF robot [8]. We have used the sample of Yamor's simulated model provided in Webots. For more details on Yamor hardware, readers are referred to [8]. We have tested on 10, 12 and 14 Yamor chain configurations (denoted by Y10, Y12, and Y14 respectively). Each Yamor module has 3 available actions: go up (+0.5 rad), go down (−1.57 rad) or stay neutral (0 rad).

We have compared our proposed approach against a random locomotion approach. Modules select a random action in every iteration and execute that. Random action

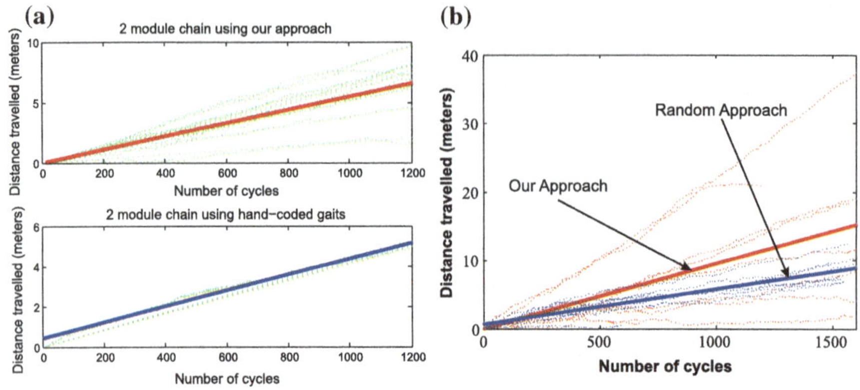

Fig. 1 Performance comparison of our proposed approach when applied on ModRED configurations: **a** 2-module chain and against hand-coded gaits **b** 5-module chain and against the random approach

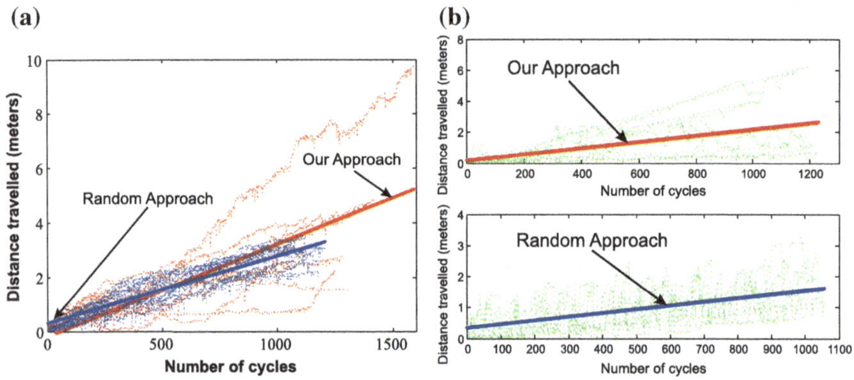

Fig. 2 Performance comparison of our proposed approach when applied on ModRED configurations: **a** 4-module and **b** 5-module chains against the random approach

strategies have been shown to provide steady performance in ATRON and M-TRAN robots [5]. The main performance metrics shown here are: distance traveled from the start point by the front module, maximum speed achieved by the configuration and number of messages passed between modules. The faint lines in the plots (Figs. 1, 2, 3 and 7) denote multiple runs and the bold red/blue lines indicate the best-fit line. Three variables in the Q-learning approach, α, ε and τ, have been set to 0.1, 0.8 and 0.9 respectively for all the tests (these values were determined experimentally). In our tests, each configuration runs Algorithm 1 for learning different types of locomotion (such as inchworm and rolling motions). Each test has been run for 30 min and 10 times each. Videos can be found here: https://youtu.be/8YiAj5xF8ag and https://youtu.be/zjKkNW_r0ZI.

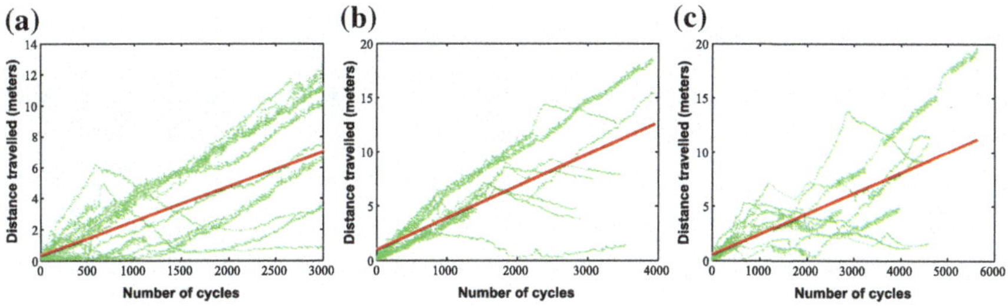

Fig. 3 Performance of our proposed approach when applied on Yamor configurations: **a** 10-module, **b** 12-module and **c** 14-module chains

5.2 Results

Inchworm Locomotion: As the hand-coded gait tables for the 2-module ModRED chain configuration are already available [6], we first compare the performance (distance metric) of our proposed algorithm applied on the 2-module ModRED chain configuration against the same using the hand-coded gaits. The result is shown in Fig. 1a. This result shows that using our approach, the ModRED configuration was able to move further than by using hand-coded gaits. The reason for this is the hand-coded gaits were built in such a way that no part of the chain is dragged along the ground, but our proposed approach learns from all available actions - does not necessarily restrict some modules being dragged. That is why our approach performed better.

Next we compare our algorithm's performance (distance metric) on M3, M4 and M5 against the random approach (Figs. 1b and 2). Although in almost all of the cases, the random approach initially performed better than our approach, but over time our algorithm was able to perform better than the random approach. In some of the plots one might notice that the distance metric (y-axis) decreases sometimes. The reason for this is we have calculated the distance from the start point - not the total distance traveled. The reward is based on distance traveled (regardless of the direction). In some cases, after traveling towards a certain direction for some time, the configurations started to move towards the start direction again. This behavior of the configurations caused the dips in distance metric in some of the plots.

Next we show the result of applying our proposed approach for locomotion learning in the Yamor modular robot [8]. As Yamor modules have fewer DOFs and each Yamor module can perform a very limited set of actions (only 3), we could test on longer Yamor chain configurations than we could do with ModRED within Webots. The performance (distance metric) of our proposed approach when applied on different Yamor configurations has been shown in Fig. 3. The results show that Yamor configurations could travel longer distances than ModRED configurations. We believe the main reason for this is the smaller size and lighter Yamor modules compared to ModRED modules. Also due to larger chain sizes, we have noticed some abrupt

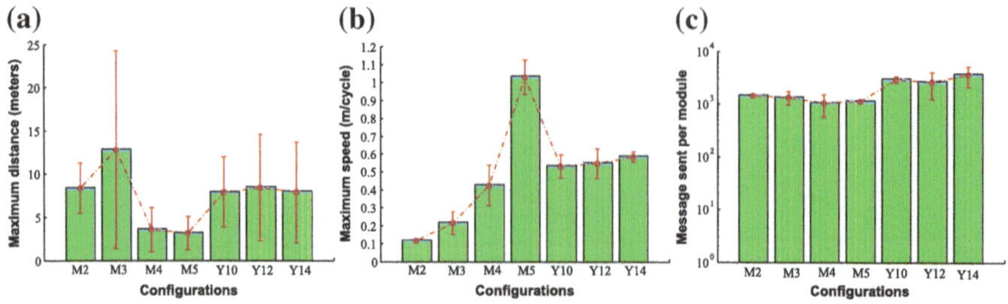

Fig. 4 **a** Maximum distance traveled in any direction from the start location by different configurations, **b** Maximum speed achieved by different configurations and **c** Average number of messages sent by each module in different configurations

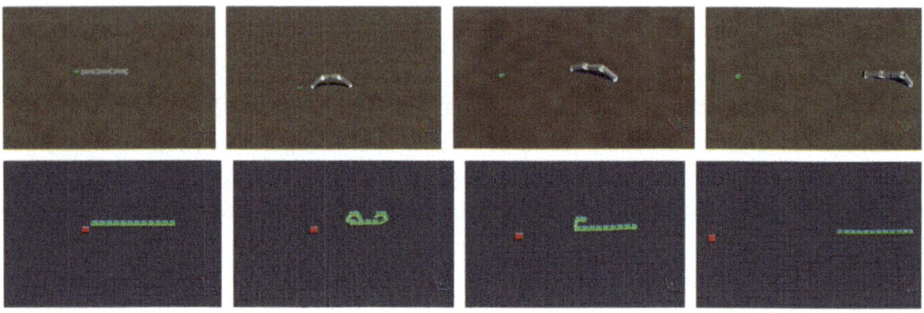

Fig. 5 Snapshot of inchworm locomotion performed by (top) ModRED and (bottom) Yamor configurations using our proposed approach

jumps during locomotion which helped the configurations to travel a longer distance in one learning cycle.

Next we summarize the results for maximum distance traveled in any direction from the start location by different configurations (Fig. 4a), maximum speed achieved (Fig. 4b) and the number of messages sent by each module in different configurations (Fig. 4c). As can be seen in Fig. 4a, maximum distance has been traveled by M3, but at the same time we can also observe a very high variance in that performance. As we have run each test for 30 min, in the case of (larger) ModRED configurations, M4 and M5, due to the slow simulation, we notice a low overall distance traveled. Although these configurations achieved the maximum speeds among all ModRED configurations (Fig. 4b). In terms of maximum achieved speed, the Yamor configurations performed very similarly, though we can notice a slight increase in maximum speed in Y14. This makes us believe that the longer chain sizes and corresponding (rare) abrupt jumps during inchworm locomotion were the reasons for larger configurations achieving higher maximum speed in both ModRED and Yamor. Figure 5 shows the snapshots of inchworm locomotion in ModRED and Yamor configurations (M3 and Y12 respectively). Green and red marks denote the start location in those pictures.

Fault Adaptation: We are also interested in understanding how our proposed approach adapts itself if some module becomes faulty during the operation.

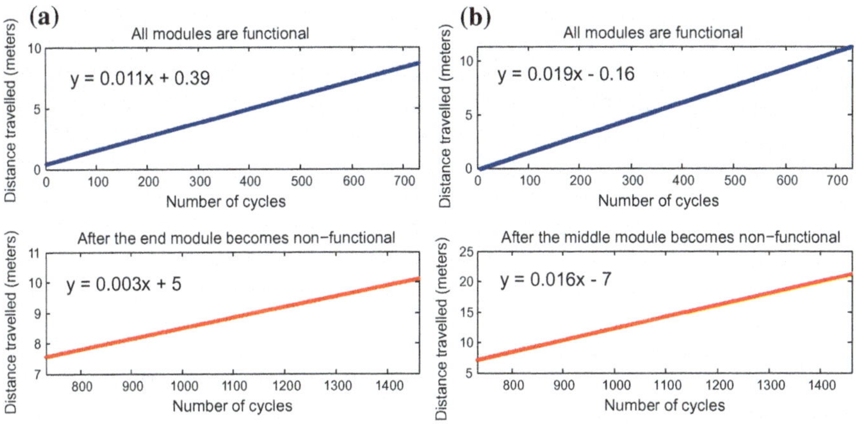

Fig. 6 Performance of our proposed algorithm after **a** the end and **b** the middle module becomes non-operational

We have tested the fault adaptability of our approach on M3 and have performed three adaptability tests: how the algorithm performs when the middle, end or last two modules stop working (i.e., stop moving and communicating). For the first 30 min in the experiment, all modules were functional, and then for the next 30 min one of the above situations occurs. Depending on the relative position of the faulty module within the configuration, the performance of the algorithm (distance metric) varies. For example, when the middle module becomes faulty, we can see a very nominal change in the distance traveled by the configuration: the slope of the best-fit lines are almost the same (Fig. 6b). But when the end module stops working (irrespective of middle module's faulty condition), then the total amount of distance traveled by the configuration drops drastically (Figs. 6a and 7a). But even with any type of fault, we can notice that our algorithm performs steadily and was able to adapt to any module's fault in the configuration.

Rolling motion of ModRED: We also present preliminary results of applying our proposed approach on a ModRED 2-module chain configuration for rolling locomotion. We compare the performance of our approach applied on the 2-module ModRED chain for rolling motion against the hand-coded gaits for ModRED proposed in [6]. The result is shown in Fig. 7b. This figure shows that using our proposed approach the ModRED configuration was able to travel almost double the distance than by using the hand-coded gaits. We also observe that using rolling actions modules were able to travel longer distances than using the inchworm locomotion. There are two main reasons for that: (1) modules have fewer actions to learn from and (2) in one rotation (clockwise/anti-clockwise) of the rotational DOF, one module travels further than any action provided for inchworm locomotion.

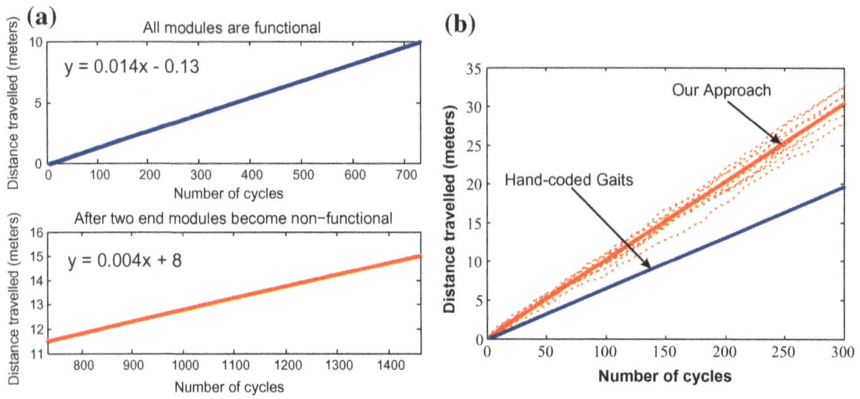

Fig. 7 **a** Performance of our proposed algorithm after two end modules become non-operational. **b** Comparison of performance of our algorithms for rolling locomotion against the same by using hand-coded gaits

6 Conclusion and Future Work

In this paper, we have proposed a Q-learning based adaptive locomotion learning strategy which learns from the module's past actions as well as from the relationship between its own actions with its neighboring modules' actions. We have empirically shown that using our proposed approach both ModRED and Yamor modular robots learn to move forward. Our approach has also been shown to be adaptive to module failure. In the future, we plan to implement this algorithm on more complex configurations built using second-generation ModRED modules. We also plan to investigate the effect of neighborhood size in learning. In this paper, each module uses action-pair data structure only for its immediate neighbors, i.e., physically connected modules. But we plan to extend this approach in which each module will learn its optimal (i.e., provides best learning behavior) neighborhood size in dynamic time. Also we plan to implement this algorithm in actual ModRED hardware.

References

1. Ahmadzadeh, H., Masehian, E.: Modular robotic systems: methods and algorithms for abstraction, planning, control, and synchronization. Artif. Intell. **223**, 27–64 (2015)
2. Baca, J., Hossain, S., Dasgupta, P., Nelson, C.A., Dutta, A.: Modred: hardware design and reconfiguration planning for a high dexterity modular self-reconfigurable robot for extra-terrestrial exploration. Robot. Auton. Syst. **62**(7), 1002–1015 (2014)
3. Baca, J., Woosley, B., Dasgupta, P., Dutta, A., Nelson, C.: Coordination of modular robots by means of topology discovery and leader election: improvement of the locomotion case. In: Distributed Autonomous Robotic Systems, pp. 447–458. Springer, Berlin (2016)

4. Castano, A., Behar, A., Will, P.M.: The conro modules for reconfigurable robots. IEEE/ASME Trans. Mech. **7**(4), 403–409 (2002)
5. Christensen, D.J., Schultz, U.P., Stoy, K.: A distributed strategy for gait adaptation in modular robots. In: 2010 IEEE International Conference on Robotics and Automation (ICRA), pp. 2765–2770. IEEE (2010)
6. Chu, K.D., Hossain, S., Nelson, C.A.: Design of a four-dof modular self-reconfigurable robot with novel gaits. In: ASME 2011 International Design Engineering Technical Conferences and Computers and Information in Engineering Conference, pp. 747–754. American Society of Mechanical Engineers (2011)
7. Cully, A., Clune, J., Tarapore, D., Mouret, J.B.: Robots that can adapt like animals. Nature **521**(7553), 503–507 (2015)
8. Dietsch, J., Moeckel, R., Jaquier, C., Drapel, K., Dittrich, E., Upegui, A., Jan Ijspeert, A.: Exploring adaptive locomotion with yamor, a novel autonomous modular robot with bluetooth interface. Ind. Robot. Int. J. **33**(4), 285–290 (2006)
9. Dutta, A., Dasgupta, P.: Simultaneous configuration formation and information collection by modular robotic systems. In: 2016 IEEE International Conference on Robotics and Automation (ICRA), pp. 5216–5221 (2016)
10. Ijspeert, A.J.: Central pattern generators for locomotion control in animals and robots: a review. Neural Netw. **21**(4), 642–653 (2008)
11. Kamimura, A., Kurokawa, H., Yoshida, E., Murata, S., Tomita, K., Kokaji, S.: Automatic locomotion design and experiments for a modular robotic system. IEEE/ASME Trans. Mech. **10**(3), 314–325 (2005)
12. Kamimura, A., Kurokawa, H., Yoshida, E., Tomita, K., Kokaji, S., Murata, S.: Distributed adaptive locomotion by a modular robotic system, m-tran ii. In: 2004 IEEE/RSJ International Conference on Intelligent Robots and Systems, 2004. (IROS 2004). Proceedings, vol. 3, pp. 2370–2377. IEEE (2004)
13. Kapetanakis, S., Kudenko, D.: Reinforcement learning of coordination in cooperative multi-agent systems. AAAI/IAAI **2002**, 326–331 (2002)
14. Nutt, J., Marsden, C., Thompson, P.: Human walking and higher-level gait disorders, particularly in the elderly. Neurology **43**(2), 268–268 (1993)
15. Shen, W.M., Salemi, B., Will, P.: Hormone-inspired adaptive communication and distributed control for conro self-reconfigurable robots. IEEE Trans. Robot. Autom. **18**(5), 700–712 (2002)
16. Shen, W.M., Will, P.: Docking in self-reconfigurable robots. In: 2001 IEEE/RSJ International Conference on Intelligent Robots and Systems, 2001. Proceedings, vol. 2, pp. 1049–1054. IEEE (2001)
17. Sproewitz, A., Moeckel, R., Maye, J., Ijspeert, A.J.: Learning to move in modular robots using central pattern generators and online optimization. Int. J. Robot. Res **27**(3–4), 423–443 (2008)
18. Sutton, R.S., Barto, A.G.: Reinforcement learning: An introduction. MIT Press, Cambridge (1998)
19. Yim, M.: Locomotion with a unit-modular reconfigurable robot. Ph.D. thesis, Citeseer (1994)
20. Yim, M., Shen, W.M., Salemi, B., Rus, D., Moll, M., Lipson, H., Klavins, E., Chirikjian, G.S.: Modular self-reconfigurable robot systems [grand challenges of robotics]. IEEE Robot. Autom. Mag. **14**(1), 43–52 (2007). https://doi.org/10.1109/MRA.2007.339623

Distributed Camouflage for Swarm Robotics and Smart Materials

Yang Li, John Klingner and Nikolaus Correll

Abstract We present a distributed algorithm for a swarm of active particles to camouflage in an environment. Each particle is equipped with sensing, computation, and communication, allowing the system to take color and gradient information from the environment and self-organize into an appropriate pattern. Current artificial camouflage systems are either limited to static patterns, which are adapted for specific environments, or rely on back-projection, which depend on the viewer's point of view. Inspired by the camouflage abilities of the cuttlefish, we propose a distributed estimation and pattern formation algorithm that allows to quickly adapt to different environments. We present convergence results both in simulation as well as on a swarm of miniature robots "Droplets" for a variety of patterns.

1 Introduction

We wish to design artificial camouflage systems that can quickly adapt to a large variety of environments. Inspired by the capabilities of cephalopods, which tightly integrate sensing, actuation (color change), neural computation, and communication, we are interested in a distributed artificial approach that mimics this tight integration [9, 21]. While animals employ camouflage mostly for escaping predators, camouflage in an engineering context is typically motivated by clandestine military operations. More broadly, everything from small robots to buildings could use these techniques to more seamlessly be a part of their environment. Nature employs a large variety of techniques to achieve these goals. For example, moths mimic patterns that they would expect in their environments, sea animals use mottle patterns to soften their

Y. Li (✉) · J. Klingner · N. Correll
Department of Computer Science, University of Colorado Boulder, Boulder, CO, USA
e-mail: yang.li-4@colorado.edu

J. Klingner
e-mail: john.klingner@colorado.edu

N. Correll
e-mail: nikolaus.correll@colorado.edu

R. Groß et al. (eds.), *Distributed Autonomous Robotic Systems*,
Springer Proceedings in Advanced Robotics 6,
https://doi.org/10.1007/978-3-319-73008-0_25

contours, and other animals decorate their body with artifacts from the environment [17]. Two animals with notable camouflage abilities are the cuttlefish and octopus, who can dramatically alter the coloration and patterning of their skin and switch between different environments in a matter of seconds [5]. These creature's camouflage behavior is not only driven by the animals' visual system (which is color-blind [11]) or brain [12], but has also been shown to rely on local sensing and control [15].

There have been multiple attempts to achieve active camouflage using a combination of cameras and projection [6, 8]. Although such systems provide "perfect" camouflage, they are highly dependent on the observer's viewpoint. Mimicking the background exactly is rarely employed in the animal kingdom, where a few simple families of patterns — mottled, striped, or simply uniform [4] — dominate. Creating such patterns requires only local coordination [10], suggesting a combination of high-level selection of appropriate motor programs [12] and self-organization [10]. Here, we are not concerned with perfectly matching the background, but rather aim to replicate the pattern matching ability of natural systems, which are able to fool sophisticated predators.

Distributing the sensing and actuation for camouflage generation makes an implementation scalable for a variety of factors, such as resolution of the camouflage pattern, the size of the area being camouflaged, and robustness against the failure of individual units. Further, a distributed camouflage system could respond to local changes in the environment, in particular when deployed on non-trivial 3D surfaces.

In this paper, we present a fully distributed approach, which we implement on a swarm of Droplets [2], each equipped with the ability to sense and emit color as well as communicate with its local neighbors. Although there exist multiple attempts to design artificial chromatophores, most work focuses on component technology, i.e. the ability to color change in a soft substrate [14, 16], but very few works articulate the systems challenges that require not only local color changes, but also local sensing and computation [21], or investigate the ability to co-locate simple signal processing with the sensors themselves [3].

Our algorithm can be broken into three phases, each described in detail below. First, we estimate a color and gradient histogram with a consensus algorithm among the particles. This information is then used to determine the parameters of a pattern formation algorithm. Finally, the pattern is formed using a reaction-diffusion process. Motivated by the mottled and striped patterns which dominate in the animal kingdom, this paper focuses on reproducing three different types of patterns: spots, horizontal stripes, and vertical stripes. The "background" in to which the swarm is trying to camouflage is projected on to the particles from above, requiring them to have color sensors. To simplify the color-identification process, the paper focuses on two-tone patterns. Finally, we arrange the particles in a grid pattern, which allows us to implement a discrete convolution operation and simplifies debugging the pattern at the low resolution that swarms in the order of tens of particles can afford.

2 Distributed Camouflage Algorithm

In this section, we describe the distributed camouflage algorithm. Figure 1 illustrates the steps of the algorithm in broad strokes. First, each robot measures the color projected on it. Then, it exchanges the measured color with neighboring robots. Once received, neighbors' color information is used to compute an estimated probability for the various pattern types (Sect. 2.1). Next, the swarm communicates their local pattern probabilities, using a weighted-average consensus algorithm to compute the most likely global pattern (Sect. 2.2). Once consensus has been achieved, the swarm reproduces the pattern collaboratively with a reaction-diffusion process (Sect. 2.3).

2.1 *Pattern Descriptor*

Once each robot has measured the local environment's color, and communicated that information, they apply a filter mask (see Fig. 2) to compute a discrete approximation of the second-order color derivative in both the horizontal and vertical directions. This is quite similar in concept to kernels used in edge detection and other computer-vision tasks [1]. Indeed, if the grid of robots is viewed as an image with each robot a pixel, these two pattern descriptors are simply the value the pixel would have after each of the two convolutions. These second-order derivatives are the *Pattern Descriptors* – denoted P_x and P_y for the horizontal and vertical directions respectively – and are used to calculate the most probable local pattern. Note that this requires either that the robots all have a common orientation, or that they are able to sense their orientation relative to other robots.

To be specific, with M denoting my local color and T, R, B, and L denoting the color of my top, right, bottom, and left neighbors, P_x and P_y are given by:

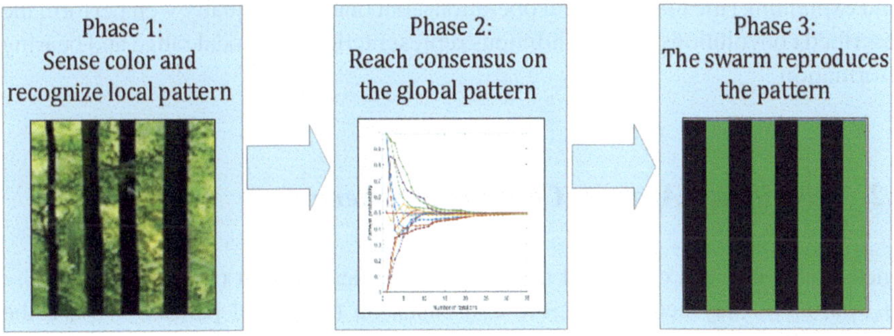

Fig. 1 Pipeline of the distributed camouflage algorithm

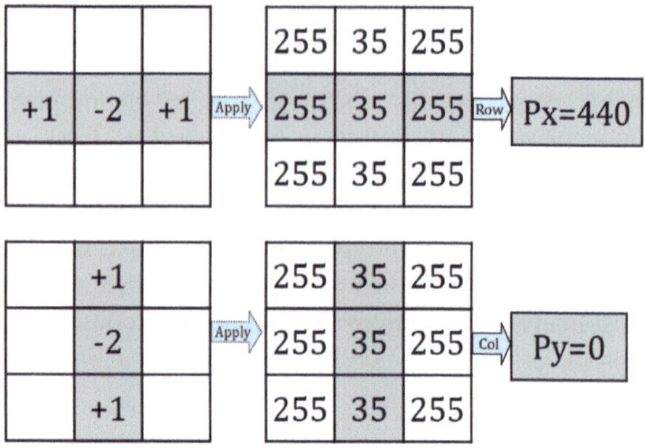

Fig. 2 Illustration of applying the two second order derivative masks

$$P_x = L + R - 2M$$
$$P_y = T + B - 2M$$

A pattern-probability array $p = [p_h, p_v, p_m]$ is used to record each robot's pattern, where p_h represents the probability of a horizontally-striped pattern, p_v the probability of a vertically-striped pattern, and p_m the probability of a mottled pattern. One pattern-type is selected and given a probability of 1 based on our local *Pattern Descriptors*, and the other probabilities are all 0. This is shown in the equation below, where T is some threshold value and $|val|$ is used to indicate abs (val).

$$p = [p_h, p_v, p_m] = \begin{cases} [1, 0, 0] & \text{if } \left|P_y\right| - |P_x| > T \\ [0, 1, 0] & \text{if } |P_x| - \left|P_y\right| > T \\ [0, 0, 1] & \text{otherwise} \end{cases} \tag{1}$$

Note that a grid representation has only been chosen for the simplicity of performing (and explaining) the mathematical operations, but one could equally well perform the described convolutions using continuous representations and local range and bearing information.

2.2 Distributed Average Consensus Scheme

Once each robot has computed the most likely local pattern (i.e., computed $p = [p_h, p_v, p_m]$), they need to achieve consensus on the global pattern. We use the distributed average consensus scheme [19] for this purpose. In each step of this scheme, the robot updates its local p to be a weighted average of its own and its neighbors'. This step is repeated many times, allowing information to diffuse through the swarm. Since the weighted average just uses local information, each step takes the same amount of time regardless of the number of robots in the swarm. The number of steps needed was determined experimentally.

The weighted-average calculation uses Metropolis weights, defined as

$$W_{i,j} = \begin{cases} \frac{1}{1+max\{d_i,d_j\}} & \text{if } (i,j) \in E, \\ 1 - \sum_{(i,k)\in E} W_{i,k} & \text{if } i = j, \\ 0 & \text{otherwise.} \end{cases} \tag{2}$$

The Metropolis weights are well-suited for distributed algorithms, since weight-calculation requires only local knowledge. Further, it is proven in [19] that Metropolis weights guarantee convergence of the average consensus provided that the infinitely occurring communication graphs are jointly connected. Once the robots have converged, the largest value in p represents the most likely global pattern. For example, $p_h > p_v$ and $p_h > p_m$ indicate that the most likely global pattern is horizontal stripes.

2.3 *Pattern Generator*

In this section, we describe the distributed pattern formation algorithm to generate a proper pattern to match the environment.

Now that a global pattern has been selected, the robots next need to generate an appropriate camouflage pattern. We use the pattern-formation algorithm presented by Young [20]: a local activator-inhibitor model. In this model, each cell (robot) is either 'on' or 'off', and can generate two kinds of morphogens: activator morphogen and inhibitor morphogen. Together, these form a "morphogenetic field". Note that the activator should be inside of the inhibitor (see left of Fig. 3). The cells (robots)

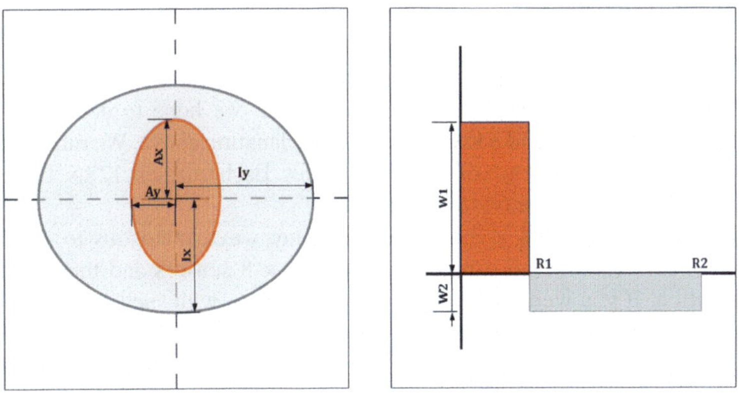

Fig. 3 Illustration of local activator-inhibitor model: on the left, the activation region (orange) is defined by A_x and A_y while the inhibition region (gray) is defined by I_x and I_y; on the right, W_1 and W_2 are the two field values. $R1$ is related to A_x and A_y and $R2$ is related to I_x and I_y

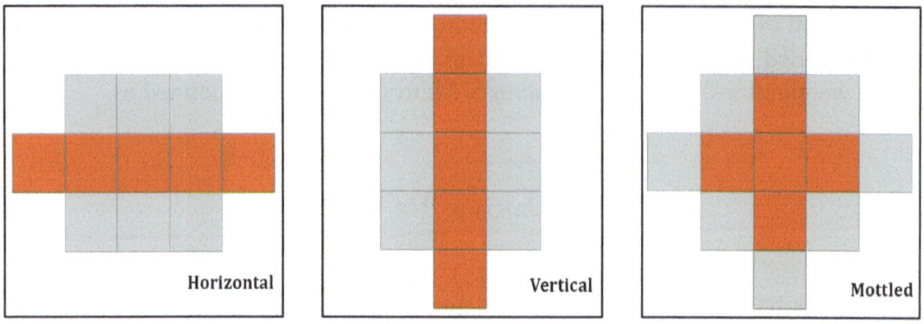

Fig. 4 Activator (orange) and inhibitor (gray) regions for each of the three patterns

in the activator morphogen contribute to stimulate change for nearby 'on' cells, and cells in the inhibitor morphogen contribute to stimulate change for nearby 'off' cells.

During each step of this algorithm, each cell changes its 'off'/'on' status based on the combined effect of all nearby morphogenetic fields. More specifically, a 'strength' is calculated with each 'on' robot contributing a positive value (W_1) if in the activator region, or contributing a negative value (W_2) if in the inhibitor region. The robot then changes its state to 'on' if the strength is greater than 0, and to 'off' otherwise. This step is repeated until the states converge to a stable pattern. In [20] the author observes that convergence typically takes around five steps. This is consistent with our observations.

In this framework, the different types of patterns are represented with differently shaped activator and inhibitor regions. The regions for each pattern are shown in Fig. 4. Note that the region sizes mean that each robot only requires information from robots within two hops of it.

3 Simulation Results

We implemented the algorithm introduced above on a centralized system for simulation. By presenting some simulated results here, we hope to demonstrate the algorithm's functionality and add clarity to the explanation above. We run these tests with three images, one for each of the pattern types. Each image is 128×128 pixels, and gray scale. We simulate 64 (8×8) robots.

Note that this grid of 8×8 robots is in many ways analogous to the sensor of a digital camera, albeit a camera with only 8×8 sensors and thus with very low resolution. If you were to recapture our test images with such a low resolution camera, many different pixels in the test image would contribute to the camera's output, resulting in a very blurry image. We therefore downsample the input image by taking the average of 16×16 pixel blocks. This blurred image is used as the color sensed by each robot for selecting the most likely pattern. For pattern generation, the initial 'on'/'off' state is determined by making the blurred image binary (i.e. white and black). Figure 5 shows the entire process for each of the three input images.

Once the Droplets calculate the local pattern based on their sensed color and that of their neighbors, they need to achieve consensus on the global pattern. As has been discussed in Sect. 2.2, convergence of this value is guaranteed.

Next, the pattern generator described above is used (Sect. 2.3) with the activator and inhibitor regions seen in Fig. 4. The activator field value of $W_1 = 1$ was used, as suggested in paper [20]. The inhibitor field value, W_2, is a parameter which gives rough control over what proportion of the robots are 'on' in the final pattern. We found that $W_2 = -0.75$ gave qualitatively good results for all three of the pattern generators.

Finally, we start pattern generation with each robot's initial state to be 'on' or 'off' status based on the sensed value. If the value is less than 127 we set it black, otherwise we set it white. The pattern generator runs for ten iterations. Robots on the image boundary use a reflection of their neighbors. A robot on the top row, for example, would count its bottom neighbor twice, as the top row is empty.

To further test the simulated algorithm, we added a simple noise model. For measurement error, instead of always assigning the appropriate color to a robot based on its position, we assign a uniformly random color with probability ρ_{meas}. For communication error, at each step in the algorithm where information from a neighbor is shared with a robot for a calculation (including the step where a robot's neighbors are calculated in the first place), a robot does not share this information with probability ρ_{comm}.

For a quantitative measure of the effects of error, we calculated the total absolute difference between the final generated pattern in the presence of error, and the final generated pattern without any error (as visible in the bottom row of Fig. 5). These results are charted in Fig. 6. Note that, with the 8×8 images used, a purely random image should give us a difference of 32, on average. The algorithm seems quite robust to errors of up to 0.15–0.2. After these thresholds, the error increases sharply. (Results shown here are for the forest image, with other images yielding similar results.) Qualitatively, we observe that even as errors started to appear, many of the resulting patterns still looked 'good', i.e., still had prominent vertical stripes. The main determining factor as the probability of error increased seemed to be in the global pattern detection. If the correct pattern (vertical stripes in this case) is selected, the resulting pattern will fit well even with large errors. Correct pattern selection grows increasingly infrequent, however.

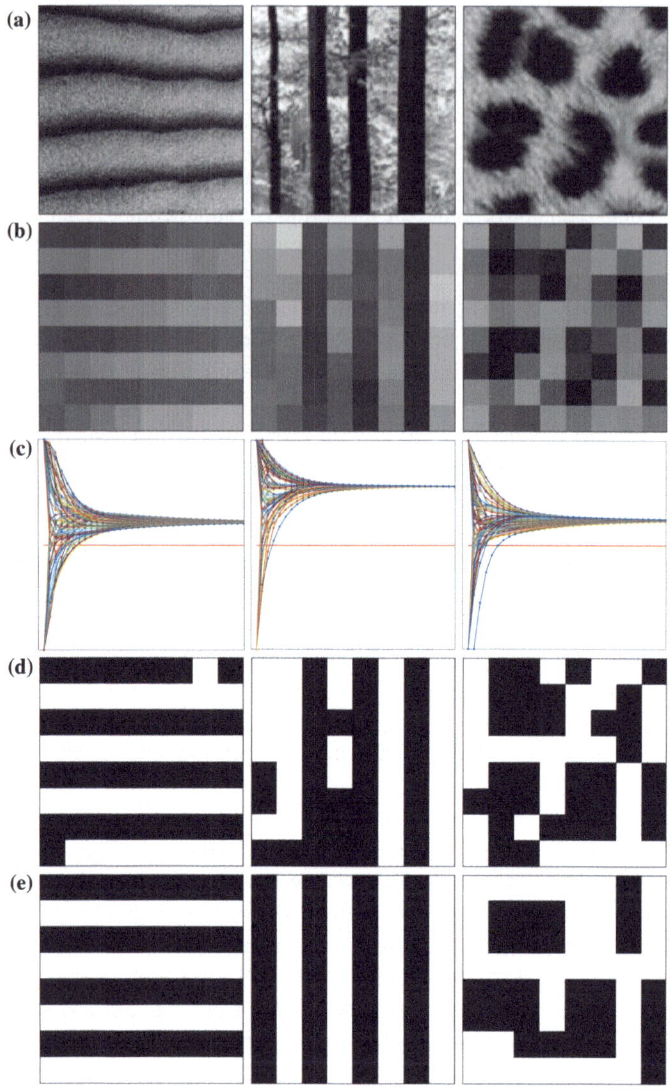

Fig. 5 The algorithm takes in the gray images (row **a**) and blurs them to images with 8×8 resolution (row **b**). These are the values sensed by each robot, and are used to calculate the pattern probabilities and choose the most probable local pattern. Row **c** shows consensus convergence for the most likely global pattern. For each of the charts in row **c**: the y axis shows pattern probability p from 0 to 1, and the x axis shows the number of steps taken from 0 to 35. The red horizontal line marks $p = 0.5$. The blurred image from row **b** is converted to binary (row **d**) to get initial states for the pattern generator, which generates the resultant pattern (row **e**)

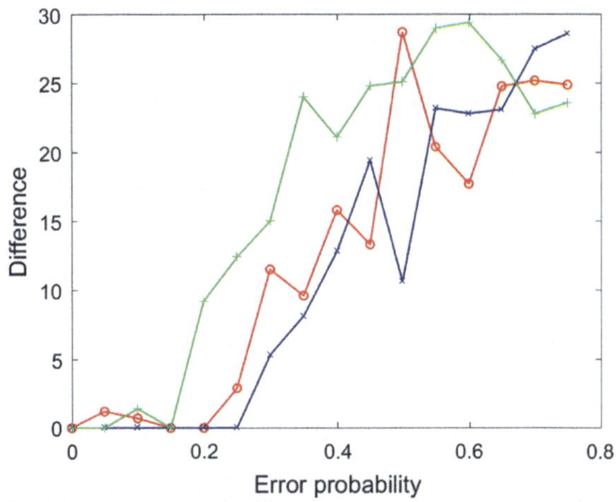

Fig. 6 The y-axis is the pixel difference from the 'correct' pattern and the x-axis is the error probability. The red line shows the effect of measurement errors (ρ_{meas}). The blue line shows the effect of communication errors (ρ_{comm}). The green line shows the effect of both measurement and communication errors ($\rho_{comm} = \rho_{meas}$). Each data point reflects the mean result over 10 trials of the forest image

4 Hardware Implementation

To validate the proposed algorithm and to understand the sorts of errors that real hardware introduces, we implemented the algorithm described above on a swarm of "Droplets" [2, 7]. The Droplets are an open-source platform, with source code and manufacturing information available online.[1] Each Droplet is roughly cylindrical with a radius of 2.2 cm and a height of 2.4 cm. The Droplets use an Atmel xMega128A3U micro-controller, and receive power via their legs through a floor with alternating strips of $+5V$ and GND. Each Droplet has six infrared emitters, receivers, and sensors, which are used for communication and for the range and bearing system [2]. The top of each board has sensors to detect the color and brightness of ambient light, and an RGB LED. Each droplet has a 16-bit unique ID.

In our implementation, each Droplet maintains an array of neighbors' IDs. Messages are labeled with phase flags and attached with Droplets' IDs. The Droplets are synchronized using a firefly synchronization algorithm [13, 18]. A simple TDMA protocol is used with 37 slots, each 350 ms long. Each frame is thus 12.95 s long. Each robot is assigned a slot based on its unique ID modulo 37. The number of slots (37) was chosen to be large enough that the probability of two adjacent robots sharing a slot is low, but small enough that the algorithm runs quickly.

In Phase 0 (neighbor identification), we initialize and configure the neighbor ID arrays which store neighboring Droplets' IDs. Range and bearing information is used to calculate positions for each Droplet's immediate neighbors, and neighbors-of-neighbors are learned by listening to the messages sent by Droplets each slot, which contain that Droplet's neighbors. The positions of Droplets and their indices in the array are illustrated in Fig. 7. Though the Droplets can sense their relative

[1] http://github.com/correlllab/cu-droplet.

Fig. 7 Neighbor array. The orange neighbors(0–3) are used for pattern recognition; the green neighbors(4–7) are used in addition to the orange for pattern consensus. All pictured neighbors (orange, green and gray) are used for pattern formation

orientations, we positioned them with a common orientation for simplicity. We allot 20 frames for Phase 0, since the neighbor information is critical to the three phases.

In Phase 1 (color sensing and recognition), each Droplet communicates the color it senses, and stores the colors its neighbors sense, as learned through communications. Once this is complete, each (non-boundary) Droplet should know the ID and position of 12 neighbors, as well as those neighbor's sensed colors. With this information, the Droplets calculate an pattern probability array p, as described in Sect. 2.1. This phase is allotted 10 frames.

In Phase 2 (pattern consensus), each Droplet communicates its pattern probability array p and receives pattern probability arrays from its neighbors. At the end of each frame, each Droplet updates its pattern probability array according to the weighted-average consensus algorithm, as described in Sect. 2.2. Each 'step' of the consensus algorithm spans one frame. This phase is allotted 35 frames.

In Phase 3 (pattern formation), each Droplet communicates its intended color for the generated pattern, and receives that information from neighboring Droplets. At the end of each frame, each Droplet updates its color in the generated pattern from corresponding Droplets. Each Droplet exchanges pattern color message with neighbors. At the end of each frame, each Droplet updates its pattern color according to the pattern generation algorithm described in Sect. 2.3. This phase is allotted 20 frames.

5 Hardware Results

A hardware implementation of the swarm camouflage algorithm is shown in Fig. 8. For this test, the projected image for the Droplets to sense is a tiger stripe pattern. The results of this test are interesting because a striped pattern is maintained, despite the failure of two units. This, in addition to the more-difficult-to-count failures in communication and color sensing.

Figure 9 shows the pattern probability convergence for a random sampling of Droplets, when run with a simple horizontal stripe pattern projected on them. The swarm reaches consensus on a horizontal pattern, converging to $p_h = 0.61$.

Fig. 8 Initial condition (left), final pattern with projected pattern (middle) and final pattern (right) for camouflaging the tiger stripe pattern

Fig. 9 Convergence of pattern probabilities of randomly chosen Droplets camouflaging tiger stripe pattern

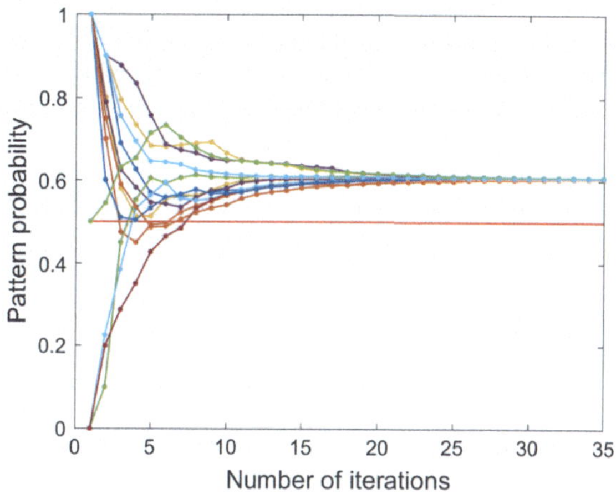

6 Conclusion

We present a distributed camouflage system, in which a robot swarm can sense the environment color, recognize the local pattern, achieve consensus on the global pattern, and generate a camouflage pattern consistent with the environment the robots are in. In our design, pattern descriptors are proposed for recognizing local patterns. A weighted-average consensus scheme is then utilized, allowing the swarm to converge to a global pattern. Finally, a pattern formation model is applied to each robot which generates a pattern appropriate for the background. This is accomplished using local communication and simple mathematical operations.

We simulated the proposed algorithm on a couple of patterns from nature: a desert, a forest, and leopard skin. After going through all the phases in the algorithm, and successfully agreeing on a global pattern, the simulation results show that robots with wrong color reading can correct themselves to match the global pattern. This is especially obvious for the horizontal and vertical patterns. We also tried to test the distributed algorithm by applying it on the Droplet swarm robotics platform. The results from the Droplets is promising since the robots can agree on the global pattern and display a proper matching pattern even if individual Droplets stop working.

We can see that the simulations are based on a regular grid of robots and the stripes in patterns are simply one pixel (or, equivalently, one robot) width. It should be possible to produce more complicated stripe patterns of varying width, for example: Young's pattern-generation model can generate patterns with stripes of different width if the parameters are carefully set. This would require more complicated pattern descriptors to correctly recognize the patterns and set appropriate corresponding parameters. Furthermore, the robots don't have to be placed regularly if the swarm is large and dense enough.

As communication on the Droplets is not perfectly reliable, the resultant patterns exhibit some random variations; they do not perfectly match simulation. Even these variations, however, will roughly follow the desired background pattern, seeming to bend or twist around the erroneous robot. In the future, we wish to test the algorithm on a more purpose-built hardware platform, which would allow for higher resolution patterns, and extend the algorithm to include consensus on the dominant colors and patterns consisting of more than two colors.

Acknowledgements This research has been supported by NSF grant #1150223.

References

1. Dalal, N., Triggs, B.: Histograms of oriented gradients for human detection. In: 2005 IEEE Computer Society Conference on Computer Vision and Pattern Recognition (CVPR'05), vol. 1, pp. 886–893. IEEE (2005)
2. Farrow, N., Klingner, J., Reishus, D., Correll, N.: Miniature six-channel range and bearing system: algorithm, analysis and experimental validation. In: 2014 IEEE International Conference on Robotics and Automation (ICRA), pp. 6180–6185. IEEE (2014)
3. Fekete, S.P., Fey, D., Komann, M., Kröller, A., Reichenbach, M., Schmidt, C.: Distributed vision with smart pixels. In: Proceedings of the Twenty-Fifth Annual Symposium on Computational Geometry, pp. 257–266. ACM (2009)
4. Hanlon, R.: Cephalopod dynamic camouflage. Curr. Biol. **17**(11), R400–R404 (2007)
5. Hanlon, R.T., Messenger, J.B.: Adaptive coloration in young cuttlefish (sepia officinalis l.): the morphology and development of body patterns and their relation to behaviour. Philos. Trans. R. Soc. Lond. B: Biol. Sci. **320**(1200), 437–487 (1988)
6. Inami, M., Kawakami, N., Tachi, S.: Optical camouflage using retro-reflective projection technology. In: Proceedings of the 2nd IEEE/ACM International Symposium on Mixed and Augmented Reality, p. 348. IEEE Computer Society (2003)
7. Klingner, J., Kanakia, A., Farrow, N., Dustin, R., Correll, N.: A stick-slip omnidirectional drive-train for low-cost swarm robotics: Mechanism, calibration, and control. In: IEEE/RSJ International Conference on Intelligent Robots and Systems (IROS) (2014)
8. Lin, H.Y., Lie, W.N., Wang, M.L.: A framework of view-dependent planar scene active camouflage. Int. J. Imaging Syst. Technol. **19**(3), 167–174 (2009)
9. McEvoy, M., Correll, N.: Materials that couple sensing, actuation, computation, and communication. Science **347**(6228), 1261689 (2015)
10. Meinhardt, H.: Models of Biological Pattern Formation, vol. 6. Academic Press, London (1982)
11. Messenger, J.B.: Evidence that octopus is colour blind. J. Exp. Biol. **70**(1), 49–55 (1977)

12. Messenger, J.B.: Cephalopod chromatophores: neurobiology and natural history. Biol. Rev. **76**(4), 473–528 (2001)
13. Mirollo, R.E., Strogatz, S.H.: Synchronization of pulse-coupled biological oscillators. SIAM J. Appl. Math. **50**(6), 1645–1662 (1990)
14. Morin, S.A., Shepherd, R.F., Kwok, S.W., Stokes, A.A., Nemiroski, A., Whitesides, G.M.: Camouflage and display for soft machines. Science **337**(6096), 828–832 (2012)
15. Ramirez, M.D., Oakley, T.H.: Eye-independent, light-activated chromatophore expansion (lace) and expression of phototransduction genes in the skin of octopus bimaculoides. J. Exp. Biol. **218**(10), 1513–1520 (2015)
16. Rossiter, J., Yap, B., Conn, A.: Biomimetic chromatophores for camouflage and soft active surfaces. Bioinspiration Biomimetics **7**(3), 036009 (2012)
17. Stevens, M., Merilaita, S.: Animal camouflage: current issues and new perspectives. Philos. Trans. R. Soc. B: Biol. Sci. **364**(1516), 423–427 (2009)
18. Werner-Allen, G., Tewari, G., Patel, A., Welsh, M., Nagpal, R.: Firefly-inspired sensor network synchronicity with realistic radio effects. In: Proceedings of the 3rd international conference on Embedded networked sensor systems, pp. 142–153. ACM (2005)
19. Xiao, L., Boyd, S., Lall, S.: A scheme for robust distributed sensor fusion based on average consensus. In: Fourth International Symposium on Information Processing in Sensor Networks, 2005. IPSN 2005, pp. 63–70. IEEE (2005)
20. Young, D.A.: A local activator-inhibitor model of vertebrate skin patterns. Math. Biosci. **72**(1), 51–58 (1984)
21. Yu, C., Li, Y., Zhang, X., Huang, X., Malyarchuk, V., Wang, S., Shi, Y., Gao, L., Su, Y., Zhang, Y., et al.: Adaptive optoelectronic camouflage systems with designs inspired by cephalopod skins. Proc. Natl. Acad. Sci. **111**(36), 12998–13003 (2014)

Evo-Bots: A Simple, Stochastic Approach to Self-assembling Artificial Organisms

Juan A. Escalera, Matthew J. Doyle, Francesco Mondada and Roderich Groß

Abstract This paper describes an alternative path towards artificial life—one by which simple modular robots with novel hybrid motion control are used to represent artificial organisms. We outline conceptually how such a system would work, and present a partial hardware implementation. The hardware, a set of self-reconfigurable modules called the evo-bots, operates on an air table. The modules use a stop-start anchor mechanism to either rest or move. In the latter case, they undergo semi-random motion. The modules can search for, harvest and exchange energy. In addition, they can self-assemble, and thereby form compound structures. Six prototypes of the evo-bot modules were built. We experimentally demonstrate their key functions, namely hybrid motion control, energy harvesting and sharing, and simple structure formation.

J. A. Escalera (✉) · M. J. Doyle · R. Groß
Department of Automatic Control and Systems Engineering,
The University of Sheffield, Sheffield, UK
e-mail: juan.antonio.escalera@gmail.com

M. J. Doyle
e-mail: matthew.doyle@sheffield.ac.uk

R. Groß
e-mail: r.gross@sheffield.ac.uk

J. A. Escalera
Department of Automatics and Systems Engineering,
University Carlos III of Madrid, Leganés, Spain

F. Mondada
Laboratoire de Systèmes Robotiques, Ecole Polytechnique
Fédérale de Lausanne, Lausanne, Switzerland
e-mail: francesco.mondada@epfl.ch

R. Groß et al. (eds.), *Distributed Autonomous Robotic Systems*,
Springer Proceedings in Advanced Robotics 6,
https://doi.org/10.1007/978-3-319-73008-0_26

1 Introduction

A long term goal of robotics is to produce not only robotic organisms, but entire robotic ecosystems. Such systems are not only a scientific curiosity in and of themselves, but could allow us to better understand the biological processes that they mimic. Although much work has been conducted on populations of virtual creatures with evolving morphology [8, 20, 23, 25], relatively few examples have been transferred into reality [3, 19]. This motivates one of the grand challenges in evolutionary robotics—to create ecosystems of physically evolving robots [9].

Fundamental to an evolving ecosystem is the ability of its members to reproduce. Jacobson [15] studied self-replicating sequences of track-bound modules. Chirikjian et al. [4] evaluated the feasibility of self-reproducing robots. Zykov et al. [27] demonstrated structured duplication using a lattice-based reconfigurable robotic system. In the Symbrion/Replicator project, a heterogeneous system of reconfigurable robots was considered [16]. The systems in these examples involved *self-propelling modules* of high complexity. While this may enable them to interact multifariously with their environments, the modules were expensive to make, limiting the number of them in practical experiment.

An alternative approach to a self-reproducing system is to use modules that require external stimulation to move. Penrose and Penrose developed a self-replicating mechanical system [21]. This was extended by Breivik [2] and Virgo et al. [24]. Griffith et al. developed self-replicating modules with programmable on-board controllers [12]. Other works with programmable modules present structure formation, but not replication [14, 17, 26]. The modules used in these examples were simple, *externally propelled devices*. While this made them relatively inexpensive to construct and therefore producible in large quantities, the modules could only interact with their environment in a limited manner.

In this paper we present the evo-bots,[1] a system of reconfigurable modules that combine the advantages of the above two classes of systems. Due to being externally propelled, the evo-bot modules are mechanically simple. Yet, they can indirectly control where to move, by using stop-start mechanisms (an approach seen only in simulation [6]). In addition, they can search for, harvest and share energy, as well as self-assemble, and thereby organize into distinct morphologies.

The paper is organized as follows. Section 2 concerns the evo-bot concept. Here we describe how simple modular robots can potentially be used to form an artificial life system. Section 3 describes our hardware, inspired by the evo-bot concept. Section 4 presents proof-of-concept experiments, showing that our hardware implements several features of the evo-bot concept. Section 5 concludes the paper.

[1] The conceptual foundation is based on preliminary work presented in [13]. This work also presents simulation results and a preliminary module implementation (described further in [7]).

2 Evo-Bots: Toward Artificial Life Systems

In nature, living beings exhibit a high degree of autonomy. Individually, they can adapt to the environment to obtain from it what they need to survive. Whilst as species, they have developed mechanisms to persist through time. These qualities should be mimicked by artificial systems to become truly autonomous.

The *evo-bots* system consists of building blocks, called *modules*, which do not move by themselves. Rather they require an external agitation apparatus, which causes the modules to undergo semi-random motion. The modules have a squared shape and four identical connectors—one per side. When two modules collide, they can connect to one another by chance. Once connected, they can exchange information and decide whether to remain connected. By disconnecting selectively, the modules can organize into linear composite structures called *polymers* (similar to [12]).

The evo-bot modules come in three types, each one endowed with unique functions:

- ***e*-module**: This module can harvest, store and provide energy.
- ***i*-module**: This module can interact with the environment. It can sense the environment and control its motion using a stop-start anchor mechanism.
- ***b*-module**: This module acts as a boundary, once it connects to a polymer, the latter stops growing and becomes an *organism*.

Each module assumes one of these types during each experiment. Organisms must accrete specific modules to perform specific functions. They need energy to sustain themselves. Therefore, they must contain at least one *e*-module. The *e*-modules share their energy with any module that is part of their compound structure. If the energy within a structure is depleted, the structure will decompose into its constituent modules.

Once a polymer accretes a *b*-module and becomes a complete organism, it can begin to replicate itself using free modules in the environment. This process is illustrated in Fig. 1. The replication of organisms that can sustain themselves and the decomposition of those that cannot is envisaged to give rise to populations of organisms well adapted to their environment [13]. Replication can be exact, resulting in a pair of identical organisms, or involve mutation.

Various definitions of life have been proposed [5, 18, 22]. We argue that the evo-bot concept encompasses at least one of these definitions—the seven pillars of life proposed by Koshland Jr. [18]. These pillars represent attributes that an organism should exhibit in order to be considered as a living being. They are program, improvisation, compartmentalization, energy, regeneration, adaptability and seclusion. We justify our claim that the evo-bots embody this concept as follows. Each organism encodes the information of its constituent modules, their functionality and how they interact (program). The modules are able to alter said program between

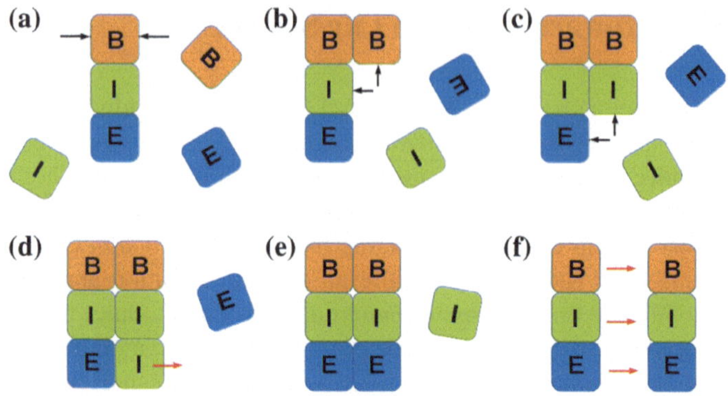

Fig. 1 Process by which an organism builds a replica of itself. **a** The organism's *b*-module activates its connectors; **b** and **c** two matching modules connect to the organism; **d** a non-matching module connects but is rejected; **e** a matching module connects; the child organism is complete and receives energy from the parent organism; **f** the parent organism repels the child organism

generations (improvisation). Each organism defines its own space, which is protected by a boundary, wherein internal processes run protected from external disturbances (compartmentalization). Organisms are able to harvest, store and distribute energy from the environment (energy). Organisms possess the capacity of self-replication (regeneration). They can vary their behavior according to external stimuli (adaptability). Finally, the information within each organism is protected from external agents (seclusion).

3 Evo-Bot Hardware

This section describes the evo-bot hardware. The physical modules do not yet represent a full implementation of the evo-bot concept, but form the basis of one. Conceptually there are three different types of evo-bot modules (*e*, *i*, *b*). However, from a mechatronic point of view, each module is identical. Each module is configured to assume only one of these types during an experiment. This generic design allows each module to execute any of the three roles, which eases fabrication and usage.

Figure 2a shows an evo-bot module. The overall dimensions of the module are $60\times60\times35$ mm. The module weighs 50 g. It has a 3-D printed ABS structure (Fig. 2a-A) with a 60 mm square base, which is surrounded by walls (faces) of 25 mm height. Viewed from above, the walls have a mildly serrated structure in order to engender self-alignment when modules connect.

The evo-bot modules float on an air table. They exploit this by using a motion control method suited to their environment. The module incorporates a stop-start mechanism, shown in Fig. 3a, b. It consists of a flanged cylinder (anchor) fitted

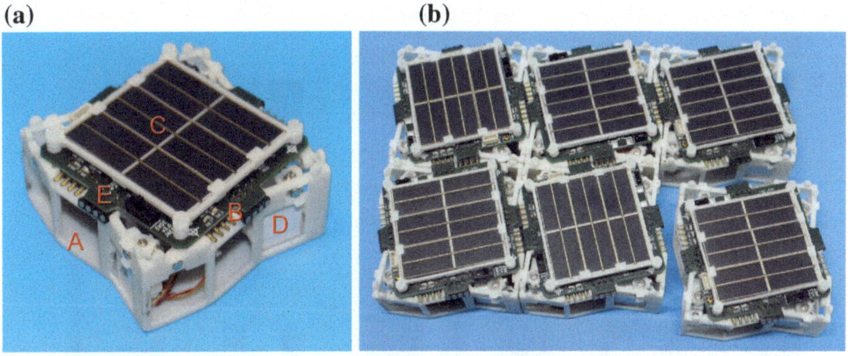

Fig. 2 **a** A single evo-bot module. **b** Six evo-bot modules, five of which are connected

vertically into the base of the module. The anchor contains one permanent magnet (A), and is positioned directly beneath a second magnet (B). Magnet B is rotatable by a servomotor. In the 'start' state magnet B is rotated to attract magnet A, pulling the anchor upwards, away from the air table surface. This allows the module to float freely. In the 'stop' state, magnet B is rotated to repel magnet A, thereby pushing the anchor down onto the air table. This causes friction between the module and the air table and renders the module incapable of floating, pinning it in place. The effectiveness of the anchor mechanism in a polymer depends on the configuration of the polymer itself. For instance, a single *i*-module activating its anchor may be unable to prevent translational motion of an entire organism.

Physical connection between modules is facilitated by a pair of permanent magnets in each face. One magnet per pair is fixed into the plastic wall. The other is mounted on a servomotor (Fig. 2a-D) and can be rotated in order to attract or repel another module [1].

The module can derive information about its environment by using its solar panel as a light sensor. This, in addition to the module's stop-start mechanism, allows it to interact with its environment.

A square, 52 mm printed circuit board (PCB) (Fig. 3c, d) is mounted on the top of the structure (Fig. 2a-E). The double-sided PCB contains the control circuitry, including the energy management system, a low power 16-bit micro-controller (the PIC24FJ128GA306) and five load switches. This provides four serial interface (UART) channels through which data can be exchanged with other modules. The micro-controller can estimate the energy balance of the module by measuring current and voltage. In addition, it can use digital outputs to switch some circuits of the module on or off.

Spring loaded contacts are mounted into each face of the PCB (Fig. 2a-B). As two modules physically connect, the contacts are brought together. This allows modules

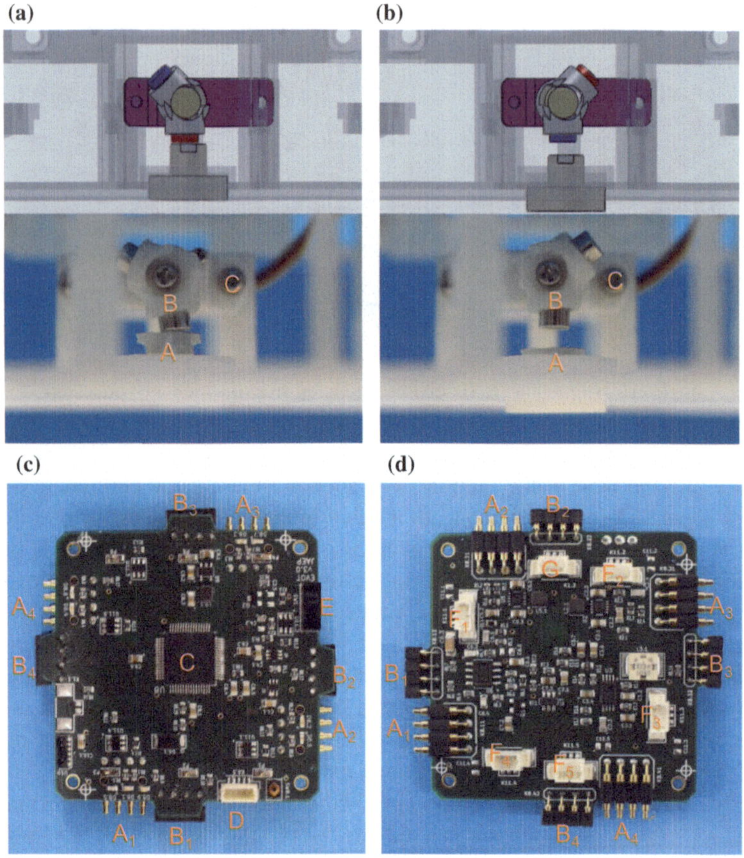

Fig. 3 **a–b** The anchor mechanism (CAD and real) showing the anchor magnet (A), the rotatable magnet (B), and the servo (C). The anchor is shown in start position **a** and stop position **b**. **c–d** The top (c) and bottom (d) layers of the PCB, showing the spring-loaded connectors (A_1 to A_4), the pad connectors (B_1 to B_4), the microprocessor (C), the programming port (D), the on/off switch (E), the servo connectors (F_1 to F_5), and the battery connector (G)

to share power and information. As shown in Fig. 4, each set of contacts passes five signals: load sharing (LS), serial transmit (TX), serial receive (RX), bus voltage (VBUS) and reference voltage (GND).

Figure 4 shows the electrical/electronic architecture used for the module to manage energy. This is carried out by two systems: power management and energy harvesting. In the following, both systems are described in detail.[2]

[2]In [11] the authors presented preliminary work about the energy management system.

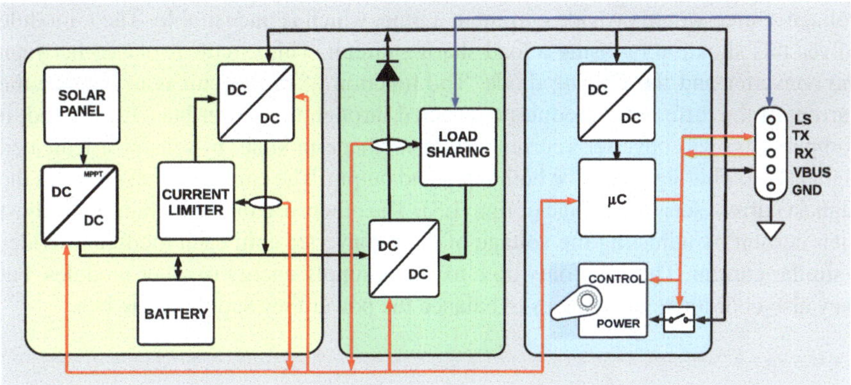

Fig. 4 Block diagram of systems within the evo-bot module. The energy management is accomplished by two systems: The energy harvesting system (yellow area) and the power management system (green area). The former comprises battery, solar charging and power bus charging. The latter includes an output voltage controlled DC/DC converter, an OR-ing diode and a load sharing IC. The remaining elements of the module (blue area) are elements for computation and actuation. The module also has four spring-loaded/pad connectors to share data and energy with other modules

3.1 Power Management System

The evo-bot module uses a 300 mAh lithium ion battery to store energy. This supplies unregulated voltage from 2.7 to 4.2 V to a high-efficient (up to 96% efficiency) buck-boost switching voltage converter that, in turn, provides 4.9 V regulated voltage output and up to 700 mA constant current. A load sharing integrated circuit (IC) varies the converter's output voltage to control the current it supplies. The output stage consists of an OR-ing circuitry that connects the regulated voltage output to the module bus voltage (line VBUS). It comprises a P-channel MOSFET controller IC and a low resistance P-channel MOSFET, which together act as a diode with negligible losses.

If a charged *e*-module is connected with other modules, it energises them. An ad hoc controlled voltage power bus emerges through the sharing lines (VBUS) and (GND). The OR-ing circuitry preserves the integrity of the power bus formed in the polymer. Within each module, the power bus supplies two current/voltage sensors, five micro servomotors along with their switches and one micro-controller plus its additional converter (4.9/3.3 V).

If a polymer comprises more than one charged *e*-module among other types, power management becomes more complex as multiple regulated sources are connected to a common power line. Although such a situation may be favourable, it implies some additional elements to consider. Namely, the aforementioned OR-ing diode prevents the whole bus from collapsing if one source fails. Furthermore, it allows hot plugging. On the other hand, diodes prevent sources with different voltage levels from providing current simultaneously to the power bus. Therefore, only the highest voltage source would provide current at a time, which is undesirable. The *e*-module solves this situation by using a load sharing circuit. This circuit is placed between the converter and the OR-ing diode. The function of this circuit is to balance the current of the different *e*-modules connected through the power bus. To that end, it compares its own converter's current to that of other *e*-modules by sensing a dedicated signal (LS). This signal can be both input and output. The converter that provides the highest current sets the voltage of line (LS). The others use this information to adjust their current by adjusting the voltage of their converter until each module provides a similar current. Thus, not only do *e*-modules supply energy to other modules, but they also coordinate themselves to balance the power they supply to the bus.

3.2 Energy Harvesting System

The *e*-module uses a solar panel to obtain energy from the environment. The panel is mounted on the top of the module in a 3-D printed ABS holder (Fig. 2a-C). It consists of three photovoltaic (PV) cells (14 × 45 mm each) that convert radiation into electrical power. Each PV cell has an efficiency of 22% and can harvest up to 270 mW. The output of the panel feeds a non-linear charger IC, which is connected to the module's battery. This charger, which uses the MPPT[3] algorithm, performs with an efficiency of up to 95%. Moreover, as the battery is connected to the power management system's converter, the solar panel charger is also connected to it. Thus, the solar panel can also supply power directly to that converter.

Energy can be transferred between *e*-modules. A low energy *e*-module can provide energy to its own battery by drawing it from a high energy *e*-module through the power bus using dedicated battery charger circuitry. This circuitry integrates the same buck-boost converter as the power management system along with a current limiter circuit. As a result, a high-efficiency constant current/constant voltage charger is implemented, which is suitable for recharging lithium ion batteries.

3.3 Fabrication and Software Loading

3-D printing a set of four chassis for the evo-bot modules took 17 h. Assembling an evo-bot module given a 3-D printed chassis and populated PCB takes 1–2 h. This involves sanding the module and attaching the magnets, servo horns, servos and PV cells. The total cost to construct a single module is approximately £175, including PCB manufacture and 3-D printing costs. The evo-bot software, written in C code, is loaded onto the modules via a suitable programmer.

[3]Maximum Power Point Tracking. Weak sources may collapse if they have to supply power that exceeds their limit. MPPT algorithms reach the maximum power point of a source and stay at this level. Therefore, the source supplies its maximum power in a safe way.

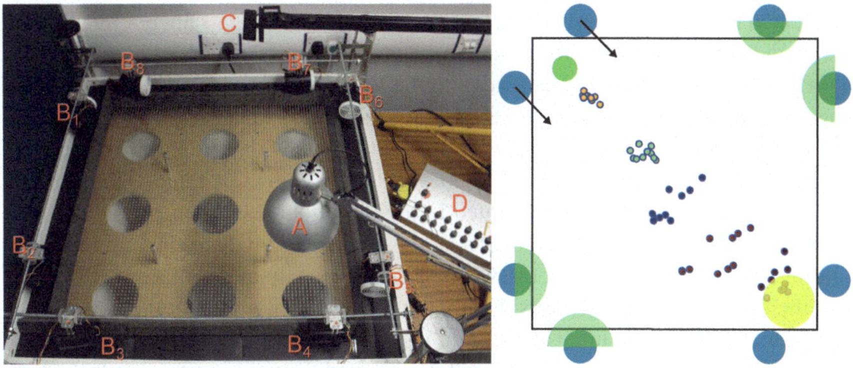

Fig. 5 Left: The evo-bot air table, showing the overhead lamp (A), the side fans (B_1 to B_8), the overhead camera (C), and the fan control box (D). Right: The light detection and motion control experiment. The side fans are indicated by large blue circles. The yellow circle in the bottom right indicates the position of the light source, and the green circle in the top left the module starting position. The colored dots represent the final positions of the modules (10 trials per color), from top left to bottom right: orange—0.3 V, teal—0.4 V, blue—0.5 V, red—0.8 V, brown—2.0 V

4 Experiments

In this section we demonstrate the key features of the evo-bots. Videos of the experiments are available [10].

The experimental environment is shown in Fig. 5 (left side). It consists of a square air table of side-length 85 cm. The surface is a 10 mm thick acrylic sheet with holes drilled in a square pattern with a spacing of 10 mm. The holes have a diameter of 2.5 mm. The main upward force is supplied by four industrial air blowers positioned below the table. The environment has foam boundaries of height 30 mm, which protect the modules from damage as they contact the borders. Eight side fans (two per side) are mounted on rails around the table, pointing inwards. These supply motive forces to the modules on the air table. Each fan can rotate in the horizontal plane up to 180°. The rotation and speed of each fan can be set via a control box, which allows the fans to be activated in preset patterns throughout the experiments. An 18.5 W LED lamp with a 25° beam angle is used as the light source for the experiments unless otherwise noted. Mounted on a lamp stand it can be positioned arbitrarily over the air table. Finally, an overhead camera is used to track the modules throughout the experiments for post-analysis.

4.1 Light Detection and Motion Control

For an organism to maintain a constant supply of energy it must be able to locate an energy source. To do this, an evo-bot module must be capable of detecting whether it is located beneath an energy (light) source, and of activating its anchor mechanism to stop and charge. We test this capability here. The specific set up is shown in Fig. 5 (right). An evo-bot module was placed in the top left corner of the table. The light source was positioned 12 cm above the surface of the table in the bottom right corner, facing downwards vertically. For this experiment a bulb with a 360° beam angle was used, in order to provide a smooth light gradient over the air table. The two side fans in the top left corner were pointed in the direction of the bottom right corner. The bottom left and top right side fans were set to oscillate over a 180° range. The bottom right side fans were not used. The fans were set up in this manner in order to push the evo-bot module towards the light source. The evo-bot module was given a certain threshold for the intensity of light incident on its solar panel (determined by measuring the voltage across the PV cells). The module was blown towards the light source and deployed its anchor once the threshold was exceeded. Its final position was recorded.

We conducted 50 trials: 10 trials each for 5 different voltage thresholds. Figure 5 (right) shows the recorded positions at the end of each trial. As can be seen, the final positions of the module are stratified depending on their thresholds—modules with higher thresholds get closer to the light source before stopping. This confirms that the module interacts with its environment as intended.

4.2 Energy Harvesting and Sharing

We examine the ability of an evo-bot module to be charged via charging station (at 4.9 V), solar panel, or another module (which we refer to as trophallaxis). In each case, the module started fully discharged. It was then charged for a certain period, and then discharged to evaluate its operating time. During the charging period only the microprocessor was running. During the discharging period, in addition, one servomotor was activated every 2 s in order to simulate the typical energy expenditure of a module during experimentation. This value was chosen based on the other experiments we conduct in this paper. However less time-sensitive experiments may not require such frequent activity, reducing energy expenditure and increasing operating time.

The charging/discharging curves are shown in Fig. 6. Via charging station the module charging period lasted until the battery was charged to 4.2 V. This took 140 min, and provided 240 min of operating time. Due to the slow rate of charge via solar panel, charging up to 4.2 V was not feasible. Instead, the charging period was set to 500 min. This charged to 3.9 V and provided 120 min of operating time. For trophallactic charging, the donor module started fully charged (4.2 V). The receiving module was then charged by the donor module until the donor module was fully discharged. The receiving module charged to 3.9 V over 160 min, subsequently providing 30 min of operating time. An artificial ecosystem could see these charging methods combined. All organisms could feed from a limited number of charging stations, but demand would be high. Those organisms with *e*-modules would be able to locate and charge from a distribution of light sources.

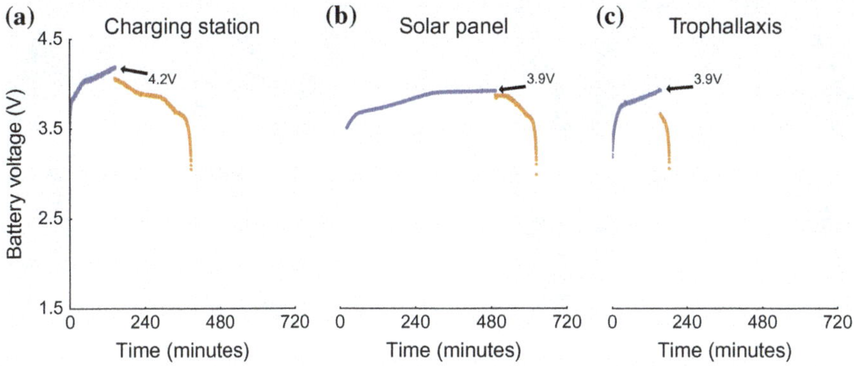

Fig. 6 Graphs showing the charging and subsequent discharging of an evo-bot module. Charging via **a** charging station, **b** solar panel, or **c** another evo-bot module

4.3 *Polymer Formation*

We examine the ability of the evo-bots to form linear polymers. The modules were set up as follows. Individual modules were able to form a connection on any of their faces. Once a module connected to another, it refused connections with new modules on faces orthogonal to its existing connection(s). In this way only linear polymers could form permanently (other configurations could form temporarily until incorrectly attached modules were rejected). Modules periodically closed and opened their connections in order to free themselves if connected in the wrong position. At the outset of each trial four evo-bot modules were positioned at the corners of a square of side-length 34 cm, concentric with the air table. The fifth module was placed at the center of the table. The rotation of the modules was arbitrary. The modules were not given any specific type designation (*e*-module, *i*-module, *b*-module) for the purpose of this experiment.

A total of ten trials were performed. The average length of time needed to form a 5-module polymer was 188 s. The longest time taken was 407 s. The shortest time taken was 62 s. Figure 7 shows stills from an example trial.

4.4 *Summary*

The experiments we have detailed in this section relate the evo-bot hardware to the concept discussed in Sect. 2. We have shown that the evo-bots can grow via module accretion, locate energy of various intensities, and harvest energy and share it with connected modules. These attributes represent the first stages of a hardware implementation of the evo-bot concept.

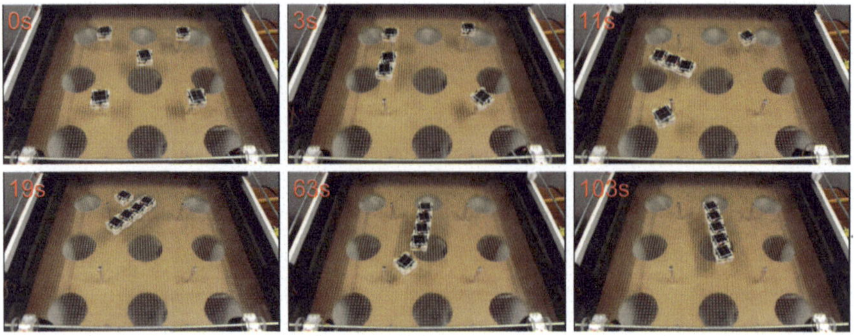

Fig. 7 A sequence of stills taken from a video of five evo-bot modules forming a linear polymer

5 Conclusion

This paper described the evo-bot concept—a simple modular system that could be used to physically implement artificial life. We have shown that a potential full implementation of the concept, in which artificial organisms are capable of growing, harvesting energy and replicating, satisfies at least one definition of life. We have developed and presented a set of physical modules inspired by the evo-bot concept. While these modules do not yet represent a full implementation of the concept, they form the basis of one. The evo-bots feature a novel stop-start motion control mechanism, which requires only a single binary actuator, but allows them control over their movement. We have performed experiments with up to five prototypes to demonstrate the system's main capabilities: hybrid motion control, light sensing, energy harvesting and trophallaxis, and self-assembly. Our next aim is to introduce polymer self-replication in order to fully implement the evo-bot concept. In addition we will conduct long-term experiments involving a large number of modules.

Acknowledgements This research was supported by a Marie Curie European Reintegration Grant within the 7th European Community Framework Programme (grant no. PERG07-GA-2010-268354). It was also funded by the Engineering and Physical Sciences Research Council (EPSRC) through scholarship support (M. Doyle) and grant no. EP/K033948/1. In addition the authors would like to thank Paul Eastwood and Michael Port for their invaluable assistance in preparing the experimental environment.

References

1. Bishop, J., Burden, S., Klavins, E., Kreisberg, R., Malone, W., Napp, N., Nguyen, T.: Programmable parts: A demonstration of the grammatical approach to self-organization. In: 2005 IEEE/RSJ International Conference on Intelligent Robots and Systems, pp. 3684–3691. IEEE (2005)

2. Breivik, J.: Self-organization of template-replicating polymers and the spontaneous rise of genetic information. Entropy **3**(4), 273–279 (2001)
3. Brodbeck, L., Hauser, S., Iida, F.: Morphological evolution of physical robots through model-free phenotype development. PLOS ONE **10**(6), e0128,444 (2015)
4. Chirikjian, G.S., Zhou, Y., Suthakorn, J.: Self-replicating robots for lunar development. IEEE/ASME Trans. Mechatron. **7**(4), 462–472 (2002)
5. Cleland, C.E., Chyba, C.F.: Defining life. Origins of Life Evol. Biosph. **32**(4), 387–393 (2002)
6. Demir, N., Açıkmeşe, B.: Probabilistic density control for swarm of decentralized on-off agents with safety constraints. In: 2015 American Control Conference (ACC), pp. 5238–5244. IEEE (2015)
7. Ding, R., Eastwood, P., Mondada, F., Groß, R.: A stochastic self-reconfigurable modular robot with mobility control. In: TAROS 2012, vol. 7229 pp. 416–417. LNCS, Springer (2012)
8. Eiben, A.: Evosphere: The world of robot evolution. In: International Conference on Theory and Practice of Natural Computing, vol. 9477, pp. 3–19. LNCS, Springer (2015)
9. Eiben, A.E.: Grand challenges for evolutionary robotics. Front. Robot. AI **1**(4), 1–2 (2014)
10. Escalera, J.A., Doyle, M.J., Mondada, F., Groß, R.: Online supplementary material (2016). http://naturalrobotics.group.shef.ac.uk/supp/2016-006/
11. Escalera, J.A., Mondada, F., Groß, R.: Evo-bots: A modular robotics platform with efficient energy sharing. In: Modular and Swarm Systems Workshop at IROS 2014 (2014). https://sites.google.com/site/iros2014mss/abstracts
12. Griffith, S., Goldwater, D., Jacobson, J.M.: Robotics: self-replication from random parts. Nature **437**(7059), 636 (2005)
13. Groß, R., Magnenat, S., Küchler, L., Massaras, V., Bonani, M., Mondada, F.: Towards an autonomous evolution of non-biological physical organisms. In: ECAL 2009, vol. 5777 pp. 173–180. LNAI, Springer (2011)
14. Haghighat, B., Droz, E., Martinoli, A.: Lily: A miniature floating robotic platform for programmable stochastic self-assembly. In: ICRA 2015, pp. 1941–1948. IEEE (2015)
15. Jacobson, H.: On models of reproduction. Am. Sci. **46**(3), 255–284 (1958)
16. Kernbach, S., Meister, E., Schlachter, F., Jebens, K., Szymanski, M., Liedke, J., Laneri, D., Winkler, L., Schmickl, T., Thenius, R., et al.: Symbiotic robot organisms: replicator and symbrion projects. In: 8th Workshop on Performance Metrics for Intelligent Systems, pp. 62–69. ACM (2008)
17. Klavins, E.: Programmable self-assembly. IEEE Control Syst. **27**(4), 43–56 (2007)
18. Koshland, D.E.: The seven pillars of life. Science **295**(5563), 2215–2216 (2002)
19. Lipson, H., Pollack, J.B.: Automatic design and manufacture of robotic lifeforms. Nature **406**(6799), 974–978 (2000)
20. Miconi, T.: Evosphere: evolutionary dynamics in a population of fighting virtual creatures. In: 2008 IEEE Congress on Evolutionary Computation, pp. 3066–3073. IEEE (2008)
21. Penrose, L.S., Penrose, R.: A self-reproducing analogue. Nature **179**(4571), 1183 (1957)
22. Ruiz-Mirazo, K., Peretó, J., Moreno, A.: A universal definition of life: autonomy and open-ended evolution. Origins Life Evol. Biosph. **34**(3), 323–346 (2004)
23. Spector, L., Klein, J., Feinstein, M.: Division blocks and the open-ended evolution of development, form, and behavior. In: 9th Annual Conference on Genetic and Evolutionary Computation, pp. 316–323. ACM (2007)
24. Virgo, N., Fernando, C., Bigge, B., Husbands, P.: Evolvable physical self-replicators. Artif. Life **18**(2), 129–142 (2012)
25. Weel, B., Crosato, E., Heinerman, J., Haasdijk, E., Eiben, A.: A robotic ecosystem with evolvable minds and bodies. In: 2014 IEEE International Conference on Evolvable Systems (ICES), pp. 165–172. IEEE (2014)
26. White, P., Kopanski, K., Lipson, H.: Stochastic self-reconfigurable cellular robotics. In: ICRA 2004, vol. 3, pp. 2888–2893. IEEE (2004)
27. Zykov, V., Mytilinaios, E., Adams, B., Lipson, H.: Robotics: self-reproducing machines. Nature **435**(7039), 163–164 (2005)

Geometrical Study of a Quasi-spherical Module for Building Programmable Matter

Benoît Piranda and Julien Bourgeois

Abstract The aim of the Claytronics project is to build spherical micro-robots, called catoms for Claytronics atoms able to stick to each other and able to move around each other. An ensemble of catoms is therefore a huge modular self-reconfigurable robot. However, the shape of these catoms have not been studied yet and remains a difficult problem as there are numerous constraints to respect. In this article, we propose a quasi-spherical catom which answers to all the constraints to build programmable matter.

1 Introduction

On a broad scope, programmable matter is a matter which can change one or several of its physical properties, most likely its shape, according to an internal or an external action. An example of a mug being created by an ensemble of micro-robots is presented in Fig. 1. Programmable matter can have different properties depending on the underlying technology chosen: evolutivity, programmability, autonomy, interactivity [2].

Only modular self-reconfigurable robots (MSR) can implement this full set of properties as they can embed computation. MSR [15, 18] also named earlier as metamorphic robotic systems [3] or as cellular robotic systems [6] are composed of individual modules able to move relatively to each other to create different configurations. There are four kinds of MSR: lattice-based when modules are aligned on a lattice, chain-type when the module are aligned but with more degree of freedom, hybrid which is a mix between lattice-based and chain-type and mobile when each module can move autonomously. The expected properties of MSR are: versatility,

B. Piranda (✉) · J. Bourgeois
University of Bourgogne Franche-Comté (UBFC), University of Franche-Comté (UFC)
FEMTO-ST Institute, Montbéliard, France
e-mail: benoit.piranda@femto-st.fr

J. Bourgeois
e-mail: julien.bourgeois@femto-st.fr

R. Groß et al. (eds.), *Distributed Autonomous Robotic Systems*,
Springer Proceedings in Advanced Robotics 6,
https://doi.org/10.1007/978-3-319-73008-0_27

Fig. 1 A simulation view of a mug and a toy car made of programmable matter and composed of micro robots

used to fulfill different tasks, robustness as a faulty module can be discarded and affordable price as the mass production of identical modules is likely reducing the overall cost [20]. MSR is an active field of research which has produced interesting hardware starting from CEBOT [5], or Polybot [17], as pioneered approaches, to SMORES [4], ATRON [11] or M-Blocks [14], for the latest ones.

This work is part of the Claytronics project. Claytronics, which stands for *Clay-Electronics* is an implementation of programmable matter initiated by Carnegie Mellon University and Intel Corporation, and then joined by FEMTO-ST Institute. In Claytronics, mm-scale robots called catoms, for Claytronics Atoms, are assembled to form larger objects. The idea is that each micro-robot has very restricted or, let us say, only strictly mandatory functionalities and as each catom is simple, hundreds of thousands can be assembled all together to create new solid objects of any shape or size. A catom should be a mass-producible, sub-mm, MEMS using computationally controlled forces for adhesion and locomotion. Each catom therefore embeds a chip for computation and for driving its actuators and communication capabilities. A first Claytronics prototype has been realized which embeds actuation and a chip for managing the movement of the micro-robots [9]. There are many challenges to solve before Claytronics could transform matter into programmable matter. These challenges and perspectives have been enumerated in [1].

In the context of programmable matter, the size of each module should be as small as possible, mm-scale seems a good resolution for approximating a shape. Scaling down the hardware is a difficult task but it also offers benefits. For example, actuation could be easier at the micro-scale by using electrostatic forces. MSR are usually composed of few modules whereas the objective of programmable matter is to have hundred thousands modules, at least, which adds even more complexity in the whole design.

To implement programmable matter as an MSR several solutions are possible. We think a lattice-based MSR, in which each module is aligned on a lattice, is the best solution. Chain-type MSR is not suitable as it is limited in the number of modules, hybrid could be used but lattice-based allows more flexibility for having a better approximation of the shape, more nodes are also able to move during self-reconfiguration and robustness is increased as any module can fail without compromising the rest.

For lattice-based MSR, different types of modules have been studied: cubic, cylindric, polyhedral and spherical. Cubic modules offer a good adhesion surface but they are difficult to move, that is why Miche [7] and Pebbles [8] have opted for self-disassembly which drastically reduces the possibilities of self-reconfiguration. Smart Blocks' [12] uses electro-permanent magnets for 2D moves but they are not strong enough to fight against gravity. The best realization so far are M-blocks [14] in which each module turns around another one using a gyroscope. However, this technology is difficult to miniaturize.

Within the Claytronics project, two kinds of cylindrical modules have been built: a macro-size vertical cylinder linked with others by electromagnetic forces [10] and a mm-scale horizontal cylinder using electrostatic forces to stick and move [9]. This latest realization has pushed the limits of miniaturization for an autonomous microrobot and is a source of inspiration for the future of Claytronics.

Designing a miniaturized sphere is complex. A macro-size example is ATRON [11]. The latching is mechanical and each quasi-spherical module is split in two and can rotate. However, as for M-blocks, the technologies employed cannot be used at the mm-scale. In the Claytronics project, a sphere has been fabricated using petals closing the shape [13]. This sphere was a first prototype of a mm-scale 3D shape although it is not autonomous, did not embed the connectors and the structure was too fragile.

The rhombic dodecahedron (RD) shape proposed by Yim et al. [19] is an interesting geometrical solution. It allows to place a set of modules that completely fills the 3D space without hole. But the main drawback is mentioned in the document, it concerns the 120 rotation around an edge of a module in order to pass from a cell to a neighbor. This motion requires an actuator technically difficult to produce.

To sum up, cubes are difficult to move in 3D, cylinder cannot move in 3D and a real sphere is difficult to built at a mm-scale from a 2D sheet of material. In this article, we present a proposition of a quasi-spherical module, called a catom in the context of the Claytronics project, able to fit all the requirements of programmable matter.

2 Objectives and Constraints

Programmable matter needs a very large set of connected small modules, called catoms, that can move to change their global shape. The goal of our work is to produce very small (about 1 mm diameter) and reliable catoms, without complex mechanical systems. At this scale, the latching and moving forces can be produced by electrostatic electrodes. The proposed system must satisfy a number of constraints. The catoms composing this system must:

1. Be combined in order to regularly fill a 3D space (following a predefined grid).
2. Have a large surface of contact (we call this area 'a connector') allow to use these surfaces for communication, presence sensors and power transfer.

3. Be free to move from one position of the grid to a neighboring free one.
4. After a motion, a module must be oriented in order to place each of its connectors in front of one connector of its neighbor modules.
5. Be physically connected to many neighbors, in order to allow power transfer and P2P communication.
6. Have a finite number of connectors that are always in the same place on the surface of the catom.
7. Be fabricated from a deformation of a flat shape.

In the next section, we study the possible organization and geometry of the catoms. In Sect. 4, we present a quasi-sphere geometry and try to solve all the constraints. In Sect. 5, we discuss the catom movements.

3 Study of Candidate Shapes and Organizations

Here, we study the shape of the catoms by examining three different solutions and then the organization of the lattice which is also of prime importance. In the rest of this article, we consider that our catoms have a diameter d and a radius r.

The first idea of shape for defining a catom consists in adapting the cubic model where each catom is placed along a regular grid oriented along $(\overrightarrow{x}; \overrightarrow{y}; \overrightarrow{z})$ axes. Rotation of catoms around a neighbor is facilitated by using rounded angles at the edges of the cubes. A face of a cube is therefore composed by a flat centered area for connectors and a curved border allowing the cube to move by turning around. But, with a rounded cube, the slip-free rolling of a cube A around another cube B does not allow to reach a neighbor cell of the grid. The distance between two connected cells is equal to the length of the cube ($2 \times r$) and each rotation produces a $\alpha \times \rho$ long motion (where α is the rotation angle of the rounded cube and ρ the radius of the curved part). Then, if a cube A has to reconnect on to a face of B, we must have $\frac{\pi}{2} \times \rho = \frac{\pi}{2} \times r$ as shown in Fig. 2. But, as $\rho < r$, rounded boxes **do not verify constraints #3**.

The second proposition for this micro-robot is a sphere, that can be naturally organized in a grid. Spheres can be placed in a regular grid that reduces the distance between center and empty areas. Spheres are placed in a regular 2D grid (along $\overrightarrow{x}$ and $\overrightarrow{y}$ and d large cells) for the first floor and the second floor is placed $\frac{\sqrt{2}}{2}r$ higher and shifted by r along $\overrightarrow{x}$ and $\overrightarrow{y}$. Under this organization each sphere has

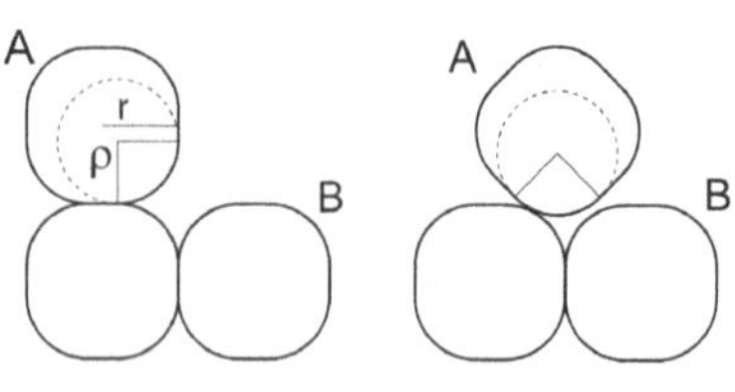

Fig. 2 Limits of the rounded cube solution: when A turns around B, it can not reach a position juxtaposing a face of A to a face of B

a maximum of 12 connected neighbors. But, two spheres only admit a single point as contact area. Moreover, manufacturing a perfect sphere is difficult. Therefore, a sphere shape **does not satisfy constraints #2 and #7**.

As the cube and the sphere do not meet all the requirements, the idea is to use a mix between them. We want to take advantage of the large surfaces of the cube for the latching, together with the easiness of rotation offered by the sphere, by designing a quasi-sphere.

4 Solving the Constraints for the Quasi-sphere

The design of the 3D catoms starts with a sphere on which we add connectors. We have to place these connectors and define their shape, size and orientation in order to verify the constraints. We will then define curves between connectors to construct paths for the rotation of a catom around one of its neighbors.

4.1 Designing Connectors for Latching

In order to answer constraint #2, connectors must be 12 planar surfaces centered in $P_0 \dots P_{11}$, where P_i are contact points of spheres in the Face-Centered Cubic lattice (FCC) as presented in [16]. Considering a sphere of radius r centered at the origin (O), these points are respectively placed at the following coordinates:

$$\begin{array}{l|l||l|l} P_0(r,0,0) & P_6(-r,0,0) & P_1(0,r,0) & P_7(0,-r,0) \\ P_2(\frac{r}{2},\frac{r}{2},\frac{r}{\sqrt{2}}) & P_8(-\frac{r}{2},-\frac{r}{2},-\frac{r}{\sqrt{2}}) & P_3(-\frac{r}{2},\frac{r}{2},\frac{r}{\sqrt{2}}) & P_9(\frac{r}{2},-\frac{r}{2},-\frac{r}{\sqrt{2}}) \\ P_4(-\frac{r}{2},-\frac{r}{2},\frac{r}{\sqrt{2}}) & P_{10}(\frac{r}{2},\frac{r}{2},-\frac{r}{\sqrt{2}}) & P_5(\frac{r}{2},-\frac{r}{2},\frac{r}{\sqrt{2}}) & P_{11}(-\frac{r}{2},\frac{r}{2},-\frac{r}{\sqrt{2}}) \end{array} \tag{1}$$

On these points, we must define a regular planar surface which will contain the connectors. As we have seen in Sect. 3, the rotation path to move from one connector to another must be equal to the distance from two connectors. The obvious choice about the connector shape is therefore a square. Furthermore, a square is a simple surface to manufacture. We then search which size (c) of the edges of the connectors allow to construct a regular 3D shape. The orientation of square is partially imposed by connection constraints #2 and #4: The squares must be placed tangentially to the sphere surface.

Figure 3 presents two c width connectors, we name corners of these connectors respectively $\{A_0, B_0, C_0, D_0\}$ and $\{A_2, B_2, C_2, D_2\}$. Constraint #4 states the distance C_0A_2 must be equal to c in order to be able to reconnect two connectors after a movement. In this section, we want to express c to define the size of the connectors. We can easily express the coordinates of A_0 and C_0, and deduce A_2 as the image of A_0 by rotation of $\frac{\pi}{4}$ around $\overrightarrow{y}$ axis and then rotation of $\frac{\pi}{4}$ around $\overrightarrow{z}$ axis (given by the matrix $\mathcal{M}$):

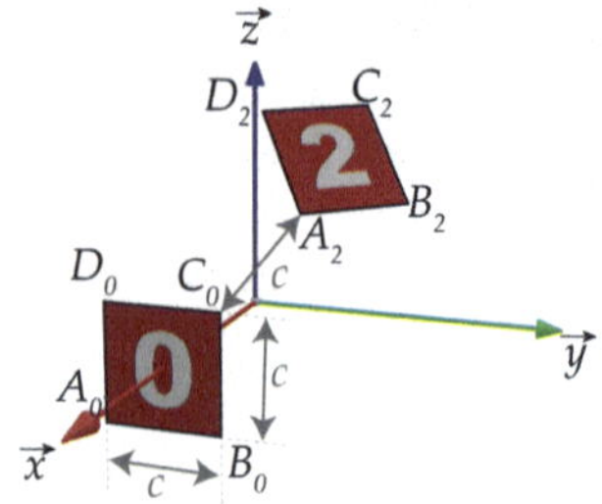

Fig. 3 Geometry used to calculate the width of square connectors c

$$A_0 = \left(r, -\frac{c}{2}, -\frac{c}{2}\right) \tag{2}$$

$$\mathcal{M} = \begin{pmatrix} \frac{1}{2} & -\frac{\sqrt{2}}{2} & -\frac{1}{2} \\ \frac{1}{2} & \frac{\sqrt{2}}{2} & -\frac{1}{2} \\ \frac{\sqrt{2}}{2} & 0 & \frac{\sqrt{2}}{2} \end{pmatrix} \tag{3}$$

$$A_2 = \mathcal{M} \times A_0 = \begin{pmatrix} \frac{r}{2} + \frac{c}{4}\left(1+\sqrt{2}\right) \\ \frac{r}{2} + \frac{c}{4}\left(1-\sqrt{2}\right) \\ \frac{\sqrt{2}}{2}r - \frac{\sqrt{2}c}{4} \end{pmatrix} \tag{4}$$

As the distance C_0A_2 must be equal to c. We can solve $\overrightarrow{C_0A_2}^2 = c^2$.

$$C_0 = \left(r, \frac{c}{2}, \frac{c}{2}\right) \tag{5}$$

$$\overrightarrow{C_0A_2} \begin{pmatrix} -\frac{r}{2} + \frac{c}{4}\left(1+\sqrt{2}\right) \\ \frac{r}{2} - \frac{c}{4}\left(1+\sqrt{2}\right) \\ \frac{\sqrt{2}}{2}r - \frac{c}{2}\left(1+\frac{\sqrt{2}}{2}\right) \end{pmatrix} \tag{6}$$

We can then express $\overrightarrow{C_0A_2}^2$:

$$\overrightarrow{C_0A_2}^2 = r^2 + \frac{c^2}{4}\left(2\sqrt{2}-1\right) - r \times c\left(\sqrt{2}+1\right) = \left(r - \frac{\sqrt{2}+1}{2}c\right)^2 \tag{7}$$

In order to obtain an expression of c, we solve:

$$r - \frac{\sqrt{2}+1}{2}c = c \tag{8}$$

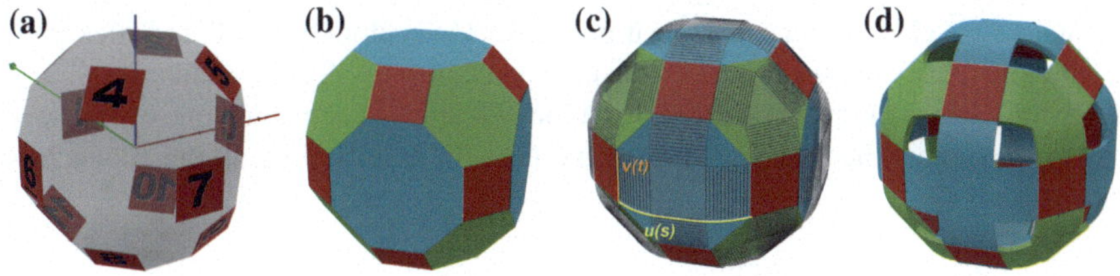

Fig. 4 Design phases of the 3D catoms: **a** Connectors are defined as squares and placed in order to be connected to neighbors. **b** The geometry of a regular filled volume linked to square connectors appears. **c** Curved actuators are placed over octagonal and hexagonal surfaces. **d** Final catoms

Finally, the connectors width c only depends on the radius r of the inscribed sphere:

$$c = \frac{2 \times r}{3 + \sqrt{2}} \approx 0.45308 \times r \tag{9}$$

This is indeed a very interesting result which shows that the size of the connectors cannot be arbitrarily fixed but must have a size only related to the size of the catom.

The geometry is therefore defined by a regular geometrical shape composed of 12 squares representing the connectors (in red, in Fig. 4b). To bind the connectors all together, 8 hexagons and 6 octagons are added (resp. in green and blue). Each connector is connected to 2 hexagons and 2 octagons. Each octagon is bound to 4 squares and 4 hexagons and each hexagon is connected to 3 squares and 3 octagons alternatively along its border. All the edges for all the shapes (squares, hexagons and octagons) have the same length c: it is an important condition to reconnect catoms after a rotation.

4.2 Designing Actuators for Allowing Movements

4.2.1 Defining the Shape of the Actuators

Curved actuators need to be placed between connectors in order to propose a path of contact during rotation of a catom around another one. The shape of this surface is relatively free, but the constraint is that the rotation must not go through any angular edges. These curved actuators need to be placed between the connectors in order to have a path of contact during rotation of a catom around another one.

We propose to define a C^1 class surface for the actuators to avoid angles. We point out that considering that a surface $S(s, t)$ is defined by two curves $S(s, t) = u(s) + v(t)$, the surface is a C^1 class surface if respectively u and v are derivable relatively to s (respectively t), and their derivative are continuous. The proposed actuators shape are made of a curved part (slice of a cylinder) and a plane, in order to obtain a C^1 class surface as shown in Fig. 4c. One catom will roll and stick onto another one between two connectors using an actuator as rotation path.

On the surface of the catom, we can define two kinds of actuators, one over octagonal faces and another one over hexagonal faces in order to link all the connectors with actuators.

4.2.2 Characterization of the Actuator on the Octagonal and Hexagonal Faces

On the octagonal faces, the actuator is formed by a square, centered on the octagon, and linked to a curved slice of a cylinder on its four sides. The square sides have the same size as the actuator width (See Fig. 4d). In order to find the radius of curvature r_1, the angle β and the center M of the blue arc, we consider two connectors C_0 and C_1 (with a side size of c, drawn in red in Fig. 5). C_0 is centered at $(r, 0, 0)$, with two corners $P_0\left(r, -\frac{c}{2}, -\frac{c}{2}\right)$ and $P_1\left(r, \frac{c}{2}, -\frac{c}{2}\right)$. C_1 is centered at $(0, r, 0)$, with two corners $P_2\left(-\frac{c}{2}, r, -\frac{c}{2}\right)$ and $P_3\left(\frac{c}{2}, r, -\frac{c}{2}\right)$. Then, we have to place a square S (drawn in orange in Fig. 5) between the two connectors C_0 and C_1. In order to respect the symmetry, the center of S is placed along the axis $(O, \overrightarrow{u})$, with $\overrightarrow{u} = \left(\frac{\sqrt{2}}{2}, \frac{\sqrt{2}}{2}, 0\right)$ and is tangent to the orthogonal vector $\overrightarrow{t} = \left(\frac{-\sqrt{2}}{2}, \frac{\sqrt{2}}{2}, 0\right)$. We can write the position of M as the intersection of the two radii MP_1 and MQ_0:

$$\begin{cases} \overrightarrow{OM} = (r - r_1)\,\overrightarrow{x} + \frac{c}{2}\,\overrightarrow{y} \\ \overrightarrow{OM} = (k - r_1)\,\overrightarrow{u} - \frac{c}{2}\,\overrightarrow{t} \end{cases} \tag{10}$$

where k is the distance between the center of S and the center of the catom. In order to obtain the position of M, we express $\overrightarrow{OM}.\overrightarrow{y}$ with the two previous relations:

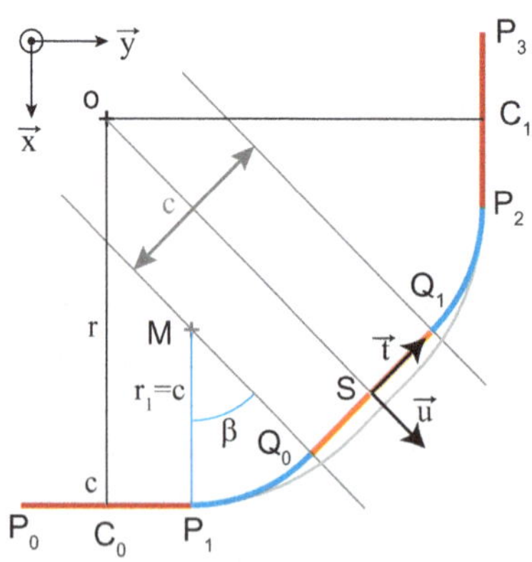

Fig. 5 Cross section of domes for octagonal area, $w = c$. Main connectors are drawn in red, flat connector in intersection of actuators are drawn in orange and curved actuator are in blue

$$\frac{c}{2} = \frac{\sqrt{2}}{2}\left(k - r_1 - \frac{c}{2}\right) \tag{11}$$

We deduce $k - r_1 = \frac{1+\sqrt{2}}{2}c$, and obtain the position of M:

$$M\left(\frac{1+\sqrt{2}}{2}c, \frac{c}{2}, -\frac{c}{2}\right) \tag{12}$$

In order to calculate r_1, we evaluate $\overrightarrow{OM}.\overrightarrow{t}$ with the relations of Eq. 10:

$$-\frac{\sqrt{2}}{2}(r - r_1) + \frac{\sqrt{2}}{2}\frac{c}{2} = -\frac{c}{2} \tag{13}$$

We obtain $r_1 = c$. Finally, as β is the angle between $\overrightarrow{x}$ and $\overrightarrow{u}$, then $\beta = 45°$ or $\frac{\pi}{4}$ rad. The cylindrical part of the actuator admits a center at M, a radius $r_1 = c$ and a height c, all the other cylinders are obtained by rotations (corresponding to combinations of matrix $\mathscr{M}$ presented in Eq. 3).

The calculation for the hexagonal faces follows the same process as for octagonal ones. The intersection of the paths are equilateral triangles of c side long. These triangles are linked to connectors by a portion of cylindrical actuator. We calculate the radius of curvature:

$$r_1' \approx 0.4670 \times r \tag{14}$$

and the angle of slice:

$$\alpha = \tan^{-1}\left(\frac{\sqrt{2}}{2}\right) \approx 0,6155 \text{ rad or } (35.26°) \tag{15}$$

Then, the length of the path is:

$$l_1' = r_1' \times \alpha \approx 0.28745 \times r \tag{16}$$

4.2.3 Adapting Geometry to Allow Movement

In the first case, that uses orthogonal domes displacements R_o (cf. in Fig. 6) are easy to apply: from a square connector face, 4 orthogonal directions are available to reach an other connector face. 90° rotations around axes are associated to each motion. In the second case, catoms use hexagonal faces to roll over another catom (following the R_h displacement in Fig. 6). From each square connector face, 4 ways are available

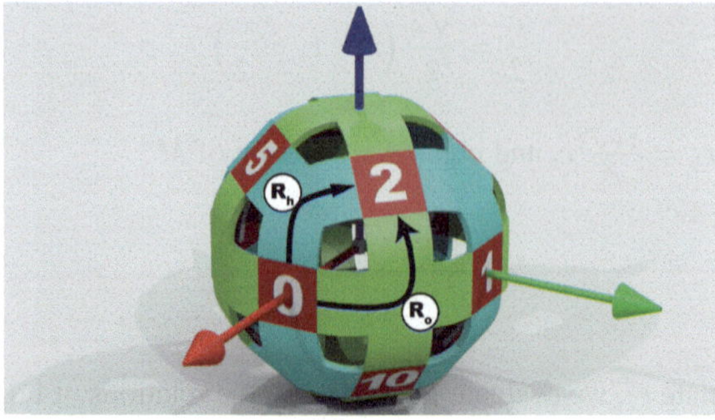

Fig. 6 A catom with two methods of rotation: R_h in the hexagonal area and R_o in the octagonal area

Fig. 7 Two cases where one of each kind of motions is possible but not the other one

too, but there are more difficult to follow. Each way is composed of two rotations of α where $\alpha = \tan^{-1}\left(\frac{\sqrt{2}}{2}\right) \approx 0.615$ rad or 35.26°.[1]

The main drawback of these two motions is the volume crossed by the moving catom. Several local configurations of neighbors prevent at least one of these motions. Images in Fig. 7 show two blocking situations, one for each motion mode. The left image shows a configuration where motion R_o is possible but not R_h, whereas the right image shows the opposite situation.

As we have found blocking situations for the first two solutions of actuation, we need to evaluate the last one using actuator on both octagonal and hexagonal faces. In Fig. 6, we can see that from the connector #0, there exists 6 other connectors that can be directly reached by rotations (#1, #2, #5, #7, #9 and #10).

More generally, we consider a mobile catom A and a fixed catom B. Graph shown in Fig. 8 details the set of rotations that can be applied from each connector. In that graph, nodes represent connectors of B (ID written in squares) and links represent rotations of α for hexagonal actuators or 45° for octagonal ones. Information in

[1] These motions are presented in a video available at the address https://youtu.be/tXbu7rB86OM.

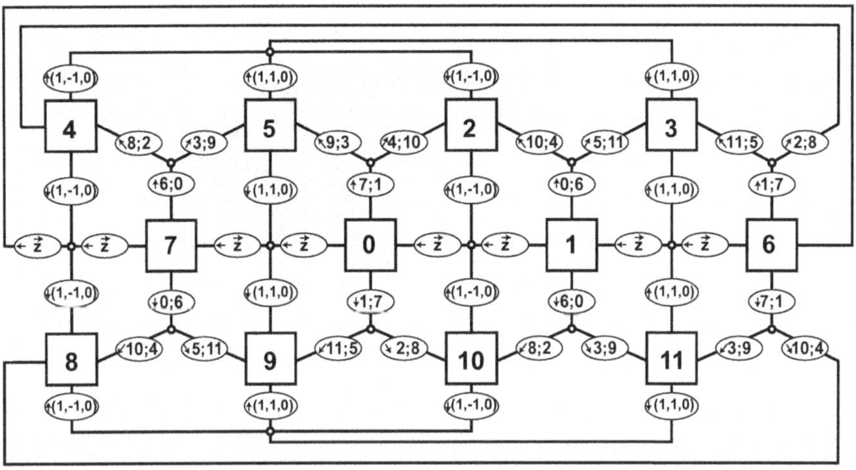

Fig. 8 Graph of possible rotations from each connector of the catom (index written in squares)

ellipses gives axes of rotation, directly specifying the rotation vector or a couple of connector centers that defines a rotation axis.[2]

5 Defining Ensemble of Catoms

In this section, we study the organization of the catoms in an ensemble.

The catoms have been designed to be organized in a FCC lattice following a hexagonal close packing placement. Figure 9 presents an example of the organization. Placing catoms on the floor on actuators (that replace one of their octagonal face), and orienting them in order that connected faces are square connectors, aligned along $\vec{x}$ and $\vec{y}$ axes, similarly to catoms shown in Fig. 6. Centers of these catoms are respectively placed at coordinates $(ix \times d; iy \times d; 0)$ where $ix \in \mathbb{Z}$ and $iy \in \mathbb{Z}$.

On the second floor, catoms are staggered over previous catoms, they center coordinates are: $(r + ix \times d; r + iy \times d; \frac{\sqrt{2}d}{2})$.

Then on the third floor, catoms are vertically placed above catoms of the first floor, 3 blue catoms in Fig. 9 are placed at: $(d; 0; \sqrt{2}d)$ $(2d; 0; \sqrt{2}d)$ $(d, d, \sqrt{2}d)$.

For programming algorithm using catoms, we need a relation between their position in space coordinates and their position in the grid. For example, during a rotation a catom turns around another one in the space coordinates, when it finishes its motion we can deduce its new position in the grid. We can write a relation between coordinates of a catom $(ix; iy; iz)$ in a regular array in memory and its position in the

[2]The video at https://youtu.be/20xRLOfoJQ4 shows a more complex succession of motions combining the two kinds of rotations as an example of the possibilities offered.

Fig. 9 Placement of 3d catoms in a regular grid

visualization space $(x; y; z)$ using $\sigma = iz\%2$, i.e. $\sigma = 0$ if iz is even and $\sigma = 1$ if iz is odd:

$$\left(x = d \times (ix + 0.5\sigma); \; y = d \times (iy + 0.5\sigma); \; z = \frac{\sqrt{2}d}{2} \times iz\right) \quad (17)$$

Similarly we can express array coordinates from space coordinates:

$$\left(ix = \left\lfloor \frac{x}{d} - 0.5\sigma \right\rfloor; iy = \left\lfloor \frac{y}{d} - 0.5\sigma \right\rfloor; iz = \left\lfloor \frac{\sqrt{2}}{d} \times z \right\rfloor\right) \quad (18)$$

Considering this approach, we express the coordinates of 12 neighbors of a catom placed at $(ix; iy; iz)$ in the regular grid.

Plane	Neighbors coordinates if iz is even	Neighbors coordinates if iz is odd
$iz + 1$	$(ix - 1; iy - 1)$ $(ix - 1; iy)$ $(ix; iy)$ $(ix; iy - 1)$	$(ix; iy)$ $(ix; iy + 1)$ $(ix + 1; iy)$ $(ix + 1; iy + 1)$
iz	$(ix - 1; iy)$ $(ix + 1; iy)$ $(ix; iy - 1)$ $(ix; iy + 1)$	$(ix - 1; iy)$ $(ix + 1; iy)$ $(ix; iy - 1)$ $(ix; iy + 1)$
$iz - 1$	$(ix - 1; iy - 1)$ $(ix - 1; iy)$ $(ix; iy)$ $(ix; iy - 1)$	$(ix; iy)$ $(ix; iy + 1)$ $(ix + 1; iy)$ $(ix + 1; iy + 1)$

6 Conclusion

We have proposed a detailed model for the realization of a quasi-spherical module for the programmable matter. This module, called catom in the context of the Claytronics project, satisfies all the constraints listed for the needs of the project. The 7th constraint has not been detailed in this paper for our catom, but an origami like unfold of our shape is possible and will be presented in future works.

We believe our quasi-spherical module will ease the movements using semi-curved shape actuators and also provide sufficiently strong connectors. Future works include a study of the physics parameters of the catom: weight, latching and actuation forces. We have already 3D printed catoms shells embedded with magnets to validate the latching and the movements. The next step is now to design the fabrication process for the MEMS/LSI integration on a flexible substratum. The major challenge will be the attachment of all the catom parts.

Acknowledgements This work has been funded by the Labex ACTION program (contract ANR-11-LABX-01-01) and ANR/RGC (contracts ANR-12-IS02-0004-01 and 3-ZG1F).

References

1. Bourgeois, J., Goldstein, S.C.: Distributed intelligent MEMS: progresses and perspectives. IEEE Syst. J. **9**(3), 1057–1068 (2015)
2. Bourgeois, J., Piranda, B., Naz, A., Lakhlef, H., Tucci, T., Mabed, H., Douthaut, D., Boillot, N.: Programmable matter as a cyber-physical conjugation. In: IEEE (ed.) IEEE International Conference on Systems, Man and Cybernetics (SMC) (2016)
3. Chirikjian, G.S.: Kinematics of a metamorphic robotic system. In: IEEE International Conference on Robotics and Automation (ICRA), pp. 449–455. IEEE (1994)
4. Davey, J., Kwok, N., Yim, M.: Emulating self-reconfigurable robots - design of the smores system. In: IEEE/RSJ International Conference on Intelligent Robots and Systems (IROS), pp. 4464–4469. Vilamoura, Algarve, Portugal (2012)
5. Fukuda, T., Kawauchi, Y.: Cellular robotic system (cebot) as one of the realization of self-organizing intelligent universal manipulator. In: IEEE International Conference on Robotics and Automation (ICRA), pp. 662–667 (1990)
6. Fukuda, T., Kawauchi, Y., Buss, M.: Communication method of cellular robotics cebot as a selforganizing robotic system. In: IEEE/RSJ International Workshop on Intelligent Robots and Systems' 89. The Autonomous Mobile Robots and Its Applications. IROS'89. Proceedings, pp. 291–296. IEEE (1989)
7. Gilpin, K., Kotay, K., Rus, D.: Miche: modular shape formation by self-dissasembly. In: Proceedings 2007 IEEE International Conference on Robotics and Automation, pp. 2241–2247 (2007)
8. Gilpin, K., Knaian, A., Rus, D.: Robot pebbles: one centimeter modules for programmable matter through self-disassembly. In: IEEE International Conference on Robotics and Automation (ICRA), pp. 2485–2492 (2010)
9. Karagozler, M.E., Thaker, A., Goldstein, S.C., Ricketts, D.S.: Electrostatic actuation and control of micro robots using a post-processed high-voltage soi cmos chip. In: IEEE International Symposium on Circuits and Systems (ISCAS) (2011)
10. Kirby, B., Campbell, J., Aksak, B., Pillai, P., Hoburg, J., Mowry, T., Goldstein, S.C.: Catoms: Moving robots without moving parts. In: Proceedings of the National Conference on Artificial Intelligence, vol. 20, p. 1730. Menlo Park, CA; Cambridge, MA; London; AAAI Press; MIT Press; 1999 (2005)
11. Østergaard, E.H., Kassow, K., Beck, R., Lund, H.H.: Design of the atron lattice-based self-reconfigurable robot. Auton. Robot. **21**(2), 165–183 (2006)
12. Piranda, B., Laurent, G.J., Bourgeois, J., Clévy, C., Le Fort-Piat, N.: A new concept of planar self-reconfigurable modular robot for conveying microparts. Mechatronics **23**(7), 906–915 (2013). https://doi.org/10.1016/j.mechatronics.2013.08.009
13. Reid, J.R., Vasilyev, V., Webster, R.T.: Building micro-robots: a path to sub-mm3 autonomous systems. In: Proceedings of nanotech (2008)

14. Romanishin, J., Gilpin, K., Rus, D.: M-blocks: momentum-driven, magnetic modular robots. In: IROS, pp. 4288–4295. IEEE (2013)
15. Stoy, K., Brandt, D., Christensen, D.J., Brandt, D.: Self-reconfigurable Robots: An Introduction. Mit Press, Cambridge (2010)
16. Vad, V., Csebfalvi, B., Rautek, P., Gr"oller, M.E.: Towards an unbiased comparison of cc, bcc, and fcc lattices in terms of prealiasing. Comput. Graph. Forum **33**(3), 81–90 (2014)
17. Yim, M., Duff, D.G., Roufas, K.D.: Polybot: a modular reconfigurable robot. In: IEEE International Conference on Robotics and Automation (ICRA) vol. 1, pp. 514–520 (2000)
18. Yim, M., Shen, W.M., Salemi, B., Rus, D., Moll, M., Lipson, H., Klavins, E., Chirikjian, G.: Modular self-reconfigurable robot systems [grand challenges of robotics]. IEEE Robot. Autom. Mag. **14**(1), 43–52 (2007)
19. Yim, M., Zhang, Y., Lamping, J., Mao, E.: Distributed control for 3d metamorphosis. Auton. Robot. **10**, 41–56 (2001)
20. Yim, M., White, P., Park, M., Sastra, J.: Modular self-reconfigurable robots. Encyclopedia of Complexity and Systems Science, pp. 5618–5631. Springer, New York (2009)

HyMod: A 3-DOF Hybrid Mobile and Self-Reconfigurable Modular Robot and its Extensions

Christopher Parrott, Tony J. Dodd and Roderich Groß

Abstract This paper presents HyMod, a hybrid self-reconfigurable modular robot, and its extensions. HyMod units feature three rotational degrees of freedom and four connectors, allowing them to move independently via differential wheels and group with other units to form arbitrary cubic lattice structures. The design is built around the high-speed genderless (HiGen) connection mechanism, allowing for single-sided disconnect and enabling units to rotate freely in place within their lattice positions. To our knowledge, HyMod is the first modular robot to combine efficient single module locomotion with free in place rotation. An analysis of HyMod is presented, as well as details of its mechanics and electronics. To augment the capabilities of HyMod, a number of extension modules are introduced. Hybrid modular robots with extensions, such as the system presented here, could see use in the areas of reconfigurable manufacturing, search and rescue, and space exploration.

1 Introduction

Modular robotics has seen numerous advances over the past decades, with the likes of M-TRAN III [9] and ATRON [14] successfully demonstrating the self-reconfiguration and collective motion of large chain and lattice structures. Each module within a modular robot is relatively simple, with typically only one or two degrees of freedom (DOF), allowing many modules to be produced at relatively low cost. Unfortunately, this means that such modules have limited or no mobility outside of a configuration. Efforts have been made to address this, with swarm systems

---

C. Parrott (✉) · T. J. Dodd · R. Groß
Department of Automatic Control and Systems Engineering, The University of Sheffield, Sheffield, UK
e-mail: c.parrott@sheffield.ac.uk

T. J. Dodd
e-mail: t.j.dodd@sheffield.ac.uk

R. Groß
e-mail: r.gross@sheffield.ac.uk

R. Groß et al. (eds.), *Distributed Autonomous Robotic Systems*, Springer Proceedings in Advanced Robotics 6,
https://doi.org/10.1007/978-3-319-73008-0_28

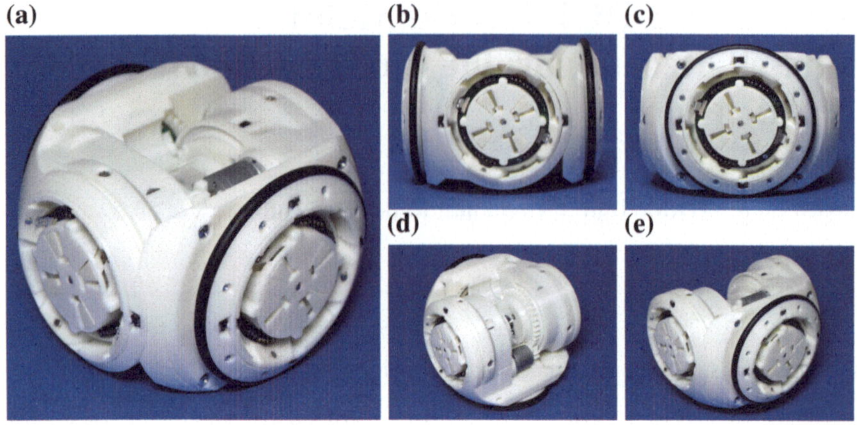

Fig. 1 HyMod: a new self-reconfigurable modular robot with three degrees of freedom (two of which form differential wheels), and four genderless connectors with single-sided disconnect. The module is shown from an **a** isometric, **b** front, and **c** side view, with its central rotational joint at zero degrees. The central joint is also shown at **d** −90, and **e** +90°

gaining the ability to self-assemble [6], and modular systems gaining dedicated drive mechanisms to provide efficient single module locomotion [11, 20]. These systems demonstrate the advantages of mobile modular robots, but compromise module self-reconfigurability in favor of individual autonomy. So far, only one modular system, to our knowledge, features efficient module mobility without sacrificing on self-reconfigurability [3, 7]; however, it lacks important features from the field such as inter-module communication and power sharing. This highlights the need for further modular robots that retain the features and reconfigurability of past successful systems, whilst also offering efficient single module locomotion.

This paper presents HyMod (Fig. 1), a self-reconfigurable modular robot that is a hybrid between chain, lattice, and mobile reconfigurable robots [23]. It extends upon preliminary work presented in [16]. Inspired by systems such as PolyBot [22] and CKbot [24], HyMod features a central rotational DOF capable of moving ±90 degrees, and is designed to form arbitrary cubic lattice structures. Two further rotational DOFs are mounted perpendicular to the central rotational joint, serving the dual purpose of emulating a spherical joint and enabling the module to drive around using a differential wheel setup. The arrangement of rotational axes shares similarities with the RobMAT [21] platform, and the use of reconfiguration joints as wheels has been explored on the iMobot [17], M^3 [10], and SMORES [3] platforms. This implementation removes the need for a separate drive mechanism for locomotion, as is the case with the modules of the Symbrion/Replicator project [8].

Connections to neighboring modules are achieved using four high-speed genderless (HiGen) connection mechanisms [15], one in each wheel and two along HyMod's central rotational axis. The connectors operate by extending hooks out of their housings to latch on to the hooks of an opposing connector, making a connection. The

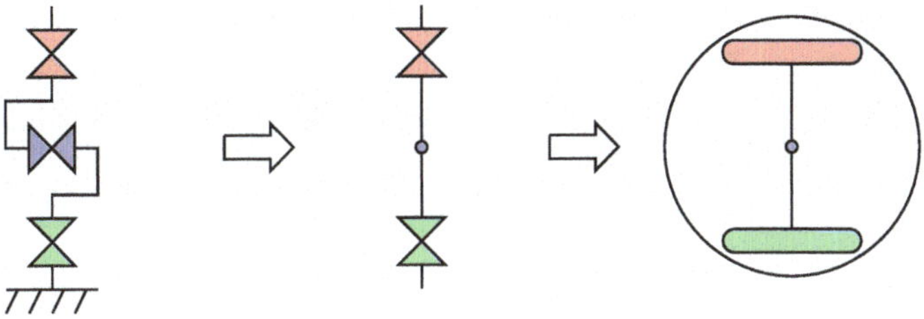

Fig. 2 The transition from the side view of a 3-DOF spherical joint (left) to a top view of a differential wheel setup (right), via an intermediate step where the middle DOF is locked at 0 degrees

use of HiGen connectors gives HyMod several advantages over other connection mechanisms, most notably the ability to independently disconnect from and produce clearance between neighboring modules. The choice of connector gave rise to the HyMod's spherical design, which allows all three of its degrees of freedom to actuate simultaneously without colliding with neighboring modules.

The remainder of this paper describes the design and implementation of HyMod (Sect. 2), and presents experiments conducted with a single unit (Sect. 3). A number of extension modules for HyMod are then explored (Sect. 4). Finally, Sect. 5 concludes the paper and discusses future work.

2 The HyMod Unit

The goal of HyMod was to create a module to address the division between mobile and self-reconfigurable systems, by integrating an efficient locomotion method that could also have a use on modules within chain or lattice structures (e.g. as a degree of freedom in a kinematic manipulator). Although the modules of systems such as M-TRAN can move independently, they are slow and have limited control over their heading when moving. A more efficient method of locomotion is that of wheels, as these can provide a constant velocity to a robot and allow for controlled turning.

To incorporate wheels into HyMod, the concept of a spherical joint was adopted (Fig. 2, left). Typically modules designed to reside in a cubic lattice have a central rotational DOF that goes from −90 to +90 degrees, allowing for a free end to move between three faces of a cube, relative to a fixed end. By adding a rotational DOF to the fixed end, the free end is able to move between five faces. Additionally, by applying a rotational DOF to the free end, any item attached to it can be oriented arbitrarily. If the central rotation axis of this spherical joint is set to zero degrees (Fig. 2, center), the remaining axes become in-line. By placing wheels on these axes a differential wheel setup is created (Fig. 2, right), granting HyMod locomotion capabilities on par with various mobile swarm robotic systems available.

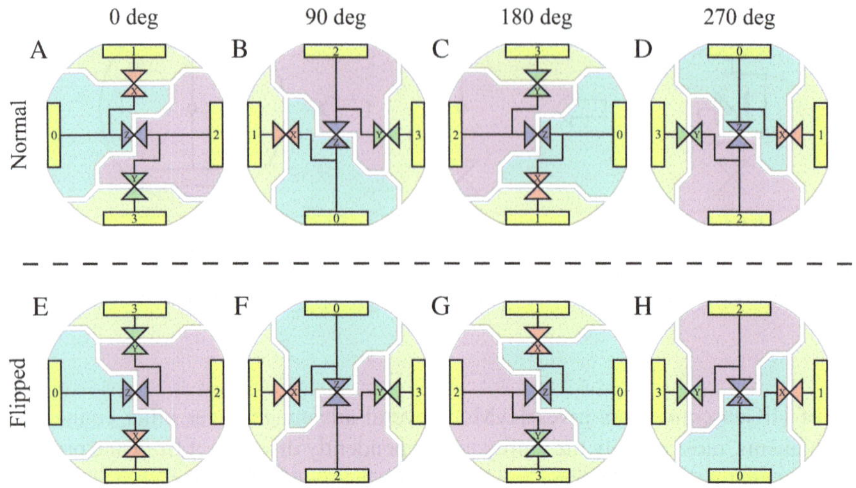

Fig. 3 The eight ways a HyMod unit can be oriented, as viewed on a 2D plane. Connectors are depicted using yellow rectangles, and are labelled 0 to 3. Rotational DOFs are depicted using connected triangles and are labelled X to Z. In this arrangement, connectors 1 and 3 can be rotated continuously, whereas connectors 0 and 2 can only be rotated ±90 degrees

2.1 Geometry Analysis

From examining the 3-DOF spherical joint (Fig. 2, left), it is apparent that an element of symmetry exists, as swapping which end is fixed can result in the same movements, provided appropriate control remapping occurs. By discovering what these symmetries are, the isomorphic configurations that can be created with a given number of HyMod units can be determined, thereby reducing the search space complexity of any self-reconfiguration algorithm that may be employed on the system.

Figure 3 shows the eight possible orientations of a HyMod unit. The orientations are depicted on a 2D plane with the central rotational joint set to zero degrees. This can either be thought of as a top-down view of the modules resting on their wheels, or a side view with the modules anchored to a surface via their bottom connectors. The module has a rotational symmetry of two, meaning that of the eight orientations shown, only four are unique. The connector and joint mapping to go between one orientation and its symmetric version are shown in Table 1. As an example, to map orientation A to C, commands that would be sent to connectors 0, 1, 2 and 3, would instead need to be send to connectors 2, 3, 0 and 1. Similarly, commands to joints X and Y would instead be sent to joints Y and X, with Z remaining unchanged.

Using the knowledge of module symmetry and the four times symmetry of HiGen connectors, the number of isomorphic configurations of two modules can be determined. By applying the mapping and discarding configurations where a connector symmetry offset (e.g. 90°) is equivalent to a wheel rotation, six isomorphic configurations are produced. These can be seen in Fig. 4. Of the six, the two

Table 1 The connector and joint index changes when mapping one HyMod orientation to another

Map				Connectors				Joints		
A ↕ C	B ↕ D	E ↕ G	F ↕ H	0 ↕ 2	1 ↕ 3	2 ↕ 0	3 ↕ 1	X ↕ Y	Y ↕ X	Z ↕ Z

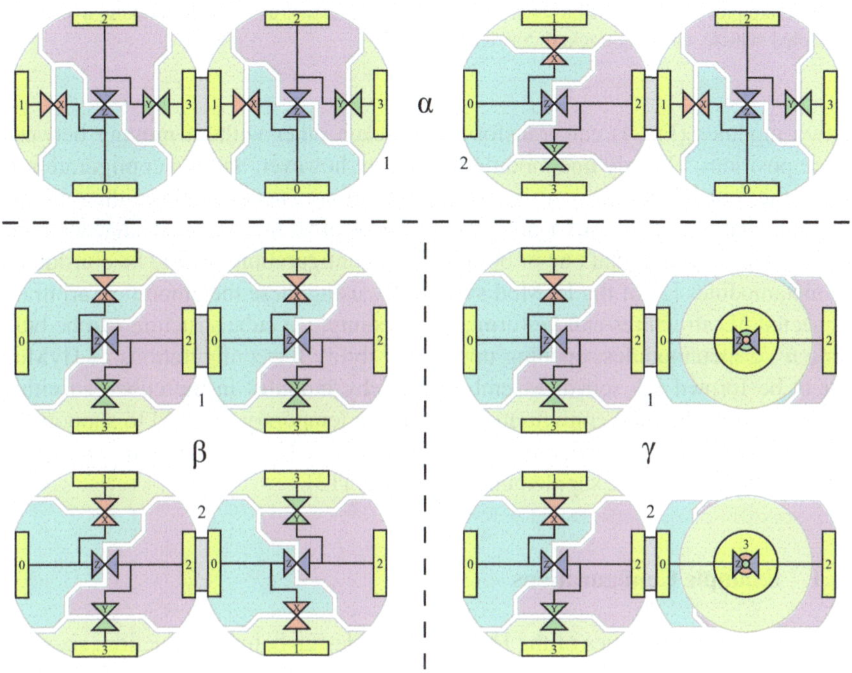

Fig. 4 All six of the isomorphic configurations that exist for two connected HyMod units

configurations labelled α offer a higher number of quantized joint angle combinations, 36 ($3 \times 4 \times 3$) versus the 9 (3×3) of the four other configurations. This is because those two configurations contain at least one continuous rotational degree of freedom between the two modules, featuring four quantized angles versus the three of the central joint. Note that rotations of wheels not connected to another module were discounted here, as they can be cancelled out by connector symmetry. Similarly, when two wheels are connected together their rotational degrees of freedom are in-line and can therefore be considered as a single joint.

As the design of HyMod is based on a spherical joint, it only occupies a single cubic lattice position. This means that in order to self-reconfigure, either four modules are needed so that a loop can be formed, or two modules and some kind of support surface (either a custom made structure or a grid of modules). By using a support surface, and provided both modules are adjacent to it, all of the isomorphic configurations

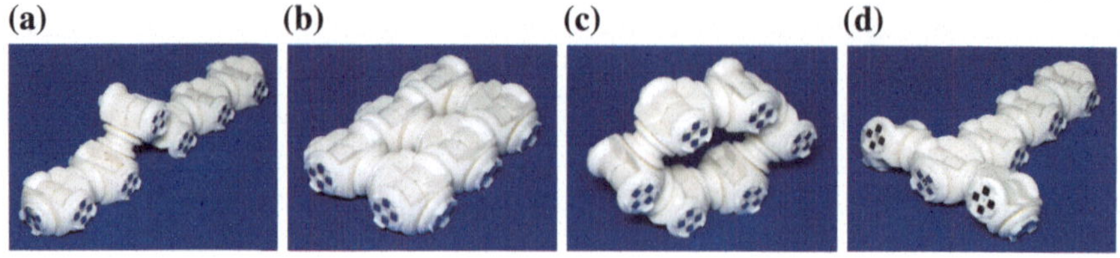

Fig. 5 Examples of four possible HyMod robot configurations, using scale models: **a** snake, **b** 6-wheeled vehicle, **c** rolling track, **d** crawler

of two modules (Fig. 4) can transform in to each other without moving between lattice positions.[1] If only one module is adjacent however, and the configuration is one of the four labelled β or γ, self-reconfiguration is not possible as there are no perpendicular rotational axes available to move the other module to be adjacent to the surface. This suggests that one or both of the α configurations should be considered the metamodules [2] of the HyMod system. By using these metamodules, arbitrary connected 3D structures can be formed. For example, a cube structure can be built using $n^3/2$ metamodules, opening up the possibility for configurations of HyMod unit to be formed via self-disassembly, whereby modules in the cube can either become part of the final configuration or act as a temporary scaffold to support the disassembly process. A render of a cube formed out of 32 HyMod unit metamodules can be seen in the online supplementary material [13].

2.1.1 Example Configurations

To explore the possibilities of the HyMod unit's design before production, six scale models were produced. They consist of four 3D printed components each and four perpendicularly mounted permanent magnets per face. This arrangement of magnets emulates the genderless property of the HiGen connector. Examples of common modular robot configurations using the scale modules are shown in Fig. 5.

2.2 Hardware Details

The module is built from two mirrored halves, forming a rotational hinge joint. This arrangement of identical halves is common with several modular robots, such as ATRON [14], Molecubes [25], UBot [19], and CoSMO [11]. Each half consists of a

[1] Note, if a surface is made up of passive HiGen connectors, that being connectors without the ability to retract their hooks, then it is not possible to self-reconfigure in to, out of, or between the two configurations labelled γ even if they are adjacent to the surface, as clearance cannot be created between the surface and the modules to allow for such a rotation.

chassis housing two HiGen connectors; one in parallel to, and the other perpendicular to the hinge axis. The parallel connector is fixed to the chassis whereas the perpendicular connector has a rotational degree of freedom through its center, forming a wheel. This gives a total of four connectors and two wheels per module.

HiGen connectors (described in more detail in [15]) operate by using a central drive motor to translate and rotate four hooks. These hooks latch on to hooks of an opposing connector, creating a genderless connection that allows for single-sided disconnect. As part of this latching process, electrical connections are made, allowing for communication and power transfer across the connectors.

Each HyMod unit consists of sixteen custom ABS plastic components (excluding the four connectors) created using 3D printing technology, fifteen custom circuit boards, two slip rings, two battery packs, and several off-the-shelf items. Four DC geared motors are used to drive the three degrees of freedom of the module (two paired together for the hinge joint), each with a ratio of 154:1 and a quoted torque of 847 mNm at 6 V. An additional 5:1 gear ratio is applied on top of each motor gearbox, increasing the torque of the rotational joints and allowing the motors to be offset from each drive axis. This setup is what facilitates the use of two motors to drive the hinge joint, enabling all motors to be identical whilst allowing the hinge joint to offer effectively twice the torque of the other degrees of freedom. This also simplifies their control because the same driver electronics can be used for each motor. The housings of the four connectors are modified from the original design to allow for extra mounting points for the wheel hubs and the addition of infrared sensors for distance sensing. Internal sensing is achieved using a potentiometer, two optical encoder setups, and an Inertial Measurement Unit (IMU). To allow for continuous rotation of the wheels whilst passing power and communication to their connectors, slip ring components are used. This is a solution adopted by past systems [14, 18].

The module measures 128 mm × 128 mm × 94 mm, when the hinge is at zero degrees. The size is governed by the dimensions of the HiGen connector, the height of the slip rings, and the chosen wheel diameter of 94 mm. This wheel diameter gives the module a 4 mm ground clearance when oriented for driving. The separation between modules in a cubic lattice is 140 mm due to the connectors extending out of their housings by 12 mm during connection. To take advantage of this ability the module is designed to fit within a spherical volume, allowing for rotation around three axes without risk of colliding with neighboring lattice modules [15, see Fig. 4]. As such the module shares visual similarity with the Roombots [18] platform, which uses its spherical design to enable the wheel-based locomotion of modules, rather than to provide clearance for self-reconfiguration. Figure 6 shows renders of the three main sections that form a complete HyMod unit. Additionally, a breakdown of the main HyMod unit properties is shown in Table 2.

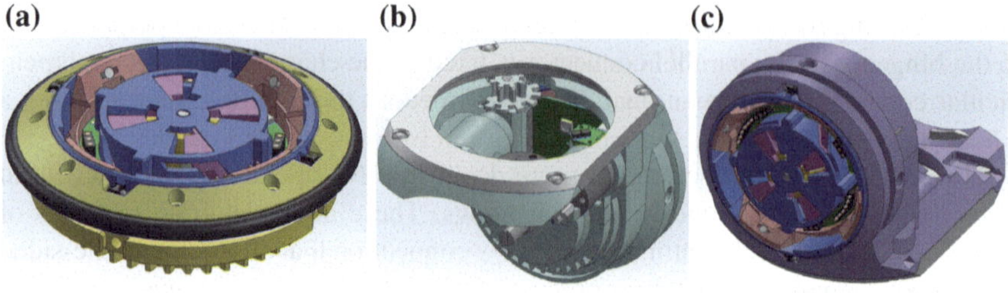

Fig. 6 3D renders of the HyMod unit's **a** wheel, **b** processing half, and **c** power half

Table 2 Properties of a HyMod unit

Property	Value
Size	128 × 128 × 94 mm
Lattice spacing	140 mm
Ground clearance	4 mm
Weight	810 g
Controllers	1 × *PJRC* Teensy 3.2
	4 × *Atmel* ATmega324P (*HiGen controller*)
Communication	1 × EGBT-046S Bluetooth modem
	1 × *NXP* fault-tolerant CAN transceiver
Sensors	1 × *Sparkfun* 9 DOF sensor stick IMU (accelerometer, gyro, magnetometer)
	12 × *Vishay* reflective optical sensor (infrared proximity)
Motors	4 × *Pololu* 154:1 metal gearmotor
	4 × *Solarbotics* 298:1 mini metal sealed gear motor
Power supply	1 × *Pololu* step-up voltage regulator (set to 9 V)
Batteries	2 × *Turnigy* 3.7 V, 750 mAh round li-po cells (total 7.4 V, 750 mAh)

2.2.1 Electronics

HyMod contains 15 custom circuit boards: 1 × *processing board*, 1 × *Bluetooth board*, 1 × *power board*, 4 × *HiGen controller*, 2 × *motor driver*, 2 × *encoder board*, and 4 × *contact ring*. The arrangement of boards is shown in Fig. 7a.

The main microcontroller for each HyMod unit is a Teensy 3.2, a 32-bit ARM Cortex-M4 based development board running at 96 MHz. This board has built-in USB, a Controller Area Network (CAN) controller, and can be programmed with the popular Arduino development environment. The Teensy is sandwiched between the *Bluetooth board* and *processing board*; the former acts as an adapter to an off-the-shelf modem, and the latter houses additional CAN components and connects to an Inertial Measurement Unit. Figure 7b shows the assembled board stack.

Each HyMod unit is powered by two 750 mAh lithium polymer (li-po) battery packs (one in each module half) connected in series to give 7.4 V. The *power board* (Fig. 7c) takes this voltage and, via a boost regulator, produces a 9 V output. This

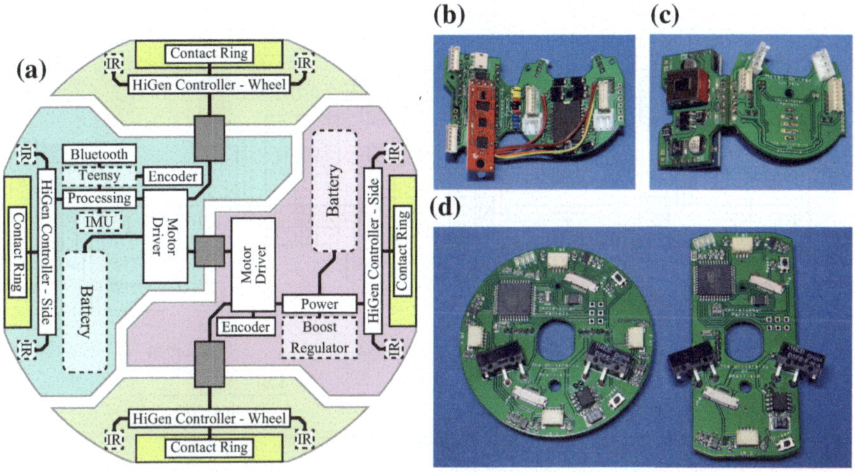

Fig. 7 **a** Block diagram showing how the circuit boards and other components within a HyMod unit connect together. White blocks are the custom boards created for this project. Assembled **b** *processing*, **c** *power* and **d** *HiGen controller* boards are also shown

output is used to power the two *motor driver* boards, which each drive two joint motors. Additionally, to enable power sharing between modules, the *power board* passes the 9 V output through an ideal diode to create a power bus. The diode prevents the current of one power supply from feeding back in to another and potentially causing damage. The power bus is then used to produce a 5 V supply for the rest of the electronics within the module.

The connectors in HyMod units are controlled using custom *HiGen controller* boards. These boards feature an ATmega324P, a motor driver, two contact switches, an analogue switch, and *contact ring* connections. The use of a separate microcontroller allows for each connector to be treated as a device on an internal communication network. Additionally, it reduces the number of connections that need to be passed through the slip rings. There are two versions of the *HiGen controller* board in each module (Fig. 7d), one for the wheel connectors and one for the side connectors. Both boards perform the same basic functions (e.g. connector actuation, infrared proximity sensing, and neighbor communication) but differ in geometry and specialized features. For instance the wheel *HiGen controller* has a grey code disc etched into it for absolute positioning of the wheel, whereas the side controller has a RGB LED for state indication and general debugging of a module.

2.2.2 Communication

Modular systems can be thought of as computer networks, where each module acts as a node, able to communicate with other nodes. There are two main ways this can be achieved, referred to as local and global communication [5]. Local communication

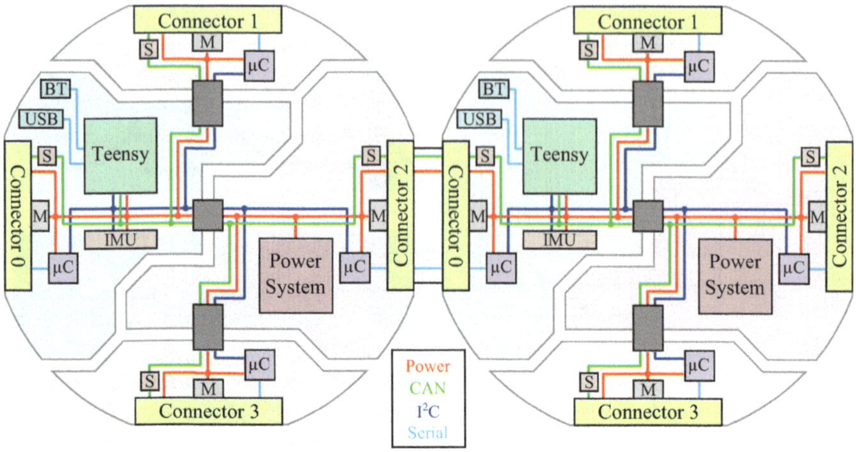

Fig. 8 The power and communication network formed between two HyMod units. BT, μC, M, and S denote Bluetooth, microcontrollers, connector motors, and bus switches, respectively

allows each module to communicate with its immediate neighbors, but requires that messages be relayed in order to reach modules other than direct neighbors. Global communication allows each module to send messages directly to any other module on the same network, but the identifier of the recipient must be known in advance. Due to the different use cases of local and global communication, both are implemented by HyMod. In addition, each unit features an internal I^2C network to communicate between components, with the Teensy acting as the master.

Local communication between two HyMod units is achieved using a serial link. Messages sent from one module to another are first sent from the Teensy over I^2C to the *HiGen controller* in question. This controller buffers the message and sends it over the serial link to the neighboring module's *HiGen controller*, which stores the message until the neighboring Teensy is ready to collect it.

Global communication between HyMod units is achieved using CAN. CAN allows for multiple connected nodes to communicate with each other by broadcasting messages on a common bus. The messages are picked up by all other networked nodes, which can then act upon the data based on an identifier. By default, CAN is designed for fixed networks where there is a single line with termination resistors at the ends. Because HyMod units are self-reconfigurable, fault-tolerant CAN was used, as this places the termination resistors at each node instead. By using digital potentiometers along with FT CAN, the network resistance can be dynamically adjusted based on the number of nodes, maintaining a stable network. Additionally, to avoid looping CAN networks that get created during self-reconfiguration, HyMod employs analogue switches at its connectors to break the network. The use of these switches also allows for hybrid networks to exist [5], whereby the global network is divided in to smaller sub-networks for task processing, with local communication being used to bridge sub-networks when necessary. Figure 8 shows both the power and communication networks produced between two HyMod units.

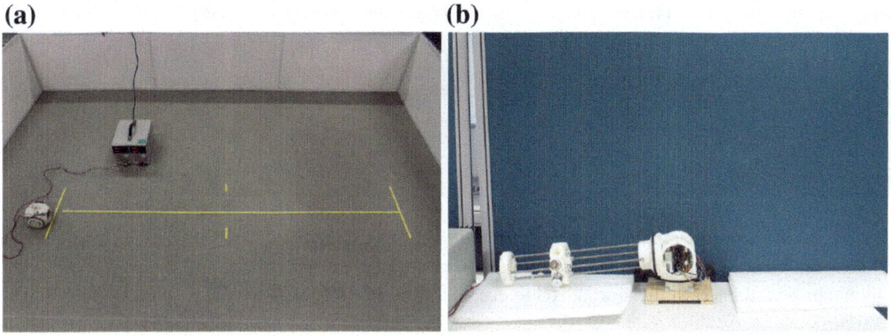

Fig. 9 The experimental setups used for conducting **a** driving and **b** lifting experiments with a single HyMod unit, tethered to a bench power supply

3 Experiments

To verify the capabilities of HyMod, a single unit was used. Three main experiments were performed using the unit, examining driving speed, lifting capability and connector actuation. For the purpose of these experiments the unit was tethered to a bench power supply set to 8.4 V (replicating the maximum battery voltage). Videos of the experiments can be found in the online supplementary material [13].

The driving speed of HyMod was determined by placing the module on the ground and timing how long it took for it to travel 2 m in a straight line. The experimental setup can be seen in Fig. 9a. The result of this experiment is that the module has a driving speed of 0.1 m s $^{-1}$.

The lifting capability of HyMod was tested using a 3D printed variable mass holder that attaches via a HiGen connector. The holder weights 520 g, and supports up to 1000 g (in 100 g increments) of additional weight. The distance from the center of the HyMod unit to the center of mass of the holder is 280 mm (two lattice spacings). Lifting tests were conducted by clamping the HyMod unit to a table and having its hinge joint rotate between −90 and +90 degrees (decelerating on the downward arc). The experimental setup used for these tests can be seen in Fig. 9b. The unit was tested lifting masses up to 1120 g, which is equivalent to lifting 1.8 modules in-line. Greater masses than 1120 g were attempted, but resulted in the failure of the 3D printed gears on the hinge joint's motors, followed by the docking hooks on the HiGen connectors themselves. If these components were constructed with stronger materials, the stated torque value of the motors suggests that higher lifting capacities would be achievable.

A final test was performed with HyMod, verifying that the two *HiGen controller* boards were able to operate the connectors as intended. Each connector was programmed to drive their motors between retracted and extended states every 2 s. The result was that both controller boards were able to successfully actuate the connectors. Further experiments involving HiGen can be found in [15].

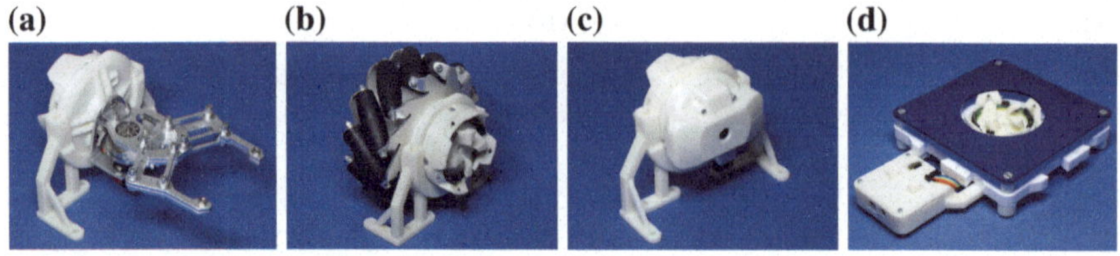

Fig. 10 Four of the extensions created for the HyMod system: **a** Gripper extension, **b** Mecanum Wheel extension, **c** Camera extension, **d** Modular Surface extension. **a**–**c** are placed on a holder, which can be attached to the side of (**d**) to create a pick-up location

4 HyMod Extensions

Unlike bespoke robotic systems made for specific tasks, modular robotic systems are intended to perform a wide variety of tasks. Some of these tasks may require specialized hardware, meaning that all modules in a homogeneous modular robot would need to feature this hardware in order for said tasks to be accomplished. This would increase the cost and complexity of each module. To overcome this problem, the HyMod system allows for extensions; modules built to add specialized capabilities to a modular robot. Past systems to employ extensions include [12, 26].

HyMod extension modules must contain processing and local communication (primarily for identification purposes), as well as at least one passive HiGen connector. A passive connector is one that is in a constant extended state, allowing for an active connector to attach to it without prior communication. This removes the need for extensions to contain their own power source. Extensions could therefore reside in known pick-up locations to be collected by modular robots when needed.

Figure 10 shows four extensions that have been developed for the HyMod system: (a) the Gripper extension uses an off-the-shelf end effector controlled by a servo motor, (b) the Mecanum Wheel extension allows for omni-directional motion when four or more are placed on a system (inspired by [1]), (c) the Camera extension uses a Raspberry Pi Zero 1.3 and Pi Cam to add video and high-powered processing to the system, and (d) the Modular Surface extension allows for square grids to be produced for modules to self-reconfigure across (inspired by [4]). The Gripper, Mecanum Wheel, and Camera weight 220 g, 560 g, and 125 g, respectively.

Figure 11 illustrates how (a) a manipulator arm and (b) an omni-directional rover can be constructed out of combinations of HyMod units and extensions. The manipulator arm takes advantage of the 3-DOF of each unit to create a 7-DOF manipulator, with the Gripper extension mounted as the end effector. The omni-directional rover takes advantage of the two continuous rotational degrees of freedom of each HyMod unit to drive Mecanum Wheel extensions to produce motion in any direction on a flat surface. A Camera extension is mounted at the front of the rover to allow for either tele-operation or autonomous operation (e.g. object following).

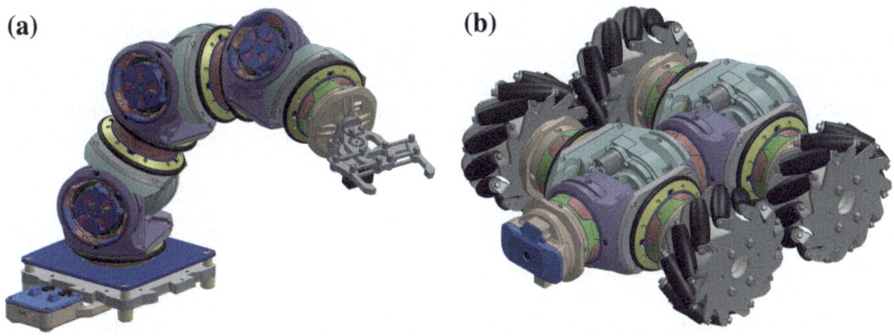

Fig. 11 Two example configurations of HyMod system modules; **a** three units and two extensions (one Gripper extension and one Modular Surface extension) forming a manipulator arm, and **b** two units and five extensions (one Camera extension and four Mecanum Wheel extensions) forming an omni-directional rover

5 Conclusions

This paper presented HyMod, a hybrid modular robot capable of self-reconfiguration and wheel-based locomotion. The module integrates the HiGen connector, allowing for single-sided disconnect and enabling units to rotate freely in place within a cubic lattice position. Details of the module were given and experiments conducted, examining the movement and lifting capabilities of a single unit. Additionally, four extensions were constructed, augmenting the capabilities of the HyMod system in the areas of manipulation, mobility, perception, and support.

Future work will involve the completion of additional HyMod units, allowing several configurations of units and extensions to be demonstrated, as well as enabling self-reconfiguration strategies to be explored.

Acknowledgements This research was funded by the Engineering and Physical Sciences Research Council (EPSRC) through scholarship support (C. Parrott) and grant no. EP/K033948/1.

References

1. Anderson, J.: RoboPlay 2014 - mecanum linkbot interlude. YouTube Video. http://www.youtube.com/watch?v=GqClygCJpKs (2016). Accessed 8 July 2016
2. Christensen, D.J., Østergaard, E.H., Lund, H.H.: Metamodule control for the ATRON self-reconfigurable robotic system. In: Proceedings, 8th Conference on Intelligent Autonomous Systems, pp. 685–692 (2004)
3. Davey, J., Kwok, N., Yim, M.: Emulating self-reconfigurable robots - design of the SMORES system. In: Proceedings, 2012 IEEE/RSJ International Conference on Intelligent Robots and Systems, pp. 4464–4469 (2012)
4. Dinh, T.K., Spröwitz, A., Bonardi, S., Ijspeert, A.: Alignment of a Roombot metamodule and extendable grid (2010) (Summer Project)
5. Garcia, R.F.M., Stoy, K., Christensen, D.J., Lyder, A.: A self-reconfigurable communication network for modular robots. In: Proceedings, 1st International Conference on Robot Communication and Coordination, pp. 23–30 (2007)
6. Groß, R., Bonani, M., Mondada, F., Dorigo, M.: Autonomous self assembly in swarm-bots. IEEE Trans. Robot. **22**(6), 1115–1130 (2006)

7. Jing, G., Tosun, T., Yim, M., Kress-Gazit, H.: An end-to-end system for accomplishing tasks with modular robots. In: Proceedings, 2016 Robotics: Science and Systems (2016)
8. Kernbach, S., Schlachter, F., Humza, R., Liedke, J., Popesku, S., Russo, S., Ranzani, T., Manfredi, L., Stefanini, C., Matthias, R.: Heterogeneity for increasing performance and reliability of self-reconfigurable multi-robot organisms. In: Proceedings, IROS 2011 Reconfigurable Modular Robotics Workshop (2011)
9. Kurokawa, H., Tomita, K., Kamimura, A., Kokaji, S., Hasuo, T., Murata, S.: Distributed self-reconfiguration of M-TRAN III modular robotic system. Int. J. Robot. Res. **27**(3–4), 373–386 (2008)
10. Kutzer, M.D.M., Moses, M.S., Brown, C.Y., Scheidt, D.H., Chirikjian, G.S., Armand, M.: Design of a new independently-mobile reconfigurable modular robot. In: Proceedings, 2010 IEEE International Conference on Robotics and Automation, pp. 2758–2764 (2010)
11. Liedke, J., Matthias, R., Winkler, L., Wörn, H.: The collective self-reconfigurable modular organism (CoSMO). In: 2013 IEEE/ASME International Conference on Advanced Intelligent Mechatronics, pp. 1–6 (2013)
12. Lyder, A., Garcia, R.F.M., Stoy, K.: Genderless connection mechanism for modular robots introducing torque transmission between modules. In: Proceedings, ICRA 2010 Workshop on Modular Robots, State of the Art, pp. 77–81 (2010)
13. Online Supplementary Material (2016). http://naturalrobotics.group.shef.ac.uk/supp/2016-005/
14. Østergaard, E.H., Kassow, K., Beck, R., Lund, H.H.: Design of the ATRON lattice-based self-reconfigurable robot. Auton. Robot. **21**(2), 165–183 (2006)
15. Parrott, C., Dodd, T.J., Groß, R.: HiGen: a high-speed genderless mechanical connection mechanism with single-sided disconnect for self-reconfigurable modular robots. In: Proceedings, 2014 IEEE/RSJ International Conference on Intelligent Robots and Systems, pp. 3926–3932 (2014)
16. Parrott, C., Dodd, T.J., Groß, R.: Towards a 3-DOF mobile and self-reconfigurable modular robot. In: IROS 2014 Modular and Swarm Systems Workshop (unpublished) (2014)
17. Ryland, G.G., Cheng, H.H.: Design of iMobot, an intelligent reconfigurable mobile robot with novel locomotion. In: Proceedings, 2010 IEEE International Conference on Robotics and Automation, pp. 60–65 (2010)
18. Spröwitz, A., Moeckel, R., Vespignani, M., Bonardi, S., Ijspeert, A.: Roombots: a hardware perspective on 3D self-reconfiguration and locomotion with a homogeneous modular robot. Robot. Auton. Syst. **62**(7), 1016–1033 (2014)
19. Tang, S., Zhu, Y., Zhao, J., Cui, X.: The UBot modules for self-reconfigurable robot. In: Proceedings, 2009 ASME/IFToMM International Conference on Reconfigurable Mechanisms and Robots, pp. 529–535 (2009)
20. Wei, H., Chen, Y., Tan, J., Wang, T.: Sambot: a self-assembly modular robot system. Proc. IEEE/ASME Trans. Mech. **16**(4), 745–757 (2011)
21. Yerpes, A., Baca, J., Escalera, J.A., Ferre, M., Aracil, R.: Modular robot based on 3 rotational DoF modules. In: Proceedings, 2008 IEEE/RSJ International Conference on Intelligent Robots and Systems, pp. 889–894 (2008)
22. Yim, M., Duff, D.G., Roufas, K.D.: Polybot: a modular reconfigurable robot. In: Proceedings, 2000 IEEE International Conference on Robotics and Automation, pp. 514–520 (2000)
23. Yim, M., m. Shen, W., Salemi, B., Rus, D., Moll, M., Lipson, H., Klavins, E., Chirikjian, G.S.: Modular self-reconfigurable robot systems [grand challenges of robotics]. IEEE Robot. Automa. Mag. **14**(1), 43–52 (2007)
24. Yim, M., Shirmohammadi, B., Sastra, J., Park, M., Dugan, M., Taylor, C.J.: Towards robotic self-reassembly after explosion. In: Proceedings, 2007 IEEE/RSJ International Conference on Intelligent Robots and Systems, pp. 2767–2772 (2007)
25. Zykov, V., Chan, A., Lipson, H.: Molecubes: an open-source modular robotics kit. In: Proceedings, IROS 2007 Self-Reconfigurable Robotics Workshop (2007)
26. Zykov, V., William, P., Lassabem, N., Lipson, H.: Molecubes extended: diversifying capabilities of open-source modular robotics. In: Proceedings, IROS 2008 Self-Reconfigurable Robotics Workshop (2008)

Network Characterization of Lattice-Based Modular Robots with Neighbor-to-Neighbor Communications

André Naz, Benoît Piranda, Thadeu Tucci, Seth Copen Goldstein and Julien Bourgeois

Abstract Modular robots form autonomous distributed systems in which modules use communications to coordinate their activities in order to achieve common goals. The complexity of distributed algorithms is generally expressed as a function of network properties, e.g., the number of nodes, the number of links and the radius/diameter of the system. In this paper, we characterize the networks of some lattice-based modular robots which use only neighbor-to-neighbor communications. We demonstrate that they form sparse and large-diameter networks. Additionally, we provide tight bounds for the radius and the diameter of these networks. We also show that, because of the huge diameter and the huge average distance of massive-scale lattice-based networks, complex distributed algorithms for programmable matter pose a significant design challenge. Indeed, communications over a large number of hops cause, for instance, latency and reliability issues.

1 Introduction

Modular robots form autonomous distributed systems in which modules communicate with each other to coordinate their activities in order to achieve common goals.

A. Naz (✉) · B. Piranda · T. Tucci · J. Bourgeois
FEMTO-ST Institute UMR CNRS 6174, University of Bourgogne
Franche-Comte (UBFC), Montbeliard, France
e-mail: andre.naz@femto-st.fr

B. Piranda
e-mail: thadeu.tucci@femto-st.fr

T. Tucci
e-mail: benoit.piranda@femto-st.fr

J. Bourgeois
e-mail: julien.bourgeois@femto-st.fr

S. Copen Goldstein
Carnegie Mellon University, Pittsburgh, PA, USA
e-mail: seth@cs.cmu.edu

R. Groß et al. (eds.), *Distributed Autonomous Robotic Systems*,
Springer Proceedings in Advanced Robotics 6,
https://doi.org/10.1007/978-3-319-73008-0_29

In this paper, we focus our attention on lattice-based modular robots composed of identical modules that communicate together using only neighbor-to-neighbor communications. In lattice-based modular robots, modules are arranged in some regular 2-dimensional or 3-dimensional lattice structure. Modules are connected to their immediate neighbors in the lattice. We consider different kinds of lattices, namely the square, the hexagonal, the simple cubic and the face-centered cubic lattices (see Fig. 1). In the neighbor-to-neighbor communication model, modules communicate only with adjacent modules. This communication model is fundamentally different than the global communication model where all modules can directly communicate together through a global bus. This approach works well in small networks but it is not scalable. Indeed, the number of hosts a bus can support is limited and packet collisions may frequently occur. Some hybrid approaches have been proposed but they are not common in modular robotics.

The considered class of modular robots captures a wide variety of existing systems, e.g., the Telecubes [28], the Miche [12] and the Distributed Flight Array [24] modular robots, some of the self-assembling systems used in [4] and the modular robotic systems developed in the Smart Blocks [26] and the Claytronics [13, 14] projects. As this work is part of the Smart Blocks and the Claytronics projects, we illustrate it using the modular robots designed in these projects, namely the Smart Blocks, the millimeter-scale 2D Catoms [18], the Blinky Blocks [19] and the 3D Catoms [25] (see Fig. 1). These modular robots are respectively arranged in the square, the hexagonal, the simple cubic, and the face-centered cubic lattice.

The Smart Blocks and the Claytronics projects propose some interesting applications based on large-scale modular robotic systems. The Smart Blocks project aims to build a large distributed modular system to convey small and fragile objects, by attaching many modules together, each one equipped with a conveyance surface. The goal of the Claytronics project is to use up to millions of modules to build programmable matter, i.e., matter that can change its physical properties in response to external and programmed events.

Communication is central to module coordination. Message and time complexities of distributed algorithms are generally expressed as a function of network properties (e.g., number of nodes, number of links, node degree, radius/diameter of the system). Many algorithms target a specific class of networks. For instance, some algorithms are more efficient in sparse networks than in dense networks (e.g., the virtual coordinate-based routing protocol in [31]). Moreover, the diameter indicates the number of hops that is required to broadcast information through the whole system. Many distributed algorithms have a worst-case time complexity that is linear to the diameter (e.g., leader election algorithms [21, 22]) or a precision that decreases with the number of hops that messages have to travel (e.g., time synchronization protocol [23]). Thus, it is crucial to take into account the network properties in order to design and choose appropriate algorithms, especially in large-scale systems.

The contribution of this paper is to characterize the network of our class of modular robots based on their lattice type and the number of modules in the system. We demonstrate that these modular robots form sparse and large-diameter networks. Moreover, we provide tight bounds on the radius and the diameter of these networks.

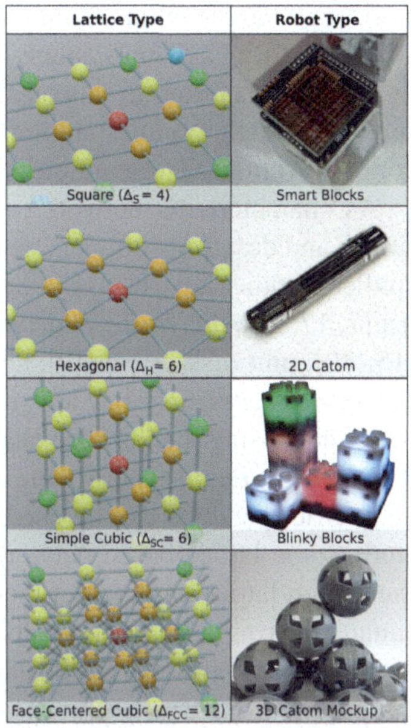

Fig. 1 The different arrangement lattices considered in this paper associated with the modular robots used to illustrate our work. For a lattice L, Δ_L denotes its coordination number, i.e., the maximum number of modules to which a module can be connected

Note that we assume perfect alignment of the modules in the lattice. However, defects in the lattice which may cause unreliable and intermittent connections, will only make the network sparser and increase both its radius and its diameter. We also discuss our results and show that efficient and effective distributed algorithms for programmable matter and more generally for massive-scale lattice-based networks may be challenging to design. Indeed, communications over a large number of hops cause, for instance, latency and reliability issues. To the best of our knowledge, this paper is the first work to characterize the networks of our class of modular robots.

The rest of this paper is organized as follows. Section 2 presents the related work. Then, Sect. 3 defines the system model and some terms. Afterwards, Sect. 4 characterizes the network density for our class of modular robots. Section 5 provides tight bounds of the radius and the diameter of the networks for our class of modular robots. Then, Sect. 6 discusses our results. Finally, Sect. 7 concludes this paper and Sect. 8 suggests future research directions.

2 Related Work

To the best of our knowledge, network characterization has attracted little attention in the modular robotic community. In [10], the authors compare the efficiency of neighbor-to-neighbor communication and global communication. Based on experimentally validated models, the authors compare the information transmission time in different scenarios for systems composed of 10–1000 modules. As mentioned in

Sect. 1, global communication through a shared medium is less scalable with system size. Since we envision systems composed of millions of units, global communication is not an option. In this paper, we focus our attention on lattice-based modular robots in which modules communicate with each other using neighbor-to-neighbor communications. These modular robots form lattice-based networks.

As characterizing network properties is crucial for choosing appropriate algorithms and designing efficient new ones, graphs and networks have been extensively studied. Studies have been conducted on various graphs and networks, e.g., the Internet [7, 17, 20], the World Wide Web [1], sensor networks [16], small-world networks [15, 30], unit disk graphs [9], and lattice-based networks [2, 3, 15]. These studies are network specific. They are either measurement based (e.g., [1, 7, 20]), or purely theoretical using the intrinsic characteristics of the network (e.g., [2, 3, 9, 16]).

Due to the regular tiling of the space in lattices, lattice-based networks obey certain geometric rules that can be used to analyze these networks. In [2, 15], the authors study some lattice-based networks, but they only consider networks embedded in the square lattice and restrict their analyze to specific network topologies, e.g., the square, the ring, etc. Their results are not generalizable to other lattices and arbitrary network topologies. In [3], the author states that the average distance between nodes in lattice networks is on the order of $n^{\frac{1}{D_L}}$, where n is the number of nodes and D_L is the dimension of the considered lattice.

In this paper, we consider lattice-based networks embedded in any of the square, hexagonal, simple-cubic and face-centered lattices. We show that these networks are sparse and large diameter. Moreover, we provide tight lower and upper bounds for the radius and the diameter of these networks.

3 System Model and Definitions

In this paper, we consider lattice-based modular robots with neighbor-to-neighbor communications. In lattice-based modular robots, modules are arranged in some regular 2-dimensional or 3-dimensional lattice L. Here, we consider the Square (S), the Hexagonal (H), the Simple Cubic (SC) and the Face-Centered Cubic (FCC) lattices. Modules can only occupy a set of discrete positions defined by L. Note that modular robots may contain holes, i.e., some positions of L may be unoccupied. Because we assume neighbor-to-neighbor communications, L also defines the module connectivity: Modules can directly communicate only with their immediate neighbors in L. D_L denotes the dimension of L and Δ_L its coordination number, i.e., the maximum number of modules to which a module can be connected.

Arbitrarily arranged modular robotic systems form lattice-based networks that can be modeled by connected, undirected, unweighted and lattice-based graphs $G = (V, E)$, where V is the set of vertices (representing the modules), E the set of edges (representing the connections), $|V| = n$, the number of vertices and $|E| = m$, the number of edges. $\delta(v_i)$ denotes v_i's degree, i.e., the number of vertices to which v_i is connected. $d(v_i, v_j)$ refers to the distance between vertices v_i and v_j, i.e., the number of edges on a shortest path between v_i and v_j. The radius, r, and the diameter, d, of G are respectively defined as $r = \min_{v_i \in V} \max_{v_j \in V} d(v_i, v_j)$ and $d = \max_{v_i \in V} \max_{v_j \in V} d(v_i, v_j)$.

Notice that we assume perfect alignment of the modules in the lattice. However, defects in the lattice which may cause unreliable and intermittent connections, will only make the network sparser and increase both its radius and its diameter.

We now define some specific graphs used in this paper. Let V_L be the infinite set of vertices representing the infinite set of positions in L. $L\text{-}Sphere(v_c, r)$ is a sphere embedded in L, where vertex v_c is the center of the sphere and $r \in \mathbb{N}$ its radius. It contains the set of vertices in V_L whose distance from v_c is equal to r:

$$L\text{-}Sphere(v_c, r) = \{v_i \in V_L \mid d(v_i, v_c) = r\} \tag{1}$$

$L\text{-}Ball(v_c, r)$ is a ball embedded in L, where v_c the center of the ball and $r \in \mathbb{N}$ its radius. It contains the set of vertices in V_L whose distance from v_c is less than or equal to r:

$$L\text{-}Ball(v_c, r) = \{v_i \in V_L \mid d(v_i, v_c) \le r\} \tag{2}$$

$$= \bigcup_{i=0}^{r} L\text{-}Sphere(v_c, i) \tag{3}$$

By abuse of notation, $L\text{-}Sphere$ and $L\text{-}Ball$ can respectively refer to sphere and ball graphs embedded in L where the connectivity between vertices is induced by the lattice structure of L. $L\text{-}Sphere(r)$ and $L\text{-}Ball(r)$ respectively refer to a sphere and a ball of radius r in the lattice L. In all the illustrations of this paper, $L\text{-}Sphere(r)$ are gradually colored from red to blue according to the value of r.

4 Network Density

In this section, we show that the networks formed by our class of modular robots are all sparse.

Corollary 1 *Let $G = (V, E)$ be the network graph of an arbitrarily arranged modular robotic system that fits the model described in Sect. 3. The vertex degree, $\delta(v_i)$, of any vertex $v_i \in V$ is bounded by:*

$$0 \le \delta(v_i) \le \Delta_L \tag{4}$$

Lemma 1 *Let $G = (V, E)$ be the network graph of an arbitrarily arranged modular robotic system that fits the model described in Sect. 3. The number of edges of G, m, is bounded as follows:*

$$n - 1 \leq m \leq n\Delta_L \quad (5)$$

Proof **Lower Bound**. A connected graph must have at least n-1 edges [15]. **Upper Bound**. Because of Corollary 1, every module cannot be connected to more than Δ_L others. Thus, the number of edges of G is upper-bounded by $n\Delta_L$. Note that a tighter upper bound can be established by considering the lattice structure of L.

Theorem 1 *Let $G = (V, E)$ be the network graph of an arbitrarily arranged modular robotic system that fits the model described in Sect. 3. If $|V| = n$ is large, then G is a sparse graph, i.e., $m \ll n^2$.*

Proof If n is large, then $\Delta_L \ll n$. Thus, we have $n\Delta_L \ll n^2$. Then, because of Lemma 1, we obtain $m \ll n^2$.

5 Network Radius and Diameter

In this section, we establish tight lower and upper bounds of the radius and the diameter of the networks of our class of modular robots.

5.1 Preliminary Materials

This section presents some preliminary results used in the computations and the demonstrations of the radius and the diameter bounds of modular robot networks. We recall that V_L is the infinite set of vertices representing the set of positions in the lattice L.

Corollary 2 $\forall v_c \in V_L$, $\forall r \in \mathbb{N}$, *L-$Ball(v_c, r)$ is centrally symmetric: The reflection v_j of every vertex v_i at distance $d(v_i, v_c) = k$ through v_c is also at distance k from v_c and $d(v_i, v_j) = 2k$.*

Proof Let L-$Ball(v_c, 1)$ be the ball of radius 1 and v_c its center. All the vertices except v_c are at distance 1 from v_c. Along every axis of the lattice L, two vertices, v_1 and v_2, are connected to v_c, one in each direction. These two vertices are symmetric through v_c, at distance 1 from v_c and at distance 2 from each other.

Let L-$Ball(v_c, r)$ be the ball of radius r and v_c its center. We assume that L-$Ball(v_c, r)$ is centrally symmetric. Let L-$Ball(v_c, r + 1)$ be the ball of radius $r + 1$ with v_c its center. By construction, L-$Ball(v_c, r + 1)$ is obtained from L-$Ball(v_c, r)$ by adding all the vertices at distance $r + 1$ from v_c. Let us consider v_3 and v_4 in L-$Ball(v_c, r)$ such that v_3 and v_4 are symmetric through v_c and $d(v_3, v_4) = 2r$. In order to construct L-$Ball(v_c, r + 1)$, we add to v_3 and v_4 two vertices v_5 and v_6 on the same axis but in the opposite direction such that $d(v_5, v_c) = d(v_6, v_c) = r + 1$. v_5 and v_6 are symmetric through v_c. Moreover, there is no shortcut between v_5 and v_6, thus, $d(v_5, v_6) = 1 + d(v_3, v_4) + 1 = 2 + 2r = 2(r + 1)$. Thus, $L\text{-}Ball(v_c, r + 1)$ is centrally symmetric.

By induction, $\forall v_c \in V_L$, $\forall r \in \mathbb{N}$, $L\text{-}Ball(v_c, r)$ is centrally symmetric.

Lemma 2 *$\forall v_c \in V_L$, $\forall r \in \mathbb{N}$, the diameter, d, of $L\text{-}Ball(v_c, r)$ is equal to $2r$.*

Proof As stated in Corollary 2, $L\text{-}Ball(v_c, r)$ is centrally symmetric. Thus, $\forall v_i \in L\text{-}Ball(v_c, r)$ such that $d(v_i, v_c) = r$, $\exists v_j \in L\text{-}Ball(v_c, r)$ with $d(v_i, v_j) = 2r$. By construction, $\nexists v_i \in L\text{-}Ball(v_c, r)$, $d(v_i, v_c) > r$. As a consequence, the diameter of $L\text{-}Ball(v_c, r)$, i.e., the largest distance between any two vertices is equal to $d = 2r$.

Corollary 3 *$\forall v_c \in V_L$, $\forall r \in \mathbb{N}$, $L\text{-}Ball(v_c, r)$ is the minimum-radius and minimum-diameter existing graph composed of $n_{L\text{-}Ball}(v_c, r) = |\, L\text{-}Ball(v_c, r)|$ vertices in L.*

Proof By construction, in $L\text{-}Ball(v_c, r)$ all the positions of the lattice L at distance less than or equal to r from v_c are occupied. Thus, if we remove a vertex v_1 and add it to an empty place adjacent to a full one (the system should remain connected) occupied by the vertex v_2, the new location of v_1 must be at distance $r + 1$ from v_c. Moreover, every vertex would be at distance $r + 1$ or more from at least one other vertex. Thus, the radius of the graph would be equal to $r + 1$. Moreover, because $L\text{-}Ball(v_c, r)$ is centrally symmetric (See Corollary 2), $\exists v_3 \in L\text{-}Ball(v_c, r)$, $d(v_2, v_3) = 2r$. Because of Lemma 2, $d(v_2, v_3)$ is the diameter of $L\text{-}Ball(v_c, r)$. Since there is no shortcut between v_1 and v_3 in its new location, $d(v_1, v_3) = d(v_2, v_3) + 1 = 2r + 1$. Thus, the diameter of the graph would be equal to $2r + 1$.

5.2 Radius and Diameter Bounds

Theorem 2 *Let $G = (V, E)$ be the network graph of an arbitrarily arranged modular robotic system that fits the model described in Sect. 3. Let $L\text{-}Ball(r_b)$ and $L\text{-}Ball(r_b + 1)$ be two ball graphs embedded in L such that the number of vertices of G, n, is between the number of vertices of these two balls, i.e., $n_{L\text{-}Ball}(r_b) \leq n < n_{L\text{-}Ball}(r_b + 1)$. The radius, r, and the diameter, d, of G are tightly bounded as follows:*

$$r_b \leq r \leq \lfloor \frac{n-1}{2} \rfloor \tag{6}$$

$$2r_b \leq d \leq n - 1 \tag{7}$$

Proof **Upper Bound**. In a connected graph, any two vertices are at most separated by all the others. In such a graph, the n vertices form a line of $n - 1$ edges. Thus, the largest distance between any two vertices, i.e., the diameter of G, is at most equal to $n - 1$ edges. The radius of G is at most equal to the half of that line, i.e., $r \leq \lfloor \frac{n-1}{2} \rfloor$.

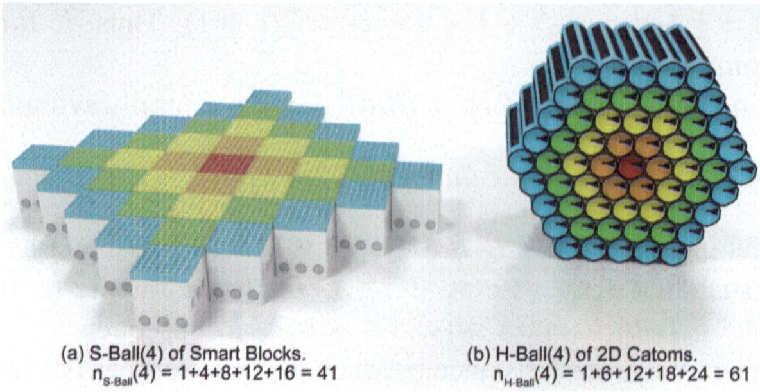

Fig. 2 An $S\text{-}Ball(4)$ and an $H\text{-}Ball(4)$ with color gradient from the center of the ball

Lower Bound. Because of Corollary 3, $L\text{-}Ball(r_b)$ is the minimum-radius and minimum-diameter graph composed of $n_{L\text{-}Ball}(r_b)$ vertices. Thus, with n vertices, G has a radius at least equal to r_b and a diameter at least equal to the diameter of $L\text{-}Ball(r_b)$, which is, because of Lemma 2, equal to $2r_b$.

In the rest of this section, we establish the formula to compute the exact radius of an $L\text{-}Ball$ according to its number of vertices in the different lattices we consider.

5.2.1 Systems in Two Dimensions: The Square and Hexagonal Lattices

In this section, we compute the exact radius of an $L\text{-}Ball$ given the number of vertices it has for the case of two-dimensional systems embedded in the Square (S) and Hexagonal (H) lattices. Figure 2 depicts an $S\text{-}Ball$ and an $H\text{-}Ball$ of radius 4, respectively composed of Smart Blocks and 2D Catoms.

Lemma 3 *In the square and the hexagonal lattices, the number of vertices in a sphere of radius $r \geq 1$, $n_{L\text{-}Sphere}(r, \Delta_L)$, can be computed by:*

$$n_{L\text{-}Sphere}(r, \Delta_L) = r\Delta_L \tag{8}$$

Proof As illustrated in Fig. 2, in the square and the hexagonal lattices, a sphere of radius $r \geq 1$ is composed of Δ_L segments of length r modules. Consequently, the number of vertices is equal to $r\Delta_L$.

Theorem 3 *In the square and the hexagonal lattices, the radius of a ball composed of $n \geq 1$ vertices, $r_{L\text{-}Ball}(n, \Delta_L)$, can be computed by:*

$$r_{L\text{-}Ball}(n, \Delta_L) = \frac{1}{2}\left(\sqrt{1 + \frac{8(n-1)}{\Delta_L}} - 1\right) \tag{9}$$

Proof By definition, $L\text{-}Ball(r)$ is the union of all the $L\text{-}Sphere(i)$ for i ranging from 0 to r. Thus, in the square and the hexagonal lattices, for $r \geq 1$, the number of vertices in an $L\text{-}Ball(r)$, $n_{L\text{-}Ball}(r, \Delta_L)$, can be computed as follows:

$$n_{L\text{-}Ball}(r, \Delta_L) = \sum_{i=0}^{r} n_{L\text{-}Sphere}(i, \Delta_L) \tag{10}$$

$$= 1 + \sum_{i=1}^{r} i\Delta_L \tag{11}$$

$$= \frac{1}{2}r^2\Delta_L + \frac{1}{2}r\Delta_L + 1 \tag{12}$$

To obtain Eq. 9, we solve Eq. 10 for r and keep only the positive root.

5.2.2 Systems in Three Dimensions: The Simple Cubic and Face-Centered Cubic Lattices

In this section, we compute the exact radius of an $L\text{-}Ball$ given the number of vertices it contains for the case of three-dimensional systems embedded in the Simple Cubic (SC) and Face-Centered Cubic (FCC) lattices. Figures 3 and 4 depict the $SC\text{-}Ball$ and the $FCC\text{-}Ball$ of radius 2, respectively composed of Blinky Blocks and 3D Catoms. Both systems can be decomposed into horizontal layers.

The Simple Cubic Lattice

Lemma 4 *In the simple cubic lattice, the number of vertices in a sphere of radius* $r \geq 1$, $n_{SC\text{-}Sphere}(r)$, *can be computed by:*

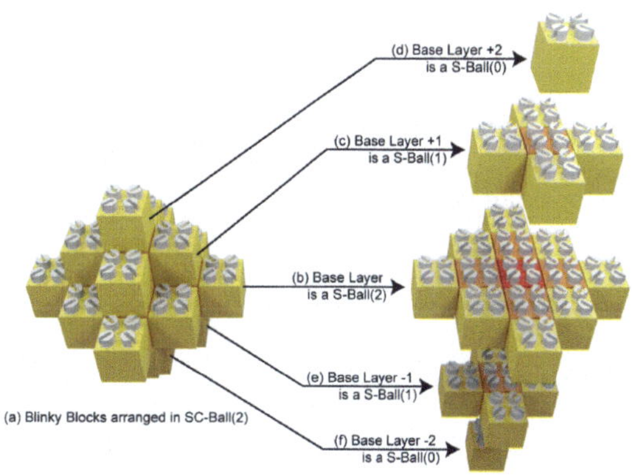

Fig. 3 An $SC\text{-}Ball(2)$ of Blinky Blocks and its decomposition in horizontal layers with color gradient from the center of the ball

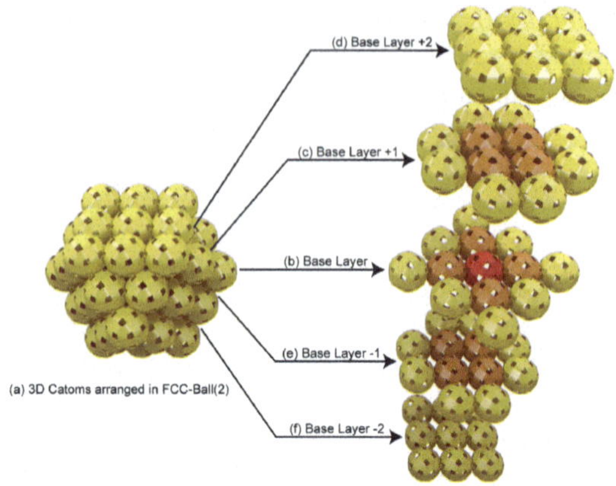

Fig. 4 An $FCC\text{-}Ball(2)$ of 3D Catoms and its decomposition in horizontal layers with color gradient from the center of the ball

$$n_{SC\text{-}Sphere}(r) = n_{S\text{-}Sphere}(r) + 2\sum_{i=0}^{r-1} n_{S\text{-}Sphere}(i) \tag{13}$$

$$= 2(2r^2 + 1) \tag{14}$$

Proof As illustrated in Fig. 3, a sphere of radius r in the simple cubic lattice can be decomposed into $2r + 1$ horizontal $S\text{-}Sphere$s of different radii. Equation 13 is obtained by summing up all the size of the $S\text{-}Sphere$s.

Theorem 4 *In the simple-cubic lattice, the radius of a ball composed of $n \geq 1$ vertices, $r_{SC\text{-}Ball}(n)$, can be computed by:*

$$r_{SC\text{-}Ball}(n) = \frac{1}{2}\left(\frac{(\sqrt{3}\sqrt{243n^2 + 125} + 27n)^{\frac{1}{3}}}{3^{\frac{2}{3}}} - \frac{5}{3^{\frac{1}{3}}(\sqrt{3}\sqrt{243n^2 + 125} + 27n)^{\frac{1}{3}}} - 1\right) \tag{15}$$

Proof By definition, $L\text{-}Ball(r)$ is the union of all the $L\text{-}Sphere(i)$ for i ranging from 0 to r. Thus, for $r \geq 1$, the number of vertices in an $SC\text{-}Ball(r)$, $n_{SC\text{-}Ball}(r)$, can be computed as follows:

$$n_{SC\text{-}Ball}(r) = \sum_{i=0}^{r} n_{SC\text{-}Sphere}(i) \tag{16}$$

$$= 1 + \sum_{i=1}^{r} 2(2i^2 + 1) \tag{17}$$

$$= \frac{4}{3}r^3 + 2r^2 + \frac{8}{3}r + 1 \tag{18}$$

To obtain Eq. 15, we solve Eq. 16 for r and keep only the real root.

The Face-Centered Cubic Lattice

Lemma 5 *In the face-centered cubic lattice, the number of vertices in a sphere of radius $r \geq 1$, $n_{FCC\text{-}Sphere}(r)$, can be computed by:*

$$n_{FCC\text{-}Sphere}(r) = 4r + 2(r+1)^2 + 2(r-1)4r \tag{19}$$

$$= 2(5r^2 + 1) \tag{20}$$

Proof As shown in Fig. 4, a sphere of radius r in the face-centered cubic lattice can be decomposed into $2r + 1$ horizontal layers. The base layer is an $S\text{-}Sphere(r)$ and contains $4r$ vertices. The bottom and the top layers both contain $(r+1)^2$ vertices. The $2(r-1)$ other layers contain $4r$ vertices each. Equation 19 is obtained by summing up the number or vertices of each layer.

Theorem 5 *In the face-centered cubic lattice, the radius of a ball of $n \geq 1$ vertices, $r_{FCC\text{-}Ball}(n)$, can be computed by:*

$$r_{FCC\text{-}Ball}(n) = \frac{1}{2}\left(\frac{(\sqrt{15}\sqrt{4860n^2+343}+270n)^{\frac{1}{3}}}{15^{\frac{2}{3}}} - \frac{7}{15^{\frac{1}{3}}(\sqrt{15}\sqrt{4860n^2+343}+270n)^{\frac{1}{3}}} - 1\right) \tag{21}$$

Proof By definition, $L\text{-}Ball(r)$ is the union of all the $L\text{-}Sphere(i)$ for i ranging from 0 to r. Thus, for $r \geq 1$, the number of vertices in an $FCC\text{-}Ball(r)$, $n_{FCC\text{-}Ball}(r)$, can be computed as follows:

$$n_{FCC\text{-}Ball}(r) = \sum_{i=0}^{r} n_{FCC\text{-}Sphere}(i) \tag{22}$$

$$= 1 + \sum_{i=1}^{r} 2(5i^2 + 1) \tag{23}$$

$$= \frac{10}{3}r^3 + 5r^2 + \frac{11}{3}r + 1 \tag{24}$$

To obtain Eq. 21, we solve Eq. 24 for r and keep only the real root.

6 Discussion

In this paper, we demonstrated some properties of networks of lattice-based modular robots with neighbor-to-neighbor communications. As shown in [5], this class of modular robots is particularly suitable to design programmable matter, i.e., matter that can change its physical properties in response to some events. In our vision, programmable matter will be composed of up to millions of modules [13, 14]. This section discusses our theoretical results and the impact on the efficiency of distributed algorithms for programmable matter and more generally for massive-scale lattice-based networks.

More precisely, we compare lattice-based networks to small-world networks [30] (e.g., the Internet network [17]) and to wireless ad-hoc networks (e.g., wireless sensor networks, multi-robot networks, etc.). Since many large real-world networks are small-world networks, it is legitimate to consider them for comparison. Wireless ad-hoc networks are highly spatially dependent, like our class of networks. Indeed, in wireless ad-hoc networks, nodes can only communicate with some neighboring nodes within some limited range. Note that wireless ad-hoc networks can fall in the class of lattice-based networks if they are deployed in a lattice structure.

We demonstrated that lattice-based networks are sparse networks (i.e., $m \ll n^2$). Because of Lemma 1 and because Δ_L is bounded by a constant for all the lattices we consider, the number of edges is $\Theta(n)$. Thus, lattice-based networks are sparser than small-world networks that have $\Omega(nlog(n))$ edges [30]. Wireless ad-hoc networks can be sparse or dense depending on the deployment environment (area/volume, obstacles, etc.), the deployment density and the node communication range.

In regular lattice networks, the typical distance between two nodes is $\sim n^{\frac{1}{D_L}}$ [3]. Thus, in lattice-based networks, i.e., lattice networks with potential holes, this distance is lower bounded by $\Omega(n^{\frac{1}{D_L}})$, while in small-world networks, this distance is $\sim log(n)$ [3]. Small-world networks have typically short distances between arbitrary pairs of nodes due to the presence of few long-range edges. As a consequence, small-world networks tend to have a small diameter. In lattice-based and sparse wireless ad-hoc networks, such long-range edges do not exist. Thus, these networks tend to have a larger average distance and a larger diameter. These phenomena are accentuated as the number of nodes in the network increases. Because programmable matter is formed of millions of modules [13, 14], the networks we consider are much larger and thus have a larger diameter than usual wireless ad-hoc networks that are typically composed of dozens of nodes to tens of thousands of nodes. We demonstrated that the radius and the diameter of lattice-based networks are lower bounded by $\Omega(\sqrt[3]{n})$ (Eqs. 9, 15 and 21 are all $\Omega(\sqrt[3]{n})$).

Studies indicate that the diameter of the Internet is around 30 hops [7, 20]. This is corroborated by the suggested values for Time-To-Live (TTL) for Internet Protocol (IP) packets. The TTL should be twice the diameter of the Internet [6] and the actual value recommended is 64 [27, 29]. As shown in Fig. 5, systems with a million 3D Catoms have a diameter of at least 132 hops, while systems with 100 million 3D Catoms have a diameter of at least 620 hops. Blinky Blocks systems have similarly large diameter, e.g., a 40,000 Blinky Blocks system has a diameter greater than 30 hops. Thus, a 40,000 Blinky Blocks system which fits in a 1.4m^3 cube, would have a diameter larger than the entire Internet that spans the whole world.

It is crucial to take into account the large diameter and large average distance to design efficient and effective distributed algorithms for large-scale modular robotic

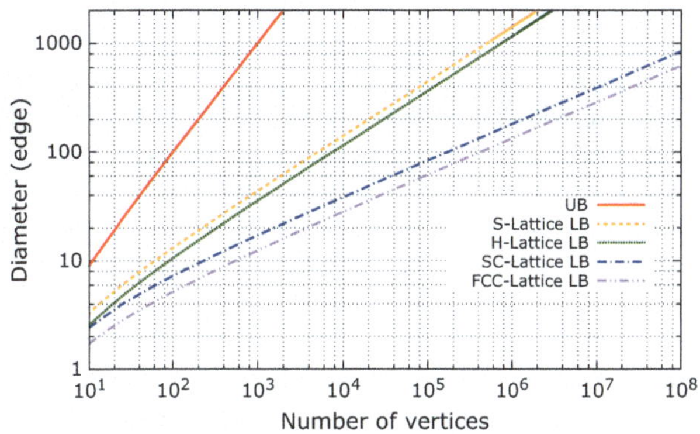

Fig. 5 Diameter bounds versus the number of vertices in the network graph for the different considered lattices. The terms "LB" and "UB" respectively stand for "lower bound" and "upper bound"

systems. For example, communication over a large number of hops causes latency and reliability issues. Assuming link faults are independent and identically distributed, the probability that a multi-hop communication fails increases exponentially with the number of hops [8]. Let us consider time synchronization and data sharing algorithms. These algorithms are required for real-time responsive programmable matter and to distribute, store and access geometry data for self-reconfiguration. However, these algorithms are challenging to design for such huge diameter and huge average distance systems. Unpredictable delays (due, for example, to queueing or retransmissions) accumulate every hop, which tends to disturb the time synchronization process and decrease the achievable synchronization precision. Moreover, in data sharing algorithms, lookup latency may be extremely long if it involves messages that have to travel a large number of hops.

7 Conclusions

In this paper, we characterize the networks of some lattice-based modular robots which use only neighbor-to-neighbor communications. We demonstrate that they form sparse and large-diameter networks. Moreover, we provide tight bounds of the radius and the diameter of these networks. Our results are generalizable to other networks embedded in the considered lattices. We also show that, it may be challenging to design efficient distributed algorithms for massive-scale lattice-based networks because of their huge diameter and their huge average distance.

8 Future Work

In future work, we will take into account the properties of huge diameter and huge average distance of massive-scale lattice-based networks in order to design efficient and effective distributed algorithms for programmable matter.

In addition, we will experimentally evaluate the practical impact of the diameter and the average distance values on the performance of some distributed algorithms executed in our class of modular robotic systems.

As previously mentioned, different communication models exist in modular robotic systems. In large-scale systems, the global communication model where all modules can directly communicate together through a global bus is not an option because the number of hosts a bus can support is limited by packet collisions. As shown in this paper, using the neighbor-to-neighbor communication model in large-scale systems implies a large diameter and a large average distance. In future work, we plan to study the network properties of modular robotic systems that use hybrid communication models in which modules communicate together through small buses, each one with a few participating modules, as proposed in [11].

Acknowledgements This work has been funded by the Labex ACTION program (contract ANR-11-LABX-01-01) and ANR/RGC (contracts ANR-12-IS02-0004-01 and 3-ZG1F) and ANR (contract ANR-2011-BS03-005).

References

1. Albert, R., Jeong, H., Barabási, A.L.: Internet: diameter of the world-wide web. Nature **401**(6749), 130–131 (1999)
2. Barrenetxea, G., Berefull-Lozano, B., Vetterli, M.: Lattice networks: capacity limits, optimal routing, and queueing behavior. IEEE/ACM Trans. Netw. (TON) **14**(3), 492–505 (2006)
3. Barthélemy, M.: Spatial networks. Phys. Rep. **499**(1), 1–101 (2011)
4. Bhalla, N., Jacob, C.: A framework for analyzing and creating self-assembling systems. In: 2007 IEEE Swarm Intelligence Symposium, pp. 281–288. IEEE (2007)
5. Bourgeois, J., Piranda, B., Naz, A., Lakhlef, H., Boillot, N., Mabed, H., Douthaut, D., Tucci, T.: Programmable matter as a cyber-physical conjugation. In: Proceedings of the IEEE International Conference on Systems, Man and Cybernetics. IEEE, Budapest, Hungary (2016)
6. Braden, R.: Requirements for Internet Hosts – Communication Layers. RFC 1122, RFC Editor (1989)
7. Cardozo, T.B., Silva, A.P.C., Vieira, A.B., Ziviani, A.: On the end-to-end connectivity evolution of the internet
8. Di Ianni, M., Gualà, L., Rossi, G.: Reducing the diameter of a unit disk graph via node addition. Inf. Process. Lett. **115**(11), 845–850 (2015)
9. Ellis, R.B., Martin, J.L., Yan, C.: Random geometric graph diameter in the unit disk with p metric. In: International Symposium on Graph Drawing, pp. 167–172. Springer (2004)
10. Garcia, R.F.M., Schultz, U.P., Stoy, K.: On the efficiency of local and global communication in modular robots. In: IEEE/RSJ International Conference on Intelligent Robots and Systems, 2009. IROS 2009, pp. 1502–1508. IEEE (2009)
11. Garcia, R.F.M., Stoy, K., Christensen, D.J., Lyder, A.: A self-reconfigurable communication network for modular robots. In: Proceedings of the 1st international conference on Robot communication and coordination, p. 23. IEEE Press (2007)

12. Gilpin, K., Kotay, K., Rus, D., Vasilescu, I.: Miche: modular shape formation by self-disassembly. Int. J. Robot. Res. **27**(3–4), 345–372 (2008)
13. Goldstein, S.C., Mowry, T.C.: Claytronics: A scalable basis for future robots. In: RoboSphere. Moffett Field, CA (2004)
14. Goldstein, S.C., Mowry, T.C.: Claytronics: An instance of programmable matter. In: Wild and Crazy Ideas Session of ASPLOS. Boston, MA (2004)
15. Hayes, B.: Graph theory in practice: Part ii. Am. Sci. **88**(2), 104–109 (2000)
16. Jennings, E.H., Okino, C.M.: On the diameter of sensor networks. In: Aerospace Conference Proceedings, 2002, vol. 3, pp. 3–1211. IEEE (2002)
17. Jin, S., Bestavros, A.: Small-world characteristics of internet topologies and implications on multicast scaling. Comput. Netw. **50**(5), 648–666 (2006)
18. Karagozler, M.E., Goldstein, S.C., Reid, J.R.: Stress-driven mems assembly+ electrostatic forces = 1mm diameter robot. In: IEEE/RSJ International Conference on Intelligent Robots and Systems, IROS 2009, pp. 2763–2769 (2009)
19. Kirby, B.T., Ashley-Rollman, M., Goldstein, S.C.: Blinky blocks: a physical ensemble programming platform. CHI '11 Extended Abstracts on Human Factors in Computing Systems. CHI EA '11, pp. 1111–1116. ACM, New York, NY, USA (2011)
20. Latapy, M., Magnien, C.: Measuring fundamental properties of real-world complex networks (2006). arXiv:cs/0609115
21. Naz, A., Piranda, B., Goldstein, S.C., Bourgeois, J.: ABC-Center: Approximate-center election in modular robots. In: IROS 2015, IEEE/RSJ International Conference on Intelligent Robots and Systems, pp. 2951–2957. Hamburg, Germany (2015)
22. Naz, A., Piranda, B., Goldstein, S.C., Bourgeois, J.: Approximate-centroid election in large-scale distributed embedded systems. In: AINA 2016, 30th IEEE International Conference on Advanced Information Networking and Applications, pp. 548–556. IEEE, Crans-Montana, Switzerland (2016)
23. Naz, A., Piranda, B., Goldstein, S.C., Bourgeois, J.: A time synchronization protocol for modular robots. In: PDP 2016, 24th Euromicro International Conference on Parallel, Distributed, and Network-Based Processing, pp. 109–118. IEEE, Heraklion Crete, Greece (2016)
24. Oung, R., DAndrea, R.: The distributed flight array. Mechatronics **21**(6), 908–917 (2011)
25. Piranda, B., Bourgeois, J.: Geometrical study of a quasi-spherical module for building programmable matter. In: DARS 2016, 13th Internatinal Symposium on Distributed Autonomous Robotic Systems. Springer (2016)
26. Piranda, B., Laurent, G.J., Bourgeois, J., Clévy, C., Möbes, S., Le Fort-Piat, N.: A new concept of planar self-reconfigurable modular robot for conveying microparts. Mechatronics **23**(7), 906–915 (2013)
27. Reynolds, J.K., Postel, J.: Assigned Numbers. RFC 1700, RFC Editor (1994)
28. Suh, J.W., Homans, S.B., Yim, M.: Telecubes: Mechanical design of a module for self-reconfigurable robotics. In: IEEE International Conference on Robotics and Automation, Proceedings. ICRA'02, vol. 4, pp. 4095–4101. IEEE (2002)
29. The Internet Assigned Numbers Authority (IANA): Internet protocol version 4 (IPv4) parameters (2016). http://www.iana.org/assignments/ip-parameters/ip-parameters.xhtml
30. Watts, D.J., Strogatz, S.H.: Collective dynamics of small-world networks. Nature **393**(6684), 440–442 (1998)
31. Zhao, Y., Chen, Y., Li, B., Zhang, Q.: Hop id: a virtual coordinate based routing for sparse mobile ad hoc networks. IEEE Trans. Mobile Comput. **6**(9), 1075–1089 (2007)

Part VI
Swarm Robotics

Decentralized Progressive Shape Formation with Robot Swarms

Carlo Pinciroli, Andrea Gasparri, Emanuele Garone and Giovanni Beltrame

Abstract We tackle the problem of achieving any given shape defined as a point cloud in a distributed manner with a swarm of robots. The contributions of this paper are (i) An algorithm that transforms a point cloud into a acyclic directed graph; (ii) A motion control law that, from the acyclic directed graph, allows a swarm of robots to achieve the target shape in a decentralized manner; and (iii) A theoretical model, which provides sufficient conditions on the convergence of the control law. The key idea of our approach is to achieve the target shape *progressively* by inducing an ordering among the robots. More precisely, we construct an acyclic directed graph so that any free robot (i.e., not part of the shape) finds its location with respect to the already placed robots. We prove that, for a 2D shape, it is sufficient for a free robot to calculate its location with respect to two already placed robots to achieve this objective. We validate our method through accurate physics-based simulations of non-holonomic robots.

1 Introduction

Swarm robotics [5] is a branch of collective robotics that focuses on decentralized approaches to the problem of coordinating large teams of robots. A common assump-

C. Pinciroli (✉)
Worcester Polytechnic Institute, Worcester, MA, USA
e-mail: cpinciroli@wpi.edu

A. Gasparri
Università Roma Tre, Rome, Italy
e-mail: gasparri@dia.uniroma3.it

E. Garone
Université Libre de Bruxelles, Bruxelles, Belgium
e-mail: egarone@ulb.ac.be

G. Beltrame
Polytechnique Montréal, Montreal, QC, Canada
e-mail: giovanni.beltrame@polymtl.ca

R. Groß et al. (eds.), *Distributed Autonomous Robotic Systems*,
Springer Proceedings in Advanced Robotics 6,
https://doi.org/10.1007/978-3-319-73008-0_30

tion in many existing approaches is that all of the robots involved in the execution of a certain algorithm are ready to take part in it. However, especially when swarm sizes involve more than a dozen robots, it is hardly conceivable that all of the robots will be available. Economical and technological constraints are major drivers in the limitation of the number of robots that can be deployed in a specific time frame. For instance, in exploratory missions of planets or deep ocean floors, it is probable that only a small number of robots can be successfully deployed in a single mission. In addition, future missions involving complex, heterogeneous robot swarms will naturally unfold in a phase-wise fashion. In a large-scale Mars exploration mission, for instance, a first wave of radiation-hardened robots might be dedicated to creating shelter and network infrastructure for future waves of cheaper, more mission-specific robots.

A new class of swarm algorithms is necessary, which focuses on swarms that are progressively deployed over time [3]. As a step in this direction, we present a progressive, decentralized algorithm that allows a robot swarm to achieve a specific shape. Shape formation is an important application for robot swarms. Typical examples include monitoring of large-scale areas, satellite displacement in orbit, and the creation of ad hoc, large-scale network infrastructure. In these applications, it is desirable that even if the final form of the target shape has not been attained yet, the already placed robots can begin performing their assigned tasks. Thus, the time taken to complete the shape is, at least to an extent, less important than the time taken to add a new robot to the shape and the precision of the positioning.

The core of our idea is to express the target shape as an acyclic directed graph, in which each robot finds and maintains its position by considering only two other robots in the shape, called the *parents*. All robots know the acyclic directed graph, but initially no robot is assigned to any specific position in the shape. The shape is constructed dynamically and in an iterative fashion—each robot joins upon being granted permission by one of the parents. Based solely on local communication, our algorithm is completely decentralized. The algorithm is also parallel, since at any time multiple robots can join different parts of the shape.

The rest of the paper is organized as follows. We discuss related work in Sect. 2. In Sect. 3, we illustrate a mathematical model that proves the convergence properties of our decentralized algorithm. In Sect. 4, we present an algorithm to construct an acyclic directed graph starting from a point cloud which represents the target shape. In Sect. 5, we describe the behavior the robots follow to achieve the target shape. We report experimental evaluation in Sect. 6. The paper is concluded by Sect. 7, in which we outline future research directions.

2 Related Work

The problem of formation control has been widely studied in the literature. Relevant, recent examples of decentralized methodologies are consensus-based approaches [11, 15, 20] and rigidity-based approaches [2, 7, 19]. The interested reader is referred to [6, 10] and references therein for a more comprehensive overview of the formation control problem.

Considering the vastness of the literature, and that we deal with the problem of achieving any given shape defined as a point cloud in a decentralized manner, we focus on approaches with a similar objective.

Spletzer and Fierro [17] consider the task of repositioning a formation of robots to a new shape while minimizing either the maximum distance that any robot travels, or the total distance traveled by the formation. Ravichandran et al. [14] present a scalable distributed reconfiguration algorithm to achieve arbitrary target configurations built on top of a distributed median consensus estimator which requires only local communication. Yu and Nagpal [21] present a theoretical study of decentralized control for sensing-based shape formation on modular multi-robot systems, where the desired shape is specified in terms of local sensor constraints between neighboring robot agents. Alonso-Mora et al. [1] consider arbitrary target patterns and detect a proper representation with an optimal robot deployment, using a method that is independent of the number of robots. Liu and Shell [8] tackle the problem of changing smoothly between formations of spatially deployed multi-robot systems, and show that this can be achieved by routing agents on a Euclidean graph that corresponds to paths computed on—and projected from—an underlying three-dimensional matching graph.

3 Mathematical Model

3.1 Problem Statement

Let us consider a team of N mobile robots moving in $\mathbb{R}^2$, each of which is assigned a label $i \in \{1, ..., N\}$. For each robot, let the vector $\mathbf{p}_i(t) = [x_i(t), y_i(t)]^T \in \mathbb{R}^2$ describe its position with respect to a fixed global reference frame $\mathcal{O}^g_{xy}$. It is assumed that each robot has a dynamics

$$\dot{\mathbf{p}}_i = \mathbf{u}_i. \tag{1}$$

We assume that each robot i does not know its absolute position $\mathbf{p}_i(t)$, but it can measure the relative distance $d_{ij}(t) = \|\mathbf{p}_i(t) - \mathbf{p}_j(t)\|$ and the relative bearing $\theta_{ij}(t)$ with respect to any neighboring robot j. We also assume that the nodes can communicate with their neighboring robots, i.e., the robots in their direct line-of-sight. Communication and distance/angle estimation can be achieved through situated communication [18] devices present on commercial robots such as the e-puck [9].

The problem statement can be defined as follows.

Problem 1 Given a desired formation expressed as a set of of desired positions $\{\bar{\mathbf{q}}_1, \bar{\mathbf{q}}_2, \ldots, \bar{\mathbf{q}}_N\}$, there exists a translation $\mathbf{r} \in \Re^2$ and a rotation $R(\theta)$ such that

$$\lim_{t\to\infty} R(\theta)\mathbf{p}_i(t) + \mathbf{r} = \bar{\mathbf{q}}_i.$$

In this problem formulation, we are not interested in having the formation converge to a specific position in the environment. Rather, we impose that the target shape can be arbitrarily rototranslated w.r.t. the fixed global reference frame. In the remainder of this section, for the sake of simplicity and without loss of generality, we will consider $\bar{\mathbf{q}}_1 = [0, 0]^T$ and $\bar{\mathbf{q}}_2 = [x_2, 0]^T$ with $x_2 > 0$. Under this assumption, we call $\mathbf{p}_i^{12}$ the coordinates of the i-th robot in the reference frame centered in $\mathbf{p}_1$. The x axis is oriented as the vector connecting $\mathbf{p}_1$ to $\mathbf{p}_2$, while the y axis is obtained through a $\pi/2$ anti-clockwise rotation of x around $\mathbf{p}_1$. We then restate Problem 1 as the following control problem:

Problem 2 Given a desired formation $\{\bar{\mathbf{q}}_1, \bar{\mathbf{q}}_2, \ldots, \bar{\mathbf{q}}_n\}$ where $\bar{\mathbf{q}}_1 = [0, 0]^T$ and $\bar{\mathbf{q}}_2 = [x_2, 0]^T$ and $x_2 > 0$, ensure that $\dot{\mathbf{p}}_i \to 0$ and $\mathbf{p}_i \to \bar{\mathbf{q}}_i \; \forall i \in \{1, \ldots, N\}$.

3.2 Proposed Solution

For robot 1, a simple solution to Problem 2 is to maintain its current position, i.e.,

$$\mathbf{u}_1 = \mathbf{0}. \tag{2}$$

For robot 2, an equally simple solution is to apply any control law of the form

$$\mathbf{u}_2 = \mathbf{f}_2(d_{1,2}(t)) \tag{3}$$

that ensures that $d_{1,2}(t)$ globally asymptotically tends to $\|\bar{\mathbf{q}}_1(t) - \bar{\mathbf{q}}_2(t)\|$. A distance-based attraction/repulsion law, such as the Coulomb potential [16], is a good choice for $\mathbf{f}_2$.

Concerning the other robots ($i > 2$), we associate to each of them two parent nodes j and k and define a continuously differentiable potential field $\Phi_i(\mathbf{p}_i, \mathbf{p}_j, \mathbf{p}_k)$, $\Phi_i : \Re^2 \times \Re^2 \times \Re^2 \to \Re$, such that, for any two bounded $\mathbf{p}_j$ and $\mathbf{p}_k$, these assumptions hold:

A1: $\Phi_i(\mathbf{p}_i, \mathbf{p}_j, \mathbf{p}_k)$ is invariant to rototranslations:

$$\Phi_i(R(\theta)\mathbf{p}_i + \mathbf{r}, R(\theta)\mathbf{p}_j + \mathbf{r}, R(\theta)\mathbf{p}_k + \mathbf{r}) = \Phi_i(\mathbf{p}_i, \mathbf{p}_j, \mathbf{p}_k).$$

A2: $\Phi_i(\mathbf{p}_i, \mathbf{p}_j, \mathbf{p}_k)$ is unbounded for unbounded $\mathbf{p}_i$:

$$\lim_{\|\mathbf{p}_i\| \to \infty} \Phi_i(\mathbf{p}_i, \mathbf{p}_j, \mathbf{p}_k) = \infty.$$

A3: If p_j and p_k are distinct, $\Phi_i(\mathbf{p}_i, \mathbf{p}_j, \mathbf{p}_k)$ admits only one stationary point, which we denote as $\bar{\mathbf{p}}_i(\mathbf{p}_j, \mathbf{p}_k)$, and that point is a minimum of the potential field. This corresponds to forcing robot i to only have one possible position to reach when communicating with robots j and k.

A4: There exist two finite scalars $\alpha_{\max}$ and $\beta_{\min}$ such that for any $\mathbf{p}_j$, $\mathbf{p}_k$

$$-\nabla\Phi_i(\mathbf{p}_i, \mathbf{p}_j, \mathbf{p}_k)^T \nabla\Phi_i(\mathbf{p}_i, \mathbf{p}_j + \mathbf{d}_j, \mathbf{p}_k + \mathbf{d}_k) \leq 0$$

for any $\|\mathbf{b}_j\|, \|\mathbf{b}_k\| \leq \alpha_{\max}$, and for any $\mathbf{p}_i = \bar{\mathbf{p}}_i(\mathbf{p}_j, \mathbf{p}_k) + \mathbf{d}_i$ with $\|\mathbf{d}_i\| \geq \beta_{\max}$. This means that, for bounded perturbations of the parents $\mathbf{p}_j$ and $\mathbf{p}_k$, if $\mathbf{p}_i$ is sufficiently far from the equilibrium, the perturbed anti-gradient is still a direction of descent of the non-perturbed potential field.

We prove the following theorem by using a control law in the form:

$$\mathbf{u}_i = -\nabla\Phi_i(\mathbf{p}_i, \mathbf{p}_j, \mathbf{p}_k), \tag{4}$$

Theorem 1 *Consider the system described by Equation* (1) *controlled by control laws in the form of Equations* (2)–(4) *that comply with assumptions **A1–A4**. Under the constraints*

1. *If j and k are parents of i, then $j < i$ and $k < i$;*
2. $\bar{\mathbf{p}}_i(\bar{\mathbf{q}}_j, \bar{\mathbf{q}}_k) = \bar{\mathbf{q}}_i \quad \forall i \in \{1, \ldots, N\}$;

Problem 2 is solved for any initial condition.

Since every position $\bar{\mathbf{q}}_i$ in the target shape is expressed with respect to two other positions $\bar{\mathbf{q}}_j$ and $\bar{\mathbf{q}}_k$, and $j < i$ and $k < i$, we can represent the target shape as an acyclic directed graph in which every node i is connected to its parents j and k. Thus, the process of positioning a specific robot i can ideally be seen as a cascade of positioning processes, which starts with robot 1 and proceeds with all other robots. If every step of the positioning dynamics is globally asymptotically stable, then the entire process converges to the desired shape. This is the core idea behind the proof in [12].

For some applications, assumptions **A3** and **A4** might prove too strong. For instance, the existence of a single stationary point topologically forbids the use of repulsion terms among robots that are usually used for collision avoidance. To fix this, consider the following relaxed condition:

A3bis: If $\mathbf{p}_j$ and $\mathbf{p}_k$ are distinct there exists $\bar{\mathbf{p}}_i(\mathbf{p}_j, \mathbf{p}_k)$ which is an isolated minimum of $\Phi_i(\mathbf{p}_i, \mathbf{p}_j, \mathbf{p}_k)$.

By using **A3bis** instead of **A3** and dropping assumption **A4**, the formation is asymptotically stable. This means that there exists a basin of attraction around the configuration of equilibrium $\dot{\mathbf{p}}_i^{1,2} = \bar{\mathbf{q}}_i$ for which Problem 2 is solved.

Theorem 2 *Consider the system described by Equation* (1) *controlled by Equations* (2)–(4) *that comply with assumptions **A1**, **A2**, and **A3bis**. Under the contraints:*

1. *If j and k are parents of i, then $j < i$ and $k < i$;*
2. $\bar{\mathbf{p}}_i(\bar{\mathbf{q}}_j, \bar{\mathbf{q}}_k) = \bar{\mathbf{q}}_i \quad \forall i \in \{1, \ldots, N\}$;

Problem 2 is solved for initial conditions in a suitable neighborhood of the equilibrium configuration of $\dot{\mathbf{p}}_i^{1,2} = \mathbf{q}_i \quad \forall i \in \{1, \ldots, N\}$.

The proof is reported in [12].

4 From Point Cloud to Acyclic Directed Graph

The target shape is expressed as a 2D point cloud. In this section, we present an algorithm that produces an acyclic directed graph starting from the 2D point cloud (see also Fig. 1). The specific frame of reference used to express the point cloud is not important, since the labeled graph must only encode the relative distances between a robot and its two predecessors.

The algorithm proceeds iteratively and collects the points in three lists:

- the *unlabeled* list contains the points for which no label was found yet;
- the *delayed* list contains the points for which a label was searched, but it was impossible to find one in the current iteration;
- the *labeled* list stores the labeled points.

The algorithm starts by storing all points as *unlabeled*, and by calculating the center of mass of the point cloud. The points are then sorted by their distance to the center of mass. The two closest points are labeled 0 and 1, 0 is stored as a predecessor of 1, and the points are moved to *labeled*. Using the center of mass allows the shape formation process to start at the center of the shape and to grow outwards, making it easier for joining robots to navigate to their target position.

Subsequent points are labeled as follows. First, a candidate point is drawn from the *unlabeled* set. Next, the viable predecessors among the points in *labeled* are

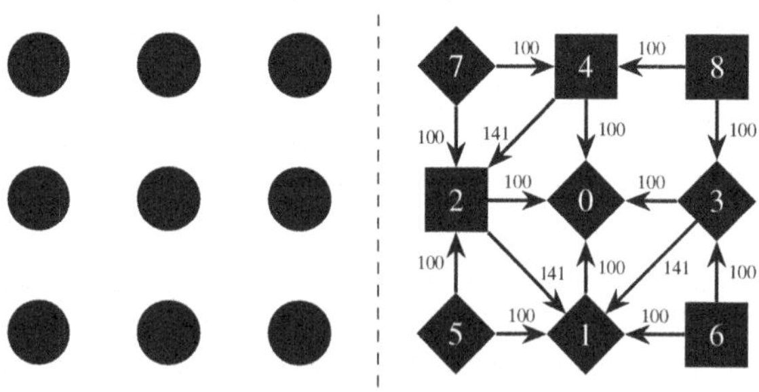

Fig. 1 Depiction of the input and output of the algorithm presented in Sect. 4. The left side shows the input, an unlabeled point cloud. The right side illustrates the labeled acyclic directed graph output by the algorithm. In the output, numerical labels identify the position in the acyclic directed graph; arrows indicate the parents of a specific shape node, and the associated numbers denote the parent-child distance in cm. Square nodes are those located on the positive side with respect to the vector joining the two parents; diamond nodes are on the negative side. Sides are calculated considering the right-hand rule

searched. Two points are viable if, with the candidate point, they form a triangle in which every robot has unobstructed line-of-sight with the other two. If no pair of predecessor can be found, the candidate is moved to *delayed* and a new candidate point is drawn. As soon as two viable predecessors are found, the candidate point is stored into *labeled* and a new graph element is added (each node stores labeled node, predecessors, and positional information—see Fig. 1, right). The *delayed* points are moved back to *unlabeled*, and a new iteration of the algorithm is performed.

The algorithm ends when no points are present in *unlabeled*. If *delayed* is an empty set, the algorithm finished successfully, as all points have been labeled; otherwise, the algorithm fails.

5 Robot Behavior

Assumptions. At time zero, no robot is part of the structure. The process that brings a robot to the decision of triggering the shape formation behavior depends on the specific experiment under study. Here, we assume that a single robot makes such decision, i.e., the process is not started twice in the same swarm. For simplicity, we also assume that every robot knows the acyclic graph of the target shape. However, this assumption could be lifted. For instance, the robot that triggers the behavior could broadcast the data of the labeled graph chosen for the task at hand. Nearby robots, upon receiving this data, could help diffuse it either unconditionally, or only when they join the shape.

Behavior Structure. The behavior is composed of four states (see Fig. 2):

FREE: A robot is this state is not part of the structure, nor actively requesting to join. In this state, a robot monitors its neighborhood searching for a label i that is not currently part of the shape nor being requested by other robots. If such label is found and both parents for it are in direct line-of-sight, the robot switches to state ASKING.

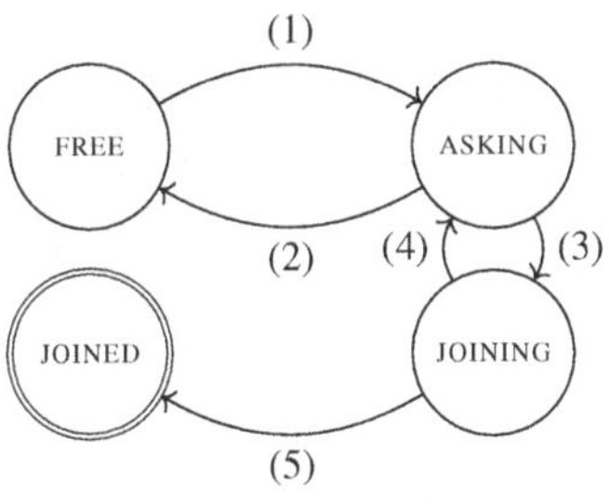

transition	condition
(1)	both parents seen for free label i
(2)	lost sight of a parent for label i
(3)	accepted to join with label i
(4)	rejected or reply timeout expired
(5)	distance to target within tolerance

Fig. 2 A finite state machine representation of the robot behavior

ASKING: A robot in this state broadcasts a request to join with label i. If the request receives no response for a predefined time, or if another robot is accepted, any rejected robot switches back to FREE. Otherwise, the robot switches to state JOINING.

JOINING: A robot in this state uses the two parents as references to join the shape. The robot navigates to the target position. As soon as its distance to the target is smaller than a tolerance ε, the robot switches to JOINED.

JOINED: A robot in this state is part of the shape and performs two activities: (i) It maintains its position with respect to the relevant robots around it—the relevant robots are its direct parents and children; and (ii) It collects requests from ASKING robots, granting access to the one in the best position to assume a specific label.

Communication. During the construction of the shape, every robot broadcasts a message to inform its neighbors of its current state. In addition, some states require extra information to be broadcast. When a robot is ASKING, it broadcasts a request structured as a tuple $\langle$ASKING,i,$z\rangle$, in which z is the request id. The latter is drawn uniformly at random between 1 and a sufficiently large number Z.[1] A robot in ASKING state is such that the parents of label i are in direct line-of-sight. Thus, both parents receive the request. The parent with the higher label m is in charge of responding to the request. This parent collects the requests received in the last time step and ranks them from the simplest to achieve (i.e., the requesting robot is close to its target position) to the hardest to achieve. The best request is granted by broadcasting a message $\langle$JOINED,m,i,$\bar{z}$, `granted`$\rangle$, where $\bar{z}$ identifies the request for label i univocally. The accepted robot switches to JOINING, while the other robots that asked for i switch back to state FREE. Rarely, it might happen that two robots choose the same request id z for the same label i. In this case, the robot in charge replies with $\langle$JOINED,m,i,`repeat`$\rangle$, which forces the conflicting robots to repeat their request, while informing other robots that their request for i was refused.

Navigation. The mathematical model described in Sect. 3 neglects aspects such as non-holonomic motion, body size, and collision avoidance to guarantee tractability. In a physical implementation of the algorithm, however, these aspects play a crucial role in system dynamics. As discussed in Sect. 3, we assume that the robots are capable of detecting their distance and angle to a number of neighbors in range. Navigation is based solely on this information—we do not require the robots to estimate their global position in the environment. Navigation is implemented following the well-known virtual physics approach [16], whereby virtual forces are calculated by overlapping multiple potential fields according to the state of the robots.

FREE: When a robot is in this state, it must keep at a safe distance from the structure being built to prevent congestion. In addition, FREE robots follow the

[1] In our experiments we set $Z = 2^{16} - 1$. With this value, 302 robots must simultaneously request the same label i to have a probability > 0.5 that at least two robots choose the same value for z. We observed that, on average, a robot has only about 8 neighbors; assuming that all of them request the same i simultaneously, the probability that at least two requests have the same value for z is $\approx 4.27 \cdot 10^{-4}$.

external perimeter of the shape and search for free labels. This is obtained by combining two virtual forces: one that imposes a sufficiently large distance from the shape, and one that pushes the robots perpendicularly to any JOINING or JOINED robot in sight.

ASKING: Robots in this state must keep in sight to the two parents of the label being requested. The virtual force is calculated by imposing bounded repulsion to FREE robots and bounded attraction to JOINED robots.

JOINING: To join the shape, a robot must follow the complex, position-dependent force field depicted in Fig. 3, which corresponds to $-\nabla\Phi_i(\mathbf{p}_i, \mathbf{p}_j, \mathbf{p}_k)$, discussed in Sect. 3.

JOINED: Robots that already joined the shape must keep their distance to their parents and direct children. In our implementation, we used a field based on Coulomb forces analogous to [16].

In addition to these primary states, we added a fifth state, ESCAPE, to deal with FREE robots that got trapped in the middle of JOINED robots. This extra behavior proved important to drop the failure rate of the shape formation process from about 30% to 0.37% (see Sect. 6). An example of the dynamics of the navigation behavior can be seen in Fig. 4.

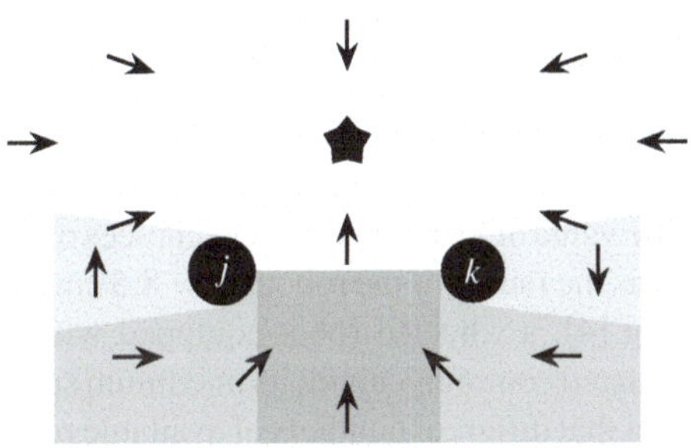

Fig. 3 Virtual force field experienced by a joining robot with respect to its parents j and k. The black star indicates the target position to be reached by the joining robot. This force field is a realization of $-\nabla\Phi_i(\mathbf{p}_i, \mathbf{p}_j, \mathbf{p}_k)$, discussed in Sect. 3

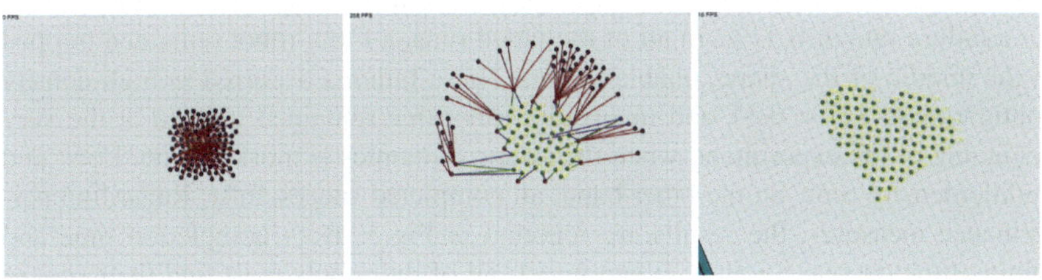

(a) $t = 10$ sec (b) $t = 400$ sec (c) $t = 1500$ sec

Fig. 4 The dynamics of the presented behavior with a heart-like shape formed by 114 robots. The black robots are FREE, and the green-yellow robots are JOINED. Screenshot taken with the ARGoS simulator. A video is available at http://carlo.pinciroli.net/DARS2016/heart.mp4

6 Experimental Evaluation

We evaluated the performance of our shape formation algorithm through a set a simulated experiments. We employed a realistic, physics-based simulator for multi-robot systems called ARGoS.[2] The models present in ARGoS have been validated, and robot controllers developed with this simulator have been succesfully ported to real platforms in many works [13].

Performance measures. We concentrated our analysis on three aspects of the shape formation process: completion time, joining time, and absolute positioning error. *Completion time* is the time necessary to form a complete shape. *Joining time* is the time elapsed between two successive join events, i.e., the moments in which the robots become part of the shape, switching to JOINED. In the context of a progressive algorithm, this measure is more interesting than completion time, because it accounts for recruitment efficiency. The *absolute positioning error* E_{ij} between robots i and j is calculated as $E_{ij} = \left|(d_{ij} - \bar{d_{ij}})/\bar{d_{ij}}\right|$ where $\bar{d_{ij}}$ is the expected distance between robots i and j (as stored in the labeled graph), and d_{ij} is the actual distance measured at the end of the experiment.

The target shape. While our algorithm can form generic shapes, to ensure comparability among different experimental conditions, the robots are tasked with the formation of a grid. We worked with three grid sizes—5×5, 10×10, and 15×15.

Experimental setup. The experimental arena is an empty square of side L, in which M robots are uniformly distributed with a certain density D. We considered three density configurations—broad ($D = 0.01$), medium ($D = 0.05$), and tight ($D = 0.1$). To enforce a specific value of D across different choices of M, we calculated $L = \sqrt{(N\pi R^2)/D}$, where R is the radius of the robot ($R = 8.5$ cm). For our experiments, we utilized the marXbot [4], a wheeled robot equipped with a range-and-bearing sensor that allows for situated communication; its maximum speed was set to 10 cm/s. We also studied the effect that different numbers of available robots have on the shape formation algorithm. To this aim, we defined N as the number of robots necessary to complete a shape (e.g., $N = 25$ for a 5×5 grid) and set $M = kN$, with $k \in \{1, 2, 3\}$. Every configuration $\langle$grid size, $D, k\rangle$ was tested 30 times, for a total of 810 runs.

Discussion. Out of the 810 runs, in only 3 of them the shape was not completed, for a failure rate of 0.37%. In all of the failed runs, a FREE robot remained trapped in the middle of the shape, unable to exit. These failures occurred in high-density configurations ($D = 0.1$), and involved a robot that remained trapped at the very beginning of the experiment, when the swarm dynamics is most chaotic. Low- and middle-density runs, on the other hand, all completed successfully. Regarding performance measures, the results are reported in Fig. 5. Both completion time and joining time decrease with k. When $k = 1$, all of the robots must find their spot in the shape. Towards the end of an experiment, however, the number of available spots diminishes, and the remaining few robots must spend a long time looking for it. For

[2] http://www.argos-sim.info/.

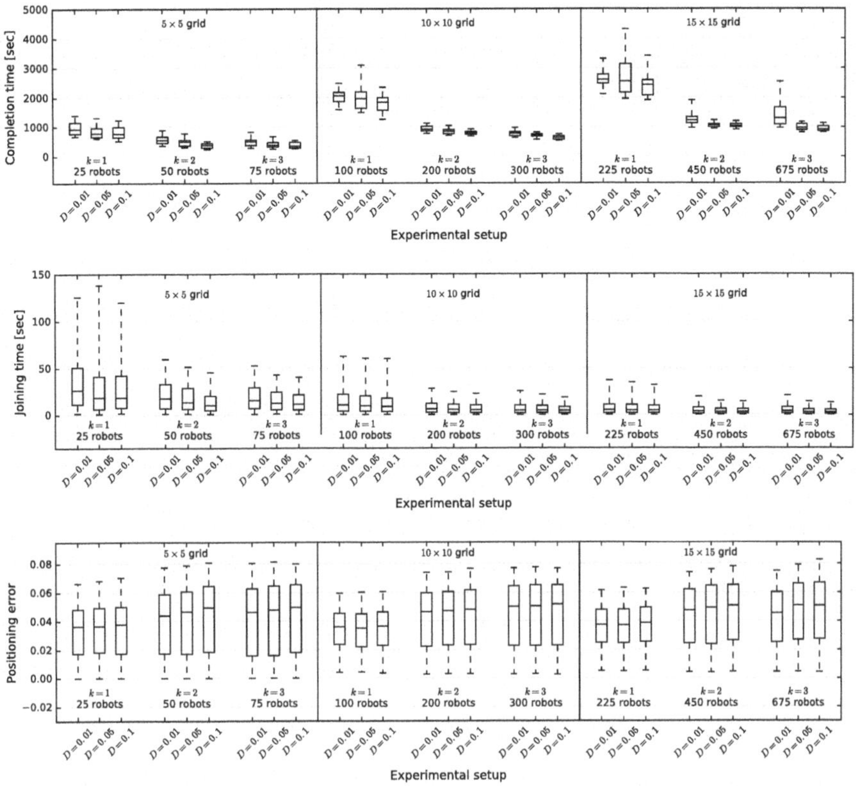

Fig. 5 Experimental evaluation of the shape formation behavior presented in Sect. 5. Each box corresponds to the distribution of a performance measure of a specific experimental configuration ⟨grid size, D, k⟩ over 30 runs (see Sect. 6). The whiskers in the box plots indicate the 5th and 95th percentile. The top plot reports the completion time in seconds; the middle plot reports the joining time in seconds; and the bottom plot reports the positioning error

higher values of k, however, the probability that a free robots finds a vacant label increases, thus lowering the joining time. The effect of the deployment density D is a slight decrease of completion time for tighter initial distributions. Although this effect is only minor, through observation we could notice that, with tighter initial distributions, the robots have to travel shorter distances and, as a consequence, tend to find vacant labels more frequently. The positioning error does not seem to be significantly affected in any configuration, remaining always lower than 8% and with a median slightly above 4%. The fact that these parameters do not affect the error is expected, because the behavior is designed to separate FREE and ASKING robots from JOINING and JOINED robots. None of the considered parameters can affect this separation. The actual value of the error is due to the parameter configuration of the force field that a robot follows to reach its place in the shape. With non-holonomic robots, it is common to be unable to reach the exact designated position due to motion limitations. For instance, a 8% error over a distance of 100 cm (i.e., the node distance in all of our experiments) corresponds to 8 cm, which is comparable to the radius of the marXbot ($R = 8.5$ cm).

7 Conclusions

Progressive deployment is a milestone towards the creation of real-world robot swarms for complex scenarios. In this paper, we presented an approach to the formation of any shape described through a 2D point cloud. Through a mathematical model, we provided sufficient conditions to ensure convergence; we proposed an algorithm to derive an acyclic directed graph from a 2D point cloud; and illustrated a decentralized robot behavior that attains the target shape represented by the acyclic directed graph. We validated our approach through a set of experiments that considered swarm size, available robots, and density of the initial robot distribution.

Future work involves the validation of our approach on real robots. In addition, joining and completion time could be improved by allowing JOINED robots to guide FREE robots to places where vacant labels are present.

References

1. Alonso-Mora, J., Breitenmoser, A., Rufli, M., Siegwart, R., Beardsley, P.: Multi-robot system for artistic pattern formation. In: 2011 IEEE International Conference on Robotics and Automation (ICRA), pp. 4512–4517 (2011)
2. Anderson, B., Yu, C., Fidan, B., Hendrickx, J.: Rigid graph control architectures for autonomous formations. IEEE Control Syst. **28**(6), 48–63 (2008)
3. Beal, J.: Functional blueprints: an approach to modularity in grown systems. Swarm Intell. **5**(3), 257–281 (2011)
4. Bonani, M., Longchamp, V., Magnenat, S., Rétornaz, P., Burnier, D., Roulet, G., Vaussard, F., Bleuler, H., Mondada, F.: The marXbot, a miniature mobile robot opening new perspectives for the collective-robotic research. In: Proceedings of the IEEE/RSJ International Conference on Intelligent Robots and Systems (IROS), pp. 4187–4193. IEEE Press, Piscataway, NJ (2010)
5. Brambilla, M., Ferrante, E., Birattari, M., Dorigo, M.: Swarm robotics: a review from the swarm engineering perspective. Swarm Intell. **7**(1), 1–41 (2013)
6. Bullo, F., Cortés, J., Martínez, S.: Distributed Control of Robotic Networks. Applied Mathematics Series. Princeton University Press (2009). http://coordinationbook.info
7. Krick, L., Broucke, M.E., Francis, B.A.: Stabilisation of infinitesimally rigid formations of multi-robot networks. Int. J. Control **82**(3), 423–439 (2009)
8. Liu, L., Shell, D.A.: Distributed autonomous robotic systems: The 11th international symposium. In: M. Ani Hsieh, G. Chirikjian (eds.) Multi-Robot Formation Morphing through a Graph Matching Problem, pp. 291–306. Springer Berlin Heidelberg, Berlin, Heidelberg (2014)
9. Mondada, F., Bonani, M., Raemy, X., Pugh, J., Cianci, C., Klaptocz, A., Zufferey, J.C., Floreano, D., Martinoli, A.: The e-puck, a Robot Designed for Education in Engineering. In: Gonçalves, P.J.S., Torres, P.J.D., Alves, C.M.O. (eds.) Proceedings of Robotica 2009–9th Conference on Autonomous Robot Systems and Competitions, vol. 1, pp. 59–65. IPCB, Castelo Branco, Portugal (2006)

10. Oh, K.K., Park, M.C., Ahn, H.S.: A survey of multi-agent formation control. Automatica **53**, 424–440 (2015)
11. Olfati-Saber, R., Fax, J., Murray, R.: Consensus and cooperation in networked multi-agent systems. Proc. IEEE **95**(1), 215–233 (2007)
12. Pinciroli, C., Gasparri, A., Garone, E., Beltrame, G.: Decentralized progressive shape formation with robot swarms: Proofs. Technical report, École Polytechnique de Montréal, Canada (2016). http://carlo.pinciroli.net/DARS2016/proofs.pdf
13. Pinciroli, C., Trianni, V., O'Grady, R., Pini, G., Brutschy, A., Brambilla, M., Mathews, N., Ferrante, E., Di Caro, G., Ducatelle, F., Birattari, M., Gambardella, L.M., Dorigo, M.: ARGoS: a modular, parallel, multi-engine simulator for multi-robot systems. Swarm Intell. **6**(4), 271–295 (2012)
14. Ravichandran, R., Gordon, G., Goldstein, S.: A scalable distributed algorithm for shape transformation in multi-robot systems. In: IEEE/RSJ International Conference on Intelligent Robots and Systems, IROS 2007, pp. 4188–4193 (2007)
15. Ren, W.: Consensus strategies for cooperative control of vehicle formations. IET Control Theory Appl. **1**(2), 505–512 (2007)
16. Spears, W.M., Spears, D.F., Hamann, J.C., Heil, R.: Distributed, physics-based control of swarms of vehicles. Auton. Robots **17**(2/3), 137–162 (2004)
17. Spletzer, J., Fierro, R.: Optimal positioning strategies for shape changes in robot teams. In: Proceedings of the 2005 IEEE International Conference on Robotics and Automation, ICRA 2005, pp. 742–747 (2005)
18. Støy, K.: Using situated communication in distributed autonomous mobile robots. Proceedings of the 7th Scandinavian Conference on Artificial Intelligence, pp. 44–52 (2001)
19. Williams, R., Gasparri, A., Priolo, A., Sukhatme, G.: Distributed combinatorial rigidity control in multi-agent networks. In: 2013 IEEE 52nd Annual Conference on Decision and Control (CDC), pp. 6061–6066 (2013)
20. Xiao, F., Wang, L., Chen, J., Gao, Y.: Finite-time formation control for multi-agent systems. Automatica **45**(11), 2605–2611 (2009)
21. Yu, C.H., Nagpal, R.: Sensing-based shape formation on modular multi-robot systems: A theoretical study. Proceedings of the 7th International Joint Conference on Autonomous Agents and Multiagent Systems -. AAMAS '08, vol. 1, pp. 71–78. International Foundation for Autonomous Agents and Multiagent Systems, Richland, SC (2008)

Discovery and Exploration of Novel Swarm Behaviors Given Limited Robot Capabilities

Daniel S. Brown, Ryan Turner, Oliver Hennigh and Steven Loscalzo

Abstract Emergent collective behaviors have long interested researchers. These behaviors often result from complex interactions between many individuals following simple rules. However, knowing what collective behaviors are possible given a limited set of capabilities is difficult. Many emergent behaviors are counter-intuitive and unexpected even if the rules each agent follows are carefully constructed. While much work in swarm robotics has studied the problem of designing sets of rules and capabilities that result in a specific collective behavior, little work has examined the problem of exploring and describing the entire set of collective behaviors that can result from a limited set of capabilities. We take what we believe is the first approach to address this problem by presenting a general framework for discovering collective emergent behaviors that result from a specific capability model. Our approach uses novelty search to explore the space of possible behaviors in an objective-agnostic manner. Given this set of explored behaviors we use dimensionality reduction and clustering techniques to discover a finite set of behaviors that form a taxonomy over the behavior space. We apply our methodology to a single, binary-sensor capability model. Using our approach we are able to re-discover cyclic pursuit and aggregation, as well as discover several behaviors previously unknown to be possible with only a single binary sensor: wall following, dispersal, and a milling behavior often displayed by ants and fish.

Research performed by all authors while at AFRL, Rome NY.

D. S. Brown (✉)
Computer Science Department, University of Texas, Austin, TX, USA
e-mail: dsbrown@cs.utexas.edu

R. Turner · O. Hennigh · S. Loscalzo
Air Force Research Laboratory, Rome, NY, USA
e-mail: ryan.turner.10@us.af.mil

O. Hennigh
e-mail: oliver.hennigh@us.af.mil

S. Loscalzo
e-mail: steven.loscalzo@ecs-federal.com

R. Groß et al. (eds.), *Distributed Autonomous Robotic Systems*,
Springer Proceedings in Advanced Robotics 6,
https://doi.org/10.1007/978-3-319-73008-0_31

1 Introduction

Biological swarms have long fascinated researchers and laymen alike. The ability of these swarms to perform complex tasks such as building temperature-controlled nests, comparing potential new nest sites, and coordinating and synchronizing flight patterns [4] have caused some observers to attribute these behaviors to supernatural abilities such as telepathy between flying birds [19] or centralized control from a queen. This notion has continued to persist in the popular media where swarm intelligence is often portrayed as many individuals controlled simultaneously by a single individual. However, despite humans' seemingly innate desire to attribute complex behaviors to higher-level, complex intelligence, researchers in robotics, biology, computer science, and physics continue to show that complex swarm behaviors are often a result of extremely simple local rules. Indeed, much of the research on swarms is focused on finding mappings between sets of specific rules and sets of specific behaviors and can be broken down into two questions: (1) Given a desired behavior, can we determine a set of rules that synthesize this behavior? and (2) Given a specific set of rules, can we determine what the resulting emergent collective behavior will be?

We propose to study a third fundamental, yet less well-defined question, that has received little attention: (3) Given a set of capabilities (i.e., computational power, number and type of sensors, communication range, etc.) what are the possible collective behaviors that can emerge?

Knowing what collective behaviors are possible given a limited set of capabilities is often quite difficult. Many emergent behaviors are counterintuitive and unexpected. Furthermore, even if the rules each agent follows are carefully constructed, it is difficult to predict what behavior will emerge. While much work has studied the problem of designing sets of rules and capabilities that result in a desired collective behavior, little work has examined the problem of characterizing the set of possible collective behaviors that can result from a limited set of rules or behaviors. We take what we believe is the first approach to address this problem.

In particular, we propose a general architecture for discovering a taxonomy of possible emergent swarm behaviors given a set of capabilities. Similar to Wolfram's work on characterizing the behaviors of simple cellular automata [23] we take the approach of a naturalist and seek a taxonomy of possible collective behaviors that can result from a set of capabilities. To form a taxonomy of the emergent behaviors that result from a given capability model, we use an approach based on novelty search [13] which allows us to explore the space of possible behaviors in an objective-agnostic manner. By optimizing novelty rather than any particular task or objective, and keeping track of novel behaviors in an archive, we are able to generate a large number of controllers that synthesize a wide variety of behaviors. Given this set of explored behaviors we use dimensionality reduction and clustering techniques to explore and categorize the space of possible behaviors.

We evaluate our architecture on a simple agent capability model that assumes only a single line-of-sight sensor that has only two possible values. Despite this parsimonious agent model, we show that there is a surprising variety of interesting collective behaviors. This approach allows us to "re-discover" previously studied computation-free circling and aggregation [7] as well as identify several behaviors previously unknown to be possible given a swarm of memory-less, single-sensor agents. These new behaviors include wall following, dispersal, and coordinated milling often found in ants and schools of fish [18, 21].

Our main contributions are summarized as follows:

- We propose the first general architecture to explore and form a taxonomy of the space of possible behaviors given a limited-capability robot model.
- We demonstrate the feasibility of using novelty search and archive clustering to generate a set of representative behaviors for a simple single-sensor robot capability model.
- We validate our approach by showing that it discovers previously known behaviors, as well as discovering several behaviors previously unknown to be possible for swarms of single-sensor robots.

2 Problem Statement

The main question this research seeks to answer can be stated as follows:

What is the set of possible emergent behaviors in a swarm of robots possessing a specific set of individual capabilities?

To formalize this problem we provide the following definitions. We define a *capability model* as a three-tuple $\langle S, M, A\rangle$ composed of sensors S, memory and computational processing resources M, and actuators A. The capability model captures what information a robot can collect from the world, how it can process that information, and how it can change its state and the state of its environment. Given a set of N agents each with capability model $c_i = \langle S_i, M_i, A_i\rangle$ for $i = 1, \ldots, N$, we define the capability model of the swarm as $\mathscr{C} = \{c_i : i = 1, \ldots, N\}$.[1]

We define an *emergent behavior* as a global pattern or structure resulting from local interactions between a collection of agents. We denote the *set of possible emergent behaviors* as $\mathscr{B}$.

We also define an *environment* $\mathscr{E}$. Given a capability model $\mathscr{C}$ and an environment $\mathscr{E}$ we desire to find a mapping from capabilities and environments to behaviors. Note that we have made no mention of how the capabilities are used by an agent with capability $c_i \in \mathscr{C}$ in environment $\mathscr{E}$. Rather than specifying the controller, we desire to find the image of a function Φ that maps from all possible controllers that can be instantiated on capability model $\mathscr{C}$ for environment $\mathscr{E}$ to the set of possible emergent behaviors $\mathscr{B}$. Thus we desire to find $\mathscr{B}$ where

[1]This formalism also captures homogeneous swarms which can be modeled by letting $c_i = c,\ \forall i$.

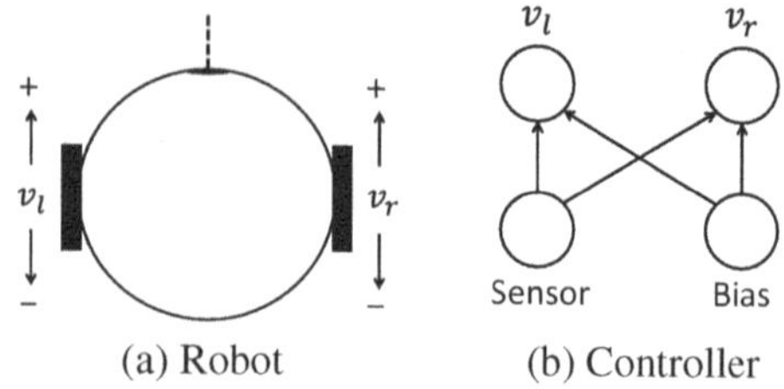

Fig. 1 Simple neural network controller for a single binary line-of-sight differential drive robot. The sensor value is input to the network which outputs the left and right wheel velocities

$$\Phi : \mathscr{U}(\mathscr{C}) \times \mathscr{E} \to \mathscr{B} \tag{1}$$

where $\mathscr{U}(\mathscr{C})$ is the controller space resulting from the capability model $\mathscr{C}$.

It is worth noting that in general the controller space is enormous. For example, consider the popular, and very simple, Kilobot robot platform [17]. A Kilobot could be modeled as a three-tuple $\langle S_k, M_k, A_k\rangle$ where S_k includes the infrared receiver and the ambient light sensor, M_k represents all programs programmable on an Atmega328 microprocessor with 32 K of memory, and A_k includes the speed of the two vibrating motors, as well as the output of the infrared LED transmitter.

As a first step towards discovering novel collective behaviors, we examine an even simpler capability model based on the e-puck robot [16], where S is a single, on/off, binary sensor, A consists of two differential drive wheels, and M is a fully connected 2 layer neural network connecting sensor inputs to wheel velocities. This capability model is depicted in Fig. 1. Even for this simple example, the controller space is $\mathbb{R}^4$, the space of all neural network weights.

Given the size of the controller-space, brute force or analytical methods for determining the mapping Φ would be incredibly difficult, if not impossible. Thus, we resort to a genetic search methodology outlined in the next section.

3 Behavior Discovery Architecture

Our proposed architecture relies heavily on novelty search as a means for exploring the space of emergent behaviors. Lehman and Stanley [13] proposed novelty search as a way to avoid getting stuck in local minima and to overcome deceptive fitness landscapes in genetic algorithms. Rather than using an objective function that rewards fitness, they show that simply trying to maximize the novelty of an evolved behavior will often generate a solution to the original problem more quickly than using pure fitness.

Rather than measuring similarity on the actual genotype, novelty measures similarity on the phenotype—the actual behavior resulting from executing the evolved controller. There are many potential ways for a user to define a behavior space; however, behavior spaces are typically represented by a vector with components that contain statistics over the state of the simulation collected periodically [10, 13]. Given a representation of a learned behavior in d dimensional behavior space, the typical measure of novelty as used in [13] is the sparseness of a point $b \in \mathbb{R}^d$ in behavior space, defined as $\texttt{Novelty}(b) = \frac{1}{k}\sum_{i=0}^{k} dist(b, \beta_i)$, where β_i is the ith nearest neighbor of b with respect to the distance metric $dist$. The nearest neighbor calculations take into consideration both individuals in the current population as well as previous members of the population that are stored in an archive that is updated each generation.

Novelty search has been used successfully to evolve many different types of single agent [6] and swarm behaviors [10]. The success of novelty search is attributed to its success in exploring the behavior space and discovering successively more complex behaviors [13]; however, to the best of our knowledge, our approach is the first to use novelty search purely for exploration without a specific task in mind.

Our approach proceeds as follows, we first start with a random population of controllers. Each controller is evaluated in our environment and a feature vector describing the resulting behavior is calculated. Given these behavior features, each policy is evaluated for novelty. Based on some archiving scheme, some or all of the policies are stored in an archive. Then, artificial evolution and mutation is used to create the next generation of controllers, where novelty is used as the fitness score. This process is repeated until it reaches some stopping criterion, at which point the discovered behaviors in the archive are clustered and representatives of each cluster are used to form an approximate taxonomy of possible emergent behaviors. The basic algorithmic outline is given in Algorithm 1.

Algorithm 1 `NovelBehaviorDiscovery`

Require: environment $\mathscr{E}$, capability model $\mathscr{C}$, and controller model $\mathscr{U}$
  $P \leftarrow$ `InitializePolicies`$(\mathscr{U}(\mathscr{C}))$ ▷ Generate initial population P_0
  archive ← `InitializeArchive`(P)
  while stopping criterion not met **do**
    for each policy p_i in population P **do**
      $f_i \leftarrow$ `ExtractFeatures`$(p_i, \mathscr{E})$ ▷ extract features by evaluating policy
      $n_i \leftarrow$ `Novelty`$(f_i,$ **archive**$)$ ▷ evaluate novelty
      if `addToArchive`$(\langle p_i, f_i\rangle)$ **then**
        archive.`add`(p_i, f_i) ▷ store individual in novelty archive
      end if
    end for
    $P \leftarrow$ `Update`(P, n) ▷ update population using a GA with fitness replaced by novelty
  end while
  $K \leftarrow$ `Cluster`(**archive**) ▷ Cluster on archive and return K representative behaviors
  return K ▷ Return cluster representatives as taxonomy

4 Implementation

4.1 Simple Capability Model

We use a homogeneous capability model based on Gauci et al.'s recently proposed single, binary-sensor, line-of-sight robots [7]. Each robot is equipped with a differential drive and a single line-of-sight sensor that provides it with one bit of information that lets the robot know whether it is facing another agent (see Fig. 1a).

Using this simple robot capability model, Gauci et al. optimized controllers to perform aggregation [7] demonstrating that highly robust aggregation was possible despite extremely limited capabilities. Subsequent research has shown that increasing the robot capability to include trinary sensors allows specific controllers to be evolved to accomplish tasks such as collecting pucks [8] and forming a perimeter, aggregating to a specific location, and foraging [11]. Our work extends previous work on simple, single-sensor swarms by examining the entire space of collective behaviors that are possible given a swarm of robots whose input is limited to a single, binary, line-of-sight sensor.

4.2 Simulation Environment

Due to the infeasibility of evaluating thousands of controllers on physical robots, we follow the common practice of using a simulator [7, 10] to allow rapid exploration of the behavior space. Following recent work on novelty search for swarms [10], we used the MASON multi-agent simulation environment [14] to simulate the physics of simple differential drive robots modeled after the e-puck robot [16]. Agent movement is simulated within a frictionless walled region of 50 by 50 units, where one unit equals one robot diameter. Each agent has two differential drive wheels. The controller for each robot is a simple neural network with one input node for the binary sensor and one output node for each wheel. The output is fixed in the range $[-1, 1]$ by a tanh function. The actual robot velocity on each wheel is then the output multiplied by the maximum speed. Figure 1 shows a representation of the robot and the controller architecture.

Following the approach used by [10], we use NeuroEvolution of Augmenting Topologies (NEAT) [20] with novelty search to optimize the weights on the neural network controller shown in Fig. 1b. We computed novelty using the 15 nearest neighbors in the archive, consistent with best practices found by Gomes et al. [9]. In our work we are interested in the full space of behaviors so we keep all individuals from each generation and add them to the novelty archive.

4.3 *Behavior Vector*

To explore the impacts of different behavior features on the discovered behaviors, we used a five element behavior vector. The five element behavior vector measures the average speed, scatter, radial scatter, angular momentum, and group rotation. *Average speed* measures the average speed of the agents in the swarm. *Scatter* [7] measures the average squared distance of the agents to the center of mass μ where $\mu = \frac{1}{N}\sum_{i=1}^{N} x_i$. *Radial variance* measures the variance of the distance of the agents to the center of mass μ. *Angular momentum* measures the true angular momentum about the center of mass of the swarm. Finally, *group rotation* measures a normalized angular momentum, ignoring the length of the moment arm [21]. R is the world radius (distance from center of world to corner in the case of a square world). The value R is used to normalize several of the features to be invariant to the size of the world. To create our final behavior vector, we used a sliding window average of each feature over the last 100 time steps. The details of these behavior vectors are shown in Table 1.

4.4 *Dimensionality Reduction and Clustering*

While it is possible to cluster in the high-dimensional behavioral space, interpreting the clusters becomes more difficult and single behaviors tend to be falsely split into multiple clusters. To reduce the dimensionality of our data we use t-distributed stochastic neighbor embedding (t-SNE) [22], a state-of-the-art dimensionality reduction technique shown to outperform other standard techniques such as PCA, Sammon mapping, and Isomap. t-SNE is especially suited for taking high-dimensional data that lies on several low-dimensional manifolds and mapping it to a 2-dimensional mapping that preserves and reveals this structure. We used t-SNE [22] to reduce the dimensionality and compare two types of clustering: k-means and hierarchical single-link (min distance) agglomerative clustering to partition the behaviors.

Table 1 Behavior vector feature descriptions

Name	Equation	Name	Equation
Average speed	$\frac{1}{N}\sum_{i=1}^{N} \lVert v_i \rVert_2$	Scatter	$\frac{1}{R^2 \cdot N}\sum_{i=1}^{N} \lVert x_i - \mu \rVert^2$
Ang. momentum	$\frac{1}{R \cdot N}\sum_{i=1}^{N} (v_i \times (x_i - \mu))$	Group rotation	$\frac{1}{N}\sum_{i=1}^{N}\left(v_i \times \frac{x_i - \mu}{\lVert x_i - \mu \rVert}\right)$
Radial variance	$\frac{1}{R^2 \cdot N}\sum_{i=1}^{N}\left(\lVert x_i - \mu \rVert - \frac{1}{N}\sum_{i=1}^{N} \lVert x_i - \mu \rVert\right)^2$		

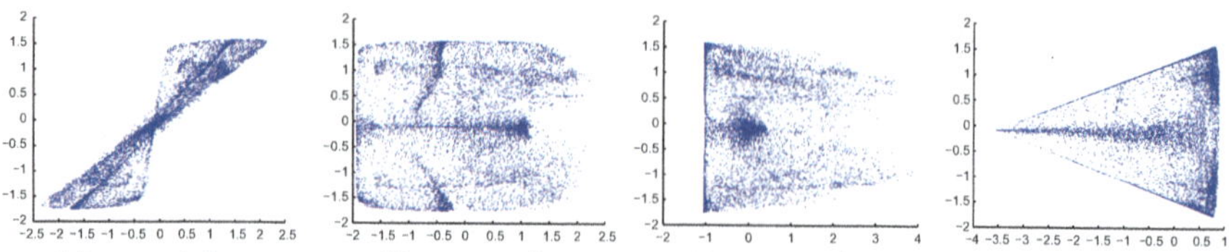

(a) group rotation vs. angular momentum (b) group rotation vs. scatter (c) group rotation vs. radial variance (d) group rotation vs. average speed

Fig. 2 Two-dimensional projections of the behavior space. The y-axis is group rotation

5 Results and Analysis

Using novelty search we ran 100 generations of 100 populations using NEAT to obtain an archive of 8020 data points[2] in 5 dimensional behavior space. Each experiment used 30 simulated robots. Figure 2 shows several 2-dimensional projections of the 5-dimensional data. Based on these results we see that there is definite structure captured by these features. We also see that group rotation and angular momentum are highly correlated, as expected, but do capture different information.

We used the MATLAB implementation of t-SNE[3] to map our 5-dimensional data to 2 dimensions. The resulting 2-dimensional data has definite structure and visible clusters as seen in Fig. 4, as opposed to the projected data shown in Fig. 2. We performed the dimensionality reduction before clustering to both reduce the computation time required for clustering and to make the results easy to visualize.

5.1 K-Medoids

k-Means is the de facto clustering algorithm to begin data exploration. We use a related clustering algorithm called k-Medoids that returns k actual data points as cluster centers. We use the medoids as the representative behaviors.

Because our goal is to discover and categorize emergent behaviors, we do not have any way of knowing the number of clusters ahead of time. Thus, we explored the resulting clusters for values of k between 2 and 10 and visually inspected the behavior of the resulting medoids for each value of k. The results are shown in Table 3a using. the abbreviations listed in Table 2. Sample trajectories of these behaviors are shown in Fig. 3.

[2]Due to the elitism feature of NEAT, the best performing policies (most novel) in one generation are kept in the population for the next generation. Thus, the algorithm explores fewer than 10,000 unique controllers.

[3]https://lvdmaaten.github.io/drtoolbox/.

Table 2 Abbreviations used to describe common behaviors

Abbreviation	Description	Abbreviation	Description
cycp	Cyclic pursuit	mill	Milling
wall	Wall slide	rand	Individual circling w/out emergence
aggr	Aggregation	cw	Clockwise motion
disp	Dispersal	ccw	Counterclockwise motion

Table 3 Results of examining centers from k-Medoids and Hierarchical clustering on the t-SNE embedded behavioral data. #x denotes that # cluster medoids were of that type

(a) k-Medoids

	cycp		wall		aggr	disp	mill		rand
k	cw	ccw	cw	ccw			cw	ccw	
2	x						x		
3				x			x	x	
4				x	x	x		x	
5						x	2x	2x	
6	x	x	x	x		x	x		
7	x	x	x		x	x	x	x	
8	x		x	2x	x	x		x	x
9	x	x	2x	x	x	x	x		
10	x	2x	2x	x	x	x		x	x

(b) Hierarchical clustering

	cycp		wall		aggr	disp	mill		rand
k	cw	ccw	cw	ccw			cw	ccw	
2	x								x
3	x	x							x
4	x	2x							x
5	x	2x			x				x
6	x	3x	x		x				
7	x	4x				x			x
8	x	4x			2x	x			
9	2x	4x	x		x	x			
10	4x	4x			x	x			

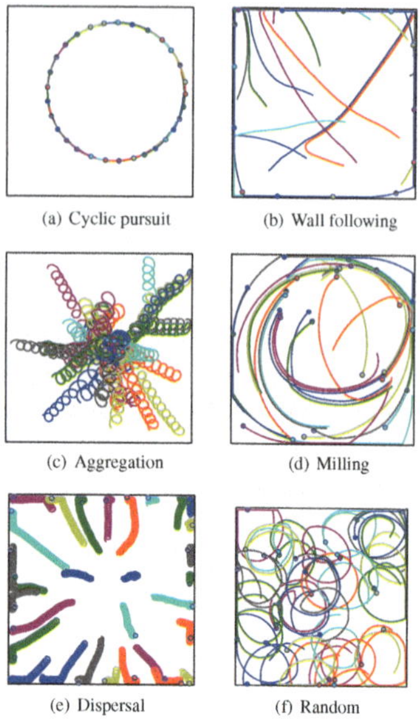

Fig. 3 Partial trajectories of swarm behaviors possible given a single, line-of-sight sensor. *Cyclic pursuit* forms a perfectly spaced, revolving circle. *Wall following* consists of agents spreading out to the boundary and then sliding along the walls. In *Aggregation*, the robots spiral into a single cluster. Robots in the *Milling* behavior constantly chase each other around in circles without ever forming a perfect circle. *Dispersal* is the opposite of the aggregation behavior, and results in agents spiraling away from each other. Finally, some behaviors were classified as *Random* due to agents never forming a coherent behavior

An example of the results is shown in Fig. 4a. We evaluated each cluster by comparing the medoids. While k-Medoids works well for forming equally sized clusters, it also ignores much of the structure in the 2-dimensional embedding. This is a common downside to k-Means and k-Medoids clustering.

5.2 *Hierarchical Single-Link Clustering*

The previous section showed that just a simple k-Medoids approach allows us to partition the behavior space into roughly equal cells. We then examined the medoids returned from each cluster to determine how many distinct behaviors were discovered. However, even a superficial examination of the 2-dimensional t-SNE embedding and the clustering shown in Fig. 4a shows that clusters in many cases do not fit well with the underlying structure. To try to remedy this we next examined hierarchical agglomerative single-link clustering.

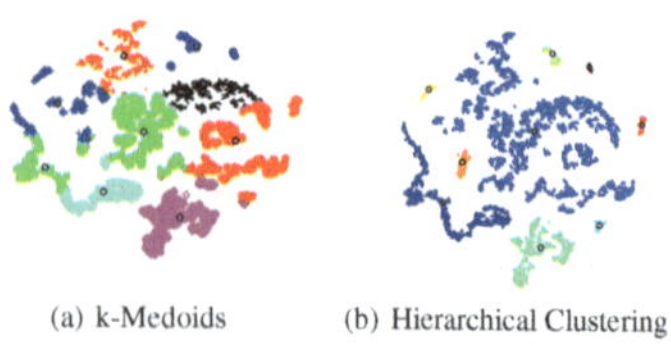

Fig. 4 An example clustering from k-medoids with k = 6. This approach partitions the t-SNE embedded behavior space into roughly equal partitions

As shown in Fig. 4b hierarchical clustering sequentially picks out isolated islands in the embedded 2-d space. However, the resulting cluster centers do not exhibit the range of behaviors found through k-Medoids. We inspected the clusters and representative behaviors and found that many of the small clusters were simply different variations of cyclic pursuit with variations in radius and speed.

6 Discussion

Our clustering results show that k-Medoids provides the most representative sampling of distinctly different behaviors, while hierarchical clustering tended towards finding different variations of cyclic pursuit while failing to find the milling behavior. One method is not clearly better than the other. If finding the largest number of clearly distinct behaviors is desired, then k-Medoids seems to perform the best. On the other hand, if a more nuanced definition of emergent behavior is desired, the hierarchical clustering seems better at uncovering the variations within behaviors.

Our clustering analysis found six possible behaviors. However, one of them, random circling, appears to not have any kind of collective behavior but is instead just a collection of robots moving in circles with no emergent properties. Thus, we focus on the five behaviors that we classify as emergent: cyclic pursuit, aggregation, wall following, dispersal, and milling. Cyclic pursuit resulting from robots with a single, binary sensor was first mentioned by Gauci et al. [7], but treated as a local minima in the search for an aggregation controller. Our method is able to "rediscover" this emergent behavior without an explicit objective. Cyclic pursuit is also well studied problem in control theory [15]; however, these problems are often solved using complex policies requiring positional and heading information, as opposed to the simple capability model we study here.

Aggregation is another behavior re-discovered by out method. Gauci et al. [7] first explored the problem of using a single-binary sensor to investigate whether they could evolve an aggregation algorithm that required no computation or memory.

Unlike cyclic pursuit or aggregation, the wall following behavior found by our algorithm is, to the best of our knowledge, a novel behavior for our capability model. While this behavior is a result of our specific environment, namely a walled environment without friction, it shows the power of our method in finding a novel behavior unknown to be possible in a swarm of memoryless single sensor robots. While it is possible to argue that wall following is simply a circle that is too big for the world size, this ignores the fact that the space of behaviors is inherently tied to the characteristics of the environment. As stated in our problem formalism, we are interested in discovering the different behaviors that are possible given both a capability model as well as an environment. Thus, for our specific environment, we argue that the wall following and cyclic pursuit are different behaviors due to their unique movement patterns and behavioral features.

The dispersal behavior is also a novel behavior that has not been previously shown to exist for single sensor swarms. Given that aggregation has previously been shown possible, it is not surprising that dispersal is also possible; however, the fact that our method finds both aggregation and dispersal shows the effectiveness of our approach.

The final emergent behavior that our method discovered in the milling, or torus behavior. The existence of this behavior is rather remarkable given the limited capability model we studied. It is well known that ants and fish form these types of milling patterns in the wild. However, we believe this is the first demonstration of these patterns shown to be possible with no memory and only a single bit of sensory information. This behavior is well-studied in the swarming community and is one of the four fundamental group types shown to emerge from the celebrated Couzin's model [5]. However, unlike Couzin's model, which makes strong assumptions about every agent being able to sense and respond to its neighbor's relative positions and velocities, we have discovered that a milling behavior is possible using only a single binary sensor.

7 Conclusions and Future Work

In this paper we formalized the problem of determining the emergent behaviors possible given a limited set of capabilities. Applying our method to a single binary sensor model, first proposed by Gauci et al. [7], we found that our method was able to rediscover a cyclic pursuit circling behavior, as well as aggregation. We also discovered three new behaviors not previously shown to be possible given our assumed capabilities: wall following, dispersal, and milling. We investigated both k-means clustering and hierarchical clustering after reducing the dimensionality of our data. We found that the centers of the k-Medoids clusters resulted in a wider variety of behaviors than the centers of clusters obtained from hierarchical clustering. Hierarchical clustering found fewer distinctly different behaviors, but was able to better select for variations within behaviors, such as speed, rotation direction, and radius.

While we believe that the problem we have studied is of fundamental and practical importance, we acknowledge the fundamental subjectivity in assigning boundaries between behavior types. Though this is an inherently subjective problem that may never admit an objective solution, we believe we have made some progress towards the goal of discriminating between qualitative behavioral groups in a principled way. While emergent behaviors will always, in some sense, be relative to the eye of the beholder, our approach allowed us to find a set of visually distinct behaviors, some of which were not previously known to exist for the single binary sensor capability model. While is it still difficult to know how well our approach will scale to more complex capability models, our proposed methodology could be useful to both scientists wanting to understand why some collective behaviors are present in a given animal species as well as engineers wishing to explore and design emergent behaviors to accomplish different tasks.

Future work should extend our method to investigate the space of possible emergent behaviors given more complex models, such as multiple sensors with more than two possible inputs, limited communication between agents, and more complex environments that include obstacles and movable items. It also remains to be seen how changes in the size and shape of the environment and number of robots affect the behaviors that are possible.

Future work should also investigate better techniques for determining what features are important for clustering. Our results have shown the difficulty in defining a behavior and in partitioning the explored space of behaviors without requiring a user to visually inspect the results and hand-tune parameters such as the number of clusters. There has been some work on using hand-crafted or learned features for classifying swarm behaviors [1, 2]; however, these methods are designed for already known behaviors, whereas we are interested in finding features that allow us to discover new behaviors. One possible avenue toward better disambiguation between behaviors would be to leverage crowd sourcing or machine learning. We hypothesize that human feedback combined with more advanced machine learning techniques such as deep convolutional neural networks [12] could allow us to better learn behavior features and similarities and improve the scalability of our approach.

Finally, we note that while discovering emergent behaviors is an interesting scientific question, there are also many open questions about how to interact with and use these behaviors. As more complex emergent behaviors are discovered, we hope there will also be research into how to use simple interactions with a swarm, either by changing the behavior of a subset of the agents [3], or even by changing the environment [11], to control and switch between different collective behaviors to accomplish interesting tasks.

References

1. Berger, M., Seversky, L.M., Brown, D.S.: Classifying swarm behavior via compressive subspace learning. In: 2016 IEEE International Conference on Robotics and Automation (ICRA), pp. 5328–5335. IEEE (2016)
2. Brown, D.S., Goodrich, M.A.: Limited bandwidth recognition of collective behaviors in bio-inspired swarms. In: Proceedings of the 2014 International Conference on Autonomous Agents and Multi-agent Systems, pp. 405–412 (2014)
3. Brown, D.S., Kerman, S.C., Goodrich, M.A.: Human-swarm interactions based on managing attractors. In: Proceedings of the 2014 ACM/IEEE International Conference on Human-Robot Interaction, pp. 90–97. ACM (2014)
4. Couzin, I.D., Krause, J.: Self-organization and collective behavior in vertebrates. Adv. Study Behav. **32**(1), 1–75 (2003)
5. Couzin, I.D., Krause, J., James, R., Ruxton, G.D., Franks, N.R.: Collective memory and spatial sorting in animal groups. J. Theor. Biol. **218**(1), 1–11 (2002)
6. Doncieux, S., Mouret, J.B.: Behavioral diversity with multiple behavioral distances. In: 2013 IEEE Congress on Evolutionary Computation (CEC), pp. 1427–1434. IEEE (2013)
7. Gauci, M., Chen, J., Dodd, T.J., Groß, R.: Evolving aggregation behaviors in multi-robot systems with binary sensors. In: Distributed Autonomous Robotic Systems, pp. 355–367. Springer, Berlin (2014)
8. Gauci, M., Chen, J., Li, W., Dodd, T.J., Groß, R.: Clustering objects with robots that do not compute. In: Proceedings of the 2014 International Conference on Autonomous Agents and Multi-agent Systems, pp. 421–428 (2014)

9. Gomes, J., Mariano, P., Christensen, A.L.: Devising effective novelty search algorithms: a comprehensive empirical study. In: Proceedings of the 2015 on Genetic and Evolutionary Computation Conference, pp. 943–950. ACM (2015)
10. Gomes, J., Urbano, P., Christensen, A.L.: Evolution of swarm robotics systems with novelty search. Swarm Intell. **7**(2–3), 115–144 (2013)
11. Johnson, M., Brown, D.S.: Evolving and controlling perimeter, rendezvous, and foraging behaviors in a computation-free robot swarm. In: International Conference on Bio-inspired Information and Communications Technologies (2015)
12. Krizhevsky, A., Sutskever, I., Hinton, G.E.: Imagenet classification with deep convolutional neural networks. In: Advances in Neural Information Processing Systems, pp. 1097–1105 (2012)
13. Lehman, J., Stanley, K.O.: Abandoning objectives: evolution through the search for novelty alone. Evol. Comput. **19**(2), 189–223 (2011)
14. Luke, S., Cioffi-Revilla, C., Panait, L., Sullivan, K., Balan, G.: Mason: a multiagent simulation environment. Simulation **81**(7), 517–527 (2005)
15. Marshall, J.A., Broucke, M.E., Francis, B.A.: Formations of vehicles in cyclic pursuit. IEEE Trans. Autom. Control **49**(11), 1963–1974 (2004)
16. Mondada, F., Bonani, M., Raemy, X., Pugh, J., Cianci, C., Klaptocz, A., Magnenat, S., Zufferey, J.C., Floreano, D., Martinoli, A.: The e-puck, a robot designed for education in engineering. In: Proceedings of the 9th Conference on Autonomous Robot Systems and Competitions, vol. 1, pp. 59–65. IPCB: Instituto Politécnico de Castelo Branco (2009)
17. Rubenstein, M., Ahler, C., Nagpal, R.: Kilobot: a low cost scalable robot system for collective behaviors. In: 2012 IEEE International Conference on Robotics and Automation (ICRA), pp. 3293–3298. IEEE (2012)
18. Schneirla, T.C.: A unique case of circular milling in ants, considered in relation to trail following and the general problem of orientation. Citeseer (1944)
19. Selous, E.: Thought transference (or what?) in birds (1931)
20. Stanley, K.O., Miikkulainen, R.: Evolving neural networks through augmenting topologies. Evol. Comput. **10**(2), 99–127 (2002)
21. Tunstrøm, K., Katz, Y., Ioannou, C.C., Huepe, C., Lutz, M.J., Couzin, I.D.: Collective states, multistability and transitional behavior in schooling fish. PLoS Comput. Biol. **9**(2), e1002,915 (2013)
22. Van der Maaten, L., Hinton, G.: Visualizing data using t-sne. J. Mach. Learn. Res. **9**(2579–2605), 85 (2008)
23. Wolfram, S.: A New Kind of Science, vol. 5. Wolfram Media, Champaign (2002)

Effects of Spatiality on Value-Sensitive Decisions Made by Robot Swarms

Andreagiovanni Reina, Thomas Bose, Vito Trianni and James A. R. Marshall

Abstract Value-sensitive decision-making is an essential task for organisms at all levels of biological complexity and consists of choosing options among a set of alternatives and being rewarded according to the quality value of the chosen option. Provided that the chosen option has an above-threshold quality value, value-sensitive decisions are particularly relevant in case not all of the possible options are available at decision time. This means that the decision-maker may refrain from deciding until a sufficient-quality option becomes available. Value-sensitive collective decisions are interesting for swarm robotics when the options are dispersed in space (e.g., resources in a foraging problem), and may be discovered at different times. However, current design methodologies for collective decision-making often assume a well-mixed system, and clever design workarounds are suggested to deal with a heterogeneous distribution of opinions within the swarm (e.g., due to spatial constraints on the interaction network). Here, we quantify the effects of spatiality in a value-sensitive decision problem involving a swarm of 150 kilobots. We present a macroscopic model of value-sensitive decision-making inspired by house-hunting honeybees, and implement a solution for both a multiagent system and a kilobot swarm. Notably, no workaround is implemented to deal with the spatial distribution of opinions within the swarm. We show how the dynamics presented by the robotic system match or depart from the model predictions in both a qualitative and quantitative way as a result of spatial constraints.

A. Reina (✉) · T. Bose · J. A. R. Marshall
Department of Computer Science, The University of Sheffield, Sheffield, UK
e-mail: a.reina@sheffield.ac.uk

T. Bose
e-mail: t.bose@sheffield.ac.uk

J. A. R. Marshall
e-mail: james.marshall@sheffield.ac.uk

V. Trianni
ISTC, Italian National Research Council, Rome, Italy
e-mail: vito.trianni@istc.cnr.it

R. Groß et al. (eds.), *Distributed Autonomous Robotic Systems*,
Springer Proceedings in Advanced Robotics 6,
https://doi.org/10.1007/978-3-319-73008-0_32

1 Introduction

Engineering large robot swarms with predictable performance is a very challenging problem, which is exacerbated by the spatiality aspects inherent to robotic systems that are widely distributed in space and that feature a highly heterogeneous and dynamic interaction network. For this reason, available design methods for system control resort to space-time models or low-dimensional abstractions [3, 9, 14]. Indeed, even with simple mobility models such as random walks [5], the system dynamics are the more difficult to predict the more the individual actions are influenced by information available only locally. If the state of a robot strongly depends on its spatial location (which in turns determines the interactions with neighbours), it is very likely that the swarm robotic system will present heterogeneities through space that may have a bearing on the macroscopic dynamics. The effects of spatiality are negligible only if the swarm is "well-mixed": in analogy with chemical systems [8], a certain robot state should be uniformly distributed within the swarm, or, in alternative, interactions between any two robots in the swarm should be equally likely. This condition is however not customary in swarm robotics, due to limited motion speed and local communication abilities that prevent sufficient mixing. As a result, the system dynamics may strongly deviate from the predictions of abstract macroscopic models [4, 25].

In collective decision-making problems, spatiality may be determinant for the system dynamics, especially when the decision is the result of the formation of spatial heterogeneities (e.g., in self-organised aggregation [1, 7]). In other cases, it can play against convergence to a coherent outcome due to the formation of spatially isolated clusters that do not sufficiently interact, resulting in a decision deadlock or in long convergence times [25]. The attentive design of the individual robot behaviour can cancel out or even exploit the effects of spatiality [15, 20, 26]. Design methodologies based on well-mixed assumptions propose clever workarounds to deal with spatial constraints, such as limiting the interaction between agents from different populations only when/where the agents populations mix, e.g., at a home location [21].

In this work, we address the design of a collective decision-making behaviour in a swarm robotics scenario characterised by spatial heterogeneities, and analyse the effects of spatiality in the system dynamics. The decision problem falls in the general class of *value sensitive decision-making* [17], that is, it requires that a decision is taken only if there is at least one option that has a sufficiently high quality (i.e., above a given threshold θ_v). In case such a high-quality option is not available, a decision should not be taken in the expectation that a supra-threshold option would become available at a later time (see Sect. 2). Due to spatiality, it may be possible that only low-quality alternatives are discovered first, due to random exploration, so that the decision should be delayed until a high-quality one is eventually found. Quantifying the effects of spatiality in such a decision problem is therefore a fundamental step toward the engineering of swarm robotics systems.

We provide an implementation for the value-sensitive decision problem following a recently introduced design pattern for decentralised decision-making [21], as detailed in Sect. 3. Although the design pattern provides some guideline to deal with spatiality—at least to a certain extent—, in this work we implement no special workaround apart from parameterising the system in a way to enhance mixing among robots. We study the system behaviour in the special case of binary symmetric decisions, that is, two options are available with the same quality $v \in [v_m, v_M]$, and we analyse when the system is able to break a decision deadlock or remain stuck at indecisions (i.e., in the expectation that a high-quality option would later appear, see Sect. 4). We provide results for an abstract multiagent system characterised by point-mass particles not physically interacting with each other, and for experiments with a swarm of 150 kilobots. We conclude in Sect. 5 by discussing the relevance of the obtained results for the engineering of large-scale swarm robotics systems.

2 Case Study: Value-Sensitive Decision-Making by a Robot Swarm

Problem description. We consider a case study in which a robot swarm must search and select among options deployed in a bidimensional environment, with an option becoming visible to a robot only in its immediate surroundings. The collective decision is taken when a large fraction x_Q of robots commits to the same option. Each option, O_i is characterised by its quality $v_i \in [v_m, v_M]$. The robots have no a priori information about the decision problem they have to solve, that is, the robots do not know: (i) how many options are available, (ii) where the options are located, and (iii) which is the quality of the options. The swarm is asked to explore the environment in order to identify all the available options and estimate their quality. Finally, the swarm must select the option with the highest quality if the quality is above a given threshold θ_v, otherwise should remain undecided.

The system property of committing or not to a decision as a function of the estimated quality of the options reflects value-sensitivity: The swarm response is *sensitive* to the value of the perceived options' quality. Value-sensitivity is relevant to engineering [21], biology [17] and also neuroscience [18], and is an advantageous property for systems that have to make decisions among an unknown number of options which have to be discovered. This is typical when the discovery of alternatives is an episodic event. In similar conditions, options may become available at various moments in time, for example, some options may need more time to be discovered or may appear later in the environment. In these cases, committing too early to the current best option may preclude the opportunity of selecting a better option that would get discovered later. This situation may be frequent in scenarios where options are deployed in a spatial environment (e.g., nest-sites in house-hunting social insects [24]). Farther options may take longer to get discovered although they might have a better quality than nearer options. This phenomenon is well-known in biology: for instance, ant colonies that change their nest are able to select the best new nest independently of the distance from the old nest [6].

Experimental setup. The robot used in this study is the kilobot [23], a small and simple robot with limited sensing and actuation capabilities. A kilobot operates at a clock frequency of about 32 Hz (which corresponds to a clock period $\tau_c \simeq 31$ms). It can move on a flat surface and control its movements through the modulation of the power applied to two vibration motors. The motion speed varies from robot to robot and also depends on the ground friction. In our experiments, a robot can move straight at a speed of 13 mm/s and rotates at 40°/s, on average. Through IR communication, a robot can exchange 9 bytes messages with neighbours in a limited range d_s which varies in relation to the reflectance of the ground and the brightness of the environment. In our experiments, the communication range was about 100 mm. Finally, each robot is equipped with a RGB LED that we use to let the robot display their current state.

The environment is a circle with a radius of 750 mm and glass ground. Each option is signalled through two static kilobot robots acting as beacons (which hereafter we call simply *beacons* to differentiate them from the robots that compose the swarm). Each beacon broadcasts every second a message with the option ID and its quality. We consider a binary decision problem characterised by two options, and we assign a unique colour to each of them (i.e., red and blue); the colours will help later to visualise each robots commitment state (see Sect. 4). We allocate two beacons for each option to allow the robots to perceive the option messages in a larger area. The two beacons are located at a distance of 150 mm from each other and thus cover an area of $A_O \simeq 0.058\,\text{m}^2$. Therefore, each robot can perceive an option in a very limited portion of space compared to the whole environment which has an area of $A_E = 1.77\,\text{m}^2$. The two options are located at a distance of about 380 mm from the environment center at diametrical opposite position. Thus, the distance between the two options is about 760 mm.

We analyse the robot experiments postprocessing the video of the experiments that we record through four overhead cameras. The cameras have overlapping fields of view and have been calibrated in order to match the coordinates of each tracked robot in a common reference frame to remove duplicate detection. We use a four-camera tracking system to maximise the probability of detecting the LED colour of every robot. In fact, multiple reduced fields of view allow a view on each area that is more orthogonal to the ground plane and reduces the occurrence of robots occluding the view of the LED with their own body.

Given the specification of the environment and of the robots communication capabilities, the swarm size (i.e., the number of robots) determines the average number of neighbours for each robot. Given N robots moving in an environment of size A_E and interacting over a range d_s, the resulting interaction network has average degree $\langle k \rangle = \pi N d_s / A_E$ [25]. We size the swarm to 150 robots which corresponds to having a network with an average interaction degree of $\langle k \rangle \simeq 2.67$. This value allows to have frequent interactions but remains below the percolation threshold $\langle k \rangle \simeq 4,51$ which determines the emergence of a single connected component in the interaction network [25].

3 Top-Down Implementation Through the Design Pattern

We implemented a swarm robotics system for decentralised decision-making following a design methodology based on the concept of design patterns, as proposed in [21]. The agent behaviour takes inspiration from a mathematical model for honeybee nest-site selection [24]. This model describes the dynamics of a honeybee swarm collectively deciding their future nesting site through a system of ODEs. In [21], the honeybee model is formalised in a design pattern supporting the design of swarm systems (e.g., robot swarms) by linking the macroscopic model parameters to the individual agent behavioural rules.

Preconditions. The decentralised decision strategy that we implemented requires a set of abilities at the individual robot level, as prescribed by the design pattern of [21]. Each robot must be able to:

- explore the environment searching for available options;
- recognise available options once found;
- individually estimate the options' quality;
- exchange with other robots the options' ID and quality.

Each robot modulates its actions as a function of the estimated quality of the opinion to which it is committed to. All these preconditions are met by the kilobot platform, hence the design pattern methodology can be applied.

Individual robot rules. The decision process works as follows. Robots can be committed to an option i (state C_i, and the total number of robots committed to i is N_i) or uncommitted (state C_U, and the total number of uncommitted robots is N_U). An uncommitted robot explores the environment and upon *discovery* of a potential option i it gets committed with probability P_{γ_i}. A robot committed to option i actively *recruits* uncommitted robots (which also become committed to i) with probability P_{ρ_i}. A robot committed to option i sends stop signals to robots committed to option j, with $j \neq i$. The robot that receives the stop signal becomes *inhibited* and reverts to an uncommitted state with probability P_{β_i}. Finally, a robot committed to option i spontaneously *abandons* its commitment and reverts to an uncommitted state with probability P_{α_i}.

Each robot, every second, broadcasts a message with information of its commitment state and, if committed to an option, the estimated option's quality. A robot updates its commitments state every $\tau_u = 400$ clock cycles through one of the four transitions: discovery, abandonment, recruitment and cross-inhibition. The mechanism of state update is described by the probabilistic finite state machine presented in Fig. 1. Some of the transitions are available only if in the latest τ_u clock cycles the robots has encountered an option (necessary for discovery) or received a message from other committed robots (necessary for recruitment and cross-inhibition).

Every robot moves in the environment through an isotropic random walk determined by the alternate sequence of straight motions for $\tau_m = 300$ clock cycles and on-place rotations in a random direction for τ_r clock cycles, which are chosen randomly from a uniform distribution $U(1, 150)$. The random walk is necessary (i) to

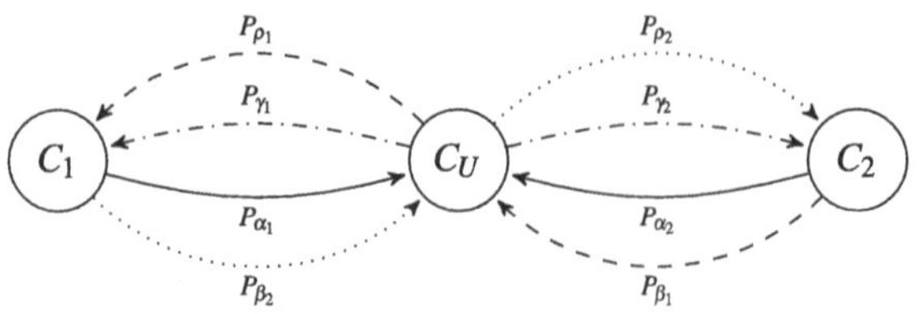

Fig. 1 Robot behaviour described as a probabilistic finite state machine. A robot updates its commitment state every τ_u clock cycles. Solid arrows are spontaneous individual transitions that can happen at any update. Dot-dashed lines are spontaneous individual transitions that can happen only if the robot has found an option in the latest τ_u clock cycles. Dashed and dotted arrows represent interactive transitions that can happen only if the latest message received by the robot in the last τ_u clock cycles is from another robot committed to option O_1 or O_2, respectively

let uncommitted robot explore the environment to discover potential options, and (ii) to allow robots to mix with each other and thus change their communication neighbourhood [5].

The robot software is available online at http://diode.group.shef.ac.uk/resources/.

Macroscopic parameterisation. The macroscopic model of the decision process can be described through a system of stochastic differential equations (SDEs) that describe the changes in the proportion of agents committed to each option. In this paper, we limit the study to binary decisions, hence the model describes the changes of the proportion of agents committed to option O_1 and O_2 as $x_1 = N_1/N$ and $x_2 = N_2/N$, where $N = N_U + N_1 + N_2$ is the total number of robots composing the swarm. The macroscopic model is:

$$\begin{cases} dx_1 = (\gamma_1 x_U - \alpha_1 x_1 + \rho_A x_1 x_U - \beta_2 x_1 x_2)dt + \sigma dW_1(t) \\ dx_2 = (\gamma_2 x_U - \alpha_2 x_2 + \rho_B x_2 x_U - \beta_1 x_1 x_2)dt + \sigma dW_2(t)\,, \\ x_U + x_1 + x_2 = 1 \end{cases} \quad (1)$$

where $\gamma_i, \alpha_i, \rho_i$ and β_i are the transition rates, respectively, of discovery, abandonment, recruitment and cross-inhibition for option i; and σdW_i is a Wiener process which represents additive noise with strength $\sigma \geq 0$. In order to implement a system able of value-sensitive decisions, we use a parameterisation similar to the one proposed in [17]. We make two modifications prescribed by the design pattern [21]: (i) The transition rate of discovery (γ_i) is changed in order to take into account the episodic nature of a discovery; (ii) All transition rates are scaled to meet the maximum speed of the process. The employed parameterisation is:

$$\gamma_i = v_i \cdot P_D, \qquad \alpha_i = v_i^{-1}, \qquad \rho_i = v_i, \qquad \beta_i = \beta, \qquad dt = d\tau/s, \qquad i \in \{1, 2\} \quad (2)$$

with P_D the episodic discovery probability, $v_i \in (1, 5]$ the option i quality and $s = 0.008$ the temporal scaling. Similarly to [13], we estimate geometrically the probability of encountering an option, $P_D = A_O/A_E \simeq 0.033$, where A_O is the area where an option can be detected by a robot and A_E is the full environment area.

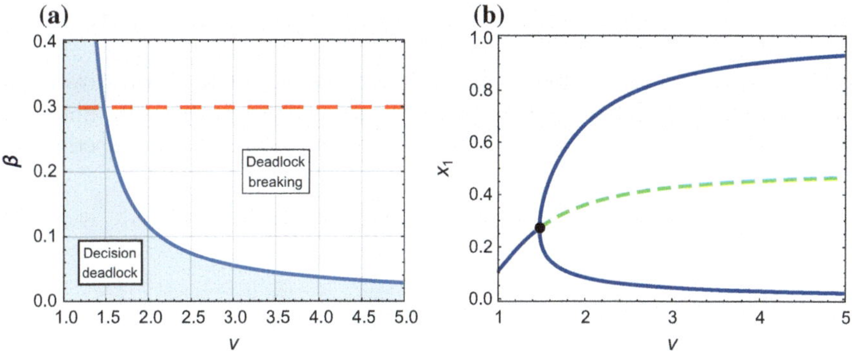

Fig. 2 Analysis of the macroscopic model of Eq. (1) with parameterisation (2) for the binary case $v = v_1 = v_2$. **b** Stability diagram in the parameter space (v, β). In the shaded area, the system has a single attractor with an equal number of committed robots to both options. In the white area, the system has two stable solutions with committed population biased for either of the two options. The horizontal red dashed line shows the selected value of $\beta = 0.3$. **b** Bifurcation diagram for $\beta = 0.3$ as a function of the option quality v. The solid blue lines are stable equilibria, the green dashed line is an unstable saddle point. As desired, the system undergoes a pitchfork bifurcation and breaks the decision deadlock for quality values greater than the threshold $\theta_v \simeq 1.5$

The macroscopic dynamics of the system in (1) can be studied analytically for varying option values [17]. A particularly interesting case is the symmetric condition in which both options have the same quality (i.e., $v_1 = v_2 = v$). In this situation, the two options are equivalent, therefore the swarm must select any of the two but only if their value is higher than θ_v. The study presented in this paper focuses on this case, as it is paradigmatic for evaluating the value-sensitivity property.

Figure 2a shows the stability diagram as a function of the options' quality v and the cross-inhibition rate β. The diagram shows that there exist two zones determining the macroscopic behaviour: In the grey shaded area, there exist a single attractor corresponding to a decision deadlock, which means that the swarm remains locked at indecision; In the white area, the system presents two stable attractors that correspond to decisions for the one or the other option, and a third unstable saddle point. This diagram can be exploited to select the system parameterisation that provides a desired value-sensitive behaviour. For instance, in order to have a value-sensitive behaviour with $\theta_v = 1.5$, it is necessary to select $\beta = 0.3$ (displayed as a horizontal red dashed line in Fig. 2a). Figure 2b shows the bifurcation diagram for the selected value of β, highlighting how the system breaks the decision deadlock when the options' quality v is greater than $\theta_v \simeq 1.5$.

Microscopic parameterisation. The design pattern of [21] explains how to convert the macroscopic parameters in the individual robot probabilities. The main difference between the macroscopic model and the agent implementation concerns the change of the temporal domain. While the macroscopic description is a continuous time model, the robots operate, as the most part of electrical devices, in discrete time (i.e., CPU clock cycles). Following the conversion rules of [21], we obtain:

$$P_{\gamma_i} = \gamma_i \cdot T, \quad P_{\alpha_i} = \alpha_i \cdot T, \quad P_{\rho_i} = \rho_i \cdot T, \quad P_{\beta_i} = \beta_i \cdot T, \quad i \in \{1, 2\} \quad (3)$$

with T the update timestep length, which is determined by the time between two commitment updates $\tau_u = 400\tau_c$ (where τ_c is the kilobot clock period) and the timescale of the macroscopic model, which is rescaled by the term s. In our experiments, $T = s\tau_u$.

4 Results

In this study, we contrast the dynamics predicted by the macroscopic models with the results obtained from the swarm robotics system which we implemented both in simulation and on real robots. At the macroscopic level, we study the system dynamics through time integration of Eq. (1) via a generalised Heun method. Using this numerical scheme the computed solution converges to the Stratonovich solution of the SDEs [10]. Starting from the SDEs model, it is possible to derive the corresponding master equation, which allows the analysis of the finite size effects where the magnitude of the stochastic fluctuations is determined by the finite number of robots composing the swarm [21]. We approximate the solution of the master equation through the simulation of the Gillespie algorithm [8].

The swarm robotics system has been implemented and analysed both through multiagent simulations and through a 150 kilobot swarm. The multiagent system is implemented in MASON [12] and simulates point-mass particles that move in a bidimensional plane. The scenario, the agent behaviour and its parameterisation are coherent with the kilobot implementation described in Sects. 2 and 3. However, in this system, noise and collisions among agents are not taken into account. The kilobot swarm implementation, instead, allows us to study the dynamics of a real physical system that includes all the aspects inherent to robotic experimentation. Figure 3 shows a screenshot of one experiment, while two videos of the robot experiments are available online[1].

A first assessment of the effects of spatiality can be obtained looking at the asymptotic dynamics of the system of Eq. (1) using the parameterisation (2), in comparison with the dynamics of the multiagent simulations. We performed a large set of simulations by extensively varying the quality value v, and Fig. 4 contrasts the macroscopic bifurcation diagram as a function of the options' quality v with the final distribution of the simulated swarm at convergence. We can appreciate a good qualitative agreement of the multiagent dynamics with the stable attractors of Eq. (1). The existence of a bifurcation is well predicted by the macroscopic model, although the multiagent simulations appear to break the symmetry for slightly higher values of v. Figure 5 shows the comparison of the temporal dynamics of the macroscopic models and of the swarm robotics systems (both real and simulated). The plots show aggregated

[1] https://www.youtube.com/watch?v=Gdy5o18y5lg and https://www.youtube.com/watch?v=EJtcpuj1Q5o.

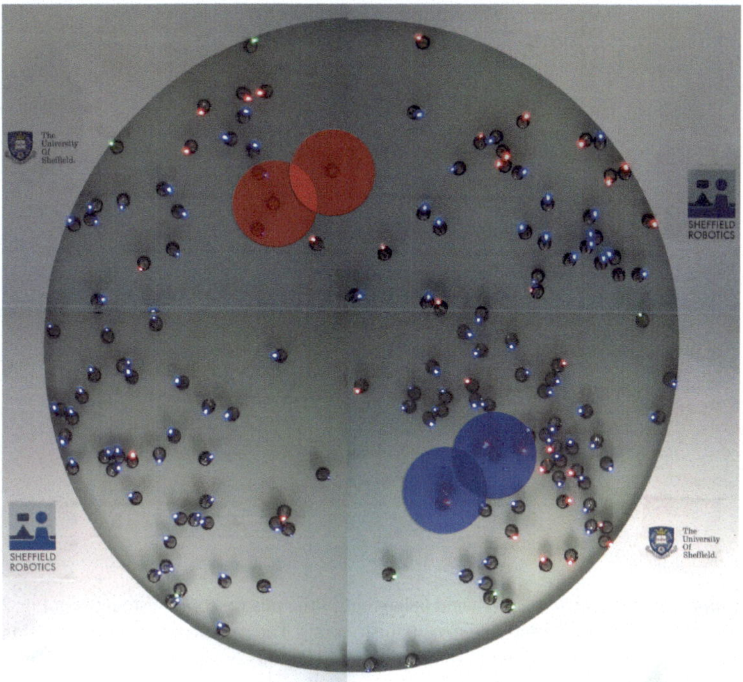

Fig. 3 Screenshot of a 150 kilobots experiment taken from the four-overhead-camera tacking system. The overlaying coloured circles show the two options localised in the environment. The robots light up their LED in a colour that corresponds to their internal commitment state: green for the uncommitted state C_U, red or blue for commitment to the option of the respective colour. This screenshot shows the final state of the swarm, after 30 min, for an experiment with $v_1 = v_2 = 5$. The swarm has a majority of robots committed to the blue option (108), only 35 robots committed to option red and 7 uncommitted robots

results of several experiments for two options' quality values: $v = 1.5$ and $v = 5$. The results show that both the multiagent simulations and the kilobot swarm have slightly different dynamics compared to the macroscopic description (SDE and Gillespie simulations). For $v = 1.5$, the system is close to the critical bifurcation point, and the dynamics are not quantitatively matched by the spatial system (see Fig. 5a). For $v = 5$ instead, the asymptotic behaviour is very similar, but the dynamics of the swarm robotics system are slower (see Fig. 5b). A slower convergence is the consequence of spatiality that leads to a drift of the system from a well-mixed condition. We observe indeed a slightly heterogeneous distribution of the commitment state among robots, with the emergence of clusters of robots with uniform commitment and a slow mixing of the two populations. Such a slowing-down effect is typical of consensus problems on regular lattices [2], and appears also with mobile agents in case the mobility pattern is not sufficient to produce an adequate mixing [25]. Additionally, collisions among kilobots represent an additional factor that slows down the

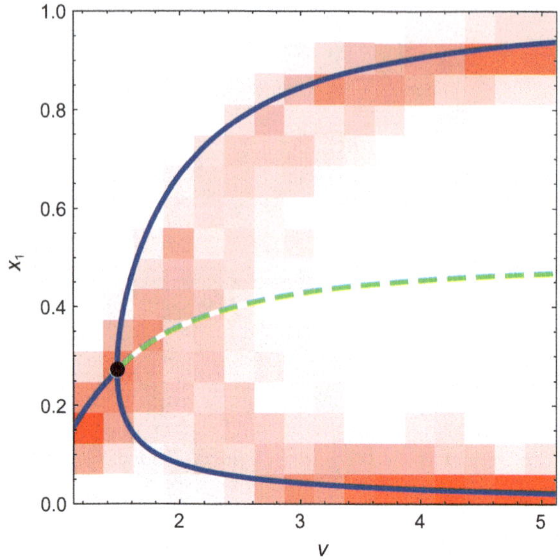

Fig. 4 Comparison between the predicted behaviour of the macroscopic system (as in Fig. 2b) and the multiagent simulations (underlying density histogram). The results of the multiagent simulation correspond to the final distribution of population x_1 in 100 runs of length 5 h

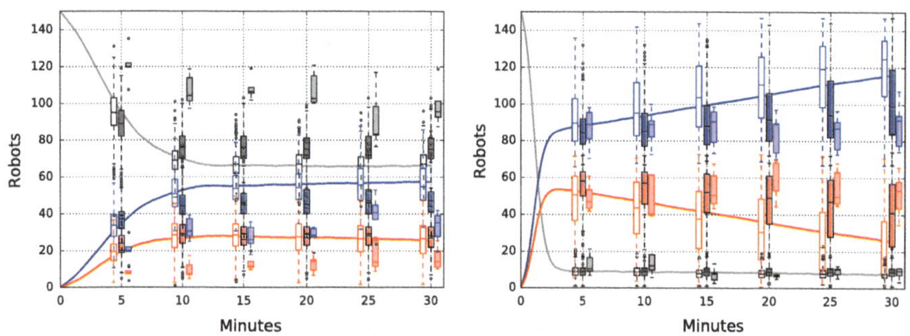

(a) $v = 1.5$ (b) $v = 5$

Fig. 5 Comparison of the temporal dynamics of the macroscopic models and the swarm robotics system. The solid lines show the dynamics of Eq. (1) with noise strength $\sigma = 0.0032$ (average over 500 runs). The empty boxplots are the dynamics of the master equation describing the process with $N = 150$ agents (the solution is approximated with 500 runs of the Gillespie algorithm). The darker boxplots are the results of the spatial multiagent simulations (500 runs). Finally, the lighter boxplots are the results of 150-robots experiments (5 runs). Colours represent the three subpopulations: gray the uncommitted robots, blue/red the committed robots. Since both options have the same quality, the results show each time as blue the selected option and as red the *discarded* option diffusion of robots in the environment, which justifies the slower dynamics detected during the kilobots experiments with respect to multiagent simulations, as shown in Fig. 5.

5 Discussions

The study we have presented highlights the complexity of dealing with spatiality in the engineering of a swarm robotics system. Although the specific decision problem we tackle seems ideal for obtaining a quantitative match between macroscopic models and experimental system—due to the agents/robots living (and mixing) in the same space—a non-negligible deviation from the model predictions is observed in both the multiagent simulations and the experiments with kilobots, especially in the transitory dynamics, which are slower for the spatial system. The design pattern for decentralised decision-making [21] is key for achieving a good qualitative match, as demonstrated with the extensive multiagent simulations we performed. Similarly to the results presented in this study, other work [27] obtained a qualitative match between spatial systems' dynamics and non-spatial mathematical models. However, a quantitative micro-macro link requires some additional workaround to better approximate a well-mixed system. For instance, the introduction of a distinction between latent and interactive agents as suggested in [21] could be key to allow a better mixing of the populations. In future work, we will implement such workaround and evaluate the extent to which the well-mixed condition is attained. More generally, effects of spatiality should be included in the macroscopic models, possibly resorting to heterogeneous mean-field approximations [8, 16]. In this way, the design pattern methodology could be enriched with tools to deal with spatially heterogeneous systems, and also for systems interacting on networks with heterogeneous topology (e.g., scale-free networks) [16]. Another possible approach to obtain a quantitative match of the system dynamics influenced by spatiality consists in including such factors into the macroscopic models [3, 9, 19].

An additional problem we recognise is given by collisions among robots, which further reduce mobility and limit mixing within the system, as already noted in previous studies [11, 25]. The kilobot platform does not provide means for efficient collision avoidance, and the high density that characterises experimentation with large groups plays against the population mixing. Indeed, besides being well-mixed, large-scale systems need also be "diluted" to ensure a good micro-macro link [8]. Efforts to provide guidelines to deal with less dilute systems will greatly benefit the engineering practice for swarm robotics systems. Indeed, we plan to investigate through further studies the effects of density on robot mobility in order to provide a model with diffusion coefficient as a function of the robot density.

The experimentation we performed in this study is limited to the symmetric binary decision case, which we deem sufficient to identify the relevant dynamics and compare with the model predictions. However, the implementation we provide is agnostic on the number of possible alternatives, and on the relative difference in quality, as already demonstrated in [21]. This means that the implemented system would work out-of-the-box also with an increasing number of alternatives. In such a best-of-n scenario, however, the macroscopic dynamics may be influenced by the number of available options, and the specific parameterisation we selected (i.e., the transitions rate strengths, as suggested by [17]) may need to be adjusted to produce value-sensitive decisions for any given number of options. Analytical studies in this direction are ongoing [22], and tests with a swarm robotics system will allow the validation of the design method beyond the binary case also for large-scale physical systems.

Acknowledgements This work was partially supported by the European Research Council through the ERC Consolidator Grant "DiODe: Distributed Algorithms for Optimal Decision-Making" (contract 647704). Vito Trianni acknowledges support from the project DICE (FP7 Marie Curie Career Integration Grant, ID: 631297). Finally, the authors thank Michael Port for the valuable help in building the infrastructure necessary to conduct the robot experiments.

References

1. Amé, J.M., Halloy, J., Rivault, C., Detrain, C., Deneubourg, J.L.: Collegial decision making based on social amplification leads to optimal group formation. Proc. Natl. Acad. Sci. **103**(15), 5835–5840 (2006)
2. Baronchelli, A., Dall'Asta, L., Barrat, A., Loreto, V.: Topology-induced coarsening in language games. Phys. Rev. E, Stat. Nonlinear, Soft Matter Phys. **73**(1), 015,102 (2006)
3. Berman, S., Kumar, V., Nagpal, R.: Design of control policies for spatially inhomogeneous robot swarms with application to commercial pollination. In: Proceedings of the 2011 IEEE International Conference on Robotics and Automation (ICRA), pp. 378–385. IEEE Press (2011)
4. Correll, N., Martinoli, A.: Collective inspection of regular structures using a swarm of miniature robots. In: The 9th International Symposium on Experimental Robotics (ISER) (Springer Tracts in Advanced Robotics), vol. 21, pp. 375–385. Springer, Berlin (2006)
5. Dimidov, C., Oriolo, G., Trianni, V.: Random walks in swarm robotics: an experiment with kilobots. In: Dorigo, M. et al. (ed.) Proceedings of the 10th International Conference on Swarm Intelligence (ANTS 2016). LNCS, vol. 9882, pp. 185–196. Springer, Berlin (2016)
6. Franks, N.R., Hardcastle, K.A., Collins, S., Smith, F.D., Sullivan, K.M., Robinson, E.J., Sendova-Franks, A.B.: Can ant colonies choose a far-and-away better nest over an in-the-way poor one? Anim. Behav. **76**(2), 323–334 (2008)
7. Garnier, S., Jost, C., Gautrais, J., Asadpour, M., Caprari, G., Jeanson, R., Grimal, A., Theraulaz, G.: The embodiment of cockroach aggregation behavior in a group of micro-robots. Artif. Life **14**(4), 387–408 (2008)
8. Gillespie, D.T., Hellander, A., Petzold, L.R.: Perspective: stochastic algorithms for chemical kinetics. J. Chem. Phys. **138**(17), 170,901–170,915 (2013)
9. Hamann, H., Wörn, H.: A framework of spacetime continuous models for algorithm design in swarm robotics. Swarm Intell. **2**(2–4), 209–239 (2008)
10. Kloeden, P.E., Platen, E.: Numerical Solution of Stochastic Differential Equations. Stochastic Modelling and Applied Probability, vol. 23. Springer, Berlin (1992)
11. Lerman, K., Galstyan, A.: Mathematical model of foraging in a group of robots: effect of interference. Auton. Robot. **13**(2), 127–141 (2002)
12. Luke, S., Cioffi-Revilla, C., Panait, L., Sullivan, K., Balan, G.: Mason: a multiagent simulation environment. Simulation **81**(7), 517–527 (2005). Transactions of the society for Modeling and Simulation International

13. Martinoli, A., Easton, K., Agassounon, W.: Modeling swarm robotic systems: a case study in collaborative distributed manipulation. Int. J. Robot. Res. **23**(4), 415–436 (2004). Special Issue on Experimental Robotics, Siciliano, B. (ed.)
14. Michael, N., Kumar, V.: Control of ensembles of aerial robots. Proc. IEEE **99**(9), 1587–1602 (2011)
15. Montes, M., Ferrante, E., Scheidler, A., Pinciroli, C., Birattari, M., Dorigo, M.: Majority-rule opinion dynamics with differential latency: a mechanism for self-organized collective decision-making. Swarm Intell. **5**(3–4), 305–327 (2010)
16. Moretti, P., Liu, S., Baronchelli, A., Pastor-Satorras, R.: Heterogenous mean-field analysis of a generalized voter-like model on networks. Eur. Phys. J. B **85**(3), 1–6 (2012)
17. Pais, D., Hogan, P.M., Schlegel, T., Franks, N.R., Leonard, N.E., Marshall, J.A.R.: A mechanism for value-sensitive decision-making. PLoS ONE **8**(9), e73,216 (2013)
18. Pirrone, A., Stafford, T., Marshall, J.A.R.: When natural selection should optimise speed-accuracy trade-offs. Front. Neurosci. **8**(73) (2014)
19. Prorok, A., Corell, N., Martinoli, A.: Multi-level spatial modeling for stochastic distributed robotic systems. Int. J. Robot. Res. **30**(5), 574–589 (2011)
20. Reina, A., Miletitch, R., Dorigo, M., Trianni, V.: A quantitative micro-macro link for collective decisions: the shortest path discovery/selection example. Swarm Intell. **9**(2–3), 75–102 (2015)
21. Reina, A., Valentini, G., Fernández-Oto, C., Dorigo, M., Trianni, V.: A design pattern for decentralised decision making. PLoS ONE **10**(10), e0140,950 (2015)
22. Reina, A., Marshall, J.A.R., Trianni, V., Bose, T.: Model of the best-of-N nest-site selection process in honeybees. Phys. Rev. E. **95**(5), 052411 (2017)
23. Rubenstein, M., Ahler, C., Hoff, N., Cabrera, A., Nagpal, R.: Kilobot: a low cost robot with scalable operations designed for collective behaviors. Robot. Auton. Syst. **62**(7), 966–975 (2014)
24. Seeley, T.D., Visscher, P.K., Schlegel, T., Hogan, P.M., Franks, N.R., Marshall, J.A.R.: Stop signals provide cross inhibition in collective decision-making by honeybee swarms. Science **335**(6064), 108–11 (2012)
25. Trianni, V., De Simone, D., Reina, A., Baronchelli, A.: Emergence of consensus in a multi-robot network: from abstract models to empirical validation. IEEE Robot. Automat. Lett. **PP**(99), 1–1 (2016)
26. Valentini, G., Ferrante, E., Hamann, H., Dorigo, M.: Collective decision with 100 kilobots: speed versus accuracy in binary discrimination problems. Auton. Agent. Multi-Agent Syst. **30**(3), 553–580 (2016)
27. Valentini, G., Hamann, H.: Time-variant feedback processes in collective decision-making systems: influence and effect of dynamic neighborhood sizes. Swarm Intell. **9**(2–3), 153–176 (2015)

Emergence and Inhibition of Synchronization in Robot Swarms

Fernando Perez-Diaz, Stefan M. Trenkwalder, Rüdiger Zillmer and Roderich Groß

Synchronization can be a key requirement to perform coordinated actions or reach consensus in multi-robot systems. We study the effect of robot speed on the time required to achieve synchronization using pulse coupled oscillators. Each robot has an internal oscillator and the completion of oscillation cycles is signaled by means of short visual pulses. These can, in turn, be detected by other robots within their cone of vision. In this way, oscillators influence each other to attain temporal synchrony. We observe in simulation and in physical robotic experiments that synchronization can be fostered or inhibited by tuning the robot speed, leading to distinct dynamical regimes. In addition, we analyze the effect of the involved parameters on the time it takes for the system to synchronize.

1 Introduction

Timing synchronization in multi-robot systems can be a prerequisite to coordinate actions or achieve consensus in a decentralized manner [9, 15]. For instance, certain approaches to box-pushing involve synchronization messages to ensure that robots

F. Perez-Diaz (✉)
Department of Computer Science, The University of Sheffield, Sheffield, UK
e-mail: fernando.perez.diaz@sheffield.ac.uk

S. M. Trenkwalder · R. Zillmer · R. Groß
Department of Automatic Control and Systems Engineering, The University of Sheffield, Sheffield, UK
e-mail: s.trenkwalder@sheffield.ac.uk

R. Groß
e-mail: r.gross@sheffield.ac.uk

R. Groß et al. (eds.), *Distributed Autonomous Robotic Systems*,
Springer Proceedings in Advanced Robotics 6,
https://doi.org/10.1007/978-3-319-73008-0_33

push an object simultaneously [9, 19]. Temporal synchronization has also been used in distributed sensing and data gathering in groups of robots [14, 23]. For example, a swarm of robots is able to track a target that moves faster than any individual robot by synchronizing the timings of the robots' observations [4]. Synchronization has been also proposed as a means to save power in distributed robotic systems and sensor networks by keeping their communication channels idle most of the time and only using them at precise times [24].

Pulse-coupled oscillators (PCOs) are one of the simplest methods for clock synchronization. They have been applied in swarm robotics [2, 10] as well as other multi-agent systems, such as sensor networks [20, 21] and ad-hoc networks [18, 22]. Each agent has an internal clock and signals the completion of a full clock cycle by means of a short pulse. This pulse is detected by the emitter agent's neighbors, which, in turn, update their clocks in an attempt to match the incoming stimulus [1]. Over time, the system can achieve global synchronization through the interactions between agents. PCOs are specially suitable in noisy or communication-limited environments because they operate at the physical layer by transmitting simple identical pulses, rather than packet messages [20–22].

Typically, communication among agents in a swarm only spans a certain local vicinity. Depending on the density of agents in the environment, they may not form a connected communication network. Therefore, agent mobility is usually needed to form a continuously evolving network that allows for synchronization to be achieved at the global scale.

Previous work in mobile pulse-coupled oscillators (MPCOs) has shown that synchronization occurs monotonically faster with higher agent speed if each agent influences others lying within a certain range [13]. At high speeds all agents interact with each other frequently, whereas at low speeds the neighborhood of any particular agent remains unchanged for long periods of time, leading to a rapid local synchrony but needing a longer time to achieve global synchronization.

In contrast, if the agents influence only their nearest neighbor a regime of intermediate speeds is observed where synchronization is retarded, while both the slow and fast regimes remain the same [12]. In a preliminary study, we observed with simulated agents that an interaction based on a cone of vision could link both monotonic and nonmonotonic behaviors by appropriately selecting the dimensions of such cone [10, 11].

This paper extends our previous work and focuses on embodied agents that signal their pulses visually using LED lights and are influenced only by others in their cone of vision. We study the parameters that influence synchronization in a simulated robotic environment and present a validation of the results with physical robot experiments.

This paper is structured as follows. Section 2 introduces the methodology, with an overview of the system as well as a description of the agents' temporal and spatial dynamics. Section 3 presents the simulation setup and results, with an analysis of the effect of the parameters involved. Section 4 describes the robotic trials and confirms experimentally the simulated results. Section 5 concludes the paper.

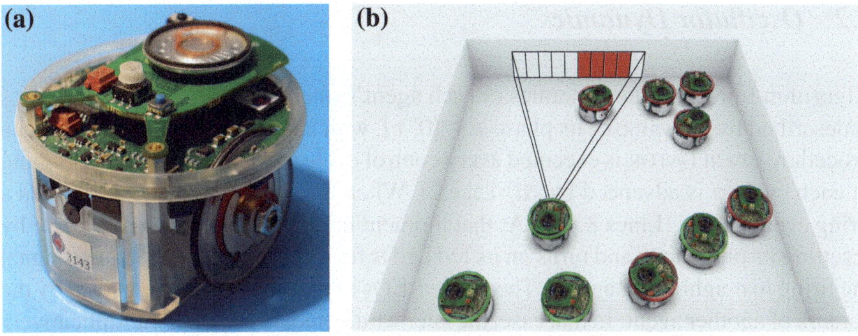

Fig. 1 **a** An e-puck robot. **b** Simulation setup in the Enki simulator. The overlaid drawing illustrates an e-puck's field of view, where a few pixels have detected a flashing robot (note that the actual amount of pixels in Enki is 60)

2 Methods

We consider a group of N robots moving in a walled square arena of length L. Every robot possesses an associated internal oscillator. The completion of a robot's oscillation cycle is signalled visually, and can be detected by neighboring robots (see details below). Thereby, the robots can influence each other in an attempt to synchronize their oscillators.

2.1 Robotic and Simulation Platforms

The robot used is the e-puck [7], which is a differential-wheeled cylindrical robot of 7.4 cm in diameter (Fig. 1a). The distance between the wheels is 5.1 cm, and their speed can be set independently within the range $[-12.8,\ 12.8]$ cm/s. The e-puck is equipped with a ring of red LED lights, eight short-range infra-red proximity sensors and a camera located at its front. The camera has a 56° horizontal viewing angle, with a resolution of 640×480 RGB pixels. Each robot signals the completion of a cycle of its internal oscillator by activating its red LED ring for a short duration. This, in turn, can be detected by another robot with its camera if the flash is in its field of view (cone of vision).

The robot simulation was performed using the open-source Enki library [6], see Fig. 1b. Enki provides a faster than real time 2-D simulation of the physics and dynamics of groups of robots, and it contains a built-in model of the e-puck. In Enki the camera capture is a single row of 60 pixels spanning the field of view of the e-puck.

2.2 Oscillator Dynamics

Algorithm 1 presents the dynamics of each agent's internal oscillator. The dynamics is described by the value of its phase $\phi \in [0, \tau]$, which is initialized at random.[1] The procedure OSCILLATOR is executed every control cycle ($\Delta t \ll 1$ s). At the beginning of each cycle ϕ is advanced by Δt (Line 2). When the threshold $\phi = \tau$ is reached a firing event occurs (Lines 8–11). At that moment the oscillator starts a new cycle by resetting its phase to 0, and turns on its red LEDs for a certain period of time to signal the firing to neighboring agents. The active LEDs of an agent can be perceived by the camera of another agent. In that case, the later would update its phase multiplicatively by a factor ε [2, 10, 12, 13] (Lines 4–6),

$$\phi \leftarrow (1 + \varepsilon)\phi \,. \tag{1}$$

The LEDs of a robot need to be active for a sufficient amount of time, $\phi \in [0, \phi_{LED}]$ (Lines 12–14) to ensure that the cameras of other robots can detect it. In addition, because of the necessary duration of the flashing signal and the delays in processing it, an oscillator could get displaced from synchrony if it detected a firing shortly after starting a new cycle. To compensate for this effect, a refractory period, $\phi \in [0, \phi_{ref}]$ was added during which an oscillator is not influenced by any other [5] (Line 3).

We found empirically that accounting for the average delay due to the frame rate ($\Delta = ½ framerate^{-1} \ll \phi_{LED}$) yielded a more stable synchronization in the real e-puck implementation. The function PHASEUPDATE in Algorithm 1 is replaced by PHASEUPDATE', defined in Algorithm 2. The basic principle is to apply the phase update due to an LED flash detected at time t to the phase at time $t - \Delta$, as if it had been perceived instantaneously (Line 2). Subsequently the phase is advanced by Δ to obtain the current value (Line 10). Note that setting $\Delta = 0$ in Algorithm 2 yields the original behavior described in Algorithm 1.

2.3 Motion Controller

Each robot moves in a straight line with speed V while there are no obstacles blocking its way. A collision avoidance algorithm is used to avoid running into walls or other robots. This algorithm implements a small turn if a robot is detected and a random reorientation upon encounter with a wall. The distinction between other robots and walls is determined heuristically by the time spent near the obstacle (see supporting material for code and detailed description [8]).

[1] All random numbers used in this paper are generated using uniform distributions.

Algorithm 1 Oscillator dynamics

Require: $\phi \leftarrow \text{random}\,(0, \tau)$ ▷ Initialize ϕ randomly within $[0, \tau)$
Require: $\phi_{\text{ref}} \in [0, \tau) \wedge \phi_{LED} \in [0, \tau) \wedge \phi_{ref} \geq \phi_{LED}$
Require: $\varepsilon > 0$
procedure OSCILLATOR
  $\phi \leftarrow \phi + \Delta t$ ▷ Linear increase of phase
  if $\phi > \phi_{ref}$ **then** ▷ If phase is outside the refractory period
    if *wasFlashDetected* **then**
      $\phi \leftarrow \text{PHASEUPDATE}(\phi) = (1 + \varepsilon)\phi$
    end if
  end if
  if $\phi \geq \tau$ **then** ▷ If a full oscillation cycle is complete
    $\phi \leftarrow 0$
    turnLEDsOn();
  end if
  if $\phi > \phi_{LED}$ **then** ▷ If the cycle is outside the flashing period
    turnLEDsOff();
  end if
end procedure

Algorithm 2 Modified phase update function to compensate for camera delay

Require: $\Delta \geq 0,\ \Delta \ll \phi_{LED}$ ▷ Average camera delay
Require: $\phi_{\text{ref}} \in [0, \tau)$
Require: $\varepsilon > 0$
function PHASEUPDATE$'(\phi)$
  $\phi \leftarrow (\phi - \Delta) \bmod \tau$ ▷ Phase at estimated time of flash
  if $\phi > \phi_{ref}$ **then** ▷ If phase at time of flash is outside the refractory period
    $\phi \leftarrow (1 + \varepsilon)\phi$
  end if
  if $\phi > \tau$ **then** ▷ If this would have triggered a new oscillation cycle
    $\phi \leftarrow 0$
    turnLEDsOn();
  end if
  $\phi \leftarrow \phi + \Delta$ ▷ Calculate current phase
  return ϕ
end function

3 Simulation

3.1 Setup

Simulations were performed for a range of values of the system parameters: size of the environment L, number of robots N, oscillator period τ and phase update factor ε. The LED and refractory periods were fixed to $\phi_{LED} = 0.075\,\tau$ and $\phi_{ref} = 0.15\,\tau$ respectively for all simulations. In addition, the effects of the dimensions of the cone of vision were studied by considering only certain fractions of the total camera pixels: from 4 to 60 pixels centered at the middle of the row. All the combinations of parameters were simulated in 200 trials for a range of 20 agent speeds V, from near stop to the maximum e-puck speed. In all trials the initial positions, orientations and phases of each robot were set at random.

In order to measure the level of synchrony, a certain robot is selected as reference. At the time of its kth firing, T_k, the complex order parameter is calculated as follows [1],

$$r(T_k)e^{\mathrm{i}\frac{2\pi\phi(T_k)}{\tau}} = \frac{1}{N}\sum_{j=1}^{N} e^{\mathrm{i}\frac{2\pi\phi_j(T_k)}{\tau}}, \tag{2}$$

where $\phi(T_k)$ is the mean phase and modulus $r(T_k)$ measures the level of synchrony, from $r(T_k) = 0$ for a totally incoherent system, to $r(T_k) = 1$ for complete synchronization. The simulations were stopped once $r(T_k)$ reaches a threshold r_{sync}, set here to $r_{sync} = 0.95$ [10]. We have observed that from this point synchronization becomes stable.[2] The number of cycles k to synchronization is recorded. We refer to this value as the synchronization time T_{sync}. This measure represents how long it takes the system to synchronize independently of the oscillation period.

3.2 Results

Figure 2 shows the synchronization time T_{sync} for a variety of parameters. We observe three clearly distinct regimes. For slow agent speed the system takes long to synchronize and T_{sync} decreases with V. For high enough speeds synchronization occurs significantly faster and T_{sync} also decreases with V. For an intermediate range of speeds the monotonically decreasing dependence observed in the other two regimes is broken, observing a local maximum. The precise position and strength of this peak, as well as the local minimum that precedes it, depend on the oscillator parameters.

We performed a least-square fitting of V_p and V_m, the speeds at which the intermediate regime peak and the preceding minimum occur respectively (Fig. 2c), with respect to each parameter and found that,

$$V_m \propto \frac{\varepsilon L}{N^{3/2}} \quad \text{and} \quad V_p \propto \frac{\varepsilon L}{N^{1/2}},$$

which agrees with the analysis by Prignano et al. [12] on point-like agents influencing their nearest neighbor. No clear dependence of either point with τ was found. In addition, we found the following relationships for the corresponding synchronization times, T_p and T_m, for both points,

$$T_{p,m} \propto \frac{Le^N}{\tau^2} + K_{p,m},$$

[2] We consider the synchronization to be stable if the system continues in coherence ($r(T_k) \gtrsim r_{sync}$) from that point in time onwards.

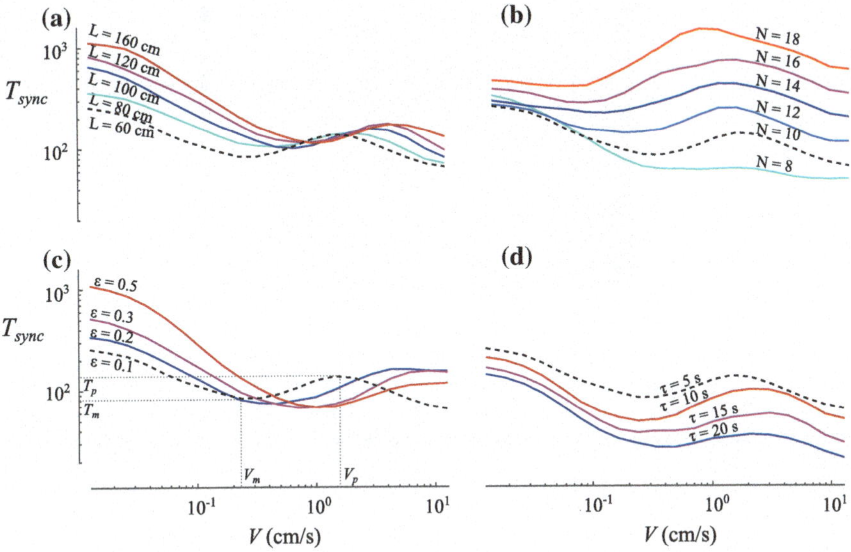

Fig. 2 Synchronization time T_{sync} (in number of cycles) as a function of agent speed V for a variety of parameters. **a** Varying the arena size L while fixing $N = 10$, $\tau = 5$ s and $\varepsilon = 0.1$. **b** Varying the number of agents N while fixing $L = 60$ cm, $\tau = 5$ s and $\varepsilon = 0.1$. **c** Varying the update factor ε while fixing $L = 60$ cm, $N = 10$ and $\tau = 5$ s. **d** Varying the oscillation period τ while fixing $L = 60$ cm, $N = 10$ and $\varepsilon = 0.1$. In all cases $\phi_{LED} = 0.075\,\tau$, $\phi_{ref} = 0.15\,\tau$ and the full image was considered (i.e. 60 pixels). The four dashed lines correspond to the exact same parameter configuration. Points (V_m, T_m) and (V_p, T_p) denote the synchronization speed and time for the local minimum and the peak in the intermediate regime

where $K_{p,m}$ are some constant offsets. No evident dependence with ε was found.

Figure 3 shows the effect of changing the dimensions of the cone of vision by considering different amounts of pixels as described in the previous section. The three synchronization regimes can be clearly identified over the whole considered range of pixels. In addition, we observe that the synchronization time increases for all speeds as the number of pixels is reduced (narrower cone of vision).

4 Physical Implementation and Experiments

4.1 Setup

Physical experiments were run in parallel with two groups of 10 e-pucks[3] in one of two adjacent $60 \times 60\,\text{cm}^2$ white-walled arenas. Five trials for 10 different robot speeds were performed, starting with random initial phases until the system achieved

[3]Hardware revision HWRev 1.2. http://www.gctronic.com/doc/index.php/E-Puck.

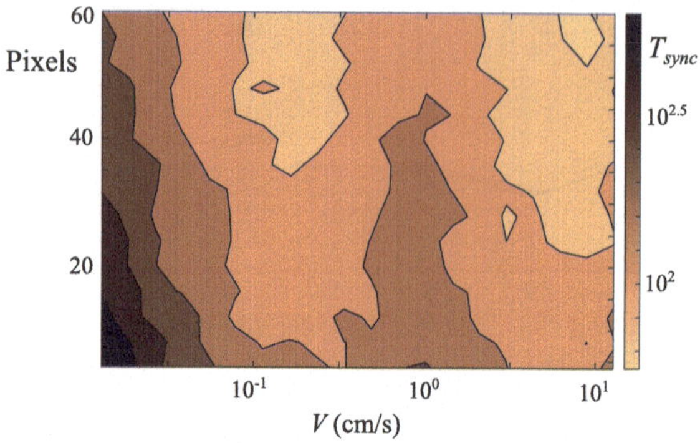

Fig. 3 Time T_{sync} (in number of cycles) required to synchronize 10 agents moving at speed V in a square arena of side 60 cm while observing different fractions of the total camera image. The oscillator parameters are the same as for the dashed lines in Fig. 2: $\tau = 5$ s, $\varepsilon = 0.1$, $\phi_{LED} = 0.075\,\tau$ and $\phi_{ref} = 0.15\,\tau$

global synchrony (see below for details). The robots' initial positions were randomly selected from a 5×5 grid with 10 cm spacing between points and equal padding to the walls. The initial orientations were randomly selected at 45° degree intervals. The experiments were performed in total darkness, and the robots only started moving once the room lights were switched off. In a few trials, a robot stopped moving and was unable to recover, or the battery ran out and accidentally restarted (resetting its phase). In such cases, the trial was discarded and repeated.

The software was implemented utilizing functions of OpenSwarm, an embedded operating system running directly on each e-puck [16]. The LED detection was performed at approximately 8 frames per second (FPS) by analyzing a single row of the acquired camera image (20 pixels with a 32× digital zoom). If two or more pixels were identified as red, a firing was considered to be detected. In addition, the LED indicating a low battery voltage was covered with tape to prevent false-positive flash detections.

A web camera was positioned at 110 cm above each arena and was used to record the trials and to measure the level of synchrony in real time. For this purpose, tracking software was developed using OpenCV that counted the number of detected red LED rings at each frame. The system was considered to be synchronized when the standard deviation of the ring counts over time was smaller than two frames, which corresponds to approximately 130 ms with our setup and ensured that synchronization was stable thenceforth. This threshold approximately corresponds to $r_{sync} = 0.95$ in accordance with the simulation setup (Sect. 3.1) (see supplementary material for a detailed calculation [8]). The time required to achieve synchrony, T_{sync}, measured in number of cycles, is recorded. Figure 4 shows a snapshot of one experiment with the detected LED rings and a histogram of the ring count over time.

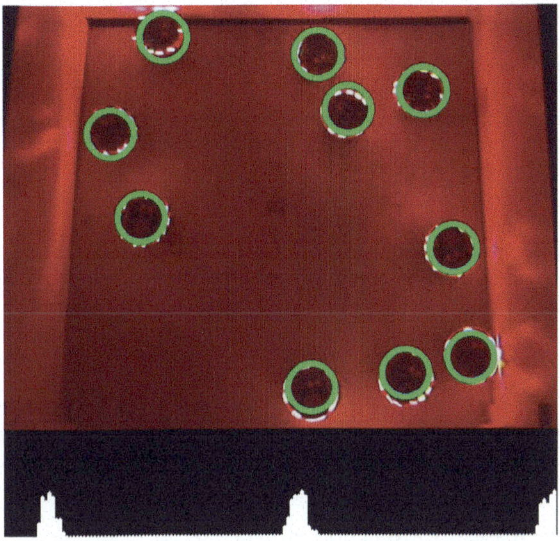

Fig. 4 Snapshot of the experimental setup with 10 physical e-puck robots in a square arena of side length $L = 60$ cm. The active LED rings are detected and superimposed as green circles. The lower strip shows an evolving histogram of the number of LED rings detected over time. The system is considered to be synchronized if the distribution of the histogram around the current peak has a standard deviation smaller than two frames

4.2 Results

Figure 5 shows the synchronization time T_{sync} (in number of cycles) required to synchronize the group of 10 e-pucks as a function of speed. The obtained curve qualitatively displays the same behavior as in simulation. The slow, intermediate and fast regimes can be clearly identified.

Observation of the trials yields visual confirmation of previous hypothesis regarding the underlying mechanisms governing each regime [10–12] (see accompanying trial videos in the online supplementary material [8]). In the slow regime an agent spends many cycles observing the same agent(s), or not seeing any other. This helps to form locally synchronized clusters. However, the whole system does not synchronize globally until a sufficient amount of neighborhood changes occur. In the fast regime, the opposite effect takes place. Each agent frequently changes the agents it sees. Therefore, global synchrony becomes much easier to attain. In the intermediate regime, the two mechanisms compete with each other. Local clusters are synchronized as in the slow regime, but they are constantly displaced by the frequent neighborhood changes.

5 Discussion

This paper studied the presence of three synchronization regimes on a system of moving embodied pulse-coupled oscillators where the influence between agents is dictated by their cone of vision. The time T_{sync} required to synchronize the system

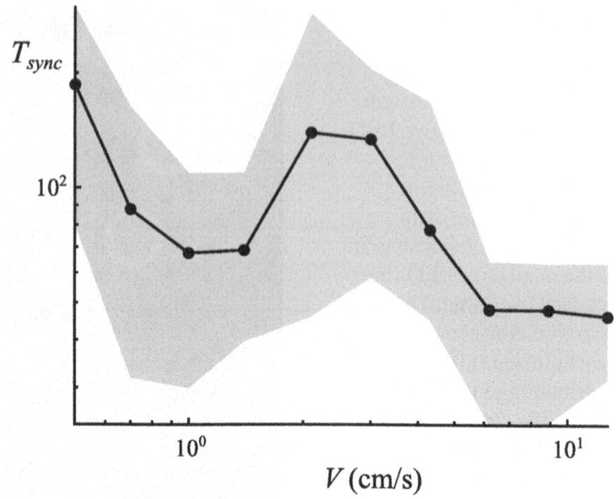

Fig. 5 Time T_{sync} (in number of cycles) needed to synchronize 10 physical e-pucks moving at speed V in a $60 \times 60\,\text{cm}^2$ arena. Black markers: average values over 5 trials. Grey contour: minimum and maximum values over the trials. The oscillator parameters are the same as for the dashed lines in Fig. 2: $\tau = 5$ s, $\varepsilon = 0.1$, $\phi_{LED} = 0.075\,\tau$ and $\phi_{ref} = 0.15\,\tau$

decreases as a function of agent speed in the slow and fast speed regimes. However for an intermediate range of speeds this dependence does not hold and a local maximum of T_{sync} is observed. Building on previous work [10, 12] we extended the understanding of the system by finding correlations between key features (namely the local minimum and maximum) and the involved parameters. In addition, the simulation results were experimentally validated in 50 trials with a group of 10 physical e-puck robots.

The quantitative differences between the simulation and the physical system may be attributed to several factors. Firstly, real robots present differences from one another: the system clocks may be slightly different or experience jitter; the onboard cameras are not equally sensitive; and the proximity sensors vary greatly, both within the same robot and between robots, which may results in robots not avoiding obstacles efficiently and getting stuck for periods of time. Secondly, real systems present other imperfections. For instance, due to the low frame rate of the robot's camera, it can miss certain flashings when turning, or when other robots are too close or far away. Lastly, the simulation does not take into account reflections of the LED flashes on the arena walls, which can affect the effective field of view.

The simulations presented here extended a preliminary study [10]. In contrast to [10], in this occasion priority was given to practical realism in order to later translate the results to the physical experiments. Firstly, in this work, the number of pixels was used to manipulate the dimensions of the cone of vision, instead of using the camera angle as in [10], which may be practically unfeasible. Secondly, obstacle avoidance and realistic reorientation at walls were implemented using the robots' infrared sensors, whereas [10] used random reorientation at walls and no obstacle avoidance mechanisms. Despite adding further constraints, this paper showed that the three dynamical regimes are still present, and that our results are translatable to a team of real robots.

One shortcoming of our experiments resides on the motion dynamics. As described in Sect. 2.3, robots move in a straight line except when avoiding other robots or reorienting at random when a wall is reached. The choice was made in order to build on the work of Prignano et al. [12] and Perez-Diaz et al. [10]. One could argue that this is not a realistic motion for robots in a swarm, which may move independently or coordinately to perform some task. In lieu of our controller, different random walks could be explored. For instance, Dimidov et al. [3] presented an in-depth study of random walks (from Brownian motion to Lévi walks) in the context of target searching in swarms of robots. It would be worth investigating the effect of the described motion models on firefly synchronization in the future.

It is worth noting that, to the best of our knowledge, this nonmonotonic dependence has only been observed elsewhere with a phase update such as the one described in Eq. 1 and with agents influencing their nearest neighbor [12].

Future work could study whether the effect of speed on synchronization observed here is translatable to other related consensus problems. For instance, Trianni et al. [17] studied the time required for a group of robots to converge to consensus in a similar setup to the one presented here. The authors studied a system of moving robots interacting with local neighbors within a certain range. As described in Sect. 1, a system of MPCOs influencing neighbors within a certain range yields a monotonically decreasing dependence of T_{sync} with agent speed [13], whereas a cone of vision leads to a nonmonotonic behavior. It would be worth investigating whether neighborhood type has similar effect on the convergence time in described consensus problem.

This work could also be extended to other robot platforms. The current method depends on LED flashing, cameras and darkness (to maximize visibility), which may not be the dominant communication method used by other robot systems. Our approach could be generalized to systems where wireless communication is the utilized signaling method.

Acknowledgements S.M. Trenkwalder is recipient of a DOC Fellowship of the Austrian Academy of Sciences.

References

1. Arenas, A., Díaz-Guilera, A., Kurths, J., Moreno, Y., Zhou, C.: Synchronization in complex networks. Phys. Rep. **469**(3), 93–153 (2008). https://doi.org/10.1016/j.physrep.2008.09.002
2. Christensen, A.L., O'Grady, R., Dorigo, M.: From fireflies to fault-tolerant swarms of robots. IEEE Trans. Evol. Comput. **13**(4), 754–766 (2009). https://doi.org/10.1109/TEVC.2009.2017516
3. Dimidov, C., Oriolo, G., Trianni, V.: Random walks in swarm robotics: An experiment with kilobots. In: International Conference on Swarm Intelligence, pp. 185–196. Springer, Berlin (2016)
4. Khaluf, Y., Mathews, E., Rammig, F.J.: Self-organized cooperation in swarm robotics. In: 14th IEEE International Symposium on Object/Component/Service-Oriented Real-Time Distributed Computing Workshops, pp. 217–226. IEEE (2011)

5. Kuramoto, Y.: Collective synchronization of pulse-coupled oscillators and excitable units. Phys. D: Nonlinear Phenom. **50**(1), 15–30 (1991)
6. Magnenat, S., Waibel, M., Beyeler, A.: Enki: the fast 2D robot simulator (2007). http://home.gna.org/enki/
7. Mondada, F., Bonani, M., Raemy, X., Pugh, J., Cianci, C., Klaptocz, A., Magnenat, S., Zufferey, J., Floreano, D., Martinoli, A.: The e-puck, a robot designed for education in engineering. In: Proceedings of the 9th Conference on Autonomous Robot Systems and Competitions, vol. 1, pp. 59–65 (2009)
8. Online supplementary material. http://naturalrobotics.group.shef.ac.uk/supp/2016-004
9. Parker, L.E.: Multiple mobile robot systems. In: Springer Handbook of Robotics, pp. 921–941. Springer, Berlin (2008)
10. Perez-Diaz, F., Zillmer, R., Groß, R.: Firefly-inspired synchronization in swarms of mobile agents. In: Proceedings of the 2015 International Conference on Autonomous Agents and Multiagent Systems, AAMAS '15, IFAAMAS, pp. 279–286 (2015). http://dl.acm.org/citation.cfm?id=2772879.2772917
11. Perez-Diaz, F., Zillmer, R., Groß, R.: Robustness of synchronization regimes in networks of mobile pulse-coupled oscillators. Phys. Rev. Appl. **7**, 054002 (2017)
12. Prignano, L., Sagarra, O., Díaz-Guilera, A.: Tuning synchronization of integrate-and-fire oscillators through mobility. Phys. Rev. Lett. **110**, 114,101 (2013). https://doi.org/10.1103/PhysRevLett.110.114101
13. Prignano, L., Sagarra, O., Gleiser, P.M., Diaz-Guilera, A.: Synchronization of moving integrate and fire oscillators. Int. J. Bifurc. Chaos **22**(07), 1250,179 (2012). https://doi.org/10.1142/S0218127412501799
14. Ranganathan, P., Morton, R., Richardson, A., Strom, J., Goeddel, R., Bulic, M., Olson, E.: Coordinating a team of robots for urban reconnaissance. In: Proceedings of the Land Warfare Conference (2010)
15. Ren, W., Beard, R.W., Atkins, E.M.: A survey of consensus problems in multi-agent coordination. In: Proceedings of the American Control Conference, 2005, vol. 3, pp. 1859–1864 (2005). https://doi.org/10.1109/ACC.2005.1470239
16. Trenkwalder, S., Lopes, Y., Kolling, A., Christensen, A., Prodan, R., Groß, R.: Openswarm: an event-driven embedded operating system for miniature robots. In: Proceedings of the 2016 IEEE/RSJ International Conference on Intelligent Robots and Systems, pp. 4483–4490 (2016)
17. Trianni, V., De Simone, D., Reina, A., Baronchelli, A.: Emergence of consensus in a multi-robot network: from abstract models to empirical validation. IEEE Robot. Autom. Lett. **1**(1), 348–353 (2016)
18. Tyrrell, A., Auer, G., Bettstetter, C.: Fireflies as role models for synchronization in ad hoc networks. In: Proceedings of the 1st International Conference on Bio Inspired Models of Network, Information and Computing Systems. ACM (2006)
19. Vig, L., Adams, J.A.: Multi-robot coalition formation. IEEE Trans. Robot. **22**(4), 637–649 (2006). https://doi.org/10.1109/TRO.2006.878948
20. Wang, Y., Núñez, F., Doyle III, F.J.: Energy-efficient pulse-coupled synchronization strategy design for wireless sensor networks through reduced idle listening. IEEE Trans. Signal Process. **60**(10), 5293–5306 (2012). https://doi.org/10.1109/TSP.2012.2205685
21. Wang, Y., Núñez, F., Doyle III, F.J.: Mobility induced network evolution speeds up synchronization of wireless sensor networks. In: Proceedings of the American Control Conference, 2014, pp. 3553–3558. IEEE (2014). https://doi.org/10.1109/ACC.2014.6858641
22. Wang, J., Xu, C., Feng, J., Chen, M.Z., Wang, X., Zhao, Y.: Synchronization in moving pulse-coupled oscillator networks. IEEE Trans. Circuits Syst. I: Regul. Pap. **62**(10), 2544–2554 (2015). https://doi.org/10.1109/TCSI.2015.2477576
23. Winfield, A.F.: Distributed sensing and data collection via broken ad hoc wireless connected networks of mobile robots. In: Distributed Autonomous Robotic Systems, vol. 4, pp. 273–282. Springer, Berlin (2000)
24. Yu, C., Werfel, J., Nagpal, R.: Collective decision-making in multi-agent systems by implicit leadership. In: Proceedings of the 2010 International Conference on Autonomous Agents and Multiagent Systems, AAMAS '10, IFAAMAS, pp. 1189–1196 (2010)

Evolving Behaviour Trees for Swarm Robotics

Simon Jones, Matthew Studley, Sabine Hauert and Alan Winfield

Abstract Controllers for swarms of robots are hard to design as swarm behaviour emerges from their interaction, and so controllers are often evolved. However, these evolved controllers are often difficult to understand, limiting our ability to predict swarm behaviour. We suggest behaviour trees are a good control architecture for swarm robotics, as they are comprehensible and promote modular reuse. We design a foraging task for kilobots and evolve a behaviour tree capable of performing that task, both in simulation and reality, and show the controller is compact and understandable.

1 Introduction

Swarm robotics is the field of robotics inspired by social insects, flocks of birds, schools of fish and other natural collective phenomena. By using many simple and cheap robots, it is hoped that goals such as pollution control, mapping and exploration, and disaster recovery could be met in ways which are resilient, scalable and decentralised [3]. The desired collective behaviour of the swarm emerges in a self-organised way from the interactions of the many individual agents that make up the swarm. Designing the controller for these agents is notoriously hard. A commonly used approach is the use of evolutionary methods to discover suitable controller designs.

Behaviour trees are widely used in the games industry to represent the decision processes of non-player characters. Recently, they have been applied to robotics,

S. Jones (✉) · M. Studley · S. Hauert · A. Winfield
Bristol Robotics Laboratory, University of the West of England, Bristol BS161QY, UK
e-mail: simon.jones@brl.ac.uk

M. Studley
e-mail: matthew2.studley@uwe.ac.uk

S. Hauert
e-mail: sabine.hauert@bristol.ac.uk

A. Winfield
e-mail: alan.winfield@uwe.ac.uk

R. Groß et al. (eds.), *Distributed Autonomous Robotic Systems*,
Springer Proceedings in Advanced Robotics 6,
https://doi.org/10.1007/978-3-319-73008-0_34

although not to our knowledge to swarm robotics. They have desirable properties that make them interesting to consider in the context of swarm robotics. They are human readable. They are hierarchical, all subtrees are themselves behaviour trees, encapsulating a complete behaviour that can exist within a larger tree, offering possibilities for modularity and building block reuse. Finally, they can be created and optimised using the techniques of Genetic Programming [26].

In this work, we design a behaviour tree controller architecture suitable for instantiation in a swarm of kilobots. We then automatically evolve behaviour trees in simulation to enable the swarm to perform a collective foraging task. The fittest behaviour tree is then evaluated in a swarm of real robots and analysed.

This paper is organised as follows; Sect. 2 gives a brief overview of swarm robotics and the kilobot platform, and introduces behaviour trees, Sect. 3 describes the experimental procedure, Sect. 4 details the results and Sect. 5 discusses results and possible further work.

2 Background and Previous Work

We work within the paradigm of swarm robotics as described by Şahin [35] taking inspiration from social insects, where many simple, homogeneous and not particularly capable robots with only local sensing and knowledge interact to produce a desired collective behaviour. There are no principled solutions to designing the controller to produce a given collective behaviour, common approaches are based on bioinspiration, evolutionary methods and gaining insight by reverse engineering the discovered controllers [19, 20, 33]. See [15] for a recent survey of the state of automatic swarm controller generation.

One problem with automatic generation of swarm controllers is that of bootstrapping; it is difficult to devise fitness functions to get complex behaviours [10, 29], the evolutionary process will often get stuck in uninteresting local maxima. Iterative approaches, with a gradually complexifying fitness function can work well, but this requires the designer to a priori specify the path to the eventual complex behaviour, lessening the likelihood of discovering novel behaviours. Hierarchical modular approaches are a promising alternative. AutoMoDe by Francesca et al. [16, 17] uses hand-designed modular and parameterised sub-behaviours which are combined within a Probabilistic Finite State Machine (PFSM), and the module parameters and PFSM topology constitute a search space over which optimisation is automatically carried out. Interestingly their automatically generated controllers have a lower reality gap compared to pure neural net approaches. Another modular approach is work by Duarte et al. [12, 13] where individual sub-behaviours are separately evolved neural net controllers which are again combined in a higher level Finite State Machine (FSM), this time hand-designed.

A behaviour tree (BT) is a hierarchical structure of nodes, with leaves that interact with the state of the world, and inner nodes that link these actions together in various conditional and sequential ways. The whole tree is evaluated at regular intervals, this is termed a *tick*. The *tick* is propagated down to the leaves and results are propagated back up according to the node types. Ogren [30] shows that all Hierarchical Dynamic Systems and therefore Finite State Machines (FSMs) can be represented by a BT, provided there are both sequence and selection type operators. With the addition of a probabilistic selector, Probabilistic Finite state Machines (PFSMs) can also be represented. Compared to an FSM or PFSM, the state transitions are implicit in the tree structure, and modular[1] structure is explicit; all subtrees are legal behaviour trees. Behaviour trees have their origins as a graphical software engineering tool before being adopted by the games industry for describing the decision processes and actions of non-player characters. Recently they have been formalised and applied to robotics [1, 2, 5, 7–9, 11, 22, 25, 27, 28, 30–32, 36, 37].

Kilobots are small cheap robots introduced by Rubenstein et al. [34]. They are capable of motion using two vibrating motors, communication with each other over a limited range using IR, distance sensing using the communication signal strength, environmental sensing with an upwards facing photo detector, and signalling with a multicolour LED. They are cheap enough to make it practical to build very large swarms and capable enough to run interesting experiments. Collective control of the kilobots in order to program and to start or stop them is achieved using a high intensity IR system using the same protocol as the inter-kilobot communication system.

3 Materials and Methods

Foraging as a collective task is often used as a benchmark for swarm systems [38]. It involves robotic agents leaving a nest region, searching for food, and returning food to the nest. Cooperative strategies are often more effective.

We designed a simple foraging experiment for a swarm of kilobots in an arena upon which we can project patterns of light to define the environment (Fig. 1). At the centre of the arena is a circular *nest* region. Surrounding this is a gap, then beyond that is the *food* region. A kilobot which moves into the food region is regarded as having picked up an item of food, a kilobot which is carrying an item of food that enters the nest region is regarded as depositing the food in the nest. Multiple kilobots are placed in the central region in a grid and all execute the same controller (homogenous swarm) for a fixed amount of time. The fitness of the swarm is related to the total amount of food returned to the nest within the test time. The maximum possible number of food items depends on the starting spatial distribution of the kilobots. Assume that the kilobots start on the edge of the *nest* region and for the duration of the test move directly back and forth between *nest* and *food* regions by the shortest distance. Let $food_{max}$ be the maximum food items, t_{test} be the test time, v_{avg} be the average linear velocity of the kilobots, n be the number of kilobots, fn_{dist} be the shortest (radial) distance between the food and nest regions:

[1]Perhaps mirroring a fundamental property of nature [6].

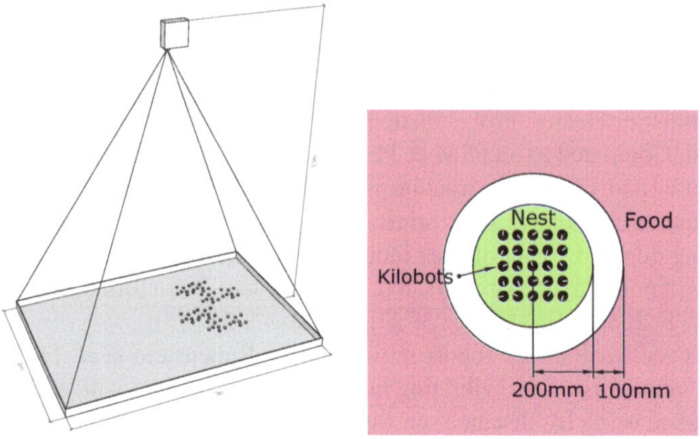

Fig. 1 Left: Kilobot arena. The arena is a 3×2 m surface upon which a projector defines the environment with patterns of light. Right: Starting configuration for kilobot foraging experiment. 25 kilobots are placed in a 5×5 grid in the centre of the nest region, with random orientations. Surrounding the nest is a 100 mm gap, then outside that is the food region

$$food_{max} = \frac{n \cdot v_{avg} \cdot t_{test}}{2 \cdot fn_{dist}} \tag{1}$$

We normalise the actual collected food items within the time of the test to give a fitness value. Let $food_{collected}$ be the total collected food items and k be a derating factor. The fitness f of the controller is given by:

$$f = k \cdot \frac{food_{collected}}{food_{max}} \tag{2}$$

The derating factor k is used to exert selection pressure towards smaller behaviour trees to ensure they will fit within the limited RAM resources of the kilobots. It is related to r_{usage} (4) in the following way: $k = 1.0$ when $r_{usage} < 0.75$ decreasing linearly to 0 when $r_{usage} = 1.0$.

Kilobots. For our experiments, we want to be able to sense whether we are within a particular region (nest or food) of the arena. Regions are delineated within the arena by using different coloured light from a video projector and detected with the upwards-facing phototransistor of the kilobots. In order to create a robust region sensing capability with a monochrome sensor, we exploited some particular characteristics of low cost DLP projectors [21].

The optical path of these type of projectors consists of a white light source, an optical modulator array, and a spinning colour wheel with multiple segments. Different full intensity primary and secondary colours produce different, quite distinct brightness modulation patterns in the light, which our eyes integrate but which we can detect easily with a series of samples from the photodetector. In our case, the projector had a wheel spinning at 120 Hz. Within each 8.3 ms period, primary colours were represented with a single pulse of about 1.2 ms, cyan and yellow with a pulse of 3.5 ms, and magenta with two pulses of 1.2 ms separated by a gap of 2 ms, giving, including black, four distinguishable patterns. We take 16 brightness samples from the phototransistor at 520 us intervals, covering one complete cycle, and classify the pattern.

The IR communication system between the kilobots has a range of about 100 mm. Twice a second, the kilobot system software sends any available outgoing message, retrying if the sending attempt collided with another sender. A kilobot receiving a valid message calls a user specified function to handle it. The message has a payload of nine bytes, and associated with the message is signal strength information to enable the distance from the sender to be calculated.

Controller. In order to control a robot with a behaviour tree, we need to define the interface between the behaviour tree action nodes and the robot, and the action nodes that act on the interface. This interface is known as the *blackboard*. Here there is a trade-off between the capabilities that we choose to hard code and those that we hope will evolve in the BT. We do not design the behaviour of the swarm but we do make assumptions about what kind of sensory capabilities might be useful for the evolutionary algorithm. This is often implicit in swarm robotics. The kilobot has no in-built directional sensors, like the range-and-bearing sensors that are common in swarm robotics experiments, so we synthesise collective sensing such that it is possible for a robot to tell if it is moving towards or away from the food or nest. We also give the capability of sensing the environment and the local density of kilobots, and of sending and receiving signals to other kilobots.

This relatively rich set of hardwired capabilities is outlined in Table 1. There are ten blackboard entries, **motors** maps to the motion control commands of the kilolib API, The **send_signal** and **receive_signal** entries allow for communication between

Table 1 Behaviour tree blackboard, defining interface between the behaviour tree and the robot

Index	Name	Access	Description
0	$motors$	W	0 = off, 1= left turn, 2 = right turn, 3 = forward
1	$scratchpad$	RW	Arbitrary state storage
2	$send\_signal$	RW	>0.5 = Send a signal flag
3	$received\_signal$	R	1 = A signal flag has been received
4	$detected\_food$	R	1 = Light sensor showing food region
5	$carrying\_food$	R	1 = Carrying food
6	$density$	R	Density of kilobots in local region
7	$\Delta density$	R	Change in density
8	$\Delta dist_{food}$	R	Change in distance to food
9	$\Delta dist_{nest}$	R	Change in distance to nest

kilobots initiated within the BT; **send_signal** is writeable from the BT. When the value is greater than 0.5, it is considered true, and a signal flag will be set in the stream of outgoing message packets. The **receive_signal** entry will be set to 1 if any message packets were received over the previous update cycle that had their signal flag set, otherwise it will remain zero. The **scratchpad** can be read and written, and has no defined meaning, it makes available some form of memory for the evolution of the BT to exploit. **Detected_food** is read-only, and is 1 if the environment sensing shows that the kilobot is in the food region, and zero otherwise, and **carrying_food** denotes whether the kilobot is considered to be carrying a food item. This entry is set to 1 if the kilobot enters the food region, and cleared to zero if the kilobot enters the nest region.

The remaining four entries are all metrics derived from the incoming stream of messages and their associated distance measurements. $density$ and $\Delta density$ are measures of the local population density and how it is changing. Each kilobot has a unique ID, which is embedded in its outgoing message packets. By tracking the number of unique IDs and the distances associated with messages from them, we can estimate the local density. Let $UID_{received}$ be the set of unique IDs received in the last update cycle, $dist_i$ be the distance in mm associated with the unique ID, the raw local density in $kilobots \cdot \mathrm{m}^{-2}$ in an update cycle d_{raw} is given by:

$$d_{raw} = \sum_{i \in UID_{received}} \frac{1}{\pi(dist_i/1000)^2} \tag{3}$$

This value is filtered with a moving average over $w = 5$ update cycles[2] to give $density(t)$ at update cycle t and $\Delta density(t) = density(t) - density(t-1)$.

The two distance metrics $\Delta dist_{food}$ and $\Delta dist_{nest}$ are calculated by tracking the minimum communication hops [18] needed to reach the respective region, illustrated in Fig. 2. For both food and nest, within the message packet are two fields, a hop count and an accumulated distance. The hop count is the minimum number of message hops to reach either the food or the nest region. The accumulated distance is the total length of those hops. Kilobots receiving messages select the lowest hop count, increment it and forward it and the new accumulated distance in the outgoing message stream. If no messages are received, we default to a distance of 0 mm if in a food or nest region, or 500 mm if not in a region. At every update cycle, we calculate two raw distance measures $dist_{food\_raw}$ and $dist_{nest\_raw}$. These are then filtered with a moving average in the same way as the $density$ value.

The behaviour tree nodes we implement are outlined in Table 2. Nodes are divided into two types; composition and action. Composition nodes are always inner nodes of the tree and combine or modify the results of subtrees in various ways. Action nodes are always leaf nodes and interface with the blackboard. Every update cycle, occurring at 2 Hz, the root node of the tree is sent the *tick* event. Each node handles the *tick* according to its function and returns *success*, *failure*, or *running*. The propagation of *tick* events down the tree and the return of the result to the root happen every cycle.

[2]Chosen in simulation as a reasonable compromise between responsiveness and stability.

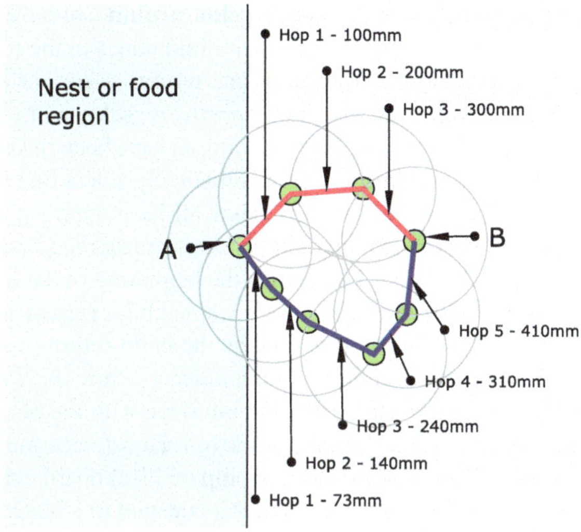

Fig. 2 Calculation of distance metrics. Kilobot 'A' is in a food or nest region, kilobot 'B' is connected to 'A' via two routes. Grey circles denote maximum communications radius. 'B' selects the message from the top route because the hop count is lowest, giving an accumulated distance along hops to the region of 300 mm

Table 2 Behaviour tree nodes. $Ch \equiv children$, $S \equiv succeeded$, $F \equiv failed$, $R \equiv running$, $N \equiv num\ children$, $I \equiv repeat\ iterations$, $r \equiv randomly\ selected\ child$, $t \equiv ticks$, $v, w \equiv blackboard\ entry$, $k \equiv contant$. Notation from [28]

Node	Size	*success* if	*failure* if	*running* if	Description
Composition nodes					
seqm 2, 3, 4	7, 9, 11	$N\ Ch\ S$	$1\ Ch\ F$	$1\ Ch\ R$	Sequence, *tick* until *failure*
selm 2, 3, 4	7, 9, 11	$1\ Ch\ S$	$N\ Ch\ F$	$1\ Ch\ R$	Selection, *tick* until *success*
probm 2, 3, 4	11, 17, 23	$Ch_r\ S$	$Ch_r\ F$	$Ch_r\ R$	Probabilistic choice
repeat	6	$I\ Ch\ S$	$1\ Ch\ F$	$Ch\ R$	Repeat subtree I times
succeed	4	$Ch\ \bar{R}$	*never*	$Ch\ R$	Always succeed subtree
failured	4	*never*	$Ch\ \bar{R}$	$Ch\ R$	Always fail subtree
Action nodes					
mf	2	$t = 1$	*never*	$t = 0$	Move forward for 1 *tick*
ml	2	$t = 1$	*never*	$t = 0$	Turn left for 1 *tick*
mr	2	$t = 1$	*never*	$t = 0$	Turn right for 1 *tick*
ifltvar	4	$v_1 < v_2$	$v_1 \geq v_2$	*never*	If $v_1 < v_2$
ifgevar	4	$v_1 \geq v_2$	$v_1 < v_2$	*never*	If $v_1 \geq v_2$
ifltcon	7	$v < k$	$v \geq k$	*never*	If $v < k$
ifgecon	7	$v \geq k$	$v < k$	*never*	If $v \geq k$
set	7	*always*	*never*	*never*	Set $w \leftarrow k$
successl	2	*always*	*never*	*never*	Always succeed
failurel	2	*never*	*always*	*never*	Always fail

The composition nodes **seqm, selm, probm** can have either 2, 3, or 4 children. On receiving a *tick* they process their child nodes in the following way: **seqm** will send *tick* to each child in turn until one returns *failure* or all children have been *tick*ed, returning *failure* or *success* respectively, **selm** will send *tick* to each child in turn until one returns *success* or all children have been *tick*ed, returning *success* or *failure* respectively, **probm** will probabilistically select one child node to send *tick* to and return what the child returns. They all have memory, that is, if a child node returns *running* the parent node will also return *running*, and the next *tick* event will start from that child node rather than the beginning of the list of child nodes. The **repeat, succeed, failured** nodes have a single child. **repeat** sends up to a constant number of *tick*s to its child for as long as the child returns *success*, **successd** and **failured** send *tick* to their child and then always return *success* or *failure* respectively. The action nodes are leaf nodes and interface with the blackboard, described in Table 1. **ml, mr, mf** turn left, right, or move forward, returning *running* for one cycle, then *success*. The various **if** nodes compare blackboard entries with each other or with a constant, and the **set** node writes a constant to a blackboard entry.

The controller runs an update cycle at 2 Hz. Message handling takes place asynchronously, and a message is always sent at each sending opportunity. Environmental sensing takes place at 8 Hz, synchronously with the update cycle, with a median filter over 7 samples to remove noise. Each cycle, the following steps take place: (1) New blackboard values are calculated based on the messages received and the environment. (2) The behaviour tree is *tick*ed, possibly reading and writing the blackboard. (3) The movement motors are activated, and the message signal flag set according to the blackboard values.

Implementation of the behaviour tree for execution on the kilobot required careful use of resources; the processor has only 2kbytes RAM, which must hold all variables, the heap, and the stack. The tree structure is directly represented in memory, with each node being a structure with type, state, and additional type-dependent data such as pointers to children. Execution of the behaviour tree involves a recursive descent following node child pointers and as such, each deeper level uses entries on the stack.

The compiled kilobot code uses about 500 bytes for all non-heap variables. We allocate 1024 bytes to the tree storage, leaving another 500 bytes for the stack and some margin for Interrupt Service Routine stack usage. Each level of tree depth uses 16 bytes of stack. Let tr_{size} be tree storage bytes and tr_{stack} be tree stack usage. The resource usage is given by:

$$r_{usage} = max\left(\frac{tr_{size}}{1024}, \frac{tr_{stack}}{500}\right) \tag{4}$$

This gives a maximum tree depth of about 30 and a maximum number of about 140 nodes at the average node size.

Evolutionary algorithm and simulator.

Behaviour trees are amenable to evolution using genetic programming techniques. Using the DEAP library [14] a primitive set of strongly typed nodes were defined to represent behaviour tree nodes and their associated allowable constants. There are several types of constants: **if** and **set** $k \in [-1.0, 1.0]$, **repeat** iterations $I \in [1..9]$, **if** blackboard index $v_i \in [1..9]$, **set** blackboard index $w \in [1..2]$, **prob** probability $p \in [0.0, 1.0]$

Evolution proceeds as follows: The population of n_{pop} is evaluated for fitness by running 10 simulations for each individual, each simulation with a different starting configuration. The starting position is always a 5×5 grid with 50 mm spacing in the centre of the nest region, but the orientation is randomly chosen from interval $(-\pi, \pi)$ radians. The simulation runs for 300 simulated seconds and fitness is as Eq. 2.

An elite of n_{elite} is transferred unchanged to the next generation. The remainder are chosen by tournament selection with size t_{size}. A tree crossover operator is applied with probability p_{xover} to all pairs of non-elite, then three different mutation operators are applied to the non-elite individuals. Firstly, with probability p_{mutu}, a node in the tree is selected at random and the subtree at that point is replaced with a randomly generated one. Next, with probability p_{muts}, a branch is chosen randomly and replaced with one of its terminals. Next, with probability p_{mutn} a node is picked at random and replaced with another node with the same argument types. Lastly, with probability p_{mute}, a constant is picked randomly and its value changed. Parameters are shown in Table 3.

We wrote a simple 2D simulator based on the games physics engine Box2D [4]. The physics engine is capable of simulating interactions between simple convex geometric shapes. We model the kilobots as disks sliding on a flat surface with motion modelled using two-wheel kinematics, with forward velocity of $8 \times 10^{-3}\text{ms}^{-1}$ and turn velocity of $0.55\,\text{rad s}^{-1}$, based on measurements of 25 kilobots. Physical collisions between kilobots, and movement into and out of communication range were handled by Box2D, with an update loop running 10 Hz. Simulator deficiencies were masked using the addition of noise [23]. Gaussian noise was added to linear

Table 3 Parameters for a single evolutionary run

Parameter	Value	Description
n_{gen}	200	Generations
t_{test}	300	Test length in seconds
n_{pop}	25	Population
n_{elite}	3	Elite
t_{size}	3	Tournament size
p_{xover}	0.8	Crossover probability
p_{mutu}	0.05	Probability of subtree replacement
p_{muts}	0.1	Probability of subtree shrink
p_{mutn}	0.5	Probability of node replacement
p_{mute}	0.5	Probability of ephemeral constant replacement

($\sigma = 1 \times 10^{-3}\text{ms}^{-1}$) and angular ($\sigma = 0.2\,\text{rad s}^{-1}$) components of motion at every simulator timestep, and each kilobot had a unique fixed linear ($\sigma = 1.3 \times 10^{-3}\text{ms}^{-1}$) and angular ($\sigma = 0.06\,\text{rads}^{-1}$) velocity bias added, to reproduce measured noise performance and variability of real kilobots. Message reception probability was fixed at 0.95. Simulation performance r_{acc}, measured using the methodology described in [24] on an iMac 3.2 GHz machine was approximately 8×10^4.

Twenty five independent evolutionary runs were conducted, each one using the parameters in Table 3. Each individual fitness evaluation was the mean over ten simulations with different starting configurations. A total of 1.1 million simulations were run.[3]

The fittest individual across the 25 separate populations was evaluated again for fitness, this time over 200 simulations with different starting configurations. This individual controller was then instantiated uniformly across a swarm of real kilobots, giving a homogenous swarm. The real kilobots were run 20 times with different starting configurations and their fitness measured.

4 Results and Discussion

The results (Fig. 3) show that we have successfully evolved a behaviour tree for use as a swarm robot controller to perform a foraging task. When instantiated in a swarm

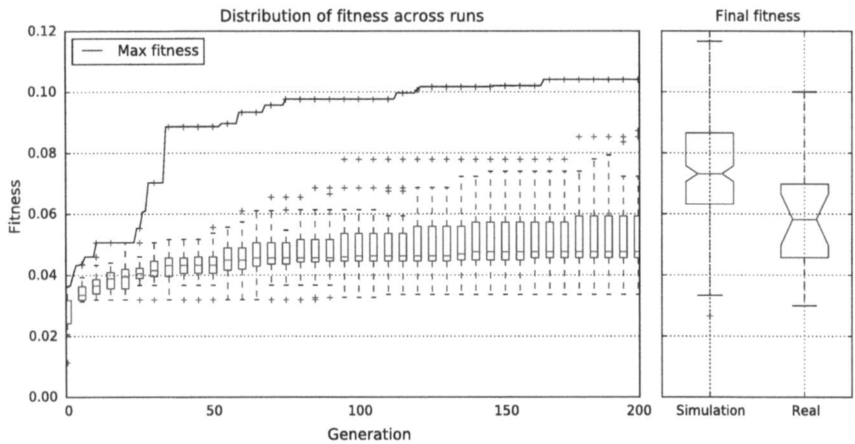

Fig. 3 Result of evolutionary runs. The left hand graph shows the maximum individual fitness across all 25 independent evolutionary runs, with a box plot every 5 generations to show the distribution. The right hand shows the distribution of fitnesses of the fittest individual, measured over 200 simulation and 20 real runs

[3]Due to the elitism policy, three individuals per generation are unchanged and need no fitness evaluation.

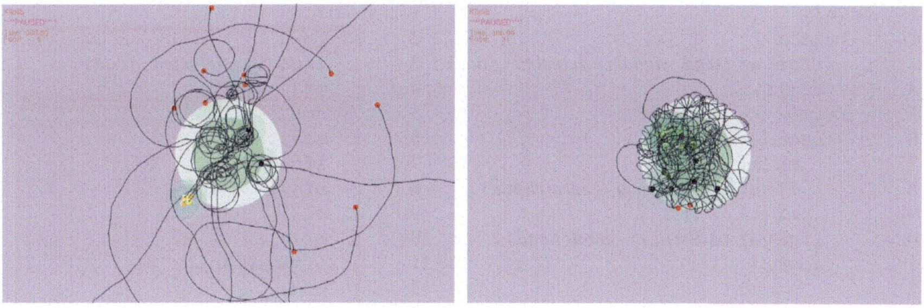

Fig. 4 Kilobot trails from simulation of the fittest controller in the first generation (left) and the 200th generation (right) of the fittest lineage

of real robots, it performs similarly to the simulation, validating the applicability of using this simulator for evolving kilobot swarm controllers. The performance is slightly lower in real life (0.058) compared to the simulated (0.075) performance, this is expected due to *reality gap* [23] effects. It is worth noting this is still a good outcome, the robots are able to effectively forage.

Fitness rises fast to about 0.03 after the first generation. This is due to the fact that an extremely simple controller that does nothing except move forward will still collect some food; because of the variability of the kilobots, some will move in large arcs that leave the nest, enter the food region and return to the nest. This type of controller is easily discovered by the evolutionary algorithm, confirmed by examining the fittest controller after one generation in the fittest lineage. The kilobot paths in simulation are shown in Fig. 4. It is noteworthy that the fittest of the 25 lineages is much fitter than the median, and the innovation seems to have been discovered around generation 30. This suggests that the evolutionary algorithm is not exploring the fitness landscape very effectively, otherwise we would expect evolution to discover similar behavioural innovations within other lineages.

We can examine the fittest BT, shown in Fig. 5, to gain insights into its workings. First of all, it is interesting to note that not all of the hardwired capabilities are used, only $detected\_food$, $\Delta dist_{food}$, and $\Delta dist_{nest}$. Both $scratchpad$ and $send\_signal$ are read but never written, so are equivalent to zero. This is not the case with all the evolved behaviour trees, see Table 4 for details of the blackboard usage of the top five fittest trees from different lineages. Between these individuals, every behaviour tree construct and blackboard entry is used. There is no obvious correlation between the features used and the fitness of the individual, perhaps indicating that there are multiple ways to solve this foraging problem.

The overall structure is a three-clause *selm*, the child trees will be *tick*ed in turn until one returns *success*. Consider a single kilobot, with no neighbours in communication with it. The first clause causes the kilobot to move forward as long as it is not in the food region. If it enters the food, the second clause comes into play, performing a series of left turns and forward movements until it moves out of the food region. Behaviour will then revert to the first clause and it will move forward

```
selm3(
  seqm2(
    ifgevar(send_signal, detected_food),
    mf()),
  seqm3(
    seqm3(
      ml(),
      ifgevar(Δdist_food, scratchpad),
      mf()),
    ifgevar(Δdist_food, scratchpad),
    seqm2(
      seqm3(
        seqm3(
          ml(),
          ifgevar(Δdist_nest, Δdist_nest),
          mf(),
        ifgevar(Δdist_food, send_signal),
        mf()),
      mf())),
  seqm3(
    ml(),
    repeat(5,
      ifltcon(Δdist_nest, −0.058530)),
    ml()))
```

```
selm3(
  seqm2(                                  Move forward until in food
    ifge(0, detected_food),
    mf()),
  seqm8(                                  Turn and forward until out of food
    ml(),
    ifge(Δdist_food, 0),
    mf(),
    ml(),
    mf(),
    ifge(Δdist_food, 0),
    mf(),
    mf()),
  seqm3(
    ml(),
    repeat(5,
      iflt(Δdist_nest, −0.058530)),
    ml()))
```

Fig. 5 Fittest behaviour tree. Left shows the code as evolved. Right shows the code with redundant lines removed by hand, the `seqm` nodes condensed, and conditionals simplified. Boxes highlight the three functional clauses

Table 4 Individuals from top five lineages and their usage of the blackboard and behaviour tree constructs. All individuals use at least the forward and one other of the motor action nodes. Usage is after redundant or unreachable nodes have been removed

Rank	Fitness	Blackboard entry									BT Nodes					
		1	2	3	4	5	6	7	8	9	SEQ	SEL	PROB	REPEAT	IF	SET
1	0.104				x				x	x	x	x		x	x	
2	0.0873		x	x	x						x	x	x		x	x
3	0.0853				x	x		x		x	x		x		x	x
4	0.0723	x	x		x	x	x	x	x	x	x	x	x	x	x	x
5	0.0710				x		x		x		x	x	x		x	

again, likely hitting the nest region. We can see that this will produce reasonable individual foraging behaviour, and this pattern is visible in the right hand trail plot in Fig. 4. The foraging behaviour will be enhanced in the presence of neighbours, since in this case the second clause will promote movement away from food generally, rather than just on the food region boundary. Finally, if the kilobot is executing the second clause, manages to leave the food then re-enters it, or moves towards it in the presence of neighbours, the third clause is triggered, which produces some additional left turning. The *repeat* sub-clause will fail on the first iteration since it is not physically possible for the kilobot to move 59 mm in one update cycle of half a second.

This evolved behaviour tree is sufficiently small that it can be analysed by hand relatively easily. It may be that greater foraging performance could be obtained by removing the selective pressure to small trees, and a larger tree would be harder to analyse. But, in contrast to evolved neural networks, which are a black box for which there are no adequate tools to predict behaviour apart from direct testing [29], it is possible at least in principle to analyse any behaviour tree, in the same way it is possible to analyse any computer program. The behaviour of each sub-tree can be analysed in isolation, descending until the size of the sub-tree is tractable, and automatic tools can simplify and prune branches which will never be entered, or will always do nothing.

Understanding the behaviour of an evolved BT does not mean that it becomes possible to predict the emergent swarm behaviour that the interaction between the kilobots will produce. However, we believe that the more easily we can understand the controller, the more likely we are to gain insights into the problem of predicting these higher-level behaviours.

5 Conclusions and Further Work

Evolved controllers for swarm robotics are generally hard to understand. We have introduced the use of behaviour trees as an architecture for evolved swarm robot controllers that are more easily human readable. A simple foraging task was designed, a behaviour tree node set and blackboard interface specified, and a population of behaviour trees were evolved for a swarm of kilobot robots. The fittest individual was tested in real robots and showed good correspondence in performance to the individual in simulation. The individual was then analysed for insight into the discovered foraging algorithm.

There are many possible avenues for exploration in the application of genetic programming to behaviour trees since little work in this area exists. Choices of the evolutionary parameter values, and the filtering of environmental signals are somewhat arbitrary and will be explored further. The choice of blackboard and action nodes is another area for further investigation. We also want to develop automatic tools for simplifying the analysis of evolved trees.

We intend to apply the evolution of behaviour trees to other collective swarm robot tasks, using a more computationally capable platform that will not be so limited in possible tree size, and will also allow the on-board adaptive co-evolution of new BT controllers in response to changing environmental conditions. We are interested in the possibility of encapsulation of various swarm behaviours such as aggregation, flocking, and dispersion. In this, we are inspired by the argument of Francesca et al. [17] that restricting the representational power of the controller allows the automatic discovery of solutions that are more resistant to reality gap effects, and feel the hierarchical structure of behaviour trees may lend themselves to tuning the bias-variance tradeoff.

Finally, we believe the increased human readability of evolved behaviour trees compared to other forms of evolved controller achieves progress towards more fully comprehending the emergence of collective behaviour from the interactions of individual agents.

References

1. Abiyev, R.H., Bektaş, Ş., Akkaya, N., Aytac, E.: Behaviour trees based decision making for soccer robots. Recent Advances in Mathematical Methods Intelligent Systems and Materials (2013)
2. Bagnell, J.A., Cavalcanti, F., Cui, L., Galluzzo, T., Hebert, M., Kazemi, M., Klingensmith, M., Libby, J., Liu, T.Y., Pollard, N., et al.: An integrated system for autonomous robotics manipulation. In: 2012 IEEE/RSJ International Conference on Intelligent Robots and Systems, pp. 2955–2962. IEEE (2012)
3. Brambilla, M., Ferrante, E., Birattari, M., Dorigo, M.: Swarm robotics: a review from the swarm engineering perspective. Swarm Intell. **7**(1), 1–41 (2013)
4. Catto, E.: Box2D: A 2D physics engine for games. World Wide Web electronic publication (2009). http://box2d.org/about
5. Champandard, A.: Behavior trees for next-gen game ai. In: Game developers conference, audio lecture (2007)
6. Clune, J., Mouret, J.B., Lipson, H.: The evolutionary origins of modularity. Proc. R. Soc. Lond. B: Biol. Sci. **280**(1755), 20122–20863 (2013)
7. Colledanchise, M., Ogren, P.: How behavior trees modularize robustness and safety in hybrid systems. In: 2014 IEEE/RSJ International Conference on Intelligent Robots and Systems (IROS 2014), pp. 1482–1488. IEEE (2014)
8. Cutumisu, M., Szafron, D.: An architecture for game behavior ai: behavior multi-queues. In: AIIDE (2009)
9. Dill, K., Martin, L.: A game ai approach to autonomous control of virtual characters. In: Interservice/Industry Training, Simulation, and Education Conference (I/ITSEC) (2011)
10. Doncieux, S., Bredeche, N., Mouret, J.B., Eiben, A.E.G.: Evolutionary robotics: what, why, and where to. Front. Robot. AI **2**, 4 (2015)
11. Dromey, R.G.: From requirements to design: formalizing the key steps. In: Proceedings of the First International Conference on Software Engineering and Formal Methods 2003, pp. 2–11. IEEE (2003)
12. Duarte, M., Gomes, J., Costa, V., Oliveira, S.M., Christensen, A.L.: Hybrid control for a real swarm robotics system in an intruder detection task. Applications of Evolutionary Computation, pp. 213–230. Springer, Cham (2016)
13. Duarte, M., Oliveira, S.M., Christensen, A.L.: Hybrid control for large swarms of aquatic drones. In: Proceedings of the 14th International Conference on the Synthesis and Simulation of Living Systems, pp. 785–792. Citeseer (2014)
14. Fortin, F.A., Rainville, D., Gardner, M.A.G., Parizeau, M., Gagné, C., et al.: DEAP: evolutionary algorithms made easy. J. Mach. Learn. Res. **13**(1), 2171–2175 (2012)
15. Francesca, G., Birattari, M.: Automatic design of robot swarms: achievements and challenges. Front. Robot. AI **3**, 29 (2016)
16. Francesca, G., Brambilla, M., Brutschy, A., Garattoni, L., Miletitch, R., Podevijn, G., Reina, A., Soleymani, T., Salvaro, M., Pinciroli, C., et al.: Automode-chocolate: automatic design of control software for robot swarms. Swarm Intell. **9**(2–3), 125–152 (2015)
17. Francesca, G., Brambilla, M., Brutschy, A., Trianni, V., Birattari, M.: AutoMoDe: a novel approach to the automatic design of control software for robot swarms. Swarm Intell. **8**(2), 89–112 (2014)

18. Hauert, S., Winkler, L., Zufferey, J.C., Floreano, D.: Ant-based swarming with positionless micro air vehicles for communication relay. Swarm Intell. **2**(2), 167–188 (2008)
19. Hauert, S., Zufferey, J.C., Floreano, D.: Evolved swarming without positioning information: an application in aerial communication relay. Auton. Robot. **26**(1), 21–32 (2009)
20. Hauert, S., Zufferey, J.C., Floreano, D.: Reverse-engineering of artificially evolved controllers for swarms of robots. In: IEEE Congress on Evolutionary Computation 2009. CEC'09, pp. 55–61. IEEE (2009)
21. Hutchison, D.C.: Introducing BrilliantColor™ Technology. Texas Instruments white paper (2005)
22. Isla, D.: Handling complexity in the halo 2 ai. In: Game Developers Conference, vol. 12 (2005)
23. Jakobi, N., Husbands, P., Harvey, I.: Noise and the reality gap: the use of simulation in evolutionary robotics. Advances in Artificial Life, pp. 704–720. Springer, Berlin (1995)
24. Jones, S., Studley, M., Winfield, A.: Mobile GPGPU acceleration of embodied robot simulation. In: Artificial Life and Intelligent Agents: First International Symposium, ALIA 2014, Bangor, UK, November 5–6, 2014. Revised Selected Papers, Communications in Computer and Information Science. Springer (2015)
25. Klöckner, A.: Interfacing behavior trees with the world using description logic. In: AIAA conference on Guidance, Navigation and Control, Boston (2013)
26. Koza, J.R.: On the programming of computers by means of natural selection. Genetic Programming, vol. 1. MIT press, Cambridge (1992)
27. Lim, C.U., Baumgarten, R., Colton, S.: Evolving behaviour trees for the commercial game defcon. Applications of Evolutionary Computation, pp. 100–110. Springer, Berlin (2010)
28. Marzinotto, A., Colledanchise, M., Smith, C., Ogren, P.: Towards a unified behavior trees framework for robot control. In: 2014 IEEE International Conference on Robotics and Automation (ICRA), pp. 5420–5427. IEEE (2014)
29. Nelson, A.L., Barlow, G.J., Doitsidis, L.: Fitness functions in evolutionary robotics: a survey and analysis. Robot. Auton. Syst. **57**(4), 345–370 (2009)
30. Ogren, P.: Increasing modularity of uav control systems using computer game behavior trees. In: AIAA Guidance, Navigation and Control Conference, Minneapolis, MN (2012)
31. Pereira, R.d.P., Engel, P.M.: A framework for constrained and adaptive behavior-based agents (2015). arXiv preprint arXiv:1506.02312
32. Perez, D., Nicolau, M., O'Neill, M., Brabazon, A.: Evolving behaviour trees for the mario ai competition using grammatical evolution. Applications of Evolutionary Computation, pp. 123–132. Springer, Berlin (2011)
33. Reynolds, C.W.: Flocks, herds and schools: a distributed behavioral model. In: ACM SIGGRAPH Computer Graphics, vol. 21, pp. 25–34. ACM (1987)
34. Rubenstein, M., Ahler, C., Nagpal, R.: Kilobot: A low cost scalable robot system for collective behaviors. In: 2012 IEEE International Conference on Robotics and Automation (ICRA), pp. 3293–3298. IEEE (2012)
35. Şahin, E.: Swarm robotics: from sources of inspiration to domains of application. Swarm Robotics, pp. 10–20. Springer, Berlin (2005)
36. Scheper, K.Y., Tijmons, S., de Visser, C.C., de Croon, G.C.: Behavior trees for evolutionary robotics. Artificial life (2015)
37. Shoulson, A., Garcia, F.M., Jones, M., Mead, R., Badler, N.I.: Parameterizing behavior trees. In: International Conference on Motion in Games, pp. 144–155. Springer (2011)
38. Winfield, A.E.: Towards an engineering science of robot foraging. Distributed Autonomous Robotic Systems 8, pp. 185–192. Springer, Berlin (2009)

Evolving Group Transport Strategies for e-Puck Robots: Moving Objects Towards a Target Area

Muhanad H. Mohammed Alkilabi, Aparajit Narayan, Chuan Lu and Elio Tuci

Abstract This paper describes a set of experiments in which a homogeneous group of simulated e-puck robots is required to coordinate their actions in order to transport cuboid objects towards a target location. The objects are heavy enough to require the coordinated effort of all the members of the group to be transported. The agents' controllers are dynamic neural networks synthesised through evolutionary computation techniques. The results of our experiments indicate that the most effective transport strategies generated by artificial evolution are those in which the robots exploit occlusion by pushing the objects across the portion of their surface, where they occlude the direct line of sight to the goal. The main contribution of this study is the analysis of the relationships between the characteristics of the object (i.e., mass and length), the morphology of the robots, and the group performance. We also test the scalability of the occlusion-based transport strategies to group larger than those used during the evolutionary design phase.

1 Introduction

This paper shows the results of a series of studies on cooperative object transport in a swarm of autonomous simulated robots required to push a cuboid object towards a

M. H. Mohammed Alkilabi (✉) · A. Narayan · C. Lu · E. Tuci
Computer Science Department, Aberystwyth University, Aberystwyth, UK
e-mail: mhm1@aber.ac.uk

A. Narayan
e-mail: apn3@aber.ac.uk

C. Lu
e-mail: cul@aber.ac.uk

E. Tuci
e-mail: elt7@aber.ac.uk

M. H. Mohammed Alkilabi
Computer Science Department, College of Science,
Kerbala University, Kerbala, Iraq

R. Groß et al. (eds.), *Distributed Autonomous Robotic Systems*,
Springer Proceedings in Advanced Robotics 6,
https://doi.org/10.1007/978-3-319-73008-0_35

target location. Cooperative transport refers to the process of transporting large item by multiple individuals simultaneously [4]. Cooperative transport is quite common in social insects, in particular in ants; the literature has documented cooperative transport in more than 40 genera of ants [10]. Ants can retrieve items/preys that are too large to be transported by single foragers without the need to dissect them on site, with the benefit of a reduced retrieval time that frees resources for other tasks, and a reduced exposure to predation [5, 14, 19]. Cooperative transport in ants is a source of inspiration for roboticists, that try to mimic the behaviour of natural swarms to design autonomous robots capable of cooperatively carry out tasks that are beyond the competencies of single individuals [2, 8, 18].

Our long term objective is to contribute to the design of more effective robotic swarms engaged in cooperative object transport tasks, by using evolutionary computation techniques to automate the design process of the individual mechanisms and rules of interactions that underpin effective group transport strategies. In [11] and more recently in [12], we showed that the individual perception of the direction of movement of the object to be transported is sufficient to allow a swarm of robots to align their pushing forces and effectively transport a heavy object. In other similar group transport studies described in the literature (e.g., [15–17]), the authors have equipped the robots with a more complex sensory apparatus which features force sensors, and an a priori defined allocation of roles (e.g., leader and follower) that is meant to facilitate the group coordination of action. Our results suggest that feedback on the movement of the object modulates the frequency with which a robot changes the point of application of its pushing forces. This modulation is sufficient for a robot to sense a quorum with respect to the direction of travel, and to break "deadlocks" in which the robots cancel each others' forces.

In this study, our objective is to extend the work describe in [11, 12] by design mechanisms to allow the robots to align their forces to initially push the object in any arbitrarily chosen direction, and, when the transport is initiated, to direct the transport toward a target location. The cooperative transport towards a target location is a challenging task in particular when the object is large enough to occlude the perception of the target (see [3]). In [3] is described an alternative group transport method which, rather than trying to overcome the limitations imposed by occlusion, it exploits occlusion. The robots are design to push the object across the portion of its surface, where it occludes the direct line of sight to the goal. The authors also provide an analytical proof of the effectiveness of the method, and results of successful empirical tests with objects of different shapes. Our study is very much related to the one described in [3], with which we share the same type of robots (i.e., the e-puck), and the use of large objects that occlude the target location. Contrary to [3], we did not explicitly program our robots to exploit occlusion to push the object toward a target. We use instead artificial evolution to integrate into a single neuro-controller the mechanisms to align the pushing forces, and to direct the transport toward a target location. Moreover, we look at scenarios in which the object is heavy enough to require the coordinated effort of all the robots of the group to be transported. The first interesting results of this study is that, as in [3], the most effective transport strategies generated by artificial evolution are those in which the robots exploit occlusion by pushing each object across the portion of its surface, where it occludes the direct line of sight to the goal. The original contribution of this study is in complementing the results shown in [3] with an analysis of the effects of object's mass and length, and of the group cardinality on the effectiveness of the occlusion-based transport strategy. We are aware that one of the criticism moved to the evolutionary approach concerns the loss of performance during transfer of the evolved solutions on real hardware [6]. Although our work is in simulation, our simulator (i.e., the robot model and the physics of the robot-world interactions) have been already used to design controllers that have being successfully ported on real e-pucks required to perform cooperative object transport tasks (see [12]). We are currently testing and comparing performances of the system in simulation and on real robots. At the time of writing, this comparative analysis has not been completed yet. However, as shown in https://www.aber.ac.uk/en/cs/research/ir/dss/#swarm-robotics, the results of few initial tests demonstrate that the best evolved solutions can be successfully ported on real robots. This paper is organized as follow: Sect. 2 describes the task and the simulation model; Sect. 3 reports the experimental results and the analysis of the operational mechanisms underpinning the single robot's behaviour; in Sect. 4 we discuss and comment on the results of this study and we point to interesting future research directions.

2 The Task and the Simulation Model

In this study, neuro-controllers are synthesised using artificial evolution to allow a homogeneous group of six autonomous simulated robots to push an elongated cuboid object (30 cm length, 6 cm width and height, 900 g mass) towards a target location (hereafter, the nest). The robots are initially positioned at 50 cm from the object centroid, facing the object. The experimental task is divided into two sub-tasks: the transport task, and the nesting task. During the transport task, the robots have to approach the object and to transport it, by collectively pushing it in an arbitrarily chosen direction. The transport task terminates when the group manages to displace the object 1 m away from its initial position. The transition from the transport to the nesting task is characterised by the appearance of the nest, towards which the object has to be moved. The nest is a circular area of 20 cm radius, whose centre is positioned 1 m away from the object centroid, at a randomly chosen angle (α) in the range $[60°, 90°]$ left or right with respect to the current object heading (see Fig. 1a). The nesting task successfully terminates when the object's centroid is inside the nest area. The object mass is set so that the coordinated effort of all six robots is required to move the object. The nest intentionally appears at the end of the transport task to avoid to provide robots with a perceptual cue that could significantly facilitate the coordination of actions required to align the pushing forces at the beginning of the trial.

The parameters of the neuro-controllers are set in a simulation environment which models kinematic and dynamic features of e-puck robots [13] operating in a boundless arena with flat terrain (see also [11] for a detailed description of the simulation

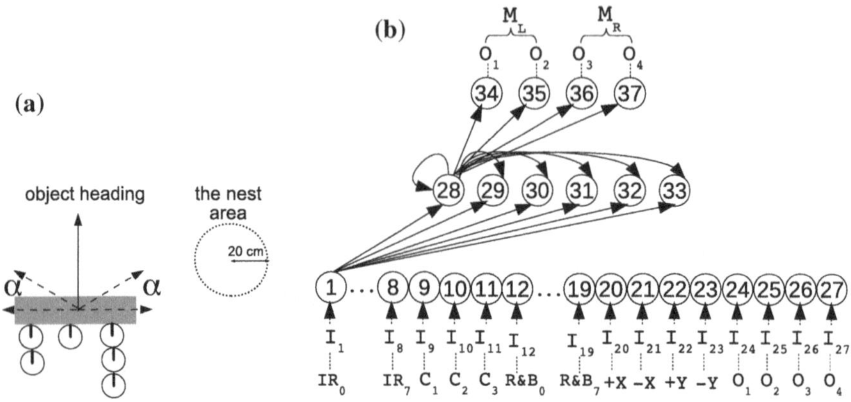

Fig. 1 **a** Experimental scenario; the empty circles refer to the robots, the grey rectangle refers to the object to be transported, the dashed arrows delimit to the angles (α) within which the nest can appear, the dotted circle refers to the nest. **b** The robot controller. The continuous line arrows indicate the efferent connections for only one neuron of each layer. Robot sensors to sensor neurons correspondence is indicated underneath the input layer, with IR_i referring to the infra-red, C_i to the camera sensors, $R\&B_i$ to the range and bearing, +X, −X, +Y, −Y to the optical flow sensor, and O_i referring to the output of the network at previous time step

environment). To take into account the dynamic aspects of this group transport scenario (e.g., forces, torque, friction, etc.), the agents and their environment have been simulated using *Bullet* physics engine. The robot model consists of three rigid bodies, a cylindrical chassis (3.55 cm radius, 6.2 cm height 200 g mass), and two motorised cylindrical wheels (2.05 cm radius, 0.2 cm height, 20 g mass) connected to the chassis with hinge joints. Both wheels can rotate forwards and backwards at a maximum speed of 8 cm/s.

The simulated robots are equipped with eight infra-red sensors (IR_i with $i = \{0, \ldots, 7\}$), a camera, the range and bearing (R&B) board (see [9]), and the optic-flow sensor. Infra-red sensors give the robot a noisy and non-linear indication of the proximity of an obstacle (e.g., the object or another robot). The IR sensor values are computed using a non-linear regression model of the sensor readings collected from the real e-puck. The camera is used to perceive coloured items (i.e., the object which is always green, or robots which are all red). The camera has a receptive field of 30°, divided in three equal sectors C_i, with $i = \{1, 2, 3\}$, each of which can return one of four possible values: 0 if no item falls within the sector's field of view; 0.4 if one or more red items are perceived; 0.7 if a green item is perceived; 1.0 if red and green items are perceived. The camera can detect coloured objects up to a distance of 50 cm. The R&B board has a circular shape, same diameter of the robot, and it is located on top of the robot. The board features 12 infra-red sensors uniformly distributed around its perimeter. Our robots can use only 8 sensors ($R\&B_i$, with $i = \{1, \ldots, 8\}$) which return 1 when they receive a signal, 0 otherwise. In our scenario, this board is used by the robots to detect the infra-red signals emitted by the nest, which emits in any direction. The position of the sensors receiving the nest signal can be used to compute the relative orientation of the nest with respect to the current robot heading. Due to physical properties and dimensions of robots and the object, the nest infra-red signals are not perceived if the object is in between the nest and the robot receiver.

The optic-flow sensor is an optical camera mounted underneath the robot chassis. It is located inside the slot originally hosting the robot battery. This sensor captures a sequence of low resolution images (i.e., 18×18 pixels) of the ground at 1500 frames per second. The images are sent to the on board DSP which, by comparing them, calculates the magnitude and the direction of movement of the robot. This information is subsequently communicated to the robot controller in the form of four normalized real values in [0, 1]: +X and −X representing the displacement on the positive and negative direction of the x axis, respectively; +Y and −Y representing the displacement on the positive and negative direction of the y axis, respectively. The optic-flow sensor generates a sensory stimulus which is a direct feedback on the consequences of the signals sent to the motors. In a collective object transport scenario multiple contingencies can result in a robot failing to execute its desired action. For example, a forward movement command may not produce the desired action if the robot is pushing a stationary object, or an object that is moving in the opposite direction due to forces exerted by other robots. The optic-flow sensor generates readings that can be used by the agents to differentiate between these circumstances and to respond accordingly. The results of the study described in [12] show that this simple feedback, generated by the optic-flow sensor, is sufficient to allow a group of real e-puck robots to coordinate their effort in order to collectively transport in an arbitrary direction an object that can not be moved by a single robot.

All robots' sensors and actuators are subject to random noise to facilitate the porting on real hardware (see [12] for details).

2.1 *The Controller and the Evolutionary Algorithm*

The robot controller is composed of a continuous time recurrent neural network (CTRNN) of 27 sensor neurons, 6 internal neurons, and 4 motor neurons (see [1] and also Fig. 1b which illustrates structure and connectivity of the network). The states of the motor neurons are used to control the speed of the left and right wheels. Sensory, internal, and motor neurons are updated using Eqs. 1–3:

$$y_i = gI_i;\ i \in \{1, \ldots, N\}; \quad \text{with } N = 27; \tag{1}$$

$$\tau_i \dot{y}_i = -y_i + \sum_{j=1}^{j=N+6} \omega_{ji}\sigma(y_i + \beta_j);\ i \in \{N{+}1, \ldots, N{+}6\}; \tag{2}$$

$$y_i = \sum_{j=N+1}^{j=N+6} \omega_{ji}\sigma(y_j + \beta_j);\ i \in \{N+7, \ldots, N+10\}; \tag{3}$$

with $\sigma(x) = (1 + e^{-x})^{-1}$. In these equations, using terms derived from an analogy with real neurons, y_i represents the cell potential, τ_i the decay constant, g is a gain factor, I_i with $i = 1, \ldots, N$ is the activation of the ith sensor neuron (see Fig. 1 for the correspondence between robots sensors and sensor neurons), ω_{ij} the strength of the synaptic connection from neuron j to neuron i, β_j the bias term, $\sigma(y_j + \beta_j)$ the firing rate f_i. All sensory neurons share the same bias (β_I), and the same holds for all motor neurons (β_O). τ_i and β_i of the internal neurons, β_I, β_O, all the network connection weights ω_{ij}, and g are genetically specified networks' parameters. At each time step, the output of the left motor is $M_L = f_{N+7} - f_{N+8}$, and the right motor is $M_R = f_{N+9} - f_{N+10}$, with M_L, $M_R \in [-1, 1]$. Cell potentials are set to 0 when the network is initialised or reset, and Eq. 2 is integrated using the forward Euler method with an integration time step T = 0.2. A simple evolutionary algorithm using roulette wheel selection is employed to set the parameters of the networks [7]. The population contains 100 genotypes. Generations following the first one are produced by a combination of selection with elitism, recombination, and mutation. For each new generation, the eight highest scoring individuals (the elite) from the previous generation are retained unchanged. The remainder of the new population is generated by fitness proportional selection from the 60 best individuals of the old population. A detailed description of the evolutionary algorithm can be found in [11].

2.2 The Fitness Function

During evolution each group undergoes a set of $E = 12$ evaluations or trials. A trial lasts 1200 simulation steps (i.e., 240 s, with 1 stimulation step corresponding to 0.2 s). A trial is terminated earlier if the group manages to transport the object inside the nest area (see Fig. 1a). As mentioned in Sect. 2, the nest intentionally appears at the end of the transportation task to avoid to provide robots with a perceptual cue that could significantly facilitate the coordination of actions required to align the pushing forces at the beginning of the trial. The nest appears with equal probability on the left and on the right of the object. In this study, we are interesting in generating controllers that allow the robots: (i) to agree on a common direction of transport solely through the interactions with the object, during the transport task; (ii) to adjust the direction of the object motion towards a specific target, during the nesting task.

At the beginning of each trial the controllers are reset, and the robots are positioned in the arena. Each trial differs from the others in the initialisation of the random number generator, which influences all the randomly defined features of the environment, such as the noise added to sensors and actuators, the robots initial position and orientation, and the relative angle between the object and the nest. The robots initial relative positions with respect to the object is an important aspect which bears upon the complexity of this task. During evolution, the robots starting positions correspond to randomly chosen points on a circle's circumference of 50 cm radius that has the object at it's centre. This circle is divided into six equals segments. Each robot is randomly placed in one part of this circle with random orientation in a way that the object can be within an angular distance of ±60° from its facing direction. These criteria should generate the required variability to develop solutions that are not sensitive to the robots initial positions.

In each trial (e), an evaluation function $F_e = f_1 + f_2 + f_3 + f_4$ rewards groups in which the robots: (i) remain close to the object to be transported (see f_1); (ii) initially transport the object 1 m from its initial position in an arbitrary direction (see f_2); (iii) adjust the direction of motion of the object towards the nest (see f_3); (iv) transport the object to the nest (see f_4). The four fitness components are computed in the following:

$$f_1 = \begin{cases} \frac{1}{TR}\sum_{t=1}^{T}\sum_{r=1}^{R}(1 - d(p_r^t, O^t)); & \text{if } d(p_r^t, O^t) > 20\text{ cm}; \\ \frac{1}{TR}\sum_{t=1}^{T}\sum_{r=1}^{R} 1; & \text{if } d(p_r^t, O^t) <= 20\text{ cm}; \end{cases}$$

$$f_2 = d(O^{t=0}, O^{t=k}); \qquad f_3 = \frac{1}{T}\sum_{t=1}^{T}\frac{(\pi - \theta(O^t, N))}{\pi}; \qquad f_4 = 1 - d(O^{t=T}, N);$$

where, $d(p_r^t, O^t)$ is the normalised Euclidean distance between the centroid of robot p_r^t and the centroid of the object O^t at time t; $d(O^{t=0}, O^{t=T})$ is the normalised Euclidean distance between the position of the object's centroid at the beginning of the trial ($t = 0$) and at the end of the transportation task ($t = k$); $d(O^{t=T}, N)$ is the normalised Euclidean distance between the centroid of the object O^t and the centroid of the nest N at the end of the trial ($t = T$); $\theta(O^t, N)$ is the smallest angular distance of the nest with respect to the object heading, chosen between the one measured clockwise and the one measured counter-clockwise. The fitness of a genotype ($\bar{F}$) is the average group evaluation score after it has been assessed $E = 12$ times.

3 Results

The primary aim of this study is to design control systems for homogeneous groups of robots required to transport objects to a target destination cooperatively. The controllers should be robust enough to deal with some variability in the object mass and length, and scalable to larger groups. During the design phase, we run 20 differently seeded evolutionary simulations, each simulation lasting 2000 generations. After evolution, in order to choose the best controller, we re-evaluated the best genotypes (i.e., groups) of the last 1000 generations of each evolutionary run. During re-evaluations, the groups are evaluated under the same experimental conditions experienced during evolution (i.e., group cardinality 6, object length 30 cm, object mass 900 g). Each group undergoes a set of 100 trials in which initial robots' positions and orientations are systematically varied. In all re-evaluation tests, we chose to measure the group performance using two binary metrics: the percentage of success in the transport task, and the percentage of success in the nesting task. Within a trial, a group succeeds in the transport task if, within the trial time limit (240 s), it manages to transport the object 1 m away from its initial position in any arbitrarily chosen direction. A group

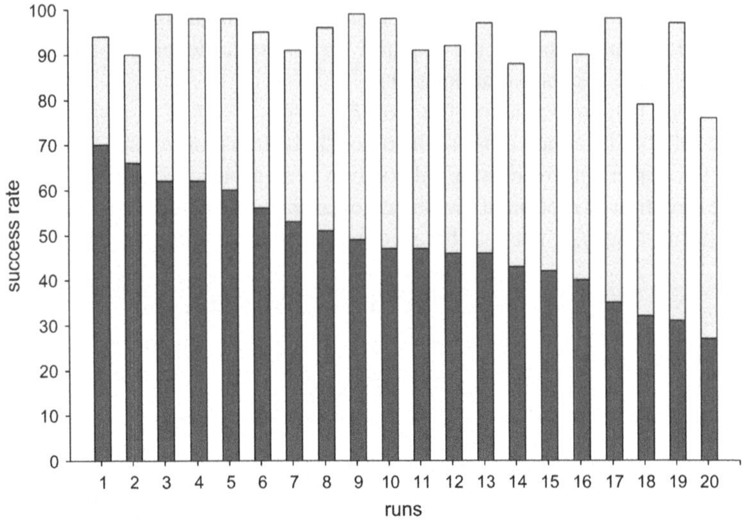

Fig. 2 Graphs showing the success rate of the best groups for each evolutionary run in the transport (grey bars) and in the nesting task (black bars). The groups are ranked from the best to the worst

succeeds in the nesting task if, within the trial time limit (240 s), it succeeds in the transport task, and it pushes the object centroid less than 20 cm away from the centre of the nest.

Figure 2 shows the performances of the best groups of each of the 20 evolutionary runs ranked in descending order with respect to success rate in the nesting task (see black bars). The graph also show the percentage of success in the transport task (see grey bars). The graph tells us that, apart from few runs, all best evolved groups are quite successful in the transport task, where the robots have to coordinate their actions and agree on a common direction of transport. The performances slightly drop for what concerns the percentage of success in the nesting task to a 70% for the very best group, and lower percentages for the others. This performance drop has to be interpreted with respect to the performance metrics used, which on the one hand it represents an easy and intuitive way to evaluate nesting behaviour. On the other hand, it punishes too severely groups that in spite of being able to accomplish the hardest part of the nesting task (i.e., the steering and simultaneous pushing the object towards the nest), they fail to take the object centroid less than 20 cm from the nest's centre within the trial time limit. In the remainder of this section, we will focus on the characteristics of behaviour of the very best evolved group, by looking at the robustness of the group strategy with respect to object length and mass, and at its scalability to larger group size.

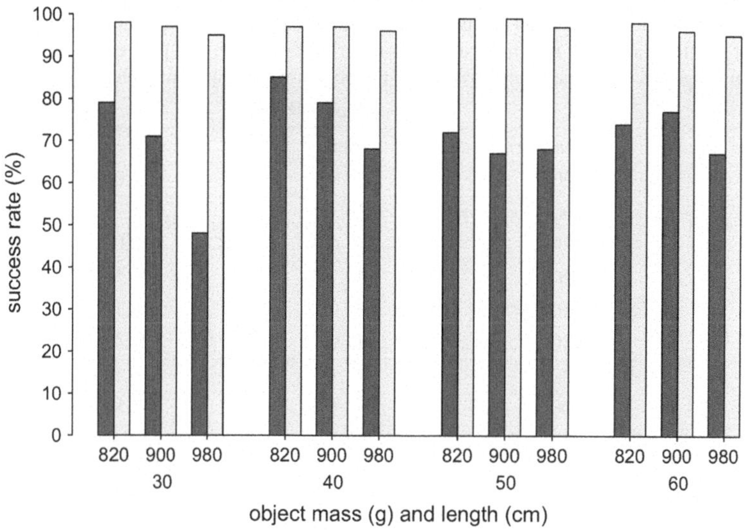

Fig. 3 Graph showing the success rate of the best evolved group for the transport (grey bars) and the nesting task (black bars) in different operational conditions, in which we varied the length of the longest object side (30, 40, 50, 60 cm) and its mass (820, 900, and 980 g)

3.1 Analysis of Best Evolved Group

For the *robustness-test*, the best evolved group was re-evaluated in 12 different operational conditions in which we varied the length of the longest object side (30, 40, 50, 60 cm) and its mass (820, 900, and 980 g). For each condition, the group underwent a set of 100 trials in which the initial robots' positions and orientations were systematically varied. The results of this *robustness-test* are shown in Fig. 3. The graph indicates that, while the initial transport is almost optimal for all conditions, the performance at the nesting task tends to drop with the increment of the object mass. There is a common performance trend for every object's length; the group does better with object of 820 g mass compared to the other two masses. This trend is particularly evident for the shorter object. The lighter the object the easier it is for the robots to transport/steer it without losing coordination. With longer objects (50 and 60 cm), the mass tends to have a smaller impact on the capability of the group to push the object toward the nest. This can be attributed to the fact that the longer the object the less momentum force required to move it. Consequently, the group can steer a heavy long object with less effort compared to shorter object with same mass. Table 1 shows the mean and the standard deviation of the distance between object and nest at the end of unsuccessful trials (i.e., trials in which the group failed to take the object centroid inside the circular nest area of 20 cm radius). Given that the initial distance between the object centroid and the centre of the nest is 1 m, the data in the Table indicates that, for all conditions, even in unsuccessful trials the group tends to

Table 1 Table showing the mean and the standard deviation of the distance between the object and the centre of the nest area at the end of unsuccessful trials during the first post-evaluation test of the best evolved group. The object-nest distance at the beginning of the nesting phase is 1 m

Object		Distance (cm)	
Length (cm)	Mass (g)	Mean	Sd
30	820	39.9	24.0
	900	34.3	25.7
	980	37.6	22.0
40	820	42.7	29.5
	900	36.9	30.1
	980	41.9	25.4
50	820	51.1	19.1
	900	53.1	15.8
	980	52.7	18.4
60	820	54.1	18.2
	900	52.0	21.7
	980	50.7	24.1

shorten the initial gap between the object and the nest, thanks to an effective pushing strategy that successfully steers the object towards the nest. Failure are primarily caused by the low speed of the transport manoeuvres with the consequence that the object does not get into the nest area within the trial time limit.

We have run a statistical analysis on the data generated by the *robustness-test*, using a logistic regression model, taking into account the random effects of the starting positions. Let p be the probability of success, and M the object mass, then the fitted model could be represented as $log(p/(1-p)) = 5.03 - 0.0046 \times M$. The model shows that the mass has a significant effect on the group success rate, indicating that the higher the mass, the lower the likelihood of success (see Fig. 4a). The analysis indicates that the object length has no significant effect on the group performance in this particular test.

We also run a further post-evaluation test, the *scalability-test*, to evaluate the scalability of the best evolved group strategy in larger group sizes, and with respect to different object lengths. The results of this test are shown in Fig. 5. Similar to the previous graphs, black bars refer to the percentage of success in the nesting task, while grey bars refers to the percentage of success in the transport task. Each condition is repeated 100 times (trials) by systematically varying the robots' initial position and orientation. The object mass does not vary within each condition. The group of 7 robots uses object of 1000 g for all 4 object's lengths. 100 g is added to the object mass for every extra robot added to the group. In each condition, all robots are required to transport the object. Results show that shorter the object, worst the group performance. This trend applies to almost all conditions for the nesting task (see Fig. 5, black bars) and to the largest groups for the transport task (see Fig. 5,

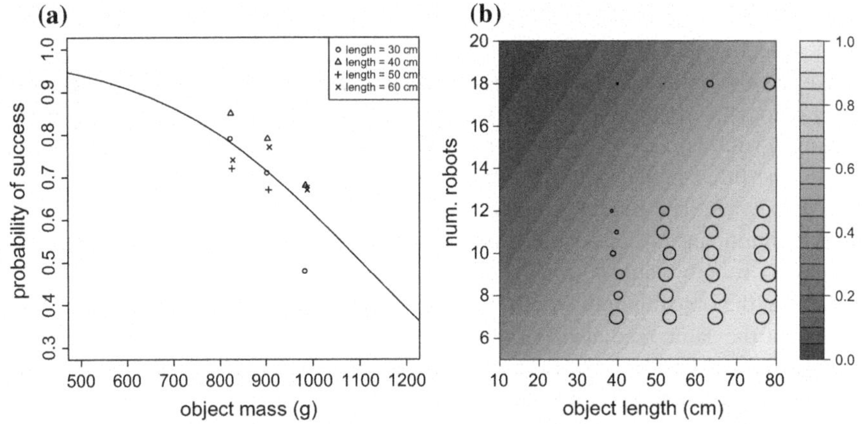

Fig. 4 **a** Graph showing mass versus probability of success at the nesting task during the *robustness-test*. The fitted logistic regression model prediction is plotted as the black curve; different symbols depict the experimental data (mass vs. empirical probability) for different sets of trials at different object lengths. **b** Graph showing the num. robots and object length versus probability of success at the nesting task during the *scalability-test*. The filled contour plot represents the fitted logistic regression model prediction; the experimental data are depicted in circles whose size are proportional to the empirical probability of success for various sets of trials

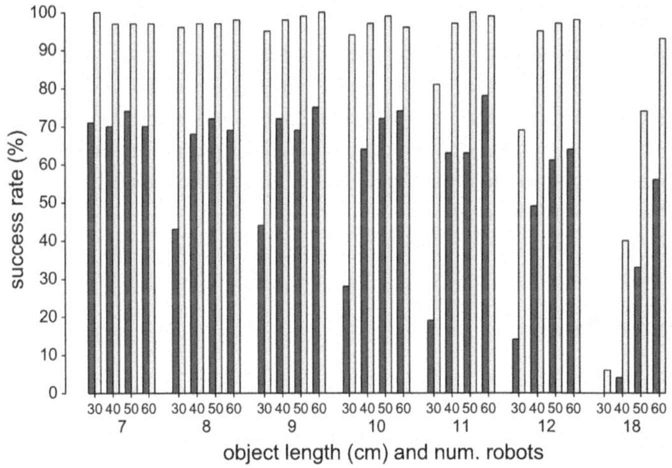

Fig. 5 Graph showing the success rate of the best evolved group for the transport (grey bars) and the nesting task (black bars) in different conditions, in which we varied the length of the longest object side (30, 40, 50, 60 cm) and the cardinality of the group (7, 8, 9, 10, 11, 12, 18)

black bars). This indicates that, in this test, the object length plays an important role on the effectiveness of the group nesting strategies.

We have run a statistical analysis on the data generated by the *scalability-test*, using a logistic regression model, taking into account the random effects of the starting positions. Let N be the number of robots in the group, and L the object length, the fitted model could be written as: $log(p/(1-p)) = -0.156 - 0.205 \times N + 0.057 \times L$. The model indicates that the number of robots and object length have a significant effect on the robots collaborative performance. Increasing the number of robots would lower the success rate for the same object length; whilst increasing the object length would increase the success rate if the number of robots were kept the same (see also Fig. 4b). These effects are mainly due to the type of strategy used by the best evolved group in steering the object toward the nest. By visually inspecting the group behaviour, we observed that, during the nesting phase, the robots tend to place the object in between themselves and the nest by moving to locations in which the infra-red signals emitted by the nest are occluded by the object. With the nest appearing either on the left or on the right side of the current direction of object motion, the tendency to hide from the nest signal let the robots to concentrate the pushing forces on the farther object side from the nest. Consequently, the object direction of motion is adjusted towards the nest.

Shorter the object, smaller is the area in which the robots are occluded from the nest signal. This implies that with shorter objects and large group sizes, the small occlusion area forces the robots to form long chains, in which progressively less robots are directly in contact with the object, while many of them push other robots. Pushing forces applied to other robots are less effective than those directly applied to the object since the robots have a cylindrical shape, and when they are in contact to each other they tend to slide one over the other. This phenomenon strongly reduces the total force applied to the object and clearly affects the capability of the group to steer the object toward the nest.

4 Conclusions

We have presented the results of a set of simulations in which a group of robots have to first transport a cuboid object 1 m away from its initial position in an arbitrary direction, and then to push it towards a nest area. The task described in this study is characterised by two distinctive features which distinguish it from similar swarm robotics studies focused on object transport, and make the task particularly challenging. First, the object is heavy enough to require the collaborative effort of all the robots of the group to be transported. Second, the initial coordination of actions and alignment of pushing forces has to be achieved through the interactions among the robots and between the robots and the object. The nest can not be used as perceptual cue to align the pushing forces since it appears only when the object is displaced 1 m away from its initial position.

The first contribution of this study is to show that, in spite of the complexity of the scenario, artificial evolution manages to integrate in a single neuro-controller the individual mechanisms underpinning the robot behaviour in all the phases of the task (e.g., during the initial alignment of pushing forces, and during the steering of the object towards the nest). Another significant contribution of this study is the analysis of the relationships between the characteristics of the object (i.e., mass and length), the morphology of the robots, and the group performance. As in [3], artificial evolution exploits an occlusion-based strategy to adjust the direction of the pushing forces, and to transport the object towards the nest. As in [3], we showed that the occlusion-based transport can be an effective cooperative strategy. We complement the results discussed in [3] by showing how the object length and mass as well as the shape of the robots relate to the effectiveness of the occlusion-based group transport strategies. In particular, with the results of our post-evaluation tests and by visually inspecting the behaviour of the best evolved group we noticed that the higher the number of e-puck robots pushing each other rather than directly the object (due to the object length and/or mass), the lower the group performance. The robots' cylindrical shape seems to be a limiting factor in the occlusion-based group transport strategy when the object surface that occludes the robots' direct line of sight to the goal is rather short. However, as discussed in [11], the same cylindrical shape facilitates the initial alignment of pushing forces by allowing the robots to move horizontally with respect to the object's side. Future work will try to overcome the limitations of the robots shape with respect to occlusion-based transport strategies without losing the beneficial effects during the alignment of pushing forces.

Acknowledgements Muhanad H. Mohammed Alkilabi thanks Iraqi Ministry of Higher Education and Scientific Research for funding his Ph.D.

References

1. Beer, R., Gallagher, J.: Evolving dynamic neural networks for adaptive behavior. Adapt. Behav. **1**(1), 91–122 (1992)
2. Berman, S., Lindsey, Q., Sakar, M., Kumar, V., Pratt, S.: Experimental study and modeling of group retrieval in ants as an approach to collective transport in swarm robotic systems. Proc. IEEE **99**(9), 1470–1481 (2011)
3. Chen, J., Gauci, M., Li, W., Kolling, A., Gross, R.: Occlusion-based cooperative transport with a swarm of miniature mobile robots. IEEE Trans. Robot. **31**(2), 307–321 (2015)
4. Czaczkes, T., Ratnieks, F.: Cooperative transport in ants (hymenoptera: Formicidae) and elsewhere. Myrmecol. News **18**, 1–11 (2013)
5. Feener, J., Donald, H., Moss, K.: Defense against parasites by hitchhikers in leaf-cutting ants: a quantitative assessment. Behav. Ecol. Sociobiol. **26**(1), 17–29 (1990)
6. Francesca, G., et al.: An experiment in automatic design of robot swarms. In: Proceedings of the 9th International Conference on Swarm Intelligence, pp. 25–37. Springer (2014)
7. Goldberg, D.E.: Genetic Algorithms in Search, Optimization and Machine Learning. Addison-Wesley, Reading (1989)
8. Groß, R., Dorigo, M.: Evolution of solitary and group transport behaviors for autonomous robots capable of self-assembling. Adapt. Behav. **16**(5), 285–305 (2008)

9. Gutiérrez, A., Campo, A., Dorigo, M., Amor, D., Magdalena, L., Monasterio-Huelin, F.: An open localization and local communication embodied sensor. Sensors **8**(11), 7545–7563 (2008)
10. Hölldobler, B., Wilson, E.: The Ants. Harvard University Press, Cambridge (1990)
11. Mohammed Alkilabi, M.H., Lu, C., Tuci, E.: Cooperative object transport using evolutionary swarm robotics methods. In: Proceedings of the European Conference on Artificial Life, vol. 1, pp. 464–471. MIT (2015)
12. Mohammed Alkilabi, M.H., Narayan, A., Tuci, E.: Design and analysis of proximate mechanisms for cooperative transport in real robots. In: Dorigo, M., et al. (ed.) Proceedings of the 10th International Conference on Swarm Intelligence (ANTS 2016). Springer (2016, in Press)
13. Mondada, F., et al.: The e-puck, a robot designed for education in engineering. In: Proceedings of the 9th International Conference on Autonomous Robot Systems and Competitions, vol. 1, pp. 59–65 (2009)
14. Tanner, C.: Resource characteristics and competition affect colony and individual foraging strategies of the wood ant formica integroides. Ecol. Entomol. **33**(1), 127–136 (2008)
15. Wang, Z., Schwager, M.: Multi-robot manipulation with no communication using only local measurements. In: Proceedings of the 54th IEEE Conference on Decision and Control (CDC), pp. 380–385. IEEE (2015)
16. Wang, Z., Schwager, M.: Kinematic multi-robot manipulation with no communication using force feedback. In: Proceedings of the IEEE International Conference on Robotics and Automation (ICRA), pp. 427–432. IEEE (2016)
17. Wang, Z., Schwager, M.: Multi-robot manipulation without communication. In: Distributed Autonomous Robotic Systems, pp. 135–149. Springer (2016)
18. Wang, Z., Takano, Y., Hirata, Y., Kosuge, K.: A pushing leader based decentralized control method for cooperative object transportation. In: Proceedings of IEEE/RSJ International Conference on Intelligent Robots and Systems, vol. 1, pp. 1035–1040. IEEE (2004)
19. Yamamoto, A., Ishihara, S., Fuminori, I.: Fragmentation or transportation: mode of large-prey retrieval in arboreal and ground nesting ants. Insect Behav. **22**, 1–11 (2009)

From Formalised State Machines to Implementations of Robotic Controllers

Wei Li, Alvaro Miyazawa, Pedro Ribeiro, Ana Cavalcanti, Jim Woodcock and Jon Timmis

Abstract Controllers for autonomous robotic systems can be specified using state machines. However, these are typically developed in an *ad hoc* manner without formal semantics, which makes it difficult to analyse the controller. Simulations are often used during the development, but a rigorous connection between the designed controller and the implementation is often overlooked. This paper presents a state-machine based notation, RoboChart, together with a tool to automatically create code from the state machines, establishing a rigorous connection between specification and implementation. In RoboChart, a robot's controller is specified either graphically or using a textual description language. The controller code for simulation is automatically generated through a direct mapping from the specification. We demonstrate our approach using two case studies (self-organized aggregation and swarm taxis) in swarm robotics. The simulations are presented using two different simulators showing the general applicability of our approach.

W. Li (✉) · J. Timmis
Department of Electronics, University of York, York, UK
e-mail: wei.li@york.ac.uk

J. Timmis
e-mail: jon.timmis@york.ac.uk

A. Miyazawa · P. Ribeiro · A. Cavalcanti · J. Woodcock
Department of Computer Science, University of York, York, UK
e-mail: alvaro.miyazawa@york.ac.uk

P. Ribeiro
e-mail: pedro.ribeiro@york.ac.uk

A. Cavalcanti
e-mail: ana.cavalcanti@york.ac.uk

J. Woodcock
e-mail: jim.woodcock@york.ac.uk

R. Groß et al. (eds.), *Distributed Autonomous Robotic Systems*,
Springer Proceedings in Advanced Robotics 6,
https://doi.org/10.1007/978-3-319-73008-0_36

1 Introduction

Safety is a major concern for autonomous robots, and the ability to provide evidence that a robotic system is safe can be demanding. Formal verification is the process of checking whether a design satisfies some requirements (properties) or that an implementation conforms to a design, and it has been used to verify a variety of robotic systems such as service robots [23] and swarming robots [20, 24].

Swarm robotics investigates how multiple robots, each with limited ability, communicate, coordinate and self-organize to accomplish certain tasks. Swarm robotics has potential in a wide range of real-world applications such as search and rescue, object transportation and environmental monitoring [4]. While using a number of simple robots to collectively perform complex tasks is desirable, designing individual controllers to guarantee the emergence of certain swarm behaviour is challenging. If swarm robotic systems are to transfer from lab-based experiments to real applications, especially those that are safety-critical, the verification of the individual controllers as well as their resulting emergent swarm behaviours needs to be conducted in a rigorous way.

Typically, the implementation of a robotic control system is conducted without establishing a strong connection between the controller code and the high-level design specifications. Here we explore the usage of a state-machine based notation, RoboChart [17], for designing robotic controllers. RoboChart has a formal semantics that allows for verification. In this paper, we extend RoboChart to support automatic code generation from the designed controllers to simulations.

Finite state machines are often adopted to design robot controllers in swarm robotics [2, 5, 10, 11, 14]. A commonly used state-machine notation is that of UML [1]. RoboChart takes inspiration from UML, and provides facilities to model timed and probabilistic systems, composed of one or more controllers.

Formal verification has been investigated in the design of controllers in swarm robotic systems [3, 7, 12, 20, 24]. In [7, 24], the authors used a temporal logic to formally specify and verify the emergent behaviour of a swarm robotic system performing aggregation. In [12], the authors used PRISM, a model checker for probabilistic automata, to formally verify the global behaviour of a foraging case scenario through exhausting all possible swarm behaviours. The analysis results were compared with those reported in [14], which used the test-driven simulation and showed a good correspondence. In these works, finite state-machine controllers were described using natural language, and there was no direct mapping from the high-level specification to low-level controller code.

In [15], the authors applied supervisory control theory to control a swarm of robots. Their approach supported automatic code generation. The controllers were specified using standard finite state machines, without any of the extra facilities for architectural modelling available, for example, in UML.

Various researchers have also explored the use of model-driven approaches to develop the high-level control of robots [6, 8, 21, 22]. The architecture analysis and design language (AADL) is a unifying component-based framework for modelling software systems with a particular focus on embedded real-time systems [8]. RoboChart could in principle be integrated into the controller component in AADL. In [22], a language was developed to program self-assembling robots. They proposed a role-based language that allowed the programmer to define the behavioural roles of each component independently from the concrete physical structure of the robots. However, in these works, the controllers of robots (e.g. state machine) were not formally specified, which makes it difficult to reason about robotic systems.

The main contribution of this paper is to reduce the gap between high-level specification and implementation of robotic controllers.

This paper is organized as follows. Section 2 briefly introduces RoboChart. This includes the elements of RoboChart and the approach to automatic code generation for simulation and deployment. Section 3 presents two case studies (self-organized aggregation [11] and swarm taxis [2]) in swarm robotics. The simulations using the automatically generated C++ code are presented. Section 4 concludes the paper and presents future work.

2 RoboChart

Figure 1 shows the RoboChart framework to combine formalised state machines and automatic implementation of robotic controllers. Once the controller is developed, code is generated automatically to be used in different simulation platforms or physical robots. Formal semantics are also automatically generated for verification. Details of the formal semantics of RoboChart can be found in [17]. In the following section, we focus on the automatic code generation for simulation and deployment.

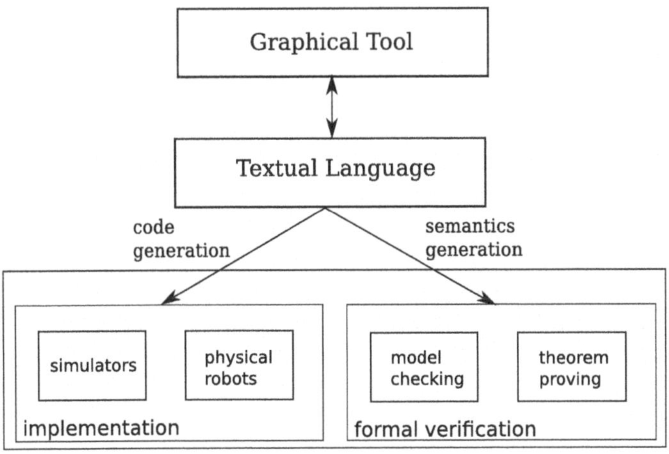

Fig. 1 The RoboChart framework for combining formalised state machines and implementation of robotic controllers

2.1 Elements of RoboChart

Central to RoboChart is a state-machine notation. RoboChart machines include states and their *entry*, *during* and *exit* operations (actions), as well as transitions possibly triggered by events. The entry operation is executed when the robot enters a state, and followed by the execution of the during operation. When a transition is triggered, the exit operation of the source state is executed. If an action is associated with the transition, it is also executed before the state machine enters the target state.

Operations and events of a state machine are described in an *interface*. A state machine can *requires* an interface. An operation can either be described without implementation or implemented by the user in a state-machine style. An operation can include a *precondition* and a *postcondition*.

Variables can be defined in a state machine, an interface or an operation. Different data types (primitive or composite) can be defined. When the behaviour is complex, multiple (potentially interacting) state machines can be used.

In addition to state machines, RoboChart also includes elements to organize specifications such as modules and robotic platforms [17]. A module defines a system, including a robotic platform and associated controllers. Each controller can be specified by one or more state machines.

RoboChart also includes time constraints. A *clock* can be defined inside a state machine to record the instant in time #T in which a transition is triggered. For example, the primitive since(T) yields the time elapsed since the most recent time instant #T. If since(T) is used as a condition (guard) on a transition with no events, then the transition will be taken immediately once the guard is true. Unless time is specified, we assume an operation takes no (or a significantly small) time.

For full details of RoboChart, refer to [17].

2.2 Simulation and Deployment

In RoboChart, the robot's controller is specified either graphically or using a textual description language. The automatically generated controller code can be imported into a wide variety of simulation platforms.

We adopt the model-view-controller (MVC) pattern in the design of simulations, where, the terms model and controller are used in a different way from that adopted in RoboChart. Figure 2 maps the RoboChart constructs to an MVC architecture. The model (M) component contains a simulation of the environment and of the RoboChart controller. We can generate a simulation of the RoboChart controller, potentially together with a simulation of the environment.[1] The controller component (C) implements the robotic platform, which corresponds to a particular robot

[1] The specification of environment is still under development. Currently the environmental stimuli are manually defined in the simulation.

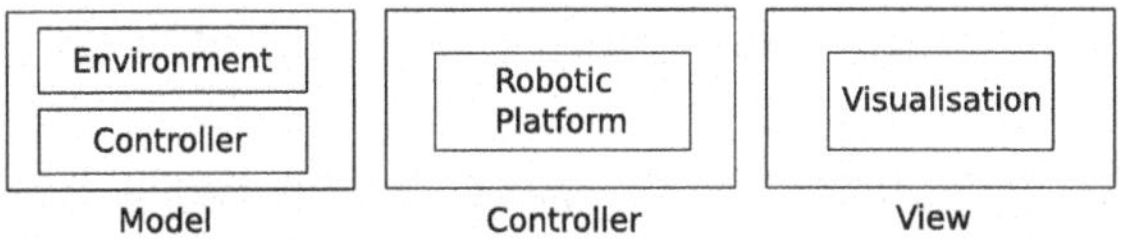

Fig. 2 RoboChart simulations pattern

RoboChart	State machine class
states	attribute of enumerated type
clocks	attribute of timer class
interfaces	inherit interface class

RoboChart	Interface class
events	attribute of enumerated type
variables	attribute
operations	methods

Fig. 3 RoboChart state machine and interface classes

in a simulation. Finally, the view component (V) defines the visualisation of the simulation.

We now describe how the controllers defined in RoboChart can be mapped into an executable language, specifically C++. Other object-oriented languages can be considered in a similar way, but are currently outside the scope of our work. The simulation of a controller is the simulation of its state machine(s). Each machine is implemented by a class. If the machine requires an interface, that interface is also implemented by a class, which is inherited by the state machine.

Figure 3 defines how constructs of a state machine and interface are mapped to elements of a class. The variables and events defined in an interface are generated as attributes of the class. The operations (*entry*, *during* and *exit*) that the robot executes in a state give rise to methods. We note that, even if an operation is specified in a state-machine style, it is generated as a method. To update the state machine, some other methods such as *MakeTransition* are also generated.

If a *clock* is defined in a state machine, a timer class is generated. It has a attribute *counter*, indicating the elapsed time, and methods such as *StartTimer* and *ResetTimer*. The state machine includes an object of the timer class as an attribute. The timer is used as a service of the state machine, which means the state machine can assess the counter. The state of the robot is updated in a cyclic manner, with the length of the cycle linked to the length of time required to capture events. The counter of the timer is updated in each control cycle.

A primitive data type is directly mapped into one in C++. For example, the type *real* corresponds to *double* in the code. A composite type is generated as a pre-defined class. For example, *vector2d* corresponds to a 2D vector class. The data-type system in RoboChart as well as its mapping are still under development.

3 Modelling Robotic Controllers Using RoboChart

To demonstrate our approach, we investigate two case studies on canonical problems in swarm robotics: aggregation [11] and swarm taxis (flocking towards a beacon) [2]. In these case studies, the robots are homogeneous. The controller of each robot is defined by a single state machine, and it is executed in the e-puck [18], which is a differential wheeled robot. It has an inter-wheel distance of 5.1 cm. The maximum speed for the left and right wheels of the e-puck is 12.8 cm/s, forward or backward.

3.1 *Case Study One: Aggregation*

3.1.1 Aggregation Behaviour

In this behaviour, each robot is equipped with a line-of-sight sensor that detects the type of item in front of it. The range of this sensor is unlimited in simulation. It gives a reading of $I = 1$ if there is a robot in the line of sight, and $I = 0$ otherwise. The environment is free of obstacles. The objective for the robots is to aggregate into a single compact cluster as fast as possible.

Each robot implements a reactive behaviour by mapping the sensor input (I) onto the outputs, that is, a pair of predefined speeds for the left and right wheels, $(v_{\ell I}, v_{rI})$, $v_{\ell I}, v_{rI} \in [-1, 1]$, where -1 and 1 correspond to the wheel rotating backwards and forwards respectively with maximum speed.

The parameters of the aggregation controller were found by performing a grid search over the space of possible combinations [11]. The controller exhibiting the highest performance was:

$$\mathbf{p} = (v_{\ell 0}, v_{r0}, v_{\ell 1}, v_{r1}) = (-0.7, -1.0, 1.0, -1.0)\,. \tag{1}$$

When $I = 0$, a robot moves backwards along a clockwise circular trajectory with a linear speed of -10.88 cm/s and an angular speed of -0.75 rad/s. When $I = 1$, a robot rotates clockwise on the spot with a linear speed of 0 and the maximum angular speed of -5.02 rad/s.

3.1.2 Modelling the Aggregation Controller in RoboChart

Figure 4 shows the diagram of the aggregation controller modelled in RoboChart. An interface, AggregationIface, declares the variables, operations and events. The state machine (AggregationFSM) requires AggregationIface. The state machine has an initial node, i, pointing to the initial state. The aggregation controller includes two states (S1 and S2), two events (*seeWall* and *seeRobot*, which correspond to $I = 0$ and $I = 1$ respectively), and two operations (*MoveClockwise* and *RotateClockwise*).

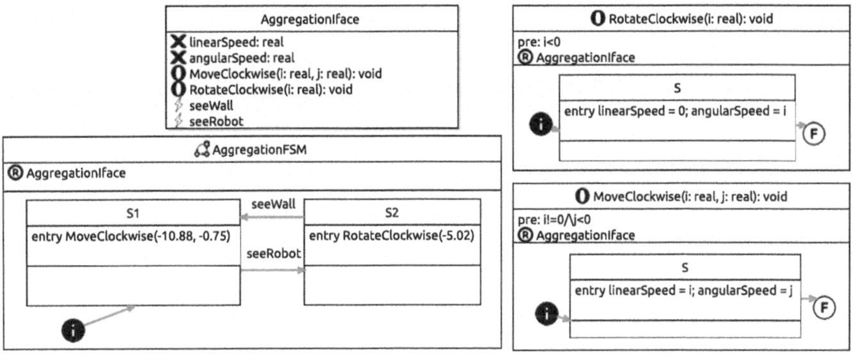

Fig. 4 Diagram of the aggregation controller modelled in Robochart

```
stm AggregationFSM {
    requires AggregationIface
    initial I
    state S1 {
        entry MoveClockwise(-10.88, -0.75)
    }
    state S2 {
        entry RotateClockwise(-5.02)
    }
    transition T1 {
        from I to S1
    }
    transition T2 {
        from S2 to S1
        trigger seeWall
    }
    transition T3 {
        from S1 to S2
        trigger seeRobot
    }
}
```

(a) aggregation controller

```
operation MoveClockwise(i: real, j:real) : void {
    precondition i != 0 /\ j < 0

    requires AggregationIface

    initial I
    final F
    state S {
        entry linearSpeed = i; angularSpeed = j
    }

    transition T1 {
        from I
        to S
    }

    transition T2 {
        from S
        to F
    }
}
```

(b) *MoveClockwise* operation

Fig. 5 Textual description of the aggregation controller and an operation in RoboChart

These operations are implemented in a state machine style with only an initial state S and final state F. Different from the AggregationFSM state machine, both operations have a final state. An operation can include precondition that must be satisfied by the caller to guarantee that the functionality of this operation is realised as specified. For example, in the *MoveClockwise* operation, the precondition requires that its first argument, an angular speed, is negative, and the second, the linear speed, is not zero. In the generated C++ code, this is realized using the *assert* function. A textual description of the AggregationFSM state machine and the *MoveClockwise* operation is shown in Fig. 5.

In the generated C++ code, two classes (AggregationIface and Aggregation-FSM) are generated. The class AggregationIface includes the attributes of two double variables (*linearSpeed* and *angularSpeed*), two methods (*MoveClockwise* and *RotateClockwise*) and two boolean events (*seeWall* and *seeRobot*). The operations are generated as virtual functions that can be overridden if necessary. The

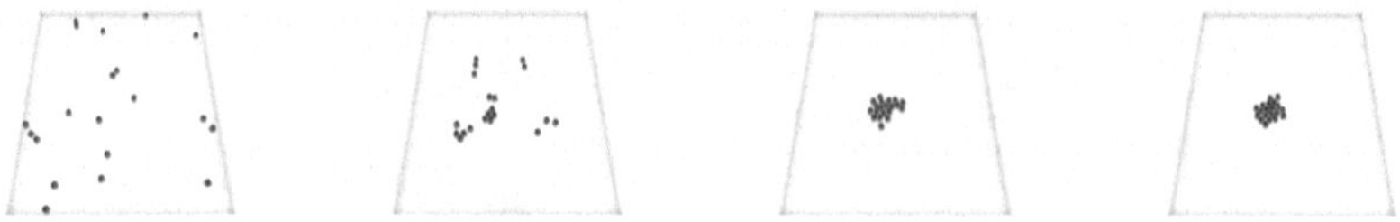

Fig. 6 Snapshots of the aggregation behaviour of 20 robots in simulation, using the automatically generated controller code from the RoboChart model

AggregationIface class is inherited by the state machine AggregationFSM class. It has the attributes of states S1 and S2, and other methods that are used to run the state machine. The generated C++ controller code can be found in the online supplementary materials [13].

3.1.3 Simulating the Aggregation Behaviour

The automatically generated code of the aggregation controller is tested in Enki [16], which has a built-in model of the e-puck robot. Enki is a 2D simulator and it can simulate swarms of robots a hundred times faster than real time. The speed of the left and right wheels of the e-puck can be set separately. The line-of-sight sensor in Enki is simulated by casting a ray from the e-puck's front and checking the first item with which it intersects (if any). The arena size is $250 \times 250\,\text{cm}^2$, and the initial position and orientation of the robots are randomly distributed. The length of the control step is set to 0.1 s, and the physics is updated every 0.01 s.

We performed 10 simulation trials with 20 robots, and in each trial the robots can aggregate into a single cluster. Figure 6 shows snapshots from a simulation trial using the automatically generated controller code from the model in RoboChart.

3.2 Case Study Two: Swarm Taxis

3.2.1 Swarm Taxis Behaviour

In the swarm taxis behaviour, the robots move towards a beacon while maintaining a coherent group. Each robot has three states: Forward, Coherence and Avoidance. The initial state is Forward. If the robot is in the Forward state for a certain number of time units without detecting any robots within avoidance radius, it enters the Coherence state. In this state, the robot turns towards the estimated center of the nearby robots. If the robot detects any robot within the avoidance radius while it is in the Forward state, it enters the Avoidance state. In this state, the robot turns away from the estimated center of the robots being avoided.

The robot can be illuminated by a beacon in the environment or shadowed by other robots (unilluminated). The avoidance radius when the robot is illuminated is larger. The avoidance radius is updated while the robot is in the Forward state. It is this mechanism that leads to the emergent swarm taxis behaviour [2].

3.2.2 Modelling the Swarm Taxis Controller in RoboChart

Figure 7 shows the diagram of the swarm taxis controller in RoboChart. The full model can be found in the online supplementary materials [13]. The interface SwarmTaxisIface defines the variables, operations and an event. A clock is defined inside the controller SwarmTaxisFSM. The initial state of the controller is Forward, where a timer T is started immediately. The timer records the time the robot stays in the state Forward. In RoboChart, an expression marked in square brackets (such as *reached* == *true* or *since(T)* < *25* in Fig. 7) is a guard for the transition. If there is no event associated with a transition, satisfaction of the condition will trigger the transition immediately. For example, once 25 time units have elapsed since the robot is in the Forward state, a transition from the Forward state to the Coherence state is triggered.

In the Forward state, the robot updates its avoidance radius through the operation *UpdateAvoidanceRadius*. The actual avoidance radius is set based on the boolean variable *illuminated* resulting from the operation *CheckIlluminationStatus*. If the robot is illuminated, the avoidance radius is set to 0.2; otherwise it is set to 0.1. As a consequence of this choice, the robots that have longer avoidance radius (are illuminated) tend to move towards the beacon and thus give rise to the beacon taxis behaviour of the whole swarm. Note that although we have declared the operation *CheckIlluminatedStatus*, we have chosen not to specify it in the RoboChart model, since it relies on the usage of the robot's sensors, which is platform dependent. If the robot detects any other robots nearby within the avoidance radius, it enters the Avoidance state, where the robot calculates the desirable turning degree

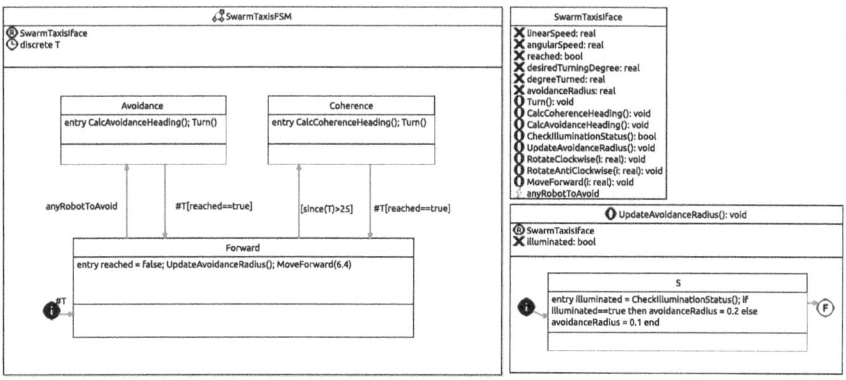

Fig. 7 Model of the swarm taxis controller in Robochart

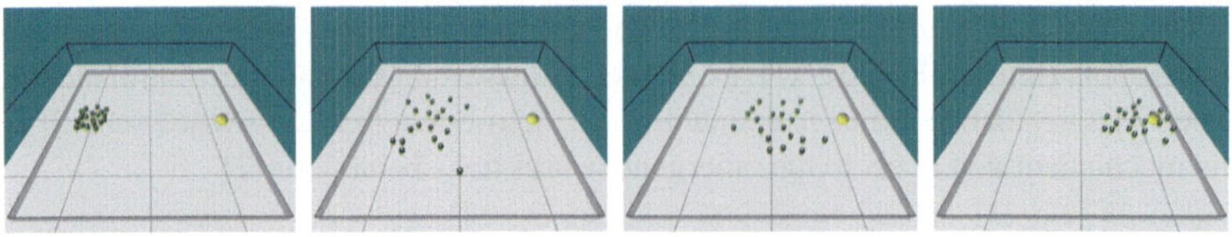

Fig. 8 Snapshots of the swarm taxis behaviour in simulation, using the automatically generated controller code from the RoboChart model. There are 20 robots (green) and one beacon (yellow)

(*desiredTurningDegree*) using the operation *CalcAvoidanceHeading* and then executes the operation *Turn*. In the operation *Turn*, the boolean variable *reached* is updated to indicate whether the robot has turned the desirable degree. Once the desirable turning degree has been achieved, the variable *reached* is set to *true*, which triggers the transition from Avoidance to Forward. Every time a transition is triggered, the timer T is started. Similar operations occur in the transition from Coherence to Forward.

For a full description of the model, refer to [13].

3.2.3 Simulating the Swarm Taxis Behaviour

The swarm taxis behaviour is simulated in ARGoS [19], which also has a built-in model of the e-puck. It is a 3D simulator. The simulated space can be divided into several sub-spaces that run different physics engines in parallel. The arena size is 400 $\times$ 400 cm^2. There is one beacon located in the right of the arena, and the robots are randomly initialized in the left region of the arena. Each robot is equipped with light sensors (to detect the beacon) and range-and-bearing sensors (to detect other robots nearby). The length of control step is set to 0.1 s. Note that in the model shown in Fig. 7 we did not attempt to optimize the value of each parameter (such as the time threshold and avoidance radius).

We performed 10 simulation trials with 20 robots, and in each trial the robots can successfully move towards the beacon while maintaining a coherent group. Figure 8 shows snapshots from a simulation trial, using the automatically generated controller code in RoboChart. A video showing the simulation of the two case studies and the automatically generated C++ code of the controllers can be found in the online supplementary materials [13].

4 Conclusion

In this paper, we have presented a state-machine based framework RoboChart for modelling the controllers of autonomous robots, combined with the automatic generation of C++ code. We believe that this is the first framework that allows for both automatic code generation for robotic simulation, deployment and formal verification. The applicability of our approach has been demonstrated through modelling two case studies (self-organized aggregation and swarm taxis) in swarm robotics. The automatically generated code of the robot's controller was run in two different simulators, which again, demonstrates the flexibility of our approach.

Our vision is to significantly reduce the gap between the high-level reasoning and low-level implementation through the use of formal methods and automatic code generation. The work presented can be seen as a first step towards the goal of verifying emergent behaviour, which is a potential application of our work to be investigated in the future. Our current focus is, however, to enrich the state machine specification of RoboChart by adding time and probability constructs, so that the framework can be applied to model a wide variety of robotic control systems. The formal semantics of RoboChart will also be enriched to make the verification feasible. In RoboChart, we focus on modelling the controller of a single robot, but we are investigating the possibility of using RoboChart models to simulate and analyse robotic swarms.

Currently, the generated controller code is a direct mapping from the elements in RoboChart to simulation. In the future, soundness of the simulation will be established by verifying the code generator. This can be realized using various software engineering techniques. In particular, we envisage that the CSP model generated from the RoboChart specification is a basis for establishing the correctness of the generated code using refinement. Practical verification can be carried out using a model checker like FDR (which also provides a facility to animate the model, and thus perform some validation), or using a theorem prover.

In this paper, we only automatically generate the code of controllers, however the simulation configurations (e.g. length of control step) in the case studies are manually defined. We intend to define simulations in an extended notation, from which the simulation configurations can also be specified. The simulation notation will be independent of specific programming languages such as C++ and Java, and of specific robotic platforms.

Possible avenue for future work is the integration of RoboChart in other tools [3, 9]. For example, in [9], an automatic design method was used to tune the free parameters of a predefined parametric architecture (e.g. probabilistic state machine) for the individual robot controller of a swarm. In this case, the controller architecture can be modelled in RoboChart, so that the obtained solution can be formally verified. In [3], a property-driven approach was proposed to design the controller of swarming robots. The designed controller can also be modelled in RoboChart to support both formal verification and code generation.

Finally, we intend to model the environmental stimuli and generate code for physical robots.

Acknowledgements The authors would like to acknowledge the support from EPSRC grant EP/M025756/1.

References

1. Bergenti, F., Poggi, A.: Exploiting uml in the design of multi-agent systems. In: Omicini, A., Tolksdorf, R., Zambonelli, F. (eds.) Engineering Societies in the Agents World: First International Workshop, pp. 106–113. Springer, Berlin, Germany (2000)
2. Bjerknes, J.D., Winfield, A.F.T.: On fault tolerance and scalability of swarm robotic systems. In: Martinoli, A., Mondada, F., Correll, N., Mermoud, G., Egerstedt, M., Hsieh, A.M., Parker, E.L., Støy, K. (eds.) Distributed Autonomous Robotic Systems: The 10th International Symposium, pp. 431–444. Springer, Berlin, Germany (2013)
3. Brambilla, M., Pinciroli, C., Birattari, M., Dorigo, M.: Property-driven design for swarm robotics. In: Proceedings of 2012 International Conference on Autonomous Agents and Multiagent Systems, pp. 139–146. IFAAMS, Richland, SC, USA (2012)
4. Brambilla, M., Ferrante, E., Birattari, M., Dorigo, M.: Swarm robotics: a review from the swarm engineering perspective. Swarm Intell. **7**(1), 1–41 (2013)
5. Chen, J., Gauci, M., Li, W., Kolling, A., Groß, R.: Occlusion-based cooperative transport with a swarm of miniature mobile robots. IEEE Trans. Robot. **31**(2), 307–321 (2015)
6. Dhouib, S., Kchir, S., Stinckwich, S., Ziadi, T., Ziane, M.: RobotML, a domain-specific language to design, simulate and deploy robotic applications. In: Noda, I., Ando, N., Brugali, D., Kuffner, J.J. (eds.) Simulation, Modeling, and Programming for Autonomous Robots, pp. 149–160. Springer, Berlin, Germany (2012)
7. Dixon, C., Winfield, A.F.T., Fisher, M., Zeng, C.: Towards temporal verification of swarm robotic systems. Robot. Auton. Syst. **60**(11), 1429–1441 (2012)
8. Feiler, P.H., Gluch, D.P.: Model-Based Engineering with AADL: An Introdution to the SAE Architecture Analysis and Design Language. Addison-Wesley, Boston (2012)
9. Francesca, G., Brambilla, M., Brutschy, A., Garattoni, L., Miletitch, R., Podevijn, G., Reina, A., Soleymani, T., Salvaro, M., Pinciroli, C., Mascia, F., Trianni, V., Birattari, M.: AutoMoDe-Chocolate: a method for the automatic design of robot swarms that outperforms humans. Swarm Intell. **9**(2–3), 125–152 (2015)
10. Gauci, M., Chen, J., Li, W., Dodd, T.J., Groß, R.: Clustering objects with robots that do not compute. In: Proceedings of 2014 International Conference on Autonomous Agents and Multiagent Systems, pp. 421–428. IFAAMS, Richland, SC, USA (2014)
11. Gauci, M., Chen, J., Li, W., Dodd, T.J., Groß, R.: Self-organized aggregation without computation. Int. J. Robot. Res. **33**(8), 1145–1161 (2014)
12. Konur, S., Dixon, C., Fisher, M.: Analysing robot swarm behaviour via probabilistic model checking. Robot. Auton. Syst. **60**(2), 199–213 (2012)
13. Li, W., Miyazawa, A., Ribeiro, P., Cavalcanti, A., Woodcock, J., Timmis, J.: Online supplementary material (2016). http://www.york.ac.uk/robot-lab/dars2016/
14. Liu, W., Winfield, A.F.T.: Modeling and optimization of adaptive foraging in swarm robotic systems. Int. J. Robot. Res. **29**(14), 1743–1760 (2010)
15. Lopes, Y.K., Trenkwalder, S.M., Leal, A.B., Dodd, T.J., Groß, R.: Supervisory control theory applied to swarm robotics. Swarm Intell. **10**(1), 65–97 (2016)
16. Magnenat, S., Waibel, M., Beyeler, A.: Enki: the fast 2D robot simulator (2011). http://home.gna.org/enki/
17. Miyazawa, A., Ribeiro, P., Li, W., Cavalcanti, A.L.C., Timmis, J., Woodcock, J.C.P.: RoboChart: a state-machine notation for modelling and verification of mobile and autonomous robots. Technical report, University of York, Department of Computer Science, York, UK (2016). www.cs.york.ac.uk/circus/publications/techreports/reports/MRLCTW16.pdf
18. Mondada, F., et al.: The e-puck, a robot designed for education in engineering. In: Proceeding of the 9th Conference on Autonomous Robot Systems and Competitions, vol. 1, pp. 59–65. IPCB: Instituto Politécnico de Castelo Branco (2009)
19. Pinciroli, C., Trianni, V., O'Grady, R., Pini, G., Brutschy, A., Brambilla, M., Mathews, N., Ferrante, E., Di Caro, G., Ducatelle, F., Birattari, M., Gambardella, L.M., Dorigo, M.: ARGoS: a modular, parallel, multi-engine simulator for multi-robot systems. Swarm Intell. **6**(4), 271–295 (2012)

20. Rouff, C.A., Hinchey, M.G., Pena, J., Ruiz-Cortes, A.: Using formal methods and agent-oriented software engineering for modeling NASA swarm-based systems. In: 2007 IEEE Swarm Intelligence Symposium, pp. 348–355. IEEE, Honolulu, Hawaii (2007)
21. Schlegel, C., Hassler, T., Lotz, A., Steck, A.: Robotic software systems: from code-driven to model-driven designs. In: Proceedings of the 14th International Conference on Advanced Robotics, pp. 1–8. IEEE, Munich, Germany (2009)
22. Schultz, U.P., Christensen, D.J., Stoy, K.: A domain-specific language for programming self-reconfigurable robots. In: Proceedings of the 2007 Workshop on Automatic Program Generation for Embedded Systems, pp. 28–36. ACM, Salzburg, Austria (2007)
23. Webster, M., Dixon, C., Fisher, M., Salem, M., Saunders, J., Koay, K.L., Dautenhahn, K., Saez-Pons, J.: Toward reliable autonomous robotic assistants through formal verification: a case study. IEEE Trans. Hum. Mach. Syst. **46**(2), 186–196 (2016)
24. Winfield, A.F.T., Sa, J., Fernandez-Gago, M.C., Dixon, C., Fisher, M.: On formal specification of emergent behaviours in swarm robotic systems. Int. J. Adv. Robot. Syst. **2**(4), 363–370 (2005)

Human Responses to Stimuli Produced by Robot Swarms - the Effect of the Reality-Gap on Psychological State

Gaëtan Podevijn, Rehan O'Grady, Carole Fantini-Hauwel and Marco Dorigo

We study the reality-gap effect (the effect of the inherent discrepancy between simulation and reality) on the human psychophysiological state, workload and reaction time in the context of a human-swarm interaction scenario. In our experiments, 37 participants perform a supervision task (i.e., the participants have to respond to visual stimuli produced by a robot swarm) with a real robot swarm and with simulated robot swarms. Our results show that the reality-gap significantly affects the human psychophysiological state, workload and reaction time. Our results also show that conducting a human-swarm interaction experiment in a virtual reality environment can be an alternative to conducting an experiment with robot swarms simulated on a computer screen. These results suggest that virtual reality can mitigate the effect of the reality-gap in human-swarm interaction experiments.

G. Podevijn (✉) · R. O'Grady · M. Dorigo
IRIDIA, Université Libre de Bruxelles, Bruxelles, Belgium
e-mail: gpodevij@gmail.com

R. O'Grady
e-mail: rogrady@ulb.ac.be

M. Dorigo
e-mail: mdorigo@ulb.ac.be

C. Fantini-Hauwel
Research Center of Clinical Psychology, Psychopathology and Psychosomatic, Université Libre de Bruxelles, Bruxelles, Belgium
e-mail: chauwel@ulb.ac.be

R. Groß et al. (eds.), *Distributed Autonomous Robotic Systems*, Springer Proceedings in Advanced Robotics 6,
https://doi.org/10.1007/978-3-319-73008-0_37

1 Introduction

There are fundamental differences between the way a human interacts with a robot swarm and the way a human interacts with a single robot. Firstly, because robots in a swarm robotics system do not have the same communication capabilities as most of the other robotics systems (e.g., text-based or voice-based communication hardware). Secondly, because even if they were augmented with communication capabilities, the large number of robots that compose a swarm robotics system would make it impractical for a human operator to interact with each individual robot—there would be too many data-points for the human operator to process.

Because of these fundamental differences, human-swarm interaction (HSI) has become an active and independent field of research. In the current HSI literature, most research tackles technical aspects of the interaction (e.g., gesture-based interaction, speech-based interaction, haptic-based interaction, leader-based robot swarm control) and only few researches are validated with user studies (see [8] for a comprehensive survey of HSI). Moreover, with the exception of [17, 19], the majority of these user studies are performed in simulation only [2, 3, 7, 12–14, 22, 23]. Though simulation is convenient for the repeatability of the experiments and for the low-cost of the infrastructure (i.e., robots and experimental room), simulation suffers from the so-called *reality-gap*—the inherent discrepancy between simulation and reality.

These user studies, performed in simulation only, have motivated us to investigate the effect of the reality-gap in the context of HSI. We are interested in understanding whether the reality-gap affects the psychology of humans interacting with a robot swarm. For instance, do human operators feel more stressed or overloaded when they interact with a real robot swarm than with a simulated robot swarm displayed on a computer screen (which is used in the majority of the human-swarm interaction user studies)? In [16], we have already shown that the reality-gap had an effect on the human psychophysiological state (i.e., the psychological state of a human assessed by physiological measures) when humans are passively interacting with a robot swarm. In this paper, we also study the effect of the reality-gap on the human psychophysiological state, however, our participants are not purely passive anymore. Even though our participants do not issue any commands to the robot swarm (because an interaction interface could influence the psychophysiological responses of the participants, thus limiting the visibility of the reality gap effect [16]), they are asked to press a button each time a robot illuminates its LEDs in red. Thanks to this task, we show in this paper that the reality-gap has also significant impacts on the human workload (i.e., the mental effort) and reaction time. In the experimental scenario designed for our experiments, 37 participants interact with a simulated robot swarm displayed on a computer screen, with a simulated robot swarm displayed in a virtual reality environment and with a real robot swarm. Figure 1 shows the three experimental scenarios.

This paper contributes to the literature in two ways. The first contribution is that our results show that our participants have stronger psychophysiological reactions and higher workload and reaction time when they interact with a real robot swarm

Fig. 1 Experimental scenario. *Left*: A participant interacts with a simulated swarm of 20 robots displayed on a computer screen. *Middle*: A participant is attached to a virtual reality head set and interacts with a simulated swarm of 20 robots. *Right:* A participant interacts with a swarm consisting of 20 real robots. The participant shown in this figure is the first author of this paper and did not take part in the experiment. The pictures shown in this figure were taken for illustration purpose

than when they interact with a simulated robot swarm displayed on a computer screen. The second contribution is that we show that our participants' workload and reaction time are higher when they interact with a robot swarm simulated in a virtual reality environment than with a robot swarm simulated on a computer screen. These results suggest that conducting simulation-based experiments in a virtual reality environment can mitigate the effect of the reality-gap in HSI and, therefore, HSI researchers can avoid to buy and maintain expensive real robots.

2 Related Literature

The effect of the reality-gap on the human psychology has been studied in human-robot interaction, and more specifically, in the context of social robotics. In social robotics, the majority of the studies show that human beings enjoy more to interact with a real robot than with a simulated one [4, 6, 10, 15, 18, 21, 24]. In these studies, the authors measured the level of enjoyment either with a self-developed questionnaire or with the game flow model—a model initially developed to measure the players enjoyment in games [20]. In HSI, though, we were the first to study the effect of the reality-gap on the human psychology [16]. We showed that due to the reality-gap, humans had stronger psychophysiological responses when they passively interacted with a real robot swarm than with a simulated robot swarm.

In HSI research, several studies had already taken into account the human psychology during the interaction with a robot swarm. In [14], the authors showed that the human workload (assessed by a subjective questionnaire) is not affected by the size of a robot swarm (i.e., the number of robots in a swarm). In [17], however, we showed that the size of a robot swarm has a significant effect on the human psychophysiological state. In this research, we studied the psychophysiological state with a combination of objective physiological measures (skin conductance and heart rate) and subjective psychological measures (via a questionnaire). The effect of two types of command propagation methods (i.e., methods to disseminate a human command among the robots of a swarm) was studied in an experiment where a human operator had to guide a swarm of robots by controlling a leader robot's velocity and heading [1, 22]. The authors showed that workload is lower with the so-called flooding method (non-leader robots all set their velocity and heading to those of the leader robot) than with the so-called consensus method (non-leader robots set their velocity and heading to the average velocity and heading of their neighbours). Workload can also be affected by different types of communication network topologies made by the robots [3]. In [19], finally, the authors showed that the mapping between the manipulability of a swarm (i.e., whether it is easy or hard to guide a robot swarm) and haptic forces also impacts the human workload. With the exception of [17, 19], all the aforementioned studies were performed in simulation. Because of the reality-gap, though, it is difficult to confirm that these results would be similar if the experiments were conducted with a real robot swarm instead of a simulated one.

This paper is different from [16] for two reasons. Firstly, because our participants are not purely passive during the interaction with a robot swarm. Secondly, because in addition to studying the reality-gap effect on the human psychophysiological state, we also study its effect on the human workload and reaction time, providing a more complete understanding of the reality-gap effect in HSI.

3 Methodology

3.1 Hypotheses

We based the experiment presented in this paper on two hypotheses:

1. The psychophysiological reactions, workload and reaction time of humans are higher when they interact with a real robot swarm than with a simulated one.
2. The psychophysiological reactions, workload and reaction time of humans are higher when they interact with a simulated robot swarm displayed in a virtual reality environment than with a simulated robot swarm displayed on a computer screen.

Confirming the first hypothesis would allow us to show that, not only the reality-gap has an effect on the human psychophysiological state, but equally importantly, that it has also an effect on the human workload and reaction time. Confirming the second hypothesis would suggest that conducting HSI experiments in a virtual reality environment could mitigate the effect of the reality-gap—the way humans would react when interacting with a robot swarm in a virtual reality environment would be more similar to the way they would react with a real robot swarm than to the way they would react with a robot swarm simulated on a computer screen.

In the next section, we present the experimental scenario used to test these two hypotheses.

3.2 Experimental Scenario

In order to test our two hypotheses, we designed an experimental scenario similar to that of [16]. In this experimental scenario, our participants' task is to supervise a swarm consisting of 20 robots for a period of 60 s. In this paper, the supervision tasks consists for our participants to watch attentively a robot swarm and press a button each time a robot of the swarm illuminates its LEDs in red.

The experimental scenario is divided into three sessions—the *Screen Simulation* session, the *Virtual Reality* session and the *Real Robots* session. In these sessions, our participants conduct the supervision task with three types of visualization interfaces. In the *Screen Simulation* session, the robot swarm is simulated in 2D and displayed on a computer screen (the participants see the robot swarm from the top view). In the *Virtual Reality* session, the robot swarm is simulated in 3D and displayed in a virtual reality environment (the participants see the robot swarm as they would see it in reality). In the *Real Robots* session, the robot swarm is composed of real robots. We decided to compare a 2D (top-view) simulation with the reality because this is how the majority of the user studies in human-swarm interaction display the robot swarm.[1] We decided to compare a 3D simulation (virtual reality) to the 2D simulation and to reality in order to test our second hypothesis. The order our participants encounter these 3 sessions is random.

3.3 Measures

In this study, we used a combination of subjective measures and objective measures. The subjective measures consist of two questionnaires—the Self-Assessment Manikin (SAM) questionnaire [9] and the NASA Task Load Index Scale (NASA-TLX) questionnaire [5]. We use the SAM questionnaire to study the participants' subjective valence (measure of pleasure-displeasure) and arousal (measure of the mental alertness and physical activity). The SAM questionnaire has 2 scales, each scale being composed of 9 pictures. In a scale, a picture represents a value of that scale (valence or arousal). We use the NASA-TLX questionnaire to study the participants' workload. In the NASA-TLX questionnaire, the workload is divided into 6 scales. In this paper, we use a simpler version of the NASA-TLX questionnaire—the NASA-RTLX, in which the result is the average of the raw score of the six scales. The objective measures consist of the physiological activity and of the reaction time of our participants. We study the physiological activity of our participants by monitoring their heart rate (number of beats per minute) and skin conductance level (slow variation of the skin conductance over time). We collect our participants' reaction

[1]Please note that this particular choice introduces the question of whether any differences observed between virtual reality and simulation are due to the different perspective. As discussed in the conclusions, this aspect will be considered in future work.

time by measuring the time taken by our participants to press a button after a robot illuminates its LEDs in red.

The baseline physiological activity of an individual (i.e., the physiological activity at rest) can be highly different from the baseline of another individual. In order to be able to compare physiological responses across our participants, we monitored our participants' physiological responses at rest and our participants' physiological responses during the experiment. To conduct our analyses, we used the difference between our participants' physiological responses during the experiment and at rest.

3.4 Physiological Data Acquisition and Robot Platform

Physiological measures were acquired by a PowerLab 26 T data acquisition system and by a supplementary GSR Amp device connected to the PowerLab 26 T system (ADInstruments). A pulse transducer was directly connected to the PowerLab 26T (for monitoring the heart rate) and two finger electrodes were connected to the GSR Amp device (for monitoring the skin conductance level). The PowerLab 26 T system was connected to a laptop computer running Mac OSX Yosemite. We used LabChart 8 to collect the physiological data.

The robotic platform used in this study is the e-puck robot platform. The e-puck robot is a robotic platform used for educational purposes [11]. In this study, we only used a subset of the e-puck's sensors and actuators. We used the proximity sensors for obstacle avoidance (i.e., to detect walls and other robots) and the wheel actuators to control the robots' motion (see Sect. 3.5).

3.5 Environment and Robot Behaviour

In our experimental scenario (see Sect. 3.2), we used a 2 m × 2 m environment, as shown in Fig. 2. The 20 robots used in this experiment are randomly placed in this

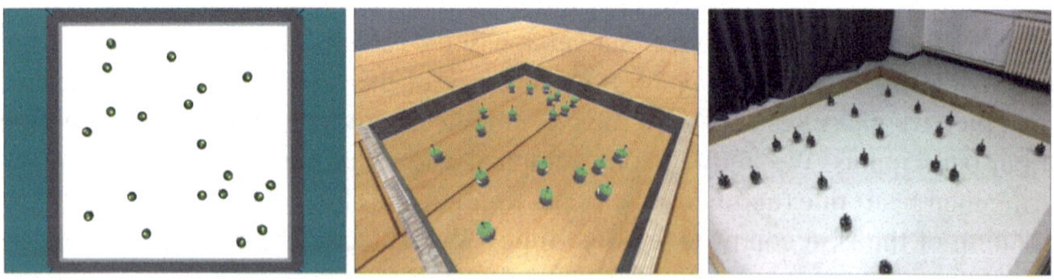

Fig. 2 Robots and environments used in the three sessions. Left: The environment displayed in the *Screen Simulation* session. Middle: The environment displayed in the *Virtual Reality* session. Right: The real environment. The views of these three environments are displayed from the participant's perspective

environment at the beginning of each of the three sessions (i.e., *Screen Simulation*, *Virtual Reality*, *Real Robots*). When a session starts, the 20 robots perform a *random walk with obstacle avoidance* behaviour that lasts 60 s. In this behaviour, each robot follows two rules. The first rule is to go straight with a constant velocity of 10 cm/s. The second rule is to change its direction when it encounters either a wall or a robot. In addition to performing a random walk with obstacle avoidance behaviour, each robot illuminates its LEDs in red with a certain probability. The probability is computed by an external software running on the experimenter's computer (there is a TCP communication link between the software and each robot). The software computes this probability as follows. Each 100 ms, with a probability of 0.02, the software randomly chooses a robot's identification number (each robot has a unique identification number). With a probability of 0.98, the software does not choose any robot's identification number. When the software selects an identification number, it sends a signal (i.e., a message via the TCP communication link) to the robot associated to that identification number. When a robot receives a signal, it illuminates its LEDs in red for 2 s. When the software chooses an identification number, it also makes sure to wait 2 additional seconds in order to prevent two robots from being illuminated at the same time.

3.6 Participants

For this experiment, we recruited 37 participants. These participants came from the campus population of the Université Libre de Bruxelles (no participant had a robotic background). They were between 17 and 30 years old with an average age of 23.2 years old ($SD = 3.54$). People with current or anterior cardiovascular problems could not participate to the experiment. Our participants had to read and sign an informed consent form explaining that we monitored their physiological activity during the experiment. We offered a 7 € financial incentive for their participation.

3.7 Experimental Procedure

The experiments took place at IRIDIA, the artificial intelligence laboratory of the Université Libre de Bruxelles. We started the experiment by explaining to the participant the supervision task (i.e., watch a swarm of robots attentively and press a button each time a robot in the swarm illuminates its LEDs in red). Then, we showed to the participant the three interfaces used in each session (the simulated robots displayed on a computer screen, the simulated robots displayed in a virtual reality headset and the real robots in the real environment). We allowed the participant to carefully look at the robots in each interface in order to get familiarised with each of them. Once familiarised with the three interfaces, we explained how to answer the SAM and the NASA-RTLX questionnaires. After the participant signed the consent form, we asked the participant to take a seat on a chair placed in a corner of the environment used in the *Real Robots* session. The participant stayed seated on the chair during the whole duration of the three sessions (we placed a computer screen in front of the participant prior to the *Screen Simulation* session and we attached a virtual reality headset to the participant prior to the *Virtual Reality* session). Once seated, we attached the physiological sensors to the participant's non-dominant hand. Prior to the first session, we collected the participant's baseline (i.e., physiological responses at rest) for five minutes. After these five minutes, we proceeded with the first session. After the first session, we administrated the SAM and the NASA-RTLX questionnaires to the participant. We pursued by collecting the participant's baseline for three minutes. This three minute baseline period allowed the participant to calm down and get back to their physiological activity at rest. We followed the same procedure for the second and third session. After the experiment, we explained the goal of the experiment to the participant and we answered their questions. The whole experiment's duration was 30 min per participant.

4 Data Analysis and Results

Due to the fact that our participants had to press a button during the experiment, the heart rate data became too noisy to be usefully analysed (the pulse transducer sensor is extremely sensible to small movements). We, therefore, decided not to analyse our participants' heart rate data. Out of the 37 participants, we had to remove the skin conductance data of 6 participants due to sensor misplacement. We also removed the SAM questionnaire data and the NASA-RTLX questionnaire data of 3 participants due to an error of the experimenter in the administration of the questionnaires. We finally removed the reaction time data of 4 participants due to a hardware problem with the button. We performed, therefore, our statistical analyses on 31 skin conductance data (17 female and 14 male), 34 SAM and NASA-RTLX questionnaire data (19 female and 15 male) and on 33 reaction time data (19 female and 14 male). We analysed our data with the R software by performing a repeated measure design analysis. We used the non-parametric Friedman test to determine whether the reality-gap has a significant effect on our participants' measures (i.e., skin conductance, arousal, valence, NASA-TLX and reaction time). In case of statistical significance of the Friedman test, we performed multiple Wilcoxon rank-signed tests with Bonferroni corrections to evaluate the significance of the differences between sessions. In Table 1, we summarise the results by giving the median and the Friedman's mean ranks and the inference statistics of the Friedman tests (i.e., p-values and χ^2).

Skin conductance level – The results of the Friedman test on the skin conductance level show a main effect of the reality-gap on our participants ($\chi^2(2) = 14$, $p < 0.001$). A Wilcoxon rank-signed test on the skin conductance level data highlights a statistically significant difference between the *Virtual Reality* session and the *Real Robots* session ($Z = 3.58$, $p < 0.001$) and between the *Screen Simulation* session and the *Real Robots* session ($Z = 3.68$, $p < 0.001$). The Wilcoxon rank-signed

Table 1 Descriptive statistics of the psychophysiological data, of the self-reported data and of the reaction time data. We report the median and the Friedman's mean rank (in parentheses) of the three sessions (*Screen Simulation, Virtual Reality, Real Robots*). We also report the inference statistics of the Friedman test (i.e., χ^2 and p value)

Dependent variable	n	Scree- simulation	Virtual-reality	Real-robots	χ^2	p
SCL	31	1.47 (1.71)	1.93 (1.74)	4.54 (2.54)	$\chi^2(2) = 14$	<**.001**
Arousal	34	3 (1.33)	5 (2.35)	5 (2.31)	$\chi^2(2) = 25.35$	<**0.001**
Valence	34	7 (1.9)	7 (1.9)	7 (2.14)	$\chi^2(2) = 1.38$	1
NASA-RTLX	34	18.33 (1.23)	35 (2.47)	27.5 (2.29)	$\chi^2(2) = 32.25$	<**0.001**
Reaction time	33	0.72 (1.39)	1.02 (2.6)	0.87 (2)	$\chi^2(2) = 24.24$	<**0.001**

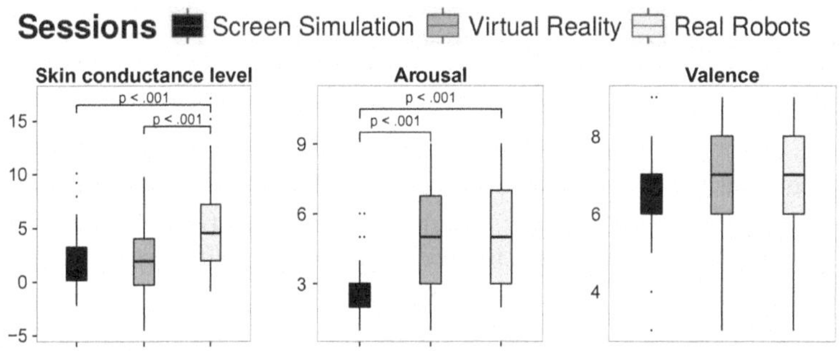

Fig. 3 Boxplots showing the skin conductance level values (left), the arousal values (middle) and the valence values (right) of all three sessions (*Screen Simulation, Virtual Reality, Real Robots*). The median value of each session is shown using the bold horizontal line in the box. Outliers are represented using dots. We also report the results of the pairwise Wilcoxon rank-signed test by connecting the boxplots of the sessions showing pairwise statistically significant differences

test does not show any statistically significant difference between the *Screen Simulation* session and the *Virtual Reality* session ($Z = 0$, $p = 1$), see Fig. 3.

SAM questionnaire – The Friedman test on the SAM questionnaire data reports a main effect of the reality-gap on our participants' arousal ($\chi^2(2) = 25.35$, $p < 0.001$). It does not show any main effect of the reality-gap on our participants' valence ($\chi^2(2) = 1.38$, $p = 1$). The Wilcoxon rank-signed test on the arousal data shows that there is a statistically significant difference between the *Screen Simulation* session and the *Real Robots* session ($Z = 4.08$, $p < 0.001$), and between the *Screen Simulation* session and the *Virtual Reality* session ($Z = 4.37$, $p < 0.001$). The Wilcoxon rank-signed test does not show any statistically significant difference between the *Virtual Reality* session and the *Real Robots* session ($Z = -0.06$, $p = 0.9$), see Fig. 3.

NASA-RTLX questionnaire – The results of the Friedman test on our participants' workload (NASA-RTLX) show a main effect of the reality-gap ($\chi^2(2) = 32.25$, $p < 0.001$). The Wilcoxon rank-signed test shows a statistically significant

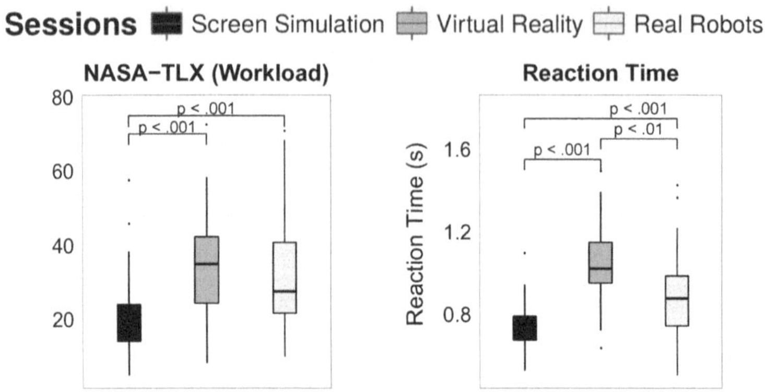

Fig. 4 Boxplots showing the workload level values (left), and the reaction time values (right) of all three sessions (*Screen Simulation*, *Virtual Reality*, *Real Robots*)

difference in the workload level of our participants between the *Screen Simulation* session and the *Real Robots* session ($Z = 4.54$, $p < 0.001$) and between the *Screen Simulation* session and the *Virtual Reality* session ($Z = 4.46$, $p < 0.001$). There was no statistical difference between the *Virtual Reality* session and the *Real Robots* session ($Z = -0.52$, $p = 0.61$), see Fig. 4.

Reaction time – Finally, the results of the Friedman test on our participants' reaction time report a main effect of the reality-gap ($\chi^2(2) = 24.24$, $p < 0.001$). The Wilcoxon signed-rank test shows a statistically significant difference between the *Real Robots* session and the *Virtual Reality* session ($Z = -2.84$, $p < 0.05$), between the *Screen Simulation* session and the *Real Robots* session ($Z = 3.42$, $p < 0.001$) and between the *Screen Simulation* session and the *Virtual Reality* session ($Z = 4.58$, $p < .001$), see Fig. 4.

5 Discussion and Conclusions

In this paper, we have shown that the reality-gap affects the psychology of humans who perform a supervision task with a robot swarm. More specifically, we have shown that the human psychophysiological state, workload and reaction time measured for the case of interaction with a real robot swarm and for the case of interaction with a simulated robot swarm displayed on a computer screen were significantly different. These results show that it is vital to take into account the reality-gap when researchers design an HSI experiment.

A solution to avoid the reality-gap effect would be to perform HSI experiments with real robots. However, real robot experiments are expensive and time consuming. Therefore, we investigated the possibility to use virtual reality as an alternative to using real robots. Our results show a difference between simulation in a virtual reality environment and simulation on a computer screen—our participants' arousal, workload and reaction time were significantly higher when they were interacting with the robot swarm in the virtual reality environment than when they were interacting with the simulated robot swarm displayed on the computer screen. These results suggest that virtual reality can mitigate the effect of the reality-gap. However, we should qualify these results because our participants' reaction time was also significantly higher when they were interacting with the robot swarm in the virtual reality environment compared to when they were interacting with the real robot swarm. Though these results do not contradict our hypothesis, we believe more research is necessary to better understand the use of virtual reality in HSI studies. In addition, we should account for the possibility that the difference of perspectives (top-view in the 2D *Screen Simulation* session and similar to the reality in the 3D *Virtual Reality* session) has also an effect on the participants' psychophysiological state, workload and reaction time. Future work should investigate whether the difference of perspectives has a significant impact on the human psychophysiological state (e.g., replicating the experiment presented in this paper by replacing the 2D top-view perspective of the *Screen Simulation* session with a 3D perspective of the robots and of the environment).

In the experimental scenario used in this paper, our participants did not issue commands to the robot swarm they interacted with. The reason behind this choice was to isolate the effect of the reality-gap—it would have been difficult to associate higher psychophysiological reactions or higher workload and reaction time to the reality-gap if our participants were, at the same time, requested to issue commands to the robot swarm. This is because the interaction interface might have some effects on our participants as well, making less clear the effect of the reality-gap. Now that we have shown the effect of the reality-gap in an experimental scenario in which human operators do not issue commands to a robot swarm, future work should focus on the effect of the reality-gap in an experimental scenario in which human operators do send commands to a robot swarm (using the results presented in this paper as a baseline).

Acknowledgements This work was partially supported by the European Research Council through the ERC Advanced Grant "E-SWARM: Engineering Swarm Intelligence Systems" (contract 246939) to Marco Dorigo. Rehan O'Grady and Marco Dorigo acknowledge support from the Belgian F.R.S.-FNRS.

References

1. Amraii, S.A., Walker, P., Lewis, M., Chakraborty, N., Sycara, K.: Explicit vs. tacit leadership in influencing the behavior of swarms. In: Proceedings of IEEE/RSJ International Conference on Robotics and Automation (ICRA), pp. 2209–2214. IEEE Press (2014)
2. Bashyal, S., Venayagamoorthy, G.: Human swarm interaction for radiation source search and localization. In: Swarm Intelligence Symposium, pp. 1–8. IEEE, St. Louis, MO, USA (2008)
3. De la Croix, J.P., Egerstedt, M.: Controllability characterizations of leader-based swarm interactions. In: AAAI Fall Symposium Series Technical Reports. AAAI Press (2012)
4. Fasola, J., Matarić, M.: A socially assistive robot exercise coach for the elderly. J. Human-Robot Interact. **2**(2), 3–32 (2013)
5. Hart, S., Staveland, L.: Development of NASA-TLX (Task Load Index): results of empirical and theoretical research. Adv. Psychol. **52**, 139–183 (1988)
6. Kidd, C., Breazeal, C.: Effect of a robot on user perceptions. In: Proceedings of the IEEE/RSJ International Conference on Intelligent Robots and System (IROS), vol. 4, pp. 3559–3564. IEEE Computer Society Press, Los Alamitos, CA (2004)
7. Kolling, A., Sycara, K., Nunnally, S., Lewis, M.: Human swarm interaction: an experimental study of two types of interaction with foraging swarms. J. Human-Robot Interact. **2**(2), 103–128 (2013)
8. Kolling, A., Walker, P., Chakraborty, N., Sycara, K., Lewis, M.: Human interaction with robot swarms: a survey. IEEE Trans. Human-Mach. Syst. **46**(1), 9–26 (2016)
9. Lang, P.J.: Behavioral treatment and bio-behavioral assessment: computer applications. In: J.B. Sidowski, J.H. Johnson, T.H. Williams (eds.) Technology in Mental Health Care Delivery Systems, pp. 119–137. Ablex (1980)
10. Leite, I., Pereira, A., Martinho, C., Paiva, A.: Are emotional robots more fun to play with? In: Proceedings of the 17th IEEE International Symposium on Robot and Human Interactive Communication (ROMAN), pp. 77–82. IEEE Press (2008)
11. Mondada, F., Bonani, M., Raemy, X., Pugh, J., Cianci, C., Klaptocz, A., Magnenat, S., Zufferey, J., Floreano, D., Martinoli, A.: The e-puck, a robot designed for education in engineering. In: Proceedings of the 9th Conference on Autonomous Robot Systems and Competitions, pp. 59–65. Instituto Politècnico de Castelo Branco, Portugal (2009)
12. Nagavalli, S., Chien, S., Lewis, M., Chakraborty, N., Sycara, K.: Bounds of neglect benevolence in input timing for human interaction with robotic swarms. In: Proceedings of ACM/IEEE International Conference on Human-Robot Interaction, pp. 197–204. ACM, New York (2015)
13. Nunnally, S., Walker, P., Lewis, M., Kolling, A., Chakraborty, N., Sycara, K.: Connectivity differences between human operators of swarms and bandwidth limitations. In: Proceedings of the Third international conference on Swarm, Evolutionary, and Memetic Computing, *Lecture Notes in Computer Science*, vol. 7677, pp. 713–720. Springer, Berlin, Germany (2012)
14. Pendleton, B., Goodrich, M.: Scalable human interaction with robotic swarms. In: Proceedings of the AIAA Infotech@Aerospace Conference, pp. 633–645. American Institute of Aeronautics and Astronautics, Va, USA (2013)
15. Pereira, A., Martinho, C., Leite, I., Paiva, A.: iCat, The chess player: the influence of embodiment in the enjoyment of a game. In: Proceedings of the 7th Iternational Joint Conference on Autonomous Agents and Multiagent Systems, pp. 1253–1256. International Foundation for Autonomous Agents and Multiagent Systems (2008)
16. Podevijn, G., O'Grady, R., Fantini-Hauwel, C., Dorigo, M.: Investigating the effect of the reality gap on the human psychophysiological state in the context of human-swarm interaction. Peer J Comput. Sci. **2**(e82) (2016)
17. Podevijn, G., O'Grady, R., Mathews, N., Gilles, A., Fantini-Hauwel, C., Dorigo, M.: Investigating the effect of increasing robot group sizes on the human psychophysiological state in the context of human-swarm interaction. Swarm Intell. **10**(3) (2016)
18. Powers, A., Kiesler, S., Fussell, S., Torrey, C.: Comparing a computer agent with a humanoid robot. In: Proceedings of the 2nd ACM/IEEE International Conference on Human-Robot Interaction (HRI), pp. 145–152. ACM, New York (2007)

19. Setter, T., Fouraker, A., Kawashima, H., Egerstedt, M.: Haptic interactions with multi-robot swarms using manipulability. J. Human-Robot Interact. **4**(1), 60–74 (2015)
20. Sweetser, P., Wyeth, P.: GameFlow: a model for evaluating player enjoyment in games. Comput. Entertain. (CIE) **3**(3), 3–3 (2005)
21. Wainer, J., Feil-Seifer, D., Shell, D., Matarić, M.: Embodiment and human-robot interaction: a task-based perspective. In: Proceedings of the 16th IEEE International Symposium on Robot and Human Interactive Communication (ROMAN), pp. 872–877. IEEE Press (2007)
22. Walker, P., Amraii, S., Lewis, M., Chakraborty, N., Sycara, K.: Human control of leader-based swarms. In: IEEE International Conference on Systems, Man, and Cybernetics (SMC), pp. 2712–2717. IEEE Press (2013)
23. Walker, P., Nunnally, S., Lewis, M., Kolling, A., Chakraborty, N., Sycara, K.: Neglect benevolence in human-swarm interaction with communication latency. In: Proceedings of the Third International Conference on Swarm, Evolutionary, and Memetic Computing, Lecture Notes in Computer Science, vol. 7677, pp. 662–669. Springer, Berlin, Germany (2012)
24. Wrobel, J., Wu, Y.H., Kerhervé, H., Kamali, L., Rigaud, A.S., Jost, C., Le Pévédic, B., Duhaut, D.: Effect of agent embodiment on the elder user enjoyment of a game. In: Proceedings of the 6th International Conference on Advances in Computer-Human Interactions (ACHI), pp. 162–167. IARIA XPS Press (2013)

Localization of Inexpensive Robots with Low-Bandwidth Sensors

Shiling Wang, Francis Colas, Ming Liu, Francesco Mondada and Stéphane Magnenat

Abstract Recent progress in electronics has allowed the construction of affordable mobile robots. This opens many new opportunities, in particular in the context of collective robotics. However, while several algorithms in this field require global localization, this capability is not yet available in low-cost robots without external electronics. In this paper, we propose a solution to this problem, using only approximate dead-reckoning and infrared sensors measuring the grayscale intensity of a known visual pattern on the ground. Our approach builds on a recursive Bayesian filter, of which we demonstrate two implementations: a dense Markov Localization and a particle-based Monte Carlo Localization. We show that both implementations allow accurate localization on a large variety of patterns, from pseudo-random black and white matrices to grayscale images. We provide a theoretical estimate and an empirical validation of the necessary traveled distance for convergence. We demonstrate the real-time localization of a Thymio II robot. These results show that our system solves the problem of absolute localization of inexpensive robots. This provides a solid base on which to build navigation or behavioral algorithms.

S. Wang (✉)
ETH Zürich, Zurich, Switzerland
e-mail: shilingwang0621@gmail.com

F. Colas
INRIA Nancy Grand Est, Nancy, France
e-mail: francis.colas@inria.fr

M. Liu
City University of Hong Kong, Hong Kong, China
e-mail: mingliu@cityu.edu.hk

F. Mondada · S. Magnenat
Mobots, LSRO, EPFL, Lausanne, Switzerland
e-mail: francesco.mondada@epfl.ch

S. Magnenat
e-mail: stephane@magnenat.net

R. Groß et al. (eds.), *Distributed Autonomous Robotic Systems*,
Springer Proceedings in Advanced Robotics 6,
https://doi.org/10.1007/978-3-319-73008-0_38

1 Introduction

Driven by consumer products, the technologies of electronics, motor and battery have made tremendous progress in the last decades. They are now widely available at prices which make affordable mobile robots a reality. This opens many new opportunities, in particular in the contexts of collective robotics.

Collective and swarm robotics focus on the scalability of systems when the number of robots increases. In that context, global localization is a challenge, whose solution depends on the environment the robots evolve in. A common approach is to measure the distance and orientation between robots [12], but this approach relies on beacons to provide absolute measurements. Yet, several state of the art algorithms require global positioning [1]. In experimental work, it is often provided by an external aid such as a visual tracker, bringing a non-scalable single point of failure into the system, and breaking the distributed aspect. Therefore, there is the need for a distributed and affordable global localization system for collective robotics experiments.

This paper answers this need by providing a distributed system using only approximate dead-reckoning and inexpensive infrared sensors measuring the grayscale intensity of the ground, without knowing the initial pose of the robots. As sensors are mounted on the robot, the localization is a local operation, which makes the approach scalable. Our solution is based on the classical Markov [6] and Monte Carlo [2] Localization frameworks that can be seen as respectively a dense and a sampling-based implementation of a recursive Bayesian filter. While these approaches are commonly used in robots with extensive sensing capabilities such as laser scanners, their implementation on extremely low bandwidth sensors is novel, and raises specific questions, such as which distance the robot must travel for the localization to converge.

In this paper, we deploy these algorithms on the Thymio II differential-wheeled mobile robot. The robot reads the intensity of the ground using two infrared sensors (Fig. 1, middle, circled red) and estimates the speed of its wheels by measuring the back electromotive force, which is less precise than encoder-based methods. For evaluation purposes, a Vicon tracking system (http://www.vicon.com/) provides the ground-truth pose (Fig. 1, right). Our first contribution is a predictive model of the necessary distance to travel to localize the robot. Our second contribution is a detailed analysis of the performances of the two implementations in comparison

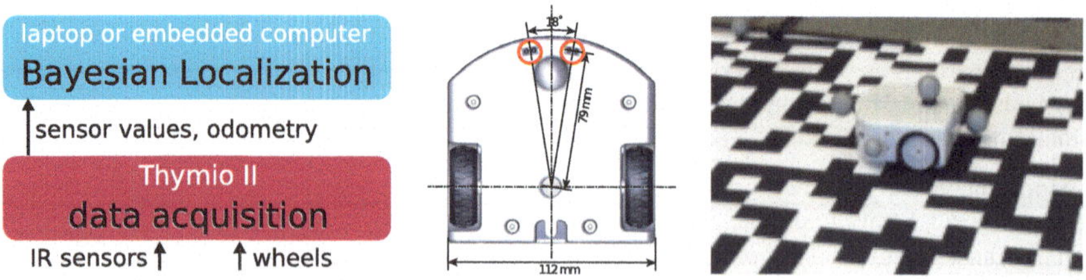

Fig. 1 The block scheme of the online system (left), the Thymio II robot with the placement of its ground sensors (middle), and markers for tracking its ground-truth pose by a Vicon system (right)

with ground-truth data. Our last contribution is an experimental validation of online, real-time localization (Fig. 1, left). The source code and all experimental data are available online[1] and the system can be seen running in a video.[2]

2 Related Work

The main challenges of solving the localization problem on affordable mobile robots are the constraints on the environment and the limited information content that inexpensive sensors can typically provide.

The work of Kurazume and Nagata [8] first raised the idea of performing inexpensive localization through the cooperative positioning of multiple robots. Prorok et al. [12] is a modern work representative of this approach. These authors have used an infrared-based range and bearing hardware along with a distributed Monte Carlo Localization approach, allowing a group of robots to localize down to a precision of 20 cm. However, this methods would require fixed robots acting as beacons to provide absolute positioning. Moreover, both radio and infrared-based range and bearing systems require complex electronics. Finally, when the cost of all robots is added, the system is far from cheap.

A cheaper approach is to use walls around an experimental arena to localize. For example, Zug et al. [13] have developed an algorithm using an array of triangulation-based infrared distance sensors. A Kalman filter algorithm is applied to localize the robot within a 2 by 1 m box. No experimental result is provided, but a simulation shows the estimated error to be within 2 cm. Dias and Ventura [3] used two horizontal lines from a VGA camera to read barcodes on the walls of an arena and localize an e-puck robot. Their system employs an Extended Kalman Filter (EKF) algorithm and reaches a precision of 1 cm and 5°. However, these systems require no obstacles between the robot and the walls, and thus are not scalable to a large number of robots.

This problem can be alleviated by detecting a known pattern visible on the ceiling and fusing this information with odometry. This approach was proven effective for localization using high-quality cameras [2]. Focusing on a low cost, Gutmann et al. [7] have developed a system using only 3 or 4 light detectors, able to localize a mobile robot in a room with an accuracy of 10 cm. However, this method requires a controlled ceiling arrangement and empty space over the robots.

Another approach is to exploit the ground for localization. Park and Hashimoto [11] proposed to localize a mobile robot over a ground equipped with randomly distributed passive RFID tags. The average localization error of this method is lower than 10 cm. However, this approach requires the ground to be equipped with tags which can become expensive and tedious to deploy when the area grows.

The system proposed in this paper builds on the idea of using information from the ground for absolute localization, with performance comparable with related work. To the best of our knowledge, existing solutions for localization of multi-robot systems

[1] https://github.com/epfl-mobots/thymio-ground-localisation.

[2] https://www.youtube.com/watch?v=70euPzixzus.

either rely on an expensive adaptation of the environment using markers or beacons, or necessitate the embedding of expensive sensing capabilities on every robot. In contrast, our system is affordable since it only requires a low number of inexpensive infrared sensors onboard the robot (which are typically used as cliff detectors), and a simple grayscale pattern on the ground that can be achieved using a printed poster.

3 Model

The generic Bayesian filter used to estimate the pose of a robot in its environment uses the following variables:

- $X_{1:t}$ 2-D pose at times $1..t$, consisting of x, y coordinates and an angle θ.
- $Z_{1:t}$ observations at times $1..t$, consisting of the output of the sensors measuring the grayscale intensity of the ground.
- $U_{1:t}$ odometry at times $1..t$, consisting of the left and right wheel speeds.

It is classically formulated as a recursive Bayesian filter with the following joint probability distribution:

$$p(X_{1:t}, Z_{1:t}, U_{1:t}) = p(Z_t|X_t)p(X_t|X_{t-1}, U_t)p(U_t)p(X_{1:t-1}, Z_{1:t-1}, U_{1:t-1}). \quad (1)$$

This filter is based on a few assumptions. First it assumes that the current observation is independent on the past observations, the past states and odometry commands conditionally to the current state. It also features the Markov assumption: the next state is independent on former states, commands and observations conditionally to the previous state and the current command. Finally, it assumes that the actions are independent from the past. All these assumptions are standard and can be found in Kalman filtering and other classical models.

This joint probability distribution allows to formulate the problem as the estimation of the pose X_t at time t given the observations $Z_{1:t}$ and the commands $U_{1:t}$:

$$p(X_t|Z_{1:t}, U_{1:t}) \propto p(Z_t|X_t) \sum_{X_{t-1}} p(X_t|X_{t-1}, U_t)p(X_{t-1}|Z_{1:t-1}, U_{1:t-1}). \quad (2)$$

This inference involves two distributions to be specified: the *observation model* $p(Z_t|X_t)$ and the *motion model* $p(X_t|X_{t-1}, U_t)$, whose parametrizations depend on the robot. In addition, we define a *self-confidence metrics* and outline the implementation.

Observation model. Our observation model $p(Z_t|X_t) = \prod_{i=0,1}^{N_{\text{sensors}}} p(Z_t^i|X_t)$ assumes all sensor noises to be independent, and the ground color to be in the range of [0, 1] (0 being black and 1 being white). We take $p(Z_t^i|X_t) \sim \mathcal{N}(v, \sigma_{\text{obs}})$, for robot pose X_t, sensor i, and a corresponding ground intensity v according to the map. The parameter σ_{obs} is selected based on the knowledge of the sensor. Thymio II has two sensors (Fig. 1, center). For binary (black and white) patterns, we chose $\sigma_{\text{obs}} = 0.5$. For grayscale images, based on measurement on a Thymio II, we set σ_{obs} to 0.15.

Motion model. Based on the model of Eliazar et al. [4], we assume that the motion has a Gaussian error model, hence $p(X_t \mid X_{t-1}, U_t) \sim \mathcal{N}(\mu_t, \Sigma_t)$. The mean μ_t is built by accumulating the estimated displacements by dead-reckoning between times $t-1$ and t. Therefore, if $\Delta x_t, \Delta y_t, \Delta \theta_t$ are the displacement between $t-1$ and t, expressed in the robot local frame at $t-1$, μ_t is:

$$\mu_t = \begin{bmatrix} x_t \\ y_t \\ \theta_t \end{bmatrix} \text{ with } \begin{aligned} \begin{bmatrix} x_t \\ y_t \end{bmatrix} &= \begin{bmatrix} x_{t-1} \\ y_{t-1} \end{bmatrix} + R(\theta_{t-1}) \begin{bmatrix} \Delta x_t \\ \Delta y_t \end{bmatrix} \\ \theta_t &= \theta_{t-1} + \Delta \theta_t \end{aligned} \tag{3}$$

where $R(\theta)$ is the 2-D rotation matrix of angle θ. The 3×3 diagonal covariance matrix Σ_t is a function of the distance traveled, the amount of rotation, and two parameters $\alpha_{\text{xy}}, \alpha_\theta$:

$$\Sigma_t = \begin{bmatrix} \sigma_{\text{xy},t}^2 & 0 & 0 \\ 0 & \sigma_{\text{xy},t}^2 & 0 \\ 0 & 0 & (\alpha_\theta |\Delta \theta_t|)^2 \end{bmatrix} \tag{4}$$

with $\sigma_{\text{xy},t} = \alpha_{\text{xy}} \sqrt{\Delta x_t^2 + \Delta y_t^2}$.

To cope with the possibility of the robot being kidnapped and therefore its pose becoming unknown, a uniform distribution with a weight p_{uniform} is added to X_t. The parameters α_{xy}, α_θ and p_{uniform} are estimated using maximum likelihood (Sect. 5).

Implementations. We compare two variants of this filter. In Markov Localization, the distributions are discretized using regular grids [6]. For our experiments, the x, y cell resolution is 1 cm and the angular resolution varies from 20° (18 discretization steps for 360°) to 5° (72 discretization steps for 360°). The estimated pose is the coordinates of the cell of maximum probability. In Monte Carlo Localization [2], the distributions are represented using samples in a particle filter. In order to extract a single pose estimate out of the many particles, we find a maximum density area in which we average the particles. It is similar to a 1-point RANSAC scheme [5]. We implemented both algorithms in Python with some Cython (http://www.cython.org) procedures used for time-critical inner loops. The algorithms run on an embedded computer or laptop (Fig. 1, left). Thymio II is programmed through the ASEBA framework [10], which connects to Python using D-Bus.

Self confidence. We define a self-confidence term that corresponds to the ratio of the probability mass of $p(X_t)$ that is within a distance d_{xy} and an angle difference d_θ to the estimated pose. In our experiments, we use $d_{\text{xy}} = 3$ cm and $d_\theta = 10°$.

4 Theoretical Analysis of Convergence

One can estimate the time required for the robot to localize itself in a given space, by comparing the information needed and the information gained while traveling.

Information need. For the Markov Localization approach, there is a given number of discrete cells. The amount of information needed to unambiguously specify one among them all is $\mathrm{H}_{\mathrm{loc}} = \log_2(N_{\mathrm{cells}})$, with N_{cells} the number of cells.

Information gain. We can estimate the information gain at each time step. Let us consider the case of a binary map and a binary sensor. This sensor ideally yields 1 bit of information per measurement. In practice, there is a loss in information due to the sensor noise, characterized by the p_{correct} probability of the sensor to be correct:

$$\mathrm{H}_{\mathrm{noise}} = \mathrm{H}_{\mathrm{b}}(1 - p_{\mathrm{correct}}), \tag{5}$$

where H_{b} is the binary entropy function: $\mathrm{H}_{\mathrm{b}}(p) = -p\log_2(p) - (1-p)\log_2(1-p)$.

We also need to take into account that the sensor measurements are not completely independent. For example, when the robot is not moving, it always observes the same place and thus cannot really gain additional information besides being sure of the intensity of the current pixel. In a discretized world, we thus need to estimate the probability of having changed cell in order to observe something new, which depends on the distance traveled and the size of the cells. This problem is equivalent to the Buffon-Laplace needle problem of finding the probability for a needle thrown randomly on a grid to actually intersect the grid[3] [9]. In our case, the probability of changing cell is given by:

$$p_{\mathrm{diff}} = \frac{4dh - d^2}{\pi h^2}, \tag{6}$$

with d the distance traveled and h the size of the cells.

We can then compute the conditional entropy for two successive ideal binary measurements O_{t-1} and O_t separated by d based on the conditional probability. There are two cases: either the robot has not moved enough to change cell (with probability $1 - p_{\mathrm{diff}}$) and the new observation is the same as the old, or the robot has changed cell (with probability p_{diff}) and the new observation has the same probability to be the same or the opposite of the old. This can be summarized by the following conditional probability distribution (b/w = black/white):

$$p(O_t = \{\mathrm{b/w}\} \mid O_{t-1} = \{\mathrm{b/w}\}) = \begin{pmatrix} (1 - p_{\mathrm{diff}}) + p_{\mathrm{diff}}/2 & p_{\mathrm{diff}}/2 \\ p_{\mathrm{diff}}/2 & (1 - p_{\mathrm{diff}}) + p_{\mathrm{diff}}/2 \end{pmatrix} \tag{7}$$

After rearranging the terms of the conditional entropy, the loss of information due to redundancy in the traveled distance is:

[3]The needle is the segment joining the start and end points of the robot movement and the grid is the borders of the cells.

$$\mathrm{H}_{\mathrm{loss,d}} = 1 - \mathrm{H}_{\mathrm{b}}(p_{\mathrm{diff}}/2). \tag{8}$$

There is also redundancy between several sensors placed on the same robot. The probability that they see the same cell based on the distance between them is exactly the same as the probability of a sensor to see the same cell after a displacement of the same distance. The information loss due to the redundancy from the sensor placement is noted $\mathrm{H}_{\mathrm{sensors}}$ and follows the same formula as $\mathrm{H}_{\mathrm{loss,d}}$, but with d being the distance between the two sensors in Eq. 6.

Finally, we can approximate the information that our robot gathers at each time step by assuming that the trajectory of the robot does not loop (no redundancy between distant time steps) and that the trajectories of the sensors are independent. The information gathered in a time step is then:

$$\begin{aligned}\mathrm{H}(O_t^1, O_t^2 \mid O_{1:t-1}^1, O_{1:t-1}^2) &= \mathrm{H}(O_t^1, O_t^2 \mid O_{t-1}^1, O_{t-1}^2)\\ &= \mathrm{H}(O_t^1 \mid O_{t-1}^1) + \mathrm{H}(O_t^2 \mid O_{t-1}^2) - \mathrm{H}(O_t^2 \mid O_t^1)\\ &= 2 \cdot (1 - \mathrm{H}_{\mathrm{noise}} - \mathrm{H}_{\mathrm{loss,d}}) - \mathrm{H}_{\mathrm{sensors}}\end{aligned} \tag{9}$$

This formula also ignores the uncertainty in the robot motion. With these assumptions, it is an upper bound on the average information gain.

Outlook. Faster localization can be achieved by moving at a greater speed to reduce redundancy in the successive measurements, with proportional increase in sampling frequency. If designing a new robot, better sensors would reduce cross-over noise. Setting the sensors apart would also reduce the redundancy between their information but, for our specific grid size, they are sufficiently separated in Thymio II.

5 Empirical Analysis of Performance

To evaluate the performance of our localization algorithms, we remotely controlled the robot and recorded datasets covering all possible robot motions:

- `trajectory 1` and `2`: The robot alternates forward and backward movements with rotations on spot, at a speed of 3–5 cm/s.
- `linear trajectory`: The robot simply goes straight ahead along the x-axis of the map, at a speed of 3–5 cm/s.
- `trajectory with kidnapping`: The robot alternates phases of forward movement, at a speed of 15 cm/s, and turning on spot. To test the algorithm's ability of recovering from catastrophic localization failures, we perform "robot kidnapping" by relocating the robot to another part of the map every minute.

The robot moves on a 150 × 150 cm ground pattern containing 50 × 50 cells of 3 × 3 cm, each randomly black or white (Fig. 1, right). The robot is connected to ROS to synchronize its sensor values and odometry information with ground-truth data from Vicon, sampled at a period of 0.3 s. We chose this period so that with basic trajectories, at maximum speed the robot travels approximately half the length of one cell between every sample.

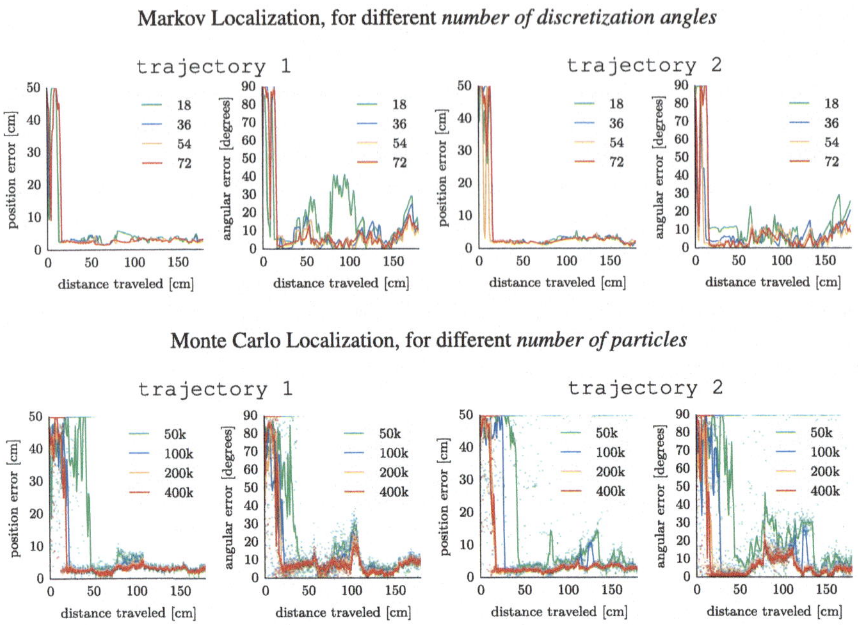

Fig. 2 The error between the estimation by the localization algorithm and the ground truth on `trajectory 1` and `trajectory 2`. For Monte Carlo Localization, the solid lines show the average over 10 trials, while the light dots show the individual trials

Parameter estimation. We estimated the noise parameters of the motion model using maximum likelihood, considering the error between the ground truth and the odometry data in the local frame between two time steps. Using `trajectory 1` and `trajectory 2`, we found the values for α_{xy} and α_θ to be in the order of 0.1. Similarly, using `trajectory with kidnapping`, we also found the value for p_{uniform} to be in the order of 0.1.

Basic localization. Figure 2 shows the error in position and orientation for the first two trajectories. In these plots, *distance traveled* represents the cumulative distance traveled by the center point between the two ground sensors of the robot.

For the Markov Localization approach, all discretization resolutions allow the robot to localize down to a precision of 3 cm and 5°. However, in `trajectory 1`, the resolution of 18 discretization steps is not enough to keep tracking the orientation at a distance traveled of 50 and 80 cm. These both correspond to the robot rotating on spot. Finer discretizations do not exhibit this problem, they are more robust and have similar precision. Therefore, we see that an angular discretization of 36 (10° resolution) is sufficient to provide accurate tracking. In `trajectory 2`, we see that an angular discretization of 54 allows for a better angular precision than 36, but 72 does not improve over 54. All discretizations provide equal position precision.

For the Monte Carlo Localization approach, we see that on `trajectory 1`, the robot localizes already with 50 k particles, but in twice the distance it takes with

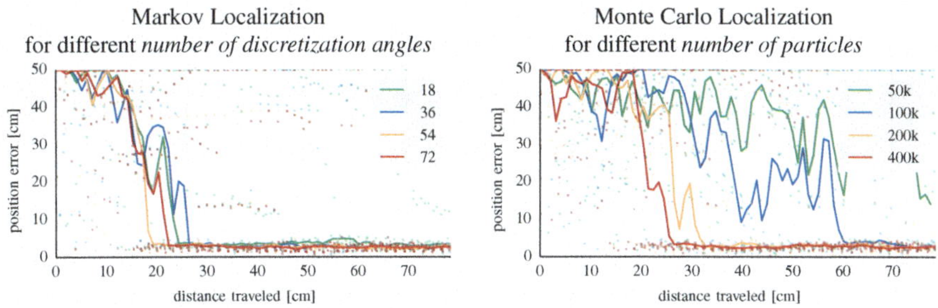

Fig. 3 The error between the estimation by the localization algorithm and the ground truth on 10 segments from the first two trajectories. The solid lines show the median while the light dots show the individual trials

100 k particles. Increasing the number of particles beyond this value only marginally decreases localization time. While 50 k particles are sufficient to localize on this run on average, in some trials, the robot loses orientation when it turns on spot. On `trajectory 2`, using 50 k particles is not enough to localize the robot. Increasing this number to 100 k leads to a good localization, except after the robot has traveled 130 cm. This corresponds to a long moment during which the robot rotates on spot, leading to less information acquisition, and therefore degraded performances.

Overall, both approaches have similar accuracy. When angular precision is critical, the Monte Carlo Localization approach might achieve better performance, as the Markov Localization approach is limited in precision by its angular discretization.

Distance to converge. Figure 3 shows the error in position for 5 different starting points in each of the first two trajectories, in which the robot moves at a speed of 3–5 cm/s. We see that, with the Markov Localization approach, the correct pose is found after about 20 cm. There are also outliers: this happens when the robot is turning on spot, in which case there is not enough information to localize the robot. With 400 k particles, the Monte Carlo Localization approach also converges after about 20 cm. Decreasing the number of particles quickly increases the distance needed for convergence, reaching 60 cm for 100 k particles. Using only 50 k particles, some trajectory segments fail to converge within 80 cm length.

It is interesting to compare these distances with theoretical estimates (see Sect. 4). One difficulty is that the theoretical model specifies the probability p_{correct} of the sensor to be correct. In our observation model, instead, we specify σ_{obs} the noise of the sensor. Therefore, we propose to compute p_{correct} from σ_{obs} by considering a sensor positioned over a black ground, and by assuming that all values below a threshold of 0.5 (0 being black and 1 being white) are correctly read:

$$p_{\text{correct}} = p(Z < 0.5 | X = 0) = \int_{-\infty}^{0.5} \mathcal{N}(0, \sigma_{\text{obs}}) \tag{10}$$

In these runs, we have assumed $\sigma_{\text{obs}} = 0.5$, leading to $p_{\text{correct}} = 0.84$ (Eq. 10) and $H_{\text{noise}} = 0.63$ (Eq. 5). Our robot moves at 3 cm/s with a time step of 0.3 s on a grid of

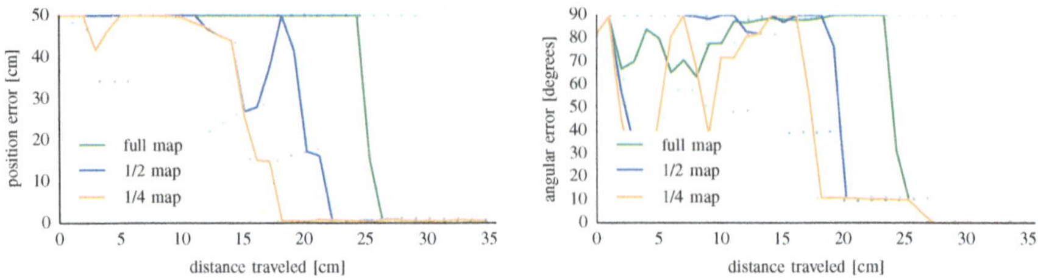

Fig. 4 The error between the estimation by the localization algorithm and the ground truth, using Markov Localization, for different map sizes on `linear trajectory`. The solid lines show the median over 3 different trajectory parts, while the light dots show the individual parts

3 cm $\times$3 cm cells in which the color is known to be similar; this yields $H_{loss,d} = 0.33$ (Eq. 8). Moreover, the sensors are 2.2 cm apart, which yields $H_{sensors} = 0.041$. As such the robot gathers on average at most 0.036 bit per time step, or 0.040 bit/cm (Eq. 9). With a 150 cm $\times$ 150 cm environment discretized with cells of 1 cm and 5° angle, the amount of information needed for the localization is $H_{loc} = 20.6$. This means that, on average, the robot should not be localized before traveling around 520 cm.[4] This distance is far larger than the observed one of about 20 cm. The reason is that using $\sigma_{obs} = 0.5$ for the binary case was an arbitrary choice and therefore the value of $p_{correct}$ does not correspond to the reality.

Computing the value of $p_{correct}$ to match the observed convergence distance by inverting the computation, we find 0.97, corresponding to $\sigma_{obs} = 0.26$. If we consider $\sigma_{obs} = 0.15$ as measured on grayscale images, then $p_{correct} = 0.99957$. This leads to a minimum theoretical localization distance of about 14.3 cm. This value is slightly lower than our experimental results, which makes sense as it is a lower bound. Moreover, this lower bound would be attained with a perfect filter observing a perfect pattern. Our filter is not perfect, because we have run it with an overestimated σ_{obs} of 0.5. Nevertheless, the filter works well, showing that it degrades gracefully with respect to imprecision in its parameters. This is important in practice, as users might not be able to provide extremely precise parameters.

Effect of map size. Figure 4 shows the effect of reducing the map size. The robot runs linearly on one quarter of the map, while the Markov Localization is performed on the whole map, half of it, and a quarter of it. We see that reducing the map size does reduce the distance traveled necessary to converge, in accordance to the theory, because less information have to be acquired to reduce the uncertainty of the pose.

Robot kidnapping. Figure 5 shows the error in position, orientation and the self confidence for the run with kidnapping. In this run, the robot is kidnapped twice, after having traveled 550 and 1000 cm. It takes the robot approximately 100 cm to re-localize, and does so successfully with both Markov and Monte Carlo Localization approaches. This difference of distance with previous runs is mostly due to the speed of the robot, which is about 5 times faster. With the Markov Localization approach,

[4] see https://github.com/epfl-mobots/thymio-ground-localisation/blob/master/theory/d_conv_from_pcorrect.py

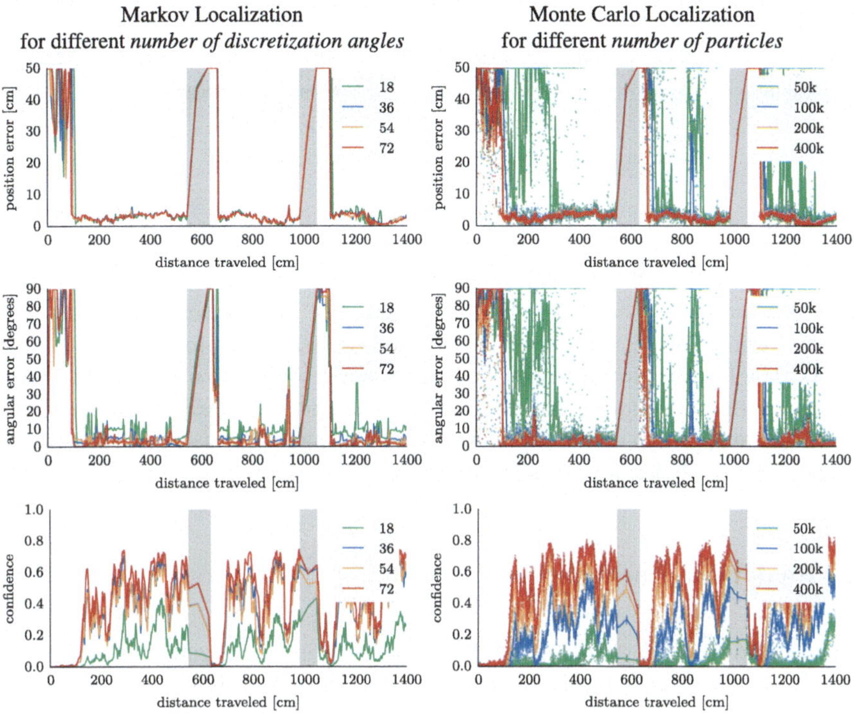

Fig. 5 The error between the estimation by the localization algorithm and the ground truth, and the self confidence of the algorithm, on the run with kidnapping. For Monte Carlo Localization, the solid lines show the median over 10 trials, while the light dots show the individual trials. The gray areas show the time during the robot is being held in the air at the occasion of kidnapping

all discretization resolutions are approximately equivalent in position performance, but 18 leads to a lower orientation precision, as well as to a lower self confidence, due to the large discretization step. Other resolutions have a confidence of about 0.5 when the robot is localized, and this value drops below 0.1 after kidnapping, clearly showing that the algorithm is able to assess its status of being lost.

With the Monte Carlo Localization approach, the robot localizes most of the time with 100 k particles, and always with 200 k particles or more. With 50 k particles, the robot eventually localizes, but this might take more than 2 m of traveled distance. We see that the self confidence increases with more particles, and, similarly to the Markov Localization approach, drops after kidnapping. This confirms the effectiveness of the self confidence measure.

Computational cost. Table 1 shows the execution duration of one step for the two algorithms with different parameters. These data were measured on a Lenovo laptop T450s with an Intel Core i7 5600U processor running at 2.6 GHz, and are averages over the two first trajectories. We see that with the Markov Localization approach, the duration scales linearly with the number of discretization angles. With the Monte Carlo Localization approach, the scaling is linear but amortized (50 k particles is

Table 1 The execution duration of one step for the two algorithms

Duration (s)	**0.64**	**1.43**	**2.15**	**2.94**	**1.97**	**2.97**	**6.08**	**12.22**
Algorithm	Markov localization				Monte Carlo localization			
Parameter	18	36	54	72	50 k	100 k	200 k	400 k
	Discretization angles				particles			

not twice faster as 100 k). This is due to the selection of pose estimate, which uses a RANSAC approach and is therefore independent of the number of particles. For similar localization accuracy, the Monte Carlo Localization approach is slower than the Markov Localization approach, and therefore we suggest to use the former with an angular discretization of 36 in practical applications. However, the Monte Carlo Localization approach might be preferred if a high angular precision is required, but at least 100 k particles are necessary for proper localization.

6 Real-Time Localization

Figure 6 shows the performance of the online, real-time localization of a Wireless Thymio II (Fig. 1, left). We tested different grayscale images of 59 × 42 cm, printed on A2 paper sheets with a resolution of 1 pixel/cm. We used Markov Localization with an angular discretization of 36 and a spatial resolution of 1 cm; the localization is performed every 0.4 s. The localization algorithm runs on a Lenovo laptop X1 3rd generation with an Intel Core i7-5500U processor running at 2.4 GHz, at a CPU load of approximately 80%. The algorithm localizes the robot globally on all images. The images with more contrast lead to a more robust and faster localization, while the ones with lower contrast lead to more imprecise trajectories. In Fig. 6, we see that the parts of trajectories in red are less precise than the ones in green. This shows that the self confidence measure is effective in assessing the quality of the localization.

Embedded use. The runtime cost is dominated by the motion model, as it involves convoluting a small window for every cell of the probability space. With the current parameters, this leads to approximately 250 floating-point multiplications per cell. For comparison, VideoCore® IV, the built-in GPU in the Raspberry Pi 3 (costing €33), has a peak processing power of 28 GFLOPS. Assuming that, due to limitations in bandwidth and dependencies between data, it can only be used at 10% of its peak floating point throughput, it could run a GPU version of our algorithm at 10 Hz on a map of 1.8 by 1.8 m with a resolution of 1 cm. Therefore, our approach can truely be used for distributed collective experiments with affordable robots.

Scalability. The computational speed of the system and the necessary distance to converge scale linearly with the size of the localized area. Also, a prolonged use of the robot might wear off the ground, slightly shifting the grayscale intensities and jeopardizing the localization. Nevertheless, modern printing options allow to use strong supports that only degrade slowly with time. Therefore, our system can be helpful for conducting collective experiments in a laboratory environment.

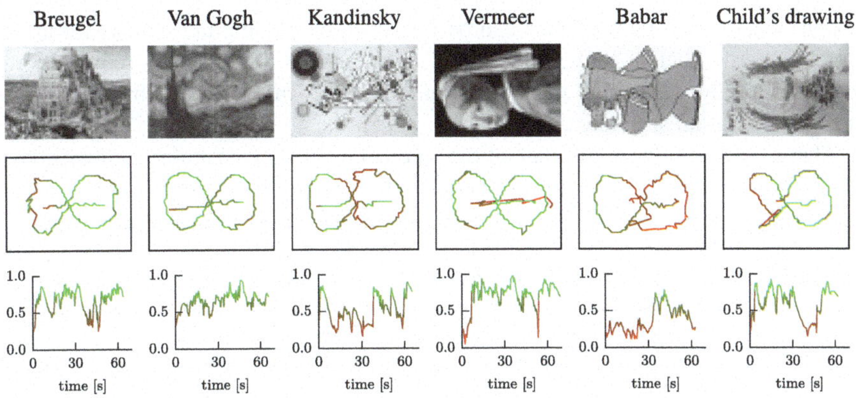

Fig. 6 The image, estimated trajectory and self confidence for different 59 × 42 cm grayscale patterns printed on A2 papers with a resolution of 1 pixel/cm. The robot is remotely controlled to draw a 8-shape figure, with similar motor commands for each image. Markov Localization with 36 discretization angles used. The color of the lines vary from red (confidence 0) to green (confidence 1). Trajectories are plotted starting from a confidence level of 0.2

Optimality. An interesting question is what is an optimal pattern. However, this question has two sides. On the one hand, a pattern could be optimal in term of quantity of information, allowing the robot to localize as fast as possible. In that case, a white-noise pattern would be ideal. On the other hand, if the robot has to interact with humans, or only be monitored by humans, such a pattern would be highly disturbing. Our experimental results with paintings and drawings show that, even with a pattern having a lot of regularities and symmetries, our system is still able to localize. Therefore, we believe it is usable with natural patterns such as photographies, drawings and grayscale schematics.

7 Conclusion

In this paper, we have implemented and empirically evaluated Markov- and Monte Carlo-based approaches for localizing mobile robots on a known ground visual pattern. Each robot requires only inexpensive infrared sensors and approximate odometry information. We have shown that both approaches allow successful localization without knowing the initial pose of the robot, and that their performances and computational requirements are of a similar order of magnitude. Real-time localization was successful with a large variety of A2 grayscale images, using the Markov Localization approach with 36 discretization steps for angle. Should larger patterns be desired, the code could be further optimized by implementing it in the GPU, allowing it to run on inexpensive boards such as the Raspberry Pi.

In addition, we have outlined, and empirically validated, a method to estimate the localization performance in function of the sensor configuration. This method provides a guide for taking decisions about the placement of sensors in a future robot design: localization performance can be improved by placing the sensors far apart on a line perpendicular to the direction of movement of the robot; moreover, more sensors allow for collecting more information, if they are separated by the size of the smallest visual structure in the map.

These contributions to the state of the art enable absolute positioning of inexpensive mobile robots costing in the €100 range. In the context of distributed collective robotics, they provide a solid base to build navigation or behavioral algorithms.

Acknowledgements The authors thank Emmanuel Eckard and Mordechai Ben-Ari for insightful comments on the manuscript and Ramiz Morina for his drawings. This research was supported by the Swiss National Center of Competence in Research "Robotics".

References

1. Breitenmoser, A., Martinoli, A.: On combining multi-robot coverage and reciprocal collision avoidance. In: Distributed Autonomous Robotic Systems, pp. 49–64. Springer (2016)
2. Dellaert, F., Fox, D., Burgard, W., Thrun, S.: Monte Carlo localization for mobile robots. In: International Conference on Robotics and Automation (ICRA), pp. 1322–1328. IEEE Press (1999)
3. Dias, D., Ventura, R.: Absolute localization for low capability robots in structured environments using barcode landmarks. J. Autom. Mob. Robot. Intell. Syst. **7**(1), 28–34 (2013)
4. Eliazar, A.I., Parr, R.: Learning probabilistic motion models for mobile robots. In: Twenty-first International Conference on Machine Learning, pp. 32–39. ACM (2004). https://doi.org/10.1145/1015330.1015413
5. Fischler, M.A., Bolles, R.C.: Random sample consensus: a paradigm for model fitting with applications to image analysis and automated cartography. Commun. ACM **24**(6), 381–395 (1981). https://doi.org/10.1145/358669.358692
6. Fox, D., Burgard, W., Thrun, S.: Markov localization for mobile robots in dynamic environments. J. Artif. Intell. Res. **11**, 391–427 (1999). https://doi.org/10.1613/jair.616
7. Gutmann, J.S., Fong, P., Chiu, L., Munich, M.: Challenges of designing a low-cost indoor localization system using active beacons. In: International Conference on Technologies for Practical Robot Applications (TePRA), pp. 1–6. IEEE Press (2013). https://doi.org/10.1109/TePRA.2013.6556348
8. Kurazume, R.Y., Nagata, S., Hirose, S.: Cooperative positioning with multiple robots. In: International Conference on Robotics and Automation (ICRA), pp. 1250–1257. IEEE Press (1994)
9. Laplace, P.S.: Théorie analytique des probabilités, 3rd rev. Veuve Courcier, Paris (1820)
10. Magnenat, S., Rétornaz, P., Bonani, M., Longchamp, V., Mondada, F.: ASEBA: a modular architecture for event-based control of complex robots. IEEE/ASME Trans. Mechatron. **16**(2), 321–329 (2010). https://doi.org/10.1109/TMECH.2010.2042722
11. Park, S., Hashimoto, S.: An approach for mobile robot navigation under randomly distributed passive rfid environment. In: International Conference on Mechatronics (ICM), pp. 1–6. IEEE Press (2009)
12. Prorok, A., Bahr, A., Martinoli, A.: Low-cost collaborative localization for large-scale multi-robot systems. In: International Conference on Robotics and Automation (ICRA), pp. 4236–4241. IEEE Press (2012)
13. Zug, S., Steup, C., Dietrich, A., Brezhnyev, K.: Design and implementation of a small size robot localization system. In: International Symposium on Robotic and Sensors Environments (ROSE), pp. 25–30. IEEE Press (2011)

Modelling Mood in Co-operative Emotional Agents

Joe Collenette, Katie Atkinson, Daan Bloembergen and Karl Tuyls

Abstract Simulating emotions based on psychological models has been a topic where work has focused on social dilemmas using simulated emotions to inform decision making within artificial agents. However human decision making is affected not only by emotions but also by other aspects of people's temperament: the mood of the person also affects their decision making, in conjunction with other factors such as inequity aversion. We propose a simulated model of mood, which is formed and validated through psychological research. We use this to inform decision making in conjunction with simulated emotions to improve the decision making within agents compared to emotions alone. We empirically evaluate our simulated model of mood in addition to emotions. We show that our mood model can be implemented in a robotic setting which can clarify aspects of multi-agent systems, such as cooperation within an agent society.

1 Introduction

We have developed an understanding of how simulated emotions and mood can be used to inform decision making in agents so as to avoid expensive computation. We propose a functional model of mood that can be used independently or in conjunction with the current work on simulated emotions. We show that our model of simulated mood can be used to allow cooperation in a social dilemma to be achieved through

J. Collenette · K. Atkinson (✉) · D. Bloembergen · K. Tuyls
University of Liverpool, Liverpool, UK
e-mail: k.m.atkinson@liverpool.ac.uk

J. Collenette
e-mail: j.m.collenette@liverpool.ac.uk

D. Bloembergen
e-mail: d.bloembergen@liverpool.ac.uk

K. Tuyls
e-mail: k.tuyls@liverpool.ac.uk

R. Groß et al. (eds.), *Distributed Autonomous Robotic Systems*,
Springer Proceedings in Advanced Robotics 6,
https://doi.org/10.1007/978-3-319-73008-0_39

choice in a multi-agent setting. With the addition of simulated mood, there is an improvement when compared to using simulated emotions exclusively. Our mood model is grounded in psychology research, using aspects of human emotions and mood to inform decision making [10, 13].

We use this developed mood model in practise to explore how cooperation flourishes within a society of agents. The resilience of cooperation growth is tested by the addition of defectors, indicating the stability of the cooperation strategy that uses our model.

Psychology research has shown emotions affect human decision making [21]. Recent work has shown that simulating these emotions within artificial agents affects the evolution of cooperation within the prisoner's dilemma game [15]. Similarly, psychology shows that mood affects decision making in humans [10]. There is a clear distinction between mood and emotion, emotions are short-term feelings that are directed towards a particular object or person [13]. Mood in contrast is a long term feeling which does not have this focus on a particular object or person [8].

Previous research has focused on simulating emotions within agents without regards for the effects that mood will have on the decision making process. [22] gives an overview of the different methods of integrating emotions into a computational model, however there has been no previous attempts to model mood. Our model for mood is integrated into previous research using a psychological background to justify the model. Whilst we recognise that emotions and mood both have physiological affects [11], we will only be considering the functional aspect where mood and emotions change the behaviour of the agents [13, 17].

We aim to provide a generic framework of mood that can be integrated into existing emotional models, which in turn provides a deeper level within the decision making captured. Within this framework interactions between agents can occur yet the agents do not need to know each others' strategies in order for cooperation to flourish. Our model of mood is grounded in psychology research. We have shown how this mood model reacts to an unknown strategy, which in our experiments is pure defection.

2 Background

We will first introduce the emotional model we will be using as part of our mood model and our experiments. Then we continue with the prisoner's dilemma game which is the setting for our experiment.

2.1 Emotional Characteristics

The simulated emotions that will be implemented in our agents are based on the Ortony, Clore and Collins model of emotions, known as the OCC model [18]. The model was developed through psychology research and has been used throughout the

Table 1 Emotional characters used

Anger threshold	Gratitude threshold	Character
1	1	Responsive
1	2	Active
1	3	Distrustful
2	1	Accepting
2	2	Impartial
2	3	Non-accepting
3	1	Trustful
3	2	Passive
3	3	Stubborn

AI community [1, 4, 15, 19]. The OCC model takes a functional view of emotions, in which emotions influence changes in behaviour. The action taken is a result of the emotional makeup of the person. The emotional makeup is a result of previous outcomes. This functional view lends itself to being a good platform for implementing emotions as the descriptions are of the outward effects of the emotions rather than how emotions are processed internally. Of the 22 emotions defined in the OCC model we will be modelling $anger$, $gratitude$ and $admiration$, so we can compare to previous work [4].

As in [4] each emotion has a threshold and a value. When that value increases past the threshold for $anger$ or $gratitude$ the action of the agent will change. When $admiration$ reaches the threshold then that agent will imitate the emotional characteristic of the agent that triggered the admiration. This is how replication is implemented in our experiment. In this paper we will be using 9 different types of emotional characters who have differing emotional makeups but they all have admiration thresholds of 3. We have chosen that value based on previous work as it gives the highest payoff in [4]. The different characteristics can been seen in Table 1.

An agent's anger increases by one when its opponent defects; gratitude increases when the opponent cooperates. For example take the two characteristics Responsive and Active. If Responsive chooses to cooperate, Active's gratitude increases to one, if Active chose to Defect then Responsive's anger increases to one. Responsive's anger level is at the anger threshold, so in the next game with that agent, Responsive will choose to defect and the anger level will return to 0.

Admiration increases when the agent believes that its opponent is performing better than itself. When a threshold is reached, the agent's behaviour changes to the emotional character that triggered the admiration emotion and the admiration value is then reset back to 0. When a mobile agent completes five games of the prisoner's dilemma, after that, the mobile agent will request the average payoff per game of its next opponent, before the game has started, and compares this value to its own average payoff. The agent will increase its admiration value towards whoever has the highest average, this will be either itself or its opponent.

Table 2 Payoff matrix of the prisoner's dilemma

	$COOP_i$	$DEFECT_i$
$COOP_j$	$3_j, 3_i$	$0_j, 5_i$
$DEFECT_j$	$5_j, 0_i$	$1_j, 1_i$

We are using average payoff, rather than total payoff which was used by [14], because we cannot be sure that each mobile agent has engaged in the same number of games as its opponent. When the admiration threshold has been reached, the agent takes on the emotional characteristics of the agent that triggered the threshold, which may be itself, so the agent will then respond to other opponents in the same way as the agent who triggered the admiration threshold. Then the admiration threshold is reset to zero. Finally, the agent plays the game with its opponent.

2.2 *Prisoner's Dilemma*

The prisoner's dilemma is a social dilemma where two players are given the choice of cooperation or defection. This choice is made simultaneously with no communication prior to the decision made. Each player then will get a payoff according to the choices made by both players. The payoff matrix is shown in Table 2.

When looking at the prisoner's dilemma outcomes, it seems in the best interest of both players to both play cooperatively since this would lead to the largest total payoff for the group as a whole. However, there is a temptation to defect as this can lead to a higher individual payoff. When both players reason this way, this then leads to the Nash equilibrium of *(DEFECT, DEFECT)*, which gives the worst outcome for the group as a whole. This highlights the dilemma of the game. Investigating methods by which self-interested agents can be incentivised to cooperate in the prisoner's dilemma has been an active area of research in the past decades, with a particular focus on the evolution of cooperation within groups of agents [2, 3, 20]. It is for this reason that we adopt this model of interaction in the current work as well.

3 Mood Model

Here we define our model of mood, with justifications for each mood state from psychological research and how this mood will affect decision making. We split the mood into three parts: negative, neutral, and positive. Our mood model only affects the decision made as we interested in what decisions are made rather than simulating how mood can affect the agent physically.

It was shown in [10, 21] that negative moods can lead to a more rational outcome in general as people tend to think more thoroughly about the action they will take. In our experiments we use low moods to lead to defection, as this is the Nash equilibrium and can be considered the more rational decision. Very low mood levels will lead to defection regardless of the emotional state of the agents.

Positive moods tend towards an ideal outcome even if that affects themselves negatively [10]. In our experiment the riskiest behaviour is cooperation as it can lead to the worst outcome for the individual agent. Cooperation is the most ideal outcome as it gives the highest payoff for the group as a whole.

For neutral moods the mood model will not affect the agents decision making. The mood will affect how agents react to unknown opponents since they do not have any emotional attachment to them. When the mood levels are extreme they will override the current emotional decision. We have done this to represent that mood levels in humans do not necessarily reflect cooperation as a whole, but affect the choice made [16].

We define the representation of each mood state as follows: a mood of below 10 is characterised as extremely low, below 30 as low, higher than 70 as high and above 90 as extremely high, and between 30 and 70 as neutral. Equation 1 shows how the agent chooses an action based on our mood model with the simulated emotions. We define an initial action, as the action an agent would take if the mood model is unable to provide an action, this is often the first interaction the agent makes. The simulated emotions in our model are defined as one of the emotional characters as described in Sect. 2.1. How and when an interaction occurs in our experiment is given in Sect. 4.

Definition 1 Let Ag be the set of all agents, with i and $j \in Ag$. Let t denote time. Let m_i^t return the mood of agent i at time t, in the range $]0, 100[$. Let $\eta_{i,j}$ return the number of interactions agent i as with agent j. Let I_i return the initial action of agent i. Let $E_{i,j}^t$ return the action that agent i would take against agent j based on i's simulated emotions, at time t.

$$Ac_{i,j}^t = \begin{cases} COOP, & \text{If } m_i^t > 90 \text{ or } (m_i^t > 70 \text{ and } \eta_{i,j} = 0) \\ DEFECT, & \text{If } m_i^t < 10 \text{ or } (m_i^t < 30 \text{ and } \eta_{i,j} = 0) \\ E_{i,j}^t, & \text{If } (30 <= m_i^t >= 70 \text{ and } \eta_{i,j} \neq 0 \\ I_i, & \text{Otherwise} \end{cases} \quad (1)$$

Our representation of positive mood values comes from psychology literature showing how people take riskier behaviour to achieve a more ideal outcome [10]. However if the mood is too positive, as it is when a person has mania, then the behaviour becomes extremely likely to hurt that person [12]. In [10, 21] it is shown that negative moods can be more likely to lead people to make a more logical and thought out choice. Research into human patients with depression shows that these people are more likely to choose defection. The research also showed that depressed patients were more critical of themselves [9]. This provides up with grounding for our choice of defection as part of our implementation of the mood model in the prisoner's dilemma, and validates how the mood values are more greatly affected when the mood is low.

The agent's mood value will go up or down based on the difference between the payoff received and their average payoff, as this represents how well the agent thinks they have done in that game [5]. Then additionally the mood value will go up or down based on how the agent feels towards inequity between the average payoffs. We will be using the inequity aversion model *Homo Egualis* to represent inequity as a value [5]. In this model we need to find an α and β, where α represents how much an agent cares when they are doing badly and β represents how much an agent cares when their opponent is doing badly. Since we want to represent an ideal solution we will take $\alpha = \beta$. This represents that an agent cares about an opponent as much as it cares about itself.

The amount the agent cares is represented by applying the mood to our α value, such that higher moods give a lower α. This results in mood changes being larger when the mood is low. If the mood is low then the agent "thinks" that they are doing poorly in the environment when compared to other agents. We do this to represent the property that humans care more about equality when doing poorly in society [5].

Definition 2 Let Ag be the set of all agents, with i and $j \in Ag$. Let t denote time. Let p_i^t return the payoff of agent i at time t. Let m_i^t return the mood of agent i at time t, in the range $]0, 100[$. Let μ_i^t denote the average payoff for agent i up to time t. Let F_i^t return the opponent of agent i at time t.

$$\alpha_i^t = (100 - m_i^{t-1})/100 \tag{2}$$

$$\Omega_{i,j}^t = \mu_i^t - \alpha_i^t \cdot \max(\mu_j^t - \mu_i^t, 0) - \alpha_i^t \cdot \max(\mu_i^t - \mu_j^t, 0) \tag{3}$$

$$m_i^t = m_i^{t-1} + (p_i^t - \mu_i^{t-1}) + \Omega_{i,j}^{t-1} \text{ where } j = F_i^t \tag{4}$$

In Eq. 2 we show how we get our α value from the current mood of an agent; this places the mood value in the range of $]0, 1[$ so it can be used as the α. For example a mood value of 75 will return an α of 0.25. Equation 3 is the simplified version of the Homo Egualis function [7], as we have only two agents in a single interaction and $\alpha = \beta$. The equation gives us a numerical representation of inequity that the agent has for that interaction. Equation 4 shows the overall implementation of mood using the previous mood value, the average payoff, the received payoff, and the Homo Egualis function to update the mood value after an interaction with another agent. This equation gives us the current mood value of an agent. The mood will increase or decrease depending on the difference in the received payoff and the average payoff, meaning that the mood will increase when this agent is doing better than expected and decrease when it is doing worse than expected. With the inclusion of Ω the amount that the mood moves can change based on how fair the agent thinks the result was for both agents.

4 Method

In this work, the agents are simulated mobile robots that drive around in an environment. The agents are given a random walk behaviour with some basic obstacle avoidance procedures. The prisoner's dilemma game is initiated whenever two agents are within close proximity, and have line of sight of each other. The game is played once, after which the agents will then continue their random walk behaviour.

The agents are placed in a random location in the environment with emotional characteristics and moods distributed randomly, uniformly and independently among the agents, given the specific proportions of each experiment. The details for the experiments conducted are given in Sect. 5.

Agents move randomly throughout their environment, while avoiding collisions with the environment or other agents. Each agent has proximity sensors located at $\{-90°, -45°, -15°, 15°, 45°, 90°\}$ w.r.t. the robot's heading. When the left sensors detect something the robot will stop and turn to the right and the reverse for the right sensors. The agents move forward at up to 10 cm/s (The speed is constant except when accelerating from stationary as built into the simulator) and can turn 45 deg/s. When there are no obstacles detected the agent moves forward with a turn speed that is between $-45°$ or $45°$ per second. A new heading is generated when the robot receives data from the sensors, resulting in a random movement pattern.

In terms of the agents' knowledge of the world they are able to differentiate between agents, but have no knowledge of the strategies others will be using. They also have no knowledge of the environment apart from the sensor data they have at that moment in time. In addition the agents have no knowledge of the payoff matrix and will purely use their mood and emotion strategy to determine whether to cooperate or defect.

5 Experiment Outline

We will be describing the experiments that we have conducted to test our mood implementation. The first experiment will explore how mood affects the evolution of cooperation. Our second experiment introduces pure defectors as an invasion force to our environment and answers the question of how resilient the cooperation is to outside defectors.

Both of the experiments will be conducted in the environment which is four corridors in a square with each corridor having a length of 5 m and a width of 1 m. This is shown in Fig. 1. The experiments will run for 10 min for each run. A run consists of a scenario and if applicable a sub-scenario, to ensure consistency of results we will be running each of the 10 min runs 10 times. In our experiments each emotional characteristic is represented equally to prevent any characteristic becoming dominate due to them having a higher initial representation. In addition the initial actions of the emotional agents will be an equal split between cooperation

Fig. 1 Environment used in this work. The environment has dimensions of 5 × 5 m with the agents themselves having a radius of 7 cm, with the traversable areas shown in white

and defection. The initial location of the agents will be generated randomly. Each of these aspects will be distributed randomly and independently of each other. For our experiments we will be simulating agents using the Player/Stage simulator [6].

5.1 Mood Experiments

The first experiment will explore how the evolution of cooperation is affected by differing initial mood levels. The initial level of mood will be categorised into three types, low, medium and high where low has a mood level of 30, medium is 50 and high is 70. There will be seven scenarios each with a different distribution of these levels among the agents which can be seen in Table 3. We refer to the outcomes given by neutral moods as medium as this better reflects the current mood value.

Each of these scenarios will be run against a number of sub-scenarios. The sub-scenarios define how many agents will be in the environment, with a range from 45 to 144 agents, the details of the scenarios can be seen in Table 4. We will be looking to see how different initial moods affect cooperation. We will explore if our mood model allows cooperation to increase in the society of agents over time.

Table 3 Mood experiment scenarios showing as a percentage the different distributions of starting mood levels for the agents

Scenario	Low mood	Medium mood	High mood
1	100	0	0
2	0	100	0
3	0	0	100
4	33	33	33
5	70	15	15
6	15	70	15
7	15	15	70

Table 4 Mood experiment sub-scenarios, showing the number of agents that will be simulated for each scenario

Sub-scenario	No. of agents
1 - Low density	45
2 - Medium density	81
3 - High density	117
4 - Very high density	144

Table 5 Resilience experiment scenarios showing the starting level of mood for the agents with our mood model and the number of pure defectors that will be added

Scenario	Mood level	No. of defectors
1	Low	43
2	Low	63
3	Low	83
4	Medium	43
5	Medium	63
6	Medium	83
7	High	43
8	High	63
9	High	83

5.2 Resilience Experiments

Our next experiment is to test the resilience of the cooperation that evolves over time. To test this we will be introducing pure defectors at the beginning of the experiment into our environment; they cannot replicate themselves but the emotional agents may take on the role of a pure defector due to their admiration emotion. Each scenario will have 63 agents whose initial mood is dictated by the scenario: the moods are categorised as high (70), medium (50) and low (30). The numbers of pure defectors are 43 (minority defectors), 63 (equal defectors and emotional agents) and 83 (majority defectors). The details of each scenario are shown in Table 5. This will show the resilience that our mood model has to these pure defectors.

6 Results

6.1 Mood Results

Figure 2 shows us the percentage of cooperation between each the number of interactions for each scenario with an extra scenario which excluded the mood model and only used the emotional strategy. The results given are quite intuitive, we see that cooperation evolves throughout the agents, and the speed at which this is achieved is directly proportional to the average level of mood. The fastest is the scenario with

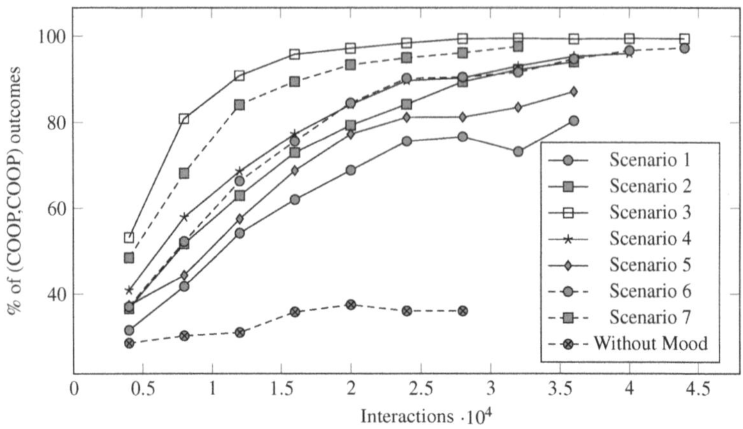

Fig. 2 Percentage (COOP, COOP) outcomes over all runs for each scenario in the mood experiment

100% of agents starting with high mood levels and the slowest is the scenario with 100% of agents having low mood levels. To attribute this to the mood model we ran the same experiments but without our mood model, where the decision making was purely based on their emotional decision making. We can attribute the rise in cooperation to the mood model as when it is removed the cooperation does not rise as quickly. The number of iterations between scenarios being uneven is due to the random nature of the agents movement.

This shows us that our mood model can support the evolution of cooperation over time and sustain cooperation; this was an expected result as when cooperation is high the mood moves very little. When two agents play the game, with one being in a high mood and one being in a low mood, the low mood will rise faster than the high mood can go down which is a property of our implementation of the egualis equation. This leads to more agents in a cooperative state raising cooperation. This effect is most apparent in the later stages of the simulation when the agents start with low moods, as the agents which are cooperating meet a group of agents which are not cooperating. This lead to a dip in cooperation followed by the continuing rise of cooperation when a large amount of agents with opposing moods meet.

To justify our claim that the speed at which cooperation is achieved is proportional to the starting level of mood we have plotted the average mood values against the number of $(COOP, COOP)$ actions, as can be seen in Fig. 3. We have shown this against scenario 1 as this is where the effect is most pronounced; we can see that cooperation between agents falls from 77% to 73% as agents who are cooperating meet larger groups of defecting agents. However the average mood level still rises, from 71.7% to 74.5%. When the cooperation rises again the standard deviation of mood levels is reduced to 26.9 from 27.5. This shows us that the mood reflects the level of cooperation, and the higher the starting level of mood the faster cooperation is achieved.

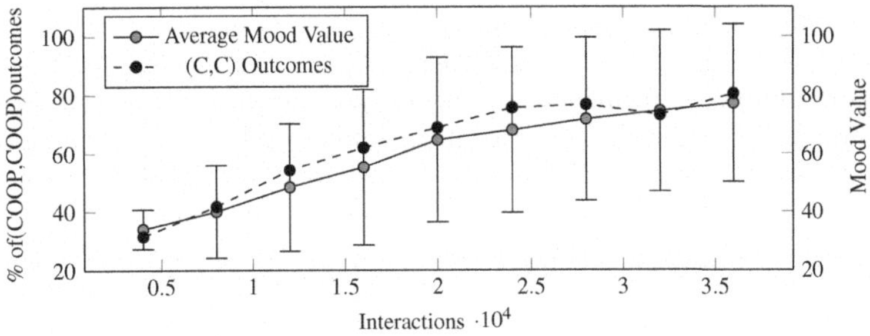

Fig. 3 Average mood value with standard deviation against percentage of $(COOP, COOP)$ outcomes in scenario 1 of the mood experiment

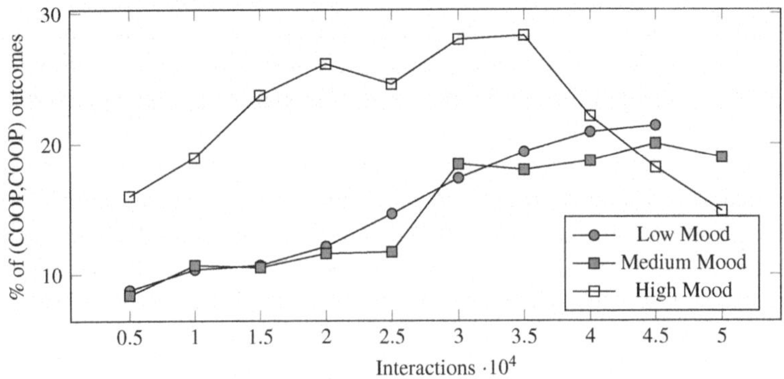

Fig. 4 Percentage of (COOP, COOP) outcomes for each initial mood level in the resilience experiment

6.2 *Resilience Results*

Figure 4 shows that when the mood is low, the emotional agents as a group are more resilient to an invading population of pure defectors. In high moods, cooperation between the emotional agents rises quickly, this in turn raises the mood of the agents as well. It therefore does not take long for the mood to increase to the point where the agents can be considered pure cooperators due to the mood level being very high. When this happens and agents are faced with the pure defectors the only outcome between an emotional agent and a pure-defector can be $(COOP, DEFECT)$. This causes the average score of the defectors to increase and the emotional agents' average to decrease. These changes in average payoffs will be affected rapidly because of the payoff difference. When replication occurs in the emotional agents they choose to become pure defectors because of this payoff difference, which leads to the collapse of cooperation as there are more pure-defectors.

In contrast when the emotional agents are in a low mood it takes longer for them to get their moods to the level where they are indistinguishable from pure cooperators; this allows them to protect themselves from the pure defectors by using their emotional choice which switches their action to defection for that particular opponent. Actions driven by emotions rather than mood are bounded to a particular opponent. This allows the agents to evolve cooperation with other emotional agents without replicating into pure defectors since the defectors have a low average as the number of $(DEFECT, DEFECT)$ actions they receive increasing over time.

These results show both expected and unexpected results. We had expected that cooperation would continue to be stable over time as the simulated moods and emotions would adapt to the invasion force, as seen in the low and medium starting moods. However the collapse of the high mood was unexpected.

To justify our hypothesis about why the mood levels have collapsed, we have shown that high moods do not adapt quickly to the pure defectors and therefore are taken advantage of. The advantage taken then leads to the emotional agents becoming pure defectors as their average score is not high enough when compared to the pure defectors. We took the difference between average score of the defectors and the average score of the emotional agents for each starting level of mood. The results showed that the difference in average score between low starting moods and the pure defectors was 0.04, for medium moods was 0.22, finally for the high moods the difference was 0.5. We can see that the high mood difference is more than double the medium mood difference. The defectors are clearing taking advantage of the high moods the most.

As the high moods are being taken advantage of the most, we expect that the payoffs for the defectors should be the highest when faced with the highest mood. The average scores of the defectors are shown in Table 6 and clearly show that the defectors do the best when faced with high moods, meaning that they will replicate the fastest in the high moods. The medium and low moods do not collapse as they adapt to the newly replicated defectors through the use of their directed emotion strategy. The high moods do not adapt, as when the mood is very high they act as pure cooperators.

Table 6 Average scores with standard deviation of the defectors, for each mood level in the resilience experiment

Starting mood level	Average	Standard deviation
Low	1.45	0.49
Medium	1.82	0.69
High	2.17	0.85

7 Conclusions

We have proposed a model of mood that can be used either independently or in conjunction with emotions. We have constructed this model using psychological research, with general cases for type of outcomes for varying mood levels. We have then applied this to an experiment with validation from psychology research for our choices made, which explores cooperation in a multi-agent setting.

For our experiments conducted we have shown that a combination of mood and emotion can support positive levels of cooperation within an agent society. We have also shown that mood levels in our agents are related to the level of cooperation that is achieved as a group. By adding an invasion force of pure defectors the cooperation between the emotional agent collapses over time when the mood levels of the emotional agents are high. In contrast when the mood is not high the cooperation over time is more stable since the agents do not give the benefit of doubt to the pure defectors, preventing the defectors from achieving a higher average. For future work we will looking into adapting our mood model to take into consideration mood fluctuations over time.

References

1. André, E., Klesen, M., Gebhard, P., Allen, S., Rist, T.: Integrating models of personality and emotions into lifelike characters. In: Paiva, A. (ed.) Affective Interactions. LNCS, vol. 1814, pp. 150–165. Springer, Berlin (2000)
2. Axelrod, R., Hamilton, W.D.: The evolution of cooperation. Science **211**(4489), 1390–1396 (1981)
3. Bloembergen, D., Ranjbar-Sahraei, B., Bou Ammar, H., Tuyls, K., Weiss, G.: Influencing social networks: an optimal control study. In: Proceedings of ECAI'14, pp. 105–110 (2014)
4. Collenette, J., Atkinson, K., Bloembergen, D., Tuyls, K.: Mobility effects on the evolution of co-operation in emotional robotic agents. In: Proceedings of ALA Workshop (2016)
5. Fehr, E., Schmidt, K.M.: A theory of fairness, competition, and cooperation. Q. J. Econ. **114**, 817–868 (1999)
6. Gerkey, B., Vaughan, R.T., Howard, A.: The player/stage project: tools for multi-robot and distributed sensor systems. In: Proceedings of ICAR'03, pp. 317–323 (2003)
7. Gintis, H.: Game Theory Evolving: A Problem-centered Introduction to Modeling Strategic Behavior. Princeton university press, Princeton (2000)
8. Gray, E.K., Watson, D., Payne, R., Cooper, C.: Emotion, mood, and temperament: similarities, differences, and a synthesis. Emotions At Work: Theory, Research and Applications for Management, pp. 21–43. Wiley, Chichester (2001)
9. Haley, W.E., Strickland, B.R.: Interpersonal betrayal and cooperation: effects on self-evaluation in depression. J. Pers. Soc. Psychol. **50**(2), 386 (1986)
10. Hertel, G., Neuhof, J., Theuer, T., Kerr, N.L.: Mood effects on cooperation in small groups: Does positive mood simply lead to more cooperation? Cogn. Emot. **14**(4), 441–472 (2000)
11. Keltner, D., Gross, J.J.: Functional accounts of emotions. Cogn. Emot. **13**(5), 467–480 (1999)
12. Leahy, R.L.: Clinical implications in the treatment of mania: Reducing risk behavior in manic patients. Cogn. Behav. Pract. **12**(1), 89–98 (2005). https://doi.org/10.1016/S1077-7229(05)80043-4
13. Levenson, R.W.: Human emotion: a functional view. Nat. Emot. Fundam. Quest. **1**, 123–126 (1994)

14. Lloyd-Kelly, M., Atkinson, K., Bench-Capon, T.: Developing co-operation through simulated emotional behaviour. In: 13th International Workshop on Multi-Agent Based Simulation (2012)
15. Lloyd-Kelly, M., Atkinson, K., Bench-Capon, T.: Fostering co-operative behaviour through social intervention. In: 2014 International Conference on Simulation and Modeling Methodologies, Technologies and Applications (SIMULTECH), pp. 578–585. IEEE (2014)
16. Lount, R.B.J.: The impact of positive mood on trust in interpersonal and intergroup interactions. J. Pers. Soc. Psychol. **98**(3), 420–433 (2010)
17. Morris, W.N.: A functional analysis of the role of mood in affective systems. Emotion (1992)
18. Ortony, A., Clore, G.L., Collins, A.: The Cognitive Structure of Emotions. Cambridge university press, Cambridge (1990)
19. Popescu, A., Broekens, J., van Someren, M.: Gamygdala: an emotion engine for games. IEEE Trans. Affect. Comput. **5**(1), 32–44 (2014). https://doi.org/10.1109/T-AFFC.2013.24
20. Santos, F.C., Santos, M.D., Pacheco, J.M.: Social diversity promotes the emergence of cooperation in public goods games. Nature **454**(7201), 213–216 (2008)
21. Schwarz, N.: Emotion, cognition, and decision making. Cogn. Emot. **14**(4), 433–440 (2000). https://doi.org/10.1080/026999300402745
22. Ventura, R.: Emotions and empathy: a bridge between nature and society? Int. J. Mach. Conscious. **2**(02), 343–361 (2010)

Programmable Self-disassembly for Shape Formation in Large-Scale Robot Collectives

Melvin Gauci, Radhika Nagpal and Michael Rubenstein

Abstract We present a method for a large-scale robot collective to autonomously form a wide range of user-specified shapes. In contrast to most existing work, our method uses a subtractive approach rather than an additive one, and is the first such method to be demonstrated on robots that operate in continuous space. An initial dense, stationary configuration of robots distributively forms a coordinate system, and each robot decides if it is part of the desired shape. Non-shape robots then remove themselves from the configuration using a single external light source as a motion guide. The subtractive approach allows for a higher degree of motion parallelism than additive approaches; it is also tolerant of much lower-precision motion. Experiments with 725 Kilobot robots allow us to compare our method against an additive one that was previously evaluated on the same platform. The subtractive method leads to higher reliability and an order-of-magnitude improvement in shape formation speed.

1 Introduction and Related Work

In nature, groups of thousands to millions of units can self-assemble into complex structures, purely through local interactions. For example, cells self-assemble into complex organisms, colonies of ants self-assemble into bridges and bivouacs made out of their own bodies, and fish self-assemble into complex schooling patterns [1, 3]. In all of these cases, a group can dramatically change its interaction with its environment through collective shape formation. In the fields of collective and modular robotics, many groups have taken inspiration from these natural examples to

M. Gauci (✉) · R. Nagpal
Harvard University, Cambridge, MA, USA
e-mail: mgauci@g.harvard.edu

R. Nagpal
e-mail: rad@eecs.harvard.edu

M. Rubenstein
Northwestern University, Evanston, IL, USA
e-mail: rubenstein@northwestern.edu

R. Groß et al. (eds.), *Distributed Autonomous Robotic Systems*,
Springer Proceedings in Advanced Robotics 6,
https://doi.org/10.1007/978-3-319-73008-0_40

create novel robotics systems that achieve higher functionality through self-assembly and shape formation. For example, researchers have designed small mobile robots that are envisioned to work together in large groups, including assembling together to achieve complex tasks [8, 9, 11, 12]. Similarly, using inspiration from multicellular development, several groups have developed modular robots that can self-reconfigure into different morphologies for grasping or locomotion tasks [16]. The ultimate goal is to shrink the size of these systems to create "programmable matter", where large numbers of tiny robotic modules can rapidly self-assemble into user-specified tools such as a wrench or a key [2, 7].

Self-assembling robot collectives and programmable matter systems pose a global-to-local algorithmic challenge: How does one program large numbers of agents to self-organize into a user-specified shape, when each agent can only interact locally with nearby agents at a scale much smaller than the global outcome? There is also the hardware challenge: How do such algorithms scale to large numbers of real robots, where individual errors can amplify into global errors?

There has been considerable progress on answering these algorithmic questions. Inspiration is often drawn from conventional engineering manufacturing processes, which can be broadly divided into two classes: additive and subtractive. In additive processes, material is added and fused to form the desired shape; examples include injection molding and 3D printing. In subtractive processes, material is removed from an initial bulk until only the desired shape remains; examples include lathing and milling. It is noteworthy that both additive and subtractive processes can also be observed in nature, such as the growth of limbs (additive) and the removal of inter-digit webs in the hand (subtractive) [15].

In additive self-assembly methods, the shape starts with a seed module that defines the origin and orientation of a coordinate system. The remaining modules then localize relative to the seed and form the shape in layers. This approach is often referred to as "directed growth" [13] and has been implemented both in mobile collectives where the robots move freely in space [8, 11, 12], and in self-reconfigurable systems where the robots are constrained to move along docking sites [16, 17]. In some cases, instead of actively mobile agents, the system uses agents that move passively in the environment, based on some source of random motion such as the mixing of a fluid or a low friction surface [2, 9, 14]. Although many additive self-assembly algorithms have been explored, most have only been validated in simulation or in at most tens of hardware agents. One recent exception is a demonstration by our group on a 1024 robot collective, namely the Kilobot system [12]. Using cooperative error correction, this approach was able to robustly self-assemble large shapes with diameters of up to 45 times the radius of local interactions without human intervention. However, it has limited parallelism in robot motion, and requires this motion to be highly precise. These factors lead to long shape formation times.

The subtractive self-assembly approach (hereafter called *self-disassembly*) has received considerably less attention so far, but a few examples do exist. In this approach, the system starts with the agents in a large, dense, stationary group. A seed module then initiates the formation of a coordinate system, and finally, non-shape agents remove themselves from the system. In the Pebbles system [7], modules are arranged in a 2D rectangular lattice, and non-shape modules release themselves and move away by exploiting surface vibrations. In the Miche system [6], larger modules are arranged into a 3D rectangular lattice, using magnetic connections to maintain the structure. The removal of non-shape modules is achieved by exploiting gravity. Both the Pebbles and the Miche systems attempt to replace active precision mobility with external actuation, in order to be able to shrink the size of modules and bring the system closer to the vision of programmable materials.

The subtractive approach has several potential advantages over the additive approach, including a high level of motion parallelism, and a low level of required motion precision, both of which become more attractive as the scale of the collective increases. The self-disassembly examples discussed above were both implemented on modular robots that can be arranged into a regular lattice structure. In this case, coordinates can be easily assigned since the modules can exploit directional communication [7, 13]. However, requiring modules to align in perfect lattices makes the hardware design more complex and any manufacturing imperfections can negatively affect the perceived locations of modules. Moreover, these systems have so far only been tested on limited numbers of modules ($\leq$30).

In this paper, we present an approach to programmable self-disassembly that extends the conceptual ideas from the Miche and Pebbles systems to a fully-autonomous mobile robot collective that operates in free space, and uses only local sensing and interactions. The robots distributively form a continuous coordinate system through distance sensing and multilateration, allowing for a larger tolerance to the localization errors of individual robots [10, 12]. The robots then use a decentralized consensus algorithm to transition from the coordinate system formation phase to the self-disassembly phase. Finally, the non-shape robots autonomously remove themselves from the shape without distorting it, using a single external light source as a motion guide. We present theoretical results to show which classes of shapes can be formed by our approach, and how the shape class determines the self-disassembly procedure. We then validate our approach on the Kilobot platform, using 725 robots, and present experimental results demonstrating high reliability and accuracy despite the noise and failures present in large-scale collectives. We also present a direct comparison of additive and subtractive algorithms, as experimentally tested on the same hardware platform, and show that while the additive approach is able to form larger shapes for a given number of robots, the subtractive approach achieves much higher motion parallelism, as well as tolerance to simpler and noisier motion. This results in a significant improvement in shape formation speed.

2 Self-disassembly Algorithm

2.1 Robot Model

Here we describe the model for a Kilobot robot; note, however, that its feature set matches closely with those of many other robots designed for large-scale collective behaviors. More details on the Kilobot platform can be found in [12]. An individual Kilobot has the following capabilities: external light sensing, computation (via a microcontroller), and non-holonomic locomotion using vibration motors. Each robot can also communicate with nearby neighbors within a radius of 3 body lengths (at most around 36 robots) via shared channel wireless broadcast; the group thus acts as a large multi-hop wireless network. Furthermore, each robot can measure its distance to its neighbors based on the communication signal strength, with a resolution of 1 mm. However, it cannot measure the bearings or headings of its neighbors.

The collective is inherently decentralized, multi-hop, and asynchronous. In addition, all aspects experience noise and manufacturing variation. The light sensor has a limited resolution and accuracy. Open-loop robot motion drifts quickly with time and the robots do not have internal odometry capabilities. Messages are communicated via a shared wireless channel which is asynchronous and experiences message loss. Distance measurements based on wireless signal strength are inherently noisy. Even after calibration, there are significant variations between robots in all aspects.

2.2 Algorithm and Implementation

The goal of the algorithm is to form a user-specified shape, starting from a configuration where all the robots are in a tightly packed group. We assume that the size of the desired final shape fits within the starting configuration. Three pre-localized *seed* robots are placed within the initial configuration, representing the origin and orientation of the coordinate system. All the robots in the configuration are given an identical program, which includes a representation of the shape to be formed and the location of a light source (with respect to the seed robots) that will serve to guide the motion of non-shape robots. The algorithm proceeds in three steps: (i) Distributed Coordinate System Formation - each robot localizes itself within the collective (with respect to a seed) and decides whether or not it is part of the desired shape; (ii) Transition Using a Consensus Algorithm: the robots collectively determine when every robot has localized in order to transition to the next step; (iii) Self-Disassembly: robots that are in the desired shape remain stationary while non-shape robots move away from the shape.

Distributed Coordinate System Formation

Since Kilobots can measure distance to and communicate with neighbors, a robot can localize itself with respect to 3 or more other localized neighbors using multilateration. If robot i has n localized neighbors with distances $d_1, \ldots, d_n$ away from it and positions $\mathbf{p}_1, \ldots, \mathbf{p}_n$, then robot i computes its best-guess position as:

$$\mathbf{p}_i^* = \arg\min_{\mathbf{p}_i} \sum_{j=1}^{n} \frac{\left| d_j - \|\mathbf{p}_i - \mathbf{p}_j\| \right|}{d_j}, \tag{1}$$

where $|\cdot|$ denotes the absolute value and $\|\cdot\|$ denotes the Euclidean norm. The d_j in the denominator serves to implement inverse variance weighting, giving less importance to larger measured distances, which tend to be less accurate. The robots localize in a 'wave' that propagates outward from the 3 pre-localized seed robots.

Although distributed coordinate system formation is a simple process at an abstracted level, it is well-known that in practice, it can be highly sensitive to initial conditions and several types of noise [10]. As such, at large scales, it requires significant additional constraints to operate reliably. As shown in the experimental section, we have developed a highly-robust distributed coordinate system formation algorithm by using the following measures.

(i) In order to account for asymmetry in distance sensing, robots exchange measured distances with each other for two-way averaging. Let d_{ij} be the distance to robot j as measured by robot i, and vice-versa for d_{ji}. Then both robots estimate the distance between them as being $\frac{1}{2}\left(d_{ij} + d_{ji}\right)$. Robot i only takes robot j into account for multilateration purposes (i.e. Eq. 1) if: (a) both d_{ij} and d_{ji} are available, (b) $\frac{1}{2}\left(d_{ij} + d_{ji}\right) \leq 3$ body lengths, and (c) $|*|\, d_{ij} - d_{ji} \leq 1$ body length. (ii) A robot only considers localizing after it has at least one set of 3 neighbors that meets the following criteria: (a) the minimum angle of the triangle formed by the positions of these 3 neighbors is larger than 35°, and (b) 2 of these neighbors are within 1.25 body lengths and the other is within 2.5 body lengths. The first criterion avoids localizing with respect to almost-collinear neighbors, in which case the best-guess position might have a flip ambiguity. The second criterion ensures that the estimated distances of the robots making up this triangle are reasonably accurate. After condition (ii) has been met, a robot waits up to 30 seconds to check if it can acquire more neighbors that satisfy condition (i). Each time a new such neighbor is acquired, this timer is reset and the waiting starts again. This ensures that each robot uses as many neighbors as possible for multilateration, increasing the amount of averaging in the process, and thus improving consistency.

Once localized, a robot determines if it is part of the shape. The shape is specified in the robots' program as a polygon with an arbitrary number of vertices ($\geq$3), and a simple point-in-polygon algorithm is used to decide if the robot's position lies inside its perimeter. At this point, if a robot is not part of the shape, it also decides based on geometric considerations which type of motion it will execute when it is its turn to start moving (this will be explained further in Sect. 2.3).

Transition using a Consensus Algorithm

Once all the robots have localized, the collective must autonomously transition to the next phase where the non-shape robots remove themselves from the configuration. This is done using a simple gossip-based consensus algorithm, as follows. Each robot transmits a consensus value. For a non-localized robot, this value is always 0. A localized robot, on the other hand, determines its value by adding 1 to the minimum of its neighbors' values (all neighbors within the communication range are used for this algorithm). As long as there is at least one non-localized robot, its 0 value will hold down the values of all the other robots; a 'gradient' pattern will form across the group, with the maximum possible value of any robot being the maximum distance in the configuration (in terms of the communication range). When there are no longer robots transmitting a value of 0, all the robots start to count up. Once a robot's value exceeds a predefined threshold (chosen to be larger than the maximum possible hop count with an unlocalized robot still present), the robot decides that consensus has been reached. This algorithm is sensitive to cases where a single robot fails to localize permanently, due to not meeting one of the conditions described above. In our experiments, this happened rarely. Nevertheless, in the future, this phase of the algorithm can be improved by using a quorum algorithm in place of a consensus algorithm, which would 'only' require a high percentage (but not 100%) of the robots to have localized.

Self-disassembly

To remove themselves from the shape, non-shape robots rely on three motion types: collision avoidance, phototaxis, and antiphototaxis. All moving robots perform collision avoidance if they are too close to their neighbors, but each robot will only perform one of phototaxis and antiphototaxis, depending on its initial position relative to the shape. After a robot has localized, its program casts a ray from its position to the light source; if this does not intersect with the shape, the robot decides on performing phototaxis; otherwise, it decides on performing antiphototaxis.

The self-disassembly process starts from the edge of the initial configuration, and proceed inwards towards the shape. In this manner, robots that are completely locked by other robots do not needlessly start moving, which could potentially distort the shape. Each robot only starts moving when it has had 4 or fewer neighbors (stationary or moving) within 1.25 body lengths for a predefined amount of time (30 s). After starting to move, robots execute the following behavior.

A robot does collision avoidance if it perceives a stationary neighbor with 2 body lengths (i.e. one that is either part of the shape, or is not but has not yet started moving). This is done to minimize the chances of moving robots colliding with the shape and distorting it, at the cost of a slower self-disassembly process. When a robot senses no stationary localized robots within 2 body lengths, it switches to phototaxis or antiphototaxis. A robot that is doing phototaxis or antiphototaxis only reverts to collision avoidance if it perceives a stationary neighbor within 1.5 (rather than 2) body lengths. This hysteresis prevents robots from continuously switching between behaviors. Note that, moving robots do not avoid each other, allowing them to move quickly and aggressively, even in the face of congestion.

The motion behaviors are implemented as follows. Collision avoidance starts with the robot turning, and looking at its distance to its closest stationary neighbor. The robot searches for a valley in the profile of this distance versus time, i.e., a transition from a negative to a positive rate of change. This indicates that the robot is facing away from its stationary neighbor, and at this point, it switches to straight motion. In phototaxis, the robot starts turning and looking at its light sensor reading. It searches for a valley in the intensity profile versus time, which indicates that it is facing towards the light source (the light sensor is on the back of the robot). At this point, the robot keeps rotating in its current direction for a small time period, then reverses its turning direction, and starts looking for a new valley. Due to the configuration of the robot's legs, this oscillatory turning causes a net motion towards the light. Antiphototaxis is implemented in the same way, but the robot now looks for peaks rather than valleys in the intensity versus time profile.

2.3 Achievable Shape Classes

In order to successfully form a desired shape, all non-shape robots must move away from the shape without disturbing it. Our current system relies on phototaxis and antiphototaxis, which have the advantage of being fast and highly tolerant of robot sensory and actuation noise. However, this imposes a constraint on the system: all (or in practice most) of the robots must be able to leave the shape using one of these two motion types. We now describe which shape types are solvable by self-disassembly under this constraint. We use polygon representations for the shapes, and divide polygons into 3 classes. An example from each class is provided in Fig. 1.

- **Class 1**: *Solvable by antiphototaxis alone*: "Star-shaped polygons" is a well-known class of polygons that can be defined as follows: There is at least one point within the polygon such that any ray cast from this point intersects the polygon's perimeter exactly once. For our application this means that, if the light source is placed at such a point, any robot outside the shape can move away from it in a straight line (i.e., antiphototaxis) without encountering the shape. Note that convex polygons are a subclass of star-shaped polygons, having the additional property that *every* point inside the polygon satisfies the light placement condition. In general, a linear-time algorithm exists for determining which points inside a given polygon, if any, satisfy this condition; if none are found, the polygon is not Class 1 [5].
- **Class 2**: *Solvable by a combination of phototaxis and antiphototaxis*. We define this class of polygons according to the following property: There is at least one point outside the polygon such that every ray cast from this point either does not intersect the polygon's perimeter, or intersects it exactly twice. If a ray does not intersect the perimeter, both phototaxis and antiphototaxis are viable from every point on this ray; we prioritize phototaxis over antiphototaxis due to its better reliability on Kilbots. If a ray intersects the perimeter twice, then phototaxis is viable from points that lie on the segment between the light source and the first intersection, while antiphototaxis is viable from points beyond the second intersection. In the supplementary material [5], we supply a linear-time algorithm that determines the

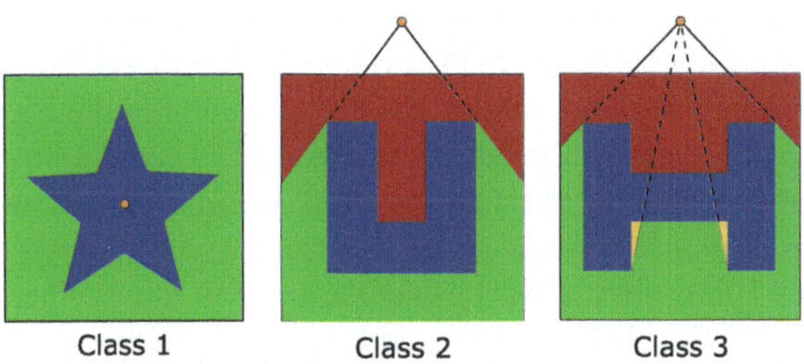

Fig. 1 Examples of shapes in the 3 classes (see the text for details). Orange dots indicate the light source location; blue indicates the shape; red and greed indicate regions from which phototaxis and antiphototaxis are possible, respectively; yellow indicates blocked regions

set of feasible light source positions, if any, for a Class 2 shape; if none exist, the polygon is not Class 2.

- **Class 3**: *Not solvable by a combination of phototaxis and antiphototaxis alone*. A polygon is in Class 3 if it is not Class 1 or Class 2. For a Class 3 polygon, wherever the light source is located, there will always be regions in the environment from which a robot can do neither phototaxis nor antiphototaxis without encountering the shape. In general, self-disassembly into Class 3 polygons is not possible under our constraints. In practice, some robots might still manage to navigate away from blocked regions using collision avoidance, but this depends on the depth of the blocked regions. It is a promising avenue for future research to investigate how our method can be augmented for such shapes.

3 Experimental Validation

3.1 Setup and Protocol

In order to validate our algorithm, we performed 3 experiments with two shapes in Class 1 and on shape in Class 2. The Class 1 shapes constituted a 5-point starfish and a wrench, both with the light source placed in the center such that all robots could perform antiphototaxis. The Class 2 shape was a U shape, with the light source placed above the U's cavity such that robots in the cavity could perform phototaxis and the rest could perform antiphototaxis. We used the experiments to look at the time efficiency and the accuracy at each stage of the algorithm.

The experimental protocol was as follows. The robots were initially arranged in an approximate hexagonal grid. For the starfish and the U shapes, this consisted of 29 rows by 25 columns (=725 robots), which yields a roughly square arrangement since $29 \times \sqrt{3}/2 \approx 25$. For the wrench shape, the initial configuration consisted of 17 rows by 42 columns (=714 robots), which yields a rectangular arrangement. In all cases, the seed robots were placed in the middle of the initial configuration.

All the robots were simultaneously programmed with the algorithm described previously. At the beginning of each experiment, the light source was placed in the appropriate location, and recording was started on an overhead camera. A run signal was then issued to all the robots, and the formation of the coordinate system was monitored by the operator based on the robots' LED colors. In order to collect data on how well the robots were estimating their own coordinates, we used an extra robot that was programmed to receive messages from localized robots and report their positions to a PC. This robot was waved over the group of robots when the coordinate system formation process had finished, and before the transition happened. During the self-disassembly phase, no manual intervention was made, including in cases of dead or stuck robots. The experiment was stopped after one hour had elapsed from the start of the self-disassembly phase.

3.2 Results and Discussion

Figure 2 shows the (a) initial and (c) final states of the robots for both experiments, as well as (b) where each robot believes its coordinates to be, with colors indicating whether it believes to be part of the shape (blue), or should leave the shape via phototaxis (red) or antiphototaxis (green). Figures 3 and 4 show snapshots of the self-disassembly process over time for the starfish shape and the U shape, respectively. The supplementary material [5] includes full-length overhead video recordings of the self-disassembly process for the starfish and the U shape experiments.

Qualitatively, it can be visually observed in Fig. 2b that in all three cases, the coordinate system formation algorithm achieved a good accuracy in all three cases. A slight rotation of the coordinate system is present in the case of the starfish shape. Note, however, that this is a global rotation, and does not cause any significant warp in the coordinate system; as such, the user-specified shape is not distorted. As a first quantitative measure of coordinate system accuracy, we look at the estimated distances between all pairs of robots that are adjacent in reality (i.e. the real distance is 1 body length = 33 mm). We looked at the distribution statistics of the error (defined as $\text{distance}_{\text{estimated}} - \text{distance}_{\text{true}}$). In the starfish experiment, the median error was +1.00 mm; the 25th and 75th percentile errors were −1.74 mm and +4.00 mm, respectively; and the extreme errors were −13.79 mm and +13.32 mm. In the wrench experiment, the median error was +1.01 mm; the 25th and 75th percentile errors were −1.94 mm and +4.05 mm, respectively; and the extreme errors were −33.00 mm

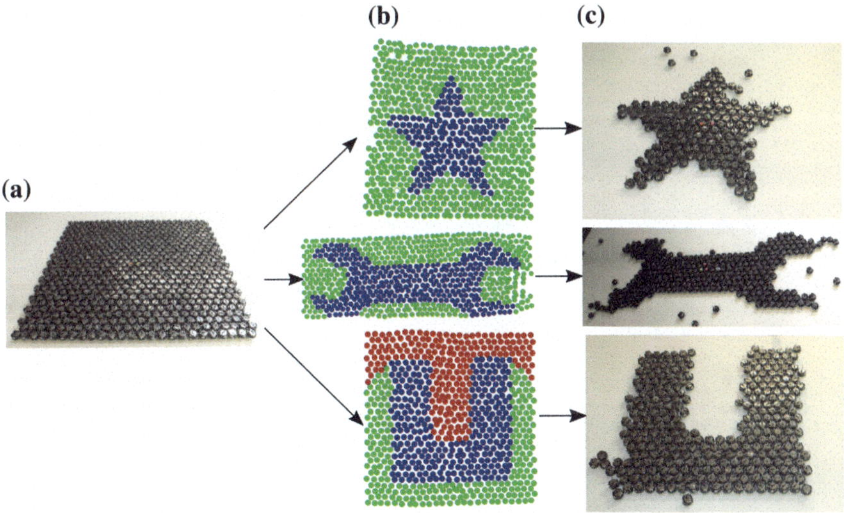

Fig. 2 Self-disassembly overview. **a** Robots start arranged in a grid. **b** A coordinate system is formed in a distributed fashion, and each robot decides if it is part of a user-specified shape (blue). If it is not, it also decides which motion type it should use to leave the configuration—phototaxis (red) or antiphototaxis (green); **c** Final state: non-shape robots have left the shape

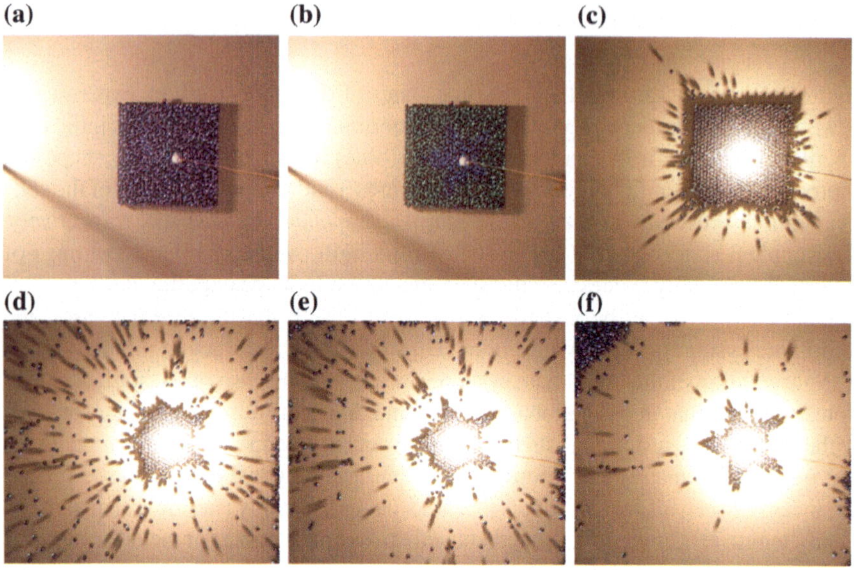

Fig. 3 Experimental self-disassembly of a starfish shape. **a** Initial configuration; only the seed robots are localized. **b** $t = 40$ min: All the robots are localized, and each robots has decided if it is part of the shape (blue) or not (green). **c** $t = 50$ min: The non-shape robots start leaving the shape. **d**, $t = 55$ min, **e** $t = 63$ min. **f** 90 min: Final state

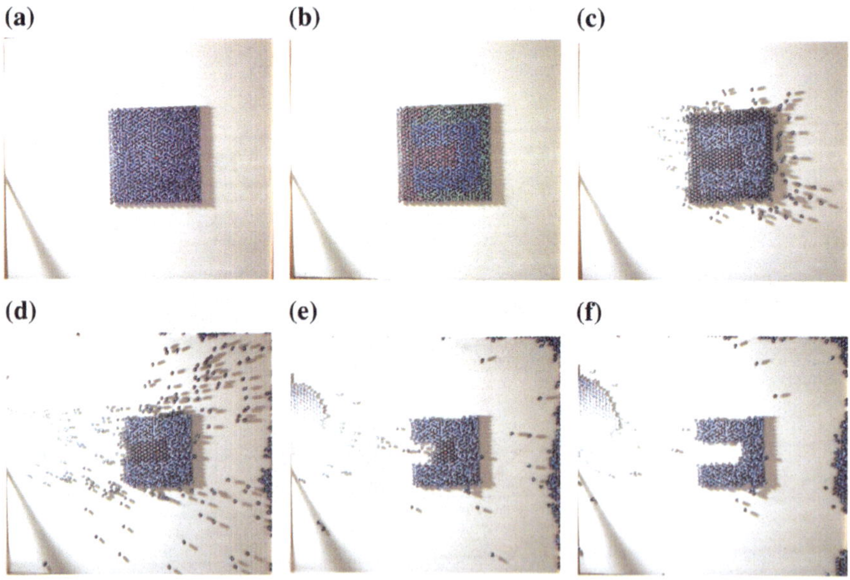

Fig. 4 Experimental self-disassembly of a U shape. **a** Initial configuration; only the seed robots are localized. **b** $t = 40$ min: All the robots are localized, and each robots has decided if it is part of the shape (blue), and if not, whether it should leave the shape using phototaxis (red) or antiphototaxis (green). **c** $t = 50$ min: The non-shape robots start leaving the shape. **d**, $t = 55$ min, **e** $t = 63$ min. **f** 90 min: Final state

and +13.96 mm. In the U shape experiment, the median error was +1.06 mm; the 25 and 75th percentile errors were −0.98 mm and +3.68 mm, respectively; and the extreme errors were −21.34 mm and +13.39 mm.

As a second quantitative measure of coordinate system accuracy, we calculated the error between the area of the estimated configuration (defined as the area enclosed by the convex hull of the estimated coordinates) and the area of the true configuration. The error is defined as $100 \times (\text{area}_{\text{estimated}} - \text{area}_{\text{true}})/\text{area}_{\text{true}}\%$. For the starfish, wrench, and U shape experiments, this error was +0.62%, −0.81%, and +2.37%, respectively.

As a third quantitative measure of coordinate system accuracy, we compared the error between the number of robots that considered themselves to be part of the shape according to the coordinate system and the theoretically-expected number (assuming a fully-packed hexagonal initial configuration). The error is defined as $100 \times (\text{number}_{\text{coordinate system}} - \text{number}_{\text{expected}})/\text{number}_{\text{expected}}\%$. In the starfish experiment, the actual and expected numbers were 164 and 181, respectively (error = −9.39%). In the wrench experiment, the actual and expected numbers were 336 and 356, respectively (error = −5.62%). In the U shape experiment, the actual and expected numbers were 284 and 298, respectively (error = −4.70%).

These three quantitative measures of coordinate system accuracy together suggest that the robots are systematically overestimating the distances to their neighbors. Note that in the first measure, the median error is positive in all three cases; if there were no systematic error in the distance sensing, a median error of 0 mm would be expected. In the second measure, the estimated area is larger than the true are in two out of three cases. In the third measure, in all three cases, the number of robots that considered themselves to be in the shape according to the coordinate system is smaller than the expected number. This is because the estimated configuration is expanded, and therefore sparser.

All three runs of the coordinate system reached steady-state after around 40 min. This time mainly reflects the strict conditions imposed on when a robot can start to localize (see Sect. 2.2). We observed that relaxing some of these conditions would cause the coordinate system to form faster, but at the cost of reduced accuracy and reliability. In the starfish experiment, 2 robots at one of the edges of the configuration failed to localize. Upon inspection, it turned out that these were neighboring a robot that had switched off due to low battery. This was exacerbated by the fact that edge robots have the fewest neighbors. They were therefore failing to meet one or more of conditions to localize described in Sect. 2.2. These two robots were manually removed to allow the consensus-based transition to take place. In the future, this problem could be addressed by: (i) slightly relaxing the conditions to localize for edge robots, and/or (ii) replacing the consensus-based transition algorithm with a quorum-based one. Both transition procedures lasted around 10 min.

The final shape formations in all three experiments shows good agreement with the formed coordinate system (see Fig. 2). In the starfish experiment, the final configuration contained 167 robots,[1] whereas 164 robots had considered themselves to be part of the shape according to the coordinate system (error = +1.82%). In the wrench experiment, the corresponding numbers were 363 and 356 (error = +1.97%), and in the U shape experiment, they were 294 and 284 (error = +3.52%). Additionally, in the wrench experiment there were four robots in the final configuration that were not connected to the shape, but still inside its convex hull. The reasons for the number of robots left in the final shape being larger than the count from the coordinate system count are twofold, and due to hardware issues. Firstly, some robots stopped working due to a low battery level, and robots close to the shape are more susceptible to this because they have to wait the longest before starting to move. Secondly, some robots suffered from miscalibration or malfunction of their light sensor and/or motors, and as such failed to perform successful phototaxis or antiphototaxis, occasionally becoming stuck to the shape. In the U shape experiment, three in-shape robots were knocked out of place by moving robots; no in-shape robots suffered this effect in the starfish experiment. These low errors and the high fidelity of the final shapes confirm that although the motion strategy of the non-shape robots is highly-parallelized and aggressive, the shape is not significantly distorted thanks to a successful collision avoidance behavior.

4 Comparison of Self-disassembly to Additive Self-assembly

The Kilobot platform not only allows us to validate complex collective behaviors at scale, but also allows for a direct comparison of different algorithmic approaches on the same hardware testbed. Here we compare two approaches to shape formation using the Kilobot platform: (i) an additive self-assembly approach, where the shape grows in layers starting from a seed [12] (similar to approaches such as [13]), and (ii) the subtractive self-disassembly approach presented here, which is similar to the Pebbles [7] and Miche [6] systems.

The self-disassembly algorithm has some disadvantages, most importantly the fact that it is more wasteful because not all robots are used in the shape. This effect is exacerbated in cases where the shapes bounding box (or convex hull) is only sparsely populated by the shape. Also, the shape classes formed by both approaches have different constraints. The self-assembly algorithm in [12] could form any simply-connected shape (i.e. no holes), and there exist self-assembly algorithms that can handle holes as well [4]. In contrast, the self-disassembly algorithm presented is currently limited to shapes classes 1 and 2, and by its nature fundamentally lacks the ability to handle holes in the shape. However, the self-disassembly algorithm can form disjointed shapes from a single seed, which is much harder to achieve with self-assembly algorithms.

[1]The number of robots in the final shape is defined as the maximal connected subgraph, with two robots being considered adjacent if they are within 1 body length apart.

At the same time, the self-disassembly algorithm has a significant advantage in terms of motion parallelism and time efficiency, both theoretically and experimentally. Theoretically, the additive self-assembly approach builds in layers that severely restricts parallelism, resulting in an $\mathcal{O}(n)$ scalability [12]. In contrast, self-disassembly has a high level of parallelism, and the time taken is bounded by the longest path that a non-shape robot takes in order to exit the convex hull, resulting in $\mathcal{O}(\sqrt{n})$ for a square or diameter of the shape in other cases. Experimentally however, this effect is even more dramatic. The wrench experiment took around 6 hours to self-assemble, and only around 40 minutes to form through self-disassembly. The high efficiency comes from the fact that self-disassembly relies on fast imprecise motion; phototaxis and antiphototaxis are simple robot behaviors that are feasible at higher speeds and are collectively robust without collision avoidance. In contrast, self-assembly requires precise edge-following motion which is reliable only at lower speeds, and collectively creates traffic lanes that allow the slowest robot's speed to dominate. Finally, the wrench experiment also demonstrates that self-disassembly algorithms can potentially achieve higher accuracy; in the previous self-assembly algorithm the final wrench coordinate system had a significant bend whereas with self-disassembly the coordinate system was close to perfect. For shapes with high-aspect ratios, small errors and drift can easily accumulate in additive algorithms whereas in a subtractive algorithm, the initial layout creates a long-range consistent localization which the shape can take advantage of.

5 Conclusion and Future Work

In this paper we present and experimentally demonstrate a self-disassembly algorithm on a large-scale robot collective (up to 725 Kilobots). Our theoretical and experimental results suggest that such a self-disassembly algorithm can achieve a wide class of shapes with high efficiency and accuracy, making it a good candidate for shape formation in modular robots and programmable materials. In the future, we intend to extend this algorithm by introducing a stochastic motion component, in order to deal with a class of shapes for which phototaxis and antiphototaxis alone are not sufficient for self-disassembly.

References

1. Anderson, C., Theraulaz, G., Deneubourg, J.L.: Self-assemblages in insect societies. Insectes Sociaux **49**(2), 99–110 (2002)
2. Bishop, J., Burden, S., Klavins, E., Kreisberg, R., Malone, W., Napp, N., Nguyen, T.: Programmable parts: a demonstration of the grammatical approach to self-organization. In: Proceedings of the 2005 International Conference on Intelligent Robots & Systems (IROS) (2005)
3. Camazine, S., Deneubourg, J.L., Franks, N.R., Sneyd, J., Theraulaz, G., Bonabeu, E.: Self-organization in Biological Systems. Princeton University Press, Princeton (2003)

4. De Rosa, M., Goldstein, S.C., Lee, P., Campbell, J., Pillai, P.: Scalable shape sculpting via hole motion: motion planning in lattice-constrained modular robots. In: Proceedings of the ICRA (2006)
5. Gauci, M., Nagpal, R., Rubenstein, M.: Programmable self-disassembly for shape formation in large-scale robot collectives: online supplementary material (2016). http://goo.gl/VKnMmk
6. Gilpin, K., Kotay, K., Rus, D.: Miche: modular shape formation by self-disassembly. Int. J. Robot. Res. **27**, 345–372 (2008)
7. Gilpin, K., Knaian, A., Rus, D.: Robot pebbles: One centimeter modules for programmable matter through self-disassembly. In: Proceedings of the 2010 International Conference on Robotics and Automation (ICRA) (2010)
8. Groß, R., Bonani, M., Mondada, F., Dorigo, M.: Autonomous self-assembly in swarm-bots. IEEE Trans. Robot. **22**, 1115–1130 (2006)
9. Haghighat, B., Platerrier, B., Waegeli, L., Martinoli, A.: Synthesizing rulesets for programmable robotic self-assembly: a case study using floating miniaturized robots. In: Proceedings of the 10th International Conference Swarm Intelligence (ANTS) (2016)
10. Moore, D., Leonard, J., Rus, D., Teller, S.: Robust distributed network localization with noisy range measurements. In: Proceedings of the 2nd International Conference on Embedded Networked Sensor Systems, pp. 50–61. ACM (2004)
11. O'Grady, R., Christensen, A.L., Dorigo, M.: Swarmorph: multirobot morphogenesis using directional self-assembly. IEEE Trans. Robot. **25**(3), 738–743 (2009)
12. Rubenstein, M., Cornejo, A., Nagpal, R.: Programmable self-assembly in a thousand-robot swarm. Science **345**, 795–799 (2014)
13. Stoy, K., Nagpal, R.: Self-reconfiguration using directed growth. In: Proceedings of the 7th International Conference on Distributed Autonomous Robotic Systems (DARS) (2004)
14. White, P., Zykov, V., Bongard, J., Lipson, H.: Three dimensional stochastic reconfiguration of modular robots. In: Proceedings of the Robotics: Science and Systems I (2005)
15. Wolpert, L., Tickle, C., Lawrence, P., Meyerowitz, E., Robertson, E., Smith, J., Jessell, T.: Principles of Development, 4th edn. Oxford University Press, Oxford (2011)
16. Yim, M., Shen, W.M., Salemi, B., Rus, D., Moll, M., Lipson, H., Klavins, E., Chirikjian, G.S.: Modular self-reconfigurable robot systems. IEEE Robot. Autom. Mag. **14**(1), 43–52 (2007)
17. Zykov, V., Mytilinaios, E., Adams, B., Lipson, H.: Robotics: self-reproducing machines. Nature **435**, 163–164 (2005)

Towards Differentially Private Aggregation of Heterogeneous Robots

Amanda Prorok and Vijay Kumar

Abstract We are interested in securing the operation of robot swarms composed of heterogeneous agents that collaborate by exploiting aggregation mechanisms. Since any given robot type plays a role that may be critical in guaranteeing continuous and failure-free operation of the system, it is beneficial to conceal individual robot types and, thus, their roles. In our work, we assume that an adversary gains access to a description of the dynamic state of the swarm in its non-transient, nominal regime. We propose a method that quantifies how easy it is for the adversary to identify the type of any of the robots, based on this observation. We draw from the theory of *differential privacy* to propose a closed-form expression of the *leakage* of the system at steady-state. Our results show how this model enables an analysis of the leakage as system parameters vary; they also indicate design rules for increasing privacy in aggregation mechanisms.

1 Introduction

To date, the issues of privacy and security remain poorly addressed within robotics at large. These issues are particularly important in heterogeneous multi-robot systems: by introducing heterogeneity by design, we impose a certain degree of uniqueness and specialization. Indeed, it is generally acknowledged that exploiting *heterogeneity* by design leads to more versatile systems [5, 19]. However, as a consequence, any given robot type may be critical to securing the system's ability to operate without failure. Hence, we must find ways of protecting the system to avoid threats that arise when the roles within the swarm can be determined by adversaries.

A. Prorok (✉) · V. Kumar
University of Pennsylvania, Philadelphia, PA, USA
e-mail: prorok@seas.upenn.edu

V. Kumar
e-mail: kumar@seas.upenn.edu

R. Groß et al. (eds.), *Distributed Autonomous Robotic Systems*,
Springer Proceedings in Advanced Robotics 6,
https://doi.org/10.1007/978-3-319-73008-0_41

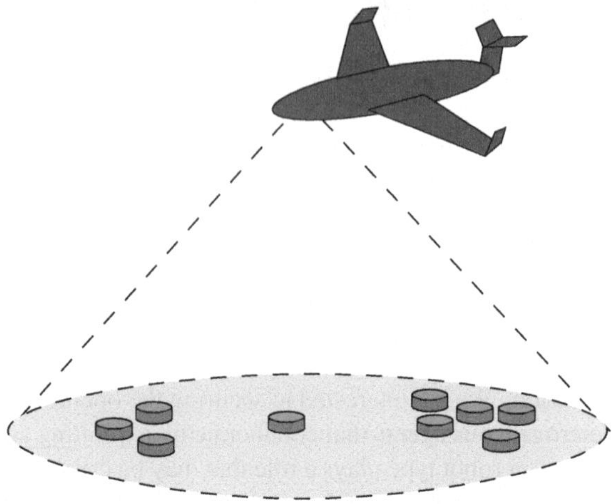

Fig. 1 Example scenario: an adversarial spy-plane can take a snapshot of a heterogeneous robot swarm. Even when the behavioral state is observable, the goal is to ensure that individual robot types/roles remain hidden

In order to collaborate and reap the benefits of their complementarity, robots create coalitions. By doing so, the robots assemble into virtual or physical formations of higher complexity and functionality. Indeed, aggregation is a mechanism that is exploited by nature to enable interactions and information exchange between biological individuals (e.g., for predator protection and collective decision-making [10, 12, 18]). Inspired by the potential of such systems, swarm roboticists have tackled the problem of engineering and analyzing aggregation behaviors [3, 4, 11, 16, 17]. The work in this paper goes beyond previous approaches by including *privacy* as a novel component, and by posing a complementary problem formulation. Let us consider the example application shown in Fig. 1. The scenario considers a heterogeneous robot swarm that is dynamically aggregating, in order to solve an underlying problem. The swarm is observable from an outside vantage point, and is visually homogeneous. By taking a snapshot of the swarm's behavioral state, an adversary gains access to observable system-level information that may allow him to infer the role of a particular robot. By exploiting this knowledge, the adversary can then scheme attacks on the system (e.g., by targeting a specific robot type that he knows is critical to the system's operation). As a consequence, we aim at answering the following question: *How easy is it for the adversary to guess the role/type of any robot in the system?* To answer this question, we develop a model that enables the analysis of privacy as a function of the parameters that define the swarm's aggregation behavior. In other words, we measure the loss of privacy caused by the observable state of the swarm. The following sections develop our models, and elaborate the interplay between privacy and swarm behavior.

2 Model of Robot System

We define a system of robots, where each robot is classified according to its capabilities, and belongs to a *species* [19]. The system is composed of N_S species $\mathscr{S} = \{1, \ldots, N_S\}$, with a total number of robots N, and $N^{(s)}$ robots per species such that $\sum_{s\in\mathscr{S}} N^{(s)} = N$. At a high level of abstraction, we model the robots' actions (and interactions) as discrete stochastic events. We denote the states that compose the system as aggregates $a^{\mathscr{I}}$, where the superscript $\mathscr{I}$ is the list of species that compose the aggregate. For example, $a^{\{1,2\}}$ is an aggregate of species 1 and 2. Aggregation mechanisms can be expressed very elegantly using the formalism of Chemical Reaction Networks (CRN) [13, 15, 17]. Indeed, CRNs are a powerful means of representing complex systems, and allow us to represent species interactions through linear as well as non-linear functions — though not a new field of research, many recent research findings that simplify the calculations are accelerating the adoption of CRNs into domains other than biology and chemistry [9]. Figure 2 shows an example of three species that aggregate.

We define our CRN as a triplet $\mathscr{N} = (\mathscr{A}, \mathscr{C}, \mathscr{R})$, where $\mathscr{A}$ is the set of aggregate states, $\mathscr{C}$ is the set of complexes, and $\mathscr{R}$ is the set of reactions.

<u>State set $\mathscr{A}$:</u> The state set encompasses all states that arise in the system, with $\mathscr{A} = \{A_1, \ldots, A_{N_A}\}$ where N_A is the number of states. States relating to a specific species s are denoted by $\mathscr{A}^{(s)}$. The set of all states is denoted

$$\mathscr{A} = \bigcup_{s=1}^{N_S} \mathscr{A}^{(s)} \text{ and } \mathscr{A}^{(s)} = \bigcup_{s\in\mathscr{I}} a^{\mathscr{I}} \tag{1}$$

We can identify the interactive (interspecific) states of an arbitrary subset of species $\tilde{\mathscr{S}} \subset \mathscr{S}$ by considering the intersection of sets $\bigcap_{i\in\tilde{\mathscr{S}}} \mathscr{A}^{(i)}$. Trivially, if $\bigcap_{i\in\tilde{\mathscr{S}}} \mathscr{A}^{(i)} = \emptyset$, then the species in $\tilde{\mathscr{S}}$ do not interact. The CRN is a population model, and allows us to keep track of the number of robots in each of the states (aggregates) in $\mathscr{A}$. Hence, we define a population vector $\mathbf{x} = [x_1, \ldots, x_{N_A}] \in \mathbb{N}_{\geq 0}^{N_A}$, where x_i corresponds to the population present in state A_i. We refer to the population vector $\mathbf{x}$ as the system-level state. In order to simplify the formulation of our case study later on, we will also use the notation $\mathbf{x}^{\mathscr{I}}$ to refer explicitly to the population in state $a^{\mathscr{I}}$.

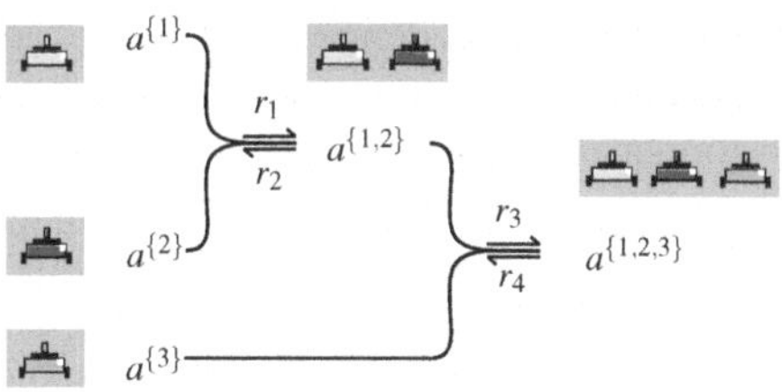

Fig. 2 Example of aggregation with three heterogeneous species that complement each other within their aggregates. Species 1 and 2 are topologically equivalent

Complex set $\mathscr{C}$: The complex set is defined as $\mathscr{C} = \{C_1, \ldots, C_{N_C}\}$, with N_C the number of complexes, and where $C_j = \sum_{i=1}^{N_A} \rho_{ij} A_i$ for $j = 1, \ldots, N_C$, with vector $\boldsymbol{\rho_j} = [\rho_{1j}, \ldots, \rho_{N_A j}]^\top \in \mathbb{N}_{\geq 0}^{N_A}$. A complex is a linear combination of states, and denotes the net input or output of a reaction. In other words, a complex denotes either (i) the states that are required for a certain reaction to take place, or (ii) the states that occur as an outcome of a certain reaction that took place. The non-negative integer terms ρ_{ij} are coefficients that represent the multiplicity of the states in the complexes.

Reaction set $\mathscr{R}$: We use complexes to formulate reactions $R_l : C_j \xrightarrow{r_l} C_k$. The reaction set is defined as $\mathscr{R} = \{R_1, \ldots, R_{N_R}\}$, with N_R the number of reactions, such that $R_l \in \{(C_j, C_k) | \exists\, C_j, C_k \text{ with } C_j \rightarrow C_k\}$ for $j, k = 1, \ldots, N_C$, and where r_l is the rate function $r_l(\mathbf{x}; \kappa_l) : \mathbb{N}_{\geq 0}^{N_A} \mapsto \mathbb{R}_{\geq 0}$ parameterized by rate constant κ_l. In this work, we use non-linear mass-action rate functions, and $r_l(\mathbf{x}; \kappa_l) = \kappa_l \prod_{i=1}^{N_A} x_i^{\rho_{ij}}$ for all $R_l = (C_j, \cdot)$. A set of complexes that is connected by reactions is termed a linkage class. The net loss and gain of each reaction is summarized in a $N_A \times N_R$ stoichiometry matrix Γ, the columns of which encode the change of population per reaction. In particular, the ith column of Γ corresponds to the ith reaction $R_i = (C_j, C_k)$ and thus, the column is equal to $\boldsymbol{\rho_k} - \boldsymbol{\rho_j}$. The elements Γ_{ji} are the so-called stoichiometric coefficients of the jth state in the ith reaction. Positive and negative coefficients denote products and reactants of the reaction, respectively.

A simple stochastic model for CRNs treats the system as a continuous time Markov chain with state $\mathbf{x} \in \mathbb{N}_{\geq 0}^{N_A}$ (i.e., the population vector), and with each reaction modeled as a possible transition for the state. Hence, the number of transitions between two neighboring states is Poisson distributed (equivalently, the time between two transitions is exponentially distributed). In order to calibrate rate constants κ_l on hand of a real system, we proceed by measuring the effective transition rates (by observing the number of transitions, assuming the number of robots is known), and using the mass-action rate functions to solve for the parameter values.

Finally, we describe the dynamics of our system with help of two functions: an execution function $f_{\mathscr{N}}$, and a query function q:

$$\begin{aligned} f_{\mathscr{N}}(\mathbf{x_0}, t) &: \mathbb{N}_{\geq 0}^{N_A} \times \mathbb{R}_{\geq 0} \mapsto \mathbb{N}_{\geq 0}^{N_A} \\ q(\mathbf{x}) &: \mathbb{N}_{\geq 0}^{N_A} \mapsto \mathbb{N}^{N_O},\ N_O \in \mathbb{N}_{>0} \end{aligned} \tag{2}$$

The execution function $f_{\mathscr{N}}$ samples a trajectory that describes the system's evolution over time, as defined by the states and reactions defined in $\mathscr{N}$, and returns the population vector $\mathbf{x}(t)$, evaluated at a fixed time t. The query function q allows us to formalize the notion of an *observable system-level* state. It takes the population vector $\mathbf{x}$ as input, and returns a vector of observable values $\mathbf{y}$. In its most basic form, the query function is the identity function, meaning that an observer is able to capture the exact (true) system-level state, and $\mathbf{x} = \mathbf{y}$. In this work, we assume that the observed values take the form of simple summations over the population vector. This assumption is well motivated when individual robots are not distinguishable from an outside vantage point, and thus, only aggregated values can be observed. I.e.,

$y_i = \sum_{j \in \Omega_i} x_j$ with $\Omega_i \subset \{1, \ldots, N_A\}$, $\cup_{i \in \{1,\ldots,N_O\}} \Omega_i = \{1, \ldots, N_A\}$, and all Ω_i disjoint. In our aggregation case-studies, we use $y_i = \sum_{\mathscr{I} \text{ s.t.} |\mathscr{I}|=i} x^{\mathscr{I}}$ for all $a^{\mathscr{I}} \in \mathscr{A}$, which counts the number of aggregates of a given size.

3 Definition of Differentially Private Swarm

In the following, we put forward a model that measures the privacy of a dynamic swarm of robots. Various measures of privacy have been proposed in the database literature so far. The early work in [1] proposes a quantification of privacy in terms of the amount of noise added to a true value, or, in other words, how closely the original value of a modified attribute can be estimated. This measure, however, omits the notion of side-information, i.e., any additional information about the underlying distribution that the adversary might own. The work in [8] extends the notion of privacy to include such prior knowledge. The proposed measure suggests a quantification of the largest difference between an adversary's a-priori to a-posterior beliefs (which corresponds to the worst-case scenario). It turns out that this model is significantly stronger, since it accounts for infrequent, but noticeable privacy breaches. In 2006, Dwork et al. introduced the notion of ε-indistinguishability, a generalization of the measure in [8], and later coined the term of *differential privacy* [7]. Today, differentially private mechanisms are enjoying tremendous success, due to their ability of dealing with arbitrary side-information (a *future-proof* quality) and worst-case scenarios [6]. For these reasons, we build our formalism on the theory of differential privacy.

In a nutshell, differential privacy is the statistical science of trying to learn as much as possible about a group while learning as little as possible about any of its individuals. It considers two key components: a *database* that holds sensitive information pertaining to individuals, and a *query* that releases information obtained from the database via a mechanism. If an observer, who can request data from the database (through a query), cannot significantly reduce the uncertainty of her prior knowledge (i.e., side information) using the requested data, the query is considered private.[1] In particular, if an individual's presence or absence in the database does not alter the distribution of the output of the query by a significant amount, regardless of the side information, then the privacy of that individual's information is assured. Our analogy applies the concepts of database and query to the context of heterogeneous swarms. First, we consider a database that represents the composition of our robot swarm, and that records the species of each of the robots. Second, we consider an adversary who 'queries the system' by taking a snapshot of its observable state. The

[1] Side information can be understood as a prior probability distribution over the database [14].

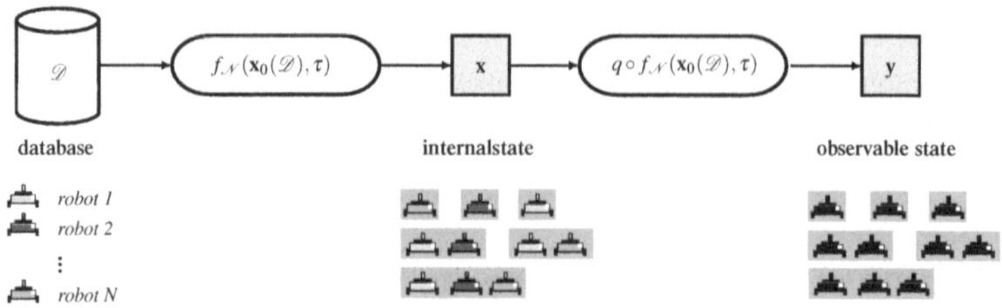

Fig. 3 The composition of the robot swarm is recorded in a database $\mathscr{D}$. The function $f_{\mathcal{N}}$ is a stochastic process that executes the swarm and returns a system-level state output $\mathbf{x}$. Query function q reads the (internal) system-level state, and returns the observable output $\mathbf{y}$. Parameter τ denotes the time at which the system is observed

adversary may also own arbitrary side-information.[2] Then, our analogous definition of privacy is the notion that the adversary cannot infer to which species individual robots belong. The composition of our robot swarm is recorded in a database $\mathscr{D} \in \mathscr{S}^N$ that consists of N entries, where each entry $\mathscr{D}_i$ denotes the species of robot i. We define an *adjacency* set Adj($\mathscr{D}$) that encompasses all databases $\mathscr{D}$' adjacent to $\mathscr{D}$. Two databases $\mathscr{D}$ and $\mathscr{D}$' are adjacent if they differ by one single entry. In other words, two robot swarms (represented by $\mathscr{D}$ and $\mathscr{D}$') are adjacent if they differ by one robot i, meaning that robot i belongs to s_i in $\mathscr{D}$ (i.e., $\mathscr{D}_i = s_i$), and to a different species $s_i' \neq s_i$ in $\mathscr{D}$' (i.e., $\mathscr{D}_i \neq s_i$). As previously described, the behavior of the robot swarm can be described by tracking the system-level states. If we let the system run, it produces a trajectory that can be evaluated at a given time τ, resulting in a snapshot of the population vector $\mathbf{x}$.[3] Our query/response model consists of a user (adversary) who is able to query this system-level state (at time τ). Hence, the query $q \circ f_{\mathcal{N}}(\mathbf{x}_0(\mathscr{D}), \tau)$ depends on the swarm composition $\mathscr{D}$, and the time at which the system is observed τ. The function $\mathbf{x}_0(\mathscr{D}) : \mathscr{S}^N \mapsto \mathbb{N}_{\geq 0}^{N_A}$ distributes the robots in $\mathscr{D}$ to their initial states. A schema of this system is shown in Fig. 3.

Our aim is to analyze the differential privacy of the observed system output. To this end, we propose a definition of differential privacy that applies to dynamic swarms [20]:

Definition 1 (ε – *indistinguishable heterogeneous swarm*) A heterogeneous swarm with dynamics defined by a system $\mathcal{N}$ is ε-indistinguishable (and gives ε-differential privacy) if for all possible swarm compositions recorded in databases $\mathscr{D}$, we have

$$\mathscr{L}(\mathscr{D}) = \max_{\mathscr{D}' \in \mathrm{Adj}(\mathscr{D})} \left| \ln \frac{\mathbb{P}[q \circ f_{\mathcal{N}}(\mathbf{x}_0(\mathscr{D}), \tau)]}{\mathbb{P}[q \circ f_{\mathcal{N}}(\mathbf{x}_0(\mathscr{D}'), \tau)]} \right| \leq \varepsilon. \tag{3}$$

[2] In our context of a robotic swarm, an example of side information could be the number of manufacturing parts ordered to build the swarm. If different robot species are made of different parts, such information can be used to construct an initial guess about the number of robots per species. Thus, one would be able to derive the probability of a robot belonging to a given species.

[3] We assume a snapshot adversary that gains system-level information at a specific time. This system-level information is a design variable, called the *observable* system-level state.

where $\mathbb{P}[\mathbf{y}]$ denotes the probability of the output $\mathbf{y}$, obtained through query q of the system-level state given by $f_{\mathcal{N}}$.

The value ε is referred to as the leakage. Intuitively, this definition states that if two swarm systems are close, in order to preserve privacy they should correspond to close distributions on their observable outputs. The above definition is stringent: for a pair of distributions whose statistical difference is arbitrarily small, the ratio may still result in an infinite value (leakage), when a point in one distribution assigns probability zero and the other non-zero. Later, in our evaluations, we use a smooth version of the leakage formula above, by adding an arbitrary, negligibly small value ν, uniformly over the support of the probability distributions. This allows us to differentiate between large and small mismatches of the output when one point of a probability distribution returns zero. Due to this addition, we are able to show continuous privacy trends as a function of the underlying parameters.

4 Complex-Balanced Swarms

The formula in Eq. (3) provides strong privacy guarantees. Yet, it requires that we have a way of specifying the probability distribution over the swarm's observable output. The choice of method for computing this probability distribution depends on the time at which the adversary takes the snapshot of the swarm. In [20], we presented a method that computes the probability distribution at a specific time (which can be during any regime, transient or stationary). Here, we present a method that computes the probability distribution at steady-state, in the swarm's operational mode. We show how this can be done very efficiently for a class of CRNs whose stationary distribution can be formulated analytically: *complex-balanced* CRNs. Later, in Sect. 5, we prove that aggregation mechanisms are complex-balanced.

4.1 Preliminaries

If a CRN is complex-balanced, it admits a single equilibrium point $\bar{\mathsf{x}} \in \mathbb{R}^{N_A}$ [21]. When modeled deterministically, the average population (rather than an exact robot count) in system states changes according to an ODE, described as follows[4]:

$$\dot{\mathsf{x}} = \mathbf{MA}\psi(\mathsf{x}), \tag{4}$$

where $\psi(\mathsf{x})$ returns a vector in $\mathbb{R}^{N_C}$ in which each entry ψ_j is the product of states in complex j (i.e., $\psi_j = \prod_{i=1}^{N_A} x_i^{\rho_{ij}}$), where $\mathbf{M} \in \mathbb{R}^{N_A \times N_C}$ is a matrix in which each entry $\mathbf{M}_{ij}$ is the coefficient of state j in complex i, and where matrix $\mathbf{A} \in \mathbb{R}^{N_C \times N_C}$ is defined

[4]The symbol $\mathbf{x}$ denotes the discrete state, whereas x denotes the average population.

as

$$\mathbf{A}_{ij} = \begin{cases} \kappa_{ji}, & \text{if } i \neq j, (C_i, C_j) \in \mathscr{R} \\ 0, & \text{if } i \neq j, (C_i, C_j) \notin \mathscr{R} \\ -\sum\limits_{(C_i, C_k) \in \mathscr{R}} \kappa_{ki}, & \text{if } i = j \end{cases}$$

If this system admits $\mathbf{A}\psi(\bar{\mathsf{x}}) = \mathbf{0}$, then the system is complex balanced, with equilibrium point $\bar{\mathsf{x}} \in \mathbb{R}^{N_A}$. Following Theorem (4.2) of [2], we can use the equilibrium point $\bar{\mathsf{x}}$ to define the stationary distribution $\bar{\pi}_{\mathscr{D}}(\mathbf{x}) = \lim_{t\to\infty} \mathbb{P}[f_{\mathscr{N}}(\mathbf{x_0}(\mathscr{D}), t)]$ of the stochastically modeled system. If the system is irreducible, this stationary distribution consists of a product of Poisson distributions and is given by

$$\bar{\pi}_{\mathscr{D}}(\mathbf{x}) = \prod_{i=1}^{N_A} \frac{\bar{\mathsf{x}}_i^{x_i}}{x_i!} e^{-\bar{\mathsf{x}}_i}, \ \mathbf{x} \in \mathscr{X}_{\mathscr{N}}(\mathscr{D}) \tag{5}$$

where $\mathscr{X}_{\mathscr{N}}(\mathscr{D})$ is the set of all possible population vectors $\mathbf{x}$ that can arise from the CRN $\mathscr{N}$ and the robot species specified by $\mathscr{D}$. We note that when the system is reducible, a similar equation exists, see [2].

4.2 Privacy

If the swarm's CRN model is complex-balanced, we are able to derive a stationary probability density function describing the steady-state of the system. We use this description to formulate a closed-form measure of the loss of privacy.

Proposition 1 *If a swarm's CRN is complex-balanced and irreducible, then the leakage at the steady-state of the corresponding swarm system defined by $\mathscr{D}$, and observed through the identity query $q_{\mathscr{N}}(\mathbf{x}) = \mathbf{x}$ is*

$$\mathscr{L}(\mathscr{D}) = \max_{\substack{\mathscr{D}' \in \mathrm{Adj}(\mathscr{D}) \\ \mathbf{x} \in \mathscr{X}_{\mathscr{N}}(\mathscr{D}) \cup \mathscr{X}_{\mathscr{N}}(\mathscr{D}')}} \left| \sum_{i=1}^{N_A} x_i \ln \frac{\bar{\mathsf{x}}_i}{\bar{\mathsf{x}}'_i} - \bar{\mathsf{x}}_i + \bar{\mathsf{x}}'_i \right| \tag{6}$$

where $\bar{\mathsf{x}}$ and $\bar{\mathsf{x}}'$ are the steady-states resulting from $\mathscr{D}$ and $\mathscr{D}'$.

Proof Starting with Eq. (3), and using query $q_{\mathscr{N}}(\mathbf{x}) = \mathbf{x}$, we have

$$\mathscr{L}(\mathscr{D}) = \max_{\mathscr{D}' \in \mathrm{Adj}(\mathscr{D})} \left| \ln \frac{\mathbb{P}[f_{\mathscr{N}}(\mathbf{x_0}(\mathscr{D}), \tau)]}{\mathbb{P}[f_{\mathscr{N}}(\mathbf{x_0}(\mathscr{D}'), \tau)]} \right|. \tag{7}$$

At steady-state we have $\lim_{\tau\to\infty} \mathbb{P}[f_{\mathscr{N}}(\mathbf{x_0}(\mathscr{D}), \tau)] = \bar{\pi}_{\mathscr{D}}(\mathbf{x})$, hence

$$\mathscr{L}(\mathscr{D}) = \max_{\substack{\mathscr{D}' \in \mathrm{Adj}(\mathscr{D}) \\ \mathbf{x} \in \mathscr{X}_{\mathscr{N}}(\mathscr{D}) \cup \mathscr{X}_{\mathscr{N}}(\mathscr{D}')}} \left| \ln \frac{\bar{\pi}_{\mathscr{D}}(\mathbf{x})}{\bar{\pi}_{\mathscr{D}'}(\mathbf{x})} \right|. \tag{8}$$

Continuing with Eq. (5) we get

$$\mathcal{L}(\mathcal{D}) = \max_{\substack{\mathcal{D}' \in \mathrm{Adj}(\mathcal{D}) \\ \mathbf{x} \in \mathcal{X}_{\mathcal{N}}(\mathcal{D}) \cup \mathcal{X}_{\mathcal{N}}(\mathcal{D}')}} \left| \ln \left(\prod_{i=1}^{N_A} \frac{\bar{\mathrm{x}}_i^{x_i}}{x_i!} e^{-\bar{\mathrm{x}}_i} \right) - \ln \left(\prod_{i=1}^{N_A} \frac{\bar{\mathrm{x}}_i'^{x_i}}{x_i!} e^{-\bar{\mathrm{x}}_i'} \right) \right|, \tag{9}$$

which yields Eq. (6). □

Corollary 1 *If a swarm's CRN is complex-balanced and irreducible, then the leakage at steady-state of the corresponding swarm system defined by $\mathcal{D}$, and observed through the query $q_{\mathcal{N}}(\mathbf{x}) = \mathbf{y}$, with $\mathbf{y} \in \mathbb{N}_{\geq 0}^{N_O}$ and each y_i of the form $\sum_{j \in \Omega_i} x_j$, with $\Omega_i \subset \{1, \ldots, N_A\}$, $\cup_{i \in \{1,\ldots,N_O\}} \Omega_i = \{1, \ldots, N_A\}$, and all Ω_i disjoint is*

$$\mathcal{L}(\mathcal{D}) = \max_{\substack{\mathcal{D}' \in \mathrm{Adj}(\mathcal{D}) \\ \mathbf{y}}} \left| \ln \frac{\sum_{\{\mathbf{x} | \mathbf{y} = q_{\mathcal{N}}(\mathbf{x}) \wedge \mathbf{x} \in \mathcal{X}_{\mathcal{N}}(\mathcal{D})\}} \bar{\pi}_{\mathcal{D}}(\mathbf{x})}{\sum_{\{\mathbf{x} | \mathbf{y} = q_{\mathcal{N}}(\mathbf{x}) \wedge \mathbf{x} \in \mathcal{X}_{\mathcal{N}}(\mathcal{D}')\}} \bar{\pi}_{\mathcal{D}'}(\mathbf{x})} \right|. \tag{10}$$

This formulation, even though less compact than Eq. (6) above, still provides a fast means of computing the leakage for complex-balanced swarms — in particular, the alternative to using this formulation is to compute the PMF via the Chemical Master Equation [20], which, in our experience, is at least one order of magnitude slower. Moreover, we note that Eq. (6) is linear in $\mathbf{x}$, and can, thus, be solved by integer linear programming (ILP) methods. In summary, the analytical formulation for the privacy of complex-balanced swarms allows us to compute the leakage efficiently. In the remainder of this work, we demonstrate the benefit of this formulation with case studies on aggregation.

5 Aggregation

To apply Corollary 1, we must first show that aggregation is a complex-balanced mechanism. In order to develop our proof, we represent the topology of the reaction networks through directed acyclic graphs (DAG) (see the example in Fig. 4).

Definition 2 (*Aggregation – DAG*) An aggregation-DAG is a topological representation of a CRN that defines an aggregation mechanism. It is a directed acyclic graph, such that each node forms a complex C_k, the sum of its in-neighbors form complex C_j, and its incoming edges form the reactions $C_j \underset{r_n}{\overset{r_m}{\rightleftharpoons}} C_k$, with $r_m, r_n > 0$. Furthermore, a node has either 0 or > 1 in-neighbors.

Proposition 2 *The aggregation of a heterogeneous swarm of robots described by a CRN is a complex-balanced mechanism if and only if the underlying CRN can be represented by an aggregation-DAG.*

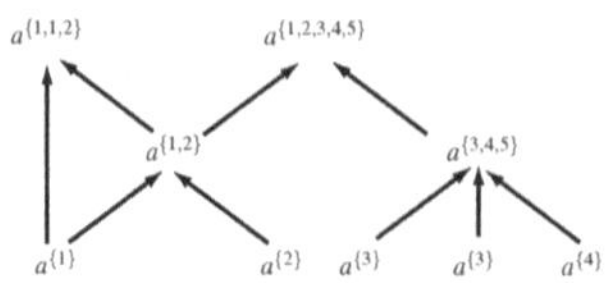

Fig. 4 Example of an aggregation mechanism that is represented as a directed acyclic graph. There are $L = 4$ linkage classes, $N_C = 8$ complexes, and $\mathrm{rank}(\Gamma) = 4$

Proof According to Theorem 4.1 of [21], a CRN is complex-balanced if (i) it is weakly reversible and (ii) it has deficiency zero. Condition (i) requires all complexes to be connected via some reaction pathway (cf. Definition 2.2 in [2]). If all aggregates can be decomposed as well as composed, this is trivially satisfied. The deficiency of a reaction network is $\delta = N_C - L - \mathrm{rank}(\Gamma)$, which is the number of complexes minus the number of linkage classes, each of which is a set of complexes connected by reactions, minus the network rank, which is the rank of the stoichiometry matrix Γ. Hence, we will show that $N_C = L + \mathrm{rank}(\Gamma)$. From Definition 2 it follows that the number of linkage classes L is equal to the number of parent nodes, and the number of complexes N_C is equal to twice the number of parent nodes. Thus, $N_C = 2L$, and it remains to be shown that $\mathrm{rank}(\Gamma) = L$. Matrix Γ is of size $N_A \times N_R$, with $N_R = 2L$ (the network is weakly reversible). Since each new linkage class includes a new parent node (i.e., aggregate state), there are L linearly independent columns in Γ, and, hence, $\mathrm{rank}(\Gamma) = L$. □

5.1 Example

We consider the example shown in Fig. 2. The system is composed of three species, $\mathscr{S} = \{1, 2, 3\}$. Aggregates are formed with one robot per species, and with species 1 and 2 aggregating prior to species 3. This behavior is formalized with the following reactions:

$$a^{\{1\}} + a^{\{2\}} \underset{r_2}{\overset{r_1}{\rightleftharpoons}} a^{\{1,2\}}, \qquad a^{\{1,2\}} + a^{\{3\}} \underset{r_4}{\overset{r_3}{\rightleftharpoons}} a^{\{1,2,3\}} \tag{11}$$

The states of this system are $\mathscr{A} = \{a^{\{1\}}, a^{\{2\}}, a^{\{3\}}, a^{\{1,2\}}, a^{\{1,2,3\}}\}$. Our population vector keeps track of the number of robots per state, and is written

$$\mathbf{x} = [x^{\{1\}}, x^{\{2\}}, x^{\{3\}}, x^{\{1,2\}}, x^{\{1,2,3\}}] \tag{12}$$

We consider an adversary who is able to observe the number of non-aggregated robots, the number of 2-aggregates, and the number of 3-aggregates. Hence, we formulate the observable data as $\mathbf{y} = [y_1, y_2, y_3]$ with

$$y_1 = x^{\{1\}} + x^{\{2\}} + x^{\{3\}}, \; y_2 = x^{\{1,2\}}, \; y_3 = x^{\{1,2,3\}}. \tag{13}$$

5.1.1 Analysis

We compute the steady-state $\bar{\mathsf{x}}$ by solving the deterministic system $\mathbf{M}\mathbf{A}_\kappa \psi(\bar{\mathsf{x}}) = \mathbf{0}$:

$$\begin{bmatrix} 1 & 0 & 0 & 0 \\ 1 & 0 & 0 & 0 \\ 0 & 1 & 1 & 0 \\ 0 & 1 & 0 & 0 \\ 0 & 0 & 0 & 1 \end{bmatrix} \cdot \begin{bmatrix} -\kappa_1 & 0 & \kappa_2 & 0 \\ 0 & -\kappa_3 & 0 & \kappa_4 \\ \kappa_1 & 0 & -\kappa_2 & 0 \\ 0 & \kappa_3 & 0 & -\kappa_4 \end{bmatrix} \cdot \begin{bmatrix} \bar{\mathsf{x}}^{\{1\}}\bar{\mathsf{x}}^{\{2\}} \\ \bar{\mathsf{x}}^{\{1,2\}}\bar{\mathsf{x}}^{\{3\}} \\ \bar{\mathsf{x}}^{\{1,2\}} \\ \bar{\mathsf{x}}^{\{1,2,3\}} \end{bmatrix} = \mathbf{0}. \tag{14}$$

Since the number of robots per species is constant, we have

$$\begin{aligned} \bar{\mathsf{x}}^{\{1\}} + \bar{\mathsf{x}}^{\{1,2\}} + \bar{\mathsf{x}}^{\{1,2,3\}} &= N^{(1)} \\ \bar{\mathsf{x}}^{\{2\}} + \bar{\mathsf{x}}^{\{1,2\}} + \bar{\mathsf{x}}^{\{1,2,3\}} &= N^{(2)} \\ \bar{\mathsf{x}}^{\{3\}} + \bar{\mathsf{x}}^{\{1,2,3\}} &= N^{(3)}. \end{aligned} \tag{15}$$

The equations above can be plugged into Eq. (14) to give a quartic equation:

$$0 = \kappa_1 \left(N^{(1)} - N^{(3)} - \bar{\mathsf{x}}^{\{1,2\}} + \frac{\kappa_4 N^{(3)}}{\kappa_3 \bar{\mathsf{x}}^{\{1,2\}} + \kappa_4} \right) \cdot \left(N^{(2)} - N^{(3)} - \bar{\mathsf{x}}^{\{1,2\}} + \frac{\kappa_4 N^{(3)}}{\kappa_3 \bar{\mathsf{x}}^{\{1,2\}} + \kappa_4} \right) - \kappa_2 \bar{\mathsf{x}}^{\{1,2\}} \tag{16}$$

which only depends on variable $\bar{\mathsf{x}}^{\{1,2\}}$. The solution to the remaining variables $\bar{\mathsf{x}}^{\{1\}}$, $\bar{\mathsf{x}}^{\{2\}}$, $\bar{\mathsf{x}}^{\{3\}}$, $\bar{\mathsf{x}}^{\{1,2,3\}}$ of this system can be found by substitution. Of the four possible solutions, there is only a single all-positive solution (which corresponds to the single equilibrium of the complex-balanced system). Finally, we can compute the leakage $\mathscr{L}(\mathscr{D})$ of the observed system according to Corollary 1.

5.1.2 Evaluation

The observable state is a function of the system-level state, and is defined by the number of robots per species $N^{(s)}$, and by reaction rates $\boldsymbol{\kappa}$. Hence, we vary these values to identify their relation to privacy. We note that this relationship is made mathematically evident in Eq. (16). We compute the leakage for two settings, shown in Figs. 5 and 6. In the first setting, we fix the reaction rates and vary the robot populations of two species, while keeping the third species fixed. In Fig. 5a, we observe a clear "valley" of minimum leakage values for an equal number of species 1 and 2. The overall minimum is at $N^{(1)} = N^{(2)} = 220$, $N^{(3)} = 200$. The plot also reveals that increasing the total number of robots increases privacy, as shown by the expansion of the valley in the upper right corner. Panel Fig. 5b shows the resulting leakage for varying species 2 and 3. We observe a sharp drop in privacy as the population $N^{(2)}$ deviates from $N^{(1)}$, with a minimum at $N^{(1)} = N^{(2)} = 220$, $N^{(3)} = 200$ (as previously observed in Fig. 5a).

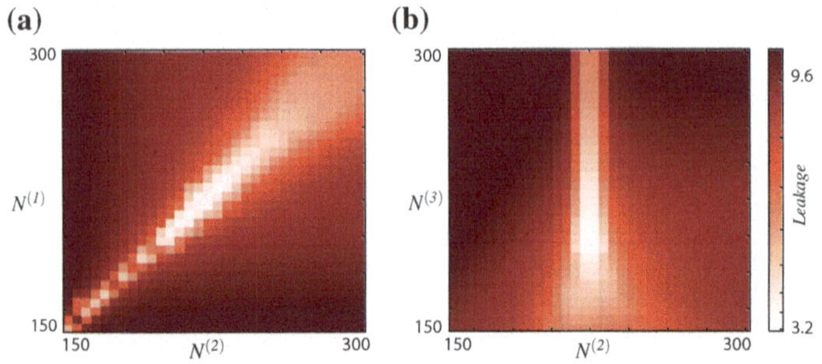

Fig. 5 Leakage for varying populations in the range [150, 300], while keeping the third species fixed, and with fixed reaction rates $\boldsymbol{\kappa} = 1$. In **a** $N^{(1)}$ and $N^{(2)}$ vary, with $N^{(3)}$ =200, and in **b** $N^{(2)}$ and $N^{(3)}$ vary, with $N^{(1)}$ =220

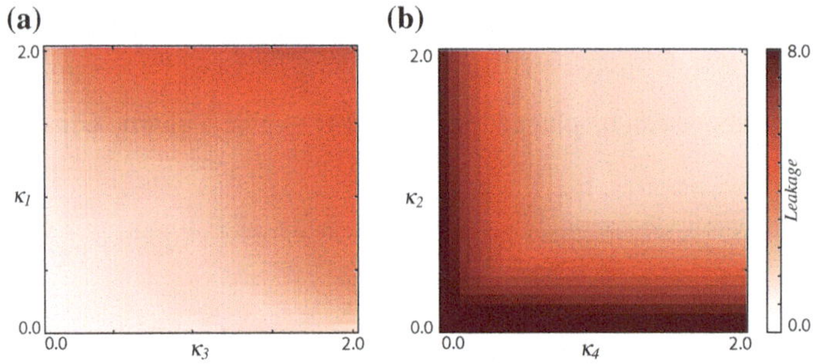

Fig. 6 Leakage for varying aggregation rates with robot populations fixed at $N^{(1)} = N^{(2)} = 220$, $N^{(3)} = 200$. In **c** we vary the rates κ_1, κ_3 at which aggregates form, while fixing $\kappa_2 = \kappa_4 = 1$. In **d** we show the reverse (varying the rates at which aggregates decompose)

We conclude that species that are topologically equivalent (species 1 and 2, as visible in Fig. 2) should have a balanced number of robots for increased privacy. In the second setting (Fig. 6), we vary the reaction rates while keeping the robot populations fixed. Figure 5b and c indicate that if we increase the probability of reaching larger sized aggregates, either by increasing the number $N^{(3)}$, or by increasing the aggregation rates, we decrease privacy. Indeed, in this setting, the 2-aggregates and 3-aggregates are unique, hence, they expose more information about the system.

5.2 Evaluating the Impact of Topology and Parameters

The results of the preceding example indicate that both the topology and parameters of the CRN affect privacy. To expose the impact of a CRN's topology on the leakage, we proceed by considering aggregation-DAGs that can be represented by binary trees. For a system composed of 16 species, we evaluate the leakage for each of the

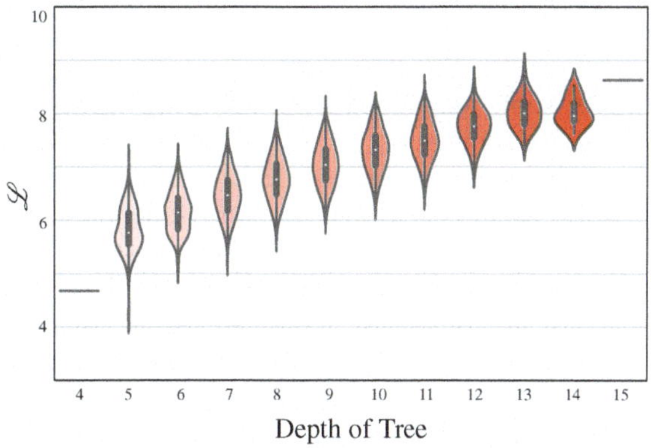

Fig. 7 Leakage for binary aggregation trees of varying topology (with 16 leaves). Trees of same depth are assembled by one violin plot that features a kernel density estimation of the underlying distribution. The Pearson correlation coefficient evaluated on this data is 0.74

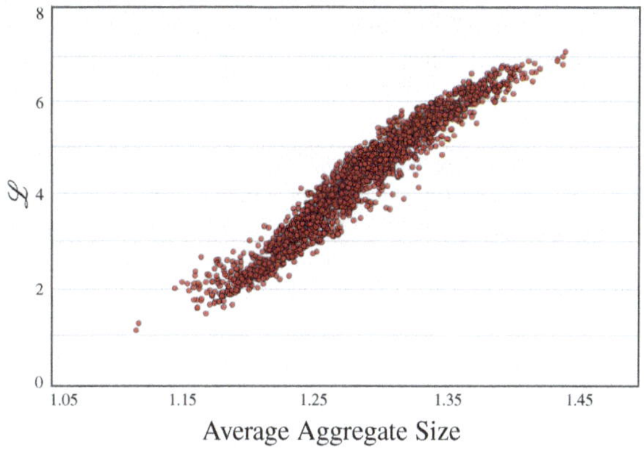

Fig. 8 Leakage for 2000 trees with identical topology (symmetric binary tree, with 8 leaves), and with all 7 aggregation rates varied uniformly and randomly in the range [0.1, 2] (decomposition rates are held constant, equal to 1). The Pearson correlation coefficient evaluated on this data is 0.97

possible 10905 unlabeled binary rooted trees with 16 leaves (which corresponds to the Wedderburn-Etherington number). Figure 7 shows the leakage as a function of the depth of the tree. We see a clear correlation between irregular, unbalanced topologies (with greater depth) and high leakage values, and between more balanced, symmetric topologies (with smaller depth) and low leakage values.

Next, to expose the impact of a CRN's parameters (i.e., aggregation rates) on the leakage, we proceed by considering a symmetric binary tree with 8 leaves (of depth 3), and we vary the aggregation rates uniformly and randomly in the range [0.1, 2], gathering 2000 datapoints. Figure 8 shows the leakage as a function of the average aggregate size (at steady-state) for each set of rates. We see that as the average size increases, so does the leakage.

These results together indicate that privacy can be increased by (i) designing aggregation mechanisms that are balanced (asymmetric aggregation mechanisms create more unique aggregates, and hence, reveal more information about the system), or by (ii) throttling the aggregation of larger aggregates (which tend to be more unique).

6 Conclusion

In this work, we showed how to analyze the privacy of aggregation mechanisms in heterogeneous robot swarms. Our main contribution consists of a closed-form expression that quantifies the leakage of dynamic swarms that can be modeled as complex-balanced reaction networks. We demonstrated that aggregation mechanisms are complex-balanced, and hence, were able to use our formula to efficiently evaluate various settings. The reported results showed how privacy levels vary, as the topology as well as the parameters of the aggregation mechanism are varied. This means that we are able to identify the relation between privacy loss at a macroscopic level, and swarm design parameters (such as reaction rates, reaction topology, and swarm composition). As a consequence, our framework paves the way for methods that control the swarm, while guaranteeing bounds on privacy loss. We intend to further this line of work by developing active privacy mechanisms that are able control the loss of privacy, while maintaining the overall performance of the underlying swarm system.

Acknowledgements The authors would like to thank the anonymous referees for their constructive feedback. We gratefully acknowledge the support of ONR grants N00014-15-1-2115 and N00014-14-1-0510, ARL grant W911NF-08-2-0004, NSF grant IIS-1426840, and TerraSwarm, one of six centers of STARnet, a Semiconductor Research Corporation program sponsored by MARCO and DARPA.

References

1. Agrawal, R., Srikant, R.: Privacy-preserving data mining. ACM Sigmod Record **29**(2), 439–450 (2000)
2. Anderson, D.F., Craciun, G., Kurtz, T.G.: Product-form stationary distributions for deficiency zero chemical reaction networks. Bull. Math. Biol. 1–23 (2010)
3. Cheng, J., Cheng, W., Nagpal, R.: Robust and self-repairing formation control for swarms of mobile agents. AAAI (2005)
4. Correll, N., Martinoli, A.: Modeling self-organized aggregation in a swarm of miniature robots. In: IEEE International Conference Robotics and Automation (ICRA) (2007)
5. Dorigo, M., Floreano, D., Gambardella, L.M., Mondada, F., Nolfi, S., et al.: Swarmanoid: a novel concept for the study of heterogeneous robotic swarms. IEEE Robot. Autom. Mag. **20**(4), 60–71 (2013)
6. Dwork, C.: Differential privacy: a survey of results. In: Theory and Applications of Models of Computation, pp. 1–19. Springer, Berlin (2008)
7. Dwork, C.: Differential privacy. In: Encyclopedia of Cryptography and Security, pp. 338–340 (2011)
8. Evfimievski, A., Gehrke, J., Srikant, R.: Limiting privacy breaches in privacy preserving data mining. In: the Twenty-second ACM SIGMOD-SIGACT-SIGART Symposium, pp. 211–222. ACM Press, New York (2003)
9. Feinberg, M.: Some recent results in chemical reaction network theory. In: Patterns and Dynamics in Reactive Media, pp. 43–70. Springer, New York (1991)
10. Garnier, S., Jost, C., Gautrais, J., Asadpour, M., Caprari, G., Jeanson, R., Grimal, A., Theraulaz, G.: The embodiment of cockroach aggregation behavior in a group of micro-robots. Artif. Life **14**(4), 387–408 (2008)

11. Groß, R., Dorigo, M.: Self-assembly at the macroscopic scale. Proc. IEEE **96**(9), 1490–1508 (2008)
12. Halloy, J., Sempo, G., Caprari, G., Rivault, C., Asadpour, M., Tache, F., Said, I., Durier, V., Canonge, S., Ame, J.M., Detrain, C., Correll, N., Martinoli, A., Mondada, F., Siegwart, R., Deneubourg, J.L.: Social integration of robots into groups of cockroaches to control self-organized choices. Science **318**(5853), 1155–1158 (2007)
13. Hosokawa, K., Shimoyama, I., Miura, H.: Dynamics of self-assembling systems: analogy with chemical kinetics. Artif. Life **1**(4), 413–427 (2010)
14. Kasiviswanathan, S.P., Smith, A.: A note on differential privacy: Defining resistance to arbitrary side information. CoRR abs. (2008)
15. Klavins, E., Burden, S., Napp, N.: Optimal rules for programmed stochastic self-assembly. In: Robotics: Science and Systems (2006)
16. Martinoli, A., Ijspeert, A.J., Gambardella, L.M.: A probabilistic model for understanding and comparing collective aggregation mechanisms. In: Advances in Artificial Life, pp. 575–584. Springer, Berlin (1999)
17. Matthey, L., Berman, S., Kumar, V.: Stochastic strategies for a swarm robotic assembly system. In: IEEE International Conference on Robotics and Automation (ICRA), pp. 1953–1958. IEEE, New York (2009)
18. Parrish, J.K., Edelstein-Keshet, L.: Complexity, pattern, and evolutionary trade-offs in animal aggregation. Science **284**(5411), 99–101 (1999)
19. Prorok, A., Hsieh, A.M., Kumar, V.: Formalizing the impact of diversity on performance in a heterogeneous swarm of robots. In: IEEE International Conference on Robotics and Automation (ICRA) (2016)
20. Prorok, A., Kumar, V.: A macroscopic privacy model for heterogeneous robot swarms. In: International Conference on Swarm Intelligence (2016)
21. Siegel, D., MacLean, D.: Global stability of complex balanced mechanisms. J. Math. Chem. **27**, 89–110 (2000)

Part VII
Multi-Robot Systems in Applications

Construction Planning for a Modularized Rail Structure: Type Selection of Rail Structure Modules and Dispatch Planning of Constructor Robots

Rui Fukui, Yuta Kato, Gen Kanayama, Ryo Takahashi and Masayuki Nakao

Abstract Remote robot operation is highly anticipated for use in hazardous environments such as nuclear accident sites. We propose an automated construction system of a modularized rail structure for working robots to have access to any operational point. The modules are delivered and constructed through the cooperation of transfer robots and a connector robot. To realize time-efficient and economical construction of the structure, it is necessary to integrate three planning procedures: path planning from a start point to the operational point, type selection planning of rail structure modules, and dispatch planning of constructor robots. This paper describes newly developed algorithms that plan the type selection of all rail structure modules using rules of thumb, and which plan the dispatch of robots to deliver or construct modules avoiding deadlock. A simulation experiment demonstrates that the geometrical constraint conditions of the structure can reduce the search space of selecting module types.

R. Fukui (✉) · Y. Kato · G. Kanayama · R. Takahashi · M. Nakao
Department of Mechanical Engineering, Graduate School of Engineering, The University of Tokyo, Tokyo, Japan
e-mail: fukui@ra-laboratory.com

Y. Kato
e-mail: kato@hnl.t.u-tokyo.ac.jp

G. Kanayama
e-mail: kanayama@hnl.t.u-tokyo.ac.jp

R. Takahashi
e-mail: r.takahashi@hnl.t.u-tokyo.ac.jp

M. Nakao
e-mail: nakao@hnl.t.u-tokyo.ac.jp

R. Groß et al. (eds.), *Distributed Autonomous Robotic Systems*,
Springer Proceedings in Advanced Robotics 6,
https://doi.org/10.1007/978-3-319-73008-0_42

1 Introduction

Robot operations are highly anticipated at the accident site of Tokyo Electric Power Company's Fukushima No. 1 Nuclear Power Plant (1F) because the high intensity of radiation makes human operations difficult and dangerous. Moreover, a secondary disaster such as collapsing rubble might occur [9]. To date, various remote-controlled robots have been developed to do specific tasks at a working destination point. Some robots are actually sent inside the reactor buildings in 1 F and conducted missions with great success, such as observations of interior spaces, monitoring of radiation levels, and acquisition of samples [3, 5, 8]. However, for the completion of decommission, it is especially necessary to bring melted debris outside 1F. To do so, multiple operational robots must locomote back and forth repeatedly to deliver the necessary resources. The robots have payload limitations for transfer. Therefore, they must locomote repeatedly. Some existing rescue robots have legs or crawlers as a locomotive mechanism. Mechanisms of these types have limited mobility, particularly in an upward direction or on terrain with rubble. Moving these robots inside 1 F forces the operational workers of these robots to undergo great stress because of the small field of vision. To use operational robots effectively, it is necessary to construct an environment for ease of locomotion and operations.

As presented in Fig. 1, Fukui et al. propose an "Automated Construction System of a Modularized Rail Structure" [1]. Robots locomote on surfaces of modularized rail structures like a gscaffold,h which means that this structure is a locomotion and operation environment for robots. Robots have a driving unit and hanging mechanism in the bottom part to prevent falling from the structure. The structure is constructed automatically and successively by specialized constructor robot groups. Therefore, robots can move upward and avoid obstacles easily if the rail structure is properly located. Up to the present, various kinds of modular robots have been developed. Terada et al. proposed an automated construction system called Automatic Modu-

Fig. 1 Conceptual sketch of "Automated Construction System of Modularized Rail Structure"

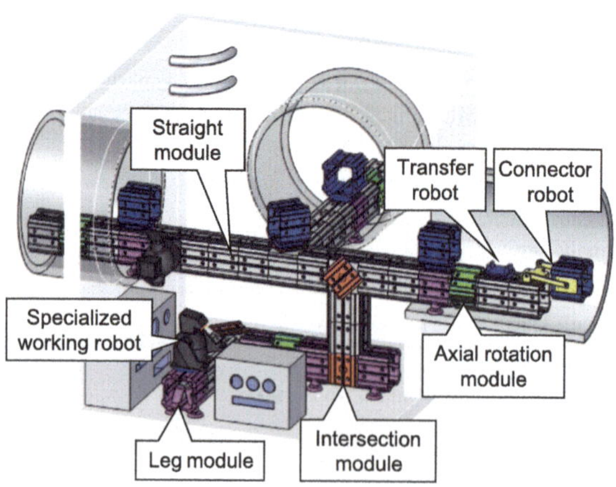

lar Assembly System (AMAS), and discussed the hardware system and distributed control method [16]. Their concept is similar to ours in the sense that the system is composed of separate structure modules and an assembler robot. However, objective of our system is to construct paths for robot locomotion and operation in hazardous environment. Therefore, the complex robot action comprising multiple DOFs (Degrees of Freedom) is not suitable for secure operation. We apply the concept of "Self-Reconfigurable Robotics" [14] to our system. Consequently, a function of changing the robot attitude is implemented in some modules. This implementation decreases the total number of actuators; then less malfunction of robot operation occurs because of high radiation and rubble.

The rail structure is modularized to be transferred easily. It has modules of four types. Each module has different functions as described below. Static modules have no actuators inside. Dynamic modules include actuators inside, so that a robot on it can be rotated. Connecting mechanisms among modules are standardized. Straight modules have a connecting mechanism of modules. The leg module mounts extensive adaptive leg mechanisms, which enable support of the whole weight of the rail structure. An axial rotation module builds in a rolling rotation mechanism. The intersection module has a function for the branch structure. Two built-in turntables can rotate in the yaw direction, independently of each other.

The following two types of constructor robots conduct construction motion of the structure. A transfer robot carries one module and moves on the rail structure to the leading end. The connector robot receives a module from a transfer robot and connects it to the leading end of the structure.

As in previous studies of this system, we verified the module delivery from a transfer robot to a connector robot in three robot attitudes. In addition, straight modules and leg modules are developed [2]. However, to apply this system to the real world, e.g. at 1F, it is necessary to develop a construction planner of rail structures that comprises three subtasks. (1) Path planning of a modularized rail structure that calculates a rail structure path from a start point to a destination while avoiding obstacles. (2) Type selection of modules that decides the attitude and type of all consisting modules. (3) The dispatch planner of robots that calculates robot behavior, as well as the total construction time from the module-assigned structure and robot numbers and specifications.

In this paper, we develop a construction planner of a modularized rail structure with a combination of the type selection of modules, which is a unique problem in this system, and the dispatch planning of constructor robots. The objective of this research is to clarify the influence on total construction time by some means to select module types and the specification of constructor robots using a developed planner.

The remainder of paper is organized as follows. In Sect. 2, we present the basic concept of the proposed planner and discuss the salient technical problems. The detailed design and implementation are described in Sect. 3. We applied the proposed planner to test the environment which imitates 1 F and conducted an experiment to generate construction planning in Sect. 4. Section 5 is the conclusion.

2 Construction Planning for a Modularized Rail Structure

2.1 Path Planning of a Modularized Rail Structure

Figure 2 presents the process flow of the proposed construction planning for the modularized rail structure. In this step, the inputs are the positions and attitudes of the starting and goal points of rail structures, and distribution of environmental obstacles. The output is a trajectory. This is the list of coordinates of sequential modules connecting the start point and goal point avoiding obstacles. To evaluate the whole system, construction planning is necessary to ascertain the total construction time and the total number of actuators in the whole structure. These two parameters are calculated using the following two steps: module type selection and dispatch planning of robots. Therefore, the suitable trajectory is obtained using the repetition of construction planning, evaluation of trajectory, and readjustment of the trajectory.

This step can be regarded as a path planning problem in a three-dimensional discrete space. Related studies have developed an algorithm to solve this problem in robotics. Many tactics have been proposed such as the roadmap method and potential method [11]. In addition, a method to recalculate and reshape the trajectory is proposed if the distribution of obstacles has changed while a robot is moving [17]. We realize our trajectory planner while applying these developed methods.

2.2 Type Selection of Modules in Trajectory

In this phase, the input is the coordinate list of all module center points from the start point to the goal point. We regard the following two conditions as constraints: (1) supporting mechanisms in leg modules can endure the force and moment of the rail structure during and after construction, and (2) robots can reach at least one surface of a module located in a goal point after locomoting on structures using dynamic modules.

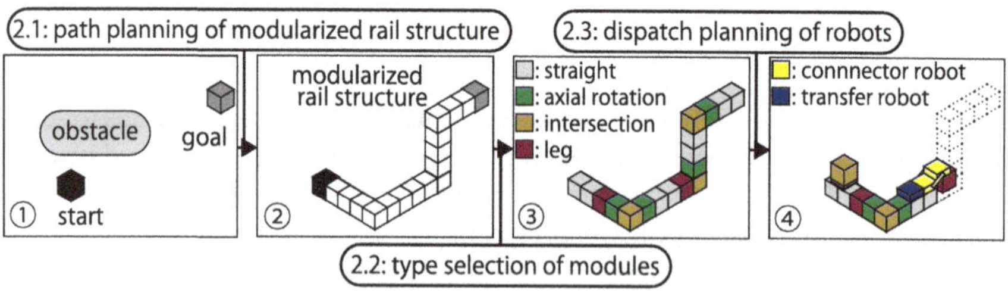

Fig. 2 Process flow of construction planning

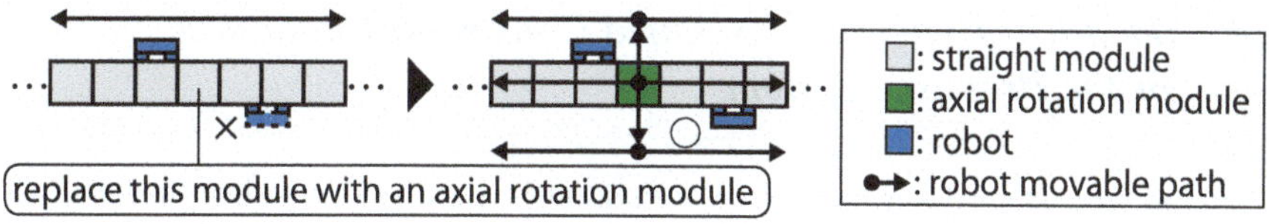

Fig. 3 Adding an axial module to the rail structure

Then the attitudes of modules and types of modules to assign (straight, leg, axial rotation and intersection) are determined. These are the output of this part. Especially, if one axial rotation module is installed to the structure, as depicted in Fig. 3, then robots can move to another surface using the additional axial rotation module. This movement changes the rail structure into a double-track pathway. It enables robots to locomote effectively and decrease the total construction time. However, the increase of axial rotation modules raises the risk of the probability of malfunction. It also raises the total cost of the module. Consequently, the number of additional axial rotation modules should be determined properly after consideration of this tradeoff.

2.3 *Dispatch Planning of Construction Robots*

In this step, the following are given as inputs: (a) position, attitude and type of all modules to be constructed, and (b) specifications of constructor robots such as numbers to be dispatched, maximal velocity, and payload. The outputs are the entire construction processes until the final module has been constructed, avoiding deadlocks and interference. The total construction time of the rail structure is calculated sequentially from these processes. Constructor robots locomote freely on the rail structure, but they cannot pass each other in the same surface. It is necessary to use another surface of the structure as a by-path.

For underground mining [4] and railways [7], studies have been conducted to optimize behaviors of vehicles according to the distribution of ore or passengers in a static locomotion environment. Otani et al. simulated and numerically calculated the construction time of a solar power plant unit in outer space, applying a multi-agent system [12]. This research is similar to that for our system because robots locomote on what is carried and constructed by themselves, which means that the locomotion environment for robots extends along with the construction. Nevertheless, because of the lower gravity, robots can easily move in any direction on hexagon-shaped solar panels. This setup poses smaller restrictions to robot locomotion.

Peasgood et al. created an algorithm and succeeded in avoiding deadlock while moving robots [13]. First, a locomotion environment for robots is interpreted as a graph structure. Then the proper motion order for robots is assigned from calculation. For this study, the authors adapt this concept to convert the modularized rail structure into a graph structure. They create decision rules based on the structure.

2.4 *Technical Issues and Assumed Constraint Conditions for the Proposed System*

From the three steps described above, we address type selection of the modules and dispatch planning of modules because in the path-planning research field includes many proposals that were produced from previous research. In addition, many available libraries can be used, such as OMPL [15]. In terms of type selection, we attempt to ascertain the influence on the total construction time of adding an axial rotation module to the rail structure. Especially for dispatch planning, we aim to clarify the effect on total construction time of an increase of the module supply ability because of the addition of transfer robots.

In this paper, we set the following as constraint conditions of proposed system to reflect the effects of technical issues on the total construction time. (a) In advance, some inspection robots can obtain available environment information. The distribution of obstacles does not change during construction. (b) The modularized rail structure has only one start and goal point. Certainly it can have branches and multiple goal points, but we suppose a longest path in possible multiple paths from the same start point. (c) There is no loop structure of modules. (d) There is no dispersion in the mobility of robots. (e) Malfunctions of robots do not occur in structure construction, but they surely have to be considered in future. (f) Module connecting motions by connector robots have a priority to the module transferring motion. (g) To avoid interference or deadlocks of constructor robots, they can wait on the structure, but not detour. (h) The robot length is less than the size of one module.

3 Design and Implementation of Construction Planning

3.1 *Type Selection of Rail Structure Modules*

Details of the process flow of module type selection are depicted in Fig. 4. The attitude of each module is calculated from the relative position of the two immediately adjacent modules. Type selection of modules consists of the five steps below.

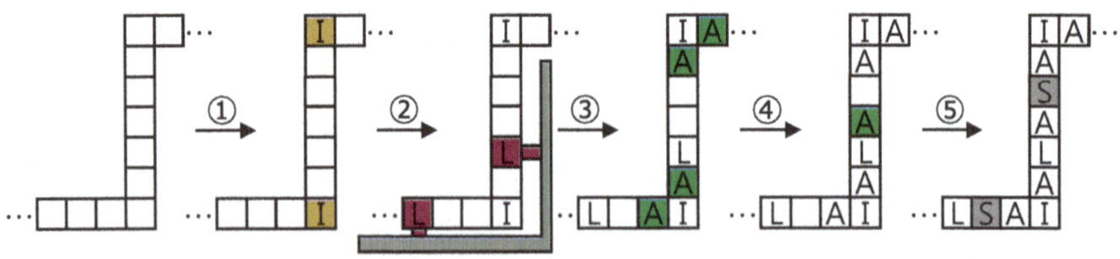

Fig. 4 Flow of function module assignment

(1) Intersection modules are assigned at the corner points of structures, which changes the extension direction of modules, to alter robots' movement direction.
(2) Leg modules are assigned after consideration of the following conditions: (a) supporting mechanisms in leg modules can support the whole weight of a structure during and after construction; also, (b) in module connection processes, mechanical position and attitude errors between a module being constructed by a connector robot and a placed module being attached to this module are smaller than the permissive tolerances. According to the load balance calculation, it is sufficiently sustainable to allocate one leg module in five modules. However, it is prohibited to assign leg modules to a position next to intersection modules. Robot access to turntables on both sides of the intersection modules is necessary to realize effective robot transfers to the goal point using a double track structure without congestion. Although this selection step does not guarantee that the total number of leg modules is minimized, it outputs candidates of type selection results satisfying the constraint conditions in a reasonable elapsed time.
(3) Mandatory axial rotation modules are assigned. Axial rotation modules are allocated in places adjacent to intersection modules to allow robots to reach and use upper and bottom turntables of intersection modules. At least one axial rotation module is put between intersection modules. This ensures that robots can locomote on any surfaces of structures using axial rotation function.
(4) Additional axial rotation modules are assigned. They are placed to limit the maximal distances between axial rotation modules in descending order. This shortens the distance from a connection place of an axial rotation module to another connection place of the next axial rotation module. After this step, the area in which robots must move on a single-track structure is decreased, which engenders the reduction of construction time.
(5) Straight modules are finally assigned to the remaining unassigned modules.

3.2 *Dispatch Planning of Constructor Robots*

The following are the process flows of the dispatch planning. The flows generate robot behavior of module transferring and connecting and total construction time of the structure.

(1) An undirected graph structure is generated from the assigned rail structure. This structure presents the connection relation of each surface of modules.
(2) Using the graph, the shortest path for robots from the start point to a goal point is calculated. A connector robot locomotes and connects all modules along the shortest path. Paths of transfer robots are determined to convey each module to the standby position of the connector robot on the locomotion path described above.
(3) Deadlock-avoiding rules are implemented in all path-determined robots. Then the state of all robots is updated sequentially. They move according to the rules. The updating ends if all modules are connected and constructed.

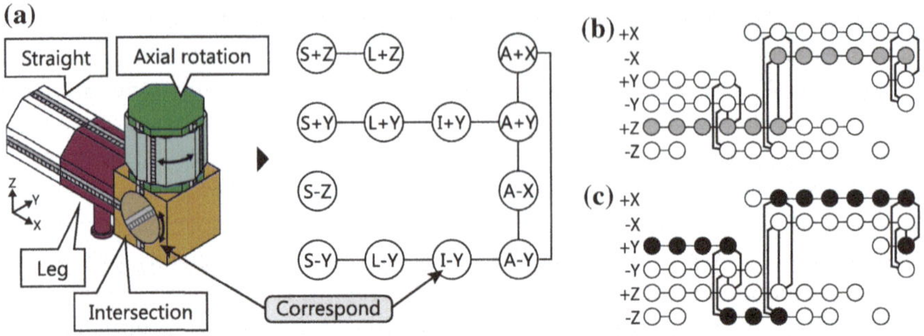

Fig. 5 **a** Converting process from a rail structure to a graph: **b** outward trajectory and **c** homeward trajectory of robots

Figure 5a presents the concept of the process, interpreting a rail structure in an undirected graph. Respective surfaces of the modules correspond to nodes of the graph. Nodes where robots can move directly are connected by edges. Axial rotation modules and intersection modules connect nodes that have different normal vectors. The weight of edges corresponds to the necessary time for robot locomotion and dynamic module operations.

From the generated graph structure, an outward and homeward trajectory to the goal point module is calculated as shown in Fig. 5b and c. The outward trajectory is used for constructor robot locomotion and module transfer to the end of the structure. After module delivery, vacant transfer robots go back to the start point along the homeward trajectory. The interference derived from crossing transfer robots in same surfaces of rail structure will be minimized by decreasing the common nodes to outward and homeward trajectory as little as possible. In the graph structure, an outward trajectory equals the shortest path from a node of start point module to a node of goal point, which is calculated using the Dijkstra algorithm. The homeward trajectory is the shortest path which includes the fewest number of nodes in the outward trajectory.

After a connector robot completes construction of a module, it moves forward to the next construction position and waits there until the arrival of another module. A transfer robot carries and transports modules to the waiting position of the connector robot. All transfer robots' moving paths are determined to go through as many nodes as possible in the homeward and outward trajectory. A state transition diagram of constructor robot updating is presented in Fig. 6. Each node can place only one robot because of the lengthy relation between the robot and module. Both module delivery motions (DELIVERY) and module connection motions (CONSTRUCTION) are conducted only when the robot exists at node. After these motions, the robot state becomes at node (ATNODE). It might block other robots' locomotion trajectories and deadlock might happen if a robot state is ATNODE and it attempts to move to next node (MOVE). In this situation, a checking algorithm must be designed and administered.

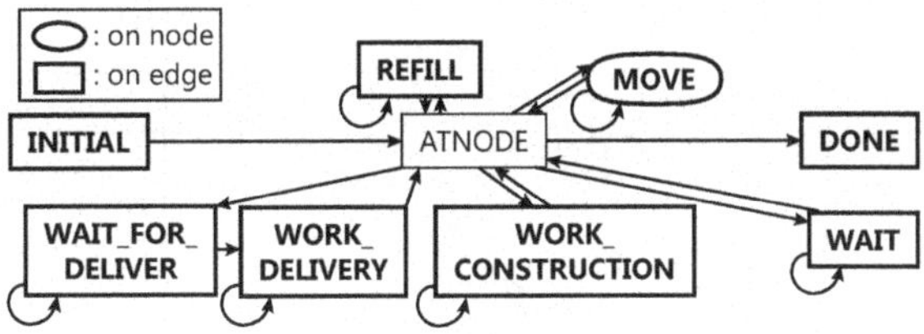

Fig. 6 Robot state transition diagram

3.3 Re-Use of the Previous Calculation Results

R_{now} is the configurations of the modularized rail structure to be simulated in its current condition of construction with total module number N. If rail structure R_{old} exists, it matches the position, attitude, and types of modules from start point to the n th module. Moreover, the input of dispatch planning is identical between R_{now} and R_{old}. Under these circumstances, the behaviors of all constructor robots are completely the same from the beginning of construction to the moment immediately before a transfer robot carrying the $n + 1$ th module is installed to the rail structure. Consequently, this part of results of updating the robots state can be reusable by copying from R_{old} to R_{now}. $S(k)$, the required updating steps to complete construction of the k th module, is proportional to the square of module total numbers if a small number of transfer robots are in operation. Reducible steps δs is up to 25% compared to the situation without re-use of $n = N/2$D

$$\delta s = S(N) - (S(N) - S(n)) = S(n) = 0.25S(N) \tag{1}$$

4 Experiment for Generating Construction Planning for a Rail Structure

Using the construction planning designed and implemented in the previous section, we conducted a generation of construction plans using an experiment imitating the situation inside 1F. The aim is to reveal the influence of the number of transfer robots and additional axial rotation modules on the construction time.

4.1 Test Trajectory

Figure 7a portrays the test trajectory. It passes a gate from the starting point, which is the surface of the 1 F reactor building. Then, it moves through the inside Reactor Pressure Vessel (RPV), and an open area of Pressure Containment Vessel (PCV). It finally reaches the goal point: the pedestal which is the understructure of RPV.

(a)

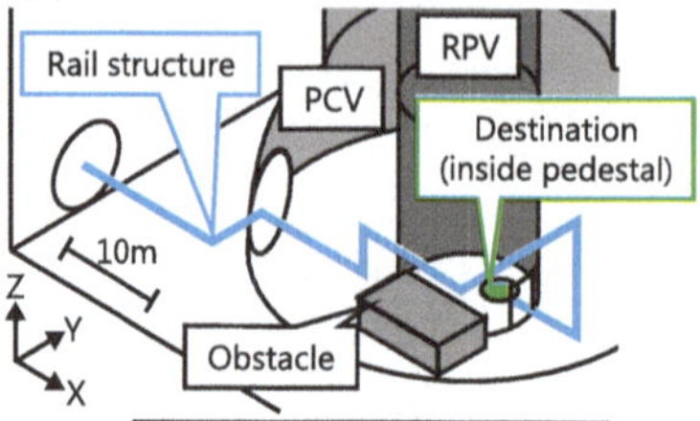

Structure length [m]	43.5
Total modules [-]	256
Number of corners[-]	7

(b)

Time for robots to reach next module [s]	16
Time for module delivery from a transfer to a connector robot [s]	112
Time for connecting one module [s]	128
Number of transfer robots [-]	1, 2, 4, 8, 16, 32
Number of additional axial rotation module [-]	0, 1, 2, 4, 8, 16, 32, 64, 128

Fig. 7 **a** Test trajectory and specifications and **b** experimental conditions

It is assumed that fuel debris, melting from a reactor core, exists in the pedestal. An opening of RPV in the pedestal is used for replacing control rods. The opening is accessible from PCV. Regarding approaches to remove fuel debris, a submerged approach is considered, but it presupposes that the submerged condition of PCV will have been preserved. Therefore, alternative approaches must be used to access through some penetrations or hatches on the side of PCV for taking humans or supplies in or out. Shape changing robot that Okada et al. [10] developed made an investigation inside 1F Unit 1, and verified the possibility of access from PCV to the pedestal. Additionally, it was revealed that numerous obstacles such as pipes and pumps exist inside PCV. Robot moving path and the information of 1F [6] are taken into consideration to design test trajectories.

4.2 Experiment of Generating Construction Planning

Using the test trajectory described before, an experiment was conducted to reveal influences that the number of additional axial rotation modules and the number of transfer robots have on the construction time. Figure 7b presents experimental conditions chosen based on the performance of transfer robots, construction robots, and straight modules developed in earlier studies. A personal computer (Core i7-4702 HQ (3.2 GHz), memory 16 GB) was used for analyses.

Figure 8a shows a simulation outcome; the relation between transfer robots, additional axial rotation modules and construction time. When the number of transfer robots is greater than 32, or when the number of additional axial rotation module is greater than 32, the decrease of construction time is saturated because, in that area, the time for connector robot to construct one module exceeds the time for a transfer robot to supply one module. We examine this point quantitatively below.

One module construction is completed by the following steps: a transfer robot transfers one module; a connector robot connects it; and the connector robot moves

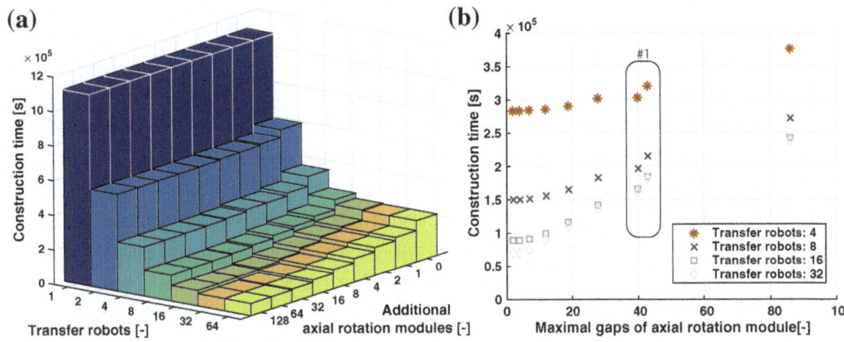

Fig. 8 **a** Relation between transfer robots, additional axial rotation modules, and construction time. **b** Relation between maximal gaps of axial rotation modules and construction time

to the next connecting point. Equation (2) shows this required time as $T_{\mathrm{connector}}$. Also, T_{del} is the time to deliver a module between robots. $T_{\mathrm{construct}}$ stands for the time for a connector robot to connect a module. T_{move} signifies the time for any robot to move by the distance of one node. However, it is necessary to consider the following two points about the supplying module time: (a) transferring module to the end of rail structure, and (b) replacing of transfer robots by passing each other near the end of rail structure. Regarding (a), the rail structure path complexity, such as a path involving a corner, is defined as α. Equation (3) shows the required time T_{transfer}, considering transfers of all M_{total} modules to the tip of the rail structure by N_{transfer} transfer robots. In terms of (b), a transfer robot recedes to the nearest axial rotation module after it completes delivery of the carried module to the connector robot. After the axial rotation module operates, a new transfer robot moves into the next delivery point. Locomoting distances of the respective transfer robots are equal to the lengths of modules until the nearest axial rotation module. This fact can be expressed as $M_{\mathrm{total}}/M_{\mathrm{axial}}$, using the number of axial rotation module M_{axial}. Therefore, T_{replace}, the necessary time with consideration of (b), is expressed as Eq. (4).

$$T_{\mathrm{connector}} = T_{\mathrm{del}} + T_{\mathrm{construct}} + T_{\mathrm{move}} \tag{2}$$

$$T_{\mathrm{transfer}} = (\alpha M_{\mathrm{total}} T_{\mathrm{move}})/N_{\mathrm{transfer}} = \alpha M_{\mathrm{total}} T_{\mathrm{move}}/N_{\mathrm{transfer}} \tag{3}$$

$$T_{\mathrm{replace}} = 2T_{\mathrm{move}} M_{\mathrm{total}}/M_{\mathrm{axial}} \tag{4}$$

A situation in which the construction performance is inferior to the module supply capability is expressed as Eq. (5). After substituting experimental conditions, Eq. (6) is made. It corresponds to experimentally obtained results. Equation (6) shows that simulations are necessary to clarify the effect of connector robots because it includes α. As Eq. (6), it is possible to the reductive effect of construction time by additional axial rotation module.

$$T_{\text{connector}} \geq T_{\text{transfer}}, \quad T_{\text{connector}} \geq T_{\text{replace}} \tag{5}$$

$$N_{\text{transfer}} \geq 16\alpha, \quad M_{\text{axial}} \geq 32 \tag{6}$$

Figure 8b portrays the relation between maximal gaps of axial rotation modules in rail structure and the construction time. The decrease of total construction time by the decrease of maximal gaps of axial rotation modules becomes smaller as the maximal gaps become smaller. When the maximal gaps of axial rotation modules are about 40 (Fig. 8b #1), an axial rotation module is added near the root of the rail structure. Therefore, it is assumed that the double track of the robot moving path is maintained from module construction of near initial position, and that the double track has an extreme effect on the decrease of construction time.

5 Conclusion

To apply "Automated Construction System of a Modularized Rail Structure" to an actual 1 F situation, it is necessary to develop the modularized structure construction planner that consists of three subtasks: path planning of a modularized rail structure, type selection of modules, and dispatch planning of robots. For this study, we developed and implemented the construction planner in combination with type selection of modules and dispatch planning of robots. Using the developed planner, we simulated the construction of a test trajectory imitating 1F Unit 3. We evaluated the influence of methods for selecting module types and the parameters of constructor robots on construction time.

We achieved the following in this research. (a) The searching space for module type selection can be reduced greatly by assigning the necessary types of modules preferentially, based on geographical constraints. (b) The single-track part of the rail structure produces a waiting queue of transfer robots, which increases the construction time. We showed quantitatively that the increase of transfer robots and axial rotation modules decreases the construction time. (c) The construction time is determined from the connecting ability of connector robots and the module supplying ability by transfer robots. (d) Adding axial rotational modules shortens the module supply interval at the end of the structure. This reduces the construction time. Installing too many axial rotation modules saturates the effect of the decreased construction time. This saturation can be calculated from the total module number and the robot moving time. (e) An increase of transfer robots enhances the module supply capability to the end of structure. Numerical simulation is necessary because the threshold of the influence depends on the structure shape.

As a subject of future work, expansion of our proposed method is necessary because of the existance of a fork in the structure to construct multiple trajectories to a goal point. Integration to path planning of rail struct-ures is also necessary. Finally, to apply this system to an actual situation, measures must be prepared to cope with high radiation and communication delays.

Acknowledgements A part of this study is the result of "HRD for Fukushima Daiichi Decommissioning based on Robotics and Nuclide Analysis" carried out under the Center of World Intelligence Project for Nuclear S&T and Human Resource Development by the Ministry of Education, Culture, Sports, Science and Technology of Japan.

References

1. Fukui, R., Kato, Y., et al.: Automated construction system of robot locomotion and operation platform for hazardous environments - basic system design and feasibility study of module transferring and connecting motions. J. Field Robot. (2015)
2. Fukui, R., Kawae, K., et al.: Rail structure supporting mechanism using foamable resin for pillar expansion, anchoring, and fixation. J. Robot. Mech. **28**(2), 129–137 (2016)
3. Furuta, T., et al.: Development of the exploring robot toward future indoor surveillance missions in the Fukushima Daiichi nuclear power plant. J. Robot. Soc. Jpn. **32**(2), 92–97 (2014). (in Japanese)
4. Haviland, D., et al.: Fundamental behaviours of production traffic in underground mine haulage ramps. Int. J. Min. Sci. Technol. **25**(1), 7–14 (2015)
5. Hirose, S.: Development of robotic systems for the decommissioning operation of Fukushima Daiichi nuclear reactor. J. Jpn. Soc. Mech. Eng. **116**, 732–735 (2013). (in Japanese)
6. IRID: Basic data of Fukushima Daiichi nuclear power plant (Summary). http://irid.or.jp/fd/?page_id=237&lang=en. Accessed 19 Feb 2016
7. Kanai, S., et al.: An optimal delay management algorithm from passengers' viewpoints considering the whole railway network. J. Rail Trans. Plan. Manag. **1**(1), 25–37 (2011). https://doi.org/10.1016/j.jrtpm.2011.09.003
8. Kawashima, M.: Unmanned anti-disaster system R & D project (NEDO). J. Robot. Soc. Jpn. **32**(2), 117–121 (2014). (in Japanese)
9. Kawatsuma, S.: Overview and issues to be solved on emergency response of robots to Fukushima NPP accidents. J. RANDEC **46**, 14–26 (2012)
10. Okada, S., et al.: Result of site test for investigation robot for inside primary containment vessel of 1F-1 -Shape changing robot-. http://irid.or.jp/_pdf/20150714_4.pdf. Accessed 31 Oct 2015
11. Ota, J., et al.: An introduction of Intelligent Robots. Corona Publishing, Tokyo (2001)
12. Otani, M., et al.: Improving recovery capability of multiple robots in different scale structure assembly. J. Adv. Comput. Intell. Intell. Inf. **15**(8), 1186–1196 (2011)
13. Peasgood, M., et al.: A complete and scalable strategy for coordinating multiple robots within roadmaps. IEEE Trans. Robot. **24**(2), 283–292 (2008). https://doi.org/10.1109/TRO.2008.918056
14. Stoy, K., et al.: Self-reconfigurable Robots: An Introduction. The MIT Press, Cambridge (2010)
15. Şucan, I., et al.: The open motion planning library. IEEE Robot. Autom. Mag. **19**(4), 72–82 (2012). https://doi.org/10.1109/MRA.2012.2205651
16. Terada, Y., Murata, S.: Automatic modular assembly system and its distributed control. Int. J. Robot. Res. **27**(3–4), 445–462 (2008)
17. van den Berg, J.P., Overmars, M.H.: Roadmap-based motion planning in dynamic environments. In: Proceedings of IEEE/RSJ International Conference on Intelligent Robots and Systems (2004)

Distributed Convolutional Neural Networks for Human Activity Recognition in Wearable Robotics

Dana Hughes and Nikolaus Correll

Abstract We investigate distributing convolutional neural networks (CNNs) for human activity recognition across computing nodes collocated with sensors at specific regions (body, arms and legs) on the wearer. We compare four CNN architectures. A distributed CNN is implemented on a network of Intel Edison nodes, demonstrating the capability of performing real-time classification. Two use a centralized, monolithic approach, and two are distributed across a number of computing nodes. While the accuracy of the distributed approaches are slightly worse than those of the monolithic CNNs, exploiting the hierarchy of the problem turns out to require much less memory — and therefore computation — than the monolithic CNNs, and only modest communication rates between nodes in the model, making the approach viable for a wide range of distributed systems ranging from wearable robots to multi-robot swarms.

1 Introduction

Human-activity recognition (HAR) from wearable sensor nodes that measure acceleration is an important problem in wearable robotics. Sensors are usually mounted at strategic body locations (Fig. 1), monitoring the dynamics of individual limbs and body parts. Interestingly, this kind of data is highly hierarchical, that is data from the feet and knees make up the group "leg", whereas data from the hand, elbows and shoulder, make up the group "arm", which then combine to upper and lower body and so on. We wish to exploit this hierarchy to distribute a classifier across the different sensing nodes and show how this approach reduces the memory and computational requirements when compared with a monolithic approach. With direct applications in the field of distributed, wearable robotics, and tracking of robots with non-traditional

D. Hughes (✉) · N. Correll
Department of Computer Science, University of Colorado, Boulder, CO, USA
e-mail: dana.hughes@colorado.edu

N. Correll
e-mail: nikolaus.correll@colorado.edu

R. Groß et al. (eds.), *Distributed Autonomous Robotic Systems*,
Springer Proceedings in Advanced Robotics 6,
https://doi.org/10.1007/978-3-319-73008-0_43

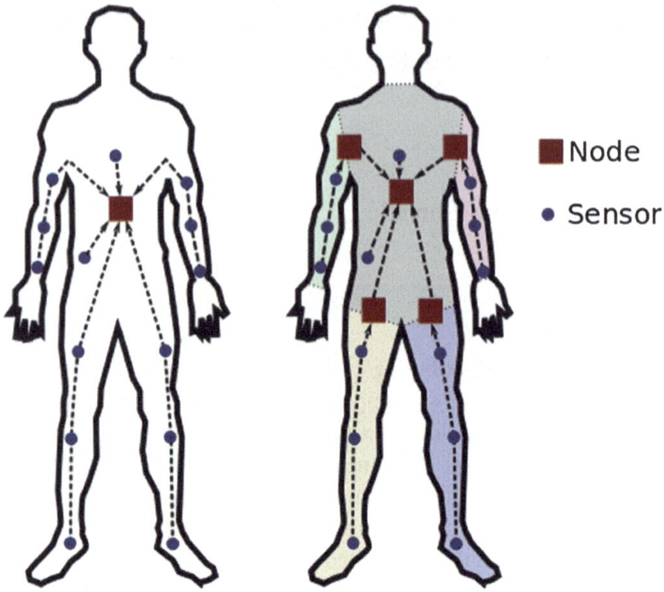

Fig. 1 Approaches to human activity recognition. Left: Data from a number of sensors are aggregated and processed at a central location. Right: Data are processed hierarchically, leading to more and more abstract representations. Note that network granularity is arbitrary

kinematics, such as soft links, or modular robots, these results also pave the way for distributing large-scale convolutional neural networks on multi-robot systems, robot swarms and"robotic materials", composites that tightly integrate sensing, actuation, computation and communication. [14, 15].

In the last few years, advances in deep learning have been applied to HAR using a very large number of sensors. These advances have resulted in the ability to automatically extract meaningful features from training data, rather than relying on hand-engineering features for specific tasks, which are robust to variations in position and scale. This paper investigates adapting deep learning approaches, which are described in more detail in [14], as a potential machine learning approach for distributed robotic systems. In this approach, each node maintains a neural network locally, and updates the state of a high-level hidden layer using local measurements and communicating state with neighboring nodes.

2 Related Work

Deep learning approaches, such as convolutional neural networks (CNNs), restricted Boltzmann machines (RBMs) and long-short term memory (LSTM), have been shown to provide a high levels of classification accuracy for HAR tasks and other classification problems. In [3], automatic feature extraction was performed on four datasets using RBMs and principle components analysis, and was suggested as a systematic way of learning meaningful features in HAR applications. Deep belief network (DBN) classifiers were built in [6] using stacked RBMs, which showed significant improvement in results when compared with shallow networks, decision trees and hidden Markov models (HMMs). CNNs were used in [4] and [5] to perform HAR on publicly available datasets. In [4], the CNNs consisted of a single convolutional and max-pooling layer, followed by a fully connected and softmax layer, while in [5], three convolutional / max-pooling layers were used. In both instances, the CNN classifier accuracy is significantly higher than classifiers based on statistical features or features automatically generated using RBMs. LSTM models were incorporated in conjunction with CNNs in [7], which provided a means of utilizing the temporal dynamics of activities to improve results. This approach showed improvement of accuracy over a baseline CNN, and resulted in predicted classes being more consistent across time. The neural network architectures used in these investigations were compared in [8] across several datasets. While neither CNNs nor LSTM models performed consistently better across all datasets, each of these architectures did show improvement over deep feed-forward neural networks.

The neural network models used in the investigations above can be very large and require powerful computing devices to processes, especially when using a large number of on-body sensors (as in [2]). Incorporating large computing elements into wearable devices or multi-robot systems is especially challenging, due to power requirements, size, and the inability to seamlessly integrate these into garments. We have begun investigating networks of inexpensive, low profile computing nodes for sound-source localization in a wearable context [16] and for classification in robotic skin [9].

Population-based learning approaches, such as particle swarm optimization (PSO), have found application for learning in swarm robotic tasks. PSO can be easily adapted for distributed settings by performing updates locally and sharing a summary of results in a local neighborhood, providing similar benefits described in this investigation. For example, a distributed PSO algorithm which employs a sampling technique known as Optimal Computing Budget Allocation is described in [11], and is used to train a recurrent neural network controller in multi-robot systems to perform tasks such as obstacle avoidance. When compared to a centralized version of the PSO algorithm, the distributed approach demonstrates similar behavior, while significantly reducing memory and communication requirements.

Machine learning has also been explored in the context of modular robots for example in the context of gait generation [13] resulting in a distributed policy, which we believe to be the inverse problem to the HAR problem. Similarly, [12] has been investigating distributing a control strategy for robot locomotion across a modular neural network. Yet, using distributing CNNs to take advantage of the computing potential of a distributed system has been so far unexplored in several contexts, including wearable robotics or HAR (see also a survey on this domain in [14]).

3 Convolutional Neural Networks

Convolutional neural networks (CNNs) are neural networks which consist of one or more convolutional layers, each of which is optionally followed by a pooling layer. For temporal data, the output of a convolutional layer, $z_i^{(l+1)}(\tau)$ is the output of the previous layer $z_j^l(\tau)$ convolved with a set of F^l feature kernels, $K_{ij}^l(\tau)$,

$$z_i^{(l+1)}(\tau) = \sigma\left(b_i^l + \sum_{j=1}^{F^l} K_{ij}^l(\tau) * z_j^l(\tau)\right) \tag{1}$$

where b_i^l is the bias of the convolutional layer, and σ is a non-linear activation function, such as a rectified linear unit (ReLU; $\sigma(x) = max(0, x)$). Convolutional layers are optionally followed by a max-pooling layer. Given a pool size and stride, the output of a max-pooling layer is simply the maximum value within the pool size, and the stride represents the number of samples between the start of each pool in a sequence.

CNNs may consist of one or more convolutional / max-pooling layers. For classification, the final layer is flattened, and used as input to a softmax layer. A softmax layer generates the probability of the input belonging to each of N classes,

$$z_i^{(l+1)} = \frac{e^{z_i^l}}{\sum_N e^{z_i^l}} \tag{2}$$

These operations are summarized in Fig. 2, which demonstrates a CNN operating on a three-channel signal with two feature kernels, a ReLU activation function and max-pooling layer.

CNNs provide several advantages over fully connected, feed forward neural networks. Primarily, they allow for automatic feature extraction from training data, which are invariant to translational and scale in the signal. Additionally, feature

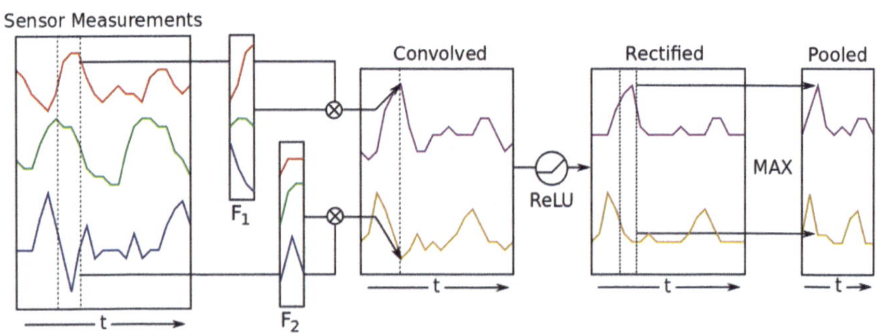

Fig. 2 Convolutional neural network consisting of two feature kernels, ReLU activation function and max-pooling layer, with three channel input

kernels provide a large amount of weight sharing in the network, reducing the total number of parameters in the network. This second point is a specific advantage for distributing neural networks across low-power nodes, as it greatly reduces the memory requirements of the overall network. Training is performed by minimizing the cross-entropy cost between the predicted class (i.e., output of the softmax layer) and the target class using any standard backpropagation algorithm (e.g., RMSProp).

4 Approach

The overall goal of this investigation is determine the potential of applying deep neural networks using a distributed set of computing nodes, rather than centrally collecting and processing all sensor measurements with a monolithic neural network. The task of classifying mid-level human activities is used to validate this approach, data that we believe to be representative for a wide range of spatio-temporal sensor data that might arise in wearable and swarm robotics. Figure 1 shows the difference between these two approaches. The network granularity for the distributed approach is arbitrary, and selected as five nodes for this paper. At an extreme, the entire body could be covered with sensors and computers, resulting in a robotic material.

4.1 Opportunity Dataset

The Opportunity Activity Recognition Dataset [2] is a publicly available benchmark dataset for classifying human activities at several levels, from low-level locomotion (e.g., walking) to complex, high-level activities (e.g., making a sandwich). Data was collected from four users performing six runs of daily living activities, and consisted of body-worn sensors as well as object and ambient sensors. For this investigation, the body-worn sensors were used for classification purposes, and the object and ambient sensors discarded, resulting in a total of 133 sensor measurements. These sensors were used to perform classification of 18 mid-level activities, using a 1 s (30 sample) sliding window with a 100 ms (3 sample) hop size (number of samples between the start of two subsequent windows). Tables 1 and 2 list the sensors and activities, respectively. The *Null* activity indicates when the subject is not performing any specific task.

To adapt this dataset for this exploration, sensors were grouped into five regions–body, left arm, right arm, left leg and right leg–as shown in Table 1. Body worn sensors either consisted of tri-axial accelerometers or IMUs (consisting of a tri-axial accelerometer, tri-axial gyroscope, tri-axial magnetometer and an estimate of orientation in quarternions), while each shoe contains both tri-axial accelerometer (near the ankle) and IMU (near the toe). By dividing the sensors into body regions, computing nodes can be associated with each region, with each node collecting and processing data from the sensors in its region.

Table 1 Location of sensors used in the opportunity dataset

Body (19[a])	Left Arm (38)	Right Arm (38)	Left Leg (16)	Right Leg (22)
Hip Acc	Upper Arm Acc	Upper Arm Acc	Shoe Acc	Upper Knee Acc
Back Acc	Lower Arm Acc	Lower Arm Acc	Shoe IMU	Lower Knee Acc
Back IMU	Hand Acc	Hand Acc		Shoe Acc
	Wrist Acc	Wrist Acc		Shoe IMU
	Upper Arm IMU	Upper Arm IMU		
	Lower Arm IMU	Lower Arm IMU		

[a]Numbers in parentheses indicate the total number of attributes (sensor values) per region

Table 2 Mid-level activities to be classified in the opportunity dataset

Open Door 1	Open Door 2	Open Drawer 1	Open Drawer 2	Open Drawer 3
Close Door 1	Close Door 2	Close Drawer 1	Close Drawer 2	Close Drawer 3
Open Fridge	Close Fridge	Open Dishwasher	Close Dishwasher	Clean Table
Drink from Cup	Toggle Switch	*Null*		

Sensor measurements were normalized to have zero mean and unit standard deviation. Using a frame size of 30 time steps (1 s) and a hop size of 3 time steps resulted in 869,387 training examples from all four subjects. The activity label for each sample was determined by selecting the most common activity in the frame. These were split into a training, validation and test set using an 80-10-10% split. Training was performed and reported separately for each subject. The resulting datasets did not have an equal number of samples for each class. Specifically, a majority of the samples belonged to the *Null* class (72.28%). This imbalance results in poorly trained networks, which typically classify *all* inputs as the dominant class (here, *Null*). To account for this bias, the cross-entropy cost of each sample was normalized by dividing the cost by the probability of selecting the target class from the dataset.

4.2 Architectures

The neural network architectures used in this investigation follow the CNN in [5]. This architecture consists of three convolutional / ReLU / max-pooling layers, followed by a softmax classification layer, and performs classification using one measurement frame. For this investigation, each of the five computing nodes implements three convolutional / ReLU / max-pooling layers. The output of each final layer is then communicated to the body node, which then predicts the activity class using a softmax layer. A comparison of these two architectures is shown in Fig. 3. The CNNs implemented on the five computing nodes in a distributed manner are referred to as "distributed CNNs" (D-CNNs) for the remainder of this paper.

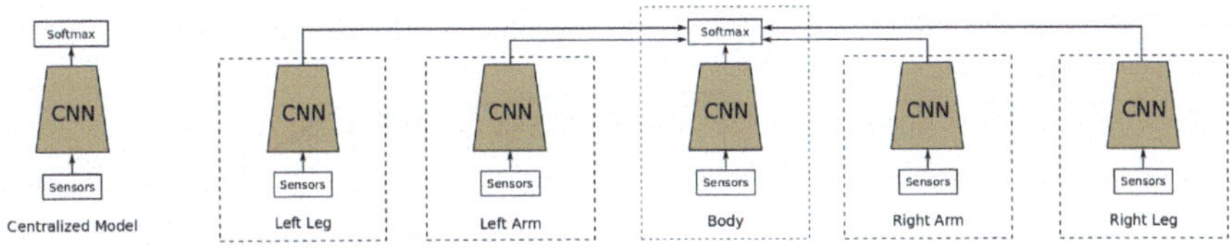

Fig. 3 Centralized architecture vs distributed architecture for human activity recognition

For this investigation, four separate architectures were considered, two centralized monolithic CNNs and two D-CNNs. All of the CNNs are based on the CNN used in [5], and consist of three convolutional / max-pooling layers, followed by a softmax layer. The filter kernels of the first two convolutional layers are five samples in width, and the width of the third layer is three samples. The max-pooling layers have a pool size of two, and a stride of two, effectively halving the size of the previous convolutional layer. The number of filter kernels were varied for each architecture, in order to explore the tradeoff between memory requirements and model accuracy. The four architectures are summarized as follows:

1. **Baseline CNN (CNN-1)**. The baseline CNN uses the same number of kernels as in [5]: 50 kernels in the first layer, 40 kernels in the second layer, and 20 kernels in the third layer.
2. **Reduced CNN (CNN-2)**. The overall number of kernels in the baseline CNN was reduced by 10% increments, and the validation accuracy calculated for each model after training. Figure 4 (solid line) provides the accuracy as a function of reduction percentage, demonstrating roughly the same level of accuracy until a reduction of 50%. The reduced CNN used the reduced kernel count at this percentage: 25 kernels in the first layer, 20 kernels in the second layer and 10 in the third layer.
3. **Baseline D-CNN (D-CNN-1)**. Using a similar approach for the reduced baseline CNN, the number of kernels used for each computing node in the D-CNN was set to match that of the baseline CNN, and reduced by 10% increments. Figure 4 (dashed line) provides the accuracy as a function of reduction percentage. The reduced D-CNN used 25 kernels in the first layer, 20 kernels in the second layer and 10 in the third layer.
4. **Region Optimized D-CNN (D-CNN-2)**. The final architecture attempts to optimize the number of kernels for the computing nodes in each region. To determine the number of kernels, a CNN with a single convolutional layer was constructed, and the validation cost plotted against the number of kernels in the layer, using 5 kernel increments to a maximum of 100 kernels. As the validation cost tends to decrease as the number of kernels increase, the number of kernels for the first layer was selected such that the validation cost was no more than 10% more than the minimum validation cost. This process was repeated for a two-layer CNN to determine the number of kernels for the second layer, setting the maximum number of kernels in the second layer to the number of kernels in the first layer. The

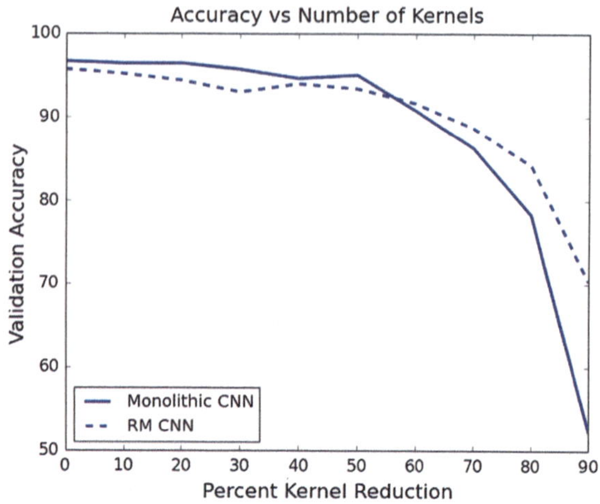

Fig. 4 Validation cost vs percentage of kernels for monolithic and distributed CNN models

Table 3 Number of kernels in each layer for the Region Specific RM-CNN

Layer	Body	Left Arm	Right Arm	Left Leg	Right Leg
First	40	40	40	25	25
Second	25	25	20	10	10
Third	10	12	10	4	4

number of kernels for the third layer was determined in a similar manner, with the exception that a 2 kernel increment was used. Table 3 provides the number of kernels used for each region's CNN.

Each model was built and trained using TensorFlow [17]. Training was performed using RMSProp with a learning rate of 0.0001 and a weight decay of 0.001, with training batch sizes of 100 samples.

5 Results

To compare the four architectures, the classification accuracy and F1 score of the test set was calculated for each subject, and is summarized in Table 4. While the monolithic CNNs provide the highest level of accuracy, the classification accuracy of the D-CNNs is comparable to the monolithic CNNs (~5 and ~2% less than CNN-1 and CNN-2, respectively). This minimal reduction in accuracy is very encouraging, especially when comparing with CNN-2, as the number of kernels used in this architecture is the same as D-CNN-1 and similar to D-CNN-2.

The communication and memory requirements are also of interest for each model, as they are indicative of a feasible implementation in a distributed, embedded system.

Table 4 Accuracy and F1 score for each architecture

	Subject 1		Subject 2		Subject 3		Subject 4	
	Accuracy (%)	F1	Accuracy (%)	F1	Accuracy (%)	F1	Accuracy (%)	F1
CNN-1	96.38	0.9647	96.39	0.9647	95.92	0.9601	97.15	0.9721
CNN-2	94.15	0.9435	93.62	0.9382	93.60	0.9381	92.79	0.9309
D-CNN-1	92.71	0.9298	92.09	0.9241	91.86	0.9222	89.59	0.9020
D-CNN-2	92.77	0.9305	91.69	0.9204	90.94	0.9136	90.74	0.9120

Table 5 Memory and communication requirements for models considered

	Model	Number of Sensors	Number Parameters	Communication Size
	Baseline CNN	133	46,138	0
	Reduced CNN	133	19,978	0
Baseline RM-CNN	Body	19	6,448	40[a]
	Left Arm	38	8,103	10
	Right Arm	38	8,103	10
	Left Leg	16	5,353	10
	Right Leg	22	6,103	10
	Total	133	34,110	40
	Average	26.6	6,822	–
Region RM-CNN	Body	19	10,363	30[a]
	Left Arm	38	13,811	12
	Right Arm	38	12,468	10
	Left Leg	16	3,499	4
	Right Leg	22	4,249	4
	Total	133	44,390	30
	Average	26.6	8,878	–

[a]Received from arm and leg nodes

The memory requirements are based on the total number of parameters (i.e., weights and biases) needed for each model; the absolute memory requirements depend on the number of bits used to represent each value (e.g., 32-bit floats). Communication requirements are based on the size of the output layer of each region's CNN–this is the number of values which need to be communicated each second to perform classification in real time. Table 5 summarizes the memory and communication requirements for each model.

Comparing the accuracy, memory and communication requirements of the four approaches demonstrates that there is a definite advantage to the D-CNN approach over a monolithic CNN for the HAR dataset. While there is minor decrease in accuracy of the D-CNNs (~2% compared to CNN-2), the amount of memory needed for each computing node in the D-CNNs is greatly reduced (~2.25–3.0x reduction). Additionally, the number of sensor measurements required per second is much less for the D-CNN nodes. Assuming fast-mode I^2C (400 kbit/s) is used to communicate with each sensor, single-byte instructions and 16-bit sensor measurements, the 133 sensors in the monolithic model can be queried at 125 Hz maximum, while the distributed approach allows for sensors to be queried at ~400 Hz–~1 kHz, depending on body region. Finally, the necessary RAM is less for the D-CNNs than the monolithic CNNs. Considering 30 samples of sensor measurements and activation layers, CNN-1 and CNN-2 have 981 and 566 units in the network (1,926 bytes and 1,132 bytes RAM for 16-bit values), respectively, while the D-CNN modules have 403–678 units (806–1,356 bytes), depending on the region and architecture.

5.1 Hardware Implementation

To analyze timing requirements and ensure the resulting distributed models can be implemented on a set of wearable sensor nodes, the D-CNN-1 model was implemented on five Intel Edison modules, which utilize a 400 MHz dual-core Intel Quark processor, using Python and Numpy to perform neural network calculations. Sensor measurements were simulated by reading the data from Run 1 of Subject 1 locally, with no frame overlap, for a total of 1,703 frames. Each module incorporates onboard WiFi, which we use for communication experiments, as it allows for simple, external monitoring of communication between nodes.

The calculation time required for the three convolutional / max-pooling layers of each region were measured for each frame–the mean computing time ranged from 177.48–181.44 ms, with a standard deviation of 0.119–1.28 ms. Calculation of the softmax classifier required a mean computing time of 0.550 ms, with a standard deviation of 15.92 μs. Communication between nodes was performed over WiFi using TCP; at each frame, the body node queries each arm and leg node for the node's current state. Each communication packet contains 80 bytes (20 32-bit floats). The mean time to communicate with the four arm and leg nodes is 62.17 ms, with a standard deviation of 36.15 ms. This time could be reduced using UDP, at the expense of possible packet loss, or through wired communication. In practice, wired communication is ideal, as microcontrollers typically implements multiple USART channels and two-wire interfaces (e.g., I^2C). For example, the hardware nodes used in [9] and [16] implement six USART channels (at up to 115,200 bps) and I^2C operating at 100 kHz or 400 kHz. Using these communication methods, the communication time required for an 80 byte packet reduces to ~5.6 ms for a USART channel, and ~1.6–6.5 ms for I^2C channels, an order of magnitude faster than WiFi communication. In addition, wired communication would require less energy than WiFi. One drawbach, however, is the potential for failure in individual wire connections.

From the above values, the total time to classify a one second frame is less than 250 ms, allowing for a classification rate of 4 Hz, which is sufficiently fast to perform real-time, on-line classification of activity.

6 Discussion

Distributing a CNN across multiple nodes (D-CNN) has lead to comparable results than using a monolithic CNN. This is encouraging, as it shows the potential for wielding the power of deep learning on distributed embedded systems ranging from wearable and modular robots to robot swarms. Although exploiting the hierarchy of the activity recognition problem does not lead to better recognition than a monolithic approach — which simply learns this hierarchy — we demonstrate significant savings in overall computation, memory and communication requirements.

The major reduction in computation, memory and communication requirements stems directly from the number of sensors each computing node needs to monitor, which has several effects. Primarily, the size of the feature kernels in the first convolutional layer are directly proportional to the dimensionality of the measurement signal. For example, the 25 feature kernels in the monolithic approach require 3,325 parameters (133 sensors x 25 kernels), whereas the left arm only requires 950 parameters (38 sensors x 25 kernels). Additionally, clustering the sensors into region-specific groups may reduce the overall number of kernels required. There is likely a high level of correlation between sensors in the same region (e.g., accelerometers in the hand and wrist), and as the number of sensor channels increases, a larger number of kernels is required to model the interactions between each pair of sensors (i.e., the "curse of dimensionality"). Ultimately, the D-CNNs allow for a high level of compression in the sensors in each region–from up to 1,140 values per second (38 sensors x 30 samples / second) to as little as 4–20 values per second.

While the individual CNNs in the D-CNN architectures could be implemented into a single hardware node with sufficient memory, implementing each CNN on a separate hardware node provides several benefits. Each node can sample its region's sensors and calculate the CNN output in parallel with other nodes, allowing for an approximate increase in sampling and processing rate by the total number of hardware nodes. Additionally, a single hardware node running multiple CNNs allows for a single point of failure; with multiple hardware nodes, activity classification could continue despite failure of one of the nodes, albeit with a likely reduction in accuracy.

As the proposed network is able to classify motion, it might also be applicable to be used to evaluate sample policies during control, for example of an exoskeleton on a human wearer, a soft robot with non-conventional kinematics and dynamics, a modular robot with previously unknown kinematics, or the trajectory of a moving source tracked by a robot swarm. While the networks used here were trained for classification of activity, adapting these to learn control policies involves shifting the learning paradigm from a supervised learning approach to a reinforcement learning approach. Reinforcement learning may be a more suitable paradigm for a hierarchical architecture such as that presented here. The error signal in a supervised learning approach must backpropagate through both multiple layers and multiple regions (e.g., from body to leg), whereas the reward signal in a reinforcement learning approach can be applied directly to each region.

7 Conclusion and Future Work

A distributed, hierarchical approach to human activity recognition using wearable sensors is presented. The approach involves training a set of five CNN modules, each associated with a major region of the body, and communicating state information to the body module for final classification. For evaluation, the distributed approach was trained to identify mid-level activities in the Opportunity dataset, and the results compared to two monolithic CNNs with structure similar to individual modules.

The resulting performance of the distributed CNNs is comparable to the monolithic CNNs, and have significantly lower requirements on memory and sensor communication. Specifically, the architectures shown here can be feasibly implemented on small, inexpensive microcontrollers that communicate via low-bandwidth channels such as serial lines, radio or infrared. The ability for the distributed architecture to perform in real-time with commercially available hardware modules and wireless communication was demonstrated.

While the results presented here show the feasibility for distributing deep neural networks in wearable devices and similar applications, there are several potential areas for future investigation. With regard to neural network architecture, the CNNs used here classified activity based on a quasi-static window of measurements. Recurrent neural networks, such as used in [7] and [8], may improve classification results, both in terms of overall accuracy, as well as consistency between measurement windows. Additionally, as shown in [7], recurrent models can be implemented with significantly fewer parameters.

The network topology explored in this investigation is a relatively simple star topology, which assigns all sensors in a particular region to a single computing node. In a robotic materials paradigm, a computing node would be collocated with each sensor [15], which, for this dataset, would result in several nodes per region and the potential for more complex networks. Exploring the effects of more complex network topologies, including deeper hierarchies and denser neighborhoods, should be explored to determine how a robotic materials paradigm may affect accuracy and convergence during training.

From a robotics perspective, activity recognition (specifically gait recognition) can be viewed as the inverse problem of gait generation for robot or exoskeleton control. As the results presented here are promising for activity identification, a similar approach to gait generation may be possible by shifting from a supervised learning paradigm to a reinforcement learning paradigm. This would allow for learning robust, low-level gait patterns independent of high-level control or planning. Finally, the architectures used here can be utilized for a variety of swarm robotic tasks, such as object tracking.

Acknowledgements This work has been supported by AFOSR grant #FA9550-15-1-0238. We are grateful for this support.

References

1. Turaga, P., Chellappa, R., Sabrahmanian, V.S., Udrea, O.: Machine recognition of human activities: a survey. IEEE Trans. Circuits Syst. Video Technol. **18**, 1473–1488 (2008)
2. Roggen, D., et al.: Collecting complex activity datasets in highly rich networked sensor environments. In: 7th International Conference on Networked Sensing Systems, 233–240 (2010)
3. Plötz, T., Hammerla, N.Y., Olivier, P.: Feature learning for activity recognition in ubiquitous computing. Int. Jt. Conf. Artif. Intell. **22**, 1729–1734 (2011)
4. Zeng, M., Nguyen, L.T., Yu, B., Mengshoel, O.J., Zhu, J., Wu, P., Zhang, J.: Convolutional Neural Networks for Human Activity Recognition using Mobile Sensors. In: 6th International Conference on Mobile Computing, Applications and Services, 197–205 (2014)
5. Yang, J.B., Nguyen, M.N., San, P.P., Li, X.L., Krishnaswamy, S.: Deep convolutional neural networks on multichannel time series for human activity recognition. Int. Jt. Conf. Artif. Intell. 25–31 (2015)
6. Alsheikh, M.A., Selim, A., Niyato, D., Doyle, L., Lin, S., Tan, H.-P.: Deep activity recognition models with triaxial accelerometers. http://arxiv.org/abs/1511.04664. Accessed 05 Jul 2016
7. Ordóñez, F.J., Roggen, D.: Deep convolutional and LSTM recurrent neural networks for multimodal wearable activity recognition. Sensors **16**, 115 (2016)
8. Hammerla, N.Y., Halloran, S., Plötz, T.: Deep, convolutional, and recurrent models for human activity recognition using wearables. http://arxiv.org/abs/1604.08880. Accessed 05 Jul 2016
9. Hughes, D., Correll, N.: Texture recognition and localization in amorphous robotic skin. Bioinspiration Biomim. **10**, 055002 (2015)
10. Ravi, N., Dandekar, N., Mysore, P., Littman, M.L.: Activity recogntion from accelerometer data. In: 17th Conference on Innovative Applications of Artificial Intelligence, 1541–1546 (2005)
11. Di Mario, E., Navarro, I., Alcherio, M.: A distributed noise-resistant particle swarm optimization algorithm for high-dimensional multi-robot learning. IEEE Int. Conf. Robot. Autom. 5970–5976 (2015)
12. Jacobs, R.A., Jordan, M.I.: Learning piecewise control strategies in a modular neural network architecture. IEEE Trans. Syst. Man Cybern. **23**, 337–345 (1993)
13. Sproewitz, A., Moeckel, R., Maye, J., Ijspeert, A.J.: Learning to move in modular robots using central pattern generators and online optimization. Int. J. Robot. Res. **27**, 423–443 (2008)
14. Hughes, D., Correll, N.: Distributed machine learning in materials that couple sensing, actuation, computation and communication. http://arxiv.org/abs/1606.03508. Accessed 05 Jul 2016
15. McEvoy, M.A., Correll, N.: Materials that couple sensing, actuation. Comput. Commun. Sci. **347**, 1261689 (2015)
16. Profita, H., Farrow, N., Correll, N.: Flutter: an exploration of an assistive garment using distributed sensing, computation and actuation. In: 9th International Conference on Tangible, Embedded, and Embodied Interaction, 359–362 (2015)
17. Abadi, M., et al.: TensorFlow: Large-scale Machine Learning on Heterogeneous Systems (2015) Software www.tensorflow.org

Formation Control of a Drifting Group of Marine Robotic Vehicles

Nicholas R. Rypkema and Henrik Schmidt

Abstract This paper presents a comparative study into three strategies for planar (2D) formation control of autonomous underwater/surface vehicles (AUVs/ASVs) in the presence of ocean or river currents. Deploying multiple AUVs in formation would provide large-scale spatial and temporal data for oceanographic research. However, multi-AUV formation control is difficult because of the communication and navigation constraints of the underwater environment. The strategies we developed to address these challenges are distributed, use relative positions of neighboring vehicles to coordinate, and leverage current to increase endurance. We present simulation results for a group of 20 AUVs operating in 4D ocean currents, and compare control strategies in terms of energy expenditure and formation quality. We validated the most promising strategy with real robot experiments using three ASVs on the Charles River, and provide an initial analysis of the data.

1 Introduction

Advances in autonomous underwater vehicle (AUV) technology have led to their wide-spread acceptance and adoption in scientific, commercial, and defense applications. At the same time, research progress in coordination and control of multi-robot systems has led to their effective deployment in the field. Applying multi-robot control concepts on AUVs would improve efficacy in missions such as tracking oceanographic processes and localizing acoustic sources, and could open up additional applications such as seismic surveying using multiple AUVs as an acoustic receiver array. Some existing missions that would be well served by multi-AUV sampling include monitoring ocean temperature, tracking algal blooms, and mapping hazardous chemical spills. However, existing multi-robot control strategies

---

N. R. Rypkema (✉) · H. Schmidt
Massachusetts Institute of Technology, Cambridge, MA, USA
e-mail: rypkema@mit.edu

H. Schmidt
e-mail: henrik@mit.edu

R. Groß et al. (eds.), *Distributed Autonomous Robotic Systems*,
Springer Proceedings in Advanced Robotics 6,
https://doi.org/10.1007/978-3-319-73008-0_44

do not take into account the unique constraints posed by AUV operations in the undersea environment; algorithms must be distributed, robust to localization error, and operable with minimal inter-vehicle communication.

We are interested in a particular multi-robot control task, known as formation control, in which a group of vehicles coordinate to establish and maintain a desired geometric pattern. Various approaches for distributed formation control have been described in previous literature, including physics-based approaches [10, 14, 15, 18], leader-follower methods [6, 13], potential-field approaches [1, 5], and virtual structure methods [2, 8, 12]. Early experimental work in underwater vehicle formation control includes the use of artificial attraction/repulsion potentials to maintain a triangular shape for three Slocum gliders [9], as well as a leader-follower method in which a simple proportional controller is used to maintain distance/velocity relative to a leading AUV [7]. More recent work [16] describes a similar leader-follower approach, in which acoustic modems mounted on three autonomous surface vehicles (ASVs) are used to determine range from a single follower to two leaders; a proportional-integral controller operating on the follower uses these ranges to maintain velocity and heading relative to the leaders to maintain a triangular formation.

The primary goal of this work is the development of planar (2D) formation control strategies allowing a group of AUVs to form and maintain a hexagonal lattice in the presence of ocean currents. Unlike previous work, we wish to leverage the prevailing current to propel the formation as a whole, with the vehicles only using their motors to maintain the desired geometry. This approach would allow the formation to drift freely with the current without breaking apart. In addition, the significant constraints of underwater communications, with data rates on the order of 100–5000 bits per second, motivate our desire to minimize communication between vehicles. We describe three behavior-based algorithms for AUVs, each of which is distributed and leaderless. Each vehicle performs formation control using relative range and bearing measurements to its neighbors within a specified radius, and does so by using these measurements to continuously calculate a vehicle target position. This is in contrast to previous approaches, in which control policies for vehicle speed and heading are explicitly derived so as to minimize error in formation position relative to leader vehicles. By specifying target positions that we wish our vehicles to track, our approach allows us to abstract away the lower-level control layers. This approach is extremely flexible and can effortlessly combine range/bearing information from an arbitrary number of neighbors, building formations that are large and leaderless. Multiple behaviors can also be combined into more complex behaviors. The cost of this flexibility is the inability to provide formal guarantees on formation convergence. In this paper, we compare the performance of three newly developed algorithms in simulation, and provide results from validating the most promising strategy in field experiments with three ASVs.

The remainder of this paper is organized as follows: Sect. 2 describes the three formation control algorithms, Sect. 3 provides a description of the experimental methods, Sect. 4 provides results of AUV simulations, and Sect. 5 provides preliminary results from ASV field experiments.

2 Formation Control Algorithms

In our approach, vehicles operate under a frontseat/backseat paradigm, in which the backseat is responsible for performing calculations for formation control, communications and autonomy using MOOS-IvP [3], a software architecture for vehicle autonomy. The backseat sends desired heading, depth and speed commands to the frontseat, which executes basic control and navigation. This behavior-based architecture allows us to separate vehicle control into two behaviors: the first to maintain constant depth, which does not use information from neighboring vehicles; and the second to maintain the desired formation using the relative positions of neighbors. In this paper we concentrate solely on the latter.

Our formation control algorithms operate by outputting a target position which represents the optimal vehicle location to maintain formation. Each control strategy is built upon a common target behavior which sends desired heading and speed commands to the frontseat, illustrated in Fig. 1. This behavior operates as follows: It begins by directing the vehicle at a heading towards the target with maximum speed. Speed is linearly decreased within a specified drifting radius, and set to zero upon location arrival (either by entering the capture radius, or if there is lack of progress within the slip radius). The vehicle is then left to drift freely until it exits the drifting radius.

The target behavior is used by vehicles to exploit the prevailing current for propulsion during times in which they are free to drift. We can trade-off formation accuracy against energy consumption by changing the drifting radius.

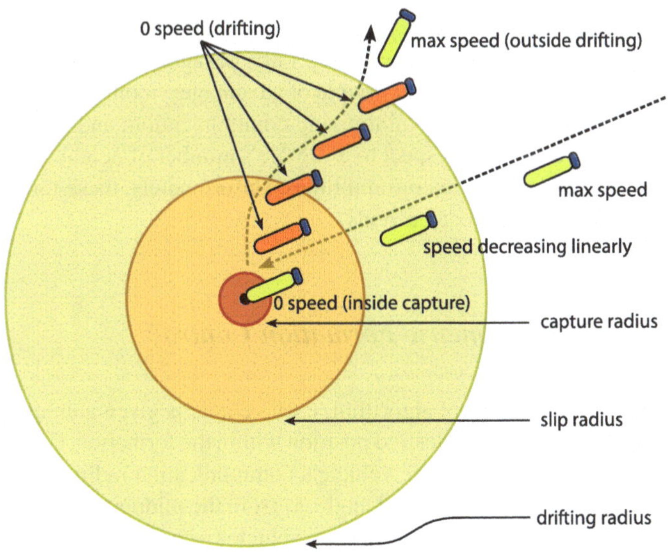

Fig. 1 Illustration of the target behaviour, which directs the vehicle to a target point and then leaves it to drift within a drifting radius

2.1 Attraction/Repulsion Formation Control

Inspired by the simplicity of physics-based approaches, our first algorithm performs formation control through the use of artificial attraction/repulsion potentials. We define our potential as:

$$f(r) = \left(\frac{s^3}{2 \cdot r^2}\right) + (r - 3 \cdot s) \tag{1}$$

This potential is placed at the positions of two neighboring vehicles, with s being the desired distance from, and r the range to, each neighbor. The first neighbor is the nearest neighbor, and the second is chosen such that the sum of edge lengths of the triangle created by the three vehicles is minimal. Equation 2 shows the minimization of Eq. 1 over two dimensions (where N_s contains the positions of the two neighbors). Solving this equation yields a minimum (x^*, y^*) representing the target position for the vehicle, which occurs at a distance s from both neighbors. Note that two minima occur, both equidistant from the two neighbors and reflected along the line joining them; the minimum closest to the vehicle is always selected as the target position by initializing our solver at the vehicle position.

$$\begin{aligned}(x^*, y^*) = \underset{(x,y)}{\operatorname{argmin}} \sum_{(x_i, y_i) \in N_s} &\left(\frac{s^3}{2 \cdot (\sqrt{(x - x_i)^2 + (y - y_i)^2})^2}\right) \\ &+ ((\sqrt{(x - x_i)^2 + (y - y_i)^2}) - 3 \cdot s)\end{aligned} \tag{2}$$

Because each vehicle attempts to position itself equidistant from two of its neighbors, this algorithm is limited to producing only hexagonal lattice formations of equilateral triangles. Even so, it has a couple of advantages: it only needs the relative positions of neighbors to operate, minimizing communication; and it is fast, with a time complexity of $O(n)$ with respect to a vehicle's number of neighbors (assuming the minimization can be done in constant time). Unfortunately, these formations are prone to breaking apart in ocean currents.

2.2 Pairwise Trigonometric Formation Control:

In our second formation control algorithm, each vehicle is given a unique ID and a user-defined plan specifying its desired position within the formation. Given this plan and the IDs of neighbors within the vehicle's communication radius, the algorithm calculates the relative distance D_d and angle ϕ_d from the midpoint between any pair of neighbors i, j to the desired position of the vehicle (where (x_{i_p}, y_{i_p}) and (x_{j_p}, y_{j_p}) are the plan-specified positions of the neighbors relative to the vehicle):

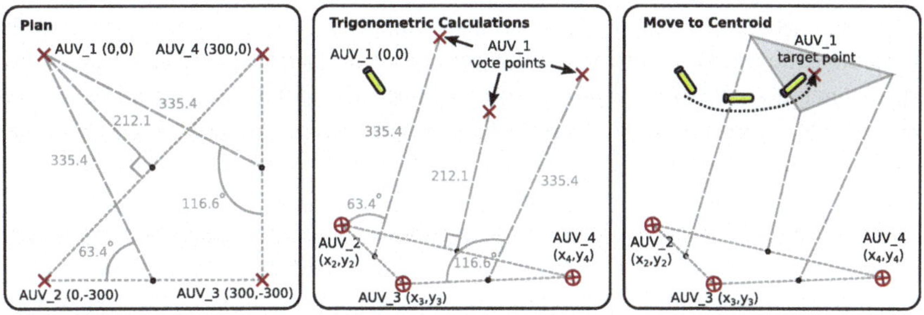

(a) Use plan to calculate distances and angles from midpoint between pairs of neighbors to desired vehicle position

(b) Apply plan-calculated distances and angles to corresponding midpoints of actual neighbor pairs

(c) Calculate vehicle target position (x^*,y^*) as centroid of votes (x^*_{ij},y^*_{ij}) from all neighbor pair combinations

Fig. 2 Illustration of the pairwise trigonometric formation control strategy

$$D_d = \sqrt{((x_{i_p} + x_{j_p})/2)^2 + ((y_{i_p} + y_{j_p})/2)^2} \tag{3}$$

$$\phi_d = atan2((y_{i_p} + y_{j_p})/2, (x_{i_p} + x_{j_p})/2) - atan2(y_{j_p} - y_{i_p}, x_{j_p} - x_{i_p}) \tag{4}$$

This plan-calculated distance and angle are then used to determine a position using the actual relative positions of these two neighbors (x_i, y_i) and (x_j, y_j):

$$(x^*_{ij}, y^*_{ij}) = ((x_i + x_j)/2, (y_i + y_j)/2) + (D_d \sin\phi_d, D_d \cos\phi_d) \tag{5}$$

The position (x^*_{ij}, y^*_{ij}) is essentially a 'vote' by this neighbor pair of where the target for the vehicle should be. The combined target position (x^*, y^*) is calculated as the centroid of all votes from every possible unique neighbor pair combination seen by the vehicle. This control strategy is visualized in Fig. 2.

Since the user can specify any formation shape in the plan, this strategy is able to form arbitrary formations; this is true as long as planned positions are within communications radius of at least two other positions (in order to calculate at least one (x^*_{ij}, y^*_{ij})). This ability, as well as its simplicity, are the main advantages of this algorithm. However, it has a few drawbacks: firstly, it has an approximate time complexity of $O(n^2)$, since for n neighbors, a vehicle must calculate distances and angles using $\frac{n(n-1)}{2}$ neighbor pairs; secondly, it has an additional overhead of communicating vehicle IDs.

2.3 *Point Set Registration Formation Control:*

Our third formation control algorithm follows from the previous in that it also requires each vehicle to have a unique ID and a corresponding position in a user-defined plan. Since a vehicle is able to detect the IDs of neighbors within its communications radius, it can look up their corresponding positions in the plan. All that remains to be done is to calculate the rigid transformation that optimally aligns desired neighbor positions to actual neighbor positions, a process broadly termed as point set registration. Unlike the previous algorithm, in which votes from neighbor pairs are used to calculate the target position, this strategy essentially treats the entire plan as a rigid structure, and transforms it to best align with neighbor positions.

Given the set of relative positions of neighbors within the vehicle's communications radius $\{(x_0, y_0) = (0, 0), (x_1, y_1), ..., (x_n, y_n)\}$, and their corresponding relative positions defined in the plan $\{(x_{0_p}, y_{0_p}) = (0, 0), (x_{1_p}, y_{1_p}), ..., (x_{n_p}, y_{n_p})\}$ (we include the vehicle's own position as (0, 0) in both sets), the optimal rigid transformation between the two point-sets is given by:

$$(R, \mathbf{t}) = \underset{R,\mathbf{t}}{\operatorname{argmin}} \sum_{i=0}^{n} ||(R \begin{bmatrix} x_{i_p} \\ y_{i_p} \end{bmatrix} + \mathbf{t}) - \begin{bmatrix} x_i \\ y_i \end{bmatrix} ||^2 \tag{6}$$

A closed form solution of this minimization is performed as in [17], resulting in an optimal rotation matrix, R, and translation vector, $\mathbf{t}$, which is applied to the vector of plan positions. As a result, a target position (x^*, y^*) for the vehicle is generated as the first point of this transformed vector. Figure 3 illustrates this algorithm.

As with the pairwise trigonometric algorithm, this strategy can create formations of arbitrary shape (since the user can define any shape in the plan), but has the disadvantage of having to communicate vehicle IDs. However, unlike the previous algorithm this strategy has a time complexity of $O(n)$ with respect to the number of neighbors; the most complex step of the closed form solution for the optimal rigid transformation requires a multiplication between an $2 \times n$ and $n \times 2$ matrix, which is $O(n)$.

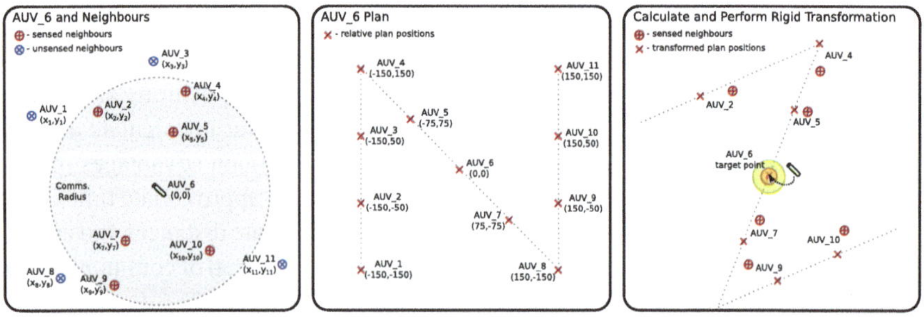

(a) AUV_6 receives pings containing IDs and relative positions of neighbors within its communication radius

(b) AUV_6 uses its plan to obtain planned relative positions of neighbors it has sensed

(c) AUV_6 applies a rigid transformation to planned positions, and as a result the plan is optimally aligned to actual neighbor positions

Fig. 3 Illustration of the point set registration formation control strategy

3 Experimental Setup

Experiments were performed both in simulation and in the field, to investigate the qualities of our three formation control algorithms. Performance of each algorithm was quantified using two measures. The first is a metric reflecting the formation quality, in which the distance between each pair of neighboring vehicles is averaged over the number of pairs, and compared to the desired inter-vehicle distance. Past literature has typically used this measure to quantify control strategy effectiveness. The second is a simplified measure of average energy expenditure, in which power consumption is calculated as the sum of three factors: integrated power consumption due to motor propulsion (which only occurs when vehicles are thrusting), integrated power consumption due to hotel load (consumption due to systems other than propulsion and acoustics), and power consumption due to acoustic pings.

Simulations were performed in the Laboratory for Autonomous Marine Sensing Systems (LAMSS) MOOS-IvP Ocean Simulation environment. This environment includes vehicle dynamics and acoustic communications, as well as physics-driven simulation of 3D time-varying oceanographic environments via MSEAS ocean models [11]. Vehicle dynamics were based on a simplified model of the Folaga AUV [4], a vehicle that has active depth and turn-in place control, and with a maximum speed of 1 m/s. We assumed that each vehicle was able to measure relative range, bearing and ID to neighbors within a radius of 550 m, with Gaussian noise with standard deviation 5° in bearing and 1.5 m in range, through the use of acoustic pings transmitted by each vehicle every 30 s. These acoustic pings were received by neighboring vehicles with a delay in reception proportional to the distance to the transmitting vehicle and assuming a 1500 m/s sound speed. The target drifting radius was set to 20 m. We performed simulation experiments using 20 AUVs, whose positions were randomly initialized within a 200 × 200 m box. The group was instructed to construct a hexagonal lattice with a desired separation distance of 300 m, and was left to drift freely in ocean currents for approximately 5 h in simulation time. Due to the fact that the attraction/repulsion algorithm can only construct hexagonal lattice formations, this formation was chosen to provide a fair comparison between each strategy.

To investigate the validity of our approach, field experiments were run with three Clearpath Kingfisher ASVs (Fig. 4). Each vehicle has a thruster in each pontoon, an IMU, compass and GPS, and has a MOOS-IvP interface to receive heading and speed commands. These ASVs are capable of a maximum speed of approximately 1.5 m/s. Commands were sent from a Raspberry Pi 2 payload computer, with our MOOS-IvP architecture running the point set registration formation control strategy. Since we did not have the hardware to detect range and bearing between vehicles, vehicle state was transmitted via 802.11 WiFi to a shoreside computer which simulated acoustic communications. Range/bearing/ID information was then transmitted back to the vehicles. Due to the limited size of the operations area (approximately 500 × 350 m), we selected a desired separation distance of 60 m, a drifting radius of 6 m, and a ping period of 20 s; we also halved the Gaussian standard deviations for simulated

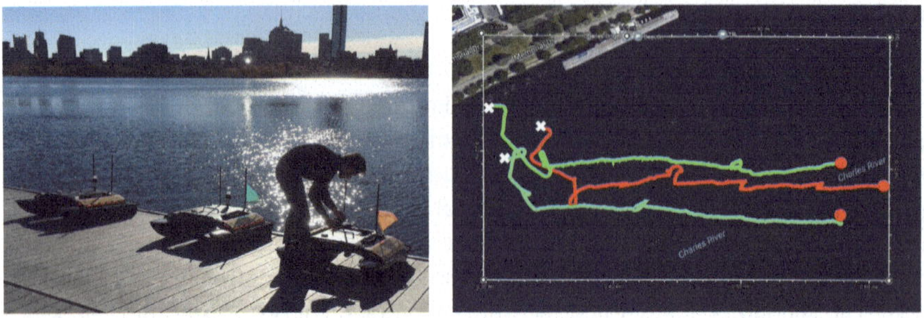

Fig. 4 *Left*: Clearpath Kingfisher ASVs used for field experiments. *Right*: Map of Charles River operating area with overlaid ASV trajectories (c/o Google Maps)

bearing and range measurements to 2.5° and 0.75 m respectively. The three vehicles were directed upstream, instructed to form an equilateral triangle, and left to drift downstream in formation.

4 Simulation Results

Simulation trials were performed for each formation control algorithm. Vehicle positions were initialized randomly in a 200×200 m area. They then constructed a hexagonal formation (and in the case of pairwise trigonometric and point set registration algorithms, each vehicle was given a formation plan for a 4×5 hexagonal lattice), and freely drifted in ocean currents modeled using an MSEAS model of the Red Sea. The maximum current velocity according to this model was approximately 0.12 m/s, about an order of magnitude lower than the maximum vehicle speed. Typical vehicle trajectories for these simulations are shown in Fig. 5.

Qualitative examination of the three algorithms indicate that although our attraction/repulsion strategy was able to construct the desired hexagonal lattice, the group eventually broke apart into three smaller groups while drifting, a consequence of continuous disturbances caused by ocean currents. Past literature concerning physics-based formation control typically demonstrated formation control in environments free of disturbances, or with few vehicles; we show here that such approaches may not be robust in disturbance-rich domains. In addition, we see that the trajectories are quite chaotic because the artificial potentials push and pull on different vehicles. These issues are partly a consequence of the fact that this algorithm only makes use of two neighbors for control. Vehicles tend to repeatedly switch their selection of which two neighbors to use for control as they travel during formation construction (neighbors are chosen to minimize the sum of edge lengths of the three vehicles); this in turn causes switching of the target position and oscillatory behavior. Two-neighbor control was used to achieve the desired inter-vehicle distance; summation of potentials from additional neighbors causes the minimum (and thus the target)

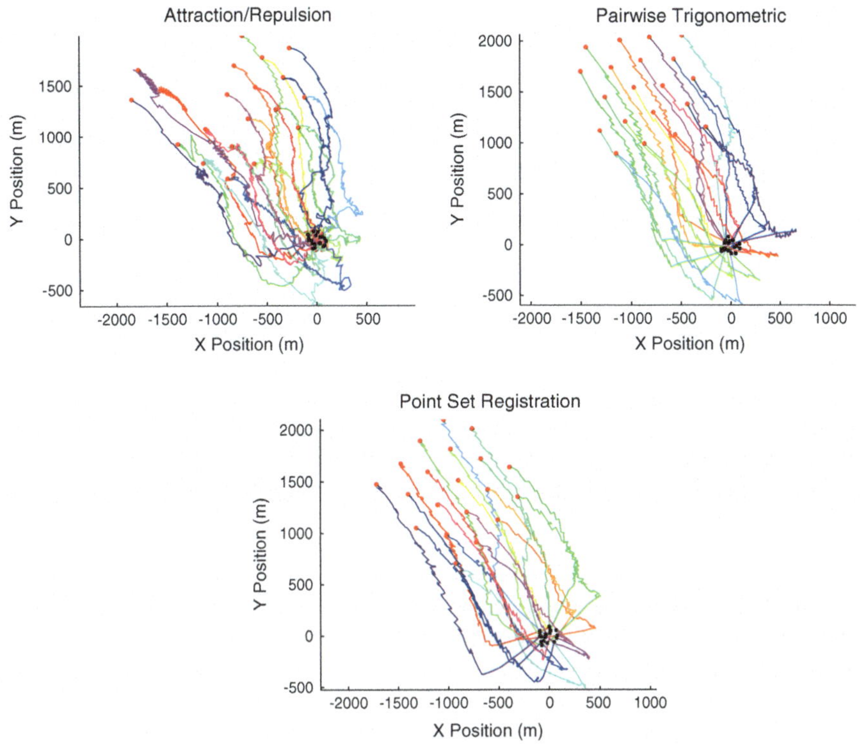

Fig. 5 Simulated trajectories of 20 AUVs for attraction/repulsion (top left), pairwise trigonometric (top right), and point set registration (bottom) formation control strategies. AUVs are randomly deployed in a 200 × 200 m area, construct a hexagonal formation, and drift in formation for 16000 s, exploiting ocean currents for propulsion. Black x's indicate vehicle starting positions, red dots indicate final positions

to shift. The formation tends to break apart when a subgroup of vehicles all select neighbors amongst themselves and the currents pull in different directions.

In contrast, both the pairwise trigonometric and point set registration strategies were able to efficiently construct and maintain the desired 4 × 5 hexagonal lattice formation, with the group using ocean currents to propel itself almost 1.8 km in 5 h. Global knowledge in the form of the user-defined formation plan has enabled each vehicle to maintain its relative position in the group, even as the group freely rotated while drifting. Vehicle trajectories possess a characteristic sawtooth movement, as vehicles alternated between drifting while in the drifting radius, and thrusting back to target.

Figure 6 illustrates the mean distance between neighboring pairs of vehicles, along with envelopes of standard deviation. This provides a measure of formation quality. Our qualitative observations of the attraction/repulsion algorithm are clearly reflected by the standard deviation envelopes of this measure. Formation break-up first occurs

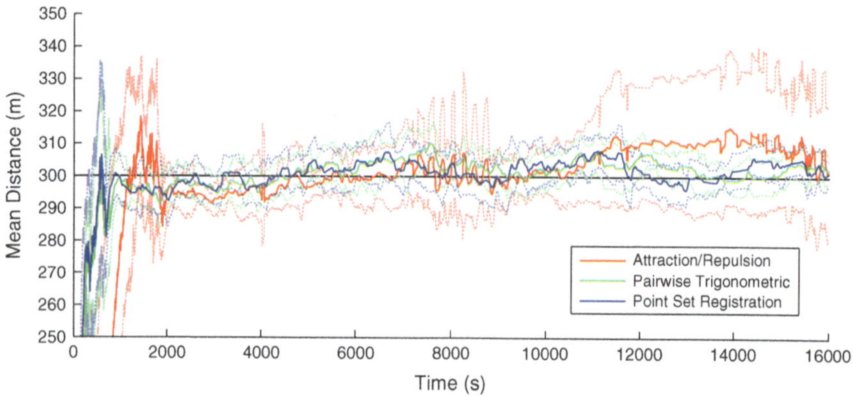

Fig. 6 Results of simulation - mean distance between vehicles in the formation; dashed lines indicate standard deviation. Experiments were conducted with 20 AUVs. AUVs regulate distance using user-specified distance and have access to relative neighbor positions and IDs within comms. radius once every 30 s

at about 7000 s, causing oscillatory movement that culminates in two distinct vehicle subgroups. A second break-up episode slowly unfolds starting at around 11000 s, causing the metric to diverge from the desired separation distance and three subgroups to emerge by the end of the mission.

Examining the formation quality metric for the pairwise trigonometric and point set registration algorithms, we see that both perform similarly. Both strategies are able to construct the formation to the desired separation distance within 1000 s. Their standard deviation envelopes are well within the set 20 m target drifting radius throughout the mission, reflecting the fact that the hexagonal lattice formation is consistently maintained to the desired accuracy.

Figure 7 illustrates vehicle mean energy expenditure with envelopes of standard deviation. Unsurprisingly, the attraction/repulsion strategy consumes the greatest amount of energy, as the artificial potentials cause vehicles to continuously jostle with their neighbors. The standard deviation envelopes demonstrate that the point set registration strategy makes use of energy more consistently across all vehicles than the pairwise trigonometric algorithm, though both have similar mean energy expenditures. The rate of energy expenditure is significantly lower once the formation has been constructed and is free to drift. These results suggest that if the prevailing current is known and along the desired trajectory, ocean current could be leveraged for formation propulsion for long-term sampling using groups of vehicles.

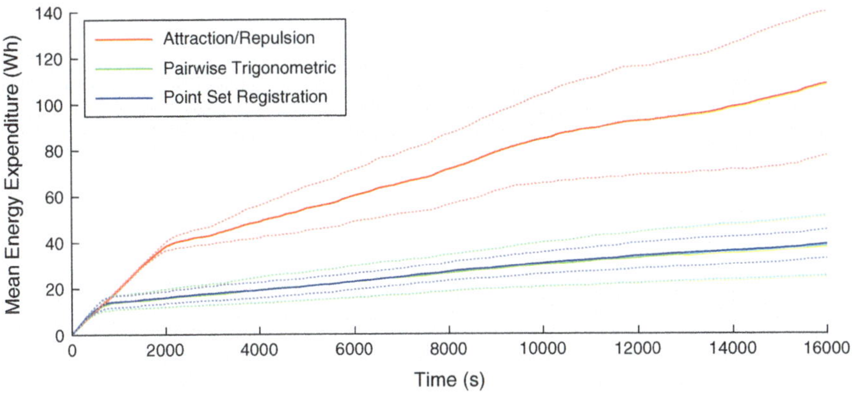

Fig. 7 Results of simulation - mean vehicle energy expenditure; dashed lines indicate standard deviation. Experiments were conducted with 20 AUVs. Expenditure is sum of integrated motor power consumption, hotel power consumption (consumption by systems excluding motor and acoustics) and acoustic transmission power consumption

5 Experimental Results

Field experiments were performed using three Kingfisher ASVs on a portion of the Charles River by the MIT Sailing Pavilion. Vehicle state was monitored on a shoreside computer, which also performed simulation of acoustic communications. The ASVs were manually directed upriver, instructed to form an equilateral triangle and left to drift downriver. Note that the ASVs could easily overcome river currents at maximum speed. Leader-following was then performed by manually directing a single vehicle upstream, causing the remaining two ASVs to naturally follow in an attempt to maintain formation. Finally, three simulated ASVs were added, easily allowing us to construct a larger equilateral triangle using three simulated and three real ASVs. All vehicles ran the point set registration algorithm in a distributed fashion. Trajectories of the ASVs during different phases are illustrated in Fig. 8.

The point set registration formation control strategy was successfully able to construct the desired triangle and maintain that formation as the group drifted downriver (Fig. 8a–c). Unlike the simulated vehicles the Kingfisher ASVs are not able to rotate in-place, but our target-based formation control strategy was still able to construct the desired formation. This highlights an advantage of our approach: formation control is achieved without needing to know exact vehicle dynamics. However, the success of the formation control algorithm in this case is partly due to the fact that the turning radius of the ASVs is much smaller than the formation inter-vehicle distance, meaning that ASV dynamics have a minimal effect on formation control performance. This would likely not be the case with smaller inter-vehicle distances or with vehicles that have a larger turning radius.

The ability to perform leader-following was a natural consequence of our formation control algorithm using only three vehicles. Figure 8d–f shows this behavior:

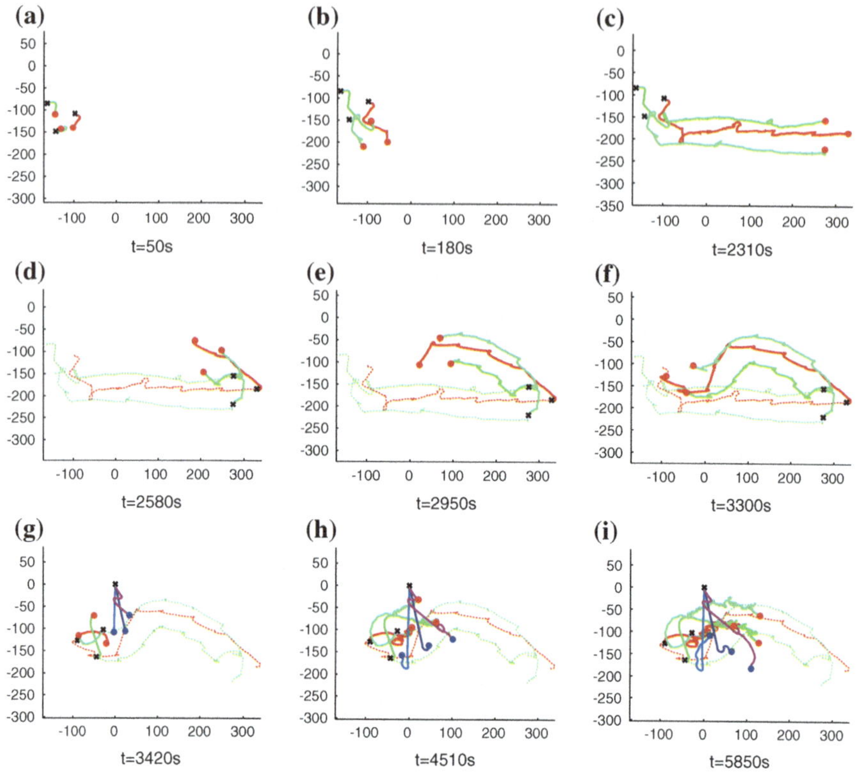

Fig. 8 Results of field experiments. Black x's indicate vehicle positions at time of preceding figure, and red dots indicate positions at current time. **a–c** Trajectories of three Kingfisher ASVs establishing formation and drifting East on the Charles River. **d–f** Trajectories of the three Kingfisher ASVs leader-following in formation, with operator-driven ASV at forward corner of triangle (red line trajectory). **g–i** Trajectories of actual Kingfisher ASVs (red) and simulated ASVs (blue) when three simulated ASVs are added and a larger formation is constructed and left to drift freely

the rightmost ASV is manually directed upstream, and the remaining two vehicles follow in an effort to maintain formation. However, due to the 20 s interval between simulated acoustic pings, the followers lag the leader and are unable to continuously maintain a perfect triangle formation.

Figure 8g–i illustrate group reformation when three simulated ASVs (blue) were added to the group. Both real and simulated vehicles can interact using the MOOS-IvP architecture. Like in the original three-vehicle case, the algorithm was able to construct the desired triangular formation. Since river currents were not simulated for the virtual vehicles, the real vehicles (red) tended to frequently reposition themselves. This caused the group to rotate, as is visible in Fig. 8h–i.

Figure 9 depicts the average distance between neighboring pairs of vehicles. The three phases of the experiment are clearly visible in this measurement, reflecting the quality of the formation. The first phase (Fig. 8a–c) occurred between 0 and 2310 s, and is reflected by a mean distance metric centered around 60 m with a standard

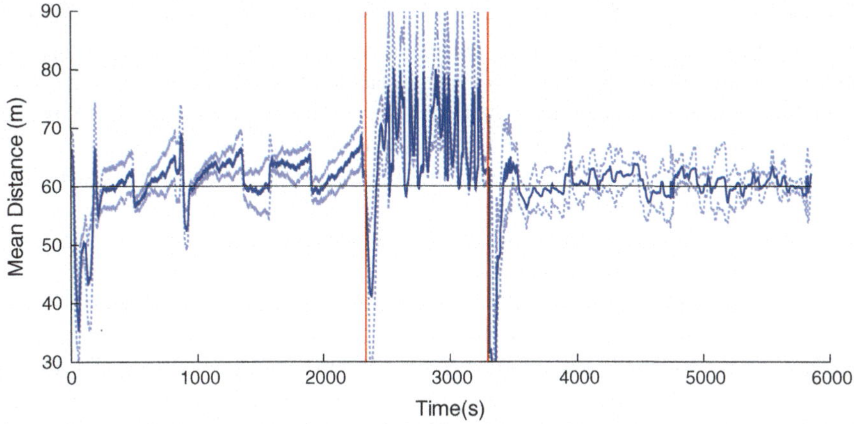

Fig. 9 Results of field experiments - mean distance between vehicles in the formation; dashed lines indicate standard deviation. Experiments were conducted with three real and three virtual ASVs. ASVs have access to neighboring vehicle information every 20 s using simulated acoustic comms. Red vertical lines separate the three experiment phases seen in Fig. 8a–c, d–f, g–i

deviation of about 4 m. The effect of the target drifting radius is clearly seen in the behavior of the metric, as it repeatedly diverges from and returns to the 60 m separation distance. The leader-following second phase (Fig. 8d–f) occurred between 2310 s and 3300 s, and is characterized by oscillations in the mean distance. This was caused by the 20 s delays between simulated acoustic pings, which resulted in lag between leader and follower movement. The leader vehicle was directed in steps to give the following vehicles time to reform, resulting in these oscillations. The third stage (Fig. 8g–i) occurred between 3300 and 5850 s when three virtual ASVs were introduced to the formation. This resulted in the mean distance metric dropping to 0 m upon vehicle introduction, and then settling again to center around 60 m once the larger formation was constructed. Interestingly, the addition of these vehicles caused the formation to more closely track the desired separation distance. This occurred because the simulated vehicles were not subject to external disturbances and because averaging over more vehicles stabilizes the metric.

6 Conclusions

We developed three algorithms for the control of a planar (2D) formation of autonomous underwater/surface vehicles in the presence of ocean and river currents. These algorithms are distributed, use low communication (where only range, bearing, and vehicle ID are needed once every 20–30 s), and exploit prevailing current for group propulsion. They were compared in simulation using formation quality and energy expenditure metrics, and it was demonstrated that two of these strategies performed well in maintaining a user-defined hexagonal formation. Simulation results suggest that exploiting ocean currents could be an advantageous strategy for sampling using multiple vehicles over large areas for extended periods of time.

Field experiments using three Kingfisher ASVs demonstrated the validity of one of the formation control algorithms tested in simulation in the presence of actual river currents and simulated acoustic communications. We also demonstrated leader-follower behavior using formation control by manually directing one of the ASVs, and dynamically added virtual vehicles to show behavior utility with more vehicles and a mixture of real and simulated ASVs.

Future work will focus on replacing the simulated acoustic communications component with actual hardware. Hydrophone arrays can be used to detect range and bearing from acoustic pingers, and unique waveforms can be used to transmit IDs. Other avenues include theoretical analysis of our Pairwise Trigonometric and Point Set Registration algorithms, including bounds on convergence, scalability, and sensitivity to parameters. Further simulation and experimental work investigating robustness of these algorithms in the event of vehicle or communications failure, and against measurement outliers would also prove useful. A final interesting possibility for future work is integrating our algorithms with optimal planning given an estimated model of ocean currents and vehicle dynamics. This could allow a group of vehicles to maintain a desired formation while following a preferred trajectory, or to avoid areas of large turbulence.

Acknowledgements The authors thank Erin Fischell for feedback and Michael Novitzky for assistance during field experiments. This material is based on work supported by APS under contract number N66001-11-C-4115 and award numbers N66001-13-C-4006 and N66001-14-C-4031.

References

1. Bachmayer, R., Leonard, N.E.: Vehicle networks for gradient descent in a sampled environment. In: 41st IEEE Conference on Decision and Control, pp. 112–117 (2002)
2. Belta, C., Kumar, V.: Motion generation for formations of robots: a geometric approach. In: IEEE International Conference on Robotics and Automation, pp. 1245–1250 (2001)
3. Benjamin, M., Schmidt, H., Newman, P., Leonard, J.: Nested autonomy for unmanned marine vehicles with MOOS-IvP. J. Field Robot. **27**(6), 834–875 (2010)
4. Caffaz, A., et al.: The hybrid glider/AUV Folaga. Robot. Autom. Mag. **17**(1), 31–44 (2010)
5. Chaimowicz, L., Michael, N., Kumar, V.: Controlling swarms of robots using interpolated implicit functions. In: IEEE International Conference on Robotics and Automation, pp. 2487–2492 (2005)
6. Desai, J.P., Ostrowski, J.P., Kumar, V.: Modeling and control of formations of nonholonomic mobile robots. IEEE Trans. Robot. Autom. **17**(6), 905–908 (2001)
7. Edwards, D.B., et al.: A leader-follower algorithm for multiple auv formations. IEEE/OES Autonomous Underwater Vehicles pp. 40–46 (2004)
8. Egerstedt, M.B., Hu, X.: Formation constrained multi-agent control. IEEE Trans. Robot. Autom. **17**(6), 947–951 (2001)
9. Fiorelli, E., et al.: Multi-auv control and adaptive sampling in monterey bay. IEEE/OES Autonomous Underwater Vehicles pp. 134–147 (2004)

10. Fujibayashi, K., et al.: Self-organizing formation algorithm for active elements. In: 21st IEEE Symposium on Reliable Distributed Systems (2002)
11. Haley, P.J., Lermusiaux, P.F.J.: Multiscale two-way embedding schemes for free-surface primitive equations in the multidisciplinary simulation, estimation and assimilation system. Ocean Dyn. **60**(6), 1497–1537 (2010)
12. Lewis, M.A., Tan, K.H.: High precision formation control of mobile robots using virtual structures. J. Auton. Robots **4**(4), 387–403 (1997)
13. Mariottini, G.J., et al.: Leader-follower formations: uncalibrated vision-based localization and control. In: IEEE International Conference on Robotics and Automation, pp. 2403–2408 (2007)
14. Prabhu, S., Li, W., McLurkin, J.: Hexagonal lattice formation in multi-robot systems. In: 11th International Conference on Autonomous Agents and Multiagent Systems (2012)
15. Shucker, B., Bennett, J.K.: Scalable control of distributed robotic macrosensors. Distrib. Auton. Robot. Syst. **6**, 379–388 (2007)
16. Soares, J.M., et al.: Joint asv/auv range-based formation control: Theory and experimental results. In: IEEE International Conference on Robotics and Automation, pp. 5579–5585 (2013)
17. Sorkine, O.: Least-squares rigid motion using svd (2007). https://igl.ethz.ch/projects/ARAP/svd_rot.pdf. Accessed 13 June 2015
18. Spears, W., et al.: Distributed, physics-based control of swarms of vehicles. Auton. Robots **17**(2–3), 137–162 (2004)

Multi-swarm Infrastructure for Swarm Versus Swarm Experimentation

Duane T. Davis, Timothy H. Chung, Michael R. Clement and Michael A. Day

Abstract This paper builds on previous Naval Postgraduate School success with large, autonomous swarms of fixed-wing unmanned aerial vehicles (UAV) to provide infrastructure for the simultaneous operation of multiple swarms. Developed in support of an event fostering swarm capability development through competition, the online referee, or Arbiter, monitors and evaluates multiple independent but interacting swarms. This Arbiter provides sensor modeling for both swarms, evaluation of inter-swarm interaction, scoring and enforcement of competition rules, and graphical display of game status. Arbiter capability is demonstrated through live-fly experiments and software-in-the-loop simulation. The Arbiter is also used to evaluate swarm behaviors that are developed for air-to-air pursuit of an opposing swarm with results provided in this paper.

1 Introduction

Improved autonomous unmanned aerial vehicle (UAV) capability has led to increased interest in multi-UAV systems, or swarms, and opened research avenues such as

T. H. Chung, M. R. Clement and M. A. Day are Contributor to this work were performed while affiliated with the Naval Postgraduate School.

D. T. Davis (✉) · M. R. Clement
Naval Postgraduate School, Monterey, CA, USA
e-mail: dtdavi1@nps.edu

M. R. Clement
e-mail: michael.clement@gmail.com

T. H. Chung
Defense Advanced Research Projects Agency, Arlington, VA, USA
e-mail: timothy.chung@darpa.mil

M. A. Day
Georgia Tech Research Institute, Atlanta, GA, USA
e-mail: michael.day@gtri.gatech.edu

R. Groß et al. (eds.), *Distributed Autonomous Robotic Systems*,
Springer Proceedings in Advanced Robotics 6,
https://doi.org/10.1007/978-3-319-73008-0_45

platforms [9, 12], communication and control [4, 11, 17], and cooperation and coordination [3, 16, 19] among others. The Naval Postgraduate School (NPS) Advanced Robotic Systems Engineering Laboratory (ARSENL) has leveraged these and other research efforts in exploring large-scale swarms of autonomous, fixed-wing UAVs [8]. As part of these efforts, the NPS ARSENL has proposed an *Aerial Combat Swarms* (ACS) Challenge to foster capability development [7]. This challenge models a capture a flag scenario in which both swarms attempt to capture an opponent's flag while defending their own by preemptively tagging opponent UAVs.

Competitive events have proven useful in fostering robotics innovation. Perhaps the most well-known example is RoboCup and its use of soccer competition to promote broad-based research [14]. RoboCup success spawned simulation leagues that further accelerated research on multi-agent teamwork, planning, recognition, and learning [1]. Additional examples include the DARPA Grand and Urban Challenges [5, 6] and annual live-fly, competitive UAV events such as the International Micro Air Vehicle Conference and Competition [21], UAV Outback Challenge [24], and AUVSI events [2]. These events typically focus on a task or mission and assess competitors based on effectiveness rather than head-to-head performance.

Live-fly experimentation with aerial swarms (which includes the proposed ACS Challenge), typically requires simultaneous monitoring of many robots. The Swarming Micro Air Vehicle Network (SMAVNET) project has utilized a UDP/IP network to monitor and control a swarm of up to 10 fixed-wing UAVs with a single workstation [13], and Vásárhelyi et al. used the same approach in outdoor experiments with 10-vehicle quadrotor swarms [25]. In both cases, decision-making is distributed to the individual vehicles, so reliance on the base station for control is minimized while operator situational awareness is maintained through periodic broadcast of vehicle state information. The NPS ARSENL has also used this approach successfully in field experiments with up to 50-UAV swarms [8]; however, the operation of multiple swarms in a competitive scenario imposes requirements beyond monitoring.

Competitive autonomous robot events require metrics or objectives by which relative success is measured by unbiased evaluators. For many competitions, this requirement is met in a fairly straight forward manner. Task-based competitions such as the Outback Challenge, for instance, do not require competing robots to interact with one another during the event. Thus, evaluators can simply determine how successful each participant is at each task [22]. Compliance with competition rules can be similarly assessed by observation of individual participants.

Of the cited examples, RoboCup most closely aligns with the proposed ACS Challenge since both call for head-to-head team competition in a common environment. RoboCup events with actual robots are not that difficult to evaluate, however, since they are typically conducted by small teams in confined arenas with easily observed objectives. In the RoboCup Standard Platform League, for instance, teams of five robots compete on a 9-meter by 6-meter field with the objective of scoring the most goals [23]. Direct observation provides a reasonable evaluation approach for these events. In addition, RoboCup intentionally requires teams to be self sufficient in order to foster the broadest-possible research [14], so no mechanism for augmenting teams' onboard perceptive capabilities is required.

Evaluation of RoboCup simulation leagues, on the other hand, imposes additional requirements that overlap with those of the ACS Challenge. These leagues require a simulation framework implementing models of the physical environment, players, and sensors and providing feedback to the individual robot agents. RoboCup simulation leagues typically utilize a client-server system such as the RoboCup Soccer Server [15, 18] to implement a virtual game environment. More recent RoboCup simulation leagues have largely maintained this client-server, simulator-dependent approach [23]. RoboCup simulation league events share many characteristics with the ACS Challenge including distributed agent decision processes and dependence on the server for agent situational awareness. However agent-to-agent communication is implemented on the server itself and a synchronous discrete event model is utilized (i.e., one action per agent per cycle) [18] making it unsuitable for field experimentation or agents that communicate directly.

In addition to requiring support for live-robot events, the proposed ACS Challenge imposes a number of constraints that are not addressed by the Soccer Server's architecture. Most obviously, the communication environment of the ACS Challenge makes the client-server model unrealistic [7]. Further, unlike RoboCup events involving actual robots, the airspace volume, coupled with the number and size of the participating UAVs, makes direct observation and evaluation of the ACS Challenge impossible.

Given the inherent challenges to judging the proposed ACS Challenge, successful execution requires a software Arbiter to serve as an autonomous referee. The Arbiter is responsible for ensuring rule compliance, assessing engagements, and providing real-time visualization [7]. This work summarizes Arbiter infrastructure and implementation in support of the ACS Challenge and also provides results and analysis of initial air-to-air swarm behaviors.

Contributions include the development of an infrastructure for evaluating air-to-air and air-to-ground capabilities of competing swarms and for assessing interactions between swarms that may not be cooperating despite operating in the same geographic area. In addition, analysis of air-to-air behaviors provides a preliminary assessment of the relative importance of swarm size and behavior robustness.

The remainder of this paper first provides a brief overview of the multi-UAV system with which this work was conducted. Section 3 then describes ACS Challenge requirements and Arbiter design. Sections 4 and 5 respectively describe and analyze currently implemented air-to-air swarm behaviors. An evaluation of Arbiter performance in experiments is also provided in Sect. 5. Finally, Sect. 6 provides a summary and proposed areas for future research.

2 The ARSENL Multi-UAV System

ARSENL experiments are conducted with the NPS-designed *Zephyr II* UAV. The *Zephyr II* leverages open-source hardware and software, hobby and commodity components, and 3D-printed parts to provide a low-cost, yet capable, UAV platform. It weighs 2.5 kg, has a nominal cruising speed of 18 meters per second and approximately 50 min of endurance [8]. Swarming behaviors are implemented on a companion computer running the open-source Robot Operating System [20]. In addition, a robust simulation environment facilitates rapid development [10].

The *Zephyr II* UAV is equipped with a wireless radio that allows participation in an ad hoc Wi-Fi network that is also utilized by ground stations. UDP broadcast is used for all swarm communications to minimize latency [8]. An application-layer protocol was developed to address ACS Challenge requirements. All messages contain a fixed-format header with typical fields such as source and destination identifiers and timestamp, while payload varies with message-type-specific fields. This protocol is used to transmit state (telemetry) at 10 Hz, navigation status at 2 Hz, and other information as required [8].

Most messages are intentionally broadcast to all participants; however, some are directed to specific recipients. Because UDP broadcast is unreliably delivered to all participants, messages must support idempotency, and participants must identify and process relevant messages and ignore irrelevant ones. Each network node is assigned a unique identifier for this purpose [8]. In addition, a number of messages exchanged between air and ground stations require reliable delivery not provided by UDP broadcast [10]. A limited reliability mechanism for point-to-point communications has been incorporated to address this requirement.

In the context of the proposed ACS Challenge, the use of UDP broadcast implicitly imposes a requirement for separate networks for participating swarms. One of the most important Arbiter capabilities, then, is to bridge these networks and exchange required information between swarms in accordance with game rules [7].

3 Multi-swarm Operations

3.1 ACS Challenge Overview

The ACS Challenge involves two competing swarms (designated blue and red) in a battle arena. Each swarm is assigned a flag location on the runway to represent the defended asset and a standby waypoint for post-launch staging as depicted in Fig. 1. All UAVs are required to remain within the geographic boundaries of the battle arena and must observe minimum and maximum altitude restrictions [7].

Successful air-to-ground engagements require landing sufficiently close to the opponent's flag, while air-to-air engagements are evaluated by the Arbiter based on notional tagging characteristics and a firing report transmitted by the firing UAV indicating the tagging and the targeted UAVs and the GPS time. Tagged UAVs are immediately ineligible to conduct further engagements but are not specifically required to depart the battle arena.

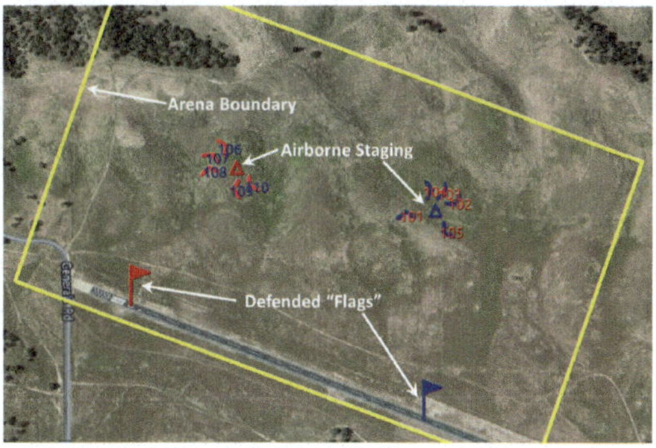

Fig. 1 Notional ACS challenge battle arena layout for McMillan airfield at Camp Roberts, CA

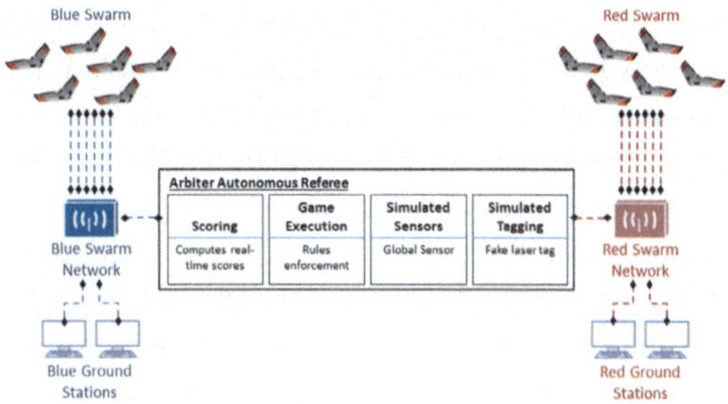

Fig. 2 Arbiter configuration for bridging red and blue networks and refereeing the ACS challenge

Each swarm is scored by the Arbiter using Eq. 1 where C_{air} and C_{ground} are constants, hit_{air} and hit_{ground} are sets of successful engagements, and f_e assesses logistical and operational effectiveness. Coefficients are adjusted to emphasize the relative importance air-to-air or air-to-ground engagements. Similarly, f_e is manipulated to incentivize factors of importance and to assess rule violation penalties. A more comprehensive description of the ACS Challenge scenario, objectives, assumptions, and rules is provided in [7].

$$Score = C_{air}\left|hit_{air}\right| + C_{ground}\left|hit_{ground}\right| + f_e \quad (1)$$

3.2 *Arbiter Implementation for Inter-swarm Interaction*

The Arbiter runs on a ground station and serves as a bridge between participating swarm networks. Since the protocol uses UDP broadcast, the Arbiter can access swarm messages by joining a swarm network, and joining both networks as depicted in Fig. 2 enables message traffic monitoring of both swarms. Arbiter situational awareness is ensured by requiring all UAVs to regularly broadcast GPS-timestamped state messages. Swarm UAVs are also required to broadcast messages to announce discrete events requiring Arbiter assessment (e.g., firing reports).

Using this pattern, the Arbiter—and more generally, the ARSENL system—is suitable for arbitrary UAV configurations so long as all simulated and live vehicles utilize the communications protocol described in Sect. 2. This flexibility allows the Arbiter to support swarms of live UAVs, simulated UAVs, and even mixed swarms of live and simulated UAVs [10]. Further, there is no requirement that UAVs be comparable to the *Zephyr II*, as both the Arbiter and the communications protocol are appropriate for fixed-wing, rotary-wing, powered, and unpowered UAVs [10].

In addition to monitoring both swarm networks, the Arbiter enforces game rules, maintains score, supplies simulated sensor data to both swarms, and simulates engagement tagging performance. Enforcement of most game rules is simply a matter of monitoring and processing UAV state information to identify violations. A breach of the battle arena boundary, for instance, is identified by comparing reported state to the battle arena geography. Maneuvering violations and many safety-of-flight issues can be similarly identified.[1] Game rule violation typically results in the removal the offending UAV from the competition [7].

Scorekeeping requires evaluation of both discrete events and continuous data that accumulates over time. In addition to engagements, discrete events such as launches, recoveries, and penalties can be incorporated into the logistical and operational effectiveness function, f_e. Continuous data, also incorporated into f_e, is useful for assessing swarm operations and might include the number of active swarm UAVs, UAV time on station, or defended area coverage among others. Given n discrete events and m accumulated continuous values, f_e is calculated using Eq. 2 where E_i is the point value of a discrete event, and T_j is a continuous value computed at time t.

$$f_e = \sum_{i=1}^{n} E_i + \sum_{j=1}^{m} \int_0^t T_j dt \tag{2}$$

Evaluation of discrete events involving multiple UAVs such as firing reports requires synchronization since alignment between affected UAV state-message and event timestamps is unlikely. This requirement is not unique to the ACS Challenge and has been addressed by the Massive Multiplayer Online Game community [26], though it is worth noting that latencies can be significantly higher in a live-fly environment using radio communications vice wired networks. Fortunately, the availability of a common time baseline from GPS time or Network Time Protocol and a level of trust afforded to challenge participants—cheating would defeat the purpose—mitigates the requirement for complicated synchronization solutions [7].

To facilitate accurate evaluation, the Arbiter maintains a scroll-like data structure (implemented as a double-ended queue) of timestamped events for each UAV. Event evaluation takes place after a predetermined delay is applied so that similarly-timed events that are received later can be properly sequenced to ensure in-order processing. When an event is evaluated, the preceding and following states are retrieved, and

[1] The Arbiter does not provide safety-of-flight services. Collision and terrain avoidance, maneuvering limits, and other hazards must be implemented on the individual UAVs.

linear interpolation is used to estimate the event-time state. For example in the case of a firing report, estimated aggressor and target states for the firing report time are used to evaluate the engagement. If an event timestamp is more recent than the last received state of an affected UAV, evaluation is deferred until a new state report is received. Run-time memory requirements are managed by periodic logging and removal of old events from UAV-specific data structures.

As described in [7], the ACS Challenge does not presume onboard sensing of opponent UAVs (organic detection can be rewarded through f_e). The Arbiter, therefore, must provide virtual sensor information to both swarms. Virtual sensor information is provided in the form of opponent-pose messages transmitted to the opposing swarm network and is based on state data maintained by the Arbiter. The Arbiter currently provides three sensor models selected independently for each swarm: an infinite range sensor (i.e., universal awareness), a finite range sensor with a swarm-wide data link (i.e., limited but swarm-wide awareness), and a finite range sensor with no data link (i.e., on-board sensing only). Noise is not currently added to virtual sensor data and sensor models are omnidirectional (more realistic and directional sensor models are a possibility for future development).

For virtual tagging the Arbiter assumes that UAVs are equipped with finite range, forward firing taggers. Requirements for successful employment (the tagging envelope) are described by minimum range r_{min}, maximum range r_{max}, maximum horizontal angle-off-the-nose α_{max}, and maximum vertical angle-off-the-nose β_{max}. An engagement is evaluated by determining the target's actual range, horizontal angle-off-the-nose, and vertical angle-off-the-nose (r, α, and β) relative to the aggressor at the time of the firing report and applying Eq. 3. A hit report is broadcast to both swarm networks for successful engagements. As with the virtual sensor, the virtual tagging model does not include noise and assumes perfect performance (i.e., if the target is in the tagging envelope then the engagement is successful).

$$hit(r, \alpha, \beta) = \begin{cases} True, & \text{if } r_{min} \leq r \leq r_{max} \wedge |\alpha| \leq \alpha_{max} \wedge |\beta| \leq \beta_{max} \\ False, & \text{otherwise} \end{cases} \tag{3}$$

The Arbiter supports operator situational awareness with map and dashboard displays. The map display provides real time position for all UAVs while the dashboard display provides the status of both swarms, echoes selected discrete events and Arbiter assessments, and displays current swarm scores.

4 Air-to-Air Algorithms

An air-to-air engagement sequence is characterized by an aggressor UAV intentionally maneuvering into a firing position against a target UAV. Once in firing position, the aggressor issues a firing report for assessment by the Arbiter. Suitable firing position can be determined using Eq. 3 based on calculated r, α, and β. Per Sect. 3.2, the

Arbiter broadcasts firing report evaluations over both swarm networks. Individual UAVs can utilize all received evaluations or only those associated with their own firing reports as implemented by their onboard behaviors.

Two air-to-air engagement behaviors were developed to test the Arbiter's usefulness in assessing swarm-versus-swarm performance. Both algorithms make target selection decisions based on Euclidean distance—the nearest target-UAV candidate is selected—and use a proportional navigation function to maneuver into firing position against the target. The selection process is obviously suboptimal in that multiple aggressors can select the same target while other candidates unengaged.

The Naive Shooter algorithm (Algorithm 1) follows a three step process: select target, maneuver into firing position, issue firing report; and continues until all opposing UAVs have been successfully engaged. This behavior is completely independent and will ignore engagement results from other UAVs. Thus, if multiple UAVs select the same target, each will continue the engagement until firing (i.e., a UAV will not disengage if another UAV issues a firing report that is assessed as a "hit"). Further, in selecting subsequent targets, UAVs that have been successfully engaged by other friendly UAVs will still be considered. The targeted UAV will be exclusively pursued until engaged even if a more attractive target emerges, and there is no break condition, so the pursuer will continue the pursuit even if the target is successfully engaged by another UAV. The behavior is intentionally rudimentary to provide a baseline against which more robust behaviors can be compared.

The Greedy Shooter algorithm (Algorithm 2) implements a simple but significant improvement. Whereas the Naive Shooter behavior ignores results of other UAVs' engagements, the Greedy Shooter algorithm incorporates these results into its own target selection and pursuit. Although the Greedy Shooter algorithm implements the same three step process as the Naive Shooter algorithm, a break condition is provided to terminate the pursuit (initiating a new target selection) if another UAV successfully engages the target. Further, successful engagements by other friendly UAVs will be factored into target selection. As with the Naive Shooter, the Greedy Shooter algorithm will continue its pursuit of the selected target until that target has been successfully engaged even if a more attractive target becomes available.

Both algorithms use the same target selection criteria and have the same communications requirements, so they differ only in one-on-one engagement termination and follow-on target selection. Over the course of an engagement, the Greedy Shooter's more efficient follow-on target selection can improve performance.

5 Experimentation

5.1 Arbiter Evaluation of Swarm Versus Swarm Interaction

ARSENL conducts regular live-fly exercises with swarms of up to 50 UAVs at Camp Roberts, California [8]. Multi-swarm experiments with the Arbiter and air-to-air behaviors described in this paper were conducted in December 2015 and May 2016

Algorithm 1 Naive Shooter

```
target_swarm ← opponent_swarm_uav_ids
hit_targets ← ∅
repeat
    target ← CHOOSE_BEST_TARGET(target_swarm)
    hit ← False
    repeat
        PURSUE(target)
        r, α, β ← GET_TARGET_PARAMETERS(target)
        if HIT(r, α, β) then
            report ← MAKE_FIRING_REPORT(time, target)
            hit ← ARBITER_EVALUATION(report)
            if hit = True then
                target_swarm ← target_swarm \ {target}
                hit_targets ← hit_targets ∪ {target}
            end if
        end if
    until target ∈ hit_targets
until target_swarm = ∅
```

Algorithm 2 Greedy Shooter

```
target_swarm ← opponent_swarm_uav_ids
hit_targets ← ∅
repeat
    target ← CHOOSE_BEST_TARGET(target_swarm)
    hit ← False
    repeat
        PURSUE(target)
        r, α, β ← GET_TARGET_PARAMETERS(target)
        if HIT(r, α, β) then
            MAKE_FIRING_REPORT(time, target)
        end if
        hit_targets ← hit_targets∪ ARBITER_EVALUATION(all)
        target_swarm ← target_swarm \ hit_targets
    until target ∈ hit_targets
until target_swarm = ∅
```

for swarm sizes of up to 15 UAVs per swarm. A visualization of a typical live-fly event involving two swarms of 10 UAVs each is provided in Fig. 3.

Field experimentation involves significant logistical investment. ARSENL field experiment success relies on an aggressive systems-engineering approach [10], a key component of which is Software-in-the-Loop (SITL) simulation. The SITL environment incorporates ground systems, accounts for environmental and communication conditions, and can include an arbitrary number of UAVs in multiple swarms [10]. Experimental results described in this paper were conducted in the SITL environment with swarms of between 10 and 25 UAVs. These studies allow for more systematic investigation of behaviors demonstrated in live-fly field tests.

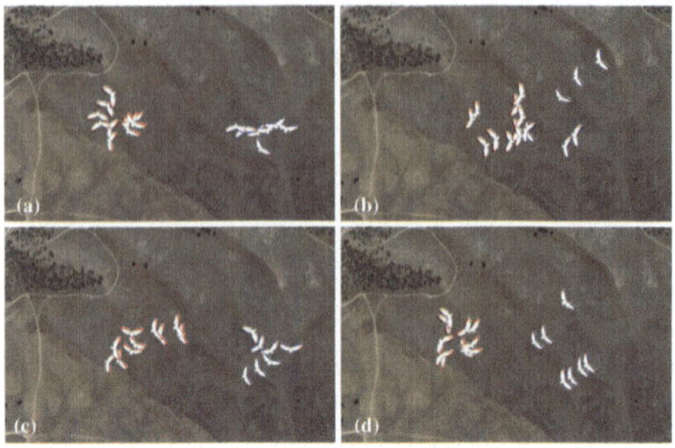

Fig. 3 Visualization in Google Earth of live-fly field experiments of 10v10 flights in December 2015. **a** Red swarm (left) and Blue swarm (right) in `swarm-ready` state; **b** Red and Blue swarms engaging using Naive Shooter algorithms; **c** Swarms disengaging for reset; **d** Blue swarm egressing at conclusion of experiment

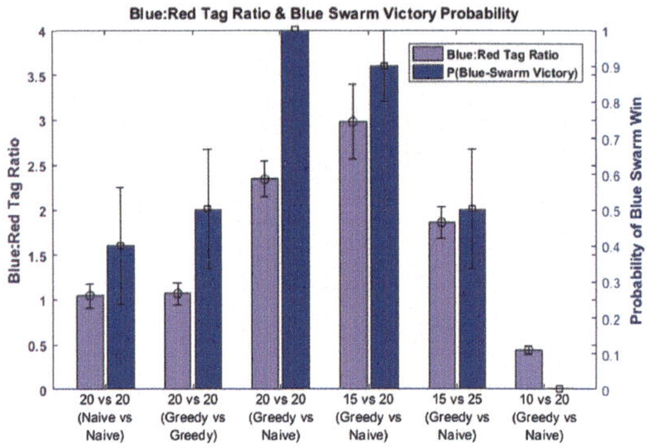

Fig. 4 Comparison of SITL experiment Naive vs. Greedy Shooter results. Blue tag ratio (left axis) describes the number of successful blue-on-red engagements per successful red-on-blue engagement. Probability of blue victory (right axis) provides the probability that the blue swarm successfully engaged all red-swarm UAVs

SITL experiments utilized the following virtual sensor and tagging parameters: infinite sensor range model, with $r_{min} = 0$ meters, $r_{max} = 150$ meters, $\alpha_{max} = 10°$, and $\beta_{max} = 90°$ (noting that this reduces engagement to a two-dimensional intercept).

Experiments were run until one swarm successfully engaged all opposing UAVs to ensure a winning swarm in every engagement. Ten iterations of each experiment were conducted with results consolidated for presentation in Fig. 4. Results are presented from the perspective of the blue swarm for consistency. Captured metrics include tag ratio (Eq. 4), and blue victory probability (i.e., the probability that all red-swarm UAVs were successfully engaged).

$$tag\_ratio = \frac{\left|hit_{air}^{blue}\right|}{\left|hit_{air}^{red}\right|} \tag{4}$$

Identical-swarm experiments were conducted with swarms of 20 UAVs to verify correct implementation and assess performance. Not surprisingly, blue:red tag ratio was close to 1:1 and blue victory probability was approximately 0.5 as the Fig. 4 "20 versus 20 (Naive vs Naive)" and "20 versus 20 (Greedy vs Greedy)" entries indicate.

Interestingly, experiments frequently stabilized with opposing UAV pairs circling one another in an unsuccessful attempt to maneuver into firing positions. When sta-

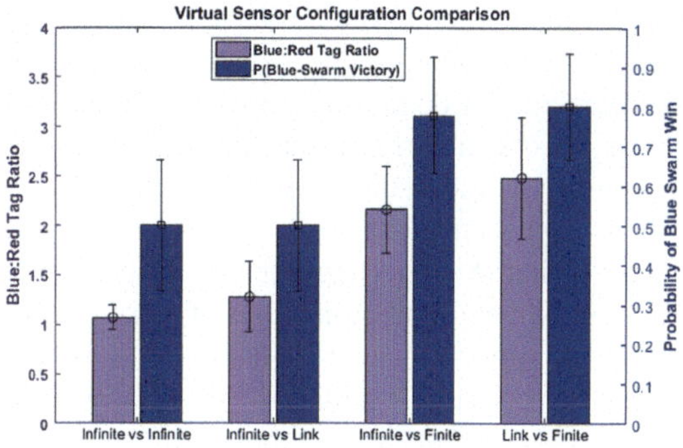

Fig. 5 Comparison of SITL experiment virtual sensor configuration results

bilization was noted, behaviors were suspended and reinitialized once the swarms separated to reset the engagement (to allow continuation until one swarm won). This phenomenon—which was also observed in live-fly experiments—results from the focused pursuit of a single opponent following target selection. That is, evenly matched UAVs will relentlessly pursue one another even if maneuvering limits prevent either from obtaining a valid firing solution.

Dissimilar-swarm experiments were conducted with the blue and red swarms utilizing the Greedy Shooter and Naive Shooter algorithms respectively. Swarm sizes were varied from evenly matched to a 2:1 advantage for the red swarm.

The blue swarm had a marked advantage in 20 versus 20 and 15 versus 20 experiments with tag ratios exceeding 2:1 and a probability of victory of nearly 1.0. Of note, the one red swarm victory resulted from a high initial volley red:blue tag ratio (i.e., red UAVs successfully engaged many blue UAVs in the initial stage). This gave the red swarm an insurmountable numerical advantage for the remainder of the experiment. In all other experiments, superior blue swarm follow-on target selection was evident.

The blue swarm maintained a tag ratio advantage in the 15 versus 25 experiments, but this did not ensure a blue swarm victory. In approximately 40% of the experiments, the red swarm successfully engaged all blue UAVs despite heavy losses (even in red swarm victories, more red UAVs were successfully engaged than blue UAVs). Once red swarm advantage was increased in the 10 versus 20 experiments, numerical superiority became decisive. Blue:red tag ratios in these experiments dropped dramatically to less than 0.5:1 and the red swarm won all engagements.

The effect of currently modeled configurations on behavior performance was also evaluated in SITL experiments with results depicted in Fig. 5. These experiments were conducted with identical 20-UAV swarms using the Greedy Shooter algorithm and a 250 m sensor range (for finite-range models). As indicated, swarms using the infinite-range sensor and finite-range sensor with a swarm-wide data link performed similarly. In head-to-head experiments these swarms were fairly evenly matched, with tag ratios near 1:1 and neither sensor model providing a noticeable advantage. Swarms using these sensor models did significantly outperform those using the finite-range model with no data link. Although preliminary, these experiments do demonstrate Arbiter usefulness in assessing sensor models.

These results demonstrate that the Greedy Shooter algorithm outperforms the Naive Shooter algorithm, but that the advantage can be overcome by numerical superiority. These results are not surprising, however, and they merely confirm fairly well-known phenomena. More importantly, these experiments demonstrate the utility of the Arbiter in supporting rigorous analysis of behavior performance in a competitive environment. Thus, outcomes of these sorts of experiments can be used not only to identify superior behaviors, but also to formally characterize the nature and limits of their superiority. In addition, the analysis itself provides an example of the sorts of comparisons that can be supported.

5.2 Evaluation of Arbiter Performance

The experiments described in Sect. 5.1 demonstrate the Arbiter's ability to support evaluation of interactive multiple swarm behavior. They also provided a mechanism for evaluating the performance of the arbiter itself and led to a number of relevant observations.

Scalability is an important consideration in any multi-agent system. Since the UDP-based ACS communications protocol does not require retransmission, network bandwidth requirements scale linearly with swarm size. Incorporation of opponent-pose messages from the Arbiter effectively doubles the bandwidth requirement (assuming same-sized swarms) on both swarm networks. Arbiter performance was successfully demonstrated with swarm sizes of up to 25 UAVs in recent field experiments, however anecdotal network performance for a single 50-UAV swarm does reveal some degradation [8]. It is reasonable to infer that communications requirements for larger multiple swarms will incur additional network performance degradation, however the actual limits in the ability of an 802.11n network to support increasing swarm sizes with this architecture need to be more formally assessed.

Synchronization was also an important factor in Arbiter performance. As noted previously, event times are based on UAV time, so out-of-order evaluation is unlikely with proper synchronization through GPS or a Network Time Server. Further, event times are explicitly encoded into the payload portion of the associated message, so transmission delays and latency are not factors. Without a Network Time Server in use, however, occasional out-of-order evaluation was noted in SITL experiments when on-UAV time differed by as little as two seconds. Evaluation of firing reports in the wrong order in these cases could lead to a UAV being evaluated as "hit", when the firing UAV should have been out of action already. No out-of-order event evaluation was noted in any of the field experiments.

Communications reliability can also factor into Arbiter performance. Use of the ACS protocol's reliability mechanism for event messages provides some assurance that these messages will be processed by the Arbiter, but the protocol does not provide for an unlimited number of retransmissions. Therefore, the possibility that a particular message will never be delivered cannot be dismissed and may require post-mission reconciliation between Arbiter and UAV logs that has not yet been implemented. Missed firing reports were not observed during field exercises with up to 25 UAVs per swarm, but were noted on occasion in SITL experiments with simulated communications loss rates of 90 percent.

Finally, UAVs that have been evaluated as "hit" can still affect game play. For safety-of-flight reasons, the Arbiter continues to transmit opponent-pose messages even after a UAV has been successfully engaged, and out-of-action UAVs are not required to immediately exit the battle arena. This means that if the aggressor's behavior does not account for out-of-action UAVs or the aggressor does not receive the hit report, it will continue to pursue the out-of-action UAV. Thus, with this Arbiter implementation, the onus for rule compliance and efficient behavior performance rests entirely upon the participant.

6 Conclusions and Future Work

This paper presented a software Arbiter for live-fly and simulation-based experimentation with multiple UAV swarms. This work extends previous ARSENL work to provide support for a proposed ACS Challenge through a multi-swarm Arbiter that was used to test and evaluate behaviors for air-to-air engagements between swarms. This analysis demonstrates the Arbiter's utility in comparative assessment of competing swarm behaviors and provides a baseline for comparing the relative values of swarm size and swarm behavior robustness. Lessons learned from this work will help to inform development and analysis of both UAV single-swarm and multi-swarm capabilities.

The efforts in this paper elucidate numerous areas for future work. Obvious Arbiter improvements include enhanced sensor and tagger models and the development of metrics for competing swarm evaluation. Future development will also benefit from more comprehensive analysis of current swarm behaviors to include experimentation with larger swarms and various sensor configurations. Further, the swarm behaviors implemented for this paper are rudimentary but provide a baseline for comparison to more robust behaviors. Future behaviors might use consensus or market-based approaches to improve target selection or more deliberative heuristics to avoid stalemate and improve survivability. Also, behavior development—for swarm-versus-swarm competition, for non-competitive interaction between swarms, and also for individual swarm implementation—is a priority. Finally, increasing the scale of experiments will facilitate more formal evaluation of the Arbiter communications architecture to include identification of practical limits in the size of the swarms that can be supported.

References

1. Asada, M., Veloso, M., Kraetzschmar, G.K., Kitano, H.: Robocup: today and tomorrow. Exp. Robot. VI **250**, 369 (1999)
2. Association for Unmanned Vehicle Systems International: 2016 Rules for AUVSI Seafarer Chapter's 14th Annual Student UAS Competition (2016)
3. Bayraktar, S., Fainekos, G.E., Pappas, G.J.: Experimental cooperative control of fixed-wing unmanned aerial vehicles. In: 43rd IEEE Conference on Decision and Control, 2004. CDC, vol. 4, pp. 4292–4298. IEEE (2004)
4. Bekmezci, I., Sahingoz, O.K., Temel, Ş.: Flying ad-hoc networks (fanets): a survey. Ad Hoc Netw. **11**(3), 1254–1270 (2013)
5. Buehler, M., Iagnemma, K., Singh, S.: The 2005 DARPA Grand Challenge: The Great Robot Race, vol. 36. Springer Science & Business Media (2007)
6. Buehler, M., Iagnemma, K., Singh, S.: The DARPA Urban Challenge: Autonomous Vehicles in City Traffic, vol. 56. Springer, Berlin (2009)
7. Chung, T.H., Jones, K.D., Day, M.A., Jones, M., Clement, M.: 50 vs. 50 by 2015: Swarm Vs. Swarm UAV Live-Fly Competition at the Naval Postgraduate School, pp. 1792–1811. AUVSI North America, Washington, DC (2013)
8. Chung, T.H., Clement, M., Day, M.A., Jones, K.D., Davis, D.T., Jones, M.: Live-fly, large-scale field experimentation for large numbers of fixed-wing UAVs. In: 2016 IEEE International Conference on Robotics and Automation. Stockholm, Sweden (2016)
9. Cole, D.T., Sukkarieh, S., Göktogan, A.H., Stone, H., Hardwick-Jones, R.: The development of a real-time modular architecture for the control of uav teams. In: Field and Service Robotics, pp. 465–476. Springer, Berlin (2006)
10. Day, M.A., Clement, M.R., Russo, J.D., Davis, D., Chung, T.H.: Multi-UAV software systems and simulation architecture. In: 2015 International Conference on Unmanned Aerial Systems, pp. 426–435. IEEE, Denver, CO (2015)
11. Gupta, L., Jain, R., Vaszkun, G.: Survey of important issues in UAV communication networks. IEEE Commun. Surv. Tutor. **18**(2), 1123–1152 (2015)
12. Han, J., Xu, Y., Di, L., Chen, Y.: Low-cost multi-UAV technologies for contour mapping of nuclear radiation field. J. Intell. Robot. Syst. **70**(1–4), 401–410 (2013)
13. Hauert, S., Leven, S., Zufferey, J.C., Floreano, D.: The swarming micro air vehicle network (SMAVNET) project (2015)
14. Kitano, H., Asada, M., Kuniyoshi, Y., Noda, I., Osawa, E.: Robocup: The robot world cup initiative. In: Proceedings of the first international conference on Autonomous agents, pp. 340–347. ACM (1997)
15. Kitano, H., Asada, M., Kuniyoshi, Y., Noda, I., Osawa, E., Matsubara, H.: Robocup: a challenge problem for AI. AI Mag. **18**(1), 73 (1997)
16. Kownacki, C., Odziej, D.: Flocking algorithm for fixed-wing unmanned aerial vehicles. In: Bordeneuve-Guibé, J., Drouin, A., Roos, C. (eds.) Advances in Aerospace Guidance, Navigation and Control SE - 24. Flocking A, pp. 415–431. Springer International Publishing (2015). http://dx.doi.org/10.1007/978-3-319-17518-8_24
17. Madey, A.G., Madey, G.R.: Design and evaluation of UAV swarm command and control strategies. In: Proceedings of the Agent-Directed Simulation Symposium, p. 7. Society for Computer Simulation International (2013)
18. Noda, I., Stone, P.: The RoboCup soccer server and CMUnited clients: implemented infrastructure for MAS research. Auton. Agents Multi-Agent Syst. **7**(1–2), 101–120 (2003)
19. Nowak, D.J., Price, I., Lamont, G.B.: Self organized UAV swarm planning optimization for search and destroy using swarmfare simulation. In: 2007 Winter Simulation Conference, pp. 1315–1323. IEEE (2007)
20. Quigley, M., Gerkey, B., Conley, K., Faust, J., Foote, T., Leibs, J., Berger, E., Wheeler, R., Ng, A.: ROS: an open-source robot operating system. In: ICRA Workshop on Open Source Software (2009)

21. Reeder, M.: Special issue on the international micro air vehicle conference and flight competition 2014 (IMAV 2014). Int. Jo. Micro Air Veh. **6**(4), i–ii (2014)
22. Roberts, J., Frousheger, D., Williams, B., Campbell, D., Walker, R.: How the outback challenge was won: the motivation for the UAV challenge outback rescue, the competition mission, and a summary of the six events (2016)
23. Stone, P.: Whats hot at robocup. In: Thirtieth AAAI Conference on Artificial Intelligence (2016)
24. UAV Challenge. https://uavchallenge.org/ (2016). Accessed: 25 June 2016
25. Vásárhelyi, G., Virágh, C., Somorjai, G., Tarcai, N., Szörényi, T., Nepusz, T., Vicsek, T.: Outdoor flocking and formation flight with autonomous aerial robots. In: 2014 IEEE/RSJ International Conference on Intelligent Robots and Systems, pp. 3866–3873. IEEE (2014)
26. Yahyavi, A., Kemme, B.: Peer-to-peer architectures for massively multiplayer online games: a survey. ACM Comput. Surv. (CSUR) **46**(1), 9 (2013)

Robust Coordinated Aerial Deployments for Theatrical Applications Given Online User Interaction via Behavior Composition

Ellen A. Cappo, Arjav Desai and Nathan Michael

Abstract We propose and evaluate a multirobot system designed to enable live theatric presentation of episodic stories through online interaction between a performer and a robot system. The proposed system translates theatric performer intent into dynamically feasible robot trajectories without requiring prior knowledge of the ordering or timing of the desired robot motions. The system enables a user to issue instructions composed of desired motion descriptors at arbitrary times to specify the motion of the robot ensemble. The system refines user motion specifications into safe and dynamically feasible trajectories thereby reducing the cognitive burden placed on the performer. We evaluate the system on a team of aerial robots (quadrotors), and show through offline simulation and online performance that the proposed system formulation translates online input into non-colliding dynamically feasible trajectories enabling the performance of an epic poem over the course of a three act performance spanning fifteen minutes of coordinated flight by a six robot team.

1 Introduction and Related Work

We wish to enable a performer to direct an ensemble of robots to act as characters during the retelling of an epic story that requires the robots to perform choreographed motions while deployed in groupings or formations. Although the choreographed motion types are known before the performance, the choice and ordering of the behaviors, their duration, and the timing between them is directed by a storyteller in an improvisational manner that may vary in tempo and tenor between performances.

E. A. Cappo (✉) · A. Desai · N. Michael
Carnegie Mellon University, Pittsburgh, PA, USA
e-mail: eacappo@cmu.edu

A. Desai
e-mail: arjavd@cmu.edu

N. Michael
e-mail: nmichael@cmu.edu

R. Groß et al. (eds.), *Distributed Autonomous Robotic Systems*,
Springer Proceedings in Advanced Robotics 6,
https://doi.org/10.1007/978-3-319-73008-0_46

To enable the theatric performer to provide online choreography direction to the multi-robot system, the performer's intent, including instructions specifying lighting, sound, or movement and timing directions, must be translated into individual robot trajectories online. This requirement of real-time intent translation is particularly challenging as the performer can issue a direction at any time and without consideration of the robots' extents and performance limits leading to violation of collision and actuator constraints. Additionally, the performer may err in the direction, leading to a motion specification that is logically invalid.

Coordinated vehicle deployments within the context of choreographed and improvisational performances are generally characterized as *scripted* or *unscripted*, respectively. Currently, most robot theatric works are scripted, fully specifying all robot trajectories before the performance, including efforts focused on choreographed aerial dance and acrobatic maneuvers [1, 9]; light shows such as the one performed at the Cannes International Festival of Creativity, Saatchi and Saatchi New Directors' Showcase 2012 Festival [6]; or the Guinness Book of World Record-holding flight of 100 Drones [7]. If present, human actors in scripted performances [8] or Cirque Du Soleil's "Sparked" [2, 3] respond to the robots' motions to invoke a sense or impression of interaction. Alternative strategies seek to blend online operator interaction with scripted robot motions that are predefined by enabling the user to dynamically trigger the start of the motions [5] as recently demonstrated by MagicLab and Rhizomatiks Research [10, 11]. Fully scripted works do not allow for any change to the choreography during the performance, and thus interaction is conveyed through the performer's reaction to the robot system. Dynamically triggered motion sequences enable the user to change elements of the performance online but do not allow the system to respond or adapt to the user intent.

We pursue a methodology that will enable a performer to direct robot motions in an improvisational (or unscripted) manner. This objective is challenging as we require a strategy to translate theatric intent into a form that can be readily specified online by the performer and yields viable motion specifications for the robots given actuator and collision constraints. We propose a framework that leverages a centralized planner that encodes user intent through predefined descriptors that capture theatric elements such as the light, sound, or spatial-temporal motion of one or multiple robots. These descriptors are combined to form *behaviors* including aesthetic behaviors such as light and sound (e.g. on or off) and motion behaviors that are composed through the combination of formation, motion, and time descriptors and manifest as individual robot trajectories as detailed in Sect. 2.1. The system ensures that the commanded behaviors are both logically valid (Sect. 2.2) and dynamically feasible through validation checks and motion plan refinement (Sect. 2.3). If the original trajectories specified by a behavior are unsafe or dynamically infeasible, the plans are refined to yield non-colliding trajectories that transition robots from their current state to the desired behavior trajectories with the requisite time-scaling to meet actuator limits. Figure 1 provides a conceptual overview of the proposed system formulation.

To evaluate coverage over the space of behavior inputs, we perform a large number of offline simulations and randomly select motion descriptor combinations given the

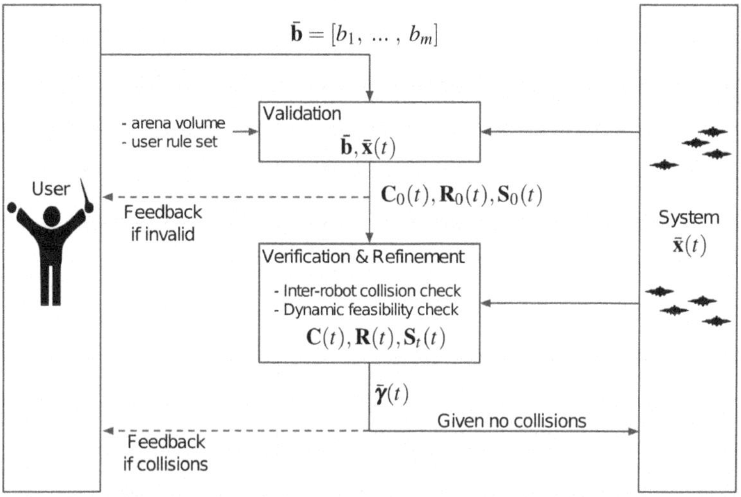

Fig. 1 The system translates user-issued input into feasible and safe behaviors. These behaviors undergo validation and verification checks and, if necessary, are modified to meet collision and actuator constraints. The resulting trajectories are distributed to the robot team. While not detailed in this work, user input is enabled via physical gesture and tablet user interfaces

set of possible start and goal locations. We confirm that when responding online to unscripted commands, resulting plans do not violate the required inter-robot clearance distance or imposed dynamic limits (Sect. 3). We validate the simulation results via experiments with six quadrotors and a performance that mirrors a subset of the coverage trials examined in simulation and confirm that the hardware performance aligns with the simulation results. We qualitatively review the outcomes of a performance with a narrative story of three acts performed by six quadrotors, where a performer directs the system online over a fifteen minute period using 100 behavior commands. Additionally, we discuss observations arising from the evaluation of the proposed system based on user interaction with the system, and conclude with directions for future system refinement (Sects. 4 and 5).

2 System Design

The proposed system design is formulated to meet the aesthetic requirements of an improvisational theatric application that leverages an aerial robot team to visually convey stories in which groups of individuals (ensembles) engage and interact as coordinated formations. Therefore, we propose a formation-based approach to enable specification of robot team motion in a manner that seeks to reduce the user interaction burden by avoiding individual robot motion specifications.

2.1 Behaviors

Performer intent is presented to the system via gesture or graphical user interfaces, as shown in Fig. 2, in the form of tokens. Tokens are composed into a behavior, $\bar{\mathbf{b}}$, consisting of m behavior descriptors, b_i. Each behavior descriptor, b_i, may take a discrete number of values. Figure 3 highlights several behavior descriptors and potential value assignments. Denoting B_i as the set of values associated with the behavior descriptor b_i, the total number of potential behaviors achievable by the system is:

$$perm(\bar{\mathbf{b}}) = \prod_{i=1}^{m} \left(\sum_{k=1}^{|B_i|} \frac{|B_i|!}{k!(|B_i| - k)!} \right). \tag{1}$$

A *motion behavior* is a collection of user specified motion behavior descriptors, of which representative descriptors are shown in Fig. 3. These descriptors specify formation type, behavior timing, and motion characteristics. A motion behavior requires the specification of the desired trajectory endpoint constraint, duration, and motion characteristics. We explain each descriptor category below as well as highlight how different descriptors contribute information to the trajectory formulation.

Behavior Duration: The starting time of the behavior, t_s, is the time at which the system receives the command from the user,[1] and concludes at t_f, the time as specified by the timing behavior descriptor.

Formation Specification: We describe a formation of robots by specifying each robot's state in a local reference frame [4], which we call the *shape* frame. The positions and headings of each robot in a local reference frame as a function of time t are $\mathbf{s}(t) = [x(t),\ y(t),\ z(t),\ \psi(t)]^{\mathrm{T}},\ \mathbf{s} \in \mathbb{R}^3 \times SO(2)$, with a vector $\mathbf{S}(t)$

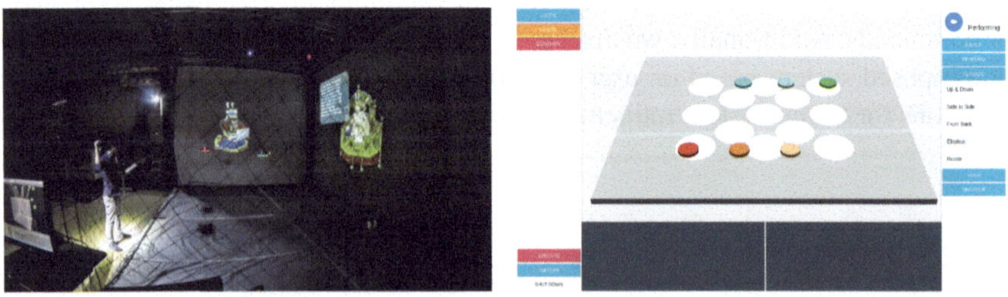

(a) A user inputs descriptors to the system via gesture.

(b) A screen shot of the graphical user interface (GUI) employed on a touch screen tablet.

Fig. 2 A user may interact with the system via gesture (**a**) or graphical user interface (**b**). The system leverages motion capture observations to identify changes in user form to recognize gesture input, while the GUI adapts selection options based on user descriptor selection

[1] In practice, t_s is set to a value slightly ahead of the instruction receipt time to account for planning computation time, allowing robots to transition between trajectories without discontinuities.

$\mathbf{C}(t), \mathbf{R}(t)$				$\mathbf{S}(t)_{xyz}$	$\mathbf{S}(t)_{\psi}$	$\mathbf{S}(t)$
robots	time	action	goal	shape	heading	manner
[1 … N]	*t* seconds	∅ hover	∅ v*dt	∅ as-is	∅ as-is	∅ fixed
		↑ takeoff	[Ax, Ay, Az]	□ polygon	*x*	variable
		↓ land	[Bx, By, Bz]	— line	*y*	drunk
		→ go-to-target	⋮	∴ filled	[Ax, Ay]	nervous
		$\overset{x}{\leftrightarrow}$ forward-rev			[Bx, By]	
		$\overset{y}{\leftrightarrow}$ side-side			⋮	
		$\overset{z}{\leftrightarrow}$ up-down				
		⊙ circle-target				
		↻ turn-in-place				

Fig. 3 Representative behavior descriptors, b_i, and descriptor sets, B_i, with associated potential values

containing all of the positions in the local reference frame of the n robots in the formation: $\mathbf{S}(t) = [\mathbf{s}_1(t), \ \ldots, \ \mathbf{s}_n(t)]$. The *shape* descriptor specifies desired starting positions, $\mathbf{s}^{xyz}(t_s)$, in the shape frame [4] and the *heading* descriptor specifies $\mathbf{s}^{\psi}(t)$ for each vehicle. A vehicle's heading can be defined relative to its current frame ("as-is") or oriented toward a target in the inertial frame (a theatrical maneuver called "spotting").

Motion Specification: Vehicle motions are defined by both the *manner* and *action* descriptors.

- The *manner* descriptor is similar to an adjective in language, giving more information about the flight characteristics that each robot should display during the behavior. Two characteristics of interest to the story are "drunk" and "nervous" mannerisms, which a robot performs by moving along a wobbly course of motion, with slower, larger motions for "drunk" and faster, smaller motions for "nervous". We represent these motions as bounded polynomial trajectories generated through randomly chosen keypoints obeying timing and distance constraints. Trajectory $\mathbf{s}_n^{xyz}(t)$ for robot n is a spline fit through k keypoints in x, y, and z [14] so that dt_{ij}, the time between each pair of consecutive waypoints i and j, is bounded ($dt_{min} \leq dt_{ij} \leq dt_{max}$) and the sum of all dt's equals the full time span: $\sum_{i=0,j=i+1}^{i=k-1} dt_{ij} = t_f - t_s$. The position of each keypoint for the jth robot lies within a ball of radius δ centered around the robot's starting position, $\mathbf{s}_j(t_s)$, $\mathbf{s}_j^{xyz}(t_s) \in \mathscr{B}_\delta(\mathbf{s}_j^{xyz}(t_s))$. The bounding values dt_{min}, dt_{max}, and δ are defined on a per-mannerism basis. The "fixed" mannerism denotes regular flight such that the vehicles hold their positions in the local frame throughout the behavior.
  All mannerisms, $\mathbf{s}(t)$, must remain within specified limits and ensured through appropriate choice of bounding values dt_{min}, dt_{max}, and δ,

$$\mathbf{s}_j^{xyz}(t) \in \mathscr{B}_\delta(\mathbf{s}_j^{xyz}(t_s)), \quad |\dot{\mathbf{s}}_j^{xyz}(t)| \leq v_{lim}, \quad |\ddot{\mathbf{s}}_j^{xyz}(t)| \leq a_{lim} \tag{2}$$

Further, the inter-robot clearance distance, d, must be respected at all times, so that for all combinations of robots in $\bar{\mathbf{b}}$:

$$|\mathbf{s}_i^{xyz}(t) - \mathbf{s}_j^{xyz}(t)| \geq d, \ \forall i, j \in \bar{\mathbf{b}}. \tag{3}$$

In general, we choose to only allow a user to specify a single mannerism descriptor. However, for appropriate choice of bounding values dt_{min}, dt_{max}, and δ, the combinations of mannerisms $\mathbf{s}(t) = \mathbf{s}_1(t) + ... + \mathbf{s}_j(t)$ will obey the constraints stated in Eqs. (2) and (3). The vehicles are therefore "nervous drunks" if required, as the length of the mannerism descriptor set is permitted to be greater than one.

- The *action* descriptor specifies the motion of the entire formation. Combined with the *goal* descriptor, we can design a trajectory that moves the local reference frame through the inertial frame. We define the state of each robot at time t by the vector $\mathbf{x}(t)$, containing position coordinates and heading of the vehicle: $\mathbf{x}(t) = [x(t),\ y(t),\ z(t),\ \psi(t)]^{\mathrm{T}},\ \mathbf{x} \in \mathbb{R}^3 \times SO(2)$. The state of an n vehicle system is given by $\bar{\mathbf{x}}(t) = [\mathbf{x}_1(t),\ \ldots,\ \mathbf{x}_n(t)]$. We design smooth trajectories for each state-space dimension via time parameterized polynomials up to an appropriate order to ensure smoothness in the trajectories and their derivatives and satisfy dynamic properties of the vehicle control model.
  The position of the origin of the local frame with respect to the inertial frame at time t is $\mathbf{C}(t) = [x(t),\ y(t),\ z(t)]^{\mathrm{T}},\ \mathbf{C}(t) \in \mathbb{R}^3$ and denote $\mathbf{R}(t) \in SO(3)$ as the time varying rotation computed from the Euler rotations around the inertial x, y, and z axes, $\mathbf{R}(t) = \mathbf{R}_z(t)\mathbf{R}_y(t)\mathbf{R}_x(t)$. To describe a smoothly varying, differentiable rotation, Euler angles are defined as polynomial trajectories [12].
  Actions such as "circle-target" or "turn-in-place" specify formation rotations, while periodic actions ("forward-rev", "side-side", and "up-down") define trajectories along the specified axis through waypoints centered on the target location. All actions are composable with all goals and timing specifications to yield valid polynomial trajectories for $\mathbf{C}(t)$ and $\mathbf{R}(t)$.

  Trajectory initial location: The starting state of a trajectory governing the motion of a formation of robots is established based on the current states of the robots at the time the instruction is specified. Upon instruction receipt, the local coordinate frame in which the formation shape is defined is established with an identity rotation and located at the mean of the specified robots' current positions and with higher order terms equal to the mean of the robots' higher order states, leading to the definition of states, $\mathbf{C}(t_s)$ and $\mathbf{R}(t_s)$.

  Trajectory ending location: The ending states, $\mathbf{C}(t_f)$ and $\mathbf{R}(t_f)$, are specified by the "goal" and "action" parameters. Goals are defined as $(x,\ y,\ z)$ locations in the inertial frame and actions specify motion primitives in relation to those locations. The initial and final states combined with the start and end times allows us to specify a polynomial trajectory for $\mathbf{C}(t)$ and $\mathbf{R}(t)$. Combining these trajectories with the motion description for a formation's local frame results in the trajectory specification for each robot in the inertial frame,

$$\boldsymbol{\gamma}_n(t) = \begin{bmatrix} \mathbf{C}(t) + \mathbf{R}(t)\mathbf{s}_n^{xyz}(t) \\ \mathbf{s}_n^{\psi}(t) \end{bmatrix}. \tag{4}$$

where $\mathbf{s}_n(t)$ is one of the n local robot trajectories as specified in $\mathbf{S}(t)$, and the superscripts xyz and ψ denote those respective elements of the local state vector. Similar to $\bar{\mathbf{x}}(t)$, the bar notation denotes a vector of robot trajectories, $\bar{\boldsymbol{\gamma}}(t)$.

2.2 Validation

There are primarily two reasons why a behavior may be invalid in an online setting. First, the current vehicle states may lead to a specified behavior colliding with flight volume boundaries. Second, a user may impose state-transition rules that limit descriptor combinations.

Therefore, we validate a behavior by first confirming that the descriptors, given the system state, do not result in rule-set violations. An allowable specification is defined as $\{\mathbf{C}_0(t), \mathbf{R}_0(t), \mathbf{S}_0(t)\}$ given the current system state and the descriptor specifications and represents the proposed desired behavior. For example, as shown in Fig. 4b, the convex hull of the behavior is confirmed to remain within the flight volume and is marked as valid.

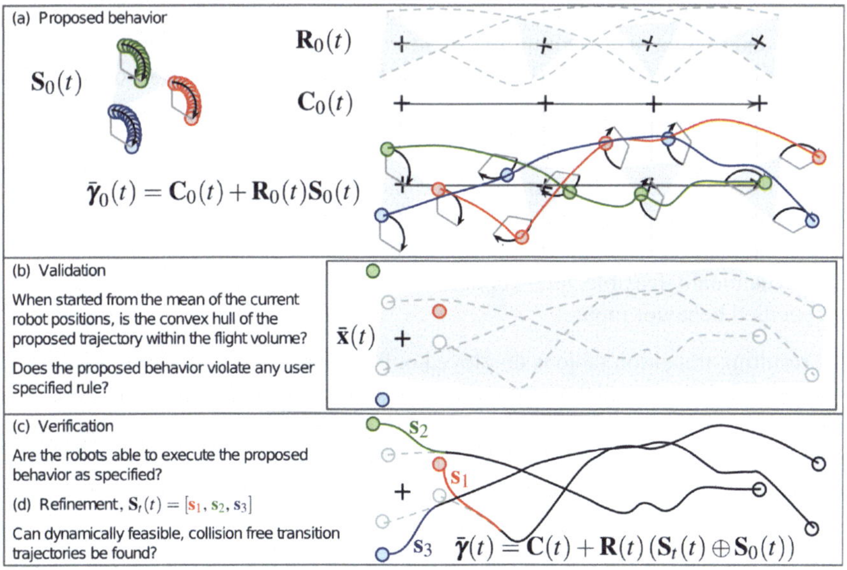

Fig. 4 Illustration of: **a** composition of a proposed behavior; **b** validation of the proposed behavior, depicted with respect to the current robot states; **c** verification given the current robot states; and **d** refined behaviors with dynamically feasible transitions

2.3 Verification and Mitigation

Given a valid desired behavior, we verify that the behavior is realizable by checking the following conditions (in order).

1. The current states of the robots specified by the behavior are sufficiently close to the starting states defined by the desired behavior, i.e., $\bar{\mathbf{x}}(t_s) \simeq \bar{\mathbf{x}}_0(t_s)$.
2. The proposed trajectory accelerations are within the specified limit, $|\ddot{\boldsymbol{\gamma}}(t)| \leq a_{lim}$.
3. The n robots in the behavior maintain an inter-robot spacing greater than or equal to the minimum clearance distance, $|\boldsymbol{\gamma}_i(t) - \boldsymbol{\gamma}_j(t)| \geq d,\ i, j \in [1 \ldots n]$.

If any condition fails, we immediately proceed to design refined trajectories [4] to mitigate the failure, leading to a dynamically feasible, inter-robot collision-free group behavior that remains within the arena volume. We now perform a final collision check across robot groups. If the desired behavior intersects with other robots, we further modify the trajectories to find collision free trajectories. However, at this time we are investigating group-based collision avoidance strategies, and so, in practice, avoid (or invalidate) behaviors that result in group collisions.

2.3.1 Offline Transition Validation and Verification

While formal methods, such as LTL [15, 16], are able to explicitly verify every behavior (at the cost of high compute time), we choose to leverage offline simulation across a large number of trials to verify all descriptor combinations, Eq. (1), assuming a discretization of the state-space of the system that approximately covers all possible starting and ending states within the flight volume. We depict the results of these offline trials in Fig. 5 and detail both the number of times a descriptor combination is tested and the number of successful behavior transitions. A behavior transition is considered successful if:

1. The descriptor combination forms a valid $\{C, R, S\}$ tuple, meaning the code implementation is error-free;
2. The descriptor combination does not violate a user specified rule; and
3. A dynamically feasible, inter-robot collision-free trajectory is generated from the specified behavior input.

The resulting transition table is employed online to assist in performing fast online validation. Behaviors with intermediate success rates frequently fail due to instruction timing. Therefore, we may choose to use this validation table as a conservative heuristic, and rather than check every online instruction, reject behavior transitions with success rates below a cutoff value.

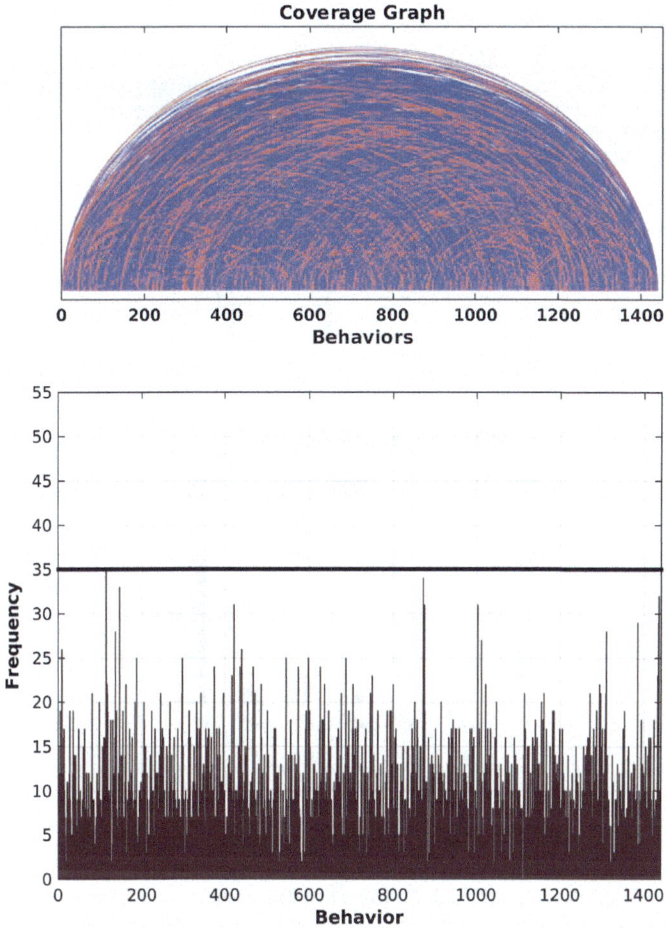

Fig. 5 Plots showing coverage over representative behavior descriptor combinations. Behaviors are validated across varying numbers of robots, with instructions issued at randomly chosen time intervals. Success indicates that the descriptor combination produces a valid behavior and the system is able to interpret, refine, and transform the behavior into a dynamically feasible, collision-free trajectory. Top plot: Arcs describe transition success rates between behaviors, where blue and red correspond to success and failure respectively. Bottom plot: Behavior validation count

3 Evaluation

We evaluate the proposed system through extensive offline simulation and online through experiments using six quadrotors, and validate that: (1) plans obey specified clearance and acceleration limits; (2) planning time does not interrupt online flight; and (3) experimental performance aligns with the simulation results to confirm that simulation may be employed as a validation mechanism.

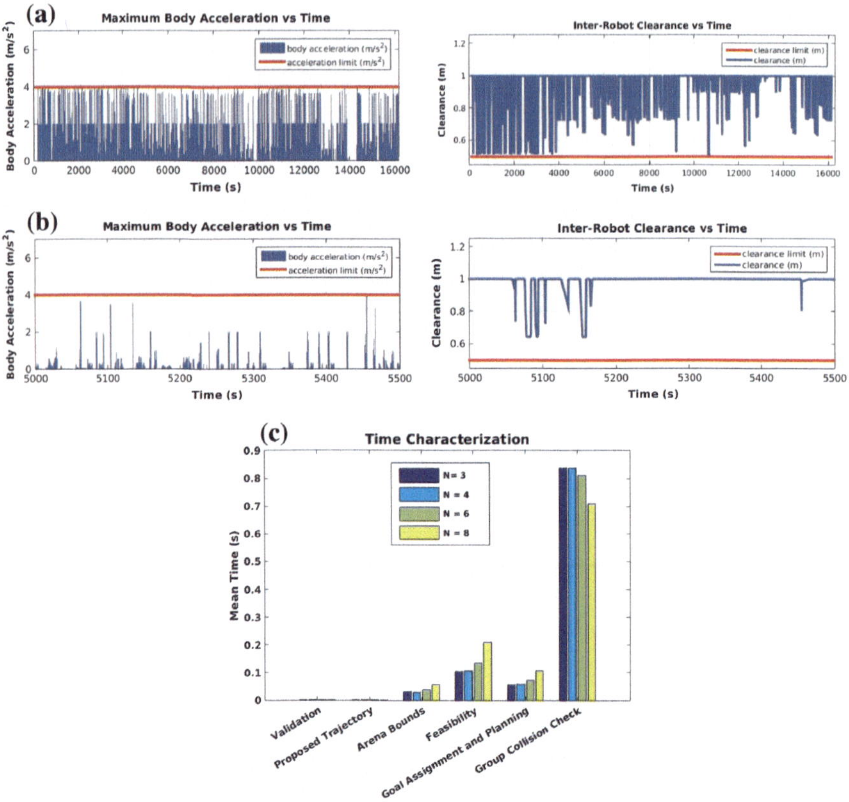

Fig. 6 Simulation evaluation. **a** Dynamic feasibility of all plans shown as continuous bounded plan accelerations; the top plot shows accelerations over four simulation hours, and the lower plot shows a close up over a smaller time period. **b** Collision avoidance, shown as minimum distance between all robots over all time. The top plot shows clearance over four simulation hours, and the lower plot shows a close up over a smaller time period. **c** Timing of system planner stage, for varying numbers of robots

As described in Sect. 2.3, we evaluate behavior descriptor combinations across varying numbers of robots through simulation, where instructions are issued at random timing intervals to simulate online user input arrival. Figure 6 confirms that a safe inter-robot clearance distance is always maintained and that all generated plans obey dynamic feasibility constraints, depicted as trajectory accelerations. Further, we display the timing across planner stages as the percentage of the total planning time per instruction, to confirm that the planning method exhibits acceptable computational complexity scaling for the number of robots considered for the application. Over 3000 descriptor combinations are tested via the representative simulation. Visualizations of transitions between representative behaviors such as single-group shape changes, group merging, and drunk behavior from simulation are shown in Fig. 7.

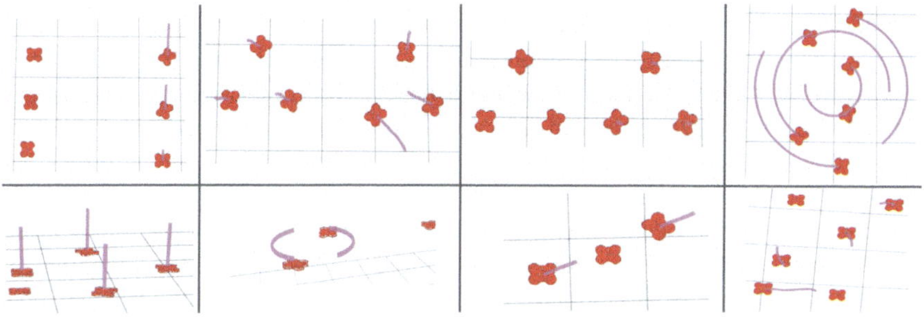

Fig. 7 Transitions between representative behaviors performed in a high fidelity dynamic simulation environment

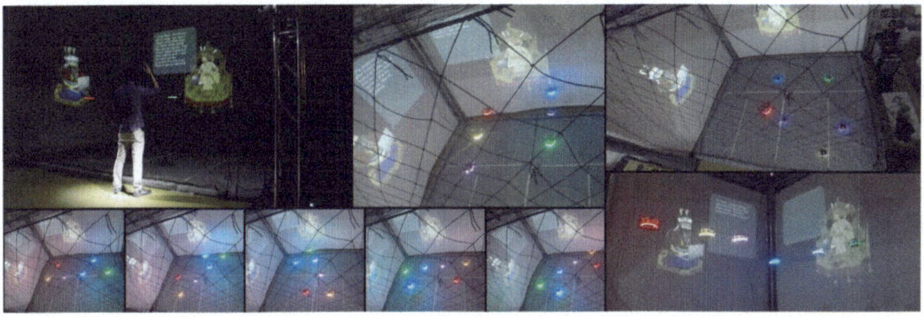

Fig. 8 Photos from a theatric performance. Clockwise, from top left: User practicing gesture-based input; two groups forming lines; one quadrotor performing a solo; red team transitions across from the blue team; bottom row: two groups of three quadrotors in triangle formations circle each other by performing a rotation maneuver as a formation of six

We also qualitatively show the performance of the system as used in a narrative storytelling. In this effort, a user directs six quadrotors through a three act performance spanning fifteen minutes of flight time, requiring 100 online-issued behavior instructions. Hardware results support the simulated evaluation, and images of the performance flight are shown in Fig. 8. A time lapsed video of the performance is available for review.[2]

4 Discussion

While we have not yet performed formal user studies, we offer observations on the design choices presented in this work and give direction for system evolution based on user interactions with the system over a time period of several months.

We find that the choice of designing behaviors in a local frame rather than attempting to design inertial motions directly aids in communicating theatrical intent with

[2]http://www.andrew.cmu.edu/user/ecappo/DARS16.mp4.

collaborators. This structure mimics the breakdown of the theatric commands, as specifications such as "Move drunkenly to Target A while spotting Target B" translate readily into "Do x in a local frame while the local frame performs y". Common to all parameterizations, the choice of polynomial trajectories and local frame representation reduces the potential design space, but by bounding position and its derivatives in a local shape frame, we are able to readily design rich motion specifications with the assurance that these trajectories will compose to form dynamically feasible motions online. Further, while we are still pursuing group collision avoidance implementations, optimization of polynomial trajectories for collision avoidance [13] show promise for use in this area.

Not unexpectedly, we find that a key factor to effective system use is the comfort and ease with which a user can specify behavior descriptor combinations. While the full lexicon of descriptors is necessary to define the many behaviors users require during a theatric performance, detailed motion specifications must be both simple and fast to execute. At times, we find that the user requires additional inputs, such as prioritizing descriptors within an instruction. For example, the system currently extends the duration of overly aggressive trajectories to render them dynamically feasible, prioritizing destination over motion. In instances where motions are meant to anthropomorphize vehicles for the sake of the narrative, however, it may be preferable to pursue fast motions to an arbitrary location, rather than slow motions to a specified target.

In contrast to specifying additional detail, we also encounter use cases where users articulate a need for input behaviors with less detail. To generate instructions quickly, we initialize all descriptors to default values to allow a user to only change a few descriptors to specify simple behaviors. While satisfactory for simple commands, more detailed behaviors still require enumerating all descriptors. Therefore, we believe exploring more advanced strategies to make the system easily accessible to the user, including using context-specific semantics or predictive behavior composition over time are important directions for tackling the problem of simplified online interaction.

5 Conclusion and Future Work

In this effort, we pursued the formulation of multi-robot trajectory generation in an online context specifically for an improvisational theatric application, using motion descriptors to allow the performer to specify theatric intent and direct robot choreographies online and using time-aware trajectory formulation methods for validation, verification, and trajectory refinement to systematically generate dynamically feasible and collision free motions. We show through evaluation that the proposed system design yields a robust approach capable of enabling online theatrical performances. In the future, we will extend the system to incorporate learning techniques to improve system performance and user interaction over many training and performance sessions. These extensions include the generation of new behaviors building on past examples and creating "macros" of recurrent behavior combinations in order to reduce performer command repetition. Further, we are investigating modeling users over repeated interactions in order to anticipate user instructions (i.e., behavior input auto-completion) with the goal of reducing latency by preemptively validating and refining potential motions.

Acknowledgements We thank James Laney and Eric Adlam for their user interface implementations and Nima Dehghani and Ali Momeni for their theatric contributions. We gratefully acknowledge support from ONR grants N00014-13-1-0821 and N00014-15-1-2929.

References

1. Augugliaro, F., Schoellig, A.P., D'Andrea, R.: Dance of the flying machines: methods for designing and executing an aerial dance choreography. IEEE Robot. Autom. Mag. **20**(4), 96–104 (2013)
2. Cirque du Soleil, ETH Zurich, Verity Studios.: Sparked: a live interaction between humans and quadcopters (2014). http://flyingmachinearena.org/sparked/. Accessed 14 July 2016
3. Coxworth, B.: Cirque du Soleil and ETH Zurich collaborate on human/drone performance. Gizmag (2014). http://www.gizmag.com/cirque-du-soleil-sparked-drone-video/33921/. Accessed 14 July 2016
4. Desai, A., Cappo, E.A., Michael, N.: Dynamically feasible and safe shape transitions for teams of aerial robots. In: Proceedings of the IEEE/RSJ International Conference on Intelligent Robots and Systems (2016)
5. Hoffman, G., Kubat, R., Breazeal, C.: A hybrid control system for puppeteering a live robotic stage actor. In: RO-MAN 17th IEEE International Symposium on Robot and Human Interactive Communication, pp. 354–359. IEEE (2008)
6. Holmes, K.: Watch this: flying quadrotor light show spectacular artist [Q&A]. The Creators Project (2012). http://thecreatorsproject.vice.com/blog/watch-this-flying-quadrotor-light-show-spectacular-artist-qa. Accessed 14 June 2016
7. Kaplan, K.: 100 dancing drones set world record. Tech Innovation (2016). http://iq.intel.com/=100-dancing-drones-set-world-record/. Accessed 14 June 2016
8. Knight, H., Gray, M.: Acting lesson with robot: Emotional gestures. In: Proceedings of the ACM/IEEE International Conference on Human-Robot Interaction, pp. 407–407. IEEE (2012)
9. Lupashin, S., Hehn, M., Mueller, M.W., Schoellig, A.P., Sherback, M., D'Andrea, R.: A platform for aerial robotics research and demonstration: the flying machine arena. Mechatronics **24**(1), 41–54 (2014)
10. MagicLab, Rhizomatiks Research: DroneMagic - Behind the Scenes (2016). https://www.youtube.com/watch?v=JEWXBEDAq60. Accessed 14 June 2016
11. MagicLab, Rhizomatiks Research: MagicLab 24 Drone Flight (2016). http://www.magiclab.nyc/research/drone_magic/. Accessed 14 June 2016
12. Mahony, R., Kumar, V., Corke, P.: Multirotor aerial vehicles: modeling, estimation, and control of quadrotor. IEEE Robot. Autom. Mag. **19**(3), 20–32 (2012)
13. Mellinger, D., Kushleyev, A., Kumar, V.: Mixed-integer quadratic program trajectory generation for heterogeneous quadrotor teams. In: Proceedings of the IEEE Interantional Conference on Robotics and Automation, pp. 477–483 (2012)
14. Richter, C., Bry, A., Roy, N.: Polynomial trajectory planning for quadrotor flight. In: Proceedings of Robotics: Science and Systems (2013)
15. Saha, I., Ramaithitima, R., Kumar, V., Pappas, G.J., Seshia, S.A.: Automated composition of motion primitives for multi-robot systems from safe LTL specifications. In: Proceedings of the

IEEE/RSJ International Conference on Intelligent Robots and Systems, pp. 1525–1532. IEEE (2014)
16. Saha, I., Ramaithitima, R., Kumar, V., Pappas, G.J., Seshia, S.A.: Implan: scalable incremental motion planning for multi-robot systems. In: ACM/IEEE 7th International Conference on Cyber-Physical Systems, pp. 1–10. IEEE (2016)

Vertex: A New Distributed Underwater Robotic Platform for Environmental Monitoring

Felix Schill, Alexander Bahr and Alcherio Martinoli

Abstract We present a new Autonomous Underwater Vehicle (AUV) system for cooperative environmental sensing. The AUV was specifically developed as a platform for distributed, cooperative sensing in lakes and coastal areas. In this paper we describe the prerequisite subsystems for a submersible multi-robot system and their interactions. In particular, we incorporate a distributed acoustic localisation system and distributed time-sliced communication systems into an agile, 5-DOF submersible robot that is small, easy to deploy and retrieve, with a modular environmental sensor payload for relevant scientific measurements. We also developed a distributed Hardware-In-the-Loop (HIL) simulation framework to facilitate early testing of algorithms in simulation while running final binary code on the actual robot hardware. To avoid communication overhead and real-time issues, the simulation of the vehicle dynamics and all proprioceptive sensors is performed on-board. Exteroceptive sensors are simulated by vehicle-to-vehicle communication where possible, supported by a central simulation supervisor where required. Finally, we present some preliminary experimental results of the system.

F. Schill (✉) · A. Bahr · A. Martinoli
Distributed Intelligent Systems and Algorithms Laboratory (DISAL),
School of Architecture, Civil and Environmental Engineering,
École Polytechnique Fédérale de Lausanne (EPFL), Lausanne, Switzerland
e-mail: felix.schill@epfl.ch

A. Martinoli
e-mail: alcherio.martinoli@epfl.ch

F. Schill · A. Bahr
Hydromea SA, Lausanne, Switzerland
e-mail: alexander.bahr@hydromea.com

R. Groß et al. (eds.), *Distributed Autonomous Robotic Systems*,
Springer Proceedings in Advanced Robotics 6,
https://doi.org/10.1007/978-3-319-73008-0_47

1 Introduction

Obtaining underwater measurements of biological, chemical and physical parameters is currently an expensive and time consuming activity. In environmental science such measurements are generally obtained manually by lowering a probe from a ship, or automatically with fixed moorings. This limits the amount of spatial data that is available to very few spots (often only the deepest point of a lake). Remote sensing techniques that are commonly used on land such as hyperspectral imaging and radar measurements do not work under water except for the surface layer, as water absorbs most of the electromagnetic spectrum with a very high attenuation. Sonar measurements can only resolve some physical features but are unsuitable for biological and chemical measurements. However, through continuous development and miniaturisation efforts there is a large range of small, precise in-situ sensors available for physical (e.g., conductivity, temperature, turbidity), chemical (e.g., pH, nitrate, chlorine, hydrocarbons) and biological parameters (e.g., fluorometers for algae, chlorophyll, dissolved organic matter). Autonomous Underwater Vehicles (AUVs) have been used for many years to carry such sensors and extend the reach of these instruments. Due to their size and very high cost, AUVs are mostly used in oceanography, but very rarely in limnology and environmental monitoring applications. Additionally, most operators can afford at most one AUV, which limits the volume of water that can be explored within a given time frame. Particularly when studying spatio-temporal phenomena such as wastewater plumes, mixing, convection and biologically active layers, it is insufficient to measure only in one spot, or with only one AUV, as the phenomenon can change drastically over time and space. A much better picture can be obtained by measuring with many sensors at many places simultaneously. Our goal is to develop a fleet of 5–10 small AUVs that can cooperatively obtain high resolution water quality measurements of natural water bodies. For practical reasons, the AUVs should be small and portable for easy deployment and retrieval, and should be designed so that they can be produced cost-effectively in large quantities. For spatially coordinated underwater operations, they have to be able to sense the relative location of nearby vehicles, and be able to communicate to their peers. Lastly, a high number of degrees of freedom is helpful as it simplifies the control of the robots. In this chapter we present the design of the Vertex AUV (Fig. 1).

Fig. 1 Left to right: The Vertex AUV with carry case, in the water at Lake Geneva, and on the ice at Lake Onego, Russia during a field campaign prior to deployment

1.1 Related Work in Distributed AUV Systems

There have been multiple efforts in the past to create an underwater swarm. In 2003 two of the authors were involved in the Serafina project by Uwe R. Zimmer at the Australian National University. An initial prototype was designed by Alexander Bahr and further built and developed by Felix Schill. The main focus of the project was to develop the prerequisites for underwater swarms, in particular scalable communication [13], and cooperative relative positioning [1, 8]. The solutions to these prerequisites were individually evaluated, but the project concluded before a large-scale swarm could be built. The CoCoRo project succeeded in building a large underwater swarm [17], but was clearly focused on indoor in-vitro experimentation [11]. The AUVs in that project have a very limited endurance and depth capability, are not suitable for outdoor use and were not designed to carry a useful payload. The optical communication system is scalable to large numbers, but has very limited range which can not easily be extended as it is dependent on water turbidity. There have been experiments with multiple underwater vehicles where communication was limited to a surface uplink [6], and an experiment with one AUV communicating underwater with a multiple stationary sensor nodes [5]. Successful formation control with three AUVs has been demonstrated recently using acoustic modems by [18], where two vehicles were at the surface using GNSS, and one vehicle followed based on acoustic range measurements. A number of AUV designs appeared in recent years with significantly reduced cost and outdoor-capable performance, such as the OceanServer Iver3, LAUV [2], Starbug [4] and COTSbot [3]; however to our knowledge they were not designed to operate in large cooperative teams, and are significantly heavier and bigger than our design presented in this paper.

1.2 Related Work in Localisation and Communication

In any cooperative multi-robot system, robots need to have some awareness of nearby other robots. In most practical systems, this means an ability to measure the relative position of other robots in proximity. Additionally, many distributed control strategies assume some means of communication with other robots. As both localisation and communication are part of control loops, they should be real-time and scalable. Local coverage of nearby robots is more important than global coverage, and often sufficient, but some distributed control strategies do require global information exchange to reach consensus. The underwater environment poses unique challenges in both domains, mainly due to the very high attenuation of electromagnetic waves. With the exception of very low radio frequencies, and the green/blue part of the visible spectrum of light, the attenuation is prohibitive even for short distances. This excludes most technologies commonly used in air, such as Global Navigation Satellite Systems (GNSS), ZigBee, WiFi or other common radio systems. Poor visibility also often precludes the use of optical channels or cameras. The most common solution is acoustic transmission for communication, localisation and remote sensing (sonar). Unfortunately, acoustic communication does not scale very well to large networks due to the high distortion and interference, as sound travels long distances but gets distorted beyond decodability [7]. Common solutions for underwater localisation are either based on external reference beacons or trackers that require fixed infrastructure (e.g., LBL, USBL), or inertial systems supported by Doppler velocity measurements taken against the sea floor (DVL) which can drift over time. For relative position sensing, such systems would still require communication and a common coordinate frame. To address these issues, we propose a combination of acoustic relative positioning [1], VLF radio communication [13] and short-range optical communication [15], GNSS (surface only) and inertial navigation (see below in Sects. 3 and 4).

2 AUV System Design

Our design goals were to develop an AUV that is small (portable in one hand), rated for outdoor operation in most inland and coastal waters, capable of carrying a useful water quality sensing payload, and designed for multi-robot operation in large groups. The final design is 70 cm long, weighs 7 kg, and is equipped with a YSI EXO2 sensor system which can be configured with up to 7 different physical, chemical and biological sensors. The maximum design depth is 300 m, which covers the majority of inland waters and coastal areas. For heavier payloads a lighter pressure hull with reduced depth rating can be used. Two battery packs provide a total of 170 Wh of energy, resulting in an estimated endurance of 6–8 h at a cruise speed of up to 1 m/s. The AUV can also operate on a single battery pack to increase payload capacity if required. The sensor payload and batteries together make up approximately 60% of the overall length of the vehicle. With the given payload we believe that the current design is close to the minimum size for such a vehicle; a further reduction in size would lead to significant performance losses (Fig. 2).

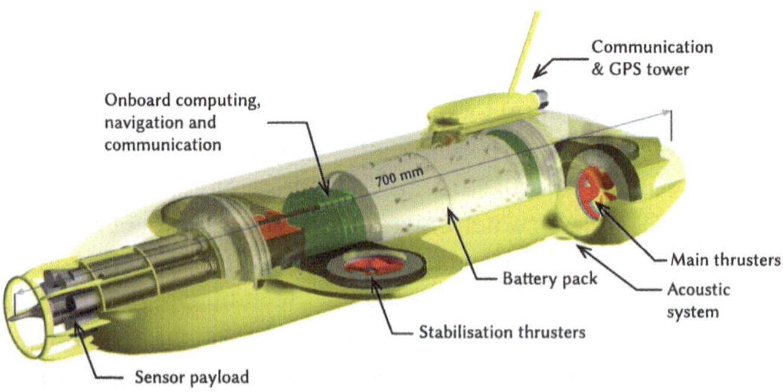

Fig. 2 CAD rendering of the Vertex AUV

2.1 Propulsion

The thrusters are a newly developed brushless rim-drive design without a central hub in the propeller. The absence of a central shaft greatly reduces the risk of entanglement with floating debris. Each thruster has an integrated motor controller, which implements 3-phase commutation and closed-loop speed control. The motor control system was optimised for fast reversal and quick response, which is important to achieve good attitude control. Five low-profile thrusters provide propulsion and attitude control to the AUV. The two lateral horizontal thrusters provide forward thrust and yaw control. Two lateral vertical thrusters mounted in the winglets in front of the center of gravity provide roll control, and contribute to pitch control and depth control together with a vertical thruster mounted centrally in the tail. The thrusters have a thickness of only 15 mm, which makes it possible to integrate them into hydrodynamically efficient winglets to reduce drag.

2.2 Hardware Architecture

The onboard embedded system consists of multiple modules built around 32-bit microcontrollers: a system management module for power management, data logging and supervision, a navigation module for inertial sensing and closed-loop control, a longwave radio module for underwater communication, and a payload module to interface to the payload sensors (Fig. 3). An external GNSS and communication

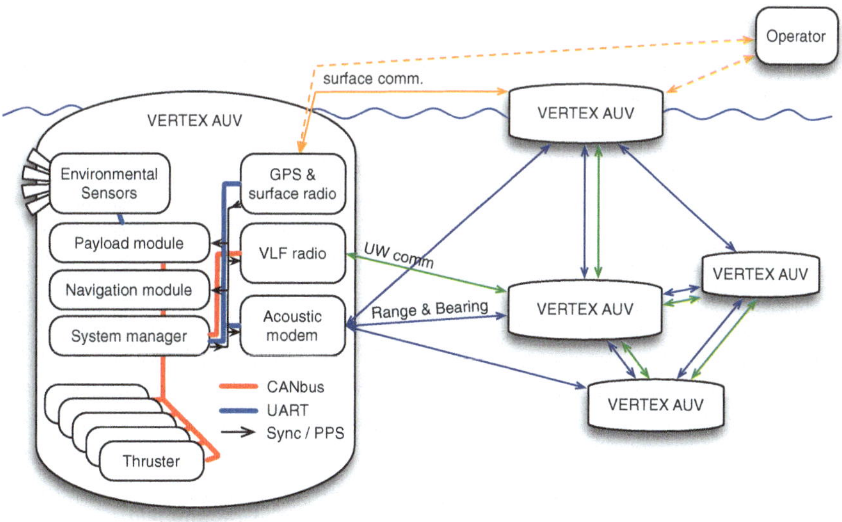

Fig. 3 System diagram of the AUV subsystems

module provides connectivity via a 868 MHz modem and GNSS positioning while at the surface. The modules are connected via a CAN bus, and dedicated UART connections. Each module can be switched on or off as needed to reduce power consumption, and all internal voltages and currents are continuously monitored. The hotel load of the complete system when fully operational without the motors running is 1.6 W without payload, and 3 W with the EXO2 payload active and sampling 5 sensors at 1 Hz.

2.3 *Synchronised Clocks*

A common time base is very useful in a distributed system for ordering events, measuring propagation delay, coordinated control, and analysing log files. The Vertex AUV has a precise reference clock that is synchronised at the surface using the GNSS time pulse. The reference clock is temperature compensated and an order of magnitude more precise than regular clock crystals used for microcontrollers (below 1 ppm). Internally, the reference clock system synthesises an electrical time pulse once per second that is distributed to all microcontroller modules, which run on regular crystals to reduce cost. Each module continuously synchronises its own clock to the time signal to within one microsecond. While submerged, without the GNSS reference, the internal clock system keeps synchrony within at most one millisecond per hour - measured drift rates are 0.3 μs/s. In most missions the AUVs can be expected to surface at least once per hour, therefore all clocks are synchronised to within one millisecond or less. We are working on a further improvement using long term drift observations against GNSS to auto-calibrate the reference clock.

2.4 *Software Framework*

A modular software framework has been developed in C which runs bare-bone on the microcontrollers. The overall software architecture was derived from the initial version of the MAV'RIC autopilot framework originally developed by Schill et al. for a swarm of quadrotors, which is now openly available in a similar, updated version [9]. For execution of subtasks at different rates, a simple fixed-priority scheduler calls the various subroutines at the specified frequency. We decided to implement subroutines with short, bounded execution times that operate on the system state, as opposed to full tasks or threads. The benefit is reduced memory usage, and lower complexity regarding task synchronisation and mutual exclusion as there are no context switches within a subroutine. Time-critical tasks are handled through interrupts and hardware events where appropriate, and concurrency is achieved by using a distributed architecture with dedicated microcontrollers. Communication between different modules (processors), as well as the telemetry link to the base station computer, is handled using the MAVlink protocol [10]. A second scheduler instance has been added for managing dynamically adjustable telemetry message update rates that can be adjusted during runtime through the operator interface. MAVlink also provides parameters that can be edited at runtime. A flexible message callback system was implemented in MAV'RIC for responding to incoming messages as they arrive. The communication channel is abstracted by a bytestream interface, which can be mapped to various interfaces such as a CAN-bus, UART, Radio or UDP. Above the hardware driver level the interface is fully transparent and hardware-independent. All internal and external message traffic is logged to an internal microSD card for post-mission analysis. A simulation layer, described later in Sect. 5, was added to simulate all relevant sensors on a low level for HIL simulation. The software framework was also ported to Linux by replacing the microcontroller hardware abstraction with a thin emulation layer. This makes it possible to compile and test almost identical code (only with a different hardware abstraction layer) on Linux, with minimal differences to onboard HIL simulation and real experiments. The internal communication streams as well as the telemetry link are transparently mapped to UDP to emulate the internal data bus as well as the telemetry uplink to the basestation software. This enables fast prototyping, debugging and testing of code on a PC of directly compatible code.

3 Inter-Vehicle Acoustic Localisation

For localisation purposes we developed a miniature acoustic modem to fit within the small scale of the Vertex AUV. Instead of relying on external, fixed infrastructure, we propose to use the AUV swarm as the reference for each vehicle, based on Bahr's previous work [1]. Each robot regularly sends a short acoustic pulse at a precisely defined time based on the synchronised reference clock. When nearby robots receive the pulse, they can immediately calculate the time of flight based on the arrival time, and derive the distance to the other robot. Additionally, together with the acoustic pulse, the sending robot also communicates its own position estimate and uncertainty. These distance measurements are then incorporated in the position state estimator, that updates the robot's own position as well as the estimated position of surrounding robots. If any of the robots is at the surface, it will use the GNSS to correct its absolute position, which will in turn improve the position estimates of the other AUVs due to the lower uncertainty of the surfaced robot. The VLF radio system will be used for communication of the position estimates and uncertainty matrices. We are currently also working on using the acoustic system also for transmission of just the position estimate in a compressed form. While acoustics have known scalability issues, it may be possible to encode a very small amount of data in short pulses that remain separated even with many nodes present. The advantage is that partial functionality of the system can be maintained over much longer ranges, and increased reliability (Fig. 4).

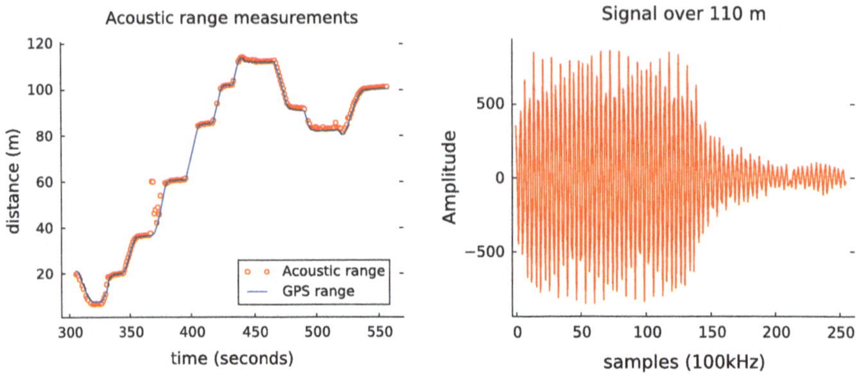

Fig. 4 Left: Acoustic range measurements (red) versus GNSS (blue). Right: Recording of acoustic pulse at maximum distance of 110 m

4 Inter-Vehicle Communication

The two challenges for an underwater communication network that is scalable to many nodes are how to transmit data through the water, and how to distribute access to the extremely limited communication channel to all nodes fairly while fully utilising the available bandwidth. The acoustic channel is not very well suited for continuous use by many nodes, due to high latency, reverberation and acoustic interference over long distances. Acoustics is therefore only suitable for sparse long-range signaling of only very small amounts of data. In previous work we developed a Very-Low-Frequency (VLF) radio system [12] that can transmit at up to 8192 bps over short to medium distances. An early prototype with 5 V transmitter drive voltage was tested in fresh water and seawater, achieving a range of approximately 4 m, and 5 m in air. We redesigned the radio for the AUV with increased power, boosting the transmitter voltage to a software-programmable value up to 35 V. Range in air improved to above 20 m (Fig. 5); range in water still has to be tested. The results shown do not use error-correction - we are planning to implement a suitable error-correcting code to reduce packet loss. Additionally, we integrated an optical communication system for short ranges up to approximately 5 m, based on [15]. Although optical links are strongly affected by water turbidity and have a very short range, they have a much higher bandwidth, in our case up to 115200 bps. While at the surface, a 868 MHz radio provides a long-range link to the operator and other vehicles that are at the surface.

Although the VLF radio offers bitrates that are higher than most acoustic modems, the available bandwidth is still very limited. To control fair access to the limited channel, we previously developed a very efficient, distributed Time Division Multiple Access (TDMA) algorithm that assigns time-slots to each node such that the channel can be fully utilised while avoiding message collisions. The algorithm is fully distributed without central coordinators; all nodes have the same behaviour. Two variants

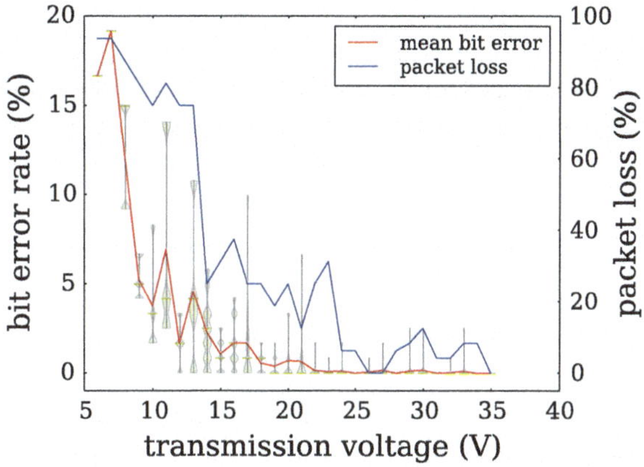

Fig. 5 Bit error rate and packet loss of VLF radio transmission over 20 m in air, for increasing transmitter drive voltage

of this algorithm where previously published: DAOS (*Distributed Ad-hoc Omnicast Scheduler*) generates fully collision-free schedules, but is less efficient for dense networks with high connectivity. The *Pruned Distributed Ad-hoc Omnicast Scheduler* (PDAOS) dynamically prunes schedules based on signal strength or distance, to give priority to nearby nodes. In simulations, PDAOS has shown excellent performance in networks with different density, and fast dynamic changes in network topology. We tested on a benchmark called "omnicast" for global information exchange, which is the time required until all nodes have received information from all other nodes, assuming bounded message size (e.g., how long it would take to reach consensus on a global maximum/minimum). A conservative theoretical upper bound of $2n - 2$ time slots for n nodes has been shown previously [16], which is clearly outperformed by the distributed algorithm, as shown in Fig. 6. In fact, global information exchange

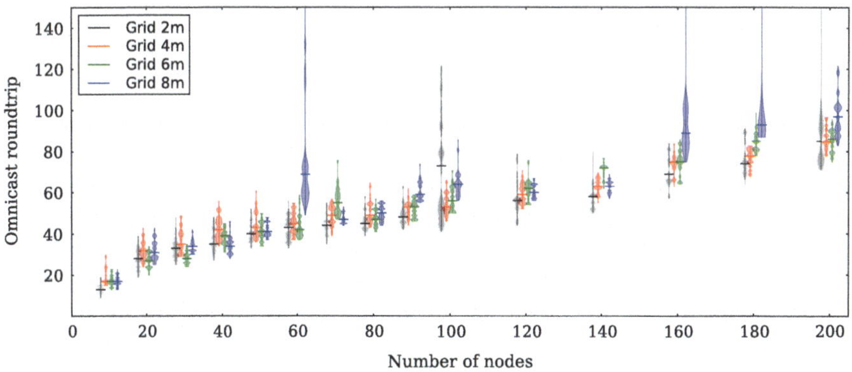

Fig. 6 Real-time simulation of the distributed PDAOS scheduling algorithm (from [14]): Communication time slots required for global information exchange (omnicast roundtrip) for varying network sizes, with nodes arranged in a 2D grid with 2, 4, 6 or 8 m distance (assuming 10 m communication range)

can be achieved in significantly less than n time slots for larger networks above 40 nodes. This is possible because nodes that are far apart can communicate in parallel; additionally, even though PDAOS deliberately allows message collisions, some nodes may still receive valid messages and further spread information despite collisions. A detailed analysis of the static and dynamic performance of PDAOS can be found in [14].

5 Simulation Framework

An important feature for testing embedded control software is HIL simulation. Generally this is done by connecting the external interfaces of a control system to a simulation computer, which reads the system outputs, simulates the system dynamics and sends emulated sensor data back to the system under test. On small, integrated embedded systems, however, it is often difficult to gain access to the sensor interfaces, as many sensors are soldered directly onto the circuit board. Also, some of these sensors are sampled at a high rate - in our case, the inertial sensors are sampled at 200 Hz. It would therefore be challenging to send actuator commands to an external computer, and send back the full set of simulated sensor data through a serial link with acceptable latency for a faithful real-time simulation.

In our system, we therefore decided to simulate the vehicle dynamics and proprioceptive sensors directly on the microcontroller itself. The cut is done at the lowest-possible level: the simulator inputs are the actuator control commands as they are sent to the motor controllers. The simulator then calculates estimated rotor speeds, propeller lift and drag, as well as thruster inertia for each thruster. The calculated forces are combined to calculate 3D torque and linear forces on the AUV body, which are then integrated to obtain angular rates, attitude, velocity and position. From these results, sensor values are simulated for the accelerometers gyroscopes, compass, GNSS and depth sensor, using the real, inverse scale factors and offsets. The remaining state estimation and control code is unchanged between simulation and operational mode - the normal attitude estimation filter, attitude/velocity control and navigation code is applied in simulation mode as it is during normal operation. The simulation update takes less than 2 ms, and runs at 200 Hz which is the same rate as the inertial sensor sampling rate during normal operation. The result is a simplified but reasonably faithful simulation of the AUV dynamics once the simulator parameters are calibrated. This allows testing of all high-level code as well as the majority of low-level code on the embedded system itself. As the project can also be compiled and run on Linux, the virtually identical code can also be tested on a PC which is often more convenient for initial development. Once the code has been successfully tested in the Linux emulation, the project is re-compiled and tested on the real hardware in the onboard simulation, which occasionally may reveal hardware-specific issues. It is then possible to switch off simulation mode, even at runtime, which activates the real actuators and sensors, and tests can continue in reality. Monitoring of the simulation occurs through the same telemetry link, using the same MAVlink message protocol, as during live experiments. This facilitates comparing results between simulation and reality, creating consistent log files and data processing pipelines.

Exteroceptive sensors can not be simulated entirely on the vehicle or vehicle processes without external inputs. We chose a distributed approach, using direct robot-to-robot communication to transmit the minimal required information, and carry out the simulation of exteroceptive sensors within the target vehicle software. This approach scales well with the number of robots in the simulation, and in many cases is very close to the real scenario. A simple example is a formation control task of multiple robots. Each robot requires range and bearing information from the other robots. This information would normally be acquired by the acoustic system of the AUV, but for tests of the higher-level formation control algorithms it can also be computed from vehicle positions. In this case, we chose to send global position messages between the simulated vehicles, as a detailed simulation of acoustics was not required. We recently showed the effectiveness of this simulation protocol in the framework of a student project with four Autonomous Surface Vehicles (ASVs) based on the same hardware and software architecture as the Vertex AUV, using a combination of radio communication and GNSS-based localisation for their inter-vehicle positioning. Experiments were carried out in PC-based and HIL simulation, as well as on Lake Geneva. This example illustrates that the flexible simulation framework can be tuned to desired abstraction levels, while carrying out meaningful simulation tests on final binary code on real hardware with multiple robots. Furthermore, to test the acoustic range and bearing measurement algorithms, the simulation was extended to synthesize acoustic samples that emulate real recorded data. In this case, the information sent between robots in simulation is their position as well as the sampled signal they transmit. Instead of sending samples through the DAC, the transmitting side sends them through the MAVlink channel to all other robots at the DAC sampling rate of 1 MHz. As other robots receive the sample messages, they look up the transmitting robot's position based on the last received position message. Based on the samples, message timestamp and position, the robots can calculate the time of flight and attenuation of the signal. The samples are then mixed into a mixing buffer which adds all received signals together. Some IIR bandpass filtering and noise is applied to emulate the frequency response and resonance of the real piezo transducers. When the receiving robot requests samples from the simulated ADC, the mixing buffer is downsampled to the ADC frequency (in this case usually 100 kHz). All remaining code beyond this mechanism is now exactly identical to the code running in reality with the real DAC, transducers and ADC. There are of course deviations between simulation and real data, but the main goal of this framework is to test code in the target language in a scenario that is realistic enough to detect bugs early, and to test algorithms under idealised conditions before subjecting them to the real world. Being able to test code early and often, initially on a PC, then on the target hardware, significantly reduces the overhead of experimental campaigns.

6 AUV Field Tests

The first AUV prototype was completed in October 2014, and tested in an indoor tank. Shortly after we were given the opportunity to participate in the field campaign "Ladoga - Life under ice" in mid-March 2015, as well as in March 2016, organised by the Limnology Center at EPFL. This campaign is a Russian-Swiss multidisciplinary collaboration, to study under-ice dynamics, aquaphysics and biology at Lake Ladoga, the largest lake in Europe, and how it is affected by climate change. Due to the unusually warm winter in 2015, the ice cover at the planned location was already disappearing. The location was therefore changed to the nearby Lake Onego, close to the city of Petrozavodsk. Ice cover at the location was 40 cm, the air temperature varied between $-15°$ and $10°$. Despite a few days of unusually high temperatures the ice cover remained safe for the duration of the campaign, but started breaking up the week after. The AUV was deployed through a hole in the ice to collect measurements horizontally in a 25–30 m radius around the hole to assist in a study of convection patterns. The payload consisted of sensors for conductivity, temperature, oxygen, pH, chlorophyll, blue-green algae, turbidity and fluorescent dissolved organic matter. During the field campaigns, the acoustic localisation system was not fully operational yet, and due to the ice cover it was impossible to surface for a GNSS fix or retrieval. A safety line was therefore attached to avoid loss of the vehicle and to aid with recovery to the ice hole. The deployment provided an opportunity for thorough system testing in harsh real-world conditions. The dry air and cold temperatures caused a significant amount of static electricity that our system was not previously exposed to, and prompted a design improvement of the electronics which has now been incorporated into the AUV. The collected sensor data (Fig. 7) was provided to the limnologists on the campaign to augment their own measurements.

More recently the Vertex AUV was also deployed at Lago Cadagno in Switzerland to investigate the spatial properties of a bacterial layer. During this campaign the AUV carried out multiple autonomous transects following a vertical zig-zag pattern in depth. An excerpt of the collected data is shown in Fig. 8. The sensor payload was similar to that of the Onego campaign, but with an additional 250 Hz fast temperature sensor for microstructure turbulence measurements. The data shown here represent an example of the possible applications of the vehicle; an in-depth discussion of the data will be published separately in collaboration with our limnology collaborators.

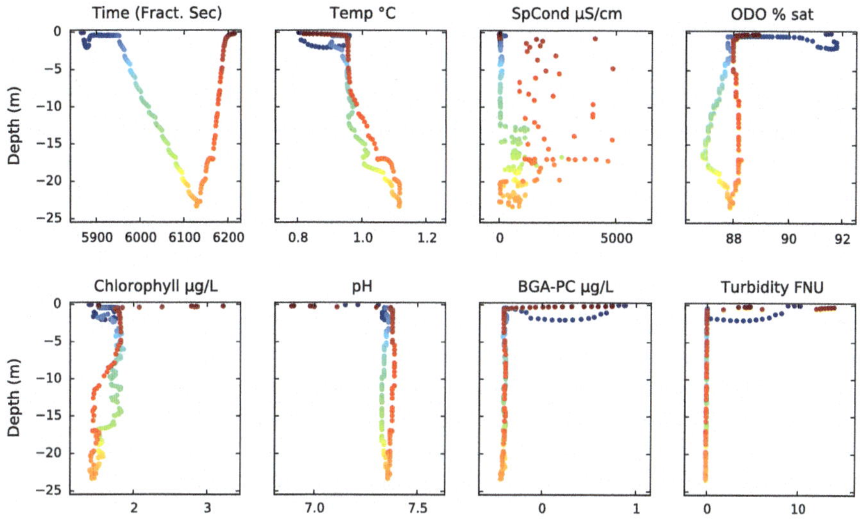

Fig. 7 Data collected during an autonomous dive at Lake Onego, 25.03.2015. The vertical axis is the depth below the surface. The color coding indicates time, starting at dark blue, through green, yellow, orange to red

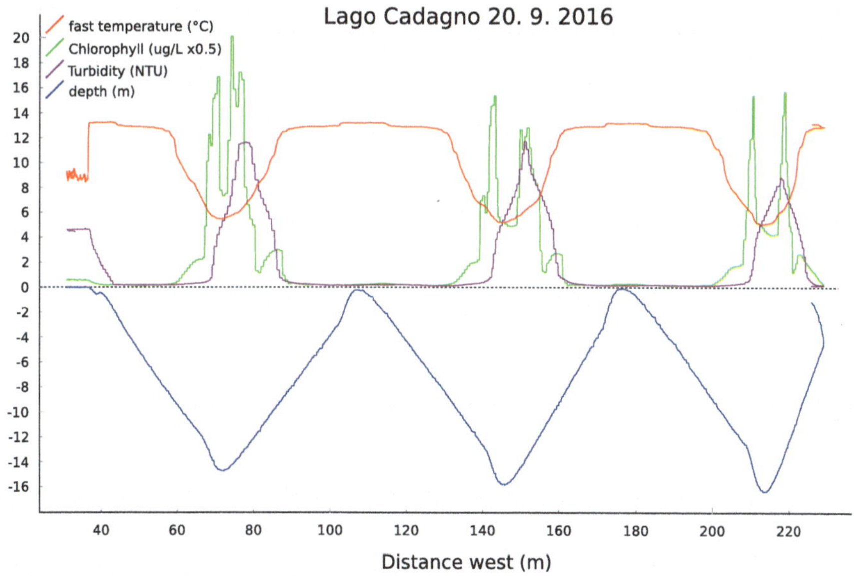

Fig. 8 Autonomous transect (east–west), excerpt from data collected at Lago Cadagno, Sept. 2016. Sensor payload: YSI EXO2: Turbidity, Chlorophyll (shown), CTD, oxygen, BGA-PE, BGA-PC (not shown); additionally a fast temperature sensor (FP07) from Hydromea measuring at 250 Hz. The AUV repeatedly traversed a distinct bacterial layer of interest at 13 m depth, detectable by a drop in Chlorophyll and sharp increase in turbidity, as well as a distinct temperature signature

7 Conclusions

We presented a new underwater robot system natively designed for distributed, cooperative aquatic sensing and environmental research. The Vertex AUV is a major effort to integrate a number of important technologies, including scalable underwater localisation and communication into one coherent system. A powerful and flexible simulation framework facilitates experimentation and research with new distributed algorithms, which can be evaluated and tested in the lab first before actual real-world experiments. We also presented preliminary experimental results of the acoustic ranging system and the VLF communication system. As a demonstration of the AUV's capabilities we also presented some results from the first field campaigns in a challenging under-ice environment as well as a mountain lake with a scientifically relevant payload. Current and future work will focus on validation of the multi-robot capabilities of the system, underwater formation control, and further field campaigns with multiple AUVs in collaboration with our partners in limnology.

Acknowledgements This work has been financially supported over multiple years by the following sponsors (in chronological order): National Competence Center in Research on Mobile Information and Communication Systems (NCCR-MICS), a center supported by the Swiss National Science Foundation under grant number 51NF40-111400, through the Spin Fund project "Serafina - Large Scale Underwater Exploration using Groups of Autonomous Underwater Vehicles"; Swiss Commission for Technology and Innovation under Grant No. 16348.1 PFES-ES; the Technology Transfer Office of EPFL, through the Enabling Grant No. TTO 6.1419; FEEL Foundation supported by Ferring Pharmaceuticals, under the project "Ladoga - Life under ice"; and the Swiss National Science Foundation under the Sinergia Grant No. CRSII2_160726/1.

References

1. Bahr, A., Leonard, J.J., Fallon, M.F.: Cooperative localization for autonomous underwater vehicles. Int. J. Robot. Res. **28**(6), 714–728 (2009)
2. da Silva, J.E., Terra, B., Martins, R., de Sousa, J.B.: Modeling and simulation of the LAUV autonomous underwater vehicle. In: 13th IEEE IFAC International Conference on Methods and Models in Automation and Robotics (2007)
3. Dayoub, F., Dunbabin, M., Corke, P.: Robotic detection and tracking of crown-of-thorns starfish. In: IEEE/RSJ International Conference on Intelligent Robots and Systems, pp. 1921–1928 (2015)
4. Dunbabin, M., Roberts, J., Usher, K., Winstanley, G., Corke, P.: A hybrid AUV design for shallow water reef navigation. In: Proceedings of the IEEE International Conference on Robotics and Automation, pp. 2105–2110 (2005)
5. Dunbabin, M., Corke, P., Vasilescu, I., Rus, D.: Experiments with cooperative control of underwater robots. Int. J. Robot. Res. **28**(6), 815–833 (2009)
6. Fiorelli, E., Leonard, N., Bhatta, P., Paley, D., Bachmayer, R., Fratantoni, D.: Multi-AUV control and adaptive sampling in Monterey Bay. IEEE J. Ocean. Eng. **31**(4), 935–948 (2006)
7. Frater, M.R., Ryan, M.J., Dunbar, R.M.: Electromagnetic communications within swarms of autonomous underwater vehicles. In: Proceedings of the 1st ACM International Workshop on Underwater Networks, pp. 64–70. ACM Press, New York, NY, USA (2006)
8. Kottege, N., Zimmer, U.R.: Underwater acoustic localization for small submersibles. J. Field Robot. **28**(1), 40–69 (2011)

9. MAVRIC autopilot project. https://github.com/lis-epfl/MAVRIC_Library
10. Meier, L., Tridgell, A., Goppert, J.: Mavlink micro air vehicle communication protocol. http://qgroundcontrol.org/mavlink/start
11. Mintchev, S., Donati, E., Marrazza, S., Stefanini, C.: Mechatronic design of a miniature underwater robot for swarm operations. In: Proceedings of the IEEE International Conference on Robotics and Automation, pp. 2938–2943 (2014)
12. Schill, F., Zimmer, U.R.: Effective communication in schools of submersibles. Proc. IEEE OCEANS **06**, 1–5 (2006)
13. Schill, F., Zimmer, U.R.: Pruning local schedules for efficient swarm communication. In: Proceedings of the International Symposium on Underwater Technology, Tokyo, Japan, pp. 594–600 (2007)
14. Schill, F., Zimmer, U.R.: A scalable electro-magnetic communication system for underwater swarms. In: 9th IFAC Conference on Manoeuvring and Control of Marine Craft, pp. 97–102 (2012)
15. Schill, F., Zimmer, U.R., Trumpf, J.: Visible spectrum optical communication and distance sensing for underwater applications. In: Proceedings of the Australasian Conference on Rotoics and Automation, pp. 1–6 (2004)
16. Schill, F., Trumpf, J., Zimmer, U.R.: Towards optimal TDMA scheduling for robotic swarm communication. In: Proceedings Towards Autonomous Robotic Systems, pp. 197–203 (2005)
17. Schmickl, T., Thenius, R., Moslinger, C., Timmis, J., Tyrrell, A., Read, M., Hilder, J., Halloy, J., Campo, A., Stefanini, C., Manfredi, L., Orofino, S., Kernbach, S., Dipper, T., Sutantyo, D.: CoCoRo – the self-aware underwater swarm. In: Proceedings of Fifth IEEE Conference on Self-Adaptive and Self-Organizing Systems Workshops, pp. 120–126 (2011)
18. Soares, J., Aguiar, A., Pascoal, A., Martinoli, A.: Joint ASV/AUV range-based formation control: theory and experimental results. In: IEEE International Conference on Robotics and Automation, pp. 5579–5585 (2013)

Author Index

A
Agmon, Noa, 163
Alkilabi, Muhanad H. Mohammed, 503
Amigoni, Francesco, 59
Arye, Idan, 31
Atkinson, Katie, 559

B
Bahr, Alexander, 679
Banfi, Jacopo, 59
Basilico, Nicola, 59
Beltrame, Giovanni, 433
Berman, Spring, 3
Bloembergen, Daan, 559
Bose, Thomas, 461
Bourgeois, Julien, 387, 415
Brown, Daniel S., 447
Bruckstein, Alfred M., 103
Bürger, Mathias, 253
Burns, Alyxander, 133

C
Campos, Mario F. M., 147
Cappo, Ellen A., 665
Cavalcanti, Ana, 517
Chaimowicz, Luiz, 313
Cho, Doo-Hyun, 269
Choi, Han-Lim, 269
Chung, Timothy H., 649
Clement, Michael R., 649
Colas, Francis, 545
Collenette, Joe, 559
Correll, Nikolaus, 359, 619
Cowley, Anthony, 45
Cruz, Patricio, 219

D
Dasgupta, Prithviraj, 345
Davis, Duane T., 649
Day, Michael A., 649
Desai, Arjav, 665
Dimarogonas, Dimos V., 253
Dodd, Tony J., 401
Dorigo, Marco., 531
Douchan, Yinon, 299
Doyle, Matthew J., 373
Dutta, Ayan, 345

E
Escalera, Juan A., 373

F
Fantini-Hauwel, Carole, 531
Fekete, Sándor P., 205
Fierro, Rafael, 219
Fukui, Rui, 605

G
Garone, Emanuele, 433
Gasparri, Andrea, 433
Gauci, Melvin, 573
Ghedini, Cinara, 89
Gillet, Denis, 75
Gini, Maria, 17
Giuggioli, Luca, 31
Goldstein, Seth Copen, 415
Groß, Roderich, 373, 401, 475

R. Groß et al. (eds.), *Distributed Autonomous Robotic Systems*,
Springer Proceedings in Advanced Robotics 6,
https://doi.org/10.1007/978-3-319-73008-0

H
Ha, Jung-Su, 269
Habibi, Golnaz, 205
Haghighat, Bahar, 329
Hauert, Sabine, 487
Heiblum Robles, Alexandro, 31
Hennigh, Oliver, 447
Hintz, Christoph, 219
Hsieh, M. Ani, 45
Hughes, Dana, 619

I
Inácio, Fabrício R., 313

J
Jensen, Elizabeth A., 17
Jones, Simon, 487

K
Kaminka, Gal A., 31, 163, 299
Kanayama, Gen, 605
Kato, Yuta, 605
Kingston, Zachary, 205
Klinger, John, 359
Kumar, Vijay, 147, 587

L
Lee, Sujin, 269
Li, Wei, 517
Li, Yang, 359
Liu, Lantao, 285
Liu, Ming, 545
Loscalzo, Steven, 447
Lowmanstone, London, 17
Lu, Chuan, 503
Lupu, Ilan, 163

M
Ma, Kai-Chieh, 285
Ma, Zhibei, 285
Macharet, Douglas G., 313
Magnenat, Stéphane, 545
Manor, Rotem, 103
Marshall, James A. R., 461
Martinoli, Alcherio, 75, 329
McLurkin, James, 205
Michael, Nathan, 665
Miyazawa, Alvaro, 517
Mondada, Francesco, 373, 545
Moon, Sunghyun, 269
Mox, Daniel, 45

N
Nagpal, Radhika, 573
Nakao, Masayuki, 605
Nam, Changjoo, 237
Narayan, Aparajit, 503
Naz, André, 415
Nelson, Carl, 345

O
O'Grady, Rehan, 531

P
Pahlajani, Chetan D., 177
Parrott, Christopher, 401
Perez-Diaz, Fernando, 475
Pinciroli, Carlo, 433
Piranda, Benoît, 387, 415
Podevijn, Gaëtan, 531
Poulakakis, Ioannis, 177
Prorok, Amanda, 147, 587

R
Ramachandran, Ragesh K., 3
Reina, Andreagiovanni, 461
Ribeiro, Carlos H. C., 89
Ribeiro, Pedro, 517
Roelofsen, Steven, 75
Romeo, Marta, 59
Rubenstein,Michael , 573
Rypkema, Nicholas R., 633

S
Sabattini, Lorenzo, 89, 191
Saldaña, David, 147
Schill, Felix, 679
Schillinger, Philipp, 253
Schmidt, Henrik, 633
Schulze, Bernd, 133
Schwager, Mac, 117
Secchi, Cristian, 191
Shell, Dylan A., 237
St. John, Audrey, 133
Studley, Matthew, 487
Su, Xuanshuo, 117, 133
Sukhatme, Gaurav S., 285

T
Takahashi, Ryo, 605
Tanner, Herbert G., 177
Taylor, C. J., 45
Timmis, Jon, 517
Trenkwalder, Stefan M., 475
Trianni, Vito, 461
Tucci, Thadeu, 415
Tuci, Elio, 503
Turner, Ryan, 447
Tuyls, Karl, 559

W
Wang, Shiling, 545
Wang, Zijian, 117
West, Jonathan, 219
Wilson, Sean, 3
Winfield, Alan, 487
Woodcock, Jim, 517

Y
Yadav, Indrajeet, 177
Yang, Guang, 117

Z
Zareh, Mehran, 191
Zillmer, Rüdiger, 475

The manufacturer's authorised representative in the EU is Springer Nature Customer Service Centre GmbH, Europaplatz 3, 69115 Heidelberg, Germany. If you have any concerns regarding our products, please contact ProductSafety@springernature.com

Printed and bound by CPI Group (UK) Ltd, Croydon, CR0 4YY
22/09/2026
02221049-0004